The Autonomous Enterprise with SAP®

SAP PRESS

Nitin Singh, Jason Porterfield, Srinivas Devarapalli,
Joffy Mathew, Cooper Walsh, Lyssa Aruda

The Autonomous Enterprise with SAP®

Apps, Data, and AI in Action

Rheinwerk
Publishing

Editor Megan Fuerst
Acquisitions Editor Emily Nicholls
Copyeditor Melinda Rankin
Cover Design Abigail Damke
Photo Credit iStock: 1332539767/© FooTToo
Layout Design Vera Brauner
Production Hannah Lane
Typesetting SatzPro, Germany
Printed and bound in the United States of America, on paper from sustainable sources

ISBN 978-1-4932-2870-6
1st edition 2026

Rheinwerk Publishing, Inc.
2 Heritage Drive, Suite 305
Quincy, MA 02171
USA
info@rheinwerk-publishing.com
+1.781.228.5070

Represented in the E. U. by:
Rheinwerk Verlag GmbH
Rheinwerkallee 4
53227 Bonn
Germany
service@rheinwerk-verlag.de
+49 (0) 228 42150-0

Library of Congress Cataloging-in-Publication Control Number: 2026027637

Contents at a Glance

Contents

Foreword by Darwin Deano

McKinsey research estimates that approximately 60 percent of all work is now automatable.[1] This is a powerful data point for the opportunity that is available to all businesses today. However, despite how much enterprises have modernized their technology landscape, they continue to face trapped working capital, supply chain disruption, slow decision cycles, and growing operational complexity. Technology investments alone are clearly not enough.

If I try to distill my three decades of large-scale enterprise transformation success in multiple industries across five continents and award-winning innovation into a couple of observed truths, they are these: that enterprises are a complex, living organism; and that realizing outsized value outcomes demands a holistic, multidimensional operating model change. Like any living organism, each enterprise has a survival instinct, an immune system that violently prefers stasis. To enable transformative change, the organism must be subjected to a systematic attack that overwhelms all defense mechanisms without being fatal. Do too little and the organism will remain unchanged, but do too much and the organism is overwhelmed, resulting in severe, undesirable, and often long-term consequences.

Transformative change for the modern enterprise will require maximum leverage of cloud, data, and AI advancements, but even together, these are not sufficient by themselves. Why? Cloud, data, and AI are powerful, but they are fundamentally abstractions of the *how* question. Cloud abstracts technology architecture and infrastructure; data abstracts business and operational outcomes; AI abstracts application programming languages and analytics. By itself, abstraction is a good thing: It enables unprecedented speed and insights. However, such abstraction has also misled business and IT executives to believe an overly simplistic notion that all change is now simple and linear. Nothing can be further from the truth when it comes to transformative change that enables outsized value outcomes.

Transformative change requires the complete multidimensional picture: not just the *how* with cloud, data, and AI, but the *why* (the business vision and strategic priorities), the *what* (the business processes and business context), the *who* (the people, AI agents, and culture), and the *when* (the time horizons of change), plus the acceptance of rapid perpetual change. Nothing is permanent, and nothing is a given. To put it simply, this complete multidimensional picture can be referred to as *operating model change*.

In partnership with the World Economic Forum, McKinsey analyzed the Global Lighthouse Network (GLN) companies, who successfully integrate advanced technologies at scale to realize significant economic and financial benefits. McKinsey found that for every two dollars spent on tech innovation, these companies spent a total of eight dollars on enabling

1 Christian Jansen, Daniele Chiarella, Kim Baroudy, Lari Hämäläinen, Lieven Van der Veken, Ruben Schaubroeck, Sébastien Lacroix, Stéphane Bout, and Guillaume Dagorret, *The Symbiotic Enterprise*, McKinsey & Company, June 18, 2026.

operating model change: three dollars in process debt reduction and five dollars for scaling and adoption.[2] This is in stark contrast to the typical AI program or ERP-enabled transformation, which tends to significantly underinvest in operating model change.

Organizations that pursue operating model change are generally those that can successfully unlock outsized value from cloud, data, and AI. That is why this book is valuable and different. This is not primarily a book about technology. It is a book about how enterprises operate, make decisions, and respond to change.

What distinguishes this book is its practical approach to a topic that is often discussed in theoretical terms. The authors connect strategy, operating model design, enterprise architecture, governance, change management, and value realization into a coherent framework. Rather than presenting autonomy as a future vision, they explain how organizations can systematically progress from today's operating models to ones that are increasingly event-driven, data-informed, and capable of acting at scale.

This book takes you inside a global consumer goods company called Korindal, where this discipline plays out in real time. You'll see how a CFO attacked trapped working capital, how supply chain improved fulfillment performance, and how IT shifted from maintaining complexity to enabling continuous innovation, not as abstractions, but as operating changes that delivered measurable value in months.

For CEOs, CFOs, COOs, CIOs, and transformation leaders, this book offers a practical perspective on one of the most important questions facing organizations today: How do we turn information into action faster, more consistently, and at greater scale?

The organizations that answer that question well will define the next generation of enterprise performance. The central argument is simple: Operating model change creates value, and technology enables it.

Darwin Deano
Distinguished Partner, McKinsey & Company

2 Dinu de Kroon, Enno de Boer, and Rahul Shahani, *Lighthouse Lessons: Four Mindsets to Make Digital Transformation Stick*, McKinsey & Company, April 23, 2025.

Preface

On a Tuesday in October, a port in Rotterdam began to slow down. Not because of a strike or a shutdown; nothing dramatic enough to make the news. A weather advisory held two vessels at anchorage; a customs backlog quietly added three hours to processing times; a crane crew rotation ran late: Each delay was minor on its own. Together, they added eighteen hours to a container shipment carrying components that a manufacturer in Stuttgart needed on Thursday morning to start a production run.

Nobody at the manufacturer knew. The shipment still showed as *in transit*.

By the time someone pieced the picture together, cross-referencing the port status, the logistics schedule, the production calendar, and the customer commitments downstream, the damage had already traveled the full length of the supply chain. The Stuttgart plant was facing a two-week production gap, a retailer in Chicago was heading into its busiest quarter short on inventory, and a CFO in Singapore was revising earnings guidance from a spreadsheet, three days after the moment that revision should have happened.

Here is the part that stings: The data existed. Every signal was there, sitting in various systems, perfectly accurate and completely disconnected. The problem was never information. It was the cost of turning information into action: the hours of human effort required to find the signals, connect them, understand what they meant together, get the right people in a room, agree on a response, and move. By the time all of that happened, the window had closed. Nobody had done anything wrong; the people involved were experienced and working hard. There were simply more signals than any team could process, more connections than any individual could hold, and less time than the situation required.

This is the defining operational challenge of the modern enterprise, and it has nothing to do with data quality, system integration, or digital transformation budgets. Most large organizations today have more data than they have ever had. They have dashboards, analytics platforms, integrated enterprise resource planning (ERP) systems, and real-time reporting. What they do not have is enough human attention to act on all of it, fast enough, every time. That constraint is what this book is about.

The premise of the Autonomous Enterprise is straightforward: AI agents can absorb the operational complexity that has always outpaced human capacity. The routine exceptions, the cross-functional coordination, the decisions that follow clear and repeatable logic—agents can handle these continuously, at scale, and without the lag that comes from routing work through human schedules and inboxes. What that frees up is not headcount. It is judgment. The experienced supply chain manager who spent three days managing the Rotterdam fallout can instead spend those three days on the supplier relationship, the strategic resourcing decisions, the questions that actually required her expertise.

The Stuttgart story, played out differently, looks like this: The moment the port began to slow, an AI agent was already working on the problem. By the time the supply chain manager arrived at her desk, the rerouting had been evaluated, the production schedule adjusted to protect the highest-priority orders, the retailer updated, and the financial impact already reflected in the working capital model. She received a summary: Here is what happened, here is what was done, here is the one decision that needs your judgment because it involves a supplier relationship the policy does not fully cover.

Meanwhile, the CFO in Singapore did not wait for the quarterly review. The system had already revised the working capital model, flagged the revenue exposure, and updated the hedge position automatically for the CFO's review, while the supply chain team was still reading the situation.

One agent, taking minutes, not days. The exception was handled, the human called in for exactly the judgment that required a human. That is the Autonomous Enterprise, and that is what this book is about.

Who This Book Is For

This book is written for executives, architects, and transformation leaders facing an operational paradox: Your organization has invested heavily in digital infrastructure, analytics, cloud platforms, and AI capabilities, yet you still struggle with the same operational constraints that have always plagued large enterprises: trapped working capital, supply chain disruptions, slow decision cycles, and decisions that arrive too late to matter.

If you are a chief information officer (CIO) or enterprise architect, this book is essential. You are tasked with building systems that can scale, innovate continuously, and integrate AI without destabilizing your core operations. The clean core principle is your mandate: Preserve the integrity of your ERP and mission-critical systems while enabling rapid innovation at the edges. But how do you actually achieve that? How do you govern AI agents, manage extensibility through SAP Business Technology Platform (SAP BTP), enable event-driven architecture, and maintain the architectural discipline that prevents technical debt from spiraling? This book provides the framework. Chapters 7 through 11 establish the platform architecture that makes autonomous enterprise technically sound, and Chapter 15 shows you how to govern and roadmap these capabilities in an integrated system, not as point solutions.

If you are a technology leader responsible for applications, integrations, or data platforms, you will find practical patterns for how SAP Cloud ERP, two-tier ERP strategies, line-of-business (LoB) applications, and SAP Business Data Cloud work together to enable real-time, intelligent operations. You will understand how to sequence these capabilities so that the foundation comes first, enabling the layers to come next, and allowing autonomous behaviors to emerge from a stable base, not from fragile, hastily assembled integrations.

If you are a CEO, COO, CFO, or chief supply chain officer (CSCO), you need to understand how operating model change, not technology investment alone, unlocks outsized value. You need practical approaches to accelerate decision cycles, reduce the human effort required to turn signals into action, and achieve competitive differentiation through operational excellence. The Korindal case study and industry-specific scenarios ground this for you.

Across all these roles, the unifying challenge is the same: Your organization has more information than it has ever had. What it lacks is the operational capacity to act on all of it, continuously, at the speed business demands. This book shows you how to solve that problem, not through more dashboards or bigger data lakes, but through a systematic approach to architecture, automation, governance, and operating model change. You will find practical frameworks, real-world case studies, and sequenced implementation guidance that treats the Autonomous Enterprise not as a future vision but as an achievable, measured transformation.

How This Book Is Structured

This book is divided into three parts: Part I: SAP Autonomous Suite; Part II: SAP Business AI Platform; and Part III: Transformation and Execution.

Part I begins your journey through SAP Autonomous Suite. It includes the following chapters:

- **Chapter 1: Introducing the Autonomous Enterprise: Apps, Data, and AI**
  This chapter introduces the Autonomous Enterprise as a unified model combining applications, data, and AI to drive intelligent, real-time business operations. It explains the shift from traditional ERP to a cloud-native, AI-infused architecture, and outlines the roles of core applications, SAP BTP, and SAP Business Data Cloud. The chapter also presents SAP Business AI as the intelligence engine.
- **Chapter 2: SAP Cloud ERP Business Capabilities**
  This chapter introduces the core business capabilities of SAP Cloud ERP across private and public cloud landscapes. It establishes SAP Cloud ERP as the digital foundation of the intelligent enterprise, highlighting how standardized best-practice processes enable operational excellence and scalability. The chapter explores key functional domains including finance, supply chain, procurement, and manufacturing.
- **Chapter 3: Two-Tier ERP Landscapes**
  This chapter explores how companies use a two-tier ERP strategy to run one main ERP system at headquarters (Tier 1) and a lighter, more flexible cloud ERP system for subsidiaries or new business units (Tier 2). It shows how this model helps organizations standardize global processes while still giving localized processes the speed and agility they need.

- **Chapter 4: LoB Apps and Integrations**
  This chapter explores how LoB applications extend the value of SAP Cloud ERP, delivering specialized capabilities for critical business domains while maintaining a clean core architecture. It provides a structured view of four major LoB areas—supply chain, finance and spend, customer experience, and human resources—and their deep integration with SAP Cloud ERP.
- **Chapter 5: Embedded AI in Business Processes**
  This chapter focuses on the intelligent technologies that elevate SAP Cloud ERP beyond transactional efficiency. It explains how embedded and generative AI, automation, predictive insights, and more augment decision-making and streamline operations, organized by finance, supply chain and operations, procurement, and other business areas.
- **Chapter 6: Industry-Specific Autonomous Scenarios**
  This chapter highlights industry-specific use cases that demonstrate how the Autonomous Enterprise drives transformation across key sectors. Each section focuses on real-world scenarios and KPIs that matter most to industry leaders.

Part II dives into the underlying platform, with a focus on the SAP Business AI capabilities that enable the Autonomous Enterprise:

- **Chapter 7: The New Platform**
  This chapter provides an overview of SAP Business AI Platform, the foundation for building, contextualizing, and governing AI across the SAP landscape. By the end of this chapter, you will understand the key components of the platform and how they work together to enable scalable, responsible, and business-driven AI.
- **Chapter 8: SAP Business Technology Platform**
  This chapter introduces SAP BTP as the foundation for the Autonomous Enterprise, enabling a clean core architecture and seamless extensibility. It explains how SAP BTP underpins application integration, data management, and AI enablement across the suite. You will learn about SAP Build, runtime choices for flexibility, and extensibility patterns that preserve system integrity while accelerating innovation. The chapter also explores platform capabilities, including integration services, event-driven architecture, and lifecycle management.
- **Chapter 9: SAP Business Data Cloud**
  This chapter introduces SAP Business Data Cloud as the data backbone of the Autonomous Enterprise, enabling trusted, governed, and AI-ready data across the enterprise. It begins by describing the architecture and explaining its role in harmonizing transactional and analytical data for real-time insights. You'll learn about setting up SAP Business Data Cloud, configuring its environments, and integrating it seamlessly with SAP Cloud ERP and LoB applications. You'll also learn about embedded AI capabilities and relevant solutions, features, and use cases.
- **Chapter 10: Extensibility and Agent Development**
  This chapter explores how organizations can extend SAP Cloud ERP while preserving a

clean core and safeguarding long-term innovation. It explains the evolution from traditional on-premise modifications to clean core extensibility models designed for cloud resiliency. This chapter offers an understanding of the extensibility spectrum, including key-user adaptation, developer extensibility, side-by-side extensions on SAP BTP, and integration patterns using APIs and events.

- **Chapter 11: Integration Capabilities**
  This chapter explains how SAP Integration Suite serves as the backbone for connecting SAP and non-SAP systems, enabling a harmonized, event-driven architecture across the enterprise. By the end, you'll understand how SAP Integration Suite and SAP Integration Suite, advanced event mesh deliver flexible, intelligent, and future-ready integration capabilities that underpin the Autonomous Enterprise.

Finally, Part III concludes the book with a look at how business transformation is achieved with SAP Autonomous Suite:

- **Chapter 12: Developing a Business Case for Your Autonomous Enterprise**
  This chapter explains how to build a compelling business case for adopting the Autonomous Enterprise. You'll learn how to quantify benefits, align investments with transformation goals, and present a CFO-ready case for executive approval.
- **Chapter 13: Leveraging Tools for Transformation**
  This chapter explains how SAP's agent-led toolchain accelerates business transformation by connecting process design, architecture, and delivery. It introduces key tools and their roles in enabling process intelligence, portfolio management, testing automation, data migration, and content governance. The chapter concludes with a look at toolchain convergence, showing how these solutions work together to provide end-to-end visibility and control for transformation programs.
- **Chapter 14: AI Governance and Responsible AI**
  This chapter addresses the governance and trust framework for AI within the Autonomous Enterprise. You'll gain practical guidance on implementing AI ethically and securely while meeting global regulatory standards.
- **Chapter 15: Operationalizing Your Vision**
  This chapter provides a practical framework for consolidating all elements of the Autonomous Enterprise into a cohesive transformation plan. It covers deployment strategies and instance planning, enterprise architecture and readiness tools, and steps to build a multiyear roadmap aligned with your business priorities. The chapter concludes with a value realization plan, ensuring organizations can measure outcomes and sustain benefits throughout their transformation journey.
- **Chapter 16: Change Management**
  This chapter emphasizes why change management is critical for successful Autonomous Enterprise adoption. It outlines strategies to prepare teams for transformation. By the end, you'll understand how structured training and proactive change management ensure faster adoption and sustained business value.

- **Chapter 17: From Analog Transactions to Autonomous Value**
  This chapter looks ahead to the evolution of the Autonomous Enterprise, from legacy tools to future functionality geared toward achieving autonomous business outcomes. You'll gain insight into emerging patterns, governance considerations, and practical steps to prepare for an autonomous future.

Acknowledgments

We extend our deep gratitude to Björn Ganzhorn, Head of Customer Advisory, SAP, for his unwavering vision and inspiration in guiding us toward bringing the Autonomous Enterprise concept to our customers in the field.

We are grateful to Sutej Kapur, Raj Singh, and Jon Sigler from SAP Value Advisory for their thoughtful review of our business case chapter (Chapter 12) and flywheel model content. Their expertise and constructive feedback strengthened both the rigor and practical relevance of this book.

We thank Muhammad Alam, Member of the Executive Board, SAP, and Suzanne Cauthorn, Chief of Staff for Product & Engineering, SAP, for their executive sponsorship and steadfast support throughout this project. Their belief in the importance of this work enabled us to bring it to completion.

We also thank Martina Bahrke and Jens Weiss from SAP for their guidance and support throughout the publishing process.

Finally, we are deeply grateful to the families of all the authors for their patience, understanding, and unwavering support during the many evenings, weekends, and hours required to bring this book to life. This book is as much a reflection of their encouragement and sacrifice as it is of our collective effort.

Conclusion

The Rotterdam scenario is not hypothetical. A port delay becomes a production disruption, and the CFO has to revise earnings guidance. This is how most enterprises operate today. The signals exist and the data is accurate, but what does not exist is the speed to translate information into action at the scale business demands.

The real constraint is not technology. It is how work moves through an organization. A decision requires sign-off from three people in different time zones; an exception is escalated, reviewed, and debated; cross-functional alignment takes weeks. By then, the moment has passed.

In many organizations, SAP already serves as the system of record for finance, supply chain, procurement, manufacturing, and workforce operations. Autonomy is most valuable where decisions affect business outcomes, and those decisions depend on business context, transactional data, and cross-functional processes. The closer intelligence moves to the systems where work is executed, the greater its potential impact.

The Autonomous Enterprise addresses this by freeing people from routine decisions. Intelligent agents handle exceptions continuously. They coordinate across domains. They act while the window is still open. Your teams focus on judgment, not firefighting.

This book explores the foundation required to make that transition possible, from SAP Cloud ERP and SAP Business Data Cloud to SAP BTP and SAP Business AI, and how these capabilities work together to enable the Autonomous Enterprise.

The question is not whether these problems can be overcome. The question is how your organization begins: Where do you start? How do you sequence the work so that the foundation comes first and each phase delivers value?

That is what this book is about, beginning in Chapter 1.

Chapter 1

Introducing the Autonomous Enterprise: Apps, Data, and AI

We'll begin our journey with a first look at the Autonomous Enterprise with SAP: the autonomous suite, the data foundation, the intelligence engine, and more.

Most enterprise transformations start with conviction and end with compromise. Systems go live; dashboards multiply; early wins are communicated confidently. Then the environment shifts. Supply chains tighten; customer expectations accelerate; a competitor moves. The organization discovers that its newly modernized systems are still organized around the same fundamental assumption: that humans will do the sensing, the deciding, and the connecting.

This book begins with a different premise. The next generation of enterprise performance will not be defined by faster transactions or cleaner data alone. It will be defined by how well an enterprise can perceive what is changing, decide what matters, and act inside its processes—continuously, at scale, and with increasing autonomy.

We call this operating pattern the sense-decide-act loop. If this loop is slow or fragmented, value erodes when it matters most. If it is tight, continuous, and AI-driven, then it compounds each cycle, producing better execution, richer data, and smarter decisions than the one before.

The Autonomous Enterprise is SAP's answer to this situation. It brings together cloud applications, a unified business data foundation, and deeply embedded AI to create an enterprise that does not merely run processes but continuously learns from them, adapts them, and, in many cases, executes them on your behalf.

This chapter lays out the foundation. It starts with the mindset shift required for autonomous operations, then introduces the flywheel that connects applications, data, and intelligence. The next sections walk through the architecture: applications, data, platform, and AI. It then shows how organizations move from augmentation to autonomy and grounds it in the Korindal case studies that continue through the book.

1.1 The Cloud and Autonomous Mindset

If mindset were irrelevant, every organization that deployed the same technology would get the same results, but they do not. Two companies can implement identical enterprise software, spend comparable budgets, and run comparable implementation programs—

and five years later, one has built a genuinely adaptive operation while the other is still wrestling with the same bottlenecks it had before. The mindset question is not "Have we adopted the latest technology?" It is something harder: "Have we changed the way we actually run?"

As shown in Figure 1.1, the path to the Autonomous Enterprise runs through a progression of mindsets, each representing a fundamentally different relationship between the organization and its technology.

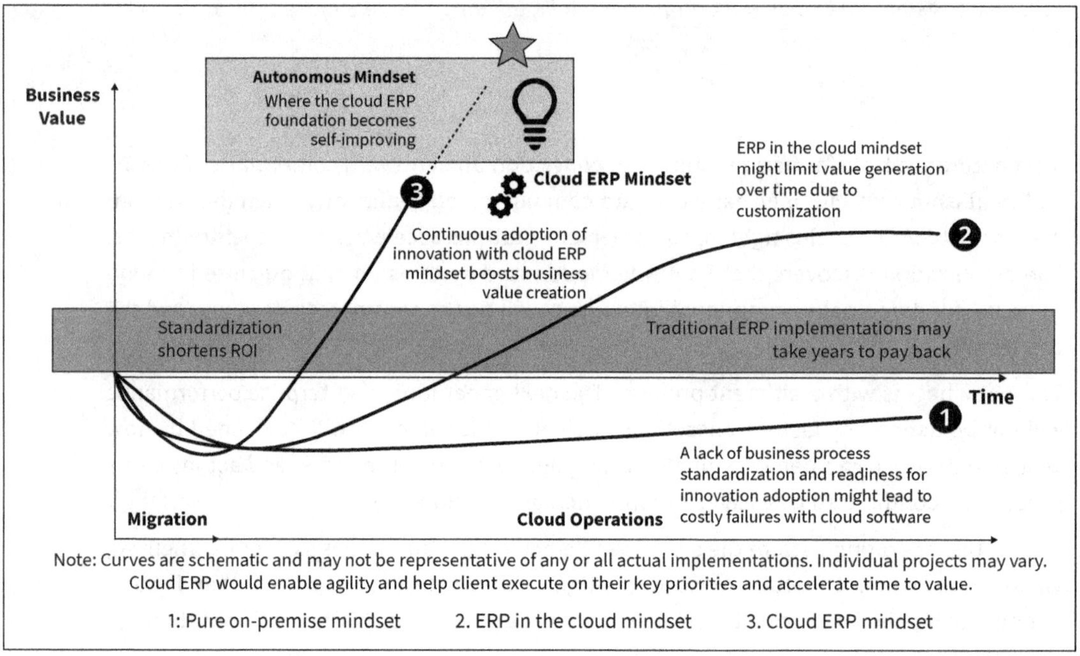

Figure 1.1: Cloud Mindset

Let's take a closer look at each mindset:

- **Purely on-premise mindset**
  The organization moves infrastructure to the cloud but replicates everything it did before inside the new environment. Technical debt relocates rather than dissolves. The cloud's most powerful capabilities go entirely untapped, and business value is slow to materialize or never arrives at all.
- **Enterprise resource planning (ERP) in the cloud mindset**
  Modern cloud platforms are adopted but heavily customized to mirror legacy workflows. Early value is captured, but each customization quietly narrows the organization's ability to absorb future innovation. Over time, the system becomes harder to update and the value curve flattens.
- **Cloud ERP mindset**
  The organization makes a genuine operating model shift: configuring rather than customizing, adopting built-in best practices, and measuring success by speed of innovation

rather than fidelity to legacy processes. This is where the value curve accelerates and the foundation for autonomous operations is established.

- **Autonomous mindset**
  The fourth mindset is the one this book is about, and it is less a technology decision than a deliberate choice about how the enterprise operates. Organizations here are not just running clean, current systems. Instead, they are actively redesigning how work is done: identifying where AI agents can handle operational complexity automatically, where human judgment adds its greatest value, and how the two work together to create an operation that continuously improves itself. The autonomous mindset does not arrive automatically after the cloud ERP mindset; it must be chosen.

What does this choice look like in practice?

Consider how most enterprises handle exceptions today: A shipment is delayed; an invoice does not match a purchase order; a production order cannot be fulfilled; a budget variance crosses an approval threshold. Each exception triggers a chain: notification, investigation, escalation, decision, communication, resolution. The chain works eventually, but it introduces lag at every link. In a competitive environment, that lag compounds.

An organization with an autonomous mindset looks at that same exception chain and asks a different question, not “How do we handle this faster?” but “Why is this being handled this way at all?” Is the resolution logic knowable? If a policy or decision rule determines the right response, then the process should run automatically, immediately, and with full auditability. Human judgment is brought in for the situations that genuinely require it: the exceptions for which context, relationships, and experience shape an outcome that no rule can fully anticipate.

This is the operating model shift at the heart of the Autonomous Enterprise, and it is available today, not at some future point when the technology matures. The organizations already making this shift are not doing it because they have bigger budgets or more advanced teams but because they made a deliberate decision to think differently about how their operations should run and then built the systems and governance to match that decision.

That combination of mindset, operating model, and technology is what separates the organizations pulling ahead from those still waiting for their transformation investment to pay off.

1.2 The Autonomous Enterprise Flywheel

Every enduring business advantage has one thing in common: It compounds. The longer an advantage runs, the harder it becomes to replicate because the gap between an organization that has it and one that does not grows with every passing cycle. Cost advantages, customer relationships, and operational learning compound.

The Autonomous Enterprise is designed around exactly this principle. Instead of a set of capabilities sitting side by side, the architecture is a flywheel, a self-reinforcing cycle in which each element continuously strengthens the others and the value produced in each cycle becomes the input that makes the next cycle better.

Understanding the flywheel is the key to understanding why the Autonomous Enterprise is not just a more efficient version of what organizations do today. It is a fundamentally different operating dynamic.

The flywheel has three elements, as shown in Figure 1.2: applications, data, and AI. None of them create the full value of the flywheel in isolation. Each depends on the other two, and the real power emerges from the way they interact.

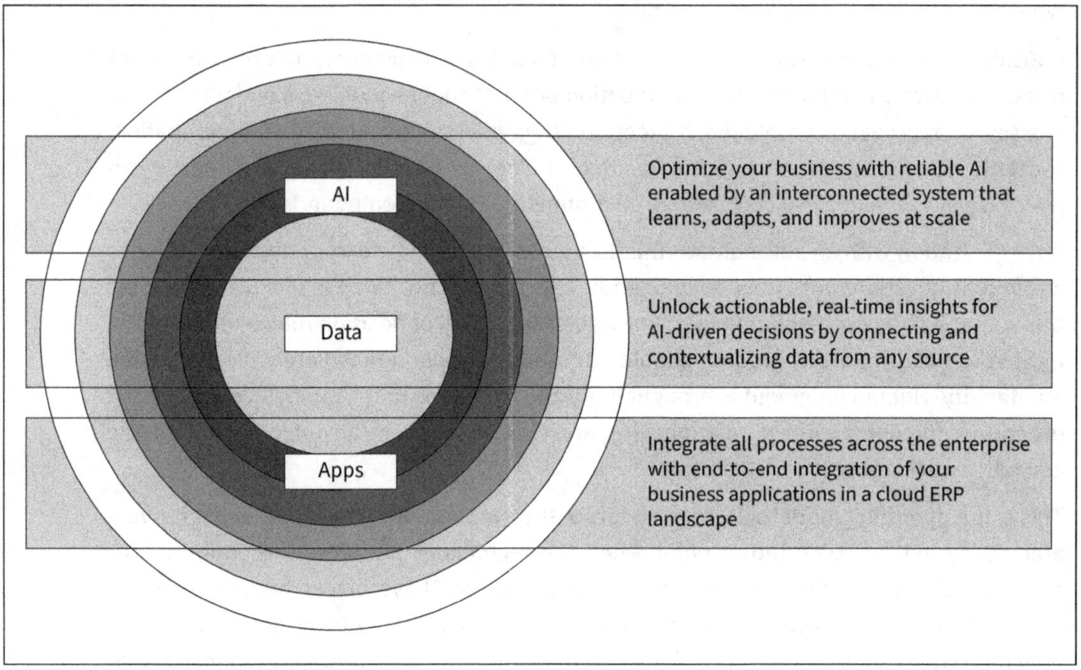

Figure 1.2: Autonomous Enterprise Flywheel: Applications, Data, and Intelligence in Continuous Self-Reinforcing Motion

Applications are where the enterprise operates: where orders are placed, invoices are processed, inventory moves, suppliers are managed, and customers are served. When those applications run on standardized, integrated processes, they do something that is easy to overlook: They generate a continuous stream of high-quality operational data as a natural by-product of execution. This is not data that has to be extracted, cleaned, and prepared before it can be used but data that is structured, consistent, and ready because the processes that produced it were designed that way.

Data is the connective layer. Data in the way most organizations have experienced it is siloed in systems, inconsistently defined, and requires months of preparation before it can support an AI initiative. In the flywheel model, data instead is unified across the enterprise and understood in the business context. A transaction is not just a number; it carries the business meaning attached to it, the process it belongs to, the entity it relates to, and the outcome it is connected to. That richness is what makes the third element possible.

Intelligence—the AI layer—takes that contextually rich operational data and does what no human team can do at scale: finds the patterns, surfaces the signals, generates the recommendations, and in many cases executes the response. Each cycle of intelligent action produces outcomes that feed back into the applications as better execution: cleaner processes, faster resolutions, and fewer exceptions. Better execution produces better data, and better data produces sharper intelligence. Thus, the loop accelerates.

How is this different from what came before?

Most enterprise architectures today are sequential, not circular: Applications run; data is extracted; analytics are produced. Humans review the analytics and decide what to do. The next execution cycle begins largely unchanged by what was learned in the previous one. Insight is always arriving after action, so learning never quite catches up with operation.

The flywheel breaks that sequence. Intelligence is not downstream of execution but embedded within it. Learning and operating happen in the same cycle, which means that every execution is informed by everything the system has learned from every execution before it. Beyond simply running its processes, the enterprise is continuously improving them.

Consider what this means over time. An organization running the flywheel for three years has accumulated three years of operational learning embedded directly into how its processes execute. Its exception rates are lower because the system has learned which patterns predict exceptions and acts before they occur. Its decisions are faster because the resolution logic has been refined through thousands of prior cycles. Its AI is more accurate because it has been trained on three years of high-quality, contextually rich operational data from its own business.

An organization that starts the same journey three years later is three years behind on technology *and* three years behind on learning. That gap does not close quickly.

The flywheel does not start itself. An organization with a purely on-premise mindset cannot run the flywheel; its processes are too fragmented and its data too inconsistent to generate the quality of input the intelligence layer needs. An organization with a heavily customized cloud system can start the flywheel but will find it turning slowly; the customizations introduce inconsistency that degrades data quality and limits what AI can reliably do with it. An organization that has made the cloud ERP mindset shift has built the foundation the flywheel requires: standardized processes, clean data, and a platform designed to absorb intelligence continuously.

An organization that has embraced the autonomous mindset does one thing more: It deliberately designs its operations around the flywheel. It asks, at every step, how the intelligence layer can be applied to improve execution, how execution can be designed to produce better data, and how that data can be used to make the intelligence sharper. It treats the flywheel as the central organizing principle of how the enterprise runs.

That decision to build around the flywheel rather than alongside it is what separates the organizations compounding value from the ones still waiting for their transformation to pay off.

Figure 1.3 shows the Autonomous Enterprise architecture, positioning the flywheel of applications, data, and intelligence as the operating core of a fully autonomous business.

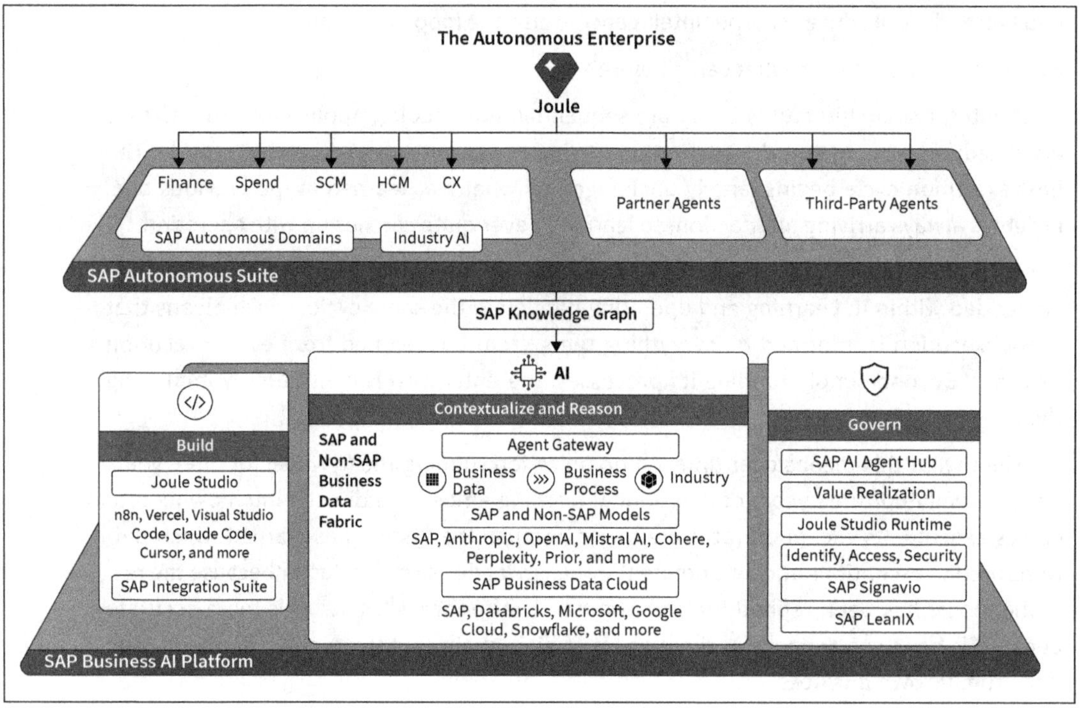

Figure 1.3: The Autonomous Enterprise

As shown in Figure 1.3, the system has three distinct layers, each with a precise role. Joule sits at the top as the engagement layer, the single place where people set intent and direct the business, with assistants and agents bringing together the right data, workflows, and actions across every system. SAP Autonomous Suite forms the operational core, organizing the enterprise into five autonomous domains: finance, spend, supply chain, human capital management (HCM), and customer experience, each running from end to end with agents, applications, and data working as one. SAP Business AI Platform sits underneath everything as the foundation: combining deep process context, unified business data, purpose-built models, and enterprise-grade governance so that the layers above it can act with both speed and trust. The following sections examine each layer in turn.

1.3 Applications: SAP Autonomous Suite

This section covers the application layer of the Autonomous Enterprise architecture. We'll first describe the SAP Cloud ERP system as the transactional foundation, then we'll explain how line-of-business (LoB) applications extend that foundation into the five autonomous domains introduced in Section 1.2. Later chapters will examine the deployment patterns and configuration choices for each domain.

Applications are the starting point of the flywheel because they are where the actual work of running a business happens—for example, orders placed, invoices processed, employees hired, and customers served. The data that feeds the intelligence layer comes primarily from these systems. Getting the application layer right and understanding how it is structured are the foundations for everything else in this book.

SAP Autonomous Suite is not a single product. It refers collectively to the SAP Cloud ERP system together with a set of LoB applications that, taken together, cover the core operational processes of most large enterprises. The two layers serve different purposes but share a common data model and a common connection to Joule. When AI agents operate across these applications, they work within a consistent environment rather than bridging isolated systems.

The five domains (autonomous finance, spend, supply chain, HCM, and customer experience) are what you get when both layers are running together with agents active across them. A *domain* is the combination of the ERP core processes, the relevant LoB applications, and the agents that coordinate work across both. Keep that distinction in mind as you read the rest of this chapter.

1.3.1 SAP Cloud ERP

SAP Cloud ERP, built on the SAP S/4HANA Cloud software, is the transactional backbone of the suite. It handles the high-volume processes that define day-to-day business operations: posting journal entries, processing purchase orders, managing inventory movements, running payroll, closing the books at period end, and so on. For most large organizations, most of their transaction volume runs through this system. Everything the AI agents do in the autonomous suite ultimately connects back to data and processes that originate here.

One reason this matters for autonomous operations is that SAP S/4HANA Cloud natively carries business context, not just data. A goods receipt in the system is not just a quantity and a date. The system knows which purchase order it closes, which supplier it came from, which plant received it, which cost center it affects, and what payment terms govern the resulting liability. An agent operating inside that environment does not need to reconstruct that context from separate tables or an analytics layer. It is already there because the transaction was modeled that way. Agents that work with extracted data must rebuild that context from scratch. At the scale of thousands of daily transactions, that difference is significant.

For agents to work reliably, the environment they work in needs to be consistent. This is the practical rationale behind what SAP calls the *clean core* principle. An SAP S/4HANA Cloud system that has been heavily modified at the code level would have custom logic (modifications to core data structures) written into standard process flows, and that introduces variability that is hard for an AI system to reason across. The same process runs slightly differently in different parts of the system. Data structures that should be identical have small inconsistencies. An agent encounters edge cases that its underlying model did not expect, because those edge cases exist only in that organization's specific customizations.

SAP Cloud ERP is available in two configurations. *SAP S/4HANA Cloud Public Edition* is fully managed by SAP, updated twice per year, and built to SAP's standard process definitions. It is the fastest path to a current, consistently maintained system and one in which SAP's agents are built to work out of the box. *SAP S/4HANA Cloud Private Edition*, delivered through RISE with SAP, allows for greater process customization and gives organizations more control over the update scheduling that is relevant for complex implementations with significant existing process investment. Both support autonomous operations; the difference is in how quickly and uniformly the AI layer can be activated across the landscape.

1.3.2 Line-of-Business Applications

SAP Cloud ERP covers the transactional core, but some business functions need more functionalities than a core system provides. Strategic procurement, talent management, advanced supply chain planning, and customer engagement each involve processes and data requirements that go beyond what SAP S/4HANA was designed to handle in depth. SAP's LoB applications fill those gaps:

- SAP Ariba for sourcing and supplier management
- SAP SuccessFactors for the full employee lifecycle
- SAP Integrated Business Planning for Supply Chain (SAP IBP) for demand sensing and inventory optimization
- SAP Customer Experience (SAP CX) suite (SAP Sales Cloud, SAP Service Cloud, SAP Commerce Cloud, and SAP Marketing Cloud) for customer-facing processes

In a conventional setup, these applications share data with SAP S/4HANA Cloud through scheduled integration: Contracts move in a defined cadence; headcount synchronizes overnight; order statuses update periodically. In the autonomous suite, they share context, and agents work across both layers continuously. A contract finalized in SAP Ariba flows into SAP S/4HANA Cloud through a standard integration, where accounts payable agents act on the resulting invoices and open items. An offer accepted in SAP SuccessFactors triggers onboarding steps that reach into SAP S/4HANA HR through a configured integration sequence. The agents operate on the data available within their domain; the integration layer that connects LoB applications to the ERP core is part of the architecture. Together, the ERP core and the relevant LoB applications form the operating environment for each autonomous domain.

This is what makes each autonomous domain coherent in practice:

- **Autonomous Finance**
  Draws on SAP S/4HANA core financials for accounting, period close, and reporting, extended by SAP applications for group consolidation, treasury management, and tax compliance. Agents handle intercompany reconciliation, cash positioning, and close tasks across the full application landscape. The Financial Close Assistant, for example, does not just work within a single module but coordinates the sequence of steps that spans the general ledger, accounts payable, accounts receivable, and consolidation, all of which may sit in overlapping SAP systems.
- **Autonomous Spend**
  Combines SAP S/4HANA Cloud procurement including requisitions, purchase orders, goods receipts, and invoice processing with SAP Ariba's sourcing and contract capabilities. The procurement cycle that previously required coordination across multiple systems is being progressively unified through agent-based automation, such as the Bid Analysis Agent, which handles the aggregation and comparison work that sourcing analysts previously did manually; the Sourcing Negotiation Agent, which conducts supplier negotiation within customer-defined parameters; and the Autonomous Sourcing Agent, which orchestrates the end-to-end sourcing sequence by coordinating the other two.
- **Autonomous Supply Chain Management**
  Extends SAP S/4HANA's planning, production, and logistics processes with SAP IBP for demand sensing and inventory optimization and with SAP Transportation Management (SAP TM) for complex logistics scenarios. When a demand signal shifts, the planning agents accelerate the decision cycle by evaluating schedule impact, proposing production adjustments, and flagging logistics implications, without waiting for the next planning run. The architecture is designed to keep planning, production, and logistics in the same decision context rather than passing information across systems sequentially.
- **Autonomous HCM**
  Brings together the SAP S/4HANA HR core with SAP SuccessFactors for talent management. A workforce planning scenario that previously required an HR analyst to cross-reference the headcount system with the talent management database becomes a query the agent resolves directly, with access to both at once. Onboarding workflows that involve HR data, system access provisioning, payroll setup, and compliance training assignment run as a coordinated agent sequence rather than a checklist tracked by a human coordinator.
- **Autonomous CX**
  Connects the SAP CX suite to the transactional reality of SAP S/4HANA. A service agent handling an escalation has the customer's order history, outstanding invoices, and contract status available in context not because someone pulled it together beforehand, but because the agents work in an environment where that information is already assembled.

This is the practical meaning of an autonomous domain: Rather than being a new system, existing systems such as the ERP core and LoB applications work together through a shared data foundation, and agents coordinate across them continuously. The data layer and AI platform that make this possible are covered in the sections that follow.

1.4 Data: The Business Data Foundation

This section describes the data layer of the Autonomous Enterprise, the second element of the flywheel. It includes SAP Business Data Cloud and the partner integrations that extend it.

In the flywheel model, applications generate data and intelligence acts on it. However, there is a step in between these actions that is easy to underestimate: Data must be usable before it is useful. A goods receipt in SAP S/4HANA, a supplier contract in SAP Ariba, a headcount record in SAP SuccessFactors: Each of these carries operational meaning in its native system. The challenge is that an AI agent trying to reason across all three does not automatically know that the supplier on the contract is the same entity as the vendor on the goods receipt, or that the headcount change affects the cost center that owns the purchase order. Without a data layer that resolves those connections, agents produce answers that are technically correct and practically incomplete.

SAP Business Data Cloud is the data foundation designed for this environment. It is not a data warehouse in the traditional sense, and it is not a data lake. It is an architecture that brings together lake storage, compute runtimes, a semantic layer, and master data governance into a single, managed environment. The structure is shown in Figure 1.4.

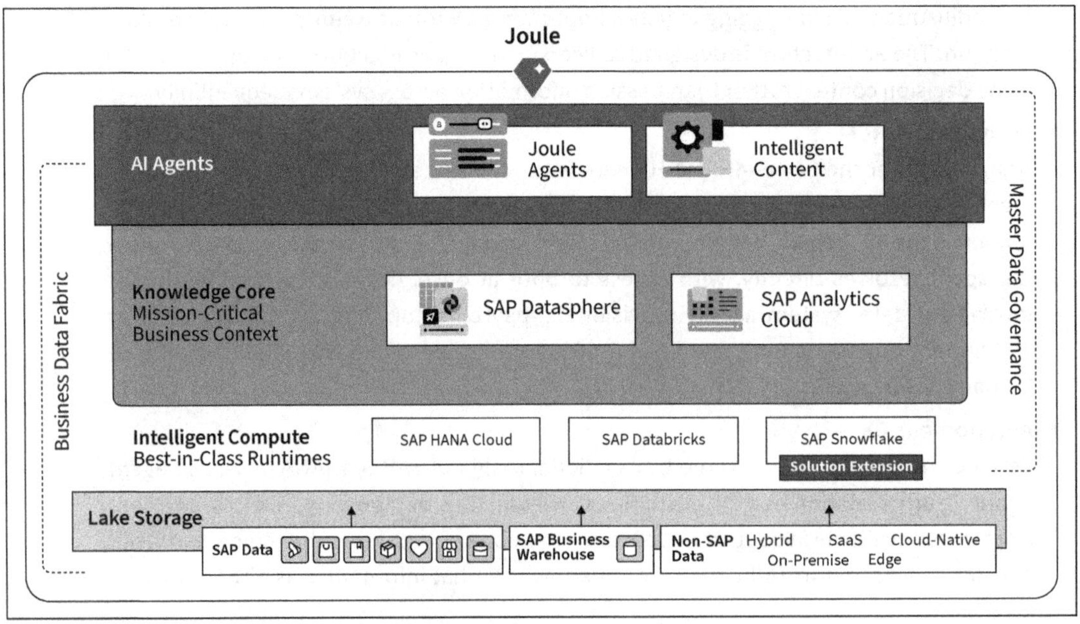

Figure 1.4: SAP Business Data Cloud: Strategy for Enterprise Data

At the base is *lake storage*, where data from SAP systems, SAP Business Warehouse (SAP BW), and non-SAP sources lands and is retained. Non-SAP data can come from any environment: hybrid, software as a service (SaaS), cloud-native, on-premise, or edge. The mechanism for bringing it in without replication is SAP Business Data Cloud Connect, which provides zero-copy data sharing while preserving the business context attached to the original source. Data does not need to be moved and reformatted before it becomes visible. It is accessed in place, governed from a central point, and made available to the layers above.

Above storage sits the *intelligent compute* layer. Different analytical and AI workloads have different runtime requirements, and SAP Business Data Cloud is designed to let each workload run on the engine best suited to it. SAP HANA Cloud handles in-memory, transactional, and real-time analytical workloads. SAP Databricks (the result of SAP's partnership with Databricks) offers machine learning (ML), large-scale data engineering, and notebook-based model development. Dremio provides high-performance *data virtualization*—that is, the ability to query across distributed sources without physically moving data, which is valuable when operational data needs to stay where it lives but still participate in cross-domain analysis. SAP Snowflake is available as a solution extension for organizations that have standardized it as their data warehouse. The common thread is that these runtimes all draw from the same underlying storage and operate under the same governance framework, so the choice of compute engine does not require rearchitecting the data products that sit above it.

The *knowledge core* is the layer that distinguishes SAP Business Data Cloud from a conventional data platform. It is built on SAP Datasphere and SAP Analytics Cloud, and its purpose is to attach business meaning to data, not just store or process it. SAP Datasphere manages the semantic layer: the business definitions, data models, active metadata, and governed data products that describe what data means in operational terms. SAP Analytics Cloud handles the analytical and planning workloads that consume that semantically enriched data. Together they form the environment in which data becomes understandable to agents not just as values in columns, but as entities with known relationships, known process context, and known business significance.

Example

When an agent queries accounts receivable aging, it does not just retrieve a list of open invoices by due date. Through the knowledge core, it understands that those invoices belong to specific customers, that those customers have credit limits and payment histories, that some have open disputes, and that their contracts contain specific collection terms. It can reason across all of that in a single pass because the semantic layer has made those relationships explicit. This is what the SAP Business Data Cloud source material describes as *universal business context*, and it is what separates an agent that can act reliably from one that generates plausible-sounding but unreliable answers.

Master data is handled separately but within the same governance boundary. Reltio, which SAP has integrated into the master data management layer, provides entity resolution and data quality management across business partners, customers, suppliers, and materials. The practical effect is that when an agent encounters a supplier record in one system and a vendor record in another, the master data layer has already resolved whether they are the same entity. Agents do not need to handle ambiguous identity matching at query time. That work is done upstream, consistently, with human-readable audit trails that support data governance requirements.

Across the right side of Figure 1.4, *master data governance* spans all four layers. This reflects an important architectural choice: Governance in SAP Business Data Cloud is not a separate tool bolted on after the fact. Data quality rules, lineage tracking, security controls, and lifecycle management apply at the storage layer, the compute layer, and the semantic layer simultaneously. An organization that has invested in data governance does not need to rebuild those controls for a new AI initiative. They carry through automatically to every workload that runs on the platform.

The result is a data foundation that can serve the full range of what the autonomous suite requires. Operational agents running in real time against SAP S/4HANA Cloud get the transactional context they need from the knowledge core. Analytical agents querying supply chain performance across systems get the cross-domain relationships that make their answers meaningful. Planning workloads in SAP Analytics Cloud run against the same governed data that operational agents use, which means that scenario models and live operations share a consistent view of the business. The data layer is not upstream of the AI layer; it runs alongside it, continuously, and in the same governed environment.

1.5 Platform: SAP Business Technology Platform as the Foundation

This section covers SAP Business Technology Platform (SAP BTP), which is the layer that connects applications, data, and AI to form a coherent operating environment. We will explain the role of SAP Integration Suite and the clean core extensibility model.

Applications and data are only as useful as the infrastructure that connects them. Most enterprise landscapes are not homogeneous: They include SAP systems running core processes; non-SAP applications managing customer relationships, logistics, or compliance; legacy platforms that are too embedded to retire quickly; and an expanding set of partner and industry tools. The question is not whether to integrate but whether integration is robust enough to support autonomous operations. Agents depend on live, reliable signals from every connected system to act without human mediation.

SAP BTP is the layer that handles this. It covers integration across all connected systems, the extensibility model that lets organizations adapt processes without modifying the core, the runtime infrastructure for AI and agents, and the shared security and governance framework that applies consistently across the landscape. These four responsibilities are not separate services that happen to share a brand name. They are designed to work together, because an Autonomous Enterprise needs all four operating in coordination.

Let's take a closer look at how integration and extensibility are supported by SAP BTP:

- **Integration at landscape scale**
  SAP Integration Suite is the integration layer within SAP BTP. It handles the exchange of data and events between SAP applications, non-SAP systems, external partners, and government-reporting endpoints. Its content library contains thousands of prebuilt business accelerators—SAP-maintained integration templates covering common scenarios across logistics, finance, HR, and procurement—and hundreds of third-party connectors for non-SAP applications. Organizations do not build those integrations from scratch; they start from tested templates and adapt where their processes diverge from the standard.

  For autonomous operations specifically, the event-driven architecture capability within SAP Integration Suite is significant. Traditional integration patterns move data on a schedule or in response to an explicit request. Event-driven integration distributes signals as they happen: a goods receipt posted, a payment cleared, a shipment delayed, a threshold breached. Agents subscribed to those events can respond immediately, without polling for updates or waiting for a batch run. The lag between signal and action that characterizes manual processes disappears when the integration layer is built around events rather than scheduled extracts. API Management within SAP Integration Suite governs how those signals are exposed: which systems can subscribe to which events and under what access controls, with full audit trails for every exchange.

- **Extensions without breaking the core**
  Every organization running SAP has processes that are specific to its business: a pricing rule that reflects a particular commercial model, a compliance check tied to a local regulation, an approval sequence that exists because of an acquisition three years ago. Historically, these were handled by modifying SAP source code. Such modifications worked until the next system update, at which point they either broke or required expensive reimplementation. The cycle repeated with each release, creating technical debt that made upgrades slower and riskier over time.

  SAP BTP addresses this through the clean core extensibility model. Custom logic runs as side-by-side extensions deployed on SAP BTP rather than as modifications to the SAP core. These extensions connect to SAP systems through stable, versioned application programming interfaces (APIs) that SAP maintains independently of the application release cycle. When SAP S/4HANA or another SAP application updates, the extension is unaffected because it does not depend on the internal code it was previously modifying. SAP Build provides the development environment for creating those extensions, with low-code tools for business-led customizations and pro-code runtimes, including Kyma and ABAP Cloud for complex scenarios that require full developer control. The practical result is that organizations can absorb SAP updates on a predictable schedule instead of deferring them because of customization risk.

The same platform also provides the runtime infrastructure for the AI layer covered in the next section, which means agents, assistants, and AI workloads run inside the same governed environment as everything else on SAP BTP, without requiring a separate trust framework.

1.6 SAP Business AI: The Intelligence Engine

The previous sections described what the Autonomous Enterprise runs on: applications organized into five autonomous domains, a unified data foundation that gives those applications shared business context, and a platform layer that connects and governs everything. This section describes what makes it intelligent.

SAP Business AI is the intelligence layer. It is not a standalone product or an add-on. It is the natively integrated suite of technologies that transforms SAP applications from systems of record into systems of action, where processes that once required human intervention at every step can now sense, reason, decide, and execute within governed boundaries.

The strategic premise is straightforward: Enterprise intelligence must live inside business processes, not beside them. When AI sits outside the operational core, organizations face the integration complexity, data disconnection, and governance gaps that keep many AI initiatives stuck in pilot mode with no path to scale. SAP starts inside the operational core and builds outward, grounding every AI capability in the semantically rich business data, deep process knowledge, and enterprise-grade governance that only the system of record can provide.

This architecture supports a dual-speed model. The first speed delivers out-of-the-box intelligence embedded directly into applications: predictive cash allocation in finance, demand sensing in supply chain, automated invoice extraction in procurement. These capabilities arrive ready to use and compound value without requiring custom development. The second speed enables competitive differentiation through custom AI built on SAP Business AI Platform, where organizations extend the foundation with proprietary models, custom agents, and industry-specific intelligence that no packaged solution can provide.

In the following section, we walk through the seven core technologies that make up the SAP Business AI intelligence layer, from generative AI and Joule at the interface through to process automation and embodied intelligence at the execution layer.

1.6.1 Intelligent Technologies

Modern enterprise efficiency is defined by the speed of data-driven action. Seven core technologies make up the SAP Business AI architecture. Combined, they form a closed-loop system that allows the enterprise to see, think, act, and learn alongside its human counterparts.

Let's walk through each:

- **Generative AI and Joule**
  Generative AI serves as both the primary interface and the orchestration backbone of the Autonomous Enterprise. Centered on Joule, SAP's natural language engagement layer, it uses large language models (LLMs) to synthesize information, generate content, interpret intent, and coordinate action. Joule balances probabilistic reasoning (LLMs for interpretation, summarization, and recommendation) with deterministic execution (structured APIs, business rules, and transactional logic). As an interface, it eliminates navigation debt. As an orchestrator, it routes user intent to the appropriate assistants and agents. This is the hierarchy in practice: People direct through Joule, assistants orchestrate, and agents execute.

  Figure 1.5 shows Joule as the engagement layer of the Autonomous Enterprise, illustrating how the AI assistant serves as the unified interface through which users set intent and the system executes across all applications.

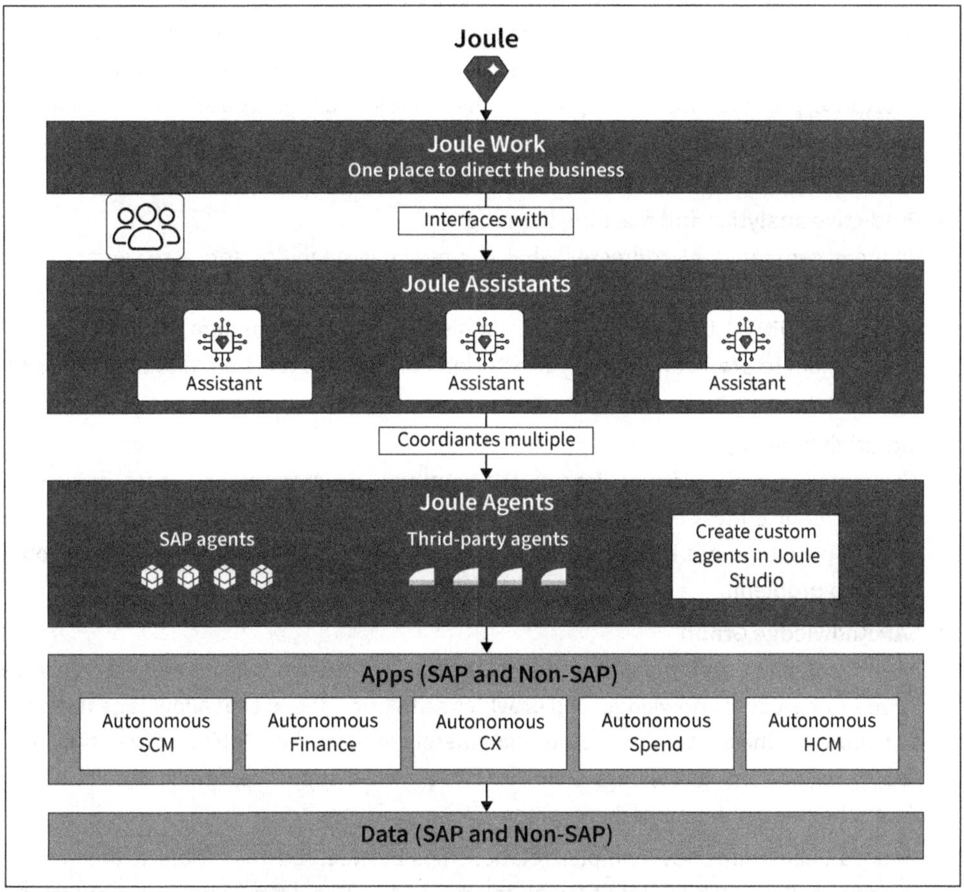

Figure 1.5: Joule: The Engagement Layer

In practice, Joule operates across a spectrum from simple retrieval to full autonomous execution. At the simpler end, users ask questions grounded in live business data. At the transactional level, Joule executes operations through natural language commands that replace multiscreen workflows. At the analytical level, it synthesizes cross-functional data into visualizations and recommendations. Finally, at the agentic level, it accepts a high-level goal, decomposes it into a plan, activates agents, and returns the completed outcome.

- **Agentic AI and assistants**
  This layer moves beyond conversation and into orchestration, transitioning the system from reactive helper to proactive partner. Joule Assistants are role-aware persona hubs, and Joule Agents are specialized digital workers that handle complex, multistep workflows using reasoning and autonomous planning. We cover these in detail in Section 1.6.2.
- **Process automation**
  This technology provides the physical execution capability. By combining robotic process automation with workflow management and AI services for processing unstructured documents, it automates the "last mile" of business processes. In the Autonomous Enterprise, process automation provides the "hands" that agents use to act on their decisions: releasing payments, creating purchase orders, updating production schedules, and so on.
- **Predictive analytics and machine learning**
  Whereas generative AI addresses what *is*, predictive analytics addresses what *will be*. This layer uses statistical algorithms, ML models, and proprietary foundation models such as SAP-RPT-1 to identify patterns in historical data. From demand sensing in the supply chain to predicting employee attrition in HR, these tools provide the confidence levels required for high-stakes planning.
- **Situation handling**
  The system's early warning framework continuously monitors business objects for exceptions or critical events. By 2026, this technology became AI-assisted. It notifies users of problems, but it also analyzes context and recommends the next best action to resolve a problem.
- **SAP Knowledge Graph**
  SAP Knowledge Graph maps complex relationships between business objects, encodes 50 years of process knowledge, and provides semantic context that allows agents to reason about business situations rather than merely process data. When a disruption hits a supply region, SAP Knowledge Graph shows affected suppliers, at-risk orders, dependent schedules, and alternative sources.

  Figure 1.6 illustrates how complex relationships between business objects are mapped and encoded in SAP Knowledge Graph to provide AI with a deep, contextual understanding of enterprise processes.

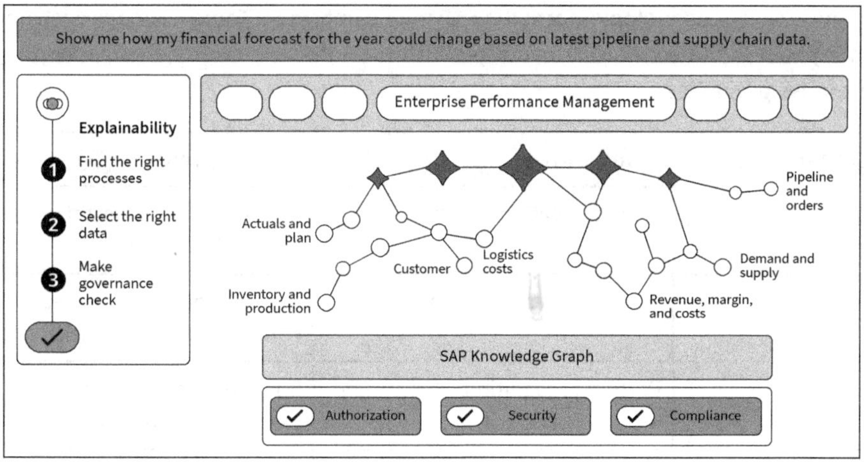

Figure 1.6: SAP Knowledge Graph

- **Embodied intelligence**
  This technology integrates reasoning into robotic hardware, extending AI from digital processes into physical operations. Autonomous mobile robots in warehouses and robotic assembly on shop floors are all guided by SAP logistics intelligence and coordinated through Joule.

  Figure 1.7 shows the integration of embodied AI into SAP Autonomous Suite, illustrating how AI reasoning is extended into robotic hardware and physical operations connected to digital business processes.

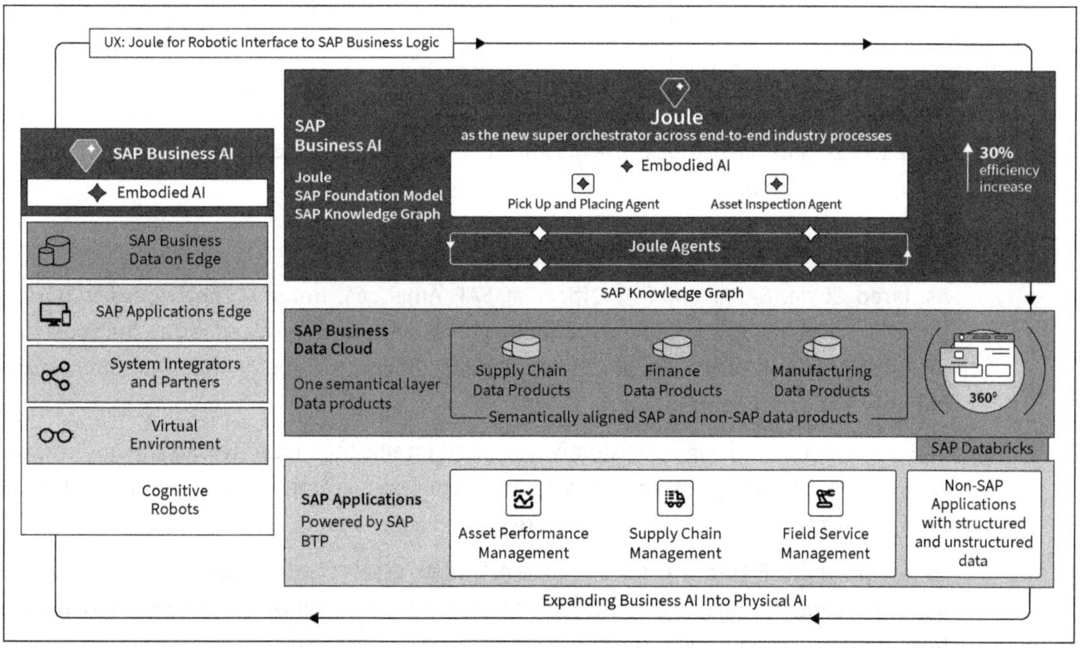

Figure 1.7: Embodied AI Integration into SAP Autonomous Suite

Together, these seven technologies form the SAP intelligent stack. Table 1.1 maps each one to its primary role in the architecture and the business value it delivers when embedded inside enterprise processes.

Intelligent Technology	Primary Role	Business Value
Generative AI	Communication and synthesis	Reduces time to insight via natural language
Agentic AI	Goal-oriented execution	Autonomously completes complex, multistep tasks
Predictive analytics	Pattern recognition	Shifts the enterprise from reactive to proactive
Situation handling	Exception management	Ensures no critical business event goes unnoticed
Process automation	Task execution	Eliminates manual data entry and "swivel chair" work
SAP Knowledge Graph	Semantic context and grounding	Eliminates AI hallucinations and provides connected construct by making explicit relationships between all SAP and non-SAP datasets
Embodied intelligence	Physical cognition and execution	Automates complex physical tasks by bridging digital intent with autonomous real-world action

Table 1.1: SAP Intelligent Stack: Primary Roles and Strategic Business Value

1.6.2 Embedded Intelligence in Applications

As Jared Coyle, SAP's Chief AI Officer at SAP Americas, noted: "The best AI is boring" (*http://s-prs.co/v633500*). The most effective AI is invisible. It surfaces inside the tools people already use and trust. By placing intelligence in context, teams maintain their muscle memory while outcomes improve.

SAP's foundational design decision is to embed intelligence directly into the applications where work happens, presenting insights as contextual, actionable elements within the user's active transaction rather than as reports in a disconnected environment.

As an example, Figure 1.8 shows an embedded asset capitalization recommendation, demonstrating how AI delivers intelligent suggestions directly within the SAP Fiori application workflow without requiring context switching.

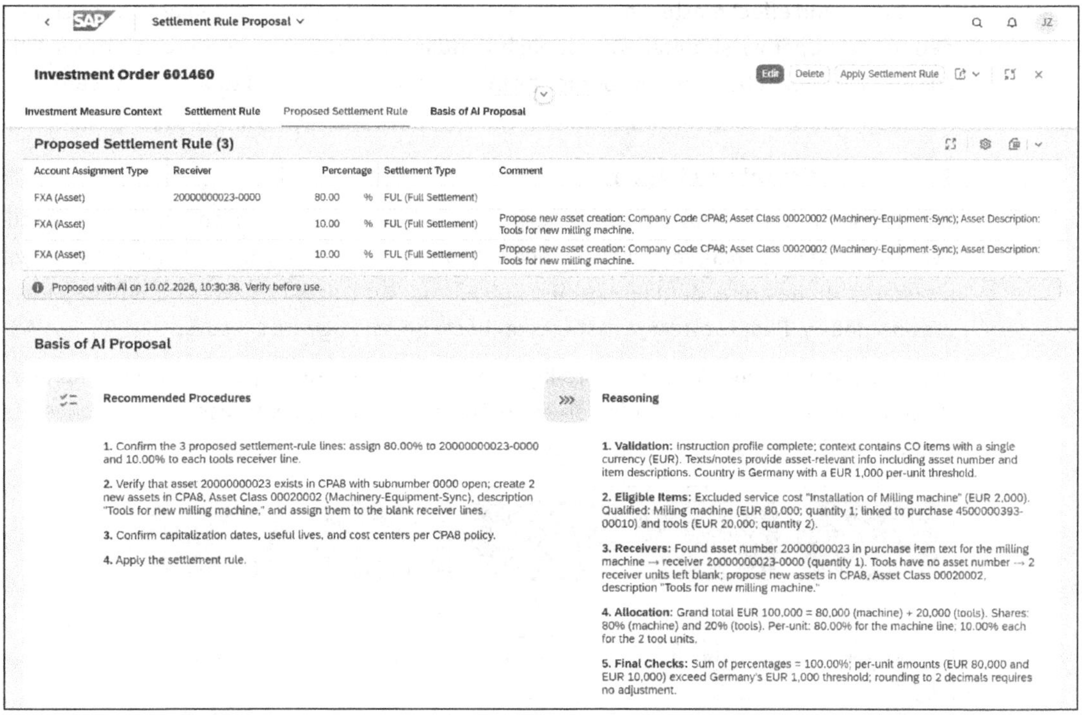

Figure 1.8: Embedded Asset Capitalization Recommendation

This shows up differently depending on the role. A procurement manager sees a supplier flagged for elevated delivery risk with a confidence bar or alternative source recommendations that can be executed without leaving the purchase order screen. A finance controller in collections sees customers with deteriorating payment patterns, predictive cash flow impact scores, and prioritized accounts with recommended strategies. A warehouse manager sees real-time zone congestion heatmaps with smart task prioritization. An account executive in SAP Sales Cloud sees opportunity scores that update dynamically with the recommended next actions.

The principle is the same: Intelligence arrives inside the workflow, not beside it. There is no context switching, query building, or interpretation gap.

Beyond the SAP Fiori layer, embedded AI operates at the process level across every function. In finance, the intelligent SAP Cash Application solution matches incoming payments by learning each customer's remittance behavior. Predictive journal entries anticipate recurring postings. Intelligent accruals estimation will project amounts before invoices arrive. In HR, SAP SuccessFactors embeds AI across the hire-to-retire lifecycle: job descriptions generated with inclusive language, skills gap analysis, attrition prediction, and so on. In procurement, intelligent sourcing evaluates suppliers against performance, risk, and capacity. Demand sensing incorporates external signals to improve forecast accuracy by 20 to 30 percent.

The compound effect matters more than any individual capability. When dozens of embedded models operate simultaneously, each handling one process with precision, then the cumulative impact transforms operations in ways no standalone AI tool can replicate.

1.6.3 AI Assistants and Agents

The transition to the Autonomous Enterprise is driven by the shift from embedded intelligence that advises to agentic intelligence that acts. Joule serves as the engagement layer, assistants orchestrate domain-specific workflows, and agents execute multistep tasks autonomously. People direct, assistants orchestrate, and agents execute.

Figure 1.9 shows how SAP scales intelligence through specialized personas, illustrating the three-tier model in which Joule serves as the engagement layer, assistants orchestrate domain workflows, and specialized agents execute multistep tasks autonomously.

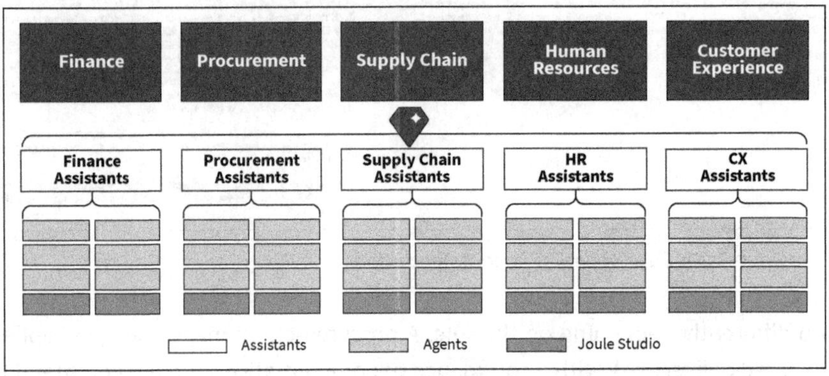

Figure 1.9: Scaling Intelligence Through Specialized Personas

Figure 1.10 shows the Autonomous Enterprise hierarchy, illustrating how the five autonomous business domains are served by AI assistants that coordinate the specialized agents operating beneath them. We'll describe them in more detail in the following sections.

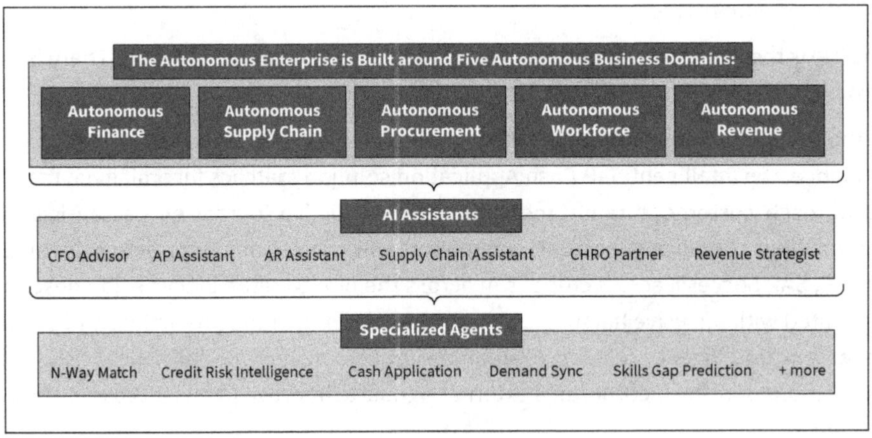

Figure 1.10: Autonomous Enterprise Hierarchy

Joule

Joule is SAP's engagement layer for the Autonomous Enterprise. People express what they want to accomplish, and the system brings together the right insights, workflows, and agents to deliver the outcome.

That experience comes to life through Joule Work, a dynamic workspace built around intent rather than application navigation. Instead of navigating between applications and manually stitching information from multiple systems, users operate from a single workspace where intent translates into outcome. Joule Work provides three interaction surfaces: *conversations* for AI-powered chat and quick task execution, *spaces* for dynamic problem-solving and analysis, and *develop* for seamless entry into Joule Studio.

The workspace is omnipresent. The Joule action bar follows users across web applications, providing one-click access to custom AI actions, reading and writing assistance, and proactive recommendations. Bidirectional integration with Microsoft 365 means that Joule reaches into Outlook, Teams, and Excel. Deep research capabilities enable multistep explorations that produce verified reports in minutes rather than hours.

As an orchestrator, Joule provides deep contextual awareness tailored to specific user roles, active transactions, and organizational context. A finance controller and a supply chain planner may ask similar questions but receive responses grounded in their respective data and decision authority. The technical architecture positions Joule as *open, but anchored*; through the Agent2Agent Protocol (A2A) and Model Context Protocol (MCP), Joule extends into external tools while maintaining identity controls, access governance, and audit trails regardless of where interactions originate.

Assistants

Joule Assistants are human-facing interfaces grounded in a specific role or process, working like expert teammates that understand a person's operational context. They support users through natural language, help them understand information and make decisions, and, critically, coordinate specialized agents to execute complex workflows from end to end.

SAP Autonomous Suite organizes over 40 assistants across the five domains. Autonomous Finance includes accounts receivable, accounts payable, billing, cash and treasury, financial closing, financial planning, expense management, sales operations, governance, tax and compliance, and enterprise architecture assistants. Autonomous Supply Chain Management (SCM) includes planning, manufacturing, logistics, product design, and asset and service assistants. Autonomous Spend includes buying, category management, invoicing, procurement contract, receiving, requisition, services procurement, sourcing, supplier management, and business network assistants. Autonomous HCM includes core HR, HR service, HR knowledge, career and talent development, compensation, recruiting, and people intelligence assistants. Finally, Autonomous CX includes deal qualification, deal closing, sales, campaign, content, case management, order management, self-service, and shopping assistants.

Each assistant orchestrates between 2 and 11 specialized agents. For example, the Accounts Receivable Assistant coordinates the Collection Insights, Collections Email Outreach, Case Preparation, Case Processing, Payment Inquiry Handling, Receivables and Payables Clearing, Dispute Creation, Dispute Resolution, Receivable Account Analysis, Dunning Insights, and Subscription Lifecycle Agents. Fine-grained specialization means that each agent masters a specific task while the assistant provides the coordination intelligence.

Agents

Joule Agents are goal-driven systems that plan, reason, and act autonomously across applications, processes, and business data to complete multistep tasks. They are goal-oriented, policy-governed entities capable of independent planning, reasoning, and action. They represent the shift from chat-based assistance to autonomous, reason-to-action capabilities.

Agents operate across a spectrum of increasing complexity, categorized into four levels:

1. Level 1 (single-shot function) agents perform targeted lookups triggered by a single request.
2. Level 2 (chaining and generic reasoning) agents introduce multistep dependency management, decomposing requests into logical sequences.
3. Level 3 (tailored, content-based) agents deliver deep domain awareness for high-value use cases like dispute resolution, invoice validation, and sourcing optimization.
4. Level 4 (optimized autonomous) agents possess advanced capabilities including continuous sensing, reflection, error self-correction, multiagent collaboration, and outcome optimization through sophisticated grounding-planning-execution-reflection loops.

Within the Autonomous Enterprise, agents operate within the governed context of SAP Autonomous Suite, accessing data through SAP Business Data Cloud, reasoning through SAP Knowledge Graph, and executing within the identity, access, and compliance controls of the govern pillar (Section 1.6.4). Every agent action is auditable, traceable, and reversible.

As the enterprise AI landscape matures, the focus shifts from isolated agents to agent-to-agent interoperability. SAP's strategy is built on the premise that no single AI solves every business problem. The goal is an enterprise in which Joule orchestrates tasks across ecosystems including Microsoft, Google Cloud, and AWS.

Figure 1.11 shows agent interoperability, illustrating the three-stage evolution from MCP connectivity through direct invocation to full agent-to-agent collaboration across SAP and third-party systems.

Three stages define this evolution:

1. MCP establishes a universal connector, enabling agents to securely access external data sources through a single interface.
2. Agent-as-tool introduces task handover. Agents identify the right peer for a job and delegate with full context.

3. A2A is the pinnacle: specialized agents collaborating autonomously, with finance, HR, and supply chain agents communicating in real time to resolve enterprise-wide challenges.

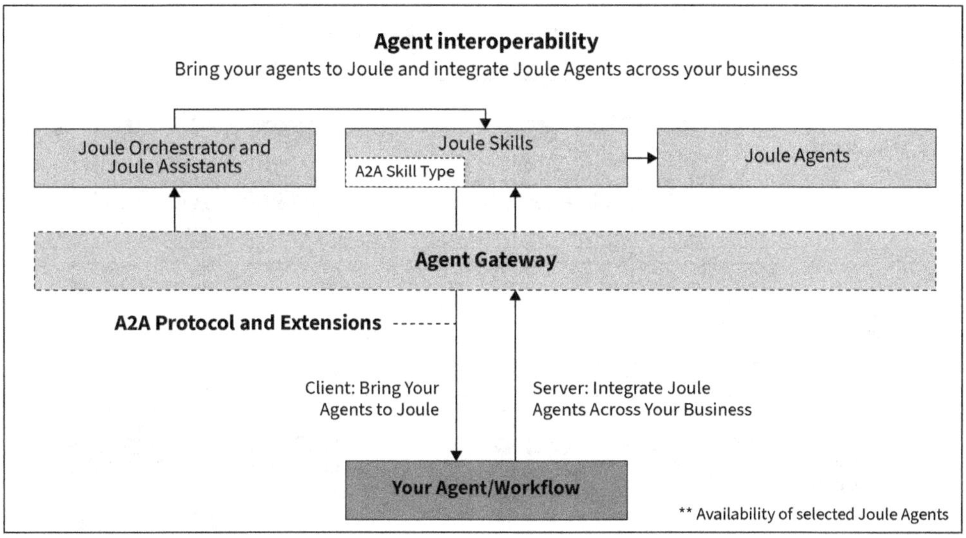

Figure 1.11: Agent Interoperability

The technical roadmap unfolds in two iterations. Direct invocation centers on connectivity through the agent gateway for trust, security, and identity propagation. The second iteration enables the Agent Gateway to invoke the Joule orchestrator itself, transitioning from isolated task execution to outcome-focused orchestration. Here, Joule independently decomposes high-level requests and coordinates the necessary agents.

Agentic Orchestration

The interplay between Joule, assistants, and agents is connected through *agentic orchestration*, the process that transforms a user request into a successfully executed business outcome. Rather than following a rigid script, the system uses a reflection-to-action loop.

Figure 1.12 shows agentic orchestration, illustrating the four-stage process by which Joule interprets user intent, contextualizes it with live business data, decomposes tasks to specialized agents, and coordinates execution.

During reflection and contextualization, the orchestrator interprets the request while applying enterprise context and user identity. Planning and task decomposition break complex intents into executable plans with optimal sequencing. Execution and monitoring activate agents with continuous oversight. Finally, synthesis and delivery consolidate outputs into a coherent response or decision context.

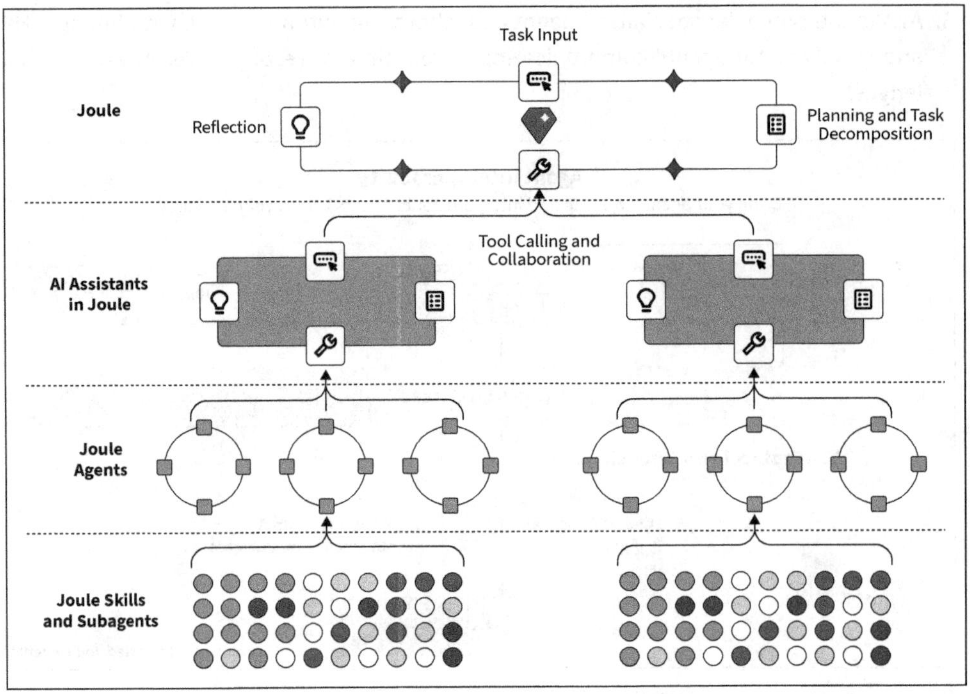

Figure 1.12: Agentic Orchestration

This hierarchy ensures that every autonomous action is grounded in live business data, compliant with enterprise policies, and traceable. As organizations expand autonomous execution, governance scales proportionally.

Table 1.2 summarizes the five patterns of AI-driven engagement, mapping each to its functional role and the primary business value it delivers as organizations progress toward full autonomy.

Layer	Role of Orchestration	Primary Responsibility
Joule (top layer)	Universal orchestrator	Handles initial task input, reflection, and high-level planning
AI assistants	Persona hubs	Coordinates specialized agents within a specific business domain (HR, finance, etc.)
Joule Agents	Specialized workers	Performs multistep reasoning and determines which tools are needed for a task
Skills and subagents	Functional blocks	Executes granular, rule-based tasks through direct API calls

Table 1.2: Agentic Orchestration Hierarchy

The capabilities we've discussed—intelligent technologies, embedded intelligence, and AI assistants and agents—do not operate independently. They run on a shared foundation that makes them trustworthy, governed, and scalable across the enterprise. In the following sections, we'll break down the three pillars of SAP Business AI.

1.6.4 The Pillars of SAP Business AI Platform

SAP Business AI Platform is the enterprise AI foundation that makes the Autonomous Enterprise a reality. It brings together SAP BTP, SAP Business Data Cloud, SAP HANA Cloud, and Business Transformation Management capabilities (which we'll discuss in future chapters) within a single, cohesive foundation. Where SAP Autonomous Suite describes what runs the business, SAP Business AI Platform describes what everything runs on.

The platform is organized around three pillars:

- **Build**
  The build pillar provides a unified path from idea to enterprise impact. Joule Studio and SAP Integration Suite together enable developers and business technologists to build AI agents, apps, and workflows grounded in business context, deploy them without operational overhead, and seamlessly integrate them across the enterprise landscape. Joule Studio is SAP's AI-first development experience. You can describe a business outcome in plain language, and the system generates an end-to-end solution structure for developers to refine. Early access customers report generating working solutions in 10 to 15 minutes that previously required three to four days. The runtime is fully SAP-managed with native connectivity and zero infrastructure burden.
- **Contextualize and reason**
  The contextualize and reason pillar anchors AI in a business data fabric that serves as the trusted knowledge core for every enterprise application and agent. SAP Business Data Cloud unifies SAP and third-party data so that every AI initiative operates from the same trusted foundation. SAP Knowledge Graph encodes process and industry intelligence as navigable relationships between business objects. Purpose-built models, including SAP-RPT-1 for structured business data prediction and SAP-ABAP-1 for enterprise code generation, arrive prebuilt on an SAP business context, while governed access to third-party models ensures the right model for every task.
- **Govern**
  The govern pillar provides the command center for enterprise-wide data and AI agent governance through SAP AI Agent Hub. This is where agent lifecycle management happens: discovery, provisioning, observation, security enforcement, optimization, and decommissioning. SAP LeanIX provides enterprise architecture context, SAP Signavio delivers process intelligence, and SAP Cloud Identity Services enforces identity and access controls. Together, these capabilities ensure that every agent operates within organizational policies, performance is continuously measured, and value realization is tracked from strategy through execution.

Seven core technologies form the intelligence layer. They embed directly into the applications where work happens, not beside them. Joule handles the interface, agents handle execution, and SAP Knowledge Graph provides the context. Predictive models surface signals before they become problems, while situation handling catches what would otherwise slip through unnoticed. Underneath all of this, SAP Business AI Platform provides the governed foundation that makes every capability trustworthy at enterprise scale: the infrastructure to build, the data fabric to contextualize, and the controls to govern agents across their full lifecycle.

The question that follows from all of this is simpler than it sounds: How does working with SAP actually feel different? The five patterns we discuss next answer that question, describing how the relationship between people and the system evolves as intelligence deepens across the organization.

1.6.5 Five Patterns of Business AI Applications

The transition to an Autonomous Enterprise follows a strategic sequencing that progresses through five patterns representing increasing autonomy. Let's walk through the five patterns of business AI engagement, sequencing the strategic progression from seamlessly embedded intelligence through assisted work, agent-driven automation, AI-native processes, and fully intent-driven autonomy:

- **Seamlessly embedded**
  The seamlessly embedded pattern delivers intelligence in context; this is the "boring AI" working invisibly within existing applications. Predictive fields, smart defaults, and automated classifications appear naturally in the workflow. This maximizes adoption because it requires no behavioral change. This is the foundation that makes Joule Work possible. When embedded models handle the routine, Joule has clean, structured signals to reason from and users have the attention to direct rather than execute.
- **Agent-driven**
  The agent-driven pattern introduces role-aware assistants that coordinate agents to carry more operational load. Users express an intent while the system orchestrates its multistep execution. This begins the shift from human *execution* to human *oversight*. Joule Work is the interface in which this shift becomes tangible. Instead of switching between applications to track status, approve exceptions, or resolve blockers, users express intent once and agents execute across the full process chain.
- **AI-native**
  AI-native pattern designs are "born intelligent"; they are built on live business graphs and continuous sensing rather than sequential workflows. These processes reimagine the flow itself around intelligence from inception. In practice, this means processes like continuous cash positioning, real-time demand sensing, and always-on compliance monitoring replace the monthly cycles and batch runs they inherited from an earlier era of enterprise software.

- **Appless**
  The appless pattern enables the traditional interface to recede. Users state what they want to accomplish, and Joule Work delivers the outcome without requiring navigation through screens. A finance controller who asks "What is our consolidated cash position and where are the risks?" receives an actionable answer drawn from live data across all entities. There is no transaction code, no dashboard, and no waiting.
- **No-apps**
  The no-apps pattern represents fully autonomous governance for routine work. The system executes within defined parameters, invoking human involvement only for genuine exceptions. This is the operating model that Chapter 5 traces through every value stream: invoice matching, goods receipt posting, supplier screening, and maintenance scheduling running continuously, with humans focused on the exceptions the system cannot resolve within governed parameters.

Table 1.3 maps each pattern to its functional role and the primary business value it delivers, providing a reference for how the relationship between people and the system shifts at each stage of the progression.

Pattern	Functional Role	Primary Business Value
Seamlessly embedded	Intuitive UX	Surfaces intelligence in context to simplify work while maintaining muscle memory
Agent-driven	Specialized automation	Coordinates multistep workflows with explicit human-in-the-loop thresholds
AI-native	Predictive insights	Reimagines workflows to lead with foresight by designing around intelligence from day one
Appless	Intent-based navigation	Intent leads directly to outcomes as the screen recedes, reducing navigation debt
No-apps	Autonomous governance	Executes routine tasks automatically; humans review exceptions, rather than everything

Table 1.3: Five Patterns of AI-Driven Engagement in Business Applications

1.7 Business AI Maturity Model: From Augmentation to Autonomy

The Rotterdam scenario from the Preface is worth revisiting: The problem that Tuesday was not a shortage of data. In fact, every signal was present: port status, vessel positions, logistics schedules, production calendars, and customer commitments downstream. The problem was the time it took to connect those signals and act on them. By the time a team assembled the full picture and issued the adjustments, the disruption had already reached Stuttgart. The response window had closed before the response began.

Better analytics would not have helped. A faster dashboard would not have closed the window. What was missing was a system that could act on the signal the moment it arrived, before the impact had time to travel, before the options had time to narrow, and before the cost of the response had time to grow. That is the line between augmentation and autonomy.

Most enterprise AI deployments today operate in augmentation mode. The system surfaces a recommendation, flags an anomaly, or summarizes a situation. A human reviews it and decides what to do. That is genuinely useful because it compresses the time between signal and human awareness. However, it does not compress the time between signal and action. The review gate remains on every transaction, every exception, and every decision. Augmentation improves the quality of human decisions, while autonomy changes what the operation actually can do. According to ASUG and SAP research from late 2025, while many organizations are actively using or planning for agentic AI, only a small percentage are in full production *(http://s-prs.co/v633501)*. The organizations that have crossed that line share one characteristic: They started with processes for which the logic was clear, the data was reliable, and the boundaries of autonomous action were precisely defined before anything went live.

Within individual functions, the results are already measurable. In finance, autonomous cash collection and invoice processing deliver 10 to 20 percent days sales outstanding (DSO) reduction and 60 to 80 percent billing productivity improvement. In the supply chain, autonomous planning agents deliver 20 to 30 percent faster planning cycles and equivalent reductions in forecast error. These are real numbers from live deployments. The more significant shift happens when the boundary between functions disappears entirely.

Single-function automation delivers value within a boundary. The Autonomous Enterprise removes the boundary. The Rotterdam disruption did not stay in the supply chain; it carried revenue exposure, working capital consequences, and treasury implications that touched finance and customer management simultaneously. In most organizations, connecting those consequences across three functions takes three teams, multiple systems, and the better part of a working day. By that point, the practical options have narrowed and the cost of the response has grown considerably.

In an enterprise running the autonomous model, that Tuesday looks different: The disruption is detected. The downstream consequences across the supply chain, finance, and customer commitments are assessed together. The response is underway across all three domains simultaneously, before the supply chain manager's morning meeting has ended.

That cross-domain response is qualitatively new, not incrementally better. It represents a class of operational capability that did not exist before—not because the data was missing, but because no human process could hold all of it together fast enough to act within the window that mattered.

The journey to the Autonomous Enterprise is a structured evolution through progressive phases. Different processes within an enterprise may operate at different maturity levels simultaneously. The following framework lays out a practical progression:

1. **Phase 1: Foundation and augmentation (months 0 to 6)**
   The initial phase establishes trust and delivers immediate value through embedded intelligence and conversational AI. Joule is activated as the engagement layer. Embedded capabilities come online: intelligent cash application, predictive demand sensing, automated document extraction. The seamlessly embedded pattern delivers productivity gains without behavioral change.

   Critically, this phase establishes the governance foundation. Identity controls, access policies, audit trails, and compliance frameworks are configured now, not added retroactively when autonomous capabilities arrive. Organizations that skip governance during augmentation face significant remediation later.

   SAP Joule for Developers transforms the software engineering lifecycle: writing optimized code, generating unit tests, explaining legacy ABAP programs, and so on. SAP Joule for Consultants streamlines implementations by navigating vast documentation libraries and recommending approaches based on industry best practices.

   Phase 1 demonstrates that SAP's AI can be trusted with routine decisions. Intelligent cash application matching at 92 percent accuracy builds confidence; predictive MRP simulation proving more accurate than manual planning builds trust.

2. **Phase 2: Orchestration and delegation (months 6 to 12)**
   Organizations transition toward agent-driven workflows. Role-aware assistants coordinate agents that execute multistep processes from end to end, with human involvement shifting from execution to oversight.

   Consider a telecommunications provider processing 35,000 customer disputes monthly. Previously, each dispute required 45 minutes of manual analysis. After deploying the Dispute Resolution Agent, routine disputes resolve autonomously: Contracts are accessed through SAP Knowledge Graph, delivery confirmations are cross-referenced, credit policies are applied. The resolution cycle time drops from 6 days to 18 hours. First-contact resolution rates improve significantly.

   Organizations that deploy agent-driven processes report that procurement contract renewal cycles are reduced 40 to 50 percent, goods receipt processing is accelerated up to 50 percent, and they experience a consistent expansion of what the organization trusts the system to handle.

3. **Phase 3: Transformation and foresight (months 12 to 18)**
   AI-native processes redesign how work is performed rather than accelerating existing workflows. Processes are built on live business graphs, continuous sensing, and predictive reasoning.

When a geopolitical event affects a supply region, the system traces its complete impact through SAP Knowledge Graph and activates agents across domains to execute mitigation before the first human escalation. For example, a multinational automotive manufacturer implementing AI-native supply chain planning reported a 43 percent reduction in production stoppages and 28 percent improvement in inventory turnover within 15 months (*http://s-prs.co/v633502*).

4. **Phase 4: Intent-driven autonomy (months 18 to 24)**
   In the fully realized Autonomous Enterprise, routine work executes automatically within governed parameters, and people focus on strategy and judgment. A global life sciences company compressed its month-end financial close from eight days to three by deploying autonomous agents across subledgers, with 35 finance professionals shifting from executing 200 discrete tasks to evaluating variances and providing strategic commentary.

The operational paradigm shifts from reactive system management to context-aware enterprise intelligence. The system resolves standard operations autonomously for governed, well-understood scenarios. It escalates only for genuine exceptions: those scenarios outside confidence thresholds, lacking precedent, or carrying risk levels reserved for human judgment. When exceptions occur, Joule Agents surface contextual information, predicted impact, alternative scenarios, and recommendations. Once a human applies judgment, agents execute the decision, update records, and adjust monitoring to handle similar situations autonomously in the future.

This is the Autonomous Enterprise: not the removal of people, but the elevation of people. AI handles the coordination and routine decision-making that previously consumed the majority of knowledge workers' time. People apply judgment to the decisions for which creativity, relationships, and strategic vision create irreplaceable value.

1.8 Case Studies

This book explores the Autonomous Enterprise through architecture, capability, and real operating detail. However, architecture without context is just theory. So before we go further, we want to introduce a company and three of the people running it, whose challenges will travel with you.

Each time a new capability is introduced, you will see it applied to a problem you have already met. Each time a solution is described, you will see what it looks like when it lands in a real operating environment with real constraints and real people who need it to work. By the final chapter, you will not just understand how the Autonomous Enterprise functions. You also will have watched it solve something. The company is fictional, but the problems are not.

1.8.1 The Company

Korindal Consumer Group is a fictitious global consumer goods manufacturer with approximately nine billion dollars in annual revenue, operating across North America, Europe, and Asia Pacific. The company manufactures and distributes branded products across multiple categories—personal care, household goods, and seasonal products—through a mix of large retail partners, regional distributors, and a growing direct-to-consumer channel that the commercial team has been building for three years without having quite the right infrastructure to support it properly. The business grew fast, mostly through acquisition.

That growth built market share, category depth, and a brand portfolio that any competitor would respect. It also built a technology landscape that nobody would have designed from scratch and that nobody has had the time, the budget, or the organizational appetite to rationalize since.

The European core business runs on SAP ERP, on-premise, with 15 years of customization layered into it by teams who had good reasons for every change they made at the time. The North American operation acquired seven years ago runs on a different ERP system entirely, with its own data model, its own chart of accounts, and its own way of defining a customer. The most recent acquisition, a regional business in Asia Pacific, brought a third system that integrates with neither of the others. On top of that foundation sits a collection of cloud point solutions assembled over the years as specific problems demanded specific fixes: a standalone customer relationship management (CRM) system, a transport management system, and a procurement platform that was introduced as a short-term measure and quietly became a permanent part of the landscape.

None of these systems share a common data model. Most of them exchange information through file-based interfaces that run overnight or through manual reconciliation processes that someone, somewhere, is maintaining in a spreadsheet. The data exist, but the connections between them largely do not.

1.8.2 The Business Problems

Korindal is not a company in distress: Revenue is solid, the leadership team is capable and commercially sharp, and the brand portfolio is strong enough that the business has been able to absorb the operational friction without the market noticing—so far. However, the operating model that got the company here is showing its limits and the margin for absorbing friction is narrowing.

Three people on the leadership team feel those limits more acutely than anyone; their names are Maria, James, and David. Their problems are different in their details, but in structure, they are the same problem, and together they are the thread that runs through this entire book. Let's walk through them.

Case Study 1: Maria, CFO: The Cash That Should Not Be Missing

Maria has been the CFO of Korindal for four years. In that time, she sat through sixteen quarterly business reviews in which DSO appeared on the agenda. It has appeared every single time. The number is running 18 days above target across the group. At Korindal's revenue scale, 18 days of DSO represents roughly four hundred million dollars in cash the company has earned and not yet collected: *four hundred million dollars*, sitting in the receivables ledger, quarter after quarter.

The cash is not missing because customers are not paying; most of them are. It is trapped in the gap between what the business knows and what its systems can act on. Invoice disputes sit unresolved for weeks because nobody has a clear, consistent view of who owns them or where they are in the resolution process. Collections follow-up varies by region because each team developed its own approach over the years and nobody has had the authority or the appetite to standardize it. Maria cannot get a single consolidated view of the full receivables position across three ERP instances without waiting for a month-end consolidation that arrives twelve days after the month has closed, by which point the numbers are already stale and the practical window for acting on them has passed.

The knowledge to fix most of this exists. The collections team knows which customers dispute most frequently. They know which invoice types generate the most exceptions. They know the resolution path for most dispute categories they encounter. That knowledge lives in spreadsheets on individual laptops, built up by experienced collectors over years of doing the job. It has never made it into a system that could apply it consistently, at scale, without depending on whether the right person happens to be available that day.

Every month, the number resets. Every quarter, it goes back on the agenda. The conversation is always the same: Acknowledge the gap, discuss the contributing factors, agree to follow up, move on.

Maria has stopped asking why DSO is high; she knows the answer. The question she has started asking instead is harder and more useful. Cash collection is a process with defined stages, known exception types, and clear resolution logic for most cases, so why does it still require this much continuous human attention to produce such inconsistent results? Why does a four-hundred-million-dollar working capital gap persist in a company running modern enterprise systems?

That is the question this book helps answer. The capabilities that address it—how AI agents handle invoice exceptions automatically, how a unified data foundation gives finance a real-time view across the full receivables position, how the autonomous finance model closes the distance between a signal and a response—are developed in the chapters on finance and intelligent operations. When we return to Maria, the numbers will look different.

Case Study 2: James, Chief Supply Chain Officer: The 14 Percent

James runs the supply chain for Korindal. The metric he measures is on-time, in-full (OTIF) delivery to customers. The current number is 86 percent, which means that 14 percent of customer orders are not arriving as committed.

As Korindal's largest retail partners enforce contractual penalty clauses for missed deliveries, that 14 percent is not just an operational shortfall. It is a direct and recurring cost: over 30 million dollars annually in penalty charges, expediting fees, and premium freight spend incurred to partially recover situations that should not have arisen in the first place, 30 million dollars a year, year after year, on problems the team fully understands and cannot consistently prevent.

The root causes are not a mystery. Demand is more volatile than the planning cycle was built to handle; the planning model was designed for a more stable environment and has never been fully recalibrated for the volatility that has become normal. Supplier confirmation reliability varies significantly across the base; some suppliers confirm accurately and deliver on time; others confirm late, revise quantities after confirmation, or go quiet entirely when a delivery is at risk. The planning system, capable as it is in steady-state conditions, is not built to sense and respond to the early signals that precede a fulfillment failure. It optimizes well, but it anticipates poorly.

The result is a consistent pattern: When conditions are stable, the operation performs well. When a demand spike arrives, a supplier short-confirms, or a logistics lane tightens unexpectedly, the system reacts after the fact rather than adjusting ahead of time. By the time the fulfillment risk appears clearly in the planning output, the low-cost response options are already gone. What remain are the expensive ones: expedite, premium freight, partial shipment, penalty.

Every week, James's team holds an exception review. It is a well-run meeting. The team is experienced, the analysis is thorough, and the decisions made in the room are generally the right ones. It is also, almost entirely, a postmortem. The team is diagnosing misses that have already happened and managing the fallout, not identifying the risks that are forming and acting before they become misses.

The signals that precede most fulfillment failures are sitting in the data days before the miss occurs: a supplier confirmation that came in three days late and was not escalated, an inventory variance that did not trigger a replanning run because it fell just below the threshold, a demand pattern that shifted outside the forecast band without updating the production schedule. The information is there, but the process is not designed to act on it when it still matters.

That gap between when a fulfillment risk becomes visible in the data and when the organization actually responds to it is what costs 30 million dollars a year. The supply chain chapters of this book address this gap directly: how continuous sensing replaces batch planning cycles, how exception handling shifts from reactive review to automated early response, and what the operation looks like when the planning process is designed to act on signals as they emerge rather than after they compound. When we return to James, the OTIF number will have moved.

Case Study 3: David, CIO: The Queue That Never Empties

David runs IT for Korindal. He has a capable organization that includes experienced architects and solid project delivery, people who know the systems deeply and work hard to support a business that makes significant demands on them. His problem is not talent but geometry.

Every improvement the business makes generates more requests than it closes. Maria's DSO initiative requires a new integration between the collections tool and the European ERP system. James's supply chain improvement needs a custom exception-routing workflow that sits outside the standard planning process. The commercial team wants the ability to make mid-cycle pricing adjustments without waiting for the next quarterly release. The sustainability team wants real-time carbon data connected to production orders. The direct-to-consumer channel needs a customer data integration that none of the existing systems were designed to support.

Each request is legitimate and makes complete sense in isolation. In aggregate, they have consumed the next three quarters of the release calendar, and the queue is not shrinking. For every item the team closes, two more arrive.

The business has started to notice. Teams that have been waiting six months for a system change have found other ways to get what they need: manual workarounds, shadow spreadsheets, point solutions procured outside the IT process. None of these are reckless decisions. They are rational responses to a system that cannot keep pace. However, they accumulate. Every workaround is a new data inconsistency. Every shadow solution is a new integration problem waiting to surface. The technical debt is growing faster than the team can work it down.

The gap David is managing is not a resourcing gap. He has made the case for additional headcount twice and received partial approval both times. Adding people to the current architecture produces diminishing returns because the constraint is not capacity but the architecture itself. Every change touches the core. Every integration is custom-built and custom-maintained. Every release requires regression testing across a fragmented landscape that was never designed to be changed at the rate the business now requires.

At a leadership offsite earlier this year, the CEO asked the question that had been in the room for a long time: Korindal has invested heavily in technology over the past decade and has a strong IT organization, so why does every change still feel harder than it should? David's answer was direct: The landscape was designed for stability, not speed. It can be maintained, but it can't evolve at the pace the business needs without a different foundation. Fixing individual requests one at a time does not solve that; the architecture itself has to change.

What that change looks like—how a clean core, a modern extensibility model, and a platform built for continuous innovation replace the current landscape without disrupting operations in the process—runs through the platform chapters of this book. When we return to David, the release calendar will look different, not because the team became bigger, but because the architecture stopped making every change harder than it needs to be.

Three Problems, One Company

Maria, James, and David do not sit together often. Finance, supply chain, and IT rarely share the same agenda, until something goes wrong badly enough to force the conversation.

What they share, without having named it this way, is the same underlying problem. In each of their worlds, the data that would enable a better outcome already exists. The signals are there. The resolution logic is largely knowable. The gap is the distance between when something becomes visible in the data and when the organization actually acts on it. In Maria's world, that distance costs four hundred million dollars in trapped cash. In James's world, it shows up as 30 million dollars in annual penalties and a weekly meeting spent on problems that have already happened. In David's world, it is a release calendar that fills faster than it empties and a business that has started quietly routing around the systems it depends on.

None of these problems are unique to Korindal. If you have worked in a large enterprise that grew through acquisition, runs more than one ERP instance, and is trying to move faster than its current architecture allows, then you have already met versions of Maria, James, or David. You may be one of them.

The chapters that follow address each problem directly—not in the abstract, but through Korindal, showing what each capability looks like when it meets a real operating environment, real constraints, and people who need it to actually work. The problems are introduced one chapter at a time and resolved the same way: incrementally, practically, with each new capability building on the one before it.

By the time the full architecture comes together, the four-hundred-million-dollar receivables gap, the thirty-million-dollar fulfillment penalty, and the release calendar that never empties will not just be familiar problems. They will be solved ones.

1.9 Summary

If we look back at the Preface, the opening story was not really about a port. It was about the gap that exists in almost every large enterprise between when something becomes visible in the data and when the organization actually responds to it. That gap is where working capital becomes trapped, fulfillment penalties accumulate, and the cost of every exception grows with each hour it goes unaddressed.

The architecture designed to close that gap is a flywheel, not a feature set. Applications generate high-quality data as a by-product of standardized execution. That data, unified through SAP Business Data Cloud, feeds the intelligence layer. Intelligence produces better execution, better execution produces richer data, and the cycle compounds. Seven intelligent technologies are embedded directly in the applications in which work happens, not beside them: from generative AI and Joule at the interface layer, through SAP Knowledge Graph, predictive analytics, and process automation at the execution layer.

The operating model is a hierarchy. People direct through Joule, assistants orchestrate domain-specific workflows with role awareness, and agents execute multistep tasks autonomously within governed boundaries. The journey to that destination is structured: Augmentation builds trust and governance, orchestration delegates execution to agents, and transformation redesigns workflows around intelligence from inception.

None of this is abstract. Maria's receivables gap, James's fulfillment penalty, and David's release calendar that fills faster than it empties exist in some form in most large enterprises today. The data that would resolve each problem already exists. The chapters that follow build the model that closes the gap, one layer at a time, through the Korindal case study, to a level of detail that makes it implementable.

PART I

SAP Autonomous Suite

Chapter 2
SAP Cloud ERP Business Capabilities

SAP Cloud ERP is the operational foundation on which standardized processes, integrated data, embedded analytics, and intelligent automation come together. That matters because autonomy in the enterprise begins when the underlying business processes are structured, connected, and reliable enough for the system to sense change, support decisions, and increasingly execute routine work within governed boundaries. This is the role of SAP S/4HANA Cloud.

This chapter introduces the core business capabilities of SAP Cloud ERP across public and private cloud landscapes. It establishes SAP Cloud ERP as the digital foundation of the Autonomous Enterprise by showing how standardized best-practice processes support operational excellence, scalability, and continuous improvement. The chapter explores the major functional domains that run the enterprise—finance, commercial sales processes, procurement, manufacturing, and supply chain.

The goal of this chapter is not to catalog every feature. Instead, the chapter highlights selected capabilities that best illustrate how SAP Cloud ERP creates business value when applications, data, and intelligence operate together inside the digital core, SAP S/4HANA Cloud.

2.1 Finance

CFOs face a paradox that has only intensified with digital transformation. Organizations have more data than ever before, yet finance leaders struggle to answer fundamental questions in real time: Where is cash actually trapped, and what is causing it? Which customers are at risk of nonpayment? What is our true profitability by customer, product, and region? Why does the month-end close still take so long despite having invested in newer systems?

The deeper issue beneath these questions is not data scarcity but timing. By the time finance teams surface insights through traditional close and reporting processes, operational decisions have already been made. Collections have been missed, disputes have aged, and margins have been surrendered. Finance ends up explaining what happened rather than steering what should happen next.

This disconnect between when issues occur and when they are surfaced is what keeps chief financial officers (CFOs) up at night. It is the gap that drives working capital inefficiency, days sales outstanding (DSO) creep, and the perpetual tension between finance needing more visibility and operations feeling constrained by finance's need for control.

SAP S/4HANA Cloud reimagines finance around a fundamentally different operating model. Instead of accumulating transactions throughout the month and then reconciling exceptions during a frantic close period, the system surfaces issues in real time as they occur. Machine learning (ML) learns what is normal and flags what is not. Routine reconciliations happen automatically. By the time the official month-end close arrives, 80 to 90 percent of the work is already complete.

The foundation for this shift is the *Universal Journal*, a single, unified ledger structure that captures all value flows simultaneously: entity and group, statutory and management, all currencies. Everything else in the finance area builds upon this foundation.

In the following sections, we'll cover the main topics relevant for CFOs and their teams. As this typically is a wide area, we'll focus on the core processes in accounting, closing, controlling, accounts receivable, and accounts payable, as well as treasury and risk management, including cash management. Finally, we'll give an overview of capabilities for SAP Risk and Assurance Management.

2.1.1 Universal Journal

The Universal Journal is a single, unified ledger structure designed to capture all value flows simultaneously. Rather than separating entity accounting from group accounting, the system manages both perspectives within the same technical environment. An entity accountant posting a transaction in their local company automatically makes that same transaction visible to the consolidation team. There is no export, no import, and no timing variance.

Similarly, the Universal Journal provides one consistent perspective on statutory and management accounting. The general ledger accountant and the business analyst work from the same source data. When they present their results to the CFO, they are not reconciling two versions of the truth but presenting different analytical lenses on the same transaction set. The profit-and-loss (P&L) statement, balance sheet, and cash flow statement presented by the statutory reporter and the management accountant naturally reconcile because they flow from the same underlying journal entries.

The system also manages multiple currencies and multiple ledgers simultaneously. An organization can run International Financial Reporting Standards (IFRS), US Generally Accepted Accounting Principles (GAAP), and local country GAAP requirements in the same environment without manual conversion between ledgers. Currency management reflects the organization's actual reporting needs: local currency for subsidiary records, group currency for consolidated results, and functional currency for hedge accounting purposes.

The Universal Journal is built on a unified ledger architecture that consolidates all accounting views in a single technical environment. The master data structure—chart of accounts, cost centers, profit centers, and segments—is defined once and used consistently across all ledgers and entities. Consolidation scope and elimination rules are embedded within the journal logic, eliminating the need for separate consolidation systems.

The system maintains complete audit trails. Every transaction is linked to its source document. Every posting is tracked with timestamp and user ID. Reversals and adjustments show the complete history. This creates a natural audit trail that satisfies regulatory requirements without requiring separate audit preparation.

2.1.2 Record-to-Report

The record-to-report process in SAP Cloud ERP is built around a simple but powerful shift. Finance is no longer organized primarily around batch-based accumulation and downstream reconciliation. Instead, accounting, closing, consolidation, and reporting are brought closer together so that execution becomes more continuous, more automated, and more transparent.

This matters because record-to-report is one of the clearest examples of how complexity accumulates in traditional enterprise systems. General ledger accountants, closing specialists, business analysts, and consolidation experts often work on related activities in partially disconnected environments, with duplicated checks, manual clean-up steps, and follow-on adjustments performed late in the cycle. SAP S/4HANA redesigns that model by giving all these personas access to the same core financial foundation through the Universal Journal. The result is the ability to make closing more exception-driven, more traceable, and less dependent on heroic effort at month end. Several capabilities illustrate this shift particularly well, as we will discuss ahead.

Groupwide steering and overall closing support in SAP S/4HANA Cloud are both vital for the finance department. SAP S/4HANA uses the Universal Journal to increase steering effectiveness and improve decision-making.

In the following sections, we'll discuss key processes within record-to-report and highlight capabilities in SAP Cloud ERP.

Consolidation and Group Reporting

Historically, consolidation was a multistep, multisystem process. Entity accountants completed their closes and submitted ledger exports to the consolidation team. The consolidation specialists loaded these exports into a separate consolidation system. They validated that totals matched and investigated mismatches. Only then could they begin actual consolidation work: eliminating intercompany transactions, recording equity adjustments, translating foreign currencies, and so on.

This approach had inherent problems. Data had to be transferred between systems, creating timing differences. Currency translations performed in the consolidation system could differ from entity-level translations. Audit trails were fragmented across systems. Changes late in the close process required re-exporting and reloading data, creating delays.

The *SAP Cloud ERP, group reporting* application eliminates the data transfer step entirely. Entity accountants post to the Universal Journal in their company codes. Consolidation specialists access the same data immediately, with no export, no import, and no timing variance. The consolidation logic runs in-database. Consolidation adjustments—intercompany eliminations, equity adjustments, consolidation entries—are calculated and applied automatically. The consolidated financial statements are available within hours of the last entity close rather than days after. Because consolidation specialists are accessing data from the same Universal Journal that entity accountants posted to, there is no reconciliation between the entity view and consolidated view. They reconcile by design. Drilldown capability allows consolidation specialists to trace consolidated numbers back to entity postings to the line-item level.

Group reporting maintains the consolidation scope (which entities are included in consolidation), elimination rules (how intercompany transactions are eliminated), and consolidation adjustments (equity transactions, noncontrolling interests, other consolidation entries). The system accesses entity data in the Universal Journal and processes it in the database.

Consolidation eliminates intercompany transactions. When Entity A recorded a sale to Entity B and Entity B recorded a purchase from Entity A, these offset each other in group reporting. The system automatically eliminates both sides. The consolidated result shows only the external transaction (Entity A's sale to the external customer).

A key SAP Fiori app for intercompany elimination is the Intercompany Matching and Reconciliation app (ID F3870). In multientity organizations, intercompany transactions create persistent reconciliation challenges. Entity A records a sale to Entity B. Entity B records a purchase from Entity A. However, the amounts might not match; the seller might record at one price and the buyer at another. Or the timing might differ; the seller records in one month and the buyer in another. Intercompany payables and receivables must clear to zero at consolidation, but manual reconciliation of hundreds of intercompany items is error-prone and time-consuming. Consolidation cannot begin until intercompany balances reconcile. This becomes a critical path item in the group close as any delays in intercompany reconciliation delay the entire consolidation process.

The Intercompany Matching and Reconciliation app maintains a repository of all intercompany transactions: sales orders, purchase orders, invoices, and payments. It applies matching logic defined by the organization: match by purchase order number, match by transaction amount, or match by transaction date range. Machine learning recognizes timing patterns. If a particular supplier consistently takes three days longer to receive goods than the buyer records receipt, the system learns this pattern and reconciles automatically rather than flagging as mismatch.

Currency translation applies to foreign subsidiaries. If Entity B operates in a foreign currency, assets and liabilities are translated at current rates, equity at historical rates, and income statement items at average rates (per IFRS or other applicable standards). The system applies these translation rules automatically and consistently.

Consolidation adjustments capture transactions that eliminate at the group level. If Entity A paid a dividend to the parent, this is eliminated in consolidation. If there was an upstream or downstream sale between entities at a profit, that profit is eliminated. The system processes these adjustments automatically based on configured rules.

Advanced Financial Close

Unlike individual accounting innovations that automate specific steps, SAP Advanced Financial Closing acts as the orchestration layer for the entire financial close. It is a cloud-based application delivered on SAP Business Technology Platform (SAP BTP) that connects systems, tasks, and stakeholders into a single, coordinated execution model. Its role is not to perform accounting postings itself, but to structure, automate, and monitor how closing activities are executed across the enterprise.

SAP Advanced Financial Closing introduces a template-driven approach to the close. Organizations define repeatable closing blueprints that include manual tasks, automated jobs, dependencies, and approval steps. These templates generate task lists for each closing cycle, ensuring consistent execution across entities, countries, and accounting standards. Instead of relying on spreadsheets, emails, or local coordination, the close becomes a system-driven process with defined sequencing, ownership, and control points. The operational impact is most visible in how the close is monitored and managed. SAP Advanced Financial Closing provides a real-time, consolidated view of closing progress across all business units. Key elements visible in the interface include the following:

- Overall task completion rate and status distribution
- Lists of open, delayed, and overdue closing tasks
- Identification of errors and bottlenecks by task list and responsible role
- Approval queues and dependencies across the closing cycle
- Completion tracking by organizational unit and fiscal period

This level of transparency represents a fundamental shift. Instead of discovering delays or inconsistencies at period end, finance teams can identify issues as they emerge, understand their dependencies, and intervene immediately. Tasks are no longer isolated activities—they become part of a coordinated execution flow, where delays, approvals, and errors are visible in context.

From an architectural perspective, SAP Advanced Financial Closing complements the core finance capabilities of SAP S/4HANA Cloud. While the Universal Journal and group reporting provide the financial data foundation, SAP Advanced Financial Closing provides the execution discipline and governance layer across the close. In distributed landscapes, it can orchestrate activities across multiple systems without requiring data replication, acting as a central control point for entity and group close processes.

There are two SAP Fiori apps worth mentioning in the financial close area:

- **Goods and Invoice Receipt Reconciliation app (ID F3302)**

  In high-volume procurement operations, goods receipts and invoice receipts frequently arrive at different times. When goods arrive before invoices, items accumulate on the goods receipt/invoice receipt (GR/IR) account. When invoices arrive before goods, they accumulate there. At month end, the GR/IR account often contains hundreds of line items that do not match. The closing team must investigate each one: Is this a legitimate timing difference? A quantity discrepancy? A pricing error? Which ones will clear naturally with follow-up transactions? Which ones require manual adjustment?

  The Goods and Invoice Receipt Reconciliation app identifies unmatched items and applies machine learning categorization. The system learns from historical clearing behavior: which items clear naturally, which require adjustment, and which have specific causes. Configuration allows each company to define tolerance thresholds for automatic clearing. Items below tolerance (say, a $100 difference on a $10,000 invoice) are automatically cleared. Items above the tolerance are routed to the appropriate reviewer based on business rules.

  SAP uses intelligent technologies to sort and filter the often-high volumes on this account (think of manufacturing and retail companies and the goods receipts they are managing daily) and to prepare the work so that the accountants can focus on the most critical differences to be cleared first, as shown in Figure 2.1.

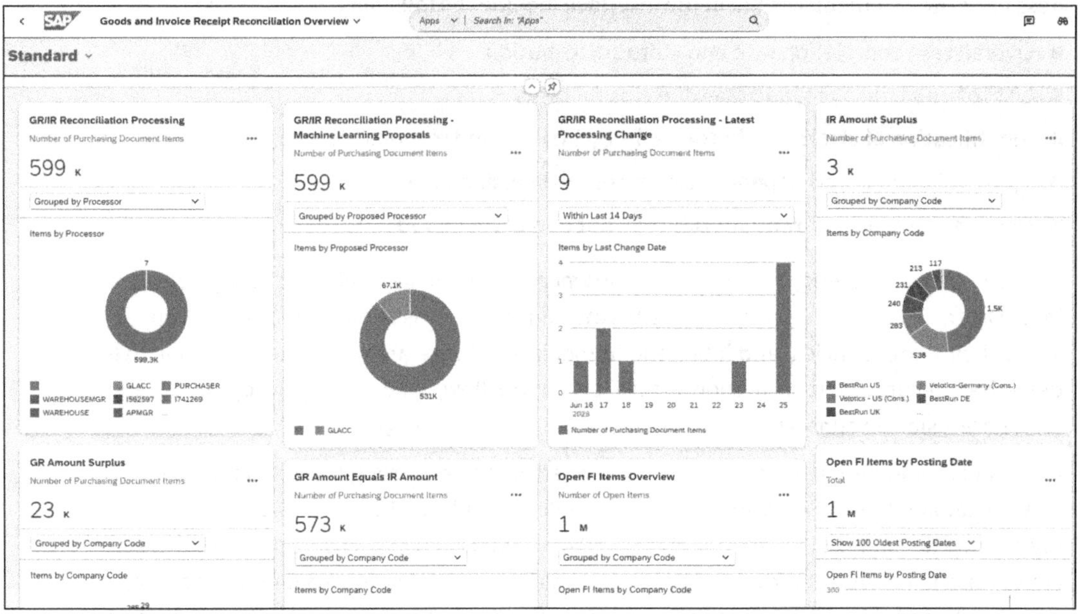

Figure 2.1: GR/IR Reconciliation

The system not only learns from what has been processed in the past but also can handle large amounts of work automatically. Over time, these capabilities can allow your accountants to shift their focus to just attending to outliers and errors—and let the machine learn and process regular tasks more regularly.

- **Accruals Management app (ID W0183)**
  At month end, open purchase orders must be accrued to reflect committed spending, but determining the right accrual amount is subjective and time-consuming. Should you accrue the full outstanding purchase order amount? This is mathematically correct and passes audit scrutiny, but it often overstates the actual obligation. Should you adjust based on historical delivery patterns? This is more realistic—some suppliers consistently deliver less than the full purchase order amount—but it requires judgment and consistency is difficult to maintain across multiple accountants. In organizations with thousands of open purchase orders at month end, determining accrual amounts can consume two to three days of the closing team's time.

  For each open purchase order requiring accrual, the system displays two proposals: the mathematical accrual (full outstanding purchase order amount) and the intelligent accrual (based on that specific supplier's historical delivery and invoice patterns for that product category). The closing specialist reviews both proposals and selects one or adjusts if needed. Configuration allows each company to set accrual triggers (accrue all open purchase orders above $X amount or apply different rules by supplier classification). Tolerance thresholds determine which accruals are automatically posted versus requiring manual review. The system captures the reasoning for each accrual decision in the audit trail.

In the context of the Autonomous Enterprise, SAP Advanced Financial Closing represents an important step beyond automation. It does not simply make tasks faster; it makes the process itself more intelligent and controllable. Closing activities can be triggered automatically based on dependencies, monitored continuously, and progressively shifted toward low-touch or no-touch execution. Finance remains fully accountable for outcomes, but it increasingly operates in an environment where the system coordinates execution, highlights exceptions, and supports timely action rather than reacting after the fact.

Figure 2.2 shows the SAP Advanced Financial Closing solution, which provides a centralized task list to manage, monitor, and accelerate the financial close process across legal entities and fiscal periods.

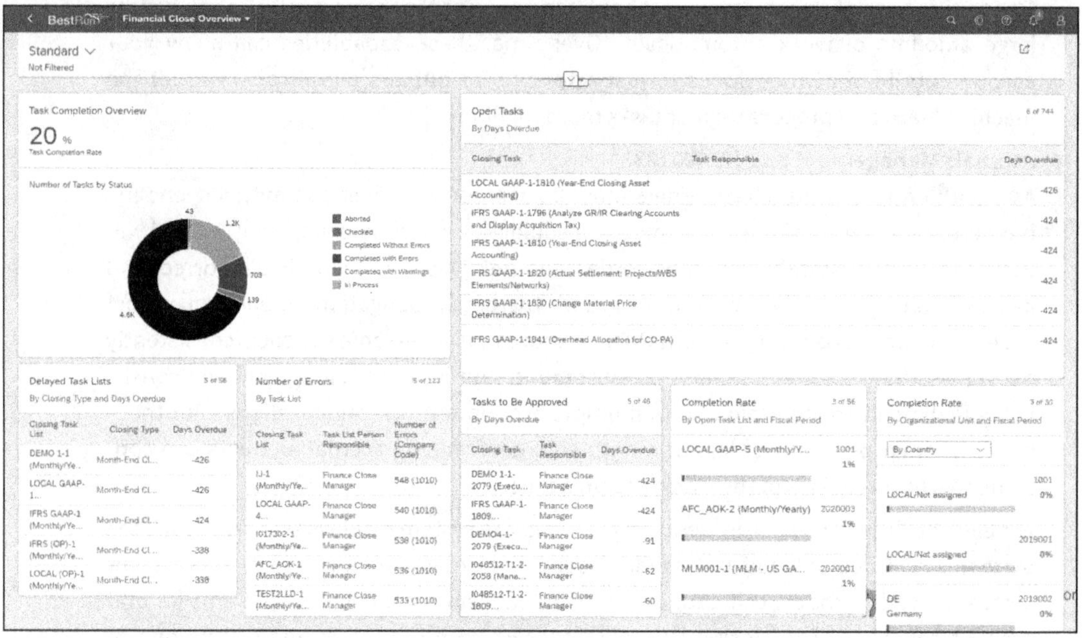

Figure 2.2: SAP Advanced Financial Closing: Managing All Closing Tasks to Achieve Faster Closing Cycles and Compliant Results

Event-Based Revenue Recognition

Event-based revenue recognition (EBRR) is a more sophisticated approach to revenue timing. Rather than waiting for milestone achievement or period-end processing, event-based recognition ties revenue directly to execution events that occur in real time. When a service is confirmed as complete, when a project task is marked finished, or when goods physically ship, those are the moments revenue should be recognized.

Traditional revenue accounting approaches introduce timing gaps. Work is done on Monday, but revenue is not recognized until the invoice is issued on Friday. By the time revenue is recognized, a few days have passed. For businesses with continuous execution and invoicing, this lag can create distorted financial visibility. This delay introduces reconciliation effort, reduces transparency during the period, and limits the ability of finance teams to influence outcomes before they are finalized.

EBRR addresses this limitation by recognizing revenue and cost directly from execution events. As business activities occur—such as service confirmations, project progress postings, goods movements, or delivery milestones—the corresponding financial impact is captured automatically. Instead of waiting for batch-driven processes, financial representation becomes continuous and aligned with real operational activity.

The impact of this shift is both operational and strategic. From an operational perspective, finance teams gain the following:

- Near-real-time visibility into revenue and margin development
- Reduced dependence on manual accruals and reconciliation steps
- A tighter connection between operational events and financial outcomes

This capability is particularly important for modern business models in which revenue is no longer tied to a single delivery point. It provides a structural foundation for the following:

- Service-centric and project-based revenue models
- Usage-based and consumption-based billing scenarios
- Integrated lead-to-cash processes in which financial outcomes must closely follow execution

There are use cases in which full IFRS 15/ASC 606 contract accounting is needed and therefore traditional revenue accounting is more suitable. Revenue recognition is both a compliance requirement and an operational reality. When revenue is recognized affects both compliance reporting (IFRS 15, ASC 606) and financial visibility. Traditional systems delay revenue recognition until month-end close or invoice generation, which creates a timing lag between when work is actually completed and when revenue is recognized in the financial statements.

SAP S/4HANA's revenue accounting and reporting functionality is supported by the Revenue Accounting app. The Revenue Accounting app monitors execution events across logistics, project management, and service management. Events such as when a service confirmation posts, when a project time entry is submitted, or when goods are picked and shipped trigger revenue recognition rules. Configuration defines how each event type maps to revenue recognition. For a services engagement, time confirmations trigger revenue recognition at the contracted rate. For a product sale, goods issue triggers revenue at the contracted price. For a subscription, the system recognizes recurring revenue on the subscription date, not at renewal or billing.

Architecturally, EBRR is deeply integrated into SAP's finance data model. Because it operates on the same foundation as the Universal Journal, it ensures that revenue recognition, margin analysis, and financial reporting all draw from a single, consistent source of truth. This eliminates the need for post hoc reconciliation between operational systems and financial results and improves both accuracy and auditability. In practical terms, finance teams interact with this capability through SAP Fiori applications that provide visibility into recognized revenue, work in progress, and profitability at a granular level. However, the value does not reside in the user interface (UI) alone. The real transformation lies in the underlying accounting logic and process integration, which allow financial outcomes to be captured automatically as part of normal business execution.

In the context of the Autonomous Enterprise, EBRR represents a critical shift from period-end reconstruction to continuous financial awareness. It enables a model in which finance does not wait for the business to complete before understanding its impact. Instead, it evolves alongside the business, providing timely, reliable, and decision-ready information at every stage of execution.

Cost Center Accounting

Management accounting relies heavily on financial analysis. Cost center managers need to understand their spend and performance. Traditionally, this analysis starts with running reports: cost center P&L statements that show actual costs versus the budget. Then comes manual investigation: Why is this line item above budget? Is this a one-time variance or a trend? What is driving the variance and what should we do about it?

The process is time-consuming. A cost center manager might spend hours running reports and investigating variances only to end up preparing a manual narrative explaining the results. For organizations with hundreds of cost centers, this monthly exercise is substantial.

The Cost Center Review Booklet app (ID F7763) transforms management accounting from fragmented reporting into a guided, system-driven financial review process. Instead of assembling insights manually from multiple reports, controllers and managers are provided with a prestructured analytical booklet that consolidates balances, plan-versus-actual comparisons, variances, and trends into a single, coherent view.

Built as an SAP Fiori application on top of SAP S/4HANA Cloud's finance data model, the booklet enables seamless navigation from high-level key performance indicators (KPIs) to detailed line-item analysis, ensuring both consistency and depth in financial reviews. Review cycles become repeatable and standardized, improving alignment across teams and reducing interpretation variability.

A key innovation is the integration of AI-assisted insights, which help summarize variances, highlight anomalies, and generate narrative explanations directly from financial data. This reduces manual effort in report preparation and allows finance teams to focus on interpretation and decision support.

2.1.3 Financial Planning and Analysis

Finance leaders are increasingly expected to answer forward-looking questions. If revenue grows 10 percent, what happens to operating margin? Which cost centers are tracking above budget? What will the cash position be next quarter? Nonetheless, most organizations rely on historical data and manual spreadsheet-based forecasting.

SAP S/4HANA Cloud integrates planning, budgeting, and forecasting directly with actual operating data through SAP Analytics Cloud, an end-to-end cloud solution that combines business intelligence, augmented analytics, predictive analytics, and enterprise planning in a single system. Plan data is maintained in the same structure as actual data, enabling real-time budget versus actual comparisons without manual harmonization. Finance leaders access the data they need through visualizations, stories, and analytic applications that are personalized to their role, such as CFO, controller, or cost center manager. Scenario modeling is built in: What if sales increase 5 percent? What if material costs increase 8 percent? Machine learning identifies anomalies and trends that humans might miss. Smart features use artificial intelligence to automatically discover meaningful insights, reveal hidden patterns in data, and forecast potential outcomes. The planning cycle itself is collaborative and iterative. Finance teams prepare the planning model, execute the planning phase with input from business units, and then finalize and analyze results, often rolling into the next quarter's forecast without starting from scratch.

SAP provides preconfigured planning content across a broad set of areas including cost centers, profit centers, balance sheet, profit and loss, liquidity, workforce-related planning, production costs, sales, and margins. Because plan data can be maintained in structures aligned with actual data, comparisons are more immediate and less dependent on manual harmonization. The planning cycle becomes shorter not because finance spends less care on planning, but because less effort is wasted translating between incompatible views of performance. The following capabilities are especially important in this area:

- **Margin analysis**
  Margin analysis, or the product profitability analysis capability, provides real-time profitability visibility. Because the Universal Journal encompasses full margin analysis capabilities, there is no separate profitability system that must be reconciled to general accounting. Business analysts access necessary detail at the line-item level, and their work harmonizes with overall accounting rather than conflicting with it.
- **Predictive accounting**
  Predictive accounting uses current data from SAP S/4HANA, integrated systems, and external sources to forecast future financial results in real time. When a sales order is created, predictive accounting generates predictive journal entries in a special prediction ledger for the expected goods issue and billing. As actual transactions occur, predictive accounting updates these forecasts by adjusting the prediction ledger amounts. Because both actual and predicted data are stored in the Universal Journal, business analysts can compare predicted versus actual results at any level, from financial statements down to individual line items. This eliminates the need for separate profitability systems or reconciliation between prediction and accounting. Finance teams gain visibility into what results will look like before the period closes, enabling faster, more informed decisions.
- **Finance-owned greenhouse gas emission planning**
  Greenhouse gas (GHG) emission planning extends financial planning to include a carbon dimension. Organizations can model trade-offs between cost and environmental impact: If we shift to a lower-cost supplier, what is the impact on our carbon footprint? If we optimize transportation routing, what is the cost savings and the emissions reduction? Sustainability considerations are embedded into core financial planning rather than treated as a separate compliance exercise.

  GHG emission planning reuses established financial planning structures, including quantity structures, cost centers, products, and allocation rules, to distribute emissions to relevant business objects. By applying emission factors to planned financial and operational data, finance teams can simulate future corporate and product-level carbon footprints and analyze how changes in volumes, cost structures, or sourcing decisions affect both financial and environmental outcomes. This approach enables sustainability metrics to be embedded into core planning and steering processes, rather than treated as a separate or downstream activity. As a result, finance organizations can incorporate sustainability considerations into their regular planning cycles, perform scenario simulations, and assess trade-offs between financial targets and emissions-reduction objectives within a single, integrated planning environment.

2.1.4 Invoice-to-Pay and Invoice-to-Cash

The invoice-to-pay and invoice-to-cash processes cover vendor- and customer-related finance processes when you pay invoices or collect cash.

The *invoice-to-pay process*, or payables management, covers the finance part of the broader procure-to-pay process that spans from vendor relations and actual procurement to payment. An example of an SAP Cloud ERP innovation for payables and cash managers with central payment management is discussed later in this section.

The *invoice-to-cash process*, or receivables management, includes all financial accounting operations related to accounts receivable, including billing customers and managing disputes, collections, and customer credit risk. Today, receivables managers must constantly balance the need to decrease DSO and bad debt write-offs while maintaining superior customer service. These managers must ensure that cash is on hand to fund business operations and new growth opportunities at low cost to deliver on margin goals.

The invoice-to-cash and invoice-to-pay processes sit close to one of the most tangible concerns of the CFO: working capital. In many companies, cash is not absent; it is delayed. It is trapped in disputes, unapplied payments, unresolved collections, blocked invoices, and process handoffs that occur outside the system or too late in the cycle to influence results. The value of cloud ERP in this area lies in bringing those activities back inside a governed operating loop.

Let's discuss a couple of key capabilities in this space:

- **SAP Cash Application**

  The final step in the invoice-to-cash process is cash application: matching incoming customer payments to outstanding invoices. This is often tedious, time-consuming work. Imagine that a customer sends a payment but does not indicate which invoices it should apply to. The payment might be a partial payment of multiple invoices, an overpayment, or an underpayment. The accounts receivable team must manually investigate: Which invoices does this payment cover? Are they applying it correctly?

  SAP Cash Application is one of the clearest examples of embedded intelligence delivering direct business value. Matching incoming payments to open invoices has always been a necessary but labor-intensive activity. Traditional rule-based automation can help, but it often becomes expensive to maintain and less effective over time. SAP's ML-based matching model improves on this by learning from past clearing behavior and using that history to propose or automate matches with greater contextual precision. Beyond productivity improvements, faster and more accurate cash application also improves accounts receivable transparency, reduces artificial inflation of DSO caused by delayed clearing, and allows finance teams to focus on true collection and dispute issues rather than processing noise.

- **Collections Worklist app**

  Collections teams traditionally work from reports exported from the accounts receivable system. Priorities are often ad hoc; collectors follow aging reports or work through customers in alphabetical order. This inconsistency means collection effort is not focused on the receivables that matter most from a cash perspective.

  The Collections Worklist app (ID F2375) generates a prioritized list of receivables that collections team members should work on each day. The list is prioritized by cash impact: The highest dollar items are listed first, then the customers at highest risk of nonpayment within that category, followed by the oldest items within that category. The list updates continuously as payments arrive and disputes are resolved. Instead of deciding what to work on each morning based on habit or intuition, collectors start with a focused, actionable list generated by the system.

  As shown in Figure 2.3, the Supervise Collections Worklist app accesses the accounts receivable subledger and applies prioritization rules. The system calculates each receivable's priority based on due date, customer credit rating, payment history, days outstanding, and any custom business rules the organization has configured. It generates a daily list of 20 to 30 receivables that the collector should work on.

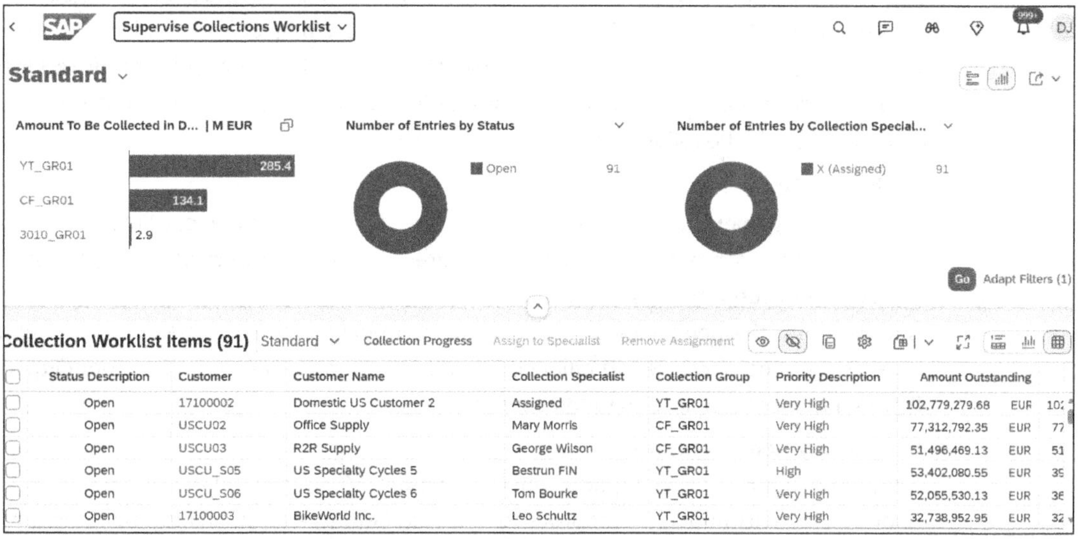

Figure 2.3: Supervise Collections Worklist

On the payable side, SAP Cloud ERP provides the same architectural advantage in reverse. Invoice-to-pay is not simply about paying vendors efficiently; it is also about controlling liabilities, reducing friction with procurement, and improving visibility across commitments and actuals. When payables processing, accruals handling, treasury visibility, and close activities are structurally connected, finance gains better control over both outbound cash and period-end accuracy. In this sense, invoice-to-pay and invoice-to-cash should not be treated as disconnected subprocesses. Together, they form the transactional edge of working capital execution.

2.1.5 Treasury and Cash Management

Treasury teams manage cash positions across multiple bank accounts and currencies, maintain banking relationships, forecast liquidity, and manage financial risks. Historically, cash position visibility was fragmented. Information came from multiple banks with delays in reporting. Manual reconciliation was required to consolidate positions across accounts and currencies.

Treasury in SAP Cloud ERP is not treated as a disconnected financial specialist domain sitting downstream of accounting. It is integrated into the same finance foundation that supports cash visibility, payment execution, exposure monitoring, and auditability. This matters because liquidity and risk are highly sensitive to timing, fragmentation, and manual handling. Treasury value increases materially when information is timely, shared, and actionable.

SAP supports treasury and risk management with capabilities for exposure visibility, hedging policy support, integration with trading environments, and automated hedge accounting. For organizations dealing with market exposure, debt instruments, and complex treasury operations, this reduces manual effort while improving control and compliance. Straight-through processing across trade handling, valuation, and accounting reduces spreadsheet dependence, improves documentation, and strengthens the audit trail. Key capabilities include the following:

- **Cash management**
  In cash management specifically, the system is designed to provide real-time analysis of bank balances, cash positions, and liquidity forecasts with drilldown capability and integrated action support. This is important not only from the perspective of the treasurer but also from the perspective of the CFO. Cash management becomes more valuable when it does not merely report balances but also supports earlier decisions about liquidity risk, payment timing, and funding needs.
- **Central payment monitoring**
  Central payment monitoring adds another important layer by addressing fragmented bank connectivity and payment processing across heterogeneous environments. By providing a more unified control point for payment monitoring and approval and by improving the visibility of cash flow updates during processing, central payment monitoring reduces operational cost, strengthens control, and lowers fraud exposure associated with manual or file-based processes.

- **Liquidity forecasting**
  As shown in Figure 2.4, the Liquidity Forecast app provides real-time visibility into key information such as current liquidity position. You can also drill down to the company level to visualize the cash flow versus the closing balance. These insights allow treasury teams and finance leaders to see risks and opportunities and take appropriate action.

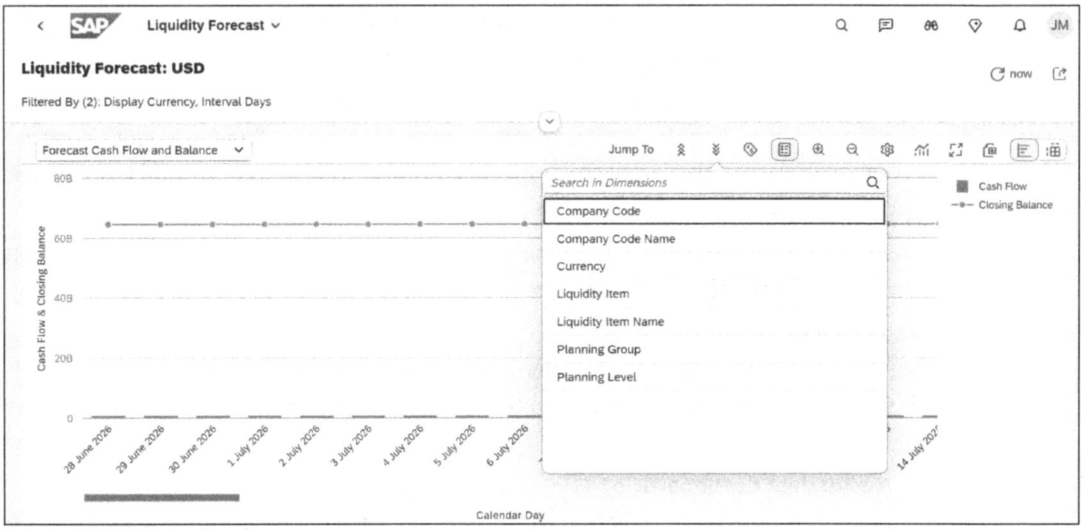

Figure 2.4: Liquidity Forecast

2.1.6 Financial Compliance Management

SAP Risk and Assurance Management, built on SAP BTP, is the next-generation control and compliance software-as-a-service (SaaS) offering integrated into SAP S/4HANA Cloud. The solution allows users to manage controls, compliance activities, and issues from compliance managers, process owners, control owners, and control performers. In addition, SAP Risk and Assurance Management provides real-time checks on inconsistencies, policy deviations, and unauthorized or fraudulent activities. Thus, it supports organizations in establishing preventive and detective measures to avoid the potential of financial misstatements, violation of national or international laws and regulations, and subsequent financial losses.

While covering the comprehensive compliance management lifecycle from documentation, execution, and issue handling to reporting and monitoring, the solution offers the following guiding features and functionalities, as shown in Figure 2.5:

- **Documentation**
  Documentation of control information and associated attributes includes the description, significance, risk level, and related organizations, processes, and regulations, accompanied by manual or automated procedures as applicable.
- **Execution of procedures**
  The control execution is supported by manual or automated procedures:

  - The *manual procedures* provide detailed step-by-step guidance to the assigned control performers, allowing them to record individual results and comments.
  - The *automated procedures* support the identification of process or policy-based exceptions or deviations using predelivered or custom-generated OData services, a standard web protocol for data exchange. The associated business logic and rules can be individually adapted to meet customer-specific requirements in identification of potential findings and subsequent issue handling.

  Automated and manual procedures can be combined to realize the utmost flexibility needs and demanding use-case scenarios. Automated scheduling for execution is available to help ensure proper control performance at all necessary times and frequencies.

- **Issue handling**
  Issues are documented and processed, supported by various classification and tracking options and meeting structured reporting and auditing objectives.
- **Reporting and monitoring**
  SAP Risk and Assurance Management offers an integrated, holistic control analytics dashboard with drilldown features in the cloud within SAP Analytics Cloud, as shown in Figure 2.5:
  - **Master data dashboard**
    Identification and remediation of potential master data inconsistencies, including missing assignments, owners, or links to processes, control, or manual/automated procedures
  - **Compliance dashboard**
    Overview of real-time compliance and issue status, including comparison features across organizations, regions, and processes

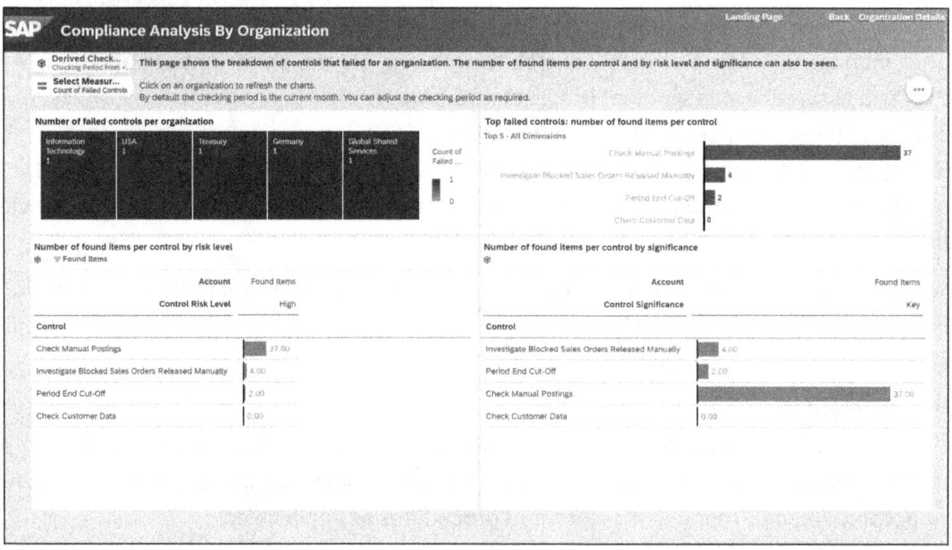

Figure 2.5: Compliance Dashboard: Review Compliance Status by Organization and Follow-Up on Failed Controls

Figure 2.5 shows the compliance dashboard, offering a real-time overview of control execution status and open issues across organizational units, enabling proactive financial compliance management.

SAP Risk and Assurance Management offers tracking and monitoring of controls and associated activities to support financial compliance for multiple regulations organization-wide while providing automation to streamline efforts and reduce costs.

Furthermore, the application is planned to allow users to deploy predefined regulation and control content, including best-practice automated and manual procedures via the Global Risk and Compliance Business Content Hub, a platform for the general and partner ecosystem that provides preconfigured, ready-to-use content.

The significance of this capability is broader than compliance alone. It shows how SAP Cloud ERP supports a central principle of the Autonomous Enterprise: More automation does not reduce the need for control; it increases the need for control to be built into the design of execution itself.

2.1.7 Outcome

In Chapter 1, Section 1.8, we introduced Maria, a CFO facing a familiar problem in Case Study 1: Cash flow was slow despite investing in the right systems and building the right teams. The issue was not visibility but timing. Cash was trapped in unresolved disputes, delayed collections, unapplied payments, and late-month handoffs. By the time issues became visible, intervention was no longer possible. The organization could only absorb the cost.

How does integration change the outcome?

The finance capabilities described in this section do not improve working capital individually. Instead, their value emerges when organized as a continuous process.

The Supervise Collections Worklist app generates a daily prioritized list of receivables ranked by cash impact and risk. Collections specialists start each day with a focused list instead of deciding what to work on. The Event-Based Revenue Recognition app (ID F2441A) recognizes revenue as work completes, allowing finance to see receivables forming during the month rather than discovering them after they have aged. The Dispute Management app (ID F0702A) tracks amounts under dispute and identifies patterns: Which customers dispute repeatedly? Which billing errors arise? Which operational issues? By fixing root causes rather than just clearing disputes, the organization prevents future problems.

SAP Cash Application removes friction at payment matching. The cash management dashboard provides continuous visibility into whether collection efforts are actually improving liquidity. Finance can intervene immediately if something changes.

The Operating Rhythm

Collections specialists work through the prioritized worklist daily. Disputes are tracked and root causes identified continuously. Management adjusts strategy based on results weekly. Finally, the close includes current working capital analysis rather than postmortem explanation monthly.

This brings us to the outcome: Cash is released more steadily. Collections work becomes focused and repeatable. Disputes are resolved faster and recur less often. The CFO's organization no longer struggles with working capital trapped in the dysfunction of reactive cash management.

The improvement does not come from new systems or new people. Existing capabilities, organized to work together and embedded into daily work, created the shift.

2.2 Sales

Sales in SAP S/4HANA Cloud represents the execution layer of the *lead-to-cash* process, in which customer demand is translated into revenue through coordinated order capture, fulfillment, billing, and financial integration. At its core, sales focuses on sales order and contract management, beginning with an inquiry or quote and concluding with billing, returns, and receivables management in finance.

What distinguishes sales in SAP S/4HANA is not simply automation of transactional processes; the ability to operate as an integrated, data-driven, and increasingly exception-based execution model is key. The system connects upstream customer engagement processes with downstream fulfillment and finance, ensuring that sales is no longer an isolated function but part of an end-to-end enterprise flow.

SAP S/4HANA Cloud redesigns sales operations around one principle: low-touch intelligent execution of high-volume routine work, with expert focus reserved for complex, relationship-driving scenarios.

Orders arrive through multiple channels—web, Electronic Data Interchange (EDI), email—but the system processes them identically. Automated checking validates credit limits, compliance requirements, and inventory availability without manual intervention. Workflows escalate only genuine exceptions, letting routine orders proceed automatically. Because sales, fulfillment, finance, and collections work from the same data, problems surface immediately rather than being discovered after they impact the customer.

Now, let's walk through the key capabilities that are part of SAP S/4HANA Sales.

2.2.1 Quote-to-Order

The quote-to-order process is where a customer's needs become an operational commitment. A customer sends an inquiry, sales responds with a quotation, the customer approves it, and the quotation becomes a sales order that drives manufacturing, procurement, and fulfillment.

Two roles drive this process. The pricing specialist governs the pricing strategy: base prices, volume discounts, promotional pricing, and customer-specific terms. The sales representative engages with customers, responds to inquiries, and prepares quotations based on this pricing framework.

We'll take a closer look at key subprocesses within quote-to-order in the following sections.

Pricing

Pricing is more dynamic than most organizations expect: Product lifecycles shorten; offerings become more complex; multiple pricing strategies apply across different markets and customer segments; market conditions change; pricing assumptions that were valid last quarter may no longer be valid this quarter.

Traditional approaches—pricing maintained in spreadsheets, pricing logic buried in email discussions—create bottlenecks. When a sales representative needs current pricing, they either wait for a response or quote based on outdated information. Customers see different prices than expected, and finance discovers pricing errors weeks after the invoice.

SAP S/4HANA Cloud centralizes pricing logic. Pricing rules are defined once and applied consistently everywhere. A salesperson enters a customer and product, and the system shows the applicable price immediately. Volume tiers and promotional pricing apply automatically, and customer-specific pricing appears automatically. If the salesperson wants to deviate, guardrails prevent margin erosion. Discounts below approved thresholds require manager approval.

The following are a few useful SAP Fiori apps for SAP S/4HANA pricing:

- **Manage Prices - Sales (ID F4111)**
  This app lets you manage all customer pricing conditions in one place, enabling flexible pricing. You can create, edit, copy, or delete multiple price records simultaneously.
- **Manage Tax Rates (ID F6972)**
  This app lets you define tax rates as condition records in your sales master data. When the system calculates tax for a business document, it uses the corresponding condition record to determine the final tax rate. Mass operations let you manage multiple tax rates efficiently.
- **Manage Sales Documents with Customer-Expected Price (ID F2713)**
  This app shows sales document items blocked by price mismatches between the net price and the customer-expected price. You have three options: Accept the customer-expected price and manually adjust the net price, decline it and release the item for processing with the original net price, or reject it to stop processing the item entirely. The latter option is useful when you cannot agree on a price and the customer no longer wants the item. You can also view customer contact information for the selected item and navigate to related apps to view or modify the corresponding sales document.

Configurable Products

For companies selling customizable or engineered products, the quoting process guides customers through options. If a customer chooses option A, which other options are still available? If they choose this configuration, what does delivery look like? How does it affect price?

The system enforces technical constraints and dependencies. A customer cannot specify an incompatible combination of options. As they make choices, the price and delivery date update in real time. By the time the quotation is prepared, both the customer and the salesperson know exactly what they are agreeing to.

The following are a few useful SAP Fiori apps for configurable products:

- **Variant Configuration Overview (ID F2940)**
  This feature displays an overview of your variant configurations. From the overview, you can quickly navigate to changed or locked configuration objects, blocked sales order items, recent object dependencies, recent simulations, and favorite configurations. You can also view a list of all variant configuration objects and open their corresponding app to modify them.
- **Simulate Configuration Models (ID F2570)**
  This feature enables you to search for configurable products and configured objects and start a configuration simulation with them. You can use the simulation to test and analyze your configuration model by configuring your object using the integrated characteristic value assignment screen and checking whether your dependencies work. You can also display the configured BOM as the result of your configuration.

From Quotation to Order

Once the customer approves the quotation, it converts directly into a sales order. The agreed-upon pricing, product configuration, and delivery terms carry forward. There is no re-entry, no renegotiation, no version confusion. The quotation becomes the binding commercial commitment that the operations team executes against.

For complex deals requiring multiple rounds of negotiation, the system maintains a version history. Customers see what changed between quotation versions. Approvals happen within the system, not via email chains.

The following are a few useful SAP Fiori apps for the quote-to-order process:

- **Quotation Conversation Rates (ID F1904)**
  This app uses machine learning to predict which quotations are likely to convert into sales orders. View your quotation conversion rates here.
- **Sales Contract Fulfillment Rates (ID F1905)**
  This analytical app displays your sales contract fulfillment rates, including the percentage of target value released in each sales contract. As a sales representative or manager, use this KPI to analyze how well contracts in your sales area are being fulfilled. It can

quickly identify contracts due for renewal and contracts underperforming against expectations.

The outcome is that quotations that would historically take days to prepare now take hours. Pricing is consistent across all customer interactions. Quotations are technically feasible before they reach operations. When a quotation converts to an order, there is no surprise about what everyone actually agreed upon.

2.2.2 Order and Contract Management

Order capture and fulfillment form the core of sales execution and are central responsibilities of the internal sales representative. In today's environment—characterized by increasing order volumes and the need for strict control of sales operating costs—efficiency and scalability are essential. Unlike traditional models, in which manual entry and tracking of orders were feasible, modern sales operations require automated order capture, low-touch processing, integrated compliance, and exception-based execution models.

SAP S/4HANA Cloud addresses these requirements by enabling the following:

- Automated and system-assisted order capture
- Integrated compliance with export, product, and trade regulations
- Embedded credit and risk management
- Workflow-driven approvals and exception handling

At the same time, contracts play a dual role: They support revenue growth by enabling flexible commercial agreements while ensuring that agreed-upon terms and service-level agreements (SLAs) are consistently enforced during fulfillment.

Let's examine the key features and types of contracts supported in SAP S/4HANA.

Manage Customer Contracts

A central responsibility in sales operations is the creation and management of customer contracts, which define the commercial framework for future transactions. SAP S/4HANA Cloud supports multiple contract types to address different business requirements:

- **Sales value contracts**
  A *value contract* defines a total monetary commitment over a specified period, allowing customers to draw products or services up to a predefined value threshold.
- **Sales quantity contracts**
  A *quantity contract* specifies the total quantity of a product to be purchased within a defined time frame, without fixing delivery dates. An important enhancement in SAP Cloud ERP is support for down payments linked to quantity contracts. This functionality enables the following:
  - Payments can be requested via billing plans
  - Delivery blocks are automatically enforced if prepayments are insufficient

  - Blocks are lifted once payments are received
  - The final billing deducts prepayments accordingly
- **Condition contracts**
  Condition contracts define pricing or settlement conditions, including rebates, commissions, and royalties. These contracts support both of the following:
  - Immediate pricing conditions applied during transactions
  - Subsequent settlements processed after fulfillment

  This enables a broad range of commercial models, including rebate programs and incentive-based selling. To support contract execution, SAP S/4HANA Cloud leverages situation handling, which proactively alerts users when contracts deviate from expected behavior—for example:
  - Contracts under- or overperforming relative to elapsed time
  - Contracts approaching expiration without follow-on agreements
  - Disproportionate fulfillment rates

Figure 2.6 shows the Manage Sales Contracts app, which is a one-stop shop to provide key insights, including the valid-from date, overall status, and valid-to date.

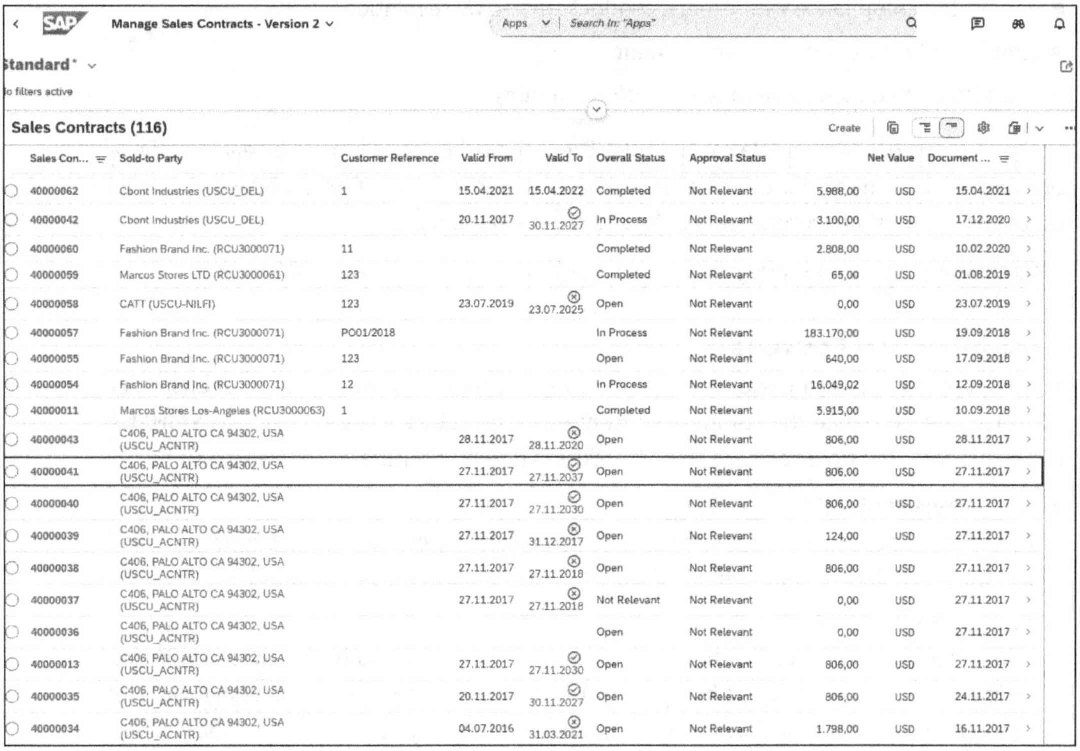

Sales Con...	Sold-to Party	Customer Reference	Valid From	Valid To	Overall Status	Approval Status	Net Value		Document ...
40000062	Cbont Industries (USCU_DEL)	1	15.04.2021	15.04.2022	Completed	Not Relevant	5.988,00	USD	15.04.2021
40000042	Cbont Industries (USCU_DEL)		20.11.2017	30.11.2027	In Process	Not Relevant	3.100,00	USD	17.12.2020
40000060	Fashion Brand Inc. (RCU3000071)	11			Completed	Not Relevant	2.808,00	USD	10.02.2020
40000059	Marcos Stores LTD (RCU3000061)	123			Completed	Not Relevant	65,00	USD	01.08.2019
40000058	CATT (USCU-NILFI)	123	23.07.2019	23.07.2025	Open	Not Relevant	0,00	USD	23.07.2019
40000057	Fashion Brand Inc. (RCU3000071)	PO01/2018			In Process	Not Relevant	183.170,00	USD	19.09.2018
40000055	Fashion Brand Inc. (RCU3000071)	123			Open	Not Relevant	640,00	USD	17.09.2018
40000054	Fashion Brand Inc. (RCU3000071)	12			In Process	Not Relevant	16.049,02	USD	12.09.2018
40000011	Marcos Stores Los-Angeles (RCU3000063)	1			Completed	Not Relevant	5.915,00	USD	10.09.2018
40000043	C406, PALO ALTO CA 94302, USA (USCU_ACNTR)		28.11.2017	28.11.2020	Open	Not Relevant	806,00	USD	28.11.2017
40000041	C406, PALO ALTO CA 94302, USA (USCU_ACNTR)		27.11.2017	27.11.2037	Open	Not Relevant	806,00	USD	27.11.2017
40000040	C406, PALO ALTO CA 94302, USA (USCU_ACNTR)		27.11.2017	27.11.2030	Open	Not Relevant	806,00	USD	27.11.2017
40000039	C406, PALO ALTO CA 94302, USA (USCU_ACNTR)		27.11.2017	31.12.2017	Open	Not Relevant	124,00	USD	27.11.2017
40000038	C406, PALO ALTO CA 94302, USA (USCU_ACNTR)		27.11.2017	27.11.2018	Open	Not Relevant	806,00	USD	27.11.2017
40000037	C406, PALO ALTO CA 94302, USA (USCU_ACNTR)		27.11.2017	27.11.2018	Not Relevant	Not Relevant	0,00	USD	27.11.2017
40000036	C406, PALO ALTO CA 94302, USA (USCU_ACNTR)				Open	Not Relevant	0,00	USD	27.11.2017
40000013	C406, PALO ALTO CA 94302, USA (USCU_ACNTR)		27.11.2017	27.11.2030	Open	Not Relevant	806,00	USD	27.11.2017
40000035	C406, PALO ALTO CA 94302, USA (USCU_ACNTR)		20.11.2017	30.11.2027	Open	Not Relevant	806,00	USD	24.11.2017
40000034	C406, PALO ALTO CA 94302, USA (USCU_ACNTR)		04.07.2016	31.03.2021	Open	Not Relevant	1.798,00	USD	16.11.2017

Figure 2.6: Manage Sales Contracts

Some additional SAP Fiori apps that are useful for managing contracts are as follows:

- **Monitor Condition Contracts (ID F2954)**
  This app provides a consolidated view of your intercompany cost distribution agreements stored as condition contracts. You have complete visibility into all contracts, the ability to search and analyze specific ones, and the capability to create new contracts.
- **Manage Sales Contracts (ID F1851)**
  As an internal sales representative, this app gives you a complete view of all sales contracts in the system. You can filter results or search for specific contracts by ID or free text. Once you locate a contract, you can navigate directly to it for detailed analysis, or you can create new contracts, modify existing ones, or create follow-on sales orders without leaving the system. The app also connects you to customer information through the 360 view, giving you the full customer context when you need it. When you need to share contract data with your team, you can export the entire list to a spreadsheet file.

Manage Customer Orders

The internal sales representative (or customer service agent) serves as the central coordination point for order management. This role requires visibility across the entire sales lifecycle, from quotation to financial posting. SAP S/4HANA Cloud provides this through a 360-degree operational view, accessible via applications such as My Sales Overview (ID F2200) (see Figure 2.7), which consolidate key information across the order-to-cash process. Key capabilities to support daily execution include the following:

- Automated sales order creation
- Workflow-based approvals and situation-driven notifications
- Flexible pricing determination
- Product recommendation and proposal management
- Exception-driven fulfillment monitoring (e.g., credit checks, compliance checks, availability checks)
- End-to-end process traceability with detailed logs

A major innovation in recent releases is the introduction of *solution orders*, which represent a shift from product-centric sales to solution-based business models. Traditionally, organizations managed separate structures for the following:

- Sales orders
- Service processes
- Projects
- Subscriptions

This led to fragmented billing, inconsistent margin visibility, and complex orchestration.

Solution orders address this by introducing a single commercial object that combines the following into one integrated structure:

- Products
- Services
- Subscriptions
- Project deliverables

This capability is foundational for organizations moving toward outcome-based and service-oriented models, where success depends on managing the entire lifecycle—from contract to revenue recognition—as a single economic unit.

Figure 2.7 shows the My Sales Overview app, which provides sales representatives with a consolidated, role-based view of their pipeline, open orders, and KPIs.

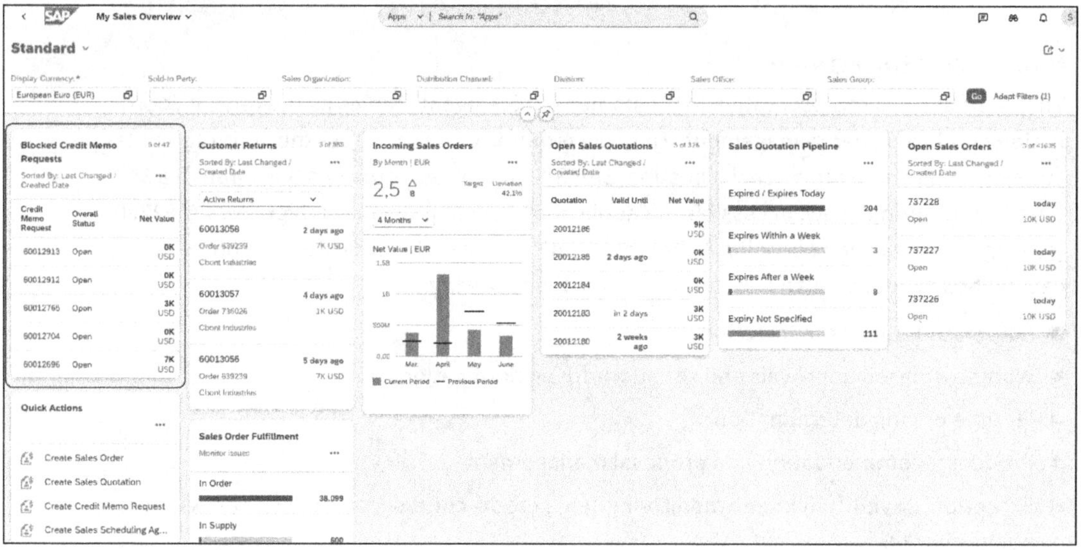

Figure 2.7: My Sales Overview App

Order Capture and Integration Scenarios

To support scalability and reduce operational costs, SAP S/4HANA Cloud enables multiple approaches for automated and system-assisted order capture:

- **Low-touch, intelligent order management**
  Orders are captured, validated, and processed with minimal manual intervention. Built-in validations (pricing, credit, compliance, availability) and workflow-based approvals ensure high accuracy and consistency. Sales representatives focus on exceptions rather than routine tasks.

- **Automated extraction from unstructured data**
  Orders received via email, PDF, or other unstructured formats can be automatically processed using machine learning and intelligent automation, thus improving efficiency and reducing errors.
- **Spreadsheet-based upload**
  Structured data uploads provide a simple mechanism for bulk order creation.
- **SAP Sales Cloud integration**
  Native integration with SAP Sales Cloud enables seamless coordination between front-office engagement and back-office execution, with full visibility into pricing, availability, and financial data.
- **SAP Commerce Cloud integration**
  Integration with SAP Commerce Cloud supports omnichannel sales by connecting web-based commerce platforms to core fulfillment processes.
- **SAP Business Network integration**
  Enables digital document exchange and collaboration between buyers and suppliers, reducing manual effort and simplifying onboarding in B2B scenarios.

Orchestrate the Fulfillment Process

Execution of sales orders extends beyond capture into fulfillment, where coordination across logistics, production, and finance is required. SAP S/4HANA Cloud supports this through real-time monitoring and exception-driven management.

Sales representatives can do the following:

- Monitor the status of orders across the entire process chain
- Identify critical issues early (e.g., delivery delays, missing documents, compliance blocks)
- Take corrective action directly within the system

Monitoring applications provide these features:

- Prioritized lists of orders requiring attention
- KPI-based alerts and visual indicators
- Drilldown capabilities across dimensions such as customer, organization, and issue type
- Integrated collaboration tools for issue resolution

The ability to navigate from high-level issues to detailed process flows ensures full transparency and control over execution.

Figure 2.8 shows the Track Sales Order app (ID F2577), which provides the fulfillment tracking view for sales documents, enabling sales and logistics teams to monitor delivery status, identify bottlenecks, and intervene where execution is at risk.

Figure 2.8: Tracking Sales Orders Fulfillment Issues

2.2.3 Customer Invoice Management

Billing represents the point at which commercial agreements are translated into financial outcomes. Billing documents serve as legally binding invoices for goods and services delivered and can be issued electronically, printed, or transmitted through digital channels. Beyond their role in customer communication, these documents also act as the interface to financial accounting, forming the basis for revenue recognition, receivables, and downstream financial reporting.

In SAP S/4HANA Cloud, billing processes are designed to address the increasing complexity of modern business models. Organizations now operate across product-based, service-based, and subscription-driven offerings, often requiring a combination of billing scenarios within a single transaction. To support this, SAP provides a flexible billing framework that combines operational efficiency with the ability to manage complex, multicomponent invoicing scenarios.

A key innovation in this area is the ability to manage preliminary billing data, allowing organizations to prepare, validate, and negotiate invoice content before final posting to financial accounting. In addition, solution billing capabilities enable the consolidation of billable data from both internal and external sources into a unified invoice, supporting evolving commercial models.

Let's explore the prominent capabilities of SAP S/4HANA Cloud in the customer invoice management domain.

Preliminary Invoicing for Billable Data

Preliminary billing introduces an additional step in the billing process that improves control and transparency, particularly in complex scenarios. Preliminary billing documents are structurally identical to final invoices, but they are not yet posted to financial accounting. Instead, they serve as a working version of the invoice that can be reviewed, adjusted, and validated.

This capability is particularly valuable when invoice details require alignment between multiple stakeholders. For example, pricing conditions, payment terms, or billing schedules may need to be confirmed with the customer before finalization. Preliminary documents can also support internal approval workflows, ensuring that billing accuracy is verified before financial impact is recorded.

Once agreement is reached or approvals are completed, these documents can be converted into final invoices and posted to financial accounting without requiring rework. This approach reduces disputes, improves invoice accuracy, and strengthens customer trust.

Invoice Processing and Execution

Billing clerks play a central role in ensuring that invoices are created accurately, processed efficiently, and delivered in a timely manner. Their responsibilities span the full billing lifecycle—from identifying billable items to resolving errors and ensuring successful payment collection.

SAP S/4HANA Cloud supports these activities through an integrated set of tools that streamline execution. Billing clerks can monitor the billing-due list, process outstanding items, and generate invoices with minimal manual intervention. Worklists and filtering capabilities allow users to focus on relevant transactions, improving both speed and accuracy.

Automation plays an increasingly important role in this process. Periodic billing activities—such as invoice creation or posting to financial accounting—can be scheduled and executed automatically, reducing operational workload. At the same time, error-handling capabilities ensure that exceptions are identified and resolved quickly, maintaining process integrity.

The integration of preliminary billing further enhances execution by allowing users to finalize invoice details before posting, reducing the likelihood of downstream corrections and disputes.

Solution Billing: Enabling Integrated Invoicing

As business models evolve, billing requirements have expanded beyond traditional scenarios. Many organizations now need to combine multiple billing types—such as one-time charges, recurring services, and usage-based fees—into a single invoice, as shown in Figure 2.9.

Solution billing in SAP S/4HANA Cloud addresses this requirement by enabling the consolidation of billing data from various sources, including both internal processes and external systems. This allows organizations to present a unified invoice to customers, even when underlying transactions originate from different channels or involve multiple business models.

Let's explore some SAP Fiori apps to support this process:

- **Manage Billing Data – Subscription Billing (ID F6803)**
  The Manage Billing Data app displays your billing data at a glance. You can also search for customer bills through the Customer Overview app and drill into bill details from there. When a subscription includes a grouping element, related bill items automatically group together in both apps, making it easier to track subscription-related charges.
- **Manage Provider Contracts for Sales Billing (ID F6404)**
  This app displays a list of all your provider contracts for sales billing. You can filter the results according to your needs by business partner, company code, or other relevant criteria in order to quickly locate the contracts you need. Click any contract in the results list to open it and view its complete details in the app.

This capability is particularly important for solution-based offerings, in which products, services, and subscriptions are delivered together. Instead of generating separate invoices for each component, organizations can create a single, integrated invoice that reflects the complete commercial relationship.

Figure 2.9 shows the solution billing capability in the Manage Provider Contracts for Sales Billing app, which consolidates charges from multiple internal and external sources into a single customer invoice for solution-based and bundled offerings.

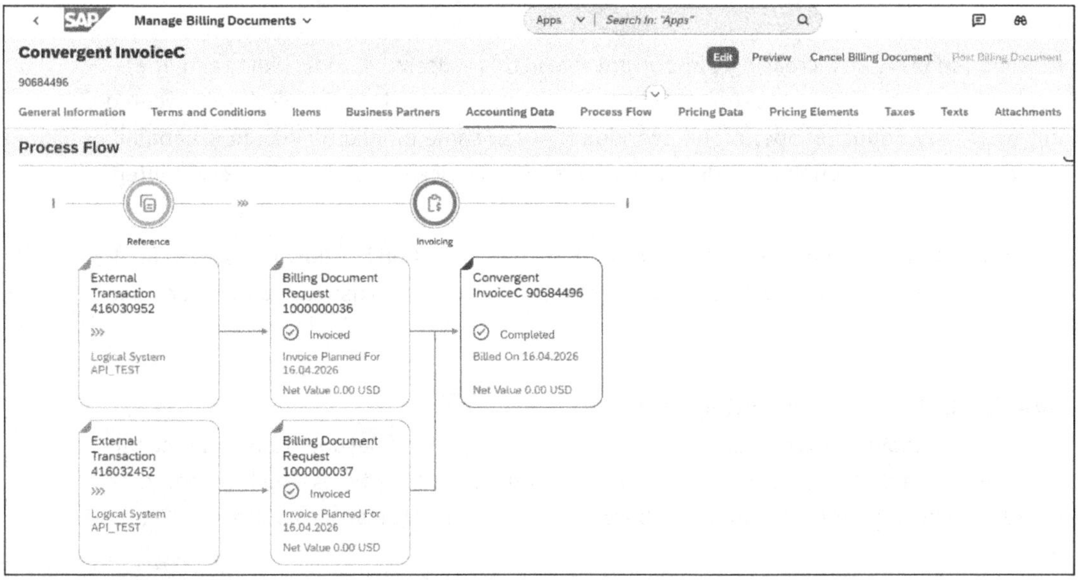

Figure 2.9: Manage Billing Documents

2.2.4 Customer Returns Management

Beyond the initial sale, organizations must manage the full lifecycle of customer engagement, including after-sales processes such as complaints handling, repairs, and returns. In today's competitive environment, customers expect returns to be handled with the same efficiency and transparency as the original sale. Returns management is therefore not only an operational necessity but also a critical driver of customer satisfaction, retention, and brand perception.

At the same time, returns processes have become more complex. Organizations must balance cost efficiency with service quality while also addressing emerging requirements such as sustainability and circular economy models. The ability to reintroduce returned goods into inventory, refurbish them, or recycle them efficiently is becoming increasingly important.

SAP S/4HANA Cloud addresses these challenges through integrated and structured returns management, enabling organizations to manage returns as a controlled, end-to-end process rather than a series of disconnected activities.

The customer returns process typically follows a sequence of coordinated steps. It begins with the initiation of a return request, where the customer's intent to return goods is captured. A return reason is recorded to support both operational handling and later analytical evaluation. The return request is then validated and authorized based on defined policies.

Once approved, the physical return is processed, and the product is received into the organization's facilities or designated returns location. Inspection follows, determining the condition of the returned item and forming the basis for subsequent actions. Depending on the outcome, the organization decides the future use of the product, which may include returning it to inventory, repairing it, scrapping it, or forwarding it to a supplier. Finally, the customer compensation is determined, resulting in a replacement, a refund, or a credit memo.

SAP S/4HANA Cloud integrates these steps into a single process flow, ensuring traceability and reducing manual coordination. The system supports a high degree of flexibility, allowing organizations to adapt their returns process to different business scenarios and product types.

The role responsible for orchestrating this process is the returns and refund clerk, whose objectives are to ensure policy compliance, deliver high-quality service, and maintain full transparency throughout the lifecycle of each return. To achieve this, the system provides a unified view of all relevant information, reduces redundant data handling, and enables full visibility into a customer's complaint history.

Embedded analytics further enhances this process by allowing organizations to identify patterns in returns, understand root causes, and optimize both operational efficiency and customer experience. As returns become more tightly integrated into the overall business process, they evolve from a cost center into a strategic capability supporting customer satisfaction and sustainability goals.

Let's discuss the key SAP Fiori app that helps you orchestrate the returns process in SAP S/4HANA Cloud: Manage Customer Returns (ID F4832). This app handles the complete returns lifecycle from creation through refund. You can create returns orders with reference to a sales order, invoice, or as standalone orders; modify partner addresses, reject items with reasons, and determine refunds; and track returns status through the process flow, manage pricing and business partners, and export data to spreadsheets or PDFs. In addition, you can navigate seamlessly to related customer and sales apps without leaving the returns workflow.

A notable trend in this area is the increasing integration of billing, predictive execution, and returns processes. Billing capabilities now support complex scenarios such as combined invoices for goods, services, and usage-based charges. Predictive analytics help identify potential issues early, enabling proactive communication with customers. Returns management, in turn, becomes a seamless extension of the lead-to-cash process, ensuring that even postsale activities are fully integrated into enterprise execution.

2.2.5 Sales Analytics

As sales volumes increase and business models become more complex, the role of analytics in sales execution becomes increasingly critical. Sales managers must continuously balance revenue growth, margin protection, and operational efficiency while managing a growing number of transactions and customer interactions.

SAP S/4HANA Cloud addresses this need by embedding analytics directly into the sales process. Rather than separating operational execution from reporting, the system enables real-time monitoring, analysis, and action within a unified environment.

Sales planning begins with the definition of targets that are distributed across organizational units such as sales regions, offices, and teams. These targets provide a framework against which actual performance can be measured throughout the period. With this structure in place, sales managers can continuously compare actual performance against the plan, enabling early identification of deviations and timely corrective action.

A key advantage of SAP S/4HANA Cloud lies in its ability to integrate analytics into operational workflows. For example, the same analytical capabilities used for reporting can also drive operational applications such as sales order fulfillment monitoring. In these scenarios, analytics identify potential issues and recommend actions directly within the process, enabling users to move from insight to action without switching contexts.

Sales analytics capabilities include multidimensional reporting on the following:

- Sales volumes and revenue trends
- Profit margins and pricing performance
- Delivery performance and fulfillment efficiency

These insights are available through flexible analysis tools that allow users to drill down from high-level KPIs to detailed transactional data. In addition, predictive capabilities support forward-looking insights such as delivery delays or product recommendations.

The value of analytics extends beyond sales alone. Because SAP S/4HANA Cloud operates on a unified data model, sales data is intrinsically linked to finance, production, and controlling. This enables a holistic view of business performance and supports enterprise-level decision-making, including integration into executive dashboards such as SAP Analytics Cloud.

For sales managers, analytics is not only about reporting but also about process control and performance optimization. Their responsibilities include defining sales strategies, managing quotas, monitoring execution, and ensuring that sales operations align with overall business objectives. Embedded analytics supports these tasks by providing timely insights and enabling proactive management of sales activities.

Ultimately, sales analytics transforms sales from a reactive function into a data-driven, proactive, and adaptive capability, enabling organizations to continuously optimize performance and respond to changing market conditions.

Figure 2.10 shows the **Sales Planning and Analytics** view for sales managers, embedding real-time KPIs and drilldown capabilities directly within the sales process to support data-driven performance management.

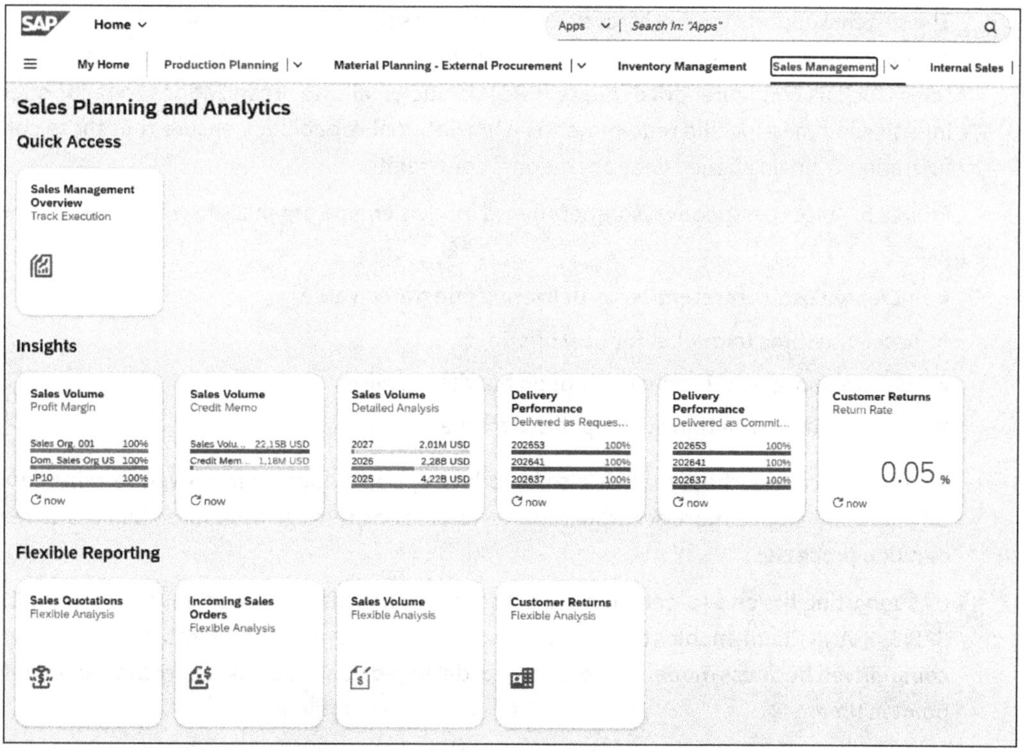

Figure 2.10: Sales Planning and Analytics View for Sales Manager

2.2.6 Solution Business Models

Traditional sales models have historically focused on delivering individual products or services to meet specific customer needs. However, as customer expectations evolve, organizations are increasingly moving toward solution-based business models, in which multiple offerings are combined into a single, integrated value proposition.

In a solution business model, organizations deliver comprehensive outcomes that may include products, services, subscriptions, and associated support activities across the entire lifecycle. For example, what was once a simple product sale may now include installation services, ongoing maintenance, digital subscriptions, and additional value-added services. These combined offerings are designed to meet broader customer needs while increasing both customer satisfaction and revenue potential.

This shift significantly expands the scope of sales operations. Organizations must now manage not only the initial sale but also the ongoing delivery and performance of the solution. This requires tight coordination across multiple process areas, including logistics, services, project management, and billing.

SAP S/4HANA Cloud supports this evolution through solution order management, which enables organizations to design, sell, and manage complex offerings within a unified framework. A solution order acts as the overarching structure that integrates all components of the offering and ensures that dependencies between them are managed consistently.

The system supports the full lifecycle of solution-based sales, from initial configuration and pricing through fulfillment, billing, and financial reporting. Integration with external systems, such as configure, price, quote (CPQ) solutions, allows organizations to tailor offerings to customer-specific requirements, while internal capabilities ensure that these configurations remain aligned with operational constraints.

From a business perspective, solution-based models enable organizations to do the following:

- Increase customer retention by delivering integrated value
- Accelerate time to market for new offerings
- Improve margin performance through bundled services
- Enhance differentiation in competitive markets

For finance leaders, these models also introduce opportunities for improved working capital management and more accurate revenue forecasting through integrated billing and recognition processes.

By supporting the end-to-end lifecycle of these complex offerings, as shown in Figure 2.11, SAP S/4HANA Cloud enables organizations to transition from product-centric selling to outcome-driven business models, where value is delivered continuously rather than at a single point in time.

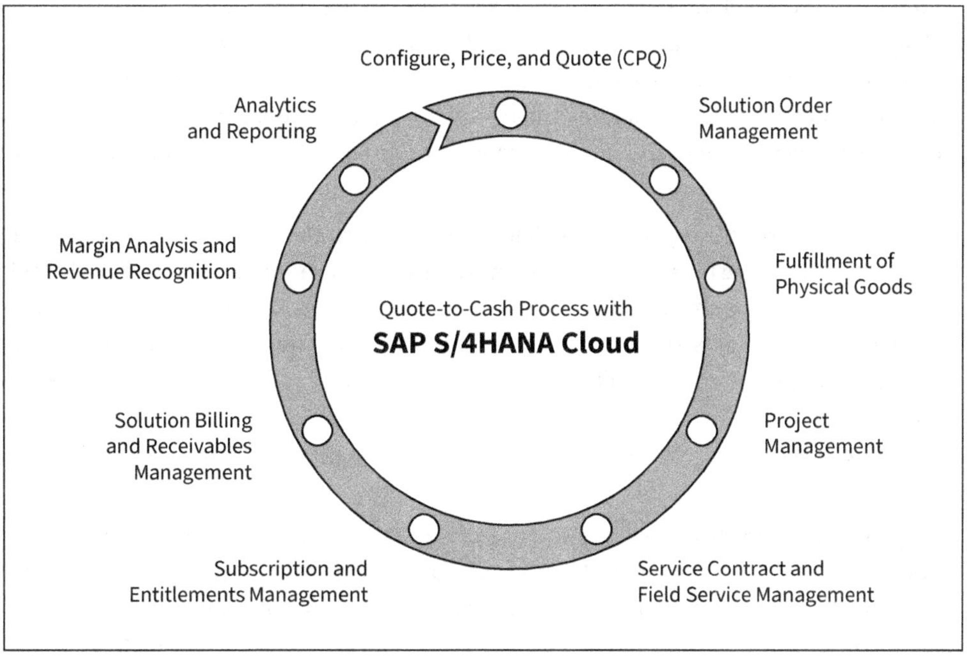

Figure 2.11: Quote-to-Cash Process with SAP S/4HANA Cloud

2.2.7 Advanced Intercompany Sales

Earlier sections focused on sales execution within a single legal entity, but many enterprises operate across multiple company codes, subsidiaries, and geographies. In such environments, customer-facing transactions often involve multiple internal entities, each responsible for different parts of the value chain. *Advanced intercompany sales* extends the standard order-to-cash process across these boundaries, enabling consistent, transparent, and automated execution across the entire enterprise.

Intercompany sales processes occur when a selling entity interacts with an external customer, but fulfillment is performed by a different entity within the same corporate group. These scenarios are common across industries, especially in organizations with centralized manufacturing, regional sales organizations, or shared service models. Historically, such processes have been difficult to manage due to fragmented systems, manual coordination, and limited visibility across entity boundaries.

SAP S/4HANA Cloud addresses these challenges through a redesigned intercompany sales model that integrates sales, procurement, logistics, finance, and compliance into a single, end-to-end process flow. At the core of this model is the concept of *value chain monitoring* (VCM), which provides a unified framework for orchestrating and tracking transactions across entities. Instead of treating each step as an isolated activity, the system manages the entire value chain as a connected sequence of events, ensuring consistency and traceability throughout the process.

In a typical advanced intercompany sales scenario, a sales organization belonging to the selling company receives an order from an external customer and creates a corresponding sales order. If the products are sourced from a plant belonging to another company code, the system automatically generates the necessary follow-on documents, including intercompany purchase orders and intercompany sales orders between the participating entities.

As the process progresses, goods are delivered directly to the customer by the delivering company. At the same time, the system manages a series of coordinated financial and logistical steps, including stock transfers, invoicing, and revenue recognition across entities. All relevant documents—customer orders, intercompany transactions, deliveries, and invoices—are generated and linked automatically, creating a fully auditable transaction chain, as highlighted in Figure 2.12. This flow illustrates how the selling company, delivering company, and end customer are coordinated across legal entities with valuated stock in transit.

A key innovation in this process is the use of valuated stock in transit. This mechanism allows the system to track ownership and control of goods as they move between entities and ultimately to the external customer. By managing transfer-of-control events explicitly, SAP S/4HANA Cloud ensures that financial postings—such as cost of goods sold (COGS) and revenue recognition—are aligned with both operational reality and regulatory requirements.

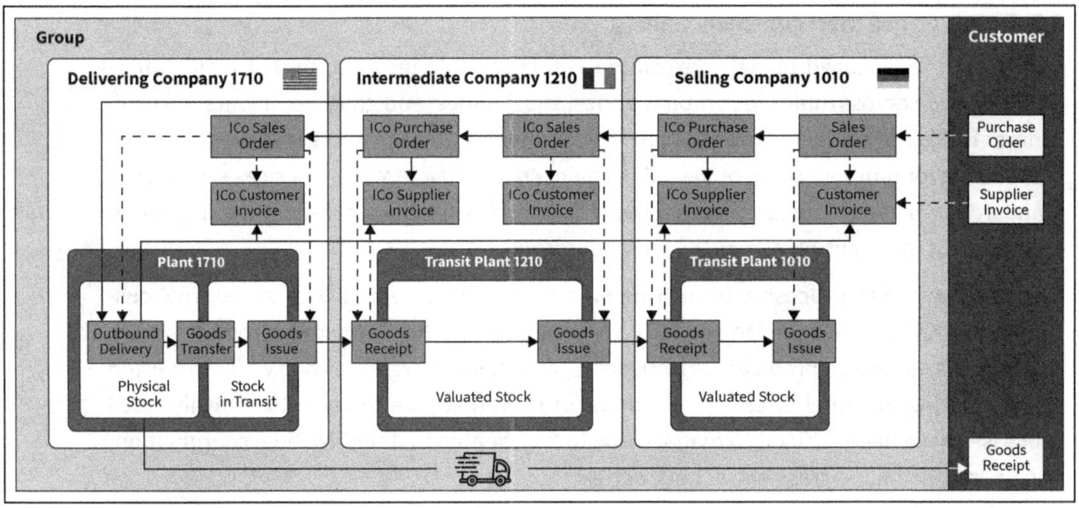

Figure 2.12: Overview of Advanced Intercompany Sales Process

Given the complexity of multientity transactions, visibility and control are critical. SAP S/4HANA Cloud provides this through the Monitor Value Chains app (ID F4854), which offers a comprehensive view of all relevant process steps across the intercompany flow, as shown in Figure 2.13.

Users can view the entire transaction either as a structured list or as a graphical flow, covering all major documents, including sales orders, intercompany orders, deliveries, material movements, and invoices. This integrated view enables users to trace the lifecycle of a transaction from the initial customer order through fulfillment and financial settlement.

Importantly, the system also supports operational interventions. In cases of errors or delays, users can repeat process steps, navigate to related applications, and resolve issues directly within the flow. This transforms intercompany processing from a reactive activity into a managed, transparent, and controllable process.

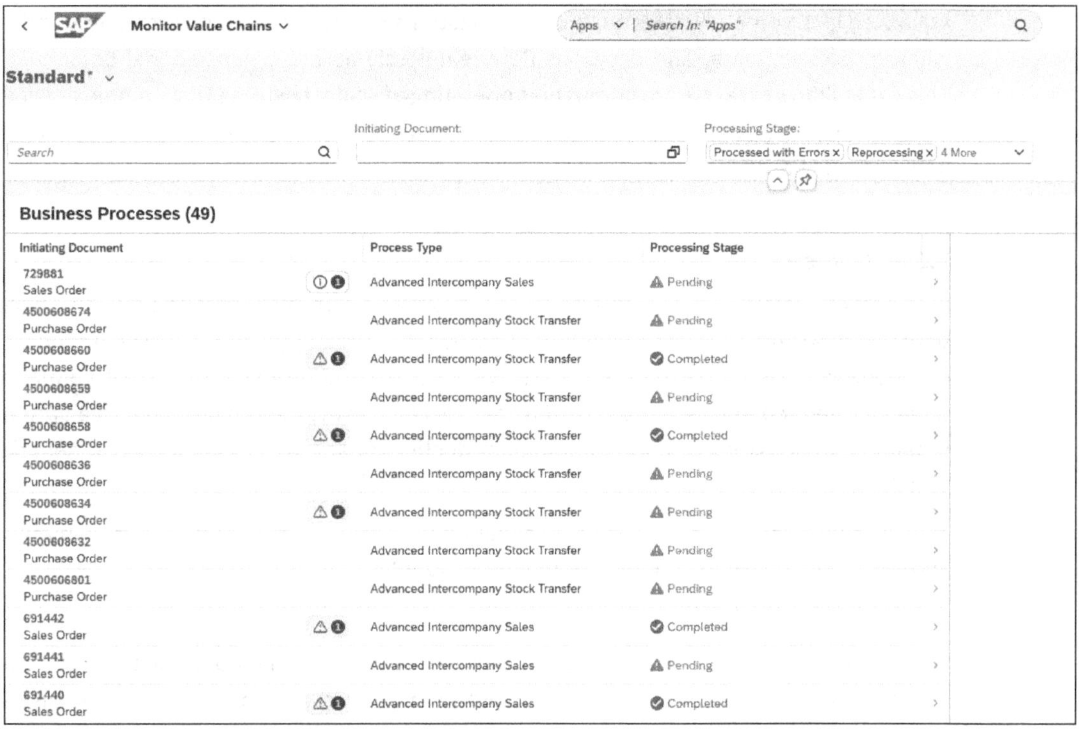

Figure 2.13: Monitor Value Chains App

Advanced intercompany sales delivers value by combining multiple capabilities into a coherent execution model. It integrates closely with the following:

- Product and trade compliance, ensuring adherence to regulatory requirements
- Transportation management, supporting logistics planning and execution
- Revenue recognition, aligning financial postings with transfer-of-control events
- Product costing and profitability reporting, enabling consistent financial analysis
- Group reporting and consolidation, ensuring seamless financial integration across entities

By embedding these capabilities directly into the process, SAP S/4HANA Cloud reduces manual effort, eliminates data inconsistencies, and improves auditability. Transactions are executed automatically based on defined rules, and data flows seamlessly across systems without the need for reconciliation.

From a business perspective, this results in the following benefits:

- Greater transparency across the entire value chain
- Faster and more reliable execution of intercompany transactions
- Improved compliance with accounting and regulatory standards
- Enhanced ability to monitor and optimize cross-entity operations

Advanced intercompany sales enables organizations to operate as a connected enterprise, where processes, data, and decisions flow seamlessly across organizational boundaries. This capability is essential for companies operating at scale, ensuring that complexity does not become a barrier to growth but instead is managed through structured, automated, and intelligent execution.

2.2.8 Outcome

In Chapter 1, Section 1.8.2, James in Case Study 2 faced a problem that haunted his operations: on-time, in-full (OTIF) delivery at 86 percent. Fourteen percent of orders arrived late or incomplete. He could see the causes—volatile demand, unreliable supplier confirmations, delayed reactions—but by the time the issues became clear, only expensive options remained, such as rush freight, expedited production, or accepting strained customer relationships.

The real problem was not lack of visibility but timing: the gap between when a risk emerged and when anyone could actually do something about it.

How does the operating model change?

SAP S/4HANA Cloud narrows that gap by pulling detection and response directly into how orders move through the system. Orders are monitored continuously as they execute. A sales order posts, and the system immediately confirms whether items are available. If not, it flags the shortage while there is still time to source from an alternative supplier or adjust the customer's delivery date. Problems do not accumulate until the weekly review; they surface the moment they start forming.

Fewer errors enter the system in the first place. Availability is checked before the order is confirmed. Pricing is validated, credit is verified, and compliance is confirmed. By the time an order reaches fulfillment, it has already passed through these gates. Fewer disruptions downstream mean fewer fires to fight.

Insights surface inside the workflow, not in a report prepared later. A fulfillment specialist sees an order flagged as at-risk delivery, with context: This customer's shipments typically go to this location, which is experiencing delays with this carrier. The specialist can adjust the route or push the delivery date proactively. The problem thus is solved before the customer notices.

The Operating Rhythm Shifts

The operational rhythm changes measurably.

Daily, teams no longer work from a list of late orders requiring recovery. They work from a prioritized view of orders at risk, allowing intervention while alternatives still exist.

Signals surface and are addressed continuously. If a supplier confirmation does not arrive by the expected time, it is flagged immediately rather than discovered during end-of-day reconciliation.

Weekly reviews shift from postmortem analysis (Why was this order late?) to forward-looking planning (How do we prevent this pattern from repeating?).

Over time, the pattern changes. Fewer orders require expediting because problems are caught earlier. Premium freight becomes an exception rather than a budget line item. Customer penalties decline because commitments are met more reliably.

OTIF improves not from better firefighting but thanks to preventing the need to fight fires in the first place. The company acts earlier, when the outcome can still be changed.

2.3 Sourcing and Procurement

Sourcing and procurement are vital functions for practically every organization. A *procurement process* is generally defined as one organization purchasing a good or service from another organization or individual. These procurement activities are classified as direct procurement or indirect procurement. With *direct procurement*, your organization is purchasing materials and services that contribute to a product being produced or delivered, while with *indirect* or *supporting procurement*, materials and services are acquired to keep the firm itself functioning.

This section provides an overview of the sourcing and procurement process in line with sourcing and procurement capabilities in SAP S/4HANA Cloud, concluding with the key innovations introduced by the latest releases of SAP S/4HANA Cloud.

Figure 2.14 shows the end-to-end sourcing and procurement process in SAP S/4HANA Cloud, from strategic sourcing and supplier management through to purchase order processing and invoice settlement.

Before purchasing a good or a service, an organization typically identifies and sources the requirements. *Sourcing* is the process of matching your organization's needs with a supplier who will provide the goods or services needed. The steps for a sourcing process include identifying and understanding the requirements, winnowing down the pool of bidders to the most viable suppliers, requesting information and bids from suppliers, awarding the purchase to the supplier with the best proposal, and then moving to the procurement process by creating a purchase order or contract.

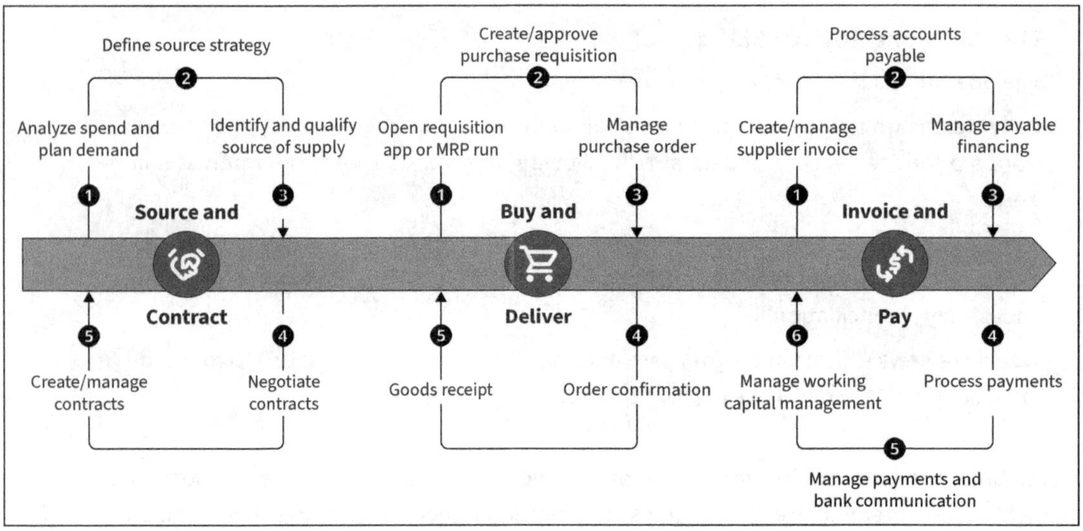

Figure 2.14: Sourcing and Procurement End-to-End Process Overview

Procurement typically takes up where sourcing leaves off, creating purchase requisitions and issuing purchase orders for goods or services to the supplier identified during the sourcing process as the source of supply. Finally, the *procure-to-pay* process arrives at payment, which includes receiving and processing receipts and invoices in follow-on processes.

Although payments are typically triggered at the end of a procurement process and at the start of a sourcing process, some procurement scenarios are common enough exceptions to undermine a one-size-fits-all sequence. Rather than insist on a sole sequence, we can more easily identify the key areas and categories of sourcing and procurement, as shown in Table 2.1.

Procurement Process Area	Description	Occurrence in Source-to-Pay Process	Main SAP S/4HANA Cloud Integration Points
Operational purchasing	Transactional procurement, core purchase order, goods receipt, service receipt, and invoices	Before invoice and payables management	Finance, inventory management, and production planning
Collaborative sourcing and contract management	Request for quote, purchase, information process, source selection through bidding or single source, and contract definition	Either after or before operational purchasing	Finance, inventory management, and production planning

Table 2.1: Procurement Process Areas

Procurement Process Area	Description	Occurrence in Source-to-Pay Process	Main SAP S/4HANA Cloud Integration Points
Invoice and payables management	Invoice submission by supplier, verification, and payment by accounts payable	After operational purchasing	Finance
Supplier management	Supplier designation, master data, and performance management	Run in parallel with the other process steps or at the end of the procurement process	SAP S/4HANA master data
Procurement analytics	Analysis of core KPIs, such as spend analysis, demand and forecast planning, and supplier performance and consolidation	Run in parallel to other process steps and/or the end of the procurement process	Finance and inventory management

Table 2.1: Procurement Process Areas (Cont.)

SAP S/4HANA Cloud addresses all the procurement areas shown in Table 2.1 in depth, both as part of the core offering native within SAP S/4HANA Cloud and through augmenting cloud solutions within SAP Ariba and SAP Business Network, SAP Fieldglass, and SAP BTP. Supporting these processes within a complex enterprise or solution stack requires integration. SAP offers both native integration from SAP S/4HANA Cloud to SAP Ariba solutions in key areas such as SAP Business Network, and extending integration layers in the form of SAP Integration Suite, managed gateway. We will discuss the SAP Ariba integration into the source-to-pay process within SAP S/4HANA Cloud in Chapter 4.

Now, let's dive into the core capabilities for sourcing and procurement in SAP S/4HANA Cloud.

2.3.1 Source-to-Contract

The source-to-contract process is where procurement earns its keep. The hard part is not cutting purchase orders; it is finding, qualifying, and contracting the right sources at the right risk level, then making those terms stick in daily buying. Most organizations leak value in three places. First, noncontracted spend creeps in because buyers cannot see approved sources or current terms at the moment of need; tail categories are bought "off system," prices drift, and savings evaporate. Second, authoring lives in email and spreadsheets while execution lives in the ERP system; renewal dates, price indexations, and service levels become manual chores, so contracts lapse or underdeliver. Third, supplier performance is fragmented; negotiations focus on price while the real costs—late deliveries, defects, expedites—show up months later in operations.

SAP S/4HANA Cloud tackles these issues by putting sourcing events, contract lifecycle, and supplier performance on one operational spine and by integrating natively with SAP Ariba Sourcing and SAP Business Network when collaborative events are required. A purchase requisition that cannot autofind a source becomes a guided sourcing request; you issue a request for quotation (RFQ) in SAP S/4HANA Cloud or run an SAP Ariba event, analyze bids with embedded scorecards, and award directly to a purchase contract or scheduling agreement. The same system that authors the contract enforces price conditions on purchase orders, tracks volumes and renewals, and alerts you before terms expire. Stakeholder surveys and supplier KPIs sit next to price during bid analysis, so awards reflect total value, not just unit cost.

The business impact is tangible: higher contract coverage and lower maverick spend, faster cycle times from request to award, fewer stockouts and expedites as sourcing sees real demand and inventory, and improved compliance because contract, price, and approval controls are embedded rather than policed. In short, sourcing and contracts in SAP Cloud ERP turns policy into practice, buying better, faster, and with less risk.

The activities involved with sourcing goods and services range from the everyday sourcing requirements to find, for instance, a supplier for office supplies to more strategic sourcing activities for projecting and forecasting demand in a core spend category for the business and negotiating the best prices with an optimal supplier mix. Sourcing is not always or solely about price. A supplier can offer the best price, but if the supplier cannot deliver on this price or is challenged by external factors, then having the best price can be an unacceptable trade-off and might result in a large-scale supply chain disruption or degradation of quality in the end product to be delivered. If your products and services are the sum of your suppliers' inputs, it follows that quality control must begin at the supplier level.

Figure 2.15 shows the Manage Sourcing Projects app (ID F4861), which enables procurement teams to create and manage competitive bidding events, compare supplier bids, and award contracts within a single application.

The sourcing cockpit in SAP S/4HANA Cloud builds on previous versions of the sourcing functionality in SAP ERP. SAP ERP has always been somewhat constrained by the collaborative requirements of strategic sourcing because opening your company's sensitive financial and HR data from behind its firewall is required to solicit bids through a RFQ. For this type of sourcing effort, the best option remains the SAP Ariba Sourcing solution.

RFQs are an important aspect of sourcing in SAP S/4HANA Cloud, and several options exist for managing this process, as well as related follow-on processes for operationalizing agreements with suppliers into contracts and purchases. You can initiate an RFQ by clicking the plus icon (+) in the Manage RFQs app (ID F2049) and creating a new RFQ, as shown in Figure 2.16.

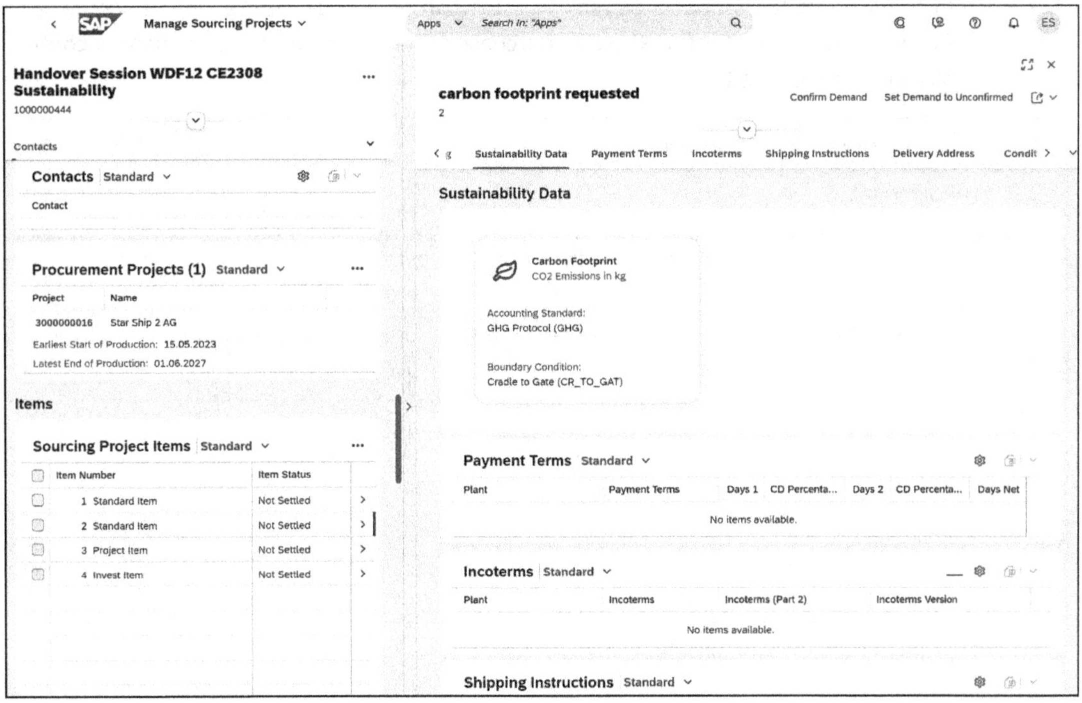

Figure 2.15: Manage Sourcing Projects App

Figure 2.16: Manage RFQs App

In the Manage RFQs app, a buyer can create an RFQ and invite suppliers, defining the follow-on document for the RFQ as a purchase order, value contract, or quantity contract, as shown in Figure 2.17.

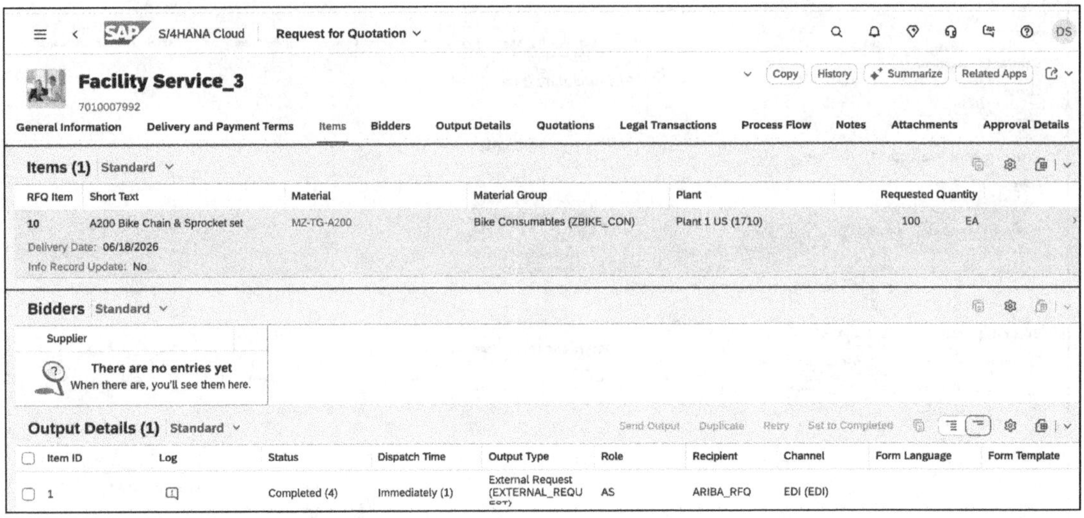

Figure 2.17: RFQ Creation

A potential supplier responds to this invitation with a bid, and the buyer can compare competing bids in the system and leverage various transactional and analytical SAP Fiori apps for suppliers in SAP S/4HANA Cloud, such as the following:

- Manage Sources of Supply
- Operational Supplier Evaluation Last 365 Days
- Overall Supplier Evaluation Last 365 Days
- Supplier Evaluation by Quantity Last 365 Days
- Supplier Evaluation by Price Last 365 Days
- Supplier Evaluation by Time Last 365 Days
- Supplier Evaluation by Questionnaire Last 365 Days
- Supplier Evaluation by User Defined Criteria Last 365 Days

Using these apps provides the buyer with a more informed view to select the best supplier to fulfill the RFQ requirements. Once a winning bid has been selected and awarded, a follow-on document is created, which leads into the operational procurement portion of the process.

2.3.2 Procure-to-Receipt

Once a sourcing event is concluded and a contract or source of supply is established, procurement moves into operational execution. The procure-to-receipt process covers everything from creating a purchase requisition through ordering, goods or service receipt, and invoice verification. It spans two primary subprocesses: service procurement for buying nonmaterial items and managed services, and operational procurement for the day-to-day purchasing of goods and services by employees and purchasing teams.

Service Procurement

In many industries, a growing trend is to move from traditional business models based on the production and selling of goods toward bundling diverse offerings combining goods, warranties, continuous maintenance, and licenses into an overall service offering. A typical example might be selling a driving experience through lease options instead of selling a car, together with a service plan, guaranteed work, and standard repairs. As a result, services increasingly are infiltrating the heart of the business world.

SAP S/4HANA Cloud provides a new feature for overall service procurement. The material master has effectively become a product master, thereby enabling a multitude of integration facets inside SAP S/4HANA Cloud and inside the overall set of SAP products. This new approach is still coming to fruition, but several valuable capabilities are available right now, including the following:

- Service procurement is integrated with service management, which enables the selling of services procured from a third-party supplier. Integration also provides goods receipt and supplier invoice information, as well as a pricing capability to determine the sales price based on the procured price plus a margin.
- Service procurement is integrated into the professional service scenario so that external supplier workers can report their spent efforts using a time sheet, which can be converted automatically into a service entry sheet.
- Service procurement is integrated with transportation management in SAP S/4HANA Cloud, enabling a much better experience during settlement of freight invoices so that the service parts on the freight invoice can be used in a three-way match. Thus, differences between the expected price and the stated price on the invoice can be identified easily to help choose the appropriate service.
- Service procurement in SAP S/4HANA Cloud is also useful when billing costs between affiliated companies (scope item 3ZB) for simple cross-company services (i.e., a service bought from one affiliate company and executed on behalf of multiple other companies). These activities include, for instance, cafeteria services, gardening, and the like, all of which can be automated when generating follow-on documents between the selling and the procuring company.

> **Note**
> SAP continuously extends service procurement integration into additional areas, including enterprise asset management and enterprise portfolio and project management. Check the SAP S/4HANA Cloud release notes for the latest scope items and integration coverage as these capabilities evolve with each quarterly release.

Operational Procurement

Operational procurement is driven primarily at the employee level, when an employee identifies a requirement and goes into the system to place an order. The three main types of procurement conducted in operational procurement are stock, consumables, and external services. *Stock procurement* is typically used for direct procurement activities and entails purchasing stock items and taking these items into inventory for management and distribution. A *consumable item* is often an indirect item that is consumed and replenished on an ongoing basis but not managed in the system as inventory. Pens, paper, coffee for the office kitchen, and other types of consumables may be ordered by the office manager or perhaps directly by an employee who notices that the office is out of a consumable good. Indirect purchasing of this nature is usually low value and high volume. The high volume makes it significant to the overall spend of an organization. High volumes also make consumable spending an ideal area to manage procurement using self-service processes in a system, allowing buyers to focus on strategic spending in your organization and enabling the requesting employee to select exactly what they require without going outside the regular approval and spend processes of the organization.

Buying services, such as building maintenance, consulting, or other tasks suitable for contingent labor, fall into the category of *external services* procurement. External services are delivered by individuals or groups who are not employees of your organization but are only involved in delivering that particular scope of work. External services can be part of direct procurement and indirect procurement activities.

SAP S/4HANA Cloud provides apps completely built on the SAP Fiori design system. You can use these apps to create and review requisitions and identify appropriate sources of supply using a consumer-grade user experience (UX). The user accesses these apps through tiles on a home screen, as shown in Figure 2.18.

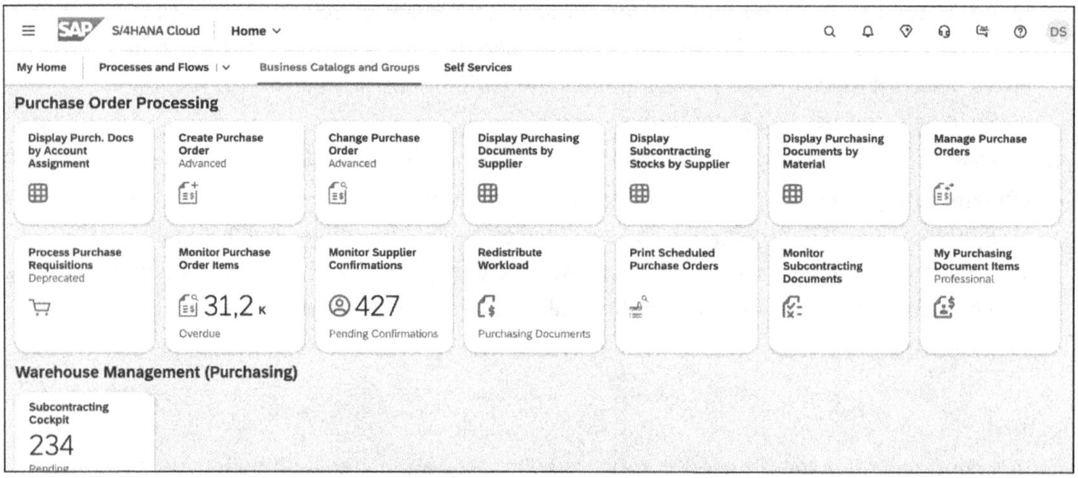

Figure 2.18: SAP S/4HANA Cloud: Sourcing and Procurement Home Page

Numerous innovations related to simplification and speed for self-service procurement in SAP S/4HANA Cloud include one-click ordering, content search, and dynamic filtering. You also have increased flexibility for *shop-on-behalf* features, via which a user can select from a group of users when creating a requisition and order something on their behalf.

Figure 2.18 shows the SAP S/4HANA Cloud sourcing and procurement home page, which provides role-based access to all procurement apps through the SAP Fiori launchpad.

The SAP S/4HANA Cloud workflow leverages the following improvements from previous SAP ERP versions (see an example in Figure 2.19):

- Executing all tasks in a single screen
- Displaying the work items list while processing a single work item
- Using mobile approval, if desired
- Filtering and sorting lists of work items
- Editing requisitions during approval
- A centralized workflow inbox

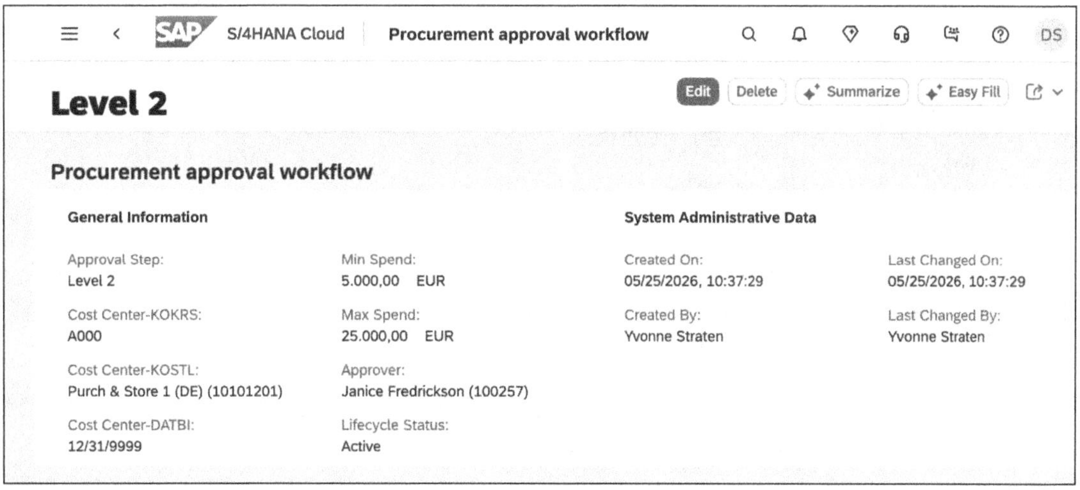

Figure 2.19: Approval Workflow

Once an order is approved, you can use SAP Business Network. Orders for suppliers not yet registered can still be routed through SAP Business Network. The supplier receives an email with the purchase order from SAP Business Network, along with instructions for registering if they wish to complete the transaction flow using SAP Business Network.

Once the purchase order has been issued, received, acknowledged, and fulfilled by the supplier, if the purchase order settings require a three-way match, you then create a goods or services confirmation. Two key apps in SAP S/4HANA Cloud support goods receipts:

- Post Goods Receipt for Inbound Delivery
- Post Goods Movement

You can also post a goods and service entry sheet using the *lean service* functionality. For example, if you have a supplier managing a conference for your company, providing both materials for both setup and service personnel to coordinate the event, you can now create and confirm a purchase order without separate confirmations for each good or service required. Lean services can be confirmed with the Manage Service Entry Sheet app (ID F2027).

The final area in a procure-to-pay or source-to-pay process is *invoicing*. For invoicing, an accounts payable user has several apps to choose from. To create a supplier invoice, click the corresponding tile (e.g., for the Create Supplier Invoice app) and then enter information into the system, as shown in Figure 2.20.

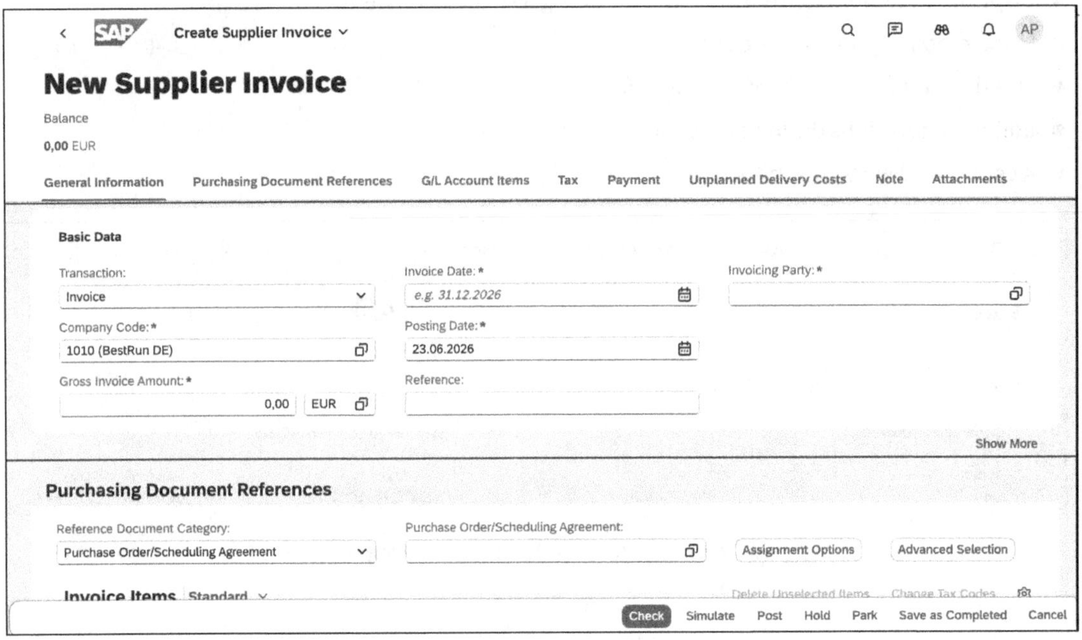

Figure 2.20: Create Supplier Invoice App

A more efficient way to enter these invoices into SAP S/4HANA Cloud for processing is through SAP Business Network. In the purchase order/invoice collaboration scenario, a supplier can "flip" a purchase order to an invoice, while following buyer-created rules for submitting an invoice. The supplier can submit an invoice that conforms to the rules of SAP S/4HANA Cloud and have this invoice automatically posted in the buyer's SAP S/4HANA Cloud system. In addition to this type of automation, SAP S/4HANA Cloud supports machine learning and other types of automation.

2.3.3 Central Procurement

Most enterprises do not buy via a single ERP system. Mergers, regional rollouts, and specialized systems leave procurement spread across SAP S/4HANA, older SAP ERP systems, and even non-SAP backends. The result is familiar: Buyers unknowingly negotiate with the same suppliers in different systems, corporate terms are applied inconsistently, category managers jump between logins to find requisitions and spend, and leadership can't see global contract consumption or risk. SAP S/4HANA Cloud for central procurement solves that problem without forcing a disruptive consolidation. It acts as a hub that connects multiple backend systems, providing one place to see and steer requisitions, purchase orders, contracts, sourcing events, and analytics while execution continues locally.

Figure 2.21 shows an overview of the central procurement process, illustrating how a central procurement hub coordinates purchasing activities across multiple connected ERP systems and legal entities.

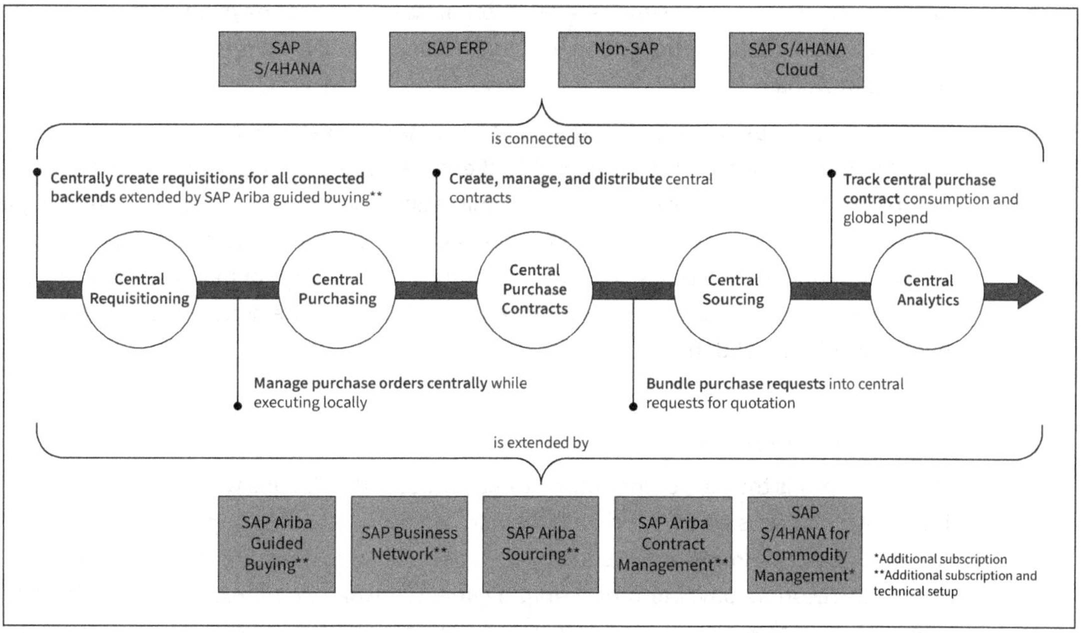

Figure 2.21: Central Procurement Process Overview

In practice, central procurement delivers five capabilities on a single spine. *Central requisitioning* lets you create and approve requests for any connected backend (and if you use SAP Ariba's guided buying feature, you can extend a consumer-grade experience to all requesters). *Central purchasing* gives lead buyers one worklist to process and monitor purchase orders created in the hub but fulfilled in local ERP systems; you can mass-change standard fields with a simulation run to catch errors before release, and an enhanced document flow shows accounting documents for posted invoices so that the payment status is visible at a glance when suppliers ask. *Central purchase contracts* allow category managers to author, distribute, and track global contracts, then push them to local systems for call-off so that negotiated prices and terms actually reach the plant. *Central sourcing* bundles demand into RFQs or sourcing events and can hand off to SAP Ariba Sourcing for large, collaborative bids, with awards flowing back to contracts or purchase orders in SAP S/4HANA. *Central analytics* closes the loop with a global view of spend, contract consumption, and compliance across all connected systems.

Two design choices make the model effective. First, orchestration rather than replacement: The hub provides policy, visibility, and control, while the connected ERP systems continue to transact, respecting local taxation, languages, and cutover realities. Second, collaboration in context: Chat-like comments on centrally managed requisitions keep conversations attached to the document, and buyers work from a single SAP Fiori inbox instead of email threads and spreadsheets. The governance benefits are immediate: better demand aggregation for negotiations, consistent use of corporate contracts and catalogs, and fewer maverick buys. Operationally, lead buyers gain a single view of requisitions, orders, status, and invoices across the estate; suppliers get faster answers because payment status and document flow are visible centrally; and executives finally see contract utilization and savings realization globally.

When should you use central procurement? Any time you need global control and transparency but must keep transacting in local backends: shared-service procurement, multi-ERP groups, postmerger landscapes, or regions migrating over time. The value story is straightforward: higher contract coverage and compliance, shorter cycle times through mass actions and one-inbox work, stronger negotiating power through demand aggregation, and lower cost to serve because you centralize the brain of procurement without paying for a big-bang ERP consolidation.

2.3.4 Invoice-to-Pay

Invoice-to-pay is the part of the source-to-pay process that begins when a supplier invoice is received and ends when the supplier is paid and the accounting impact is fully recorded. In SAP S/4HANA Cloud Public Edition, this area combines invoice capture, verification, approval, accounts payable processing, payments, bank communication, liquidity management, and analytics into an integrated flow designed to reduce manual effort and improve working capital control.

From an SAP perspective, invoice-to-pay should be understood not as a single transaction but as a controlled sequence of operational and financial activities. The business objective is simple: Pay suppliers accurately and on time while preserving compliance, transparency, and cash discipline.

The process typically begins after procurement has created the commercial commitment and a supplier invoice arrives for settlement. In SAP S/4HANA Cloud, the invoice stage is tightly linked to accounts payable, payment execution, bank communication, and liquidity management so that finance teams can manage both operational control and cash.

Invoice-to-pay is a value driver for finance and procurement because it helps organizations plan and run payments across multiple suppliers and invoices, improve cash forecast accuracy, control cash discounts, and optimize payable processing through automation. SAP also highlights end-to-end visibility into bank processes, real-time bank communication, payment monitoring, working capital management, and embedded analytics for accounts payable.

In practical terms, this means the process is not limited to posting an invoice. It supports broader finance outcomes such as better supplier relationships, fewer payment inquiries, more reliable closing, stronger liquidity forecasting, and faster exception handling when something goes wrong in bank communication or invoice approval.

Figure 2.22 shows the invoice-to-pay process overview, covering all steps from supplier invoice receipt and validation through to payment execution and cash flow planning.

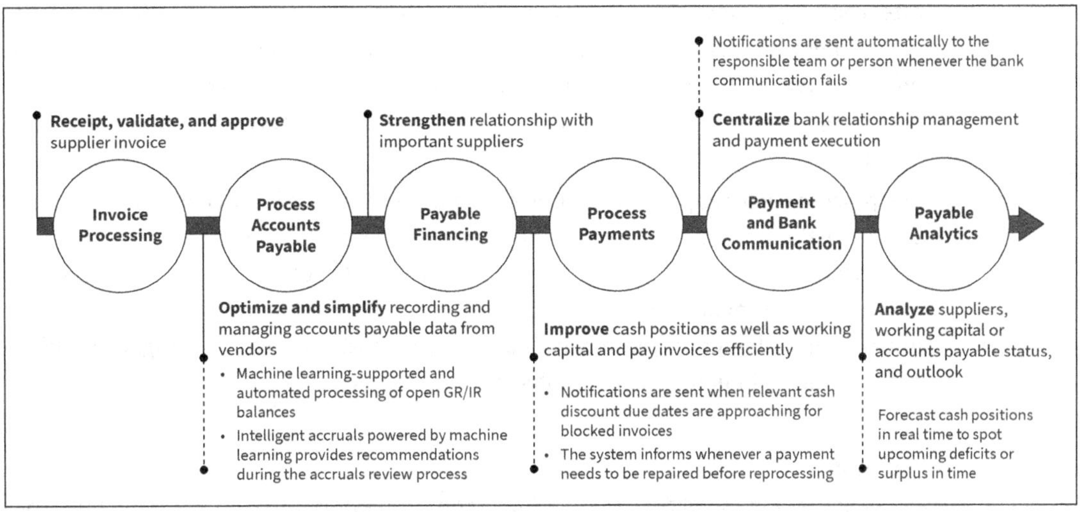

Figure 2.22: Invoice-to-Pay Process Overview

Let's walk through these steps in more detail:

1. **Receive and capture the supplier invoice**

   The process starts when the company receives a supplier invoice, either with a purchase order reference or without one. SAP supports invoice creation or upload, and the material shows that invoices can be digitized from paper into electronic form using machine learning and optical character recognition (OCR), while bots can scan incoming email attachments such as PDF, PNG, and JPEG files and upload them automatically.

   This step is especially important in low-touch invoicing scenarios. Instead of relying on manual keying by an accounts payable clerk, the system can centralize invoice intake and reduce repetitive administrative work. SAP Ariba Invoicing extends this further by fully digitizing invoice capture and processing, using embedded OCR, intelligent text processing, and rule-based general ledger account determination for non-purchase order invoices.

Example

A manufacturing company receives 800 supplier invoices per week. About 70 percent arrive as PDF attachments by email and 20 percent are scanned paper invoices from smaller vendors. With low-touch invoicing and OCR-based extraction, the accounts payable team can automatically create invoice records; identify key attributes such as the supplier, invoice number, amount, tax, and due date; and route them into the next verification step with far less manual typing.

2. **Verify invoice correctness**
   After capture, the invoice must pass a verification check to confirm that the document is correct before it can move forward. In SAP terms, this means validating master data, amounts, tax information, account assignment, and—where applicable—matching the invoice against related procurement documents such as purchase orders and goods receipts.

   The process becomes particularly powerful when the invoice is purchase-order-based. In that case, the system can compare the supplier invoice against expected quantities and values from the purchase order and the goods receipt. For organizations with large GR/IR balances, SAP provides GR/IR reconciliation capabilities that automate matching, minimize manual interaction, use machine learning for matching logic, and trigger workflows or follow-up actions such as write-offs where appropriate.

Example

A chemical company orders 1,000 valves for a turnaround project. The goods receipt confirms that 980 valves were received. If the supplier invoice is submitted for the full 1,000 units, verification detects the variance. Depending on tolerance settings and business rules, the invoice may be blocked, routed for review, or adjusted before posting.

3. **Approve the invoice**
   Once validated, the invoice moves into ML-supported automation for approval processes and show invoice approval as a distinct step before formal submission to accounts payable. This is critical when invoices exceed threshold amounts, contain exceptions, or require sign-off from cost center owners, budget holders, or procurement managers.

   Approval workflows are also a way to balance control with speed. Rather than sending approval requests by email outside the system, SAP S/4HANA Cloud embeds workflow routing and status handling inside the business process. Suppliers can also benefit indirectly because real-time visibility into invoice approval and payment status reduces inquiries and improves collaboration.

Example

A services invoice for external maintenance work is posted against a cost center rather than a purchase order. Because it exceeds the company's approval threshold, the system routes it automatically to the maintenance manager and then to finance. Once both approvers release it, the invoice proceeds to accounts payable posting without manual handoffs.

4. **Post to accounts payable**
   After approval, the invoice is submitted to accounts payable. At this point, the supplier liability is recorded, the relevant expense or inventory postings are created, and the open item becomes part of the payable population to be managed, monitored, and eventually settled.

   This is also where broader payable operations come into view. The material highlights controlling open items, embedded analytics in accounts payable, and optimization of payable processing with high automation. In other words, the posted invoice is no longer just a document; it becomes a managed financial obligation that affects working capital, supplier aging, cash forecast, and payment planning.

Example

A life sciences company receives a validated invoice for laboratory supplies worth $48,000. After posting, the invoice appears in the accounts payable open items list, contributes to the supplier balance, and feeds dashboards such as the accounts payable overview and days payables outstanding for finance monitoring.

5. **Plan payments**
   Once invoices are in accounts payable, the organization plans payment execution. SAP states that users can set up and configure payment proposals to plan payment runs containing multiple payments, business partners, or invoices. This allows treasury and accounts payable teams to balance due dates, payment terms, available liquidity, discount opportunities, and bank execution capacity.

   Planning is not merely administrative. It is a working capital decision. The system supports liquidity optimization, working capital management, and improved cash forecast accuracy through seamless integration with cash management. This helps finance decide whether to pay early for discounts, pay on the due date to preserve liquidity, or group payments for operational efficiency.

Example

At month end, a consumer products company has 1,500 approved invoices due over the next 10 days. The accounts payable team creates a payment proposal that groups domestic automated clearing house (ACH) payments separately from international wire transfers, prioritizes invoices eligible for early-payment discount, and defers non-critical invoices to their exact due dates in line with treasury guidance.

6. **Execute payments and bank communication**
   After planning comes payment execution. SAP describes support for domestic and international payments across multiple business partners, payment methods, and invoices, along with central approval processes and bank transfer management for SAP and non-SAP systems through advanced payment management. The process also includes end-to-end control of bank communication, real-time bank channels, and visibility into bank accounts and related communication.

   Bank communication is a crucial part of operational reliability. The system can automatically notify responsible teams when bank communication fails, and it can also alert users when a payment needs repair before reprocessing. These capabilities reduce the risk of silent failures, duplicate payments, or missed due dates.

Example

A multinational company launches a payment run covering suppliers in the United States, Germany, and India. Some payments go through local bank channels, while others use SWIFT. If one payment file is rejected because of invalid bank data, the system flags the exception so that accounts payable can repair and resubmit the payment instead of discovering the issue days later.

7. **Manage special scenarios**
   Invoice-to-pay also includes advanced scenarios beyond standard invoice settlement. SAP highlights *payable financing*, also referred to as *reverse factoring* or *supply chain financing*, which helps secure uninterrupted supply, strengthen supplier relationships, and optimize working capital for both the buyer and supplier through an automated financing process via bank interface.

   Authorized personnel can review, verify, approve, reject, or send back a submitted supplier down payment request using dedicated SAP Fiori apps such as Manage Teams and Responsibilities (ID F2412), Manage Workflows (ID F5343), and Verify Supplier Down Payment Requests (ID F7103 or F-47). This is valuable when suppliers require advance funding before delivery, but the company still needs strong internal control.

Example

An automotive supplier asks for a 20 percent down payment before starting production of a custom tool. The accounts payable manager reviews the request, confirms that the contractual terms are valid, and routes it through the down payment verification workflow before payment is released.

8. **Analyze and improve performance**
   The final step is continuous monitoring. SAP points to payable analytics, supplier analysis, accounts payable status and outlook, real-time cash forecasting, working capital analysis, and embedded dashboards such as Accounts Payable Overview, Bank Relationship Overview, and Days Payables Outstanding. These analytics turn the process into a management capability rather than a back-office routine.

The business value is significant. Teams can identify bottlenecks in approval, overdue invoices, discount opportunities, recurring bank issues, and supplier concentration risks. They can also improve close quality through automated purchase order accruals, which defer costs to the correct periods, shorten financial closing, improve external reporting accuracy, and reduce manual effort.

Example
A CFO reviews accounts payable analytics and notices that a cluster of blocked invoices is nearing cash discount deadlines. Because the system sends notifications for approaching due dates on blocked invoices, the team can resolve exceptions quickly and preserve discount savings that would otherwise be lost.

In day-to-day invoice-to-pay, work really lives in SAP Fiori. Accounts payable clerks and approvers move through invoice capture and verification, open-item analysis, and payment runs without leaving the role-based launchpad. For supplier down payments, the flow begins in Manage Teams and Responsibilities – Supplier Down Payment Requests and Manage Workflows – Supplier Down Payment Requests, which establish who is allowed to request and who must approve. Requests are raised and validated in Verify Supplier Down Payment Requests, and approvers act directly from their inbox and outbox—Verify Supplier Down Payments – Approver Inbox and Verify Supplier Down Payments – Approver Outbox—so governance is built into the motion rather than added after the fact. Operational invoices follow the same rhythm with Create Supplier Invoice for direct entry, Upload Supplier Invoice for OCR-assisted capture, and Manage Invoices to resolve holds, match to purchase orders or service entries, and post. Accounts Payable Overview anchors the day, surfacing aging, discounts at risk, blocked items, and duplicate checks; when liquidity timing matters, payment execution apps plan and run the payments, simulate cash impact, and reconcile exceptions.

2.3.5 Insight-to-Action

Chapter 5 covers the embedded AI and analytics capabilities available with SAP S/4HANA Cloud in depth, including AI-driven procurement insights. The procurement analytics capabilities described in this section provide the operational foundation that those AI layers build on. With procurement embedded analytics, you can perform the following tasks:

- Perform real-time operational reporting and monitoring of purchasing document items
- Gain spend visibility into actual and upcoming purchase order and scheduling agreement spend
- Control purchasing spend by monitoring off-contract and nonmanaged spend
- Evaluate supplier performance on various hard facts such as price, quantity, quality, time, and user-defined criteria and provide output options to share the scores with the suppliers
- Monitor process efficiency
- Analyze global spend, central purchase contract consumption, and the price variance in central purchase contract items for SAP S/4HANA Cloud for central procurement

Let's look at some analytical capabilities in procurement:

- **Operational supplier evaluation**
  Real-time KPIs for operational supplier evaluation, with the **Trend by Supplier** view selected, as shown in Figure 2.23. This KPI aggregates the individual scores from the different criteria—that is, price, time, quantity, and quality—by different dimensions. The corresponding weightings are applied, and the overall score for a given supplier is calculated and can be visualized over time.

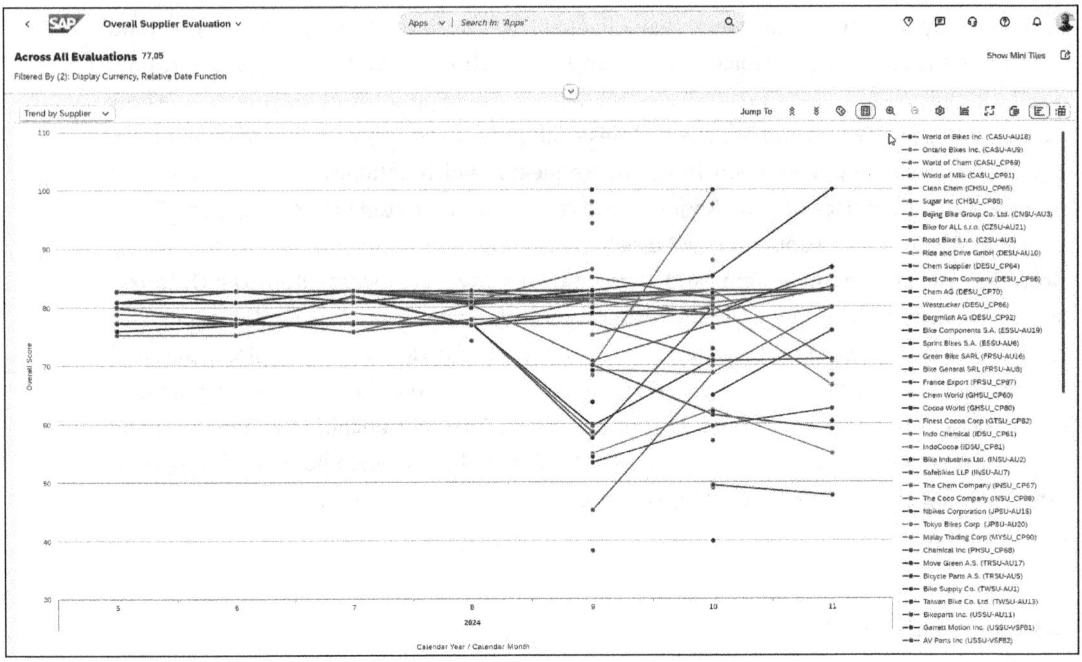

Figure 2.23: Real-Time KPIs for Operational Supplier Evaluation

- **Monitor purchase order items**
  Monitor purchase order items, to control overdue items and next deliveries. With this functionality, you can display purchase order items and monitor their status using filter criteria such as purchase order, material group, material, supplier, plant, and delivery forecast. The solution also predicts the delivery date of purchase order items and determines whether the delivery will happen earlier than scheduled, on time, or with a delay. Based on the predicted delivery date, a business user can take the necessary actions.

  Figure 2.24 shows the Monitor Purchase Order Items app, enabling buyers to track open purchase orders, identify overdue deliveries, and monitor next expected delivery dates.

Figure 2.24: Monitor Purchase Order Items App

- **Purchasing spend dashboard**
  The purchasing spend dashboard (SAP Analytics Cloud, embedded edition dashboard in SAP S/4HANA Cloud), as shown in Figure 2.25, offers your users an overview of purchase spend and off-contract spend in the organization. Prebuilt dashboards support real-time decision-making within the business process. These dashboards can be customized to increase the efficiency and effectiveness of management decisions.

More than 50 analytical procurement apps are predelivered with SAP S/4HANA Cloud and address tasks from the beginning, from posting the first business document to all the transactional applications also accessible from the SAP Fiori launchpad. These apps can be adjusted according to your company's and your end users' needs. Individual selection criteria, thresholds, chart types, or default views are just some examples. Moreover, many analytical views are ready to use in the Custom Analytical Queries app, where you can even adjust the view based on your requirements.

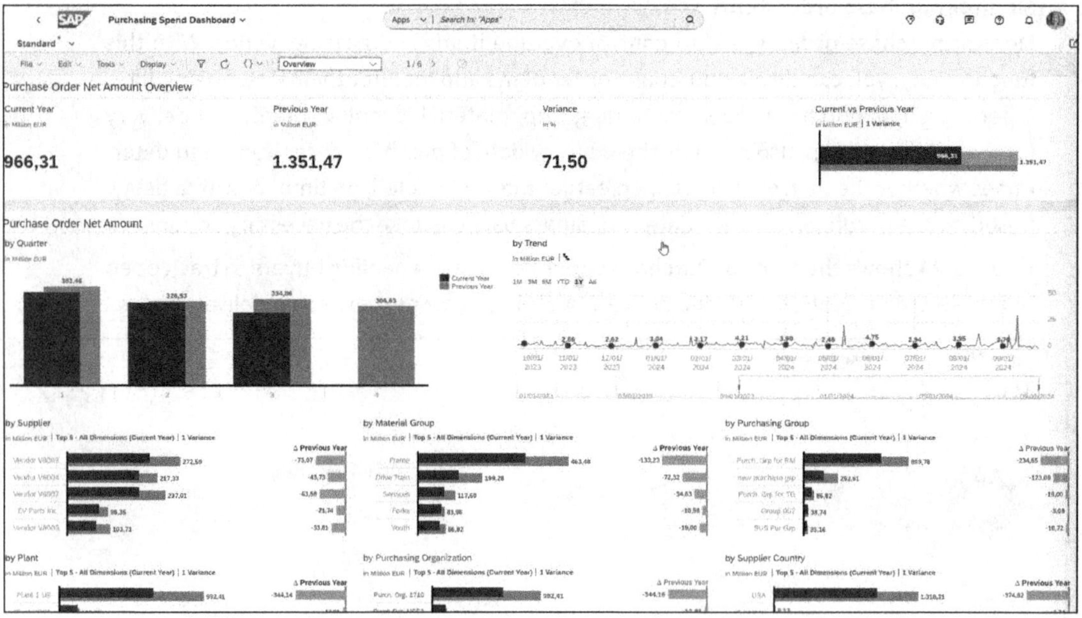

Figure 2.25: Purchasing Spend Dashboard

2.3.6 Public Cloud Versus Private Cloud

A common question from procurement stakeholders is what can be done in one edition that cannot be done in the other. For source-to-pay, the answer is that functional overlap is substantially larger than the gap. Both editions share the same core ABAP code line, and SAP delivers the same process pillars in both: sourcing, procurement contract management, operational procurement, and procurement collaboration. The scope items that power day-to-day procurement, consumable purchasing (BNX), procurement of direct materials (J45), procurement of services (22Z), subcontracting, requisitioning, scheduling agreements in procurement (BMR), supplier evaluation and performance monitoring (SL4), supplier classification and segmentation (19E), consignment AND pipeline procurement, central purchasing (2XT), central sourcing (3ZF), and supplier financing (4AV), are available in both editions. Integration with SAP Ariba Sourcing, SAP Business Network, SAP Ariba Contracts, SAP Ariba Invoicing, and guided buying for central procurement also spans both editions.

That said, meaningful differences do exist. Table 2.2 captures the primary functional divergences identified across the source-to-pay process areas, and the section that follows explains where those differences matter in practice.

Capability	Public Cloud Edition	Private Cloud Edition
Automated invoice settlement (scope item 2LH) Touchless, three-way-match-based invoice posting. The system autosettles invoices against purchase orders and goods receipts without accounts payable clerk intervention, driving straight-through processing rates and reducing the cost per invoice.	Available as a dedicated scope item (2LH) and an automotive variant (5L6). Configured through standard scope item activation; no custom development required.	Not available as a standalone scope item. Invoice processing follows the standard Create Supplier Invoice workflow. Touchless rates are achieved through integration with SAP Ariba Invoicing or the SAP Business Network invoice flip, which require separate setup.
Service entry sheet-based accruals (scope item 63L) Automated accrual posting for unapproved service entry sheets at period-end, with automatic reversal on approval. Eliminates manual accrual entries for open service purchase orders in closing cycles.	Available as a dedicated scope item.	Not available as a scope item version for current on-premise releases. Period-end service accruals are handled through manual or custom accrual processes.
Service and material procurement for project-based services (scope item J13) Integrated procurement of external resources and materials within a professional services project context, linking project demand directly to purchase requisitions and purchase orders.	Available as a cloud-native scope item confirmed in current releases.	Not confirmed as a scope item version for current private cloud/on-premise releases. Project-based procurement is handled through alternative scope items and configuration paths.
SAP Procurement Intelligence Advanced analytics, spend intelligence, and AI-driven procurement insights beyond standard embedded reporting.	Not available as a separate product. Analytics are delivered through more than 50 predelivered procurement SAP Fiori apps and embedded SAP Analytics Cloud dashboards (e.g., Purchasing Spend Dashboard, Supplier Evaluation KPIs).	Available as a dedicated add-on product: SAP S/4HANA Cloud Private Edition, Procurement Intelligence. Provides advanced spend analytics and AI-driven procurement insights as an optional licensed add-on.

Table 2.2: Functional Differences in Source-to-Pay: Public Cloud Edition Versus Private Cloud Edition

Capability	Public Cloud Edition	Private Cloud Edition
User interface Access to procurement transactions.	SAP Fiori exclusively. No access to SAP GUI transaction codes (Transactions ME21N, ME51N, MIRO, ME23N, MRBR, and similar). All procurement—purchase order creation, goods receipt, invoice entry, and approval—runs through SAP Fiori apps.	SAP Fiori plus full SAP GUI. Procurement users can use classic transactions alongside SAP Fiori apps. This supports organizations midway through a UX migration from SAP ERP.
Custom procurement logic. Ability to extend or modify standard procurement behavior.	Restricted to standard extensibility framework: custom fields on purchasing documents via the Custom Fields and Custom Logic apps, released OData/SOAP APIs, and side-by-side extensions on SAP BTP. Custom ABAP modifications to the procurement code path are not permitted.	Full ABAP environment available. Custom pricing procedures, bespoke approval workflows beyond standard configuration, user exits, and CDS-based custom procurement reports can be implemented. SAP recommends ABAP Cloud (restricted language version) to protect upgrade stability, but classic custom code remains an option where released APIs do not yet cover the requirement.

Table 2.2: Functional Differences in Source-to-Pay: Public Cloud Edition Versus Private Cloud Edition (Cont.)

For organizations running standard procurement processes, creating purchase orders, processing invoices with three-way match, managing contracts, running sourcing events through SAP Ariba, and collaborating with suppliers via SAP Business Network, both editions deliver functionally equivalent outcomes. The choice between editions is rarely driven by a missing procurement transaction.

The differences matter most in two scenarios. First, organizations that require high-volume, touchless invoice processing will find the public cloud edition more immediately capable through the Automated Invoice Settlement scope items (2LH, 5L6), which activate straight-through processing with no custom development. Private cloud customers achieve comparable results through SAP Central Invoice Management integration or the SAP Business Network invoice flip, but with additional setup effort. Second, organizations with complex custom procurement logic, such as nonstandard pricing, industry-specific approval chains, or proprietary supplier rating algorithms inherited from a prior SAP ERP landscape, require the private cloud edition as it permits the custom ABAP development that the public cloud edition does not allow. The availability of SAP Procurement Intelligence as a licensed add-on is an additional factor to consider for private cloud customers who need analytics capabilities beyond what the standard embedded apps provide.

2.3.7 Outcome

James, Korindal's chief supply chain officer (CSCO), carries a $30 million annual penalty from missed deliveries. Most of that cost does not originate in manufacturing or logistics. It originates in procurement: The wrong supplier confirmed the wrong delivery date, or a sourcing event ran three weeks because bids were compared in spreadsheets instead of in the system. The capabilities in this section address that gap directly. When sourcing events, contracts, and operational procurement run on a unified spine, the gap between the negotiated price and actual spend closes. Maverick buying drops because guided purchasing surfaces the right catalog item or approved supplier before a requisition leaves the system. Three-way match removes the manual reconciliation that delays payment and strains supplier relationships. For James, the most visible change is in supplier delivery predictability. AI-assisted supplier evaluation surfaces performance trends before they become delivery failures. Predictive lead-time models flag delayed deliveries three to five days in advance, giving the procurement team time to expedite or arrange alternatives. Production schedule disruptions from late supplier receipts fall by approximately 30 percent within the first planning cycle after deployment. The OTIF number begins to move because procurement stops feeding exceptions into the fulfillment chain.

2.4 Production Planning and Manufacturing

Plan-to-fulfill is the end-to-end orchestration that turns commercial intent into reliable promise and execution. In SAP S/4HANA Cloud, it spans three linked motions: plan and schedule what must be produced and procured, manufacture and control on the line with quality built in, and store and transport finished goods to customers. When this loop runs well, service levels rise, the need to expedite falls, and inventory stops growing just to protect against uncertainty.

Figure 2.26 shows the end-to-end plan-to-fulfill flow in SAP Cloud ERP across three motions: plan and schedule, manufacture and control, and store and transport. It starts with strategy and demand capture, runs material requirements planning (MRP), procures components, and schedules production; then stages components, builds subassemblies, and builds final assemblies with confirmation and inspection; and finishes by planning transportation, creating the outbound delivery, picking/packing/loading, posting goods issue, monitoring the shipment, and settling charges. In short, it visualizes how SAP S/4HANA Cloud turns policy and demand into finished goods delivered to the customer in one integrated loop.

At the front of the loop, planners translate strategy into practical rules. Product and customer segments receive clear policies for make or buy, sourcing mix, and where buffers belong. In SAP S/4HANA Cloud, these choices become parameters: plants and MRP areas, source lists and quota arrangements, special procurement for stock transfers or subcontracting, planning calendars, and realistic lead times. Demand baselines and consumption rules drive replenishment for steady items, while demand-driven replenishment protects volatile flows at decoupling points. MRP Live provides the heartbeat, turning plans and actuals into executable proposals (planned orders, purchase requisitions, stock transfers), with exception messages that focus attention. Predictive MRP (pMRP) offers rapid what-if simulation when demand spikes or capacity dips, and advanced available-to-promise (ATP) sets the guardrails for fair and feasible customer commitments using product allocation, supply protection, substitution, and backorder processing.

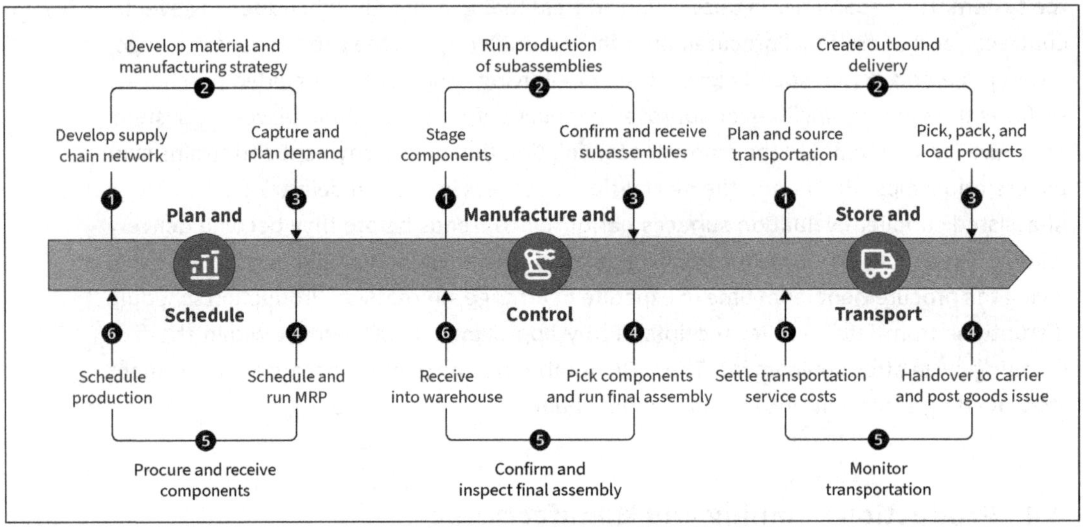

Figure 2.26: Plan-to-Fulfill with SAP Cloud ERP

The middle of the loop is where planning meets factory reality. Critical resources are scheduled finitely with production planning and detailed scheduling (PP/DS) to respect sequence and setup; the many remaining materials run steadily with MRP and sensible time fences. Component assignment and staging protect critical operations before release. Subcontracting steps are planned with component provisioning and realistic offsets. Inspection characteristics flow from routings so that "make-to-inspect" becomes one motion: Confirm yields and times, record quality, and feed facts back to the planner that they can act on. This is the core of plan to make, which we cover in detail later in this chapter: how to sequence and stage, when to firm and freeze, and how to keep the shop floor calm and predictable.

The final motion moves finished goods to customers. Delivery documents trigger picking, packing, and goods issue; warehouse teams receive and store assemblies; transportation demand is consolidated in freight units, planned into freight orders, executed with carriers, monitored in transit, and settled to close the loop. These capabilities live in the logistics footprint of SAP S/4HANA Cloud: warehouse management for bin-level control and integration to production, and transportation management for planning, execution, charge calculation, and settlement. We will cover storage and transport fully in Section 2.5, including the handoffs from outbound delivery to carrier tendering and the analytics that keep OTIF on track.

Across all three motions, the promise is kept consistent. Advanced ATP ensures that what we confirm is aligned with material and capacity realities; PP/DS finite checks prevent overcommitment on constrained lines; and exception handling—shortage monitors, backorder cockpit, and inventory analytics—gives teams a controlled way to recover when conditions change. Governance matters as much as algorithms; the plan we run should be the plan we publish, with clear rationale for any change.

In the following sections, we'll break down the core capabilities of the plan-to-make process: how the published plan becomes a finite, executable schedule, how components are staged, and how inspection feedback stabilizes the line. Later, in Section 2.5, we complete the picture with storage and transport: how deliveries are created, picked and packed, handed to carriers, monitored, and settled. Together these form a single operating rhythm—plan, make, move—that lets your enterprise keep promises confidently and efficiently.

2.4.1 Plan-to-Optimize Fulfillment Strategy

Plan-to-optimize fulfillment is the strategy spine of plan-to-fulfill. It decides how the company will keep customer promises without excess inventory or last-minute expedites. We'll use three loops to explain the rhythm. The first loop plans: You choose policies, create a believable demand baseline, and size buffers so that supply can run with confidence. The second loop promises: You confirm orders intelligently with available to promise that respects business priorities. The third loop recovers: When reality moves, you repromise and replan in a controlled way. Keep these loops in mind while reading this section; they connect strategy to daily execution.

Effective plan-to-fulfill execution depends on a small set of high-impact decisions made in the right sequence. The following operational steps guide teams from a demand signal to a published, executable plan:

- **Develop a supply chain strategy**
  A strategy is a small set of choices that you can encode in SAP S/4HANA Cloud and govern over time. For example, you can segment customers and products by service need and profitability. You can decide to make or buy, choose single or dual source, and pick where to place decoupling points. Turn those choices into parameters that planners use every day: which plants and MRP areas own planning, which source lists and quota arrangements express supplier policy, which special procurement keys indicate stock transfers or subcontracting, and which calendars and lead times reflect reality.

  For variable flows, use demand-driven replenishment at the right decoupling points so that buffers protect service. For steady items, keep it simple with forecast-based or reorder planning. The best policy grid fits on one page. "A" category items may use demand-driven MRP (DDMRP) and dual source. "B" category items can run forecast-based planning with a clear service target. "C" category items can run with a sensible reorder rule.

Spare parts can have explicit protection for field service. If a planner cannot explain the policy for a segment in one sentence, simplify it until they can.

Where network design and target setting across echelons is required, SAP Integrated Business Planning for Supply Chain (SAP IBP) provides a wide-angle view. We will discuss SAP IBP in detail in Chapter 4, Section 4.1. In this chapter, the focus is on SAP S/4HANA Cloud, where strategies become day-to-day rules that drive replenishment and promise.

- **Develop a manufacturing strategy**
  A manufacturing strategy connects the customer promise with the plant's reality. Most plants have a few resources that constrain everything else. Those need finite and sequence-aware schedules, plus protection from late changes. SAP S/4HANA Cloud gives both modes. PP/DS handles the critical resources with pegging, heuristics, capacity leveling, and setup-aware sequences. Classic MRP with sensible time fences covers the rest.

  Planning strategies set the boundaries between make-to-stock, make-to-order, and configure-to-order. Routings and work centers carry the cycle and setup data that PP/DS respects. Where the flow is stable, rate-based planning can anchor the cadence. The true test of a good strategy is a calm shop floor. Stable sequences, predictable changeovers, and very few late orders are the signs you are there.

  Longer horizon balancing across plants and months is a task for SAP IBP rough cut capacity planning. We will discuss that in Chapter 4, when we cover sales and operations planning and supply planning. Near-term feasibility remains in SAP S/4HANA and is protected with time fences and firming rules.
- **Develop a service fulfillment strategy**
  A service promise depends on part availability and technician time. The principle is simple: Protect what matters and avoid hoarding. In SAP S/4HANA Cloud, spare parts MRP sets replenishment rules and service levels. Advanced ATP becomes the safety net when orders arrive. Supply protection keeps inventory for premium contracts or field technicians. Product allocation meters scarce parts so that the first order does not consume the entire week's stock. Backorder processing reprioritizes confirmations when a critical repair appears or when supply moves. If you repair and refurbish, model the loop so that returns are converted to availability quickly.

  For multilevel service networks, where stock sits at a central distribution center, depots, and even technician trunks, SAP IBP handles service parts planning and multiechelon targets. We will cover that in Chapter 4. Here, we build a reliable first-time fix rate in SAP S/4HANA Cloud without oversizing inventory.
- **Plan demand**
  Operational demand planning is about a believable baseline that triggers the right replenishment behavior. For stable items, use forecast-based planning with modest time series methods and seasonal profiles. For volatile items, keep the forecast coarse and let DDMRP signals guide replenishment. Planned independent requirements carry the baseline. Order consumption settings control how incoming sales orders reduce it. New items

deserve careful ramp-up profiles and honest uncertainty. ABC and XYZ classification help focus attention on the few items where human judgment matters most.

Consensus forecasting, short-term demand sensing, and promotion uplift belong to SAP IBP. In SAP S/4HANA, the goal is a baseline that the supply side believes and can execute against.

Demand-driven replenishment helps production planners plan and manage supply chains efficiently based on customer demand, rather than through traditional MRP procedures. It forms the basis for a reliable material flow by defining buffers at strategically important points and adjusting these buffers regularly.

- **Plan inventory and supply**
  This is the heartbeat of SAP S/4HANA Cloud. MRP Live turns the plan and the current situation into executable proposals, such as planned orders, purchase requisitions, and stock transfers. Planners live in exceptions rather than totals, and the system supports this with clear messages for shortages and rescheduling. For the constrained few, PP/DS adds finite schedules and visible peg chains. pMRP is the quick what-if knob. If a large order lands or a line loses capacity, run a coarse simulation and decide how to shape promises.

  Buffers and safety stock handle variability. Lot sizes reduce transactional noise. Subcontracting plans align external steps with component provisioning and realistic lead times. Multiplant replenishment uses stock transfers to balance the network. When this setup is right, it feels ordinary, which is the highest praise. It is predictable, steady, and free of drama.

- **Align demand, supply, and financial plans**
  Before execution, demand, supply, and finance should tell one story. Use pMRP or PP/DS scenarios to confirm feasibility, then decide how to promise under constraint. Advanced ATP is the steering wheel. Product allocation enforces fair share when items are scarce. Alternative-based confirmation allows substitutions for a product, plant, or source to save commitments. Release for delivery checks ensure the warehouse only receives what it can ship. Backorder processing provides the recovery logic when plans move. (For more on ATP processes, see Section 2.5.4.)

  Do a quick commercial sense check. There is little value in meeting every date if margin and cash suffer. When you publish the plan, remember that governance is part of operations. The plan you run should be the plan you published, and any change needs a visible rationale.

- **Monitor and optimize supply chain performance**
  Plans drift. Leading teams are relentless about a short list of metrics and the parameter changes that make those metrics move. The service level and OTIF metric show what customers feel. Plan adherence and capacity utilization reveal factory stability. Inventory turns and slow-mover analysis show buffer health. Expedite spend and exception closure time reveal the cost of indecision.

SAP S/4HANA Cloud supports this discipline with coverage and shortage monitors, the advanced ATP backorder cockpit, and straightforward inventory analytics. Mass maintenance tools make parameter tuning repeatable at scale. Table 2.3 summarizes the key planning parameters and recommended settings for the plan-to-optimize process, providing a practical reference for implementation and ongoing tuning.

Capability Area	What It Enables	Key Features	Typical Outcome
Supply chain strategy	Segmented policies and clear decoupling choices expressed in the system	Plants and MRP areas, source lists and quota, special procurement, calendars and lead times, DDMRP, parameter governance	Higher service with less inventory because policy matches reality
Manufacturing strategy	Feasible and stable schedules on critical resources	Planning strategies for make to stock, make to order, configure to order; routings and work centers; PP/DS finite scheduling and pegging; rate-based planning; time fences and firming	Fewer changeovers and expedites, predictable throughput
Service fulfillment strategy	Protected spares and reliable appointments	Spare parts MRP, advanced ATP supply protection, product allocation, repair or refurb flows, advanced ATP backorder processing	First-time fix improves without hoarding parts
Demand planning (operational)	Believable baselines and sensible consumption	Forecast-based planning, planned independent requirements (PIRs), consumption rules, ABC or XYZ classification, ramp-up profiles, DDMRP demand signals	Coverage improves, nervousness drops, planners focus where it matters
Inventory and supply planning	Executable proposals and healthy buffers	MRP Live, safety stock and reorder rules, DDMRP buffers, PP/DS on constraints, pMRP simulations, subcontracting, stock transfers	Shortages and expedites fall, planners work exceptions rather than noise
Promise management	Smart confirmations and graceful recovery	Advanced ATP product allocation, alternative based confirmation, release for delivery checks, backorder processing	Better first-commit accuracy, fair share under constraint, faster recovery

Table 2.3: Key Planning Parameters for Plan-to-Optimize in SAP S/4HANA Cloud

Capability Area	What It Enables	Key Features	Typical Outcome
Monitoring and improvement	Parameter stewardship and rapid correction	Exception monitors, backorder processing cockpit, inventory analytics, mass maintenance of planning parameters	OTIF climbs, inventory turns improve, fewer firefights

Table 2.3: Key Planning Parameters for Plan-to-Optimize in SAP S/4HANA Cloud (Cont.)

2.4.2 Plan and Schedule

Plan and schedule is the frontend of plan-to-fulfill. Here you turn strategy and demand into a published plan the factory can honor, with the right materials arriving at the right time and a clear capacity envelope for production. In SAP S/4HANA Cloud, this motion is continuous: Capture and shape demand, translate it through MRP into executable proposals, resolve exceptions, and schedule work against real capacity before you release orders.

Figure 2.27 summarizes the plan and schedule flow in SAP Cloud ERP: Capture and optimize demand, run real-time MRP to generate supply proposals, resolve exceptions, then schedule and release production with interactive capacity views. It shows the closed loop from a believable demand signal to executable shop-floor orders and monitoring. We'll walk through each step in the process flow in more detail in the following sections.

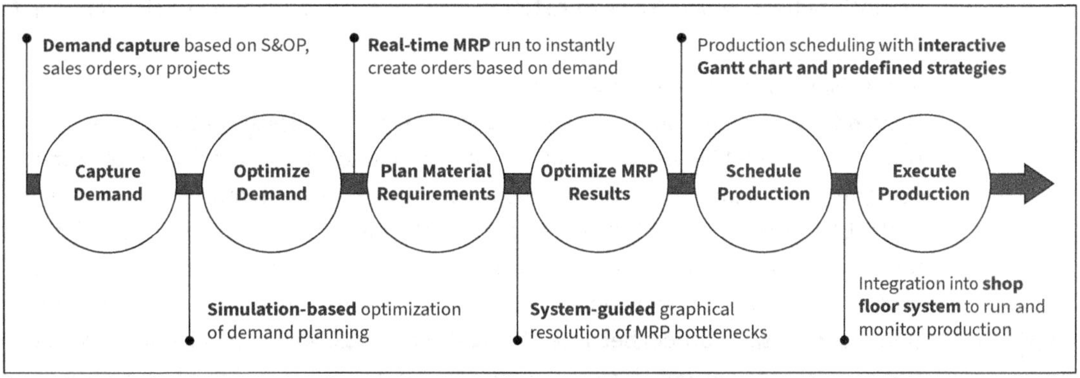

Figure 2.27: Plan and Schedule

Capture and Optimize Demand

This step creates the operational demand signal that planning can trust. Start by shaping a baseline that reflects how the business really sells, then tune it so it is both believable and economical. In SAP S/4HANA Cloud, planners maintain a planned program and let incoming sales or project orders consume it with clear rules. New items get explicit ramp-up profiles and, where seasonality matters, simple patterns rather than complex models. Classify the portfolio so that human attention lands on the few materials for which judgment moves the needle.

Optimization is the second pass. Run quick simulations to see the impact of promotions, large orders, or outages before you publish the plan. Use buffer logic at the right decoupling points so that variability does not spill across the network. The result is a demand picture that drives the right replenishment behavior without creating nervousness downstream. Longer-horizon consensus forecasting and demand sensing belong in SAP IBP and are covered later; here the focus is the daily signal inside SAP Cloud ERP.

The solution for demand-driven replenishment in SAP S/4HANA Cloud is comprised of several capabilities, as shown in Figure 2.28:

1. **Buffer positioning and sizing**
   Products at all bill of materials (BOM) levels are classified based on average demand, lead time, BOM usage, and variability to identify whether they are relevant for demand-driven replenishment. A well-balanced buffer level is proposed for products relevant to demand-driven replenishment to help ensure that they are sufficiently stocked to meet average demand but in low enough quantities to prevent excessive storage costs or losses due to expiration. An app helps planners manage safety stock, reorder points, and maximum stock through the buffer level proposals. To support the calculation of buffer levels, you can deploy pMRP.

   Demand-driven replenishment also delivers an ML service that supports the prediction of lead times for stock transfer materials based on historic lead times and their business context. This service improves the estimation of stock transfer lead times and the sizing of stock buffers, which leads to better service levels and potentially also less capital bound in inventory.
2. **Replenishment planning and execution**
   Demand-driven replenishment is fully integrated into SAP S/4HANA Cloud with the MRP run—that is, MRP Live or classic MRP. You can monitor demand-driven replenishment using an application that provides dedicated views for replenishment planning and replenishment execution. With replenishment planning, you have an overview of the buffer information sorted by planning priority. Using this information, you can trigger replenishment on time to avoid stock shortages and possible delays in fulfilling production orders or sales orders because of insufficient stock. With replenishment execution, you can view the current stock situation for each product and check that sufficient on-hand stock is available for production, and you can expedite replenishment directly through the application. You can also use the available contact details and contact the supplier of a purchase order or the production supervisor of a component production order.
3. **Demand-driven analytics**
   Demand-driven analytics helps you analyze the effects and gains of demand-driven replenishment based on KPIs such as service levels, lost sales, or inventory over time. With this information, you might decide to set new buffers or eliminate existing ones within your supply chain to further improve your KPIs.

Figure 2.28 shows the demand-driven replenishment solution in SAP S/4HANA Cloud, including buffer positioning, replenishment planning, and integration with MRP to reduce stock-out risk.

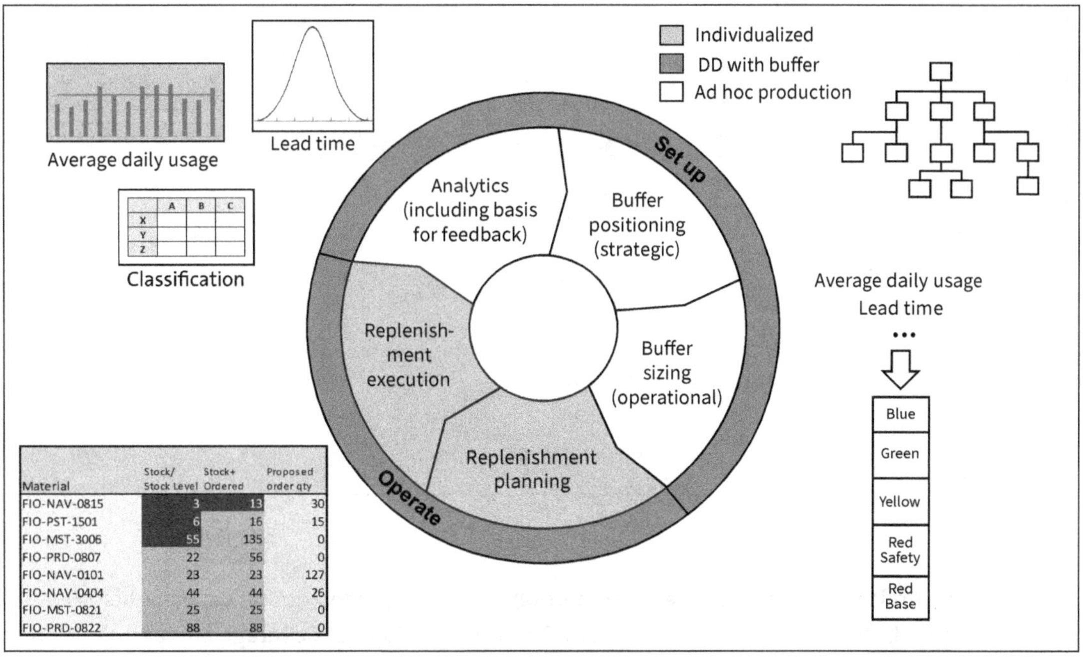

Figure 2.28: Demand-Driven Replenishment Supported by SAP S/4HANA Cloud

Plant Material Requirements and Optimize MRP Results

This is the heartbeat of planning and scheduling: Translate a believable demand signal into executable proposals and close gaps before anything reaches the line. In SAP S/4HANA Cloud, a real-time MRP run instantly creates the right orders based on demand—planned orders for make, purchase requisitions for buy, and stock transfers to move—while honoring lot sizes, lead times, safety stock targets, and subcontracting offsets. Planners stay in the interactive MRP cockpit rather than spreadsheets: they can rerun immediately when assumptions change, drill multilevel pegging to see which receipts cover which requirements, and view exception messages that spotlight true shortages and date conflicts.

During MRP results review, the Manage Material Coverage app highlights a time-phased shortage and its drivers, combining a bar chart of demands and receipts with an exception list located below the chart. Planners can simulate and apply resolution options (e.g., expedite purchase orders, reschedule receipts) directly from the **Proposed Changes** column to close the gap.

Figure 2.29 shows the Manage Material Coverage app, highlighting time-phased material shortages and their root causes to help planners resolve gaps identified during MRP results review.

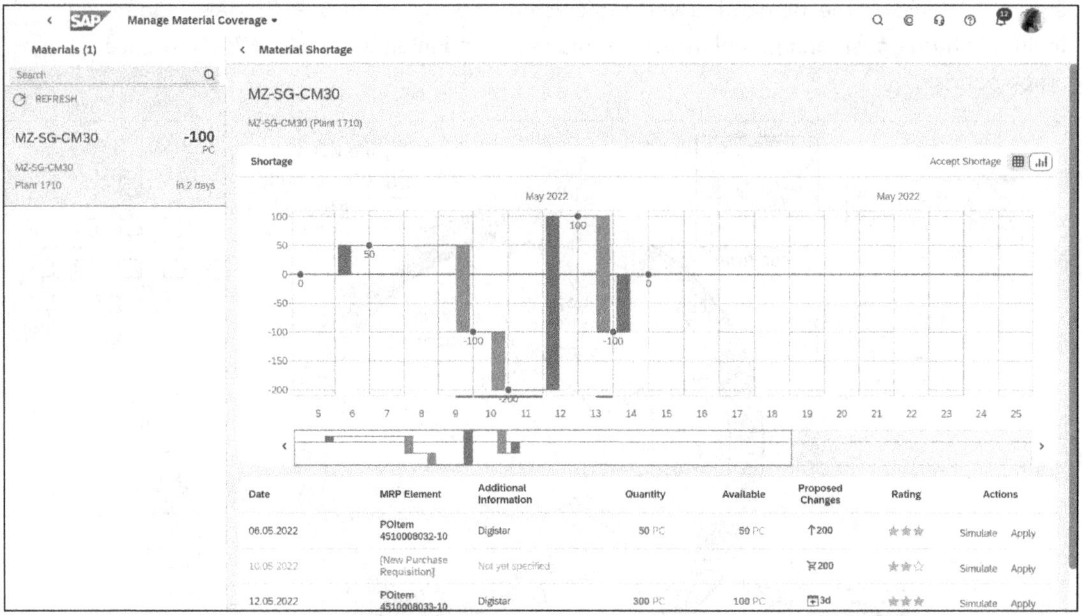

Figure 2.29: Material Coverage View

Optimization is part of the same motion. A system-guided, graphical view helps resolve MRP bottlenecks without guesswork. The Manage Material Coverage app and the Monitor Production Orders app show where a particular sales, production, or process order is at risk across all BOM levels; planners can simulate rescheduling, preproduction, or a change of source and then raise a change request to suppliers when collaboration is needed. Where variability drives the noise, buffer levels can be reviewed and resized so the next MRP run is steadier. The net result is a clean, explainable set of proposals with a short, visible list of issues that is ready for capacity scheduling and release.

Table 2.4 provides a reference for the key MRP and production scheduling parameters available in SAP S/4HANA Cloud, including planning strategies, lot-sizing procedures, and scheduling settings. Use this table alongside the Manage Material Coverage app to align your system configuration with your planning objectives.

Capability Area	What It Does	Key SAP S/4HANA Cloud Features	Expected Outcome
Real-time MRP run	Instantly converts the demand signal into executable proposals across make, buy, and move	MRP Live; lot-sizing and lead-time rules; static/time-dependent safety stock; subcontracting with component provisioning; stock transfer proposals	Early, explainable proposals with fewer last-minute surprises

Table 2.4: Key MRP and Production Scheduling Parameters in SAP S/4HANA Cloud

Capability Area	What It Does	Key SAP S/4HANA Cloud Features	Expected Outcome
System-guided bottleneck resolution	Highlights and resolves shortages/date conflicts with multilevel visibility	Interactive MRP cockpit; multilevel pegging; material coverage; order monitoring; graphical rescheduling (preproduction, reslotting, alt source)	Short, actionable exception list and faster closure
Collaborative change with suppliers	Aligns external dates/quantities when supply is the constraint	Supplier change requests integrated with SAP Business Network; confirmation tracking	Agreed reschedules without email churn; reduced mismatch risk
Stabilization for variability (optional)	Protects flow where demand/supply is noisy	DDMRP: buffer positioning and sizing, net monitoring, buffer health analytics	Higher service with right sized inventory at decoupling points
Rapid replanning and governance	Keeps the plan current as assumptions change, with traceable decisions	One-click run in the MRP cockpit; simulation and inline edits; audit of parameter changes	Up-to-date plan without offline rework; transparent decisions

Table 2.4: Key MRP and Production Scheduling Parameters in SAP S/4HANA Cloud (Cont.)

Integrated Production Scheduling

Before we cover detailed scheduling, let's discuss synchronized planning. *Synchronized planning* is a relay in which ownership of one plan shifts smoothly from a long-horizon strategy to executable detail and, finally, to verified reality on the line. SAP IBP sets the guardrails by aligning demand, inventory targets, and network response over months and weeks, then hands a single, approved plan to SAP S/4HANA PP/DS, where uncertainty narrows and the same plan is converted into a finite, feasible sequence on pacemaker resources. Pacemaker resources are the key production work centers that set the pace for the entire value stream. In SAP S/4HANA PP/DS, orders are finitely sequenced on this constraint (often the true bottleneck), and upstream/downstream operations are synchronized to its schedule, converting an approved mid-term plan into a feasible, execution-ready sequence.

Figure 2.30 shows the synchronized planning process, illustrating how demand, supply, and capacity plans are aligned across planning horizons to create a coherent and executable production schedule.

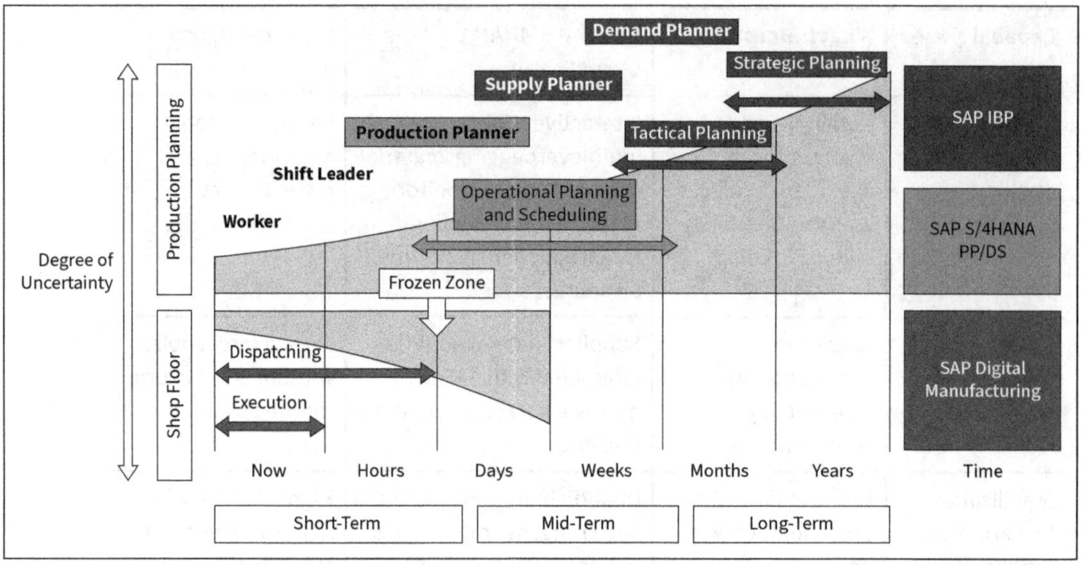

Figure 2.30: Synchronized Planning Process

Detailed scheduling in PP/DS protects a short-frozen zone the shop can trust, levels the load across work centers, and generates material-aware start and finish times that respect routings, setups, and constraints; in parallel, it feeds staging needs and promise confirmations back to order management. That schedule flows to SAP Digital Manufacturing for dispatching and execution, where operators confirm yields and times, record exceptions, and send progress signals upstream in real time. When demand changes or a disruption occurs, the update propagates as a controlled handoff—SAP IBP rebalances at the network level, PP/DS relevels the finite sequence within policy, and the shop floor executes the revised plan—so each role acts at the right cadence based on a shared truth, with detailed scheduling providing the precise bridge between planning intent and reliable production.

The process of production scheduling represents a central part of preparing for manufacturing components. Capacity planning in SAP S/4HANA Cloud is fully integrated, which helps create operational production schedules. With this capability, a production planner is equipped to design and fine-tune the most suitable production schedule for the manufacturing process.

The basic assumption for building this capability is that each order has one (known) bottleneck resource. The critical operation of an order is finitely scheduled onto the bottleneck resource, and all other (preceding and succeeding) operations are scheduled in an infinite way around the critical resource, using midpoint scheduling.

Beyond the proper setup of master data, no additional system configuration is required to enable this capability. You can use production schedules in both variants—discrete manufacturing and process manufacturing—and you can individually pick the relevant application settings.

For the calculation of required capacities for an order or operation, and of the capacities of a work center, capacity planning uses the capacity data found in the work center master data. A *work center* can be a place, a machine, or labor where production operations and activities are performed. In addition, a work center is used in routing operations and orders.

Figure 2.31 shows a real-life shop floor production-scheduling setup, demonstrating how finite capacity scheduling is applied to work centers and operations to reflect actual factory constraints.

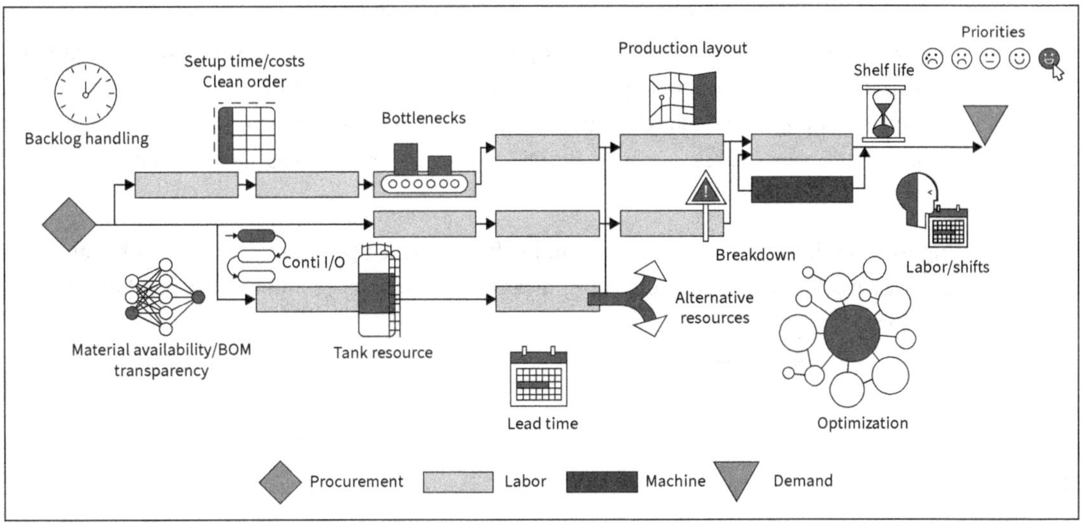

Figure 2.31: Real-Life Shop Floor Production-Scheduling Setup

Real production environments rarely match a clean schedule. A shift begins with a backlog of open orders—some with time-sensitive components, others tied to high-priority customer commitments. Continuous production lines, such as tanks or kilns, cannot be stopped and started freely, while discrete assembly lines compete for the same workforce and tooling. The capacity scheduling board addresses this complexity by giving production planners an interactive, graphical view of work center load, allowing them to drag, sequence, and level orders against real capacity constraints rather than relying on static MRP output.

The result is not a perfect plan but a reliable one: fewer rinses and changeovers, stable campaigns on continuous equipment, realistic starts that material and labor can support, and a sequence on the constraint that the shop can actually run. This is where synchronized planning meets the ground: the same plan from SAP IBP and MRP, rendered in PP/DS as minute-by-minute decisions that absorb real-life complexity without losing sight of the promise to the customer.

The main outcome of capacity planning is the production schedule. Because several different factors influence the schedule creation process, this capability is comprised of the following three subareas:

- The subarea of *maintenance* enables a planner to quickly apply changes to available capacities in order to meet the production requirements. This subarea covers actions such as the (re)definition of shifts or changes to capacity utilization.
- The subarea for *evaluation* delivers comprehensive insights into KPIs on load and backlog situations for capacities within the area of the planner's responsibility. Available capacities and capacity requirements can be selected according to various criteria. With this information, a production planner can compare available and required capacities and make informed decisions during the creation of production schedules.
- To enable the schedule creation, the *planning* subarea of the solution offers the ability of capacity leveling. The main goal is to allocate critical operations onto available and suitable resources in a desired sequence. The solution offers tabular and graphical views of production schedules with edit functionalities in order to enable fine-tuning of created schedules.

Figure 2.32 shows the capacity scheduling board, which provides planners with a graphical view of work center load, enabling interactive capacity levelling and production order sequencing.

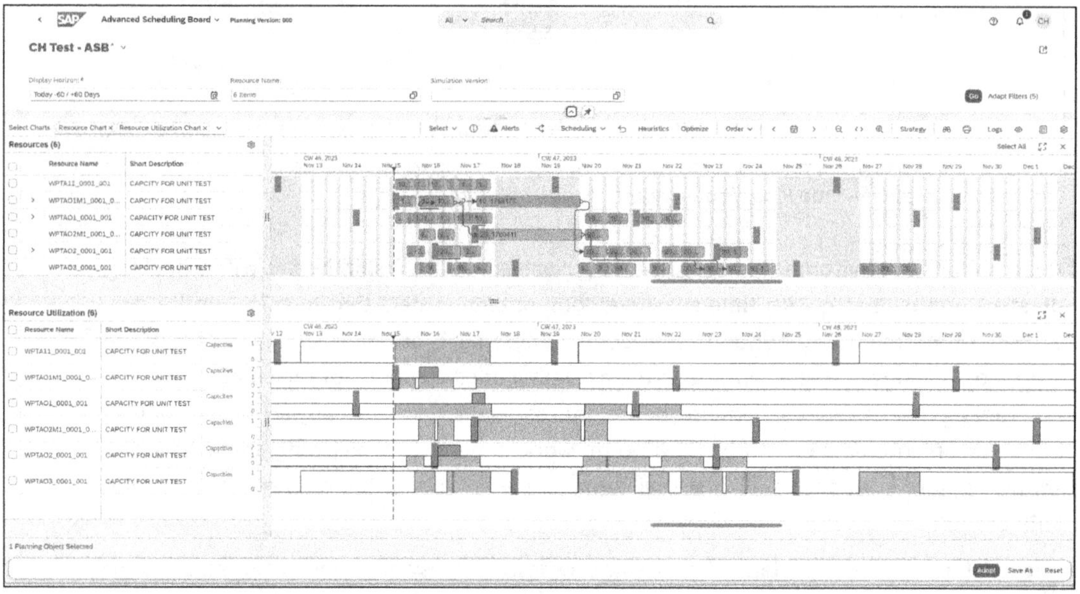

Figure 2.32: Capacity Scheduling Board

Because scheduling is integrated with MRP and order processing, handoffs are clean. Once leveled, orders are firmed and released; component staging follows the sequence; and confirmations and in-process inspections feed progress back to the planner. The result is a schedule that production trusts and a feedback loop that keeps it honest.

Best-Practice Parameters

- Keep the finite horizon to the real constraint; use time fences elsewhere to avoid overcontrol.
- Dispatch from the pacemaker resource and let upstream/downstream points follow midpoint logic.
- Maintain a short-frozen window that operations can rely on; make changes outside that window unless safety or customer impact demands otherwise.
- Treat capacity edits as governed data with validity dates and reasons so that you can learn from interventions.

2.4.3 Manufacturing

In this section, we discuss how SAP S/4HANA Cloud enables you to run your manufacturing processes and the connected logistics in a structured and well-organized way. This section covers the following processes:

- Support of production modes such as discrete manufacturing, process manufacturing, or repetitive manufacturing to satisfy industry needs
- Demand-driven production planning to help ensure the optimal use of manufacturing resources
- Integrated production scheduling to create production schedules on a detailed level
- Transparent production operations with insights into the degree of completion of production activities
- Embedded production execution and integrated production execution with integration into manufacturing execution systems (MESs) to connect the logistical view of production with the management of production machinery and its digitalized environment
- Repetitive manufacturing for planning and controlling production activities
- Transparent material flows using the Kanban lean manufacturing principle
- Support of outsourced manufacturing to reduce costs by transferring production tasks from internal staff to outside contractors
- Quality management, which is relevant for all these processes to minimize the cost of rejection

Overall, SAP S/4HANA Cloud handles the movement of input and output materials to and from the manufacturing processes. This capability allows for an up-to-date count of inventory in your warehouses and supports the performance of material valuation to control how production impacts account balances, which is needed for financial administration and for calculating the company's profit margins.

Supported Production Modes

Manufacturing in SAP S/4HANA Cloud is tailored to address the process requirements of a variety of industries and is constantly enhancing its capabilities to meet industry needs better. The different modes of production supported by SAP S/4HANA Cloud can be categorized into the following areas:

- **Discrete manufacturing**
  Discrete manufacturing is one of the most common modes of manufacturing adopted by companies using production orders. In this mode of manufacturing, a product can be assembled or disassembled and reworked if required and is generally manufactured across different work center operations. The stages of manufacturing between one operation and the next can be clearly distinguished. Manufacturing is based on a production order that is created to fulfill demand emerging from forecasts in a make-to-stock scenario or from sales orders in a make-to-order scenario. A production order is so central to this kind of manufacturing that direct materials, indirect materials, labor and overhead, and scheduling information are all captured within the production order to enable a transparent view of the actual time, material consumption, and costs of production. Master data objects such as work centers, BOMs, routings, and production versions enable discrete manufacturing. Industries that use this mode of manufacturing include high-tech manufacturers, industrial machinery and components, consumer products, and so on.

  Chapter 4 covers the extended line of business (LoB) applications for manufacturing, including SAP Digital Manufacturing for shop floor execution and SAP IBP for advanced supply chain planning.

- **Process manufacturing**
  Process manufacturing is a prominent mode of manufacturing designed for the chemicals, consumer products (food/beverages and home care products), and pharmaceuticals industries. Process manufacturing integrates different stages of production within these industries, from filling and mixing silos or tanks with inbound materials (the actual production) to putting the finished product into tanks or containers for transportation. This type of manufacturing also considers materials that are fed into the process through pipelines and therefore covers a large range of process possibilities. Process manufacturing usually occurs in phases and is spread across resources, but distinguishing one phase from another is not always possible. Most of these industries use batch-managed manufacturing driven through a process order, which also caters to both make-to-stock and make-to-order scenarios. Similar to production orders in discrete manufacturing, process orders play a central role in process manufacturing—for example, by providing a view of the actual factors of production.

  Master data objects, such as resources, BOMs, master recipes, and production versions, enable process manufacturing.

- **Repetitive manufacturing**
  Repetitive manufacturing, as the nomenclature suggests, is a mode of production that produces similar products repeatedly and continuously in a set of work centers. The major distinction from discrete and process manufacturing is that repetitive manufacturing is not driven by orders but is based on times and quantities in production. The rate at which each product is produced is an important measure in this mode of production. Planned orders are generated from an MRP run that considers both make-to-stock and make-to-order scenarios. These orders determine the run schedule, based on which production quantities and times are ascertained. Material consumption is normally done retrospectively, through periodic backflushing. The costs associated with production are periodically calculated and settled through a product cost collector depending on the material being produced. Planned order quantities are periodically reduced based on the repetitive manufacturing profile that is created for a material.

  The master data used in this type of manufacturing includes BOMs, production lines, production versions, product cost collectors, and repetitive manufacturing profiles.

In addition to these modes of manufacturing, SAP S/4HANA Cloud also enables lean manufacturing or just-in-time (JIT) manufacturing—primarily through Kanban, as described in more detail later in this section.

SAP S/4HANA Cloud offers relevant data and functionality supporting all these modes of manufacturing. Specific business roles can execute the processes in any of these modes of production through UIs for both occasional users and expert users. The actual terminology used within your organization may vary, but within the solution the primary roles are for production planners, production supervisors, production operators, quality managers, and so on.

Shop Floor Visibility and Execution Control: Production Supervisors and Operators

The capabilities of production operations within SAP S/4HANA Cloud are primarily targeted for production supervisors and production operators.

Production supervisors oversee the end-to-end manufacturing processes. They control the process based on the following data levers: material availability, capacity availability, schedule, and cost. The business object that empowers these processes is the manufacturing order. Native SAP Fiori apps such as the Manage Production Orders app, the Manage Production Operations app, and the Manage Process Orders app are windows into transparent decision-making based on these objects.

These applications are intuitive and have prominent visual aids that highlight pressing issues needing a supervisor's attention through visual cues. These applications prominently display crucial indicators and statuses that a supervisor might need to make decisions. Based on the availability of materials, on capacity, and on the schedule for fulfilling a requirement from an internal or external customer, supervisors decide when an order can be released for production. They can prioritize the allocation of several factors related to production. The exact nature and amount of materials, as well as machines, labor, production resources, and tools, are required to fulfill an order and are visible to the supervisors.

Production supervisors can also use a status to determine progress in production operations. The order changes its status through the different stages and carries information on some crucial stages in the manufacturing process. Decisions are made depending on the following criteria:

- The order is costed, and a settlement rule has been created to settle all the expenses that may go into execution of that order.
- The order has been released for production.
- Material shortages exist.
- Material has been issued for the order to begin production.
- Production has started, and the execution of operations in terms of the order's yield is confirmed either partially or fully.
- The manufactured goods have been delivered to the intended recipient storage location.
- The order is set aside as being technically complete.

Output management features within SAP S/4HANA Cloud allow supervisors to preview and print PDF files with data on the orders, the materials to be picked for the order and their storage locations, and the operations and their work centers.

Applications in SAP S/4HANA Cloud offer a smooth experience from insight to action. These applications use color-coded graphs and charts to visualize availability and operational progress so that a supervisor spends less time analyzing and more time addressing demand. These applications also visually highlight any delay in fulfilling demand and the cause of the delay. Supervisors can study any deviation in the planned and actual consumption of any of the factors and thus gain insights into the progress of the order and any deviations in cost.

To make the lives of supervisors easier, all these capabilities are available both at an individual order level and at a multiple selection level. Supervisors can choose to view order data based on their area of responsibility. SAP Fiori apps such as the Mass Processing Production Orders app (Transaction COHV) and the Monitor Production/Planned Orders app (Transaction COOIS) allow supervisors to view details from and apply actions across multiple orders.

In addition to providing insights into the execution of operations, other applications can reduce the time spent on performing transactional jobs. The Schedule Order Release Run app, the Schedule Order Closing Run app, and the Schedule Order Technical Completion Run app are SAP Fiori apps that allow supervisors to perform those actions in bulk, periodically and in the background.

Supervisors are also equipped with applications to highlight or correct any issues that may have occurred during manufacturing, such as quality issues and incorrect or incomplete material movements. These users have visibility into errors in goods movements within the system that are due to incorrect or insufficient information provided during original processing. Supervisors can reprocess failed material movements and correct errors in actual costs. Process support for performing inspection and rework is included in SAP S/4HANA Cloud as well.

Although the production supervisor controls the levers of the manufacturing process, the actual job of executing the operations that result in manufacturing is performed by production operators. SAP S/4HANA Cloud provides production operators with all information they need to perform the intended operations. They'll see details about the materials that need to be picked to execute an operation and about the storage location in which the materials are located. Production operators are provided with the operations they need to perform and the production resources and tools to be used to complete the process. Data informs these users about how much is to be produced, where to perform the operation, and what time to begin production.

Production operators also perform the crucial task of confirming that an operation is complete and providing information on the yield of the operation and time taken to complete the operation. SAP S/4HANA Cloud offers applications that allow the confirmation of both production orders and process orders at the operation/phase level or at the order-header level. The important data that a supervisor depends on to track the progress of execution is captured by the operator through these applications.

In many cases, production operators do their work with an MES that is integrated into SAP S/4HANA Cloud, or they use the manufacturing execution capabilities of SAP Digital Manufacturing, which are integrated into SAP S/4HANA Cloud. More details and other integration options are described in the next section.

In a future where manufacturing is much more connected and warrants real-time decision-making, enhancements to analytics in SAP S/4HANA Cloud will likely be a key focus area. These capabilities could provide insights into operational parameters that measure the efficiency of the manufacturing processes and learn ways to improve the processes. Creating a feedback loop that extends back to intelligently refining master data based on previous insights into the production execution is another direction in which the solution could move.

SAP S/4HANA Cloud is also constantly upgrading its integration into other systems so that the transparency of the operations happening in MESs is also enhanced. More on this subject is discussed in upcoming sections.

Embedded Production Execution and Integrated Production Execution

Production execution is a process with a high, variable degree of complexity at manufacturing companies. Based on the manufacturing products, the process varies from straightforward production execution without complex machine integration to an automated MES.

In the case of simple products with simple production processes, a production order or process order can be executed with the order confirmation, goods issue, and goods receipt functionality within SAP S/4HANA Cloud. However, a more complex production process requires a dedicated system for manufacturing execution.

In this section, we'll provide an overview of the various integration scenarios supported with the manufacturing functionality of SAP S/4HANA Cloud. Integration between SAP S/4HANA Cloud and MESs is supported with both a classic interface, using both blue-listed application programming interfaces (APIs), like Business Application Programming Interface (BAPI) and IDoc, and new APIs, like Simple Object Access Protocol (SOAP) and Open Data Protocol (OData).

> **What Is a Blue-Listed API?**
>
> OData and SOAP APIs are externally available APIs. These APIs are called *whitelisted APIs*. You can find whitelisted APIs via the SAP Business Accelerator Hub (see Chapter 11, Section 11.2).
>
> If no whitelisted API is appropriate for your organization or for partners integrating SAP S/4HANA Cloud with third-party solutions, you can send individual requests to SAP to request the ability to use specific BAPIs or IDocs on a temporary basis. These BAPIs/IDocs are called *blue-listed APIs*.

Support for the following integration scenarios is delivered as part of preconfigured best-practice support. Master data such as the material master, work center/resource, BOM, routing/master recipe, and production version replication from SAP S/4HANA Cloud to external MESs is a prerequisite for the following processes:

- **Production operation with manufacturing execution system (discrete manufacturing)**
  This best-practice support provides integration with an external MES through the classic interface using blue-listed APIs (BAPI and IDoc). The integration to the external MES is triggered once the production order is released on SAP S/4HANA Cloud. On completion of production execution (partially or fully), the MES sends back confirmation details along with a goods movement to SAP S/4HANA Cloud, where the corresponding posting is triggered automatically.
- **Third-party shop floor execution (discrete manufacturing)**
  This best-practice support provides integration with an external MES through new APIs (SOAP and OData). The execution process is similar to the first scenario with production orders.

- **Production operations with the SAP Manufacturing Execution application (discrete manufacturing)**
  This best-practice support provides integration with SAP Manufacturing Execution through the classic interface via blue-listed APIs (BAPI and IDoc). This scenario orchestrates highly automated production processes with production orders, enabling product variants with a lot size of one and their production sequences to be changed automatically, quickly, and easily. Users can access real-time transparency into production operations and KPIs, plus complete traceability based on as-built records.
- **Third-party shop floor execution (process manufacturing)**
  This best-practice support provides integration with an external MES through new APIs (SOAP and OData). The execution process is similar to the first scenario with process orders.
- **Execution with SAP Digital Manufacturing (process manufacturing)**
  This best-practice support provides a comprehensive integrated process between SAP S/4HANA Cloud, SAP IBP, SAP Asset Performance Management, logistics, and SAP Digital Manufacturing, as shown in Figure 2.33.

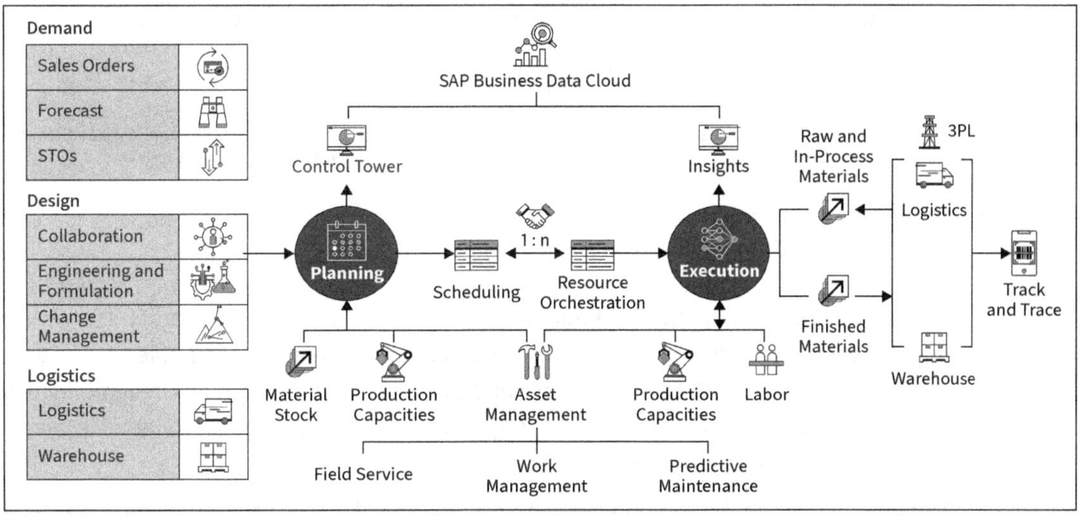

Figure 2.33: Integrated Supply Chain with SAP S/4HANA Cloud and SAP Digital Manufacturing

In this case, the integration is based on new APIs (SOAP and OData). The process starts with the creation and release of a process order from SAP S/4HANA Cloud, which sends the order to SAP Digital Manufacturing for execution. The execution process starts with the release of a shop order. Once production is completed, the shop floor operation confirms the yield, scrap quantities, activities, and goods movement using a dedicated UI, the Production Operator Dashboard. Posting activities in SAP Digital Manufacturing trigger corresponding postings in SAP S/4HANA Cloud. The leftover components are picked and issued against the production order if they are not issued through the MES.

Repetitive Manufacturing

Repetitive manufacturing is used for planning and controlling production in a repetitive manufacturing environment. The system supports the following activities:

- Planning and monitoring the material flow at a much higher level of detail than when collecting and analyzing costs
- Creating and revising production quantities on period and quantity bases
- Reducing production control effort and simpler backflushing

Figure 2.34 shows the repetitive manufacturing process overview, illustrating how production quantities are planned and backflushed against run schedules without individual production order control.

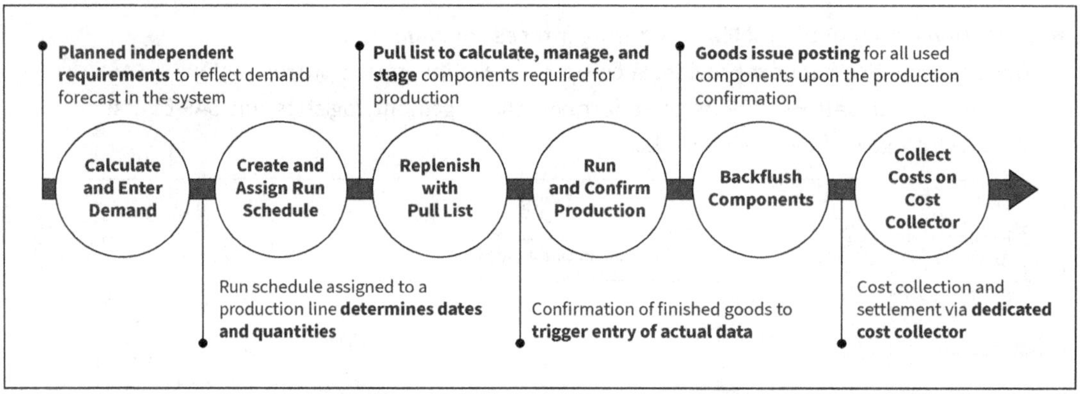

Figure 2.34: Repetitive Manufacturing Overview

In SAP S/4HANA Cloud, repetitive manufacturing supports the make-to-stock scenario. In this scenario, production is controlled without any direct reference to a sales order. Sales order quantities are delivered from stock and consume the PIR quantities in demand management, according to the planning strategy you select. A product cost collector is used to collect actual data and to settle costs.

SAP plans to extend the solution to support make-to-order scenarios, enabling more repetitive functions and configurations and adding certain aspects of flow manufacturing to the solution. *Flow manufacturing*, sometimes called continuous flow or single-piece flow, is a production method used instead of traditional mass production. In mass production, materials move through the manufacturing process in batches, which can be highly inefficient. In flow manufacturing, one product is completed before the next one is started—an approach typically used in lean organizations that help businesses achieve JIT production.

Transparent Material Flow with Kanban

In its most basic form, *Kanban* is a decentralized material flow control based on the pull principle and on actual consumption, in contrast to the classic push principle.

In Kanban, a material flow is organized using containers that are kept directly at the appropriate work centers in production. Each container contains the quantity of material required by the personnel at that work center for a certain period. As soon as a container is emptied at the demand source, replenishment is initiated. The Kanban signal for the delivery of a material can be triggered by a demand source sending a card to a supply source. This card describes which material should be delivered where and in what quantity. You can print Kanban cards containing barcodes from the system. Scanning the barcode with a standard barcode scanner suffices to transmit the data needed for procurement and to post the goods receipt upon receipt of the material. The supply source for the required material can be another place in production, an external supplier, or a warehouse. The demand source can use material from other containers until the filled container returns.

Figure 2.35 shows the Kanban process overview, illustrating the pull-based material flow control cycle, from consumption signal through container replenishment to goods receipt at the point of use.

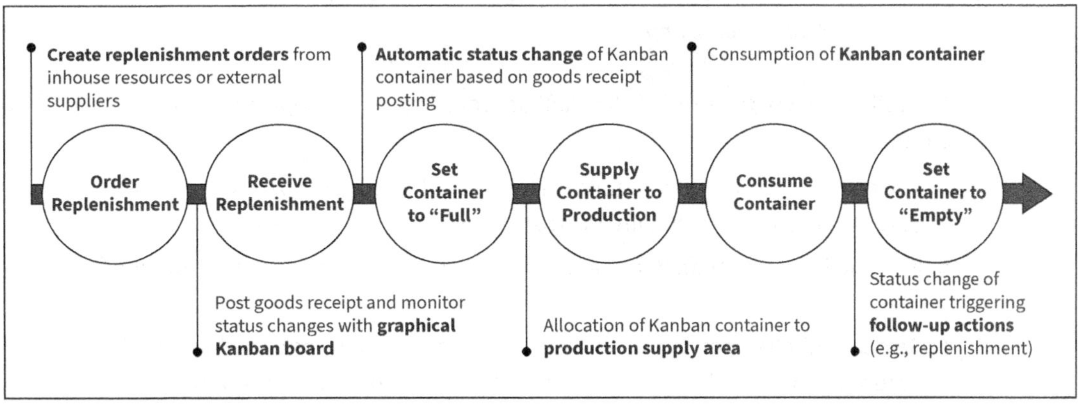

Figure 2.35: Kanban Process Overview

The aim is that the production process is controlled by production itself so that the employee must perform fewer manual postings. This self-management process and the fact that replenishment elements are created close to the time they are actually consumed means that stocks are reduced and lead times are shortened. (Replenishment is only triggered when a material is actually required, and not before then.)

The graphical Kanban Board app, shown in Figure 2.36, is the main tool for monitoring Kanban containers as they circulate in the lean manufacturing environment. A production planner or supervisor can check the actual situation in the Kanban Board app, either in the plant itself or remotely.

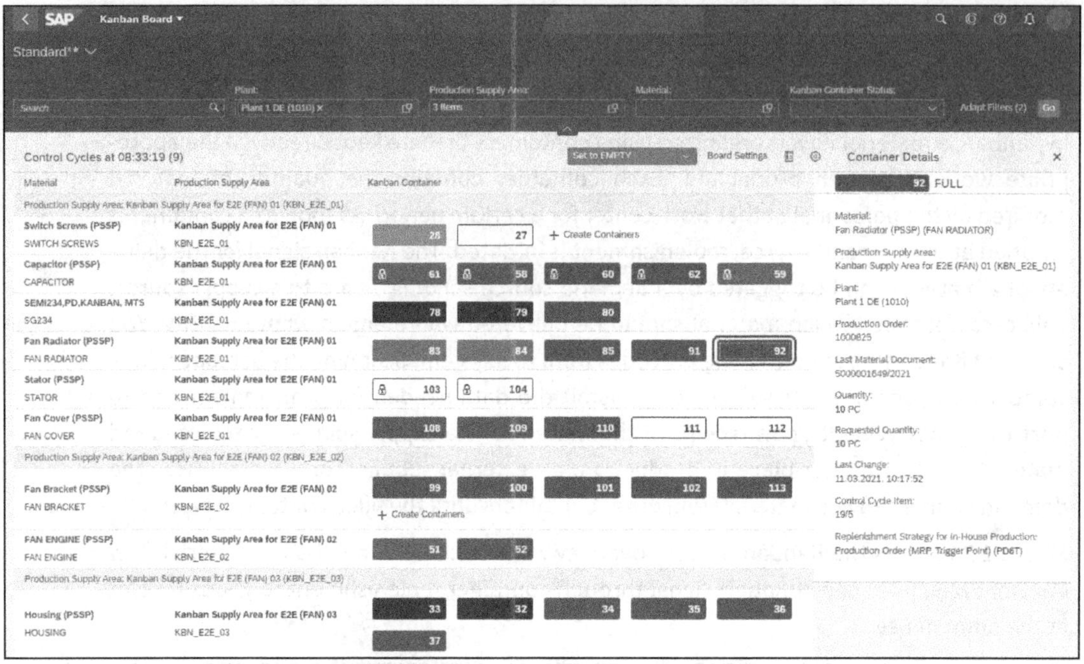

Figure 2.36: Kanban Board App

The following key features of the Kanban Board app are supported within SAP S/4HANA Cloud:

- You can use this app to monitor the status of circulating Kanban containers—particularly to identify containers that are in error status.
- Personalize the Kanban Board app view by using the filter, sort, and grouping functionalities.
- Search and filter options allow you to list Kanban control cycles and their containers according to your criteria—for example, by specific production supply areas or by locked Kanban containers.
- You can see details about a selected Kanban container, such as the actual quantity or the created documents.
- Kanban container statuses can be changed intuitively from the Kanban Board app.
- You can create event-driven Kanban containers to request replenishment.

Besides the Kanban Board app as the central lean manufacturing tool, SAP S/4HANA Cloud also supports all common replenishment strategies and offers the flexibility to adopt these strategies to meet your company's needs for the relevant classic and event-driven control cycles. Through predefined or self-customized calculation profiles, you can know that the optimal number of containers is circulating and that important lean manufacturing principles are being followed to continuously optimize production supply in a modern factory.

Furthermore, SAP aims to support a radio frequency identification (RFID) Kanban process in the cloud, based on integrated sensor and Internet of Things (IoT) data, to further automate the replenishment process and track containers within plant facilities.

Just-in-Time Manufacturing

JIT is the operating model for making and delivering only what a customer needs, in the exact quantity and at the exact time. In SAP S/4HANA Cloud it is implemented from end to end, from customer scheduling agreements to production, packing, outbound delivery, and billing. A forecast delivery schedule establishes a horizon and triggers material readiness; the real execution is driven by JIT calls (or sequenced JIT calls for just-in-sequence [JIS]) that arrive via modern APIs from the customer. Each call is grouped into packing groups, released to production, built and confirmed in the required order, packed into handling units (HUs), and shipped as an outbound delivery in sequence. Delivery confirmation flows back by API and can trigger self-billing. Because JIT is embedded in the core, planners and customer service see one set of dates and quantities across MRP, the shop floor, logistics, and invoicing.

Figure 2.37 shows the JIT manufacturing process overview, illustrating how production and delivery are synchronized with customer demand with minimal intermediate inventory.

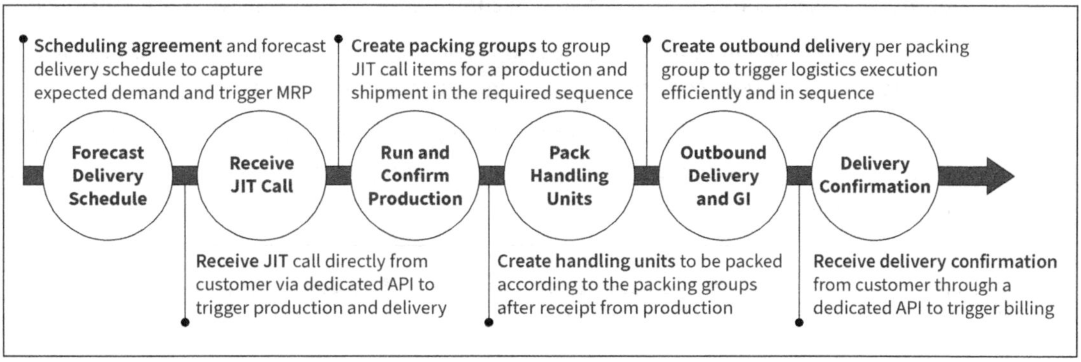

Figure 2.37: Just-in-Time Manufacturing Process Overview

What JIT changes on the shop floor is not only timing but control. The pacemaker resource is scheduled to the customer's sequence, upstream operations follow with minimal work in process (WIP), and material is staged in small lots. Exceptions are handled by reslotting the sequence, adjusting packing groups, or arranging short-interval transports rather than building inventory "just in case." The payoff is shorter lead time, lower working capital, and earlier visibility of real problems (quality, setup, or supplier reliability) so they can be fixed.

Key differences between JIT and other processes include the following:

- **JIT versus traditional push planning**
  Pushing produces to a plan and stocks finished goods; JIT produces to an external call and ships immediately, with the sequence and timing dictated by the customer.

- **JIT versus Kanban**
  JIT/JIS is a customer-driven, often intercompany process synchronized to a vehicle or line sequence, while Kanban is an internal (or short-loop) pull method that authorizes replenishment when a bin is emptied. Many plants run JIT to the customer and Kanban between work centers.
- **JIT versus plain MRP fulfillment**
  MRP ensures component readiness across the horizon; JIT consumes that readiness in final sequence and time buckets, with API-driven calls, packing groups, and delivery confirmation.

Outsourced Manufacturing

Outsourced manufacturing is an increasingly popular process in which a company contracts with another company to manage services. Outsourcing allows you to reduce costs by transferring production tasks from internal staff to outside contractors for reasons that include the use of high-cost equipment, the availability of low-cost personnel, avoiding liabilities and investments into production resources, and allowing management to focus on the core business. Outsourced manufacturing spans from single outsourced tasks to the completely outsourced manufacturing of the finished product. Multilevel outsourcing of production steps throughout a network of contractors is gaining increasing importance.

In the area of manufacturing, businesses often subcontract finishing processes, such as electroplating, powder coating and painting, and mounting and assembly tasks.

Figure 2.38 shows the outsourced manufacturing process overview, covering the subcontracting cycle from purchase order creation and component transfer through to finished goods receipt.

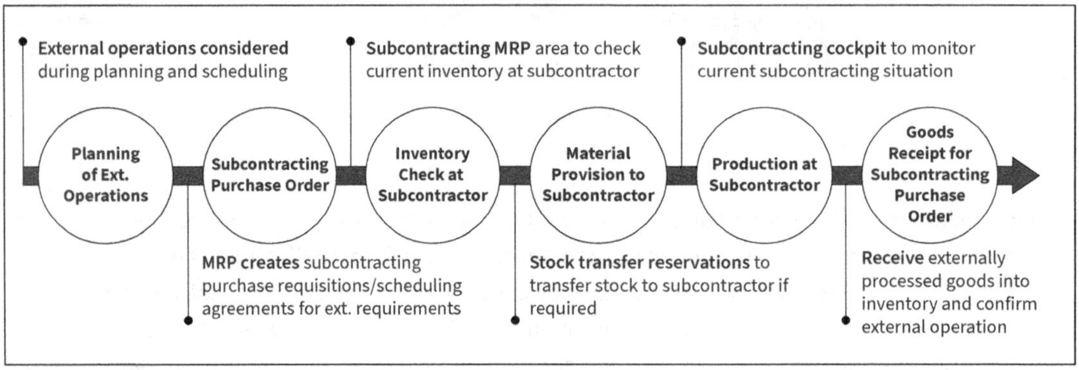

Figure 2.38: Outsourced Manufacturing Process Overview

In SAP S/4HANA Cloud, for all subcontracting variants, from single-task execution up to the production of a finished good, a subcontracting order is processed. The subcontractor receives materials like raw materials, components, or subassemblies, with which it produces the end product.

In general, the following functionalities support subcontracting in production processes:

- **Subcontracting/outsourced manufacturing in planning**
  Production planners are supported by SAP S/4HANA Cloud to plan the materials that are produced by a subcontractor. In a planning run, the system creates subcontracting purchase requisitions or schedule lines for a material made by the subcontractor, explodes the BOM of these materials, and creates subcontracting requirements.

  For subcontracting the same product at multiple contractors, SAP S/4HANA Cloud supports the creation of several production versions to cover the needs of different subcontractors. The purchasing info record of the subcontractor references the appropriate production version.

  During planning, existing inventory at subcontractor site is checked to determine whether parts need to be provided. If needed, stock transfer reservations are generated. The process is monitored automatically by the system, and the planner is notified when exceptions and errors arise.

- **Transparency in the process and in inventory**
  The outsourcing process requires special transparency. Stock is managed as part of the total valuated stock on the plant level and is available for MRP. Physical inventory of the stock of material provided to subcontractors is supported.

- **Triggering the provisioning of materials to be provided to subcontractor**
  Involved persons, like the production planner, purchaser, and shipment specialist, are provided with a comprehensive overview of all relevant information in the subcontracting cockpit. This tool provides a single entry point to gather all information on the subcontracting process and capabilities to trigger execution.

  For each purchase order item, an outbound delivery can be created. This delivery is displayed in the purchase order history for the corresponding item. In addition, purchase orders, purchase requisitions, outbound deliveries with open goods issues, reservations, and external deliveries (subcontracting components prepared by a third party) can be processed directly via the subcontracting cockpit.

- **Goods receipt for parts made by the subcontractor**
  A goods receipt for the subcontracting purchase order/schedule line triggers updates to inventory and purchasing statistics. In addition, the system backflushes components that were provided to the subcontractor.

- **External processing as a supported variant of subcontracting**
  External processing provides capabilities to instruct a supplier or subcontractor to process individual production steps such as operations or suboperations. The external processing of production order operations is frequently used for standardized process steps that cannot be performed in your own factory, such as galvanizing. In these cases, you may have environmental reasons for outsourcing a particular step to a subcontractor.

SAP S/4HANA Cloud supports all involved personas like the production planner, purchaser, shipping specialist, or inventory manager in an end-to-end subcontracting process with the relevant functionality to ensure flawless, optimized planning and execution.

2.4.4 Quality Management

To ensure quality throughout the product lifecycle, enterprises often embed multiple quality management steps in their processes to fulfill quality characteristics. A highly embedded solution guarantees integration into all relevant processes. In general, distinctions can be made among three scenarios in quality management:

- Quality management in procurement
- Internal quality management
- Quality management in sales

In all cases, a quality planner defines the characteristics, scope, method, and context of an inspection within inspection plans or routings. Copying existing plans into new contexts is a practice that improves efficiency and reduces manual defects. Sampling and dynamic modification enable enterprises to minimize effort during inspections and at the same time uphold regulations and product safety. A method provides guidance to a quality technician about what needs to be done during inspection. By defining the usage of a plan, different contexts can be covered based on a trigger point.

Figure 2.39 shows the quality management process overview, illustrating how inspection plans, result recording, and usage decisions are applied across procurement, production, and sales processes.

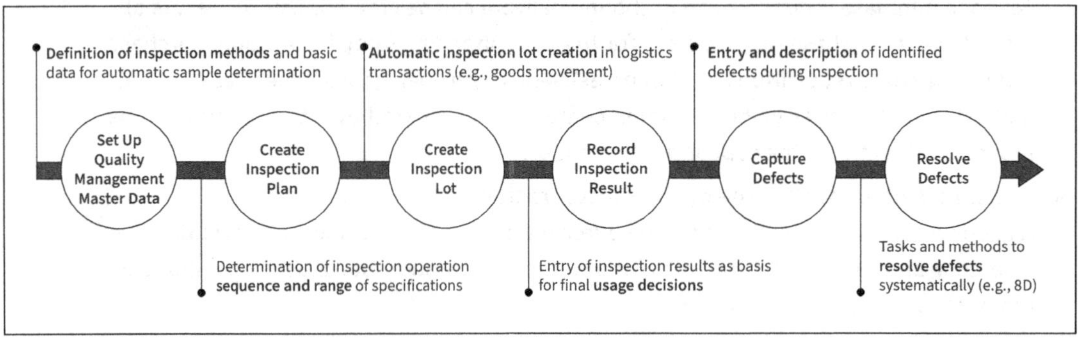

Figure 2.39: Quality Management Process Overview

Within quality management in procurement, a planner defines what relationships with a material are allowed within a plant. A certain quantity can be released for purchasing, or for a certain time. A quality planner is also authorized to block purchasing for a dedicated combination. If needed, a production part approval process (PPAP) can be defined as relevant and triggered. A supplier can then insert documentation and test results against a PPAP in accordance with the required information. An internal first article inspection can be marked as relevant based on a purchase order item. This new scenario provides further flexibility within the development phase of a product integrating with quality management.

Based on different trigger points, like a goods receipt for a purchase order, the release of a production order, a goods receipt toward a production order, an expired material in stock, or a outbound delivery creation, an inspection lot can be created based on the previous planning. With the help of the overview page, a quality technician can easily identify relevant inspection lots and start recording results based on serial numbers in a flat data format or in a table form. Within the new table form, results can be recorded more efficiently as recording is possible for multiple inspection lots at the same time. To evaluate a recording, a technician can also verify the historical results of a characteristic. A recording also can be made from other systems, using interfaces to connect other systems or measurement machines and to record defects. A quality technician needs to trigger a generic defect or a production defect for any findings during production so that issues can be corrected. Results based on batches or inspection lots can be collected to generate a certificate of analysis at delivery creation.

Figure 2.40 shows the Record Inspection Results app, in which quality inspectors record measurement values and characteristic results against an active inspection lot.

Figure 2.40: Inspection Results Recording

After an issue has been identified or a result recording finalized, the quality engineer can further identify and proceed to address the defect. The engineer either processes the defect or executes a usage decision. Both can be identified easily in an overview of the most relevant information for a quality engineer, including open usage decisions or most defective parts. The usage decision is the final step of a successfully executed inspection; it indicates the usage of a material and evaluates the result. Logistically, the affected parts will be removed from quality stock and processed into either unrestricted or blocked stock. Mixed decisions based on sorting or stock are common. Any issue identified within the process can be solved by using a task and assigning a responsible person for the execution.

If further analysis of a defect is desired, a *problem-solving process* can be started, which follows an eight-dimension approach. You can start either an internal problem-solving process or one related to a supplier. A supplier-related problem-solving process allows for collaboration with suppliers to interact on open issues.

A quality engineer is responsible not only for defect handling and usage decision-making but also for other insights to steer a process. Based on analytics, the engineer can adjust quality levels to change dynamic modification. This user can also change quality info record settings and thus block or limit releases for a material within a plant. Adjustments may be crucial to both save resources and deliver the right quality to customers, thus fostering relationships and increasing the company's reputation with its customers.

Quality management within SAP S/4HANA Cloud provides the key elements of product lifecycle quality management within an embedded environment while also providing numerous interfaces and supporting features.

2.4.5 Public Cloud Versus Private Cloud

The manufacturing capabilities in SAP S/4HANA Cloud follow a pattern similar to that of procurement; the common ground between the two editions is substantial. Discrete manufacturing, process manufacturing, and repetitive manufacturing are supported in both editions, along with MRP Live, production scheduling with capacity leveling (PP/DS), demand-driven replenishment, Kanban, production subcontracting, quality management in discrete and process manufacturing, and integration with SAP Digital Manufacturing and SAP Manufacturing Execution. BOM management, engineering change management, and production analytics are also shared.

The key functional divergences are summarized in Table 2.5.

Capability	Public Cloud Edition	Private Cloud Edition
SAP Fiori analytical apps for production planning (2JD) Dedicated SAP Fiori–based analytical applications for production planners.	Available as a dedicated scope item.	Not confirmed as a standalone scope item version. Production planning analytics are available through standard embedded analytics and CDS-based reporting.

Table 2.5: Functional Differences in Plan-to-Make, Public Cloud Edition Versus Private Cloud Edition

Capability	Public Cloud Edition	Private Cloud Edition
MRP change request integration with SAP Business Network (65D) Collaborative supply planning that propagates MRP-driven schedule changes to suppliers through SAP Business Network.	Available as a dedicated scope item.	Not confirmed as a scope item version. Supplier collaboration on schedule changes relies on standard purchase order change communication and SAP Business Network integration.
Production volume capture (4AG) Capture of actual production volumes at reporting points for process manufacturing environments.	Available as a dedicated scope item.	Not confirmed as a scope item version. Production volume tracking is typically handled through standard process order confirmations.
Warehouse-integrated manufacturing Scope items for Kanban supply into production (3VT), goods receipt from repetitive manufacturing (41U), and warehouse posting changes (3BU) that tightly couple integrated warehouse management with manufacturing execution.	Available as dedicated scope items providing native warehouse-to-production integration.	Not confirmed as standalone scope item versions. Private cloud customers address these scenarios through decentralized extended warehouse management (EWM) integration or standard inventory management postings.
Decentralized EWM—quality management (4RT) and production integration (4RU) Decentralized EWM scenarios for quality-managed warehousing and production supply.	Not confirmed as scope item versions. The public cloud edition uses the integrated warehouse management approach with EWM integration scope items (2VL, 2VN, 2VO, 2YL).	Available as dedicated scope items. Private cloud supports full decentralized EWM deployment as a separate system, which is the standard architecture for large-scale warehouse operations.
User interface and custom manufacturing logic	SAP Fiori exclusively. Extensions through released APIs and key user extensibility only.	SAP Fiori plus SAP GUI (CO01, CO02, COR1, etc.). Full ABAP available for custom production planning logic, scheduling heuristics, and MES integration adapters.

Table 2.5: Functional Differences in Plan-to-Make, Public Cloud Edition Versus Private Cloud Edition (Cont.)

In practice, both editions cover the complete manufacturing lifecycle from planning through execution and confirmation. SAP S/4HANA Cloud Public Edition provides tighter out-of-the-box integration between warehouse management and manufacturing, along with collaborative MRP through SAP Business Network. SAP S/4HANA Cloud Private Edition offers decentralized EWM for organizations running large-scale warehouse operations as a separate system, and it retains access to SAP GUI manufacturing transactions and custom ABAP for bespoke scheduling or shop floor integration requirements.

2.4.6 Outcome

James's OTIF problem is not just a procurement problem. The 14 percent of orders that miss commitment have a second cause: The production schedule responds too slowly when the reality deviates from the plan. A supplier delivers short, a machine goes unplanned down, a demand spike arrives, and by the time the exception review meeting convenes, the miss has already propagated downstream. The production planning and manufacturing capabilities in this section change the response time. MRP runs against updated signals within minutes rather than overnight. The capacity scheduling board reveals conflicts that were previously invisible until the shop floor stalled. Work center utilization improves as interactive scheduling replaces manual sequencing. Overall equipment effectiveness (OEE) monitoring identifies which losses are structural and which are recoverable. At Korindal's scale, with three plants producing roughly 2,200 production orders per week, a 10 to 15 percent OEE improvement translates directly into available capacity that does not require capital investment. Production supervisors shift from spending 30 percent of their time on exception investigation to focusing on forward-looking scheduling decisions. The manufacturing layer stops generating the exceptions that cascade into fulfillment misses, and OTIF improves as a result.

2.5 Logistics

Logistics in SAP S/4HANA Cloud covers the full store and transport lifecycle, offering comprehensive functionality for executing efficient logistics processes at all stages of the supply chain, from procuring raw materials, monitoring inventories, and managing warehouses to distributing and transporting finished products. Logistics is deeply connected with other functional areas, such as sales and distribution, procurement, and finance and controlling. As you'll see in this section, the provided capabilities help you perform the following tasks:

- Manage inventories on a quantity and value basis, engage in planning, and manage the entry and documentation of all goods movements including physical inventory
- Plan, process, and monitor inbound and outbound shipments through logistics execution (LE)

- Manage stock at the bin level, fully integrated into all other logistic components handling inbound/outbound and internal processes, returns, and physical inventory with warehouse management
- Calculate the availability of a specific product in a specific plant for order confirmations
- Consolidate transportation demands and execute the plan accompanied by integrated charges and settlement
- Manage packing-based logistics structures with handling unit management
- Track and trace material and finished goods through batch and serial number management

2.5.1 Inventory Management

With supply chains becoming more global and more complex, inventory management has had to modernize and respond to this evolution. SAP S/4HANA Cloud provides both strategic inventory planning and operational inventory management designed for high-volume transactions and tight integration. It provides real-time integration with numerous processes in sales, procurement, supply chain management, delivery manufacturing, warehouse management, and financial accounting. Inventory management is responsible for all potential stock movements and for managing different stock categories and special stocks, including consignment stock, project stock, vendor-managed stock, batches, and serialization.

In the following sections, we'll first provide an overview of inventory management in SAP S/4HANA Cloud and then dive into two of its most important processes: physical inventory and stock transfers.

Process Overview

In general, inventory management in SAP S/4HANA Cloud undertakes the following tasks:

- Planning, entry, and documentation of all the goods movement
- Managing material stock on a quantity and value basis
- Carrying out physical inventory

In inventory management, mapping of physical stock is managed in real time by logging all stock-moving transactions, which enables real-time stock updates at any given point in time. With each goods movement, stock quantities and values are updated, and related general ledger accounts also are updated for financial accounting by means of automatic account determination. This functionality helps an inventory manager view the current stock situation for a specific material and manage many of its own and external special stock forms. It further provides inventory managers with more flexibility and allows them to quickly react to any unforeseen events.

Goods movements are managed with the central entity in inventory management—the *material document*. For each goods movement, a material document is created and stored in the database to help ensure smooth documentation of a company's stock situation. Goods movement is usually tracked or posted by one of the following methods:

- Goods receipt
- Goods issues
- Physical stock transfers
- Transfer postings

A goods receipt always leads to an increase in warehouse stock and is typically created either when goods are received from a vendor and received against a purchase order or when goods are produced in house and received against a production or process order. Similarly, a goods issue leads to a decrease in warehouse stock, typically when goods are withdrawn for production or shipped and sold to a customer. Goods movements do not only occur in the form of goods receipts and goods issues. Materials often undergo movements within a company or across companies. Usually, these movements happen by physical transfer—for example, goods movement between two storage locations or plants. If goods are not physically transferred, then a logical transfer can still occur; for example, a material may be moved to another material number or the stock type may change.

Inventory management in SAP S/4HANA Cloud offers several applications to monitor and analyze goods movements, helping ensure inventory forecasts and accuracy. One such example is the Slow or Non-Moving Materials app (ID F2137), as shown in Figure 2.41.

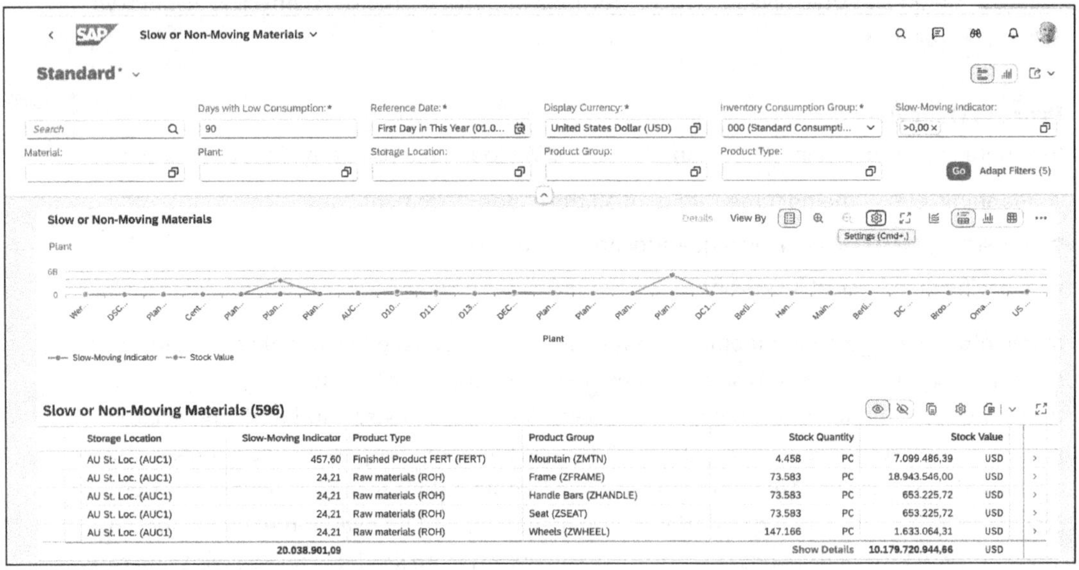

Figure 2.41: Slow or Non-Moving Materials App

This application helps you monitor and make time-dependent investigations into slow-moving or nonmoving materials. Based on these results, inventory managers can react immediately with follow-on activities, such as scrapping or stock transfers. This application provides customer-specific consumption definitions through filters and even makes it possible to receive forecast information for the predicted slow-moving indicator for materials using predictive analytics.

The prediction of the slow-moving indicator helps you gain additional insights on the future development of a material in the product lifecycle. Another functionality of this application is the Active BOM Usage Chart, which provides an overview of all valid BOMs for the reference date.

Similarly, inventory management also provides additional analytical apps such as the Stock – Multiple Materials app, the Inventory Analysis Overview app, and the Inventory KPI Analysis app to monitor critical KPIs and identify outliers at the stock identifier level in the compared time periods.

Physical Inventory

An essential building block in inventory management, physical inventory allows you to perform the periodic process of making necessary adjustments to stock on hand after a physical count.

A large variety of physical inventory procedures for legal and management accounting are supported in SAP S/4HANA Cloud, including periodic inventory, continuous inventory, and cycle counting. The process begins with the generation of the required inventory count sheets. Materials can be blocked for posting during the physical inventory process. Once inventory sheets are printed out, the actual physical inventory count is conducted for the given materials. Afterward, the count result is entered into the system, and then any discrepancies against the system quantities are reviewed. The inventory may be recounted until final counts are accepted and inventory differences are posted.

To support physical inventory procedures, inventory management in SAP S/4HANA Cloud provides various applications for the creation, counting, and evaluation phases. One such example is the Cycle Counting Classification app, which helps you classify materials according to ABC analysis or at the user's discretion. Further applications, such as the Schedule Physical Inventory Documents app, can automate the creation of physical inventory documents to reduce manual effort and improve operational efficiency for warehouse operations.

In addition, inventory management provides the Intelligent Situation Automation service, which uses the rules engine on SAP BTP to automatically notify responsible persons about counting and posting reminders in the physical inventory process. Further, the rules engine automates situation handling by making decisions or taking actions based on the user's discretion. As a result, the service checks whether recounts should be triggered or if inventory differences can be accepted automatically.

Stock Transfer

In inventory management, the transfer of goods can be performed in multiple ways, as described earlier. The most commonly used stock transfers are within a plant and across plants. The transfer that happens within a plant usually involves goods moving between storage locations. This kind of transfer can be performed in a one-step or two-step procedure. Because the stock value remains unchanged, this value is not relevant for accounting. When transferring goods across plants, different flavors of transfers exist. Stock transfers can take place within the same legal entity (different plants under the same company code) and across two different legal entities (across company codes); the transfer affects both accounting and materials planning. Stock transfers can occur in a one-step procedure, in a two-step procedure, as a stock transfer order without delivery, and as a stock transfer order with delivery.

To illustrate these concepts, let's look at a stock transfer with delivery between plants for the same legal entity (company code) next.

The stock transfer process begins with a requirement to transfer a material from one plant to another within the same legal entity. A stock transfer requisition is created in the procuring plant automatically by MRP or manually by a buyer. The buyer converts a stock transfer requisition to a corresponding stock transfer order. Based on the configuration, a confirmation key is maintained for the stock transfer order to enable the inbound delivery creation. Once a delivery is created, shipping activities, such as picking or delivery scheduling, are initiated. A shipping specialist gathers the materials and confirms the picked quantities. The delivery quantities are issued, appropriate documentation is generated, and the goods are shipped from the shipping plant.

Tracking is available in the Overdue Materials app for materials are scheduled to arrive in a receiving plant, and no goods receipts are posted for a stock transport order. This approach helps inventory managers track and plan material movements more reliably. At the receiving plant, with an inbound delivery, receiving specialists will have all necessary information beforehand because the supplying plant notifies them about the delivery ahead of time. Further, the receiving specialist can confirm the put-away of the goods received and receive goods into stock, referencing the inbound delivery.

Inventory management in SAP S/4HANA Cloud provides numerous built-in process automations, such as ML-based delivery predictions for stock in transit, as shown in Figure 2.42. Thus, inventory managers can make specific predictions based on your company's empirical data and identify that a goods receipt can be successfully completed in time.

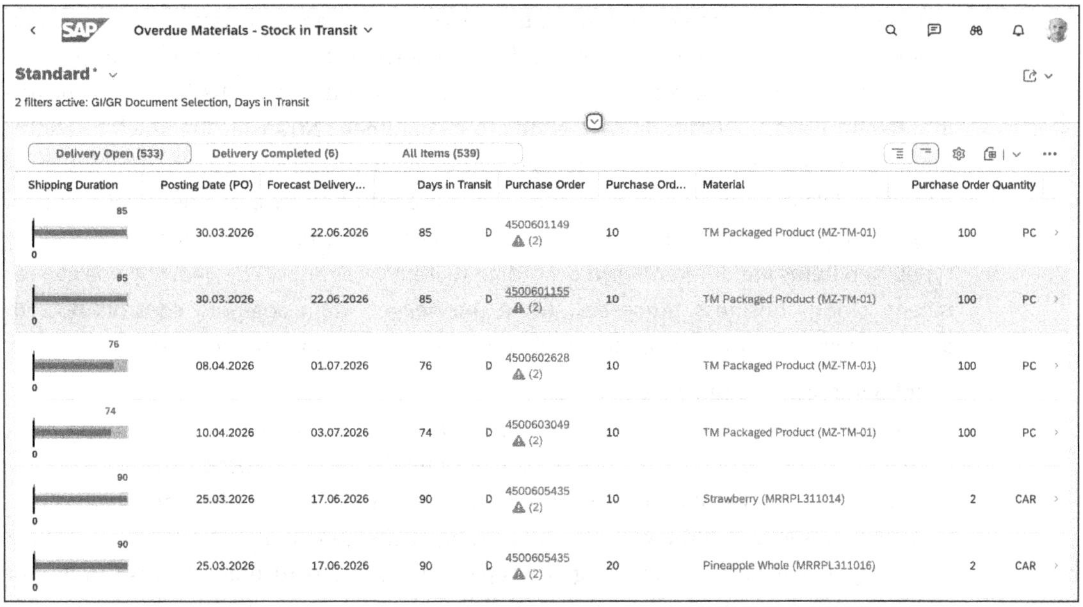

Figure 2.42: Overdue Materials – Stock-in-Transit App

2.5.2 Logistics Execution

LE offers comprehensive functionalities required for mapping efficient logistics processes at all stages of the supply chain, from procuring raw materials to distributing finished products. This capability is highly integrated into core, end-to-end logistics business processes and connected to main functional areas, such as sales and distribution, materials management, warehouse management, inventory management, and transportation management. The provided capabilities help you improve order fulfillment and on-time delivery performance with integrated LE across pick, pack, and ship processes. We'll discuss the LE functionality within SAP S/4HANA Cloud in the following sections.

Delivery Management

Delivery management is the main part of LE and involves picking, packing, and shipping processes to help ensure the right products are delivered to the right place at the right time and that incoming goods are received into storage in an efficient manner. The result is a largely automatic shipping and receiving process in which manual changes are necessary only under certain circumstances. In addition, the functionality allows you to manage stakeholder communication—for example, by sending and receiving advance shipping notifications through modern SOAP interfaces and printing and distributing necessary documents, such as delivery notes and pick lists, with the help of the output management functionality in SAP S/4HANA Cloud.

The central business object within LE, the *delivery document*, consists of a header and a multitude of items. The header contains data that applies to the entire document. The individual items contain information about the material to be delivered. Deliveries can be used in different ways, such as shipping goods to a customer or transferring stock to another plant (outbound delivery and replenishment delivery), receiving goods from a supplier (inbound delivery), or managing products that are returned by customers (returns delivery). To map these functions, delivery documents are differentiated according to delivery types, and items are differentiated according to item categories. The *delivery type* can represent various business processes, while the *delivery item category* controls how the affected delivery items are handled and processed in the following scenarios:

- **Inbound processing with deliveries**
  The goods receipt process for inbound deliveries is an essential part of the supply chain. In a typical procure-to-pay scenario, an inbound delivery is created with reference to a purchase order. The inbound delivery then serves as the basis for further activities, such as the subsequent put-away and the posting of the goods receipt posting for ordered goods. One significant advantage to depicting the goods receipt process through the inbound delivery function is that you can execute many activities in advance, even before the actual goods receipt posting takes place. You have all the necessary information beforehand because the supplier notifies you of the inbound delivery ahead of time. The inbound delivery describes exactly which materials or pallets can be received on what date and at what time.
- **Outbound processing with deliveries**
  In its role as a central object of the goods issue process, the outbound delivery enables a shipping specialist to perform all shipping activities, including picking, packing, and goods issue. During the outbound delivery process, shipping information is recorded, the status of shipping activities is monitored, and data accumulated during shipping processing is documented. When the outbound delivery is created, the shipping activities, such as picking or delivery scheduling, are initiated, and data that is generated during shipping processing is included in the delivery. You can create a single outbound delivery with or without reference to a preceding business document (such as a sales order) or multiple deliveries by manually starting a collective run or by scheduling a job to run in the background.

SAP Extended Warehouse Management Integration

SAP S/4HANA Cloud supports a predefined integration scenario for LE processes with an external, on-premise warehouse management system (WMS): SAP Extended Warehouse Management (SAP EWM) 9.5 Feature Pack 02. This integration helps you extend the scope of your core business processes with warehouse management functionality. You can manage your stock on the bin level and thus increase visibility into inventory by using best-practice process configurations for inbound and outbound processes, stock transfers, production integration, batch management, customer returns, and selected internal warehouse processes.

Logistics with Third-Party Warehouse Management

This integration scenario offers the ability to broaden the scope of SAP S/4HANA Cloud by connecting a third-party WMS. The decentralized third-party WMS runs all processes required for the physical goods movements within the warehouse complex. The non-warehouse-related parts of the processes—including sales order creation, purchase order creation, and quantity-based inventory management—are performed in SAP S/4HANA Cloud. The WMS is notified about planned goods movements by SAP S/4HANA Cloud, with the help of inbound and outbound deliveries. Once the actual delivery processing on the warehouse level has been concluded, the WMS informs SAP S/4HANA Cloud about the detailed results through confirmation messages and eventually initiates inventory changes in SAP S/4HANA Cloud on a storage-location level. The integration is covered by asynchronous SOAP APIs or IDoc technology.

2.5.3 Integrated Warehouse Management

The warehouse management functionality in SAP S/4HANA Cloud provides support with real-time transparency into managing and processing material movements flexibly in a warehouse with its own stock. Warehouse management allows you to map your warehouse in detail in the system, down to the storage bin level. Not only does this give you an overview of the total quantity of a product in the warehouse, but you can also see exactly where a specific product is at any time in your warehouse. Warehouse management allows you to plan and perform movements within the warehouse, such as the following:

- Warehouse inbound processing (GR)
- Warehouse outbound processing (goods issue)
- Internal processes (e.g., replenishment and physical inventory)

With warehouse inbound processing, products that are delivered with the intent to be stored in the warehouse are received and moved to a storage bin. Warehouse management allows different variants of inbound processing, including inbound movements from suppliers, other storage locations, and other plants.

Warehouse outbound processing moves products in the opposite direction: Goods are picked, packed, and sent from your warehouse. This process supports different scenarios. For example, you can send ordered products to external customers, return products to suppliers, and supply products to internal customers.

With warehouse internal processes, users can move products to different storage bins in your warehouse and keep track of where they are. An example of an internal movement is the replenishment process, in which you can plan and automate the creation of replenishment warehouse tasks to help ensure the right inventory levels for your products in the areas where picking is done. Also worth mentioning is physical inventory, which allows users to plan, conduct, and confirm regular counts of actual product quantities in the warehouse and compare the physical stock to the data in the system.

With warehouse management in SAP S/4HANA Cloud, you can also integrate the supply of products to production and the receipt of finished products from production into your warehouse. This integration between warehouse operations and manufacturing execution allows users to manage stock in production on a storage bin level and overall enhances the visibility and control of warehouse processes.

Warehouse management supports serial number–managed and batch-managed products. Furthermore, you can process packed goods with handling units, which are physical units consisting of packaging materials (load carriers and packing material) and the goods contained on or in these packaging materials. For better supply chain visibility, warehouse management is integrated with transportation management to plan outbound and document inbound operations.

With warehouse management in SAP S/4HANA Cloud, you have a preconfigured set of SAP GUI for HTML mobile RF screens for processing activities in a warehouse. You can run warehouse processes on mobile RF devices to carry out warehouse activities such as putaway, picking, packing, and replenishment. The system guides warehouse operators through their daily tasks.

However, managing end-to-end warehouse processes isn't the only thing warehouse management is capable of. SAP has expanded the warehouse management offering with APIs and analytics. APIs allow you to build your own applications to consume digital content from SAP and elsewhere. API connectivity is critical for enabling the building and integration of newly delivered capabilities for partners and organizations. SAP Business Accelerator Hub is where standard, interconnected APIs can be discovered and consumed. The following are some examples of APIs available for warehouse management:

- The Process Warehouse Tasks API allows you, for example, to create warehouse tasks, read the details of a warehouse task (such as source and destination bin or product and quantity to be moved), and confirm warehouse tasks using a mobile device.
- The Warehouse Outbound Delivery Order API provides, for example, an action to trigger a goods issue from a mobile device and to print the loading list.
- The Warehouse Physical Inventory API lets you enter your physical inventory count results in real time into the system using a mobile device.

One final highlight to mention relating to warehouse management is SAP Warehouse Robotics, which comes as an add-on for SAP EWM and includes prebuilt scenarios. This solution reduces the need for specialized system integrators, promotes reuse, and helps ensure compliant IT security and management. SAP Warehouse Robotics is intuitive to implement and requires no additional licensing. With it, your company can access an easier, low-risk entry into agile warehouse robotics. The solution runs and optimizes heterogenous fleets and material flows and provides a new degree of ad hoc scalability. To meet peak demand, users can integrate additional robots within minutes.

2.5.4 Order Promising: Available-to-Promise

The *ATP check* is a service that calculates the availability of a specific product in a specific plant. The most prominent use cases are sales order processing (including delivery creation and goods issue) and production planning (in particular, the components check during the conversion of planned orders into production orders). SAP S/4HANA Cloud supports both basic ATP (checking stock and inbound supply against demand) and advanced ATP, which extends this with rules-based availability checking, backorder processing, and multilevel product allocation for more complex supply scenarios.

Basic ATP checks whether a requested quantity can be confirmed against available stock, planned receipts, and scheduled supply. It answers a simple question: Is the material available on the requested date, and if not, when can we deliver? Advanced ATP adds a layer of business intelligence on top of that availability check. It considers product allocation to ensure fair distribution across customers and channels, supply protection to reserve inventory for priority contracts or service commitments, and backorder processing to reprioritize confirmations when supply shifts or a higher-priority order arrives. Where basic ATP treats every order equally on a first-come, first-served basis, advanced ATP applies configurable business rules that reflect a commercial strategy: which customers matter most, which products are constrained, and how scarce supply should be distributed when demand exceeds availability. The result is a promise that is not only feasible but commercially intelligent.

The result of an ATP check is a confirmation. In the best case, this confirmation is just verifying that the requested quantity is available at the requested date. More generally, it may consist of a time series in the form of quantity and date pairs. For example, if 10 pieces of product A are requested in plant 0001 by next Monday, the ATP check might confirm five pieces right on time (next Monday) and three pieces slightly delayed (next Wednesday), while the remaining two pieces can be promised only two weeks later.

SAP S/4HANA Cloud uses two different algorithms for calculating a confirmation for a given requirement in an ATP check:

- The product availability check (PAC) algorithm
- The product allocation algorithm

The PAC algorithm calculates product availability based on existing supply and demand elements, which are represented by business objects stored in the database, such as stocks, sales orders, or production orders. The algorithm aggregates all relevant stocks and supply elements in daily buckets in real time and compares them with the aggregated demand elements. If the cumulated supply exceeds the cumulated demand, the resulting positive ATP quantity is used to confirm the requirement that is being checked. Working with cumulated quantities rather than dedicated supply and demand elements helps ensure that the confirmations calculated by the PAC algorithm are robust and nearly immune to the minor changes that often impact transactional business objects. This degree of stability makes an ATP confirmation a reliable statement with a high business value.

In contrast to the PAC algorithm, the product allocation algorithm is based on planning figures rather than real supply elements and can consider almost every additional attribute assigned to a demand element. For example, a certain group of products may have restrictions for export to certain countries. Time series data with appropriate granularity—for example, per month—can be created to reflect that restriction (e.g., 100 pieces can be shipped each month). In addition to the previous sales-driven example, you can also model capacity restrictions, such as transport capacity or production capacity. The planning figures in product allocation could, for example, represent the weekly transport capacity of a ship to a specific country or the capacity of a machine that restricts quantities that can be produced each day.

One ATP check can include both the PAC and product allocation algorithms. Thus, both allocation types—that is, sales and capacity product allocation—can be checked in a single ATP check. The sales product allocation check is always performed before the PAC. The capacity product allocation check is performed after the PAC. The requested quantities and dates of the current check are based on the confirmation results of the preceding check.

In addition to the two basic ATP check methods, other features and functions can be used to manage and optimize the confirmation capabilities of ATP, which can be triggered by the order fulfillment manager:

- Supply protection
- Backorder processing
- Alternative-based confirmation

Supply protection is integrated into the PAC algorithm. Whereas PAC compares cumulated supply with aggregated demand elements, supply protection works with virtual demand elements. The supply protection object defines the required (virtual) safety stock. If supply is short, then supply protection prevents the confirmation of requirements that do not match the supply protection object. For requirements matching with a supply protection object, ATP can consume its virtual safety stock and therefore calculate a positive ATP quantity, which leads to a corresponding positive confirmation.

From a business perspective, supply protection helps retain supply for high-priority groups even in times of high demand from lower-priority groups. This functionality protects minimum quantities, whereas the PAC, in combination with product allocation, defines the upper limit.

The ATP confirmation of a single order is triggered whenever an order is created or changed. Each ATP check always respects existing confirmations of all other orders that have already been created or that are currently in process—even before their corresponding business processes have finished (e.g., before the underlying sales order documents were saved). The ATP confirmation, therefore, follows a strict first-come, first-served principle, which can be softened by the intelligent use of product allocation and supply protection. In addition, the ATP check offers mass processes, such as backorder processing or release-for-delivery, which in their individual contexts can be used to further optimize the confirmation situation and work for multiple orders simultaneously.

Backorder processing is a batch job that recomputes the confirmations of the orders quickly and updates the orders efficiently based on parallelization. It can be configured flexibly according to business strategy. Users can select the orders to be included in the backorder processing run, rank them by priority, and define their individual confirmation behavior. All this depends on the attributes of each individual requirement.

Release for delivery can be used to change the confirmations for multiple orders manually, with the goal of directly creating deliveries for those orders. The manipulation of the confirmed dates and quantities in the release for delivery is possible for users bearing the role of the order fulfillment specialist.

Alternative-based confirmation can be used to search for viable alternative plants, storage locations, or products. This is especially useful if the availability of the requested product in the requested plant is insufficient to fully confirm a given sales order requirement in time. Alternative-based confirmation can select the allowed alternatives and identify the most suitable combination of material and plant (and storage location) based on their current availability situation. Choosing the best alternatives can be influenced by three predefined strategies or by implementing a business add-in (BAdI). The allowed alternatives are modeled in dedicated substitution master data. The activation of alternative-based confirmation and the identification of the allowed alternatives can be influenced by the attributes of the underlying sales order.

2.5.5 Transportation Management

Transportation management in SAP S/4HANA Cloud supports you in activities connected with the physical transportation of goods from one location to another. A transportation manager can use transportation management to consolidate transportation demand and execute the plan accompanied with integrated charges and settlement.

SAP S/4HANA Cloud supports predefined integration scenarios for transportation management planning and execution processes as follows:

- **Transportation demand**
  The freight unit that represents the cargo being transported through the entire supply chain is the object needed for follow-on processes. You can automatically create freight units out of predecessor documents such as purchase orders, sales orders, or deliveries. The freight unit can be built according to configurable rules via which you can take weight and volume into consideration.
- **Transportation planning**
  You can use the freight units that have been created in the planning process to build your freight order, which represents the consolidated load. The UI is the transportation cockpit, where you select the freight units and assign them to an existing or newly created

freight order. The freight order is used for follow-on processes such as loading or unloading and subcontracting to carriers.

- **Transportation execution**
  The freight execution is based on the freight order that you have handed over to your carrier.
- **Freight charges and settlement**
  Agreements with your carriers contain the negotiated and agreed-upon rates over a period of time. With this information, you can calculate the planned and actual charges from the carrier, which can serve as the basis for subsequent accrual, settlement, and invoice verification processes.

With transportation management integrated into the end-to-end logistics flow, you have full transparency of inbound and outbound logistics. You can combine inbound and outbound logistics to better leverage existing capacity and further help reduce the carbon footprint. With standardized message interfaces, you're connected to the outside world, which enables you to set up integration scenarios with your partners.

2.5.6 Handling Unit Management

Handling unit (HU) management is related to the complete logistics chain, where material is moved between organizational structures—for instance, the customer, supplier, plant, storage location, and storage bin. This capability can be used in inbound delivery, inventory, production, and outbound delivery management to track the movements of entire HUs and the materials they contain rather than tracking each material individually as an HU. An HU is a physical unit consisting of packaging materials (load carriers and packing material) and the goods contained on or in it. HUs can be *nested*, which means they can have several levels of packaging that structure the outer handling unit. You can nest HUs by simply defining new HUs inside or under an existing one. All needed packaging materials can be added to build up the real packaging of a complete packed pallet, for example. HUs can contain both serialized materials and batch-managed ones. An HU has a unique scannable identifier that can be developed according to standards such as European Article Number (EAN) 128 or Serial Shipping Container Code (SSCC). Labels for HUs can be printed to make them identifiable wherever they are.

In HU-managed storage locations, materials can be managed in HUs only. Mixed stock (stock made up of packed and nonpacked materials within the same storage location) cannot. HUs can also be managed in interim storage types. Unpacking a material from an HU means that the stock of the material is posted to a storage location that is not HU-managed. This approach makes inventory management and HU content consistent.

The business benefit is a complete material identification of quantity and status and, if used, of serial numbers and batches for warehouse clerks, production operators, and shipping specialists by scanning labeled barcodes. This fast and error-minimizing process for handling materials stock can cover a complete logistics chain. Users only need a printer for the labels and a barcode scanner to handle stock that is packed in an HU. By shipping an HU to your customers, HU numbers can be sent in the advanced shipping notification, and your customer can consume the HU without printing new labels, if the format is aligned with your customer's systems. For example, SSCC and Global Transport Label (GTL) numbering are supported.

HU-related master data is included in the packing instruction and the determination record, where the packing instruction serves as a template for the creation of an HU. In a packing instruction, you'll define the materials and packaging materials to be packed and their quantities in an HU. This approach enables you to reproduce packaging requirements for specific customers in the system. You can also automate the packing process by creating a proposal for an HU, based on a packing instruction.

The *determination record* is used to find the relevant packing instruction by a multiple selection parameter. The selectable parameters can include, for example, material or ship-to party. Search for materials that lack a packing instruction or a determination record and simply create these records from scratch or copy an existing record.

For analytics, you can use the Handling Unit Monitor app for an overview of HUs, depending on the search you perform.

2.5.7 Public Cloud Versus Private Cloud

Logistics capabilities in SAP S/4HANA Cloud, inventory management, warehouse management, logistics execution, ATP, transportation management, and HU management are broadly shared across both editions. Core inventory management (BMC), physical inventory (BML), batch management (BLF), serial number management (BLL), warehouse inbound and outbound processing (3BR, 3BS), warehouse physical inventory (3BX), EWM integration, available-to-promise (basic and advanced), intercompany stock transfer and transportation, and returnables processing are all confirmed in both editions.

The functional differences identified are summarized in Table 2.6.

Capability	Public Cloud Edition	Private Cloud Edition
Handling unit management (4MM) Packing-based logistics structures for managing materials in handling units across the logistics chain.	Available as a dedicated scope item.	Not confirmed as a stand-alone scope item version. HU management functionality is available through integrated warehouse management and EWM scenarios.

Table 2.6: Functional Differences in Logistics: Public Cloud Edition Versus Private Cloud Edition

Capability	Public Cloud Edition	Private Cloud Edition
Warehouse inbound processing with synchronous goods receipt (5HN) Real-time synchronous goods receipt posting during warehouse inbound processing, eliminating the delay between warehouse confirmation and inventory update.	Available as a dedicated scope item.	Not confirmed as a scope item version. Goods receipt processing follows the standard asynchronous pattern or is handled through decentralized EWM.
Warehouse ad hoc goods issue (3BT) Unplanned goods issue directly from warehouse stock without a preceding outbound delivery.	Available as a dedicated scope item.	Not confirmed as a scope item version. Ad hoc goods issues are handled through standard goods movement postings.
Field logistics planning and execution (40G, 5OM) Planning and execution of field logistics operations for service and direct procurement scenarios.	Available as dedicated scope items.	Not confirmed as scope item versions. Field logistics scenarios are addressed through standard delivery and transportation management.
Decentralized EWM replenishment (4RS) Replenishment processing in a decentralized EWM deployment.	Not confirmed as a scope item version. Warehouse replenishment is handled through integrated warehouse management.	Available as a dedicated scope item. Supports full decentralized EWM replenishment scenarios for large-scale warehouse operations.
User interface and custom logistics logic	SAP Fiori exclusively. Extensions through released APIs and key user extensibility only.	SAP Fiori plus SAP GUI (Transactions VL01N, VL02N, MIGO, MB1A, etc.). Full ABAP available for custom inventory management rules, delivery determination logic, and third-party WMS integration.

Table 2.6: Functional Differences in Logistics: Public Cloud Edition Versus Private Cloud Edition (Cont.)

The logistics differences reflect a design philosophy: The public cloud edition favors tightly integrated warehouse management with synchronous processing and field logistics as native scope items, while the private cloud edition favors decentralized EWM for organizations that run warehouse management as an independent system at scale. For organizations with standard warehouse requirements, both editions are functionally equivalent. For organizations with complex, high-volume warehouse operations requiring decentralized EWM, the private cloud edition provides the necessary deployment architecture.

2.5.8 Outcome

The logistics layer is where fulfillment commitments either hold or break. An order can be planned correctly and manufactured on schedule but still arrive late because warehouse picking was slow, a carrier was not confirmed in time, or a shipment exception was not caught until the customer called. James sees this pattern regularly: Exceptions that could have been resolved the day before become escalations on the day of delivery. Inventory management gives the team a real-time view of what is available, where, and in what condition. ATP connects that view to order commitments, so delivery dates reflect actual supply rather than system defaults. Transportation management closes the loop between the warehouse and the carrier, automating freight tendering and surfacing shipment exceptions before they become delivery misses. For Korindal, the cumulative logistics improvement supports the OTIF target by addressing the last-mile causes of failure. Better warehouse resource allocation means faster order processing. Proactive transportation visibility means fewer delivery surprises. Automated freight audit recovers overcharges and provides the data to hold carriers accountable to contracted service levels. The logistics capabilities do not add complexity; they remove the friction that converts a well-planned order into a late shipment.

2.6 Asset Management

Performing maintenance on your physical assets helps you to keep them at a maximum availability, at the lowest cost, while ensuring safe operations. Given the importance of maintenance management across 25 industries—and especially for asset-centric businesses—SAP has offered designated solutions for maintenance management for decades. Allowing businesses to realize real, comprehensive processes and a smooth integration into other domains like finance, sourcing and procurement, or workforce management is SAP's strength. The availability of resources, such as technicians, spare parts, and tools, in the right time and at the right place, is equally important, as is proper financial planning and settlement of costs.

Companies that rely on physical assets for business operations may seek to enhance maintenance management practices to help improve profitability and productivity and pursue environmental and social outcomes. This all aims to serve the needs of the end customer (e.g., the production manager or site manager), who is usually responsible for operating an often diverse and complex set of physical assets under sharply defined outcome performance indicators. These roles are most often concerned about overall efficiency and ultimately need to ensure an *outcome-based operation*.

Advanced technologies such as AI, ML, and IoT are changing the domain and the way maintenance is conducted. Assets have become increasingly intelligent; they are connected and can help answer questions like “When is maintenance needed?” and “What is needed?” to mitigate potential failures or even avoid breakdowns. Such technologies enable a more diverse set of maintenance strategies—from reactive to prescriptive—and require both a holistic approach and well-defined maintenance programs.

At the same time, a shift has occurred in the roles of organizations involved in keeping assets at their maximum performance. New business models for assets need a true collaborative approach with a single source of truth for data among business partners. Manufacturers and third-party service providers extend their businesses by offering to monitor and maintain assets under equipment-as-a-service (EaaS) models.

To cope with these trends and support organizations undergoing a digital transformation in their maintenance management business, SAP provides a broad set of solutions beyond maintenance management in SAP S/4HANA Cloud, fostering support of next-generation end-to-end business processes. In this section, we'll highlight key business capabilities along modular business processes for the following asset-related activities:

- Managing assets
- Plan-to-optimize assets
- Operate-to-maintain assets

We'll also look at how to make use of the overall SAP Intelligent Asset Management solutions through the lens of the maintenance management functionality in SAP S/4HANA Cloud, pointing to other solution offerings that extend core business capabilities.

2.6.1 Intelligent Asset Management

To achieve goals in maintenance management, intelligent enterprises integrate maintenance operations across the enterprise, apply advanced technologies, merge asset performance management and maintenance operations, and adopt collaborative maintenance processes with equipment suppliers and maintenance service providers.

To embrace the two latter tasks, the maintenance management functionality in SAP S/4HANA Cloud establishes a foundation for adoption into other, strategic solution areas, including the following:

- The SAP Asset Performance Management application, to perform risk and reliability assessments, including recommendation management paired with IoT-enabled, data-driven approaches of prescriptive maintenance
- SAP Business Network for Asset Management, to manage new collaborative asset management processes

Extending Maintenance Management with Asset Performance Management

Maintenance management has been an intrinsic part of ERP offerings from SAP for the past several decades, including in SAP S/4HANA Cloud. Capabilities for managing maintenance operations—including planned and unplanned maintenance, scheduling work and resources, work order management and reporting, mobile support for field engineers, and health and safety—are delivered in an integrated way with other domains like finance, sourcing and procurement, or workforce management.

To realize a more holistic approach, asset performance management capabilities have been added as extensions and are available on SAP BTP with SAP Asset Performance Management.

With SAP Asset Performance Management, you can focus on critical assets with detailed risk assessments, maintenance strategy definitions, and recommendations for implementations. By using standardized methods such as failure mode and effect analysis (FMEA) and reliability-centered maintenance (RCM), you can define an optimal maintenance strategy for your assets and derive a maintenance program out of that strategy.

With the help of new technologies for real-time asset condition and health monitoring, your maintenance strategy can be more diverse and asset-specific. This specificity paves the way to optimizing and individualizing maintenance programs for assets while being much more prescriptive in terms of understanding of potential asset failure patterns (failure mode-centric) and maintenance strategies. This gives you the option to shift from reactive (after failure) or time-based maintenance toward more predictive- and prescriptive-oriented maintenance, as shown in Figure 2.43.

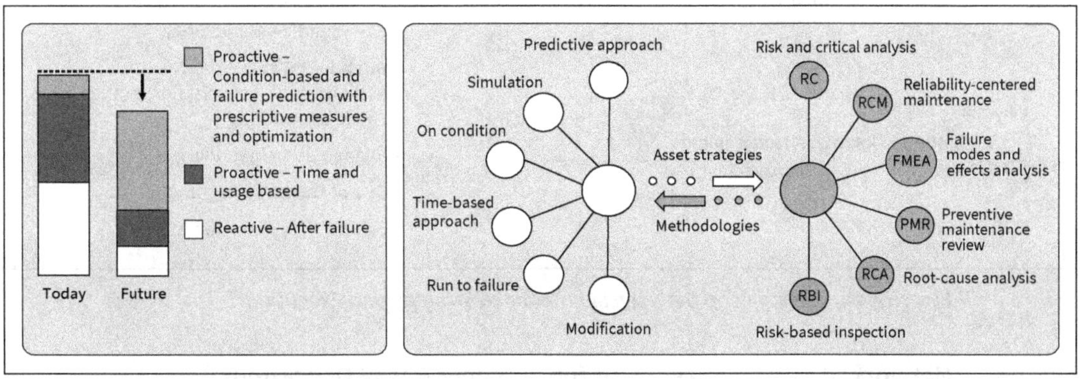

Figure 2.43: Trend to Shift from Reactive to Reliability-Centered and Prescriptive Maintenance

With SAP Asset Performance Management, you can implement maintenance strategies using condition-based, predictive, and prescriptive maintenance approaches for your asset.

Putting the asset at the center of your activities with an aligned and harmonized understanding between maintenance management and asset performance management, this solution helps you improve asset reliability and achieve operational excellence; access accurate, real-time information from your assets; predict potential operational problems before they occur by using rules and algorithms; and drive excellence by optimizing maintenance time windows and work scopes.

By extending maintenance management with asset performance management into a full set of integrated capabilities, as shown in Figure 2.44, SAP closes the loop between the maintenance strategies and performance levels and the execution of maintenance activities to improve the effectiveness of maintenance programs.

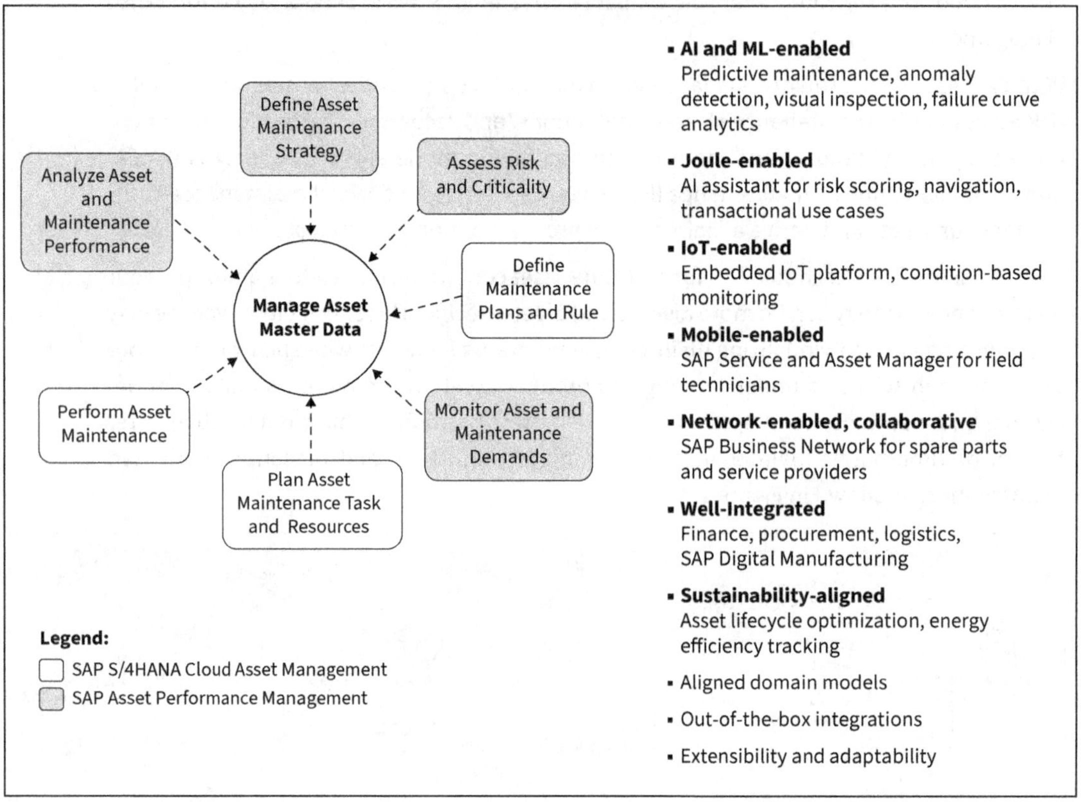

Figure 2.44: Closing Loop Between Maintenance Strategy and Execution

Networked Asset Management for Outcome-Based Operations

Another shift in the roles of organizations involved in operating and maintaining assets has occurred in the following ways:

- Assets are not always maintained by the operator. Manufacturers or third parties offer to monitor and maintain assets under EaaS models.
- Increasingly, complex assets are maintained by third parties focused on outcome-based service models, making the network dimension of asset management more important than ever.

These market trends increase the need for collaborative business services and "network" concepts. Sharing data in terms of asset data in a secured and standardized way is essential to supporting peers in the maintenance and service businesses. SAP Business Network for Asset Management provides integration capabilities for electronic data handovers and transactional collaboration for component suppliers, asset manufacturers, engineering providers, operators, and service providers.

2.6.2 Plan-to-Optimize Assets

To improve your maintenance processes and improve asset performance, you'll want to continuously assess and improve the effectiveness of your maintenance programs to support safe, reliable, and efficient operation of your assets. As a reliability engineer, you can perform risk assessments of your assets using best-practice methodologies such as reliability-centered maintenance to determine the best maintenance and service strategies that minimize maintenance costs and reduce the risk of failure for critical assets. Next-generation asset performance management refines and extends such long-standing methods in this area by making them data-driven processes.

SAP segments plan-to-optimize assets into (1) asset strategy and planning and (2) asset and maintenance optimization. This section will focus only on the latter and will point out several cross-functional relationships between these assets and other business capabilities in operate-to-maintain assets.

Analyzing Assets and Maintenance Performance

The process of analyzing assets, their performance, and related measures are supported with a wide range of insights combined with flexible tools to support users in their roles and day-to-day jobs. Along with the processes and activities, there are dedicated insight-to-action patterns with analytical insights to focus on the most relevant information.

For a broader analysis, dedicated capabilities are provided to analyze what's going on, such as the following:

- **Actual cost analysis**
  This functionality supports you in monitoring and evaluating actual costs resulting from current maintenance orders. For example, you can analyze the actual costs for materials and labor in maintenance or compare the total corrective and preventive actual maintenance costs during a selected time period. In addition, you can easily identify which maintenance activities lead to the highest costs or which parts of an asset were particularly costly on inspection. Altogether, this analytical list page offers you multiple options for evaluating actual maintenance costs stored in the Universal Journal entry (Section 2.1.1).

- **Technical object breakdowns**
  With this capability, you can analyze the causes of a breakdown more closely and calculate the distribution of the durations of the various breakdowns or repairs. You can also find the cause behind a short or long breakdown period or the period between two consecutive breakdowns.
- **Technical object damages**
  With this functionality, you can see which damages occur frequently and find the technical object parts that cause them. You can find related maintenance notifications in a list based on maintenance items. You can filter these tables and charts and navigate to the related applications.

Beyond the built-in analytical capability, rich content is provided to extract data from the maintenance management functionality of SAP S/4HANA Cloud and to consume it in custom-built analytics—for example, by using SAP Analytics Cloud. The respective core data services (CDS) are provided for key business objects in terms of master data, such as equipment and function locations, and transactional data, such as maintenance notifications and orders.

In addition to the insights and capabilities provided directly in SAP S/4HANA Cloud, you can make use of advanced analytics to bring business data and sensor data together. With SAP Predictive Asset Insights, you can use the explorer functionality to analyze your fleet of equipment, navigate to the details, and see sensor data next to business data.

Defining Asset Maintenance Strategies

Reliability engineers define the best maintenance strategies for assets. To concentrate on the most important assets, in SAP Asset Performance Management you can perform an asset risk and criticality assessment to derive respective scores for your own defined impact categories.

With detailed knowledge of potential failure patterns for the asset, you can choose from among the typical maintenance strategies: reactive maintenance (also called *run to failure*); time-, usage-, or condition-based maintenance; or predictive or prescriptive maintenance. The recommended maintenance strategy is not limited to dedicated repair work to keep the asset functioning. To mitigate the likelihood of potential failures, recommendations can also include the definition for inspection needs, design changes, and the like.

With SAP S/4HANA Cloud providing a view of the maintenance history for a particular asset or SAP Asset Performance Management providing the latest health and condition insights, a reliability engineer can decide to revise a maintenance strategy for better business results. Well-defined maintenance strategies, outlined in SAP Asset Performance Management, need to be brought into action. In this context, a maintenance planner can review and understand the recommended strategy and measures defined by the reliability engineer. In a collaborative process, the planner can make use of these definitions and implement them. For classic time- or usage-based maintenance, as well as run-to-failure and one-time measures, you can work directly in SAP S/4HANA Cloud. To set up more sophisticated maintenance rules for condition-based, predictive, or prescriptive maintenance and trigger maintenance demands, you can use SAP Asset Performance Management.

2.6.3 Operate-to-Maintain Assets

The maintenance management capabilities in SAP S/4HANA Cloud allow you to demand, plan, and execute maintenance in entirely redesigned processes to better meet the needs of a next-generation taskforce. End-to-end processes in maintenance execution typically span multiple phases. An industry best-practice scenario covers the following phases: initiation by raising a work request, screening and approval of the work request, the handover to detailed work-scope planning, order approval and release with subsequent scheduling and dispatching, the execution of work, work confirmation, and closure.

To best support various roles and to meet the aspirations of next-generation users, insight-to-action patterns have guided the redesign of existing UIs and the development of new UIs. This section will explain how to use the provided business capabilities to best operate and maintain your company's physical assets. We'll focus on key capabilities covered by the four available scope items:

- Reactive maintenance
- Proactive maintenance
- Improvement maintenance
- Operations and overhead maintenance

We'll refer to the bigger picture in overall intelligent asset management solutions. In addition, the listed scope items are supplemented with enhanced capabilities to better schedule maintenance work and resources and to execute work on a mobile device:

- SAP S/4HANA Asset Management for resource scheduling
- SAP Maintenance Assistant (mobile maintenance application)

The business capabilities for operate-to-maintain assets can be segmented into asset maintenance planning and asset maintenance execution. Like other capabilities, several cross-functional relationships exist between these capabilities and other domains (e.g., procurement, materials management) because of their end-to-end nature.

Asset Maintenance Planning

The maintenance planning capability in the maintenance management functionality of SAP S/4HANA Cloud is the process performed before maintenance execution. Maintenance planning crosses different maintenance approaches—for example, reactive or proactive maintenance.

Proactive maintenance in traditional terms is scheduled for equipment before it can fall out of tolerance or break, preventing costlier scheduled repairs or emergency repairs and downtime. With maintenance plans, task lists, and inspection templates in SAP S/4HANA Cloud, a maintenance planner can plan, in advance, the work scope and time needed for time- or usage-based measures such as inspections and maintenance work. The capability provides a set of parameters to fine-tune schedules for maintenance work—for example, once a month or every 300 operating hours.

Moving forward from a one-size-fits-all approach with classic time- or performance-based maintenance plans, you can take advantage of the built-in integration with SAP Asset Performance Management for condition-based and predictive or prescriptive maintenance. With SAP Asset Performance Management, you can set up rules and benefit from frequent time-series data being sent from, say, asset sensors and their interpretation. The system calculates the degradation of asset health or its condition and triggers the generation of a maintenance demand—as well as inspection demands—if threshold violations occur or remaining useful life is becoming critical. Thus, you can shift to a much more individualized maintenance program for assets, with optimized time windows and work scopes, for better asset performance and availability.

In a holistic approach, the need for an asset to be covered by proactive measures can be prescribed and recommended by a maintenance strategy defined in SAP Asset Performance Management.

Monitoring the maintenance demand for assets holistically is a first key step to preparing a maintenance backlog in an optimized way. For the demand generation itself, you can benefit in different ways with the following features:

- **Manual work request creation by the user**
  In SAP S/4HANA Cloud, maintenance demands in the form of work requests can be manually created by any employee or maintenance technician, as shown in Figure 2.45. Using a new UI, you're guided to provide the essential information to create and submit a work request in an intuitive way—for example, the identification of the asset, supported through a recommendation to classify with failure and malfunction data, or text suggestions from other historical events. A risk- and criticality-aware prioritization capability is provided to harmonize and standardize the prioritization of maintenance requests via consequence and likelihood schemas. For maintenance technicians, the SAP Maintenance Assistant mobile application can be used to raise and process such demands on a mobile device in a timely manner—either online or offline.
- **Maintenance demand generation by maintenance plans and rules**
  Maintenance demands can also be generated automatically from previously defined maintenance plans in SAP S/4HANA Cloud or by the setup of rules in SAP Asset Performance Management. In addition, you can make use of an alert monitor and subsequently decide to convert alerts into qualified notifications in SAP S/4HANA Cloud.

  Figure 2.45 shows the Create Maintenance Request screen, where technicians or system rules generate a work request to document a maintenance demand or asset failure notification.

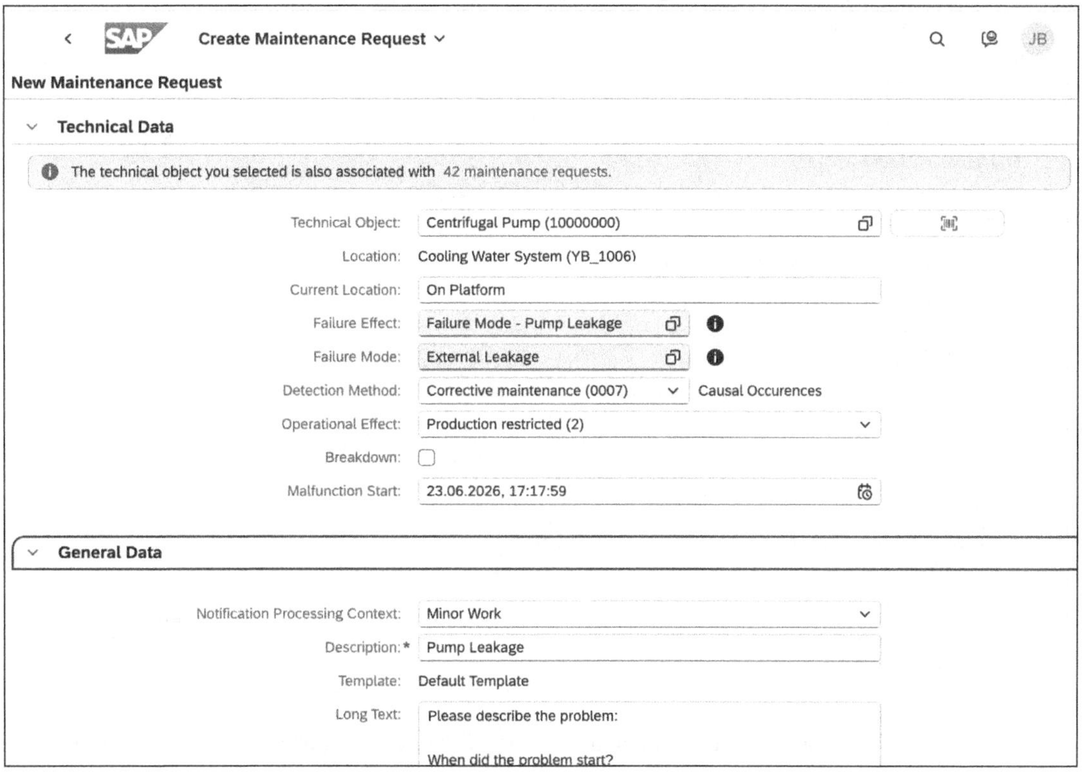

Figure 2.45: Create Maintenance Request (Notification)

- **Screen maintenance demands**

  The central screening functionality for maintenance requests in SAP S/4HANA Cloud, shown in Figure 2.46, gives a screener or maintenance supervisor, for example, control over incoming requests. By using embedded intelligence, the functionality brings key data to the user's immediate attention to help them efficiently screen and decide on the demand, such as asset criticality, demand priority, maintenance history, and others, including open notifications and orders for the same asset. An automated ranking can be created that covers the risk and criticality of the asset and the event.

  With clarity into maintenance demands, their related failures or malfunctions, and the consequences of failure, a screener can finalize prioritization, determine a final due date for the measure, and pave the way for more detailed work planning.

- **Plan asset maintenance (including resource scheduling)**

  The planning of work scope happens in terms of order management in the maintenance management functionality of SAP S/4HANA Cloud. To best organize the day, a maintenance planner can use the Maintenance Planning Overview app, shown in Figure 2.47, to access an overview of what is going on and to focus on what to do next.

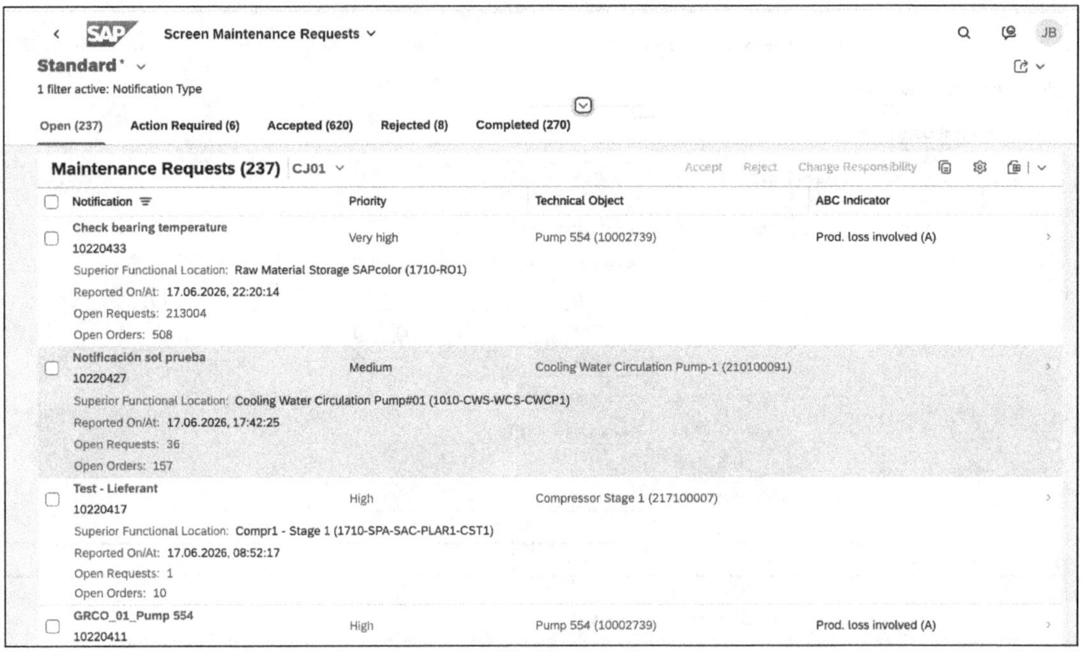

Figure 2.46: Screen Maintenance Request

Figure 2.47 shows the Maintenance Planning Overview app, which consolidates open work orders, planned dates, and resource requirements to support structured maintenance work planning.

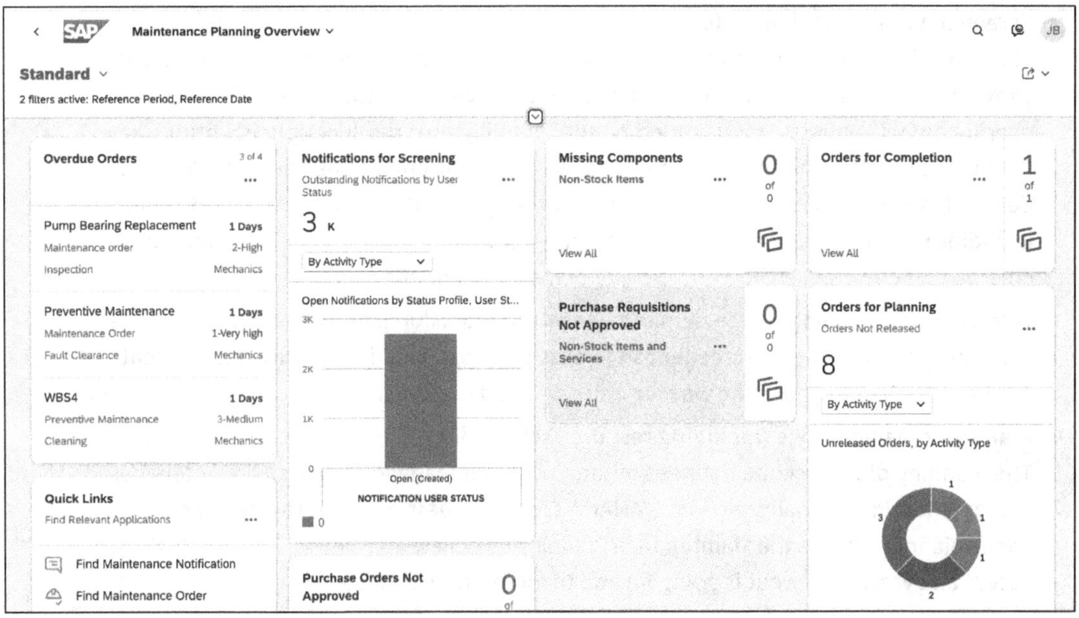

Figure 2.47: Maintenance Planning Overview

The trigger for planning activities is often an "approved" maintenance notification representing the maintenance demand in terms of manually created maintenance requests or proactively scheduled demands. Additional work demand for the maintenance crew also can be raised by investments and capital work. Digging more deeply into a maintenance order, you can define the right activities to be performed, including the necessary resources, such as skilled technicians (either internal or external workforce), needed spare parts and tools, and documentation, such as permissions needed. You can benefit from this capability because of its tight integration into other domains such as finance (for costing purposes) and sourcing and procurement (for supply purposes). To plan most efficiently, you can make use of best practices predefined in task lists. For even better work sequencing, individual activities can be categorized as prework, main work, and postwork. With the main work completed, an asset can be handed back to operations while postwork is still going on.

To organize a whole backlog of maintenance work for streamlined planning, you can make use of planning buckets. With planning buckets, you can cluster work along aligned maintenance windows for recurring or one-time events, as well as for other planning elements such as plants, planner groups, work centers, functional locations, and equipment. With insights from the Manage Maintenance Backlog app, as shown in Figure 2.48, you'll see the readiness of your planned orders with regard to, for example, the supply of material (spare parts) or services.

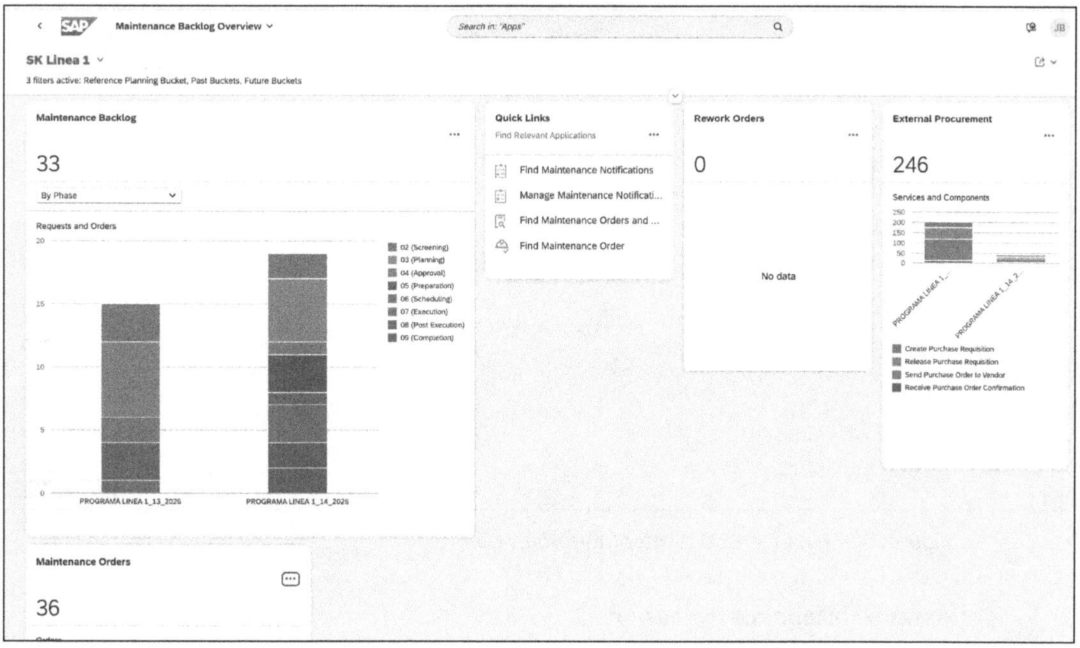

Figure 2.48: Maintenance Backlog Overview

After financial approval, you can release the complete work order or individual operations and trigger the supply of needed resources. By getting full insights into resource availability for the rough-cut plan, you can finalize planning with dedicated scheduling capabilities.

With SAP S/4HANA Asset Management for resource scheduling, as a maintenance scheduler, you can extend the digital core with more advanced scheduling capabilities. As a maintenance planner, you can use relevant information to determine, in an automated way, what work can be done with what resources and by when on a detailed level. The capabilities in the tool span monitoring and analyzing maintenance workloads, identifying capacity bottlenecks through graphical views or lists, creating and collaborating (e.g., with production) on schedules, taking actions, and properly distributing the work to dispatch with a binding ("frozen") schedule. The Manage Work Center Utilization app shown in Figure 2.49 provides an overview on an operational level across orders and shows the utilization of the work centers you're responsible for. As a result, you can easily balance technicians' loads to meet defined targets with only a few clicks.

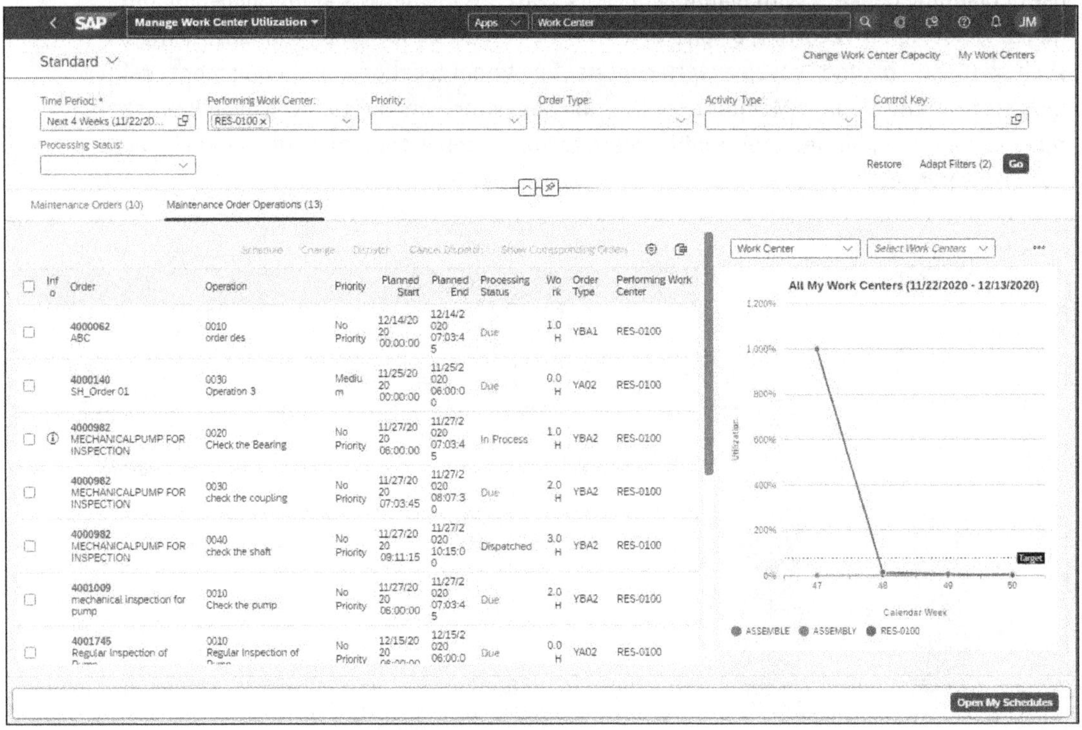

Figure 2.49: Manage Work Center Utilization App

Asset Maintenance Execution

After proper maintenance planning is done, it's time to execute the plan—to keep assets functioning or restore them. Let's briefly explore the main solution capabilities relevant for maintenance technicians performing work.

When planning and scheduling is done, you can hand over the work to maintenance technicians or to the maintenance team. With dedicated assignments, a technician can see open jobs and can start working on them. Assuming the needed tools (e.g., cranes and scaffolds) or spare parts have been supplied and are available, all permits are granted, and safety measures are in place, a maintenance technician can now perform the outlined activities. The technician can manage all system-related interactions while performing the work, such as partial- or final-time confirmations, using delivered desktop applications.

To be the most effective, maintenance technicians can use SAP Maintenance Assistant, shown in Figure 2.50, to have all asset insights and maintenance-event-related data at hand, whether online or offline. Technicians can enhance their tool use and deliver efficiencies over paper-based systems. Performing adjustments in work, timely reporting of work progress, and confirming final time and materials are key to providing information, especially for integrated asset management in terms of time bookings, material consumptions, and settlement in financials, for example. To close the loop of asset performance management, technicians can record maintenance history with failure and malfunction data.

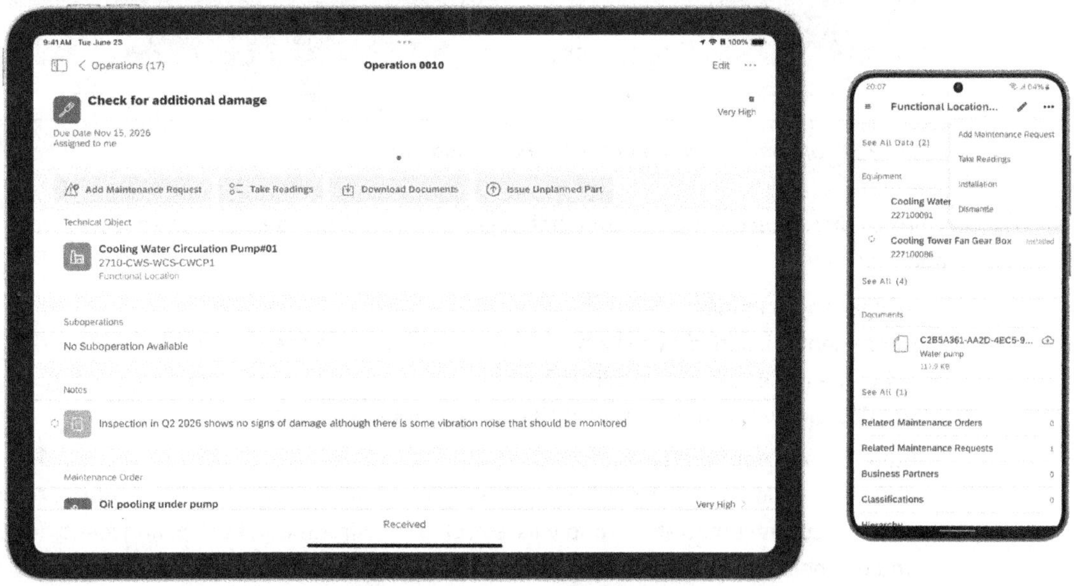

Figure 2.50: Asset Management on Mobile Device Using SAP Maintenance Assistant

2.6.4 Offboard-to-Decommission

Offboard-to-decommission is the end-of-life capability that retires physical assets and related properties safely, compliantly, and with financial accuracy. It is typically the fifth movement in the acquire-to-decommission value chain after operations and maintenance. The purpose is straightforward: make an asset safe, remove it from service, dispose of it responsibly, and close the books with a clean audit trail. Although often spotlighted in asset-intensive industries such as oil and gas, utilities, mining, and engineering/construction/operations, the capability is cross-industry and applies to any enterprise managing significant fleets of equipment, facilities, or infrastructure.

Figure 2.51 shows the offboard-to-decommission process map, covering all steps from the business decision to retire an asset through safe disposal, financial settlement, and audit closure.

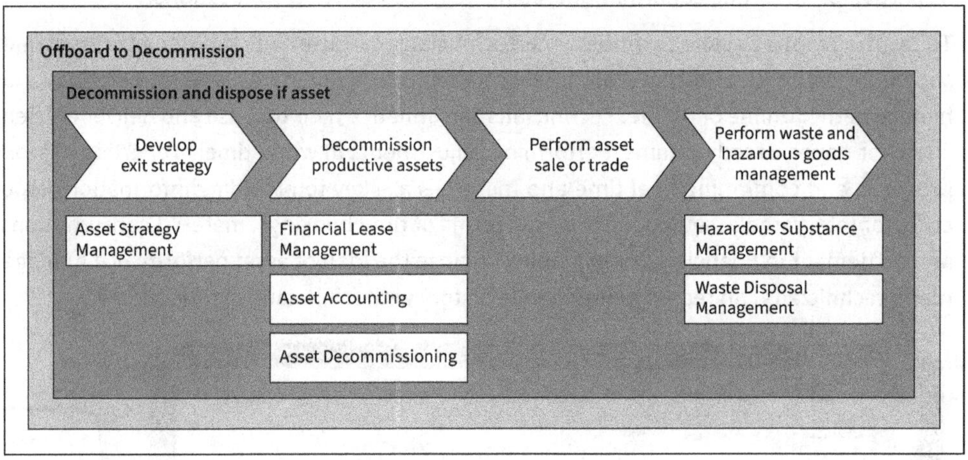

Figure 2.51: Offboard-to-Decommission Process Map

Think of six connected movements that begin with a business decision and end with auditable closure:

1. **Trigger and scoping**
   Retirement is justified by reason of safety, regulatory change, or economics. The organization identifies the scope across technical objects, hazardous inventories, connected utilities, land rights or property entitlements, and active contracts.
2. **Planning and approvals**
   A decommissioning plan defines method statements, isolation and lock-out/tag-out, permits, contractor strategy, and waste profiles. Environment, health, and safety (EHS) and engineering assess risk; finance validates expected financial outcomes; and regulatory approvals and community notifications are secured.
3. **Execution and dismantling**
   Work packages are released, isolation is verified, dismantling proceeds under permit, and remediation or site restoration is completed. Materials are routed to the best-value outcome—reuse, resale, recycling—or hazardous disposal with a full chain of custody.
4. **Master data and records**
   Technical objects move to a retired/decommissioned status, equipment hierarchies are frozen, and closure evidence—permits, manifests, certificates—is stored with legal retention.

5. **Financial close-out**
   Stop depreciation; calculate net book value and proceeds; postretirement with or without revenue; recognize gain/loss; derecognize AUC/WBS where applicable; reconcile subledger to ledger; and prepare disclosures.
6. **Compliance and reporting**
   Complete environmental closure and land/lease terminations; consolidate sustainability metrics (waste diversion, recycling rates, hazardous waste compliance); and update the portfolio baseline and capture lessons learned.

Table 2.7 maps the six decommissioning movements to the SAP Cloud ERP capabilities that support them, along with optional AI enhancements and the master data and document controls required at each stage. Use this table as a practical implementation reference when configuring the offboard-to-decommission process for your asset portfolio.

SAP Cloud ERP Capability	What It Enables	Key Cloud Features	Example KPIs
Decommissioning planning and approvals	Scope retirement, structure tasks, route approvals	Maintenance orders/operations; WBS planning; workflow and approvals	Plan cycle time; plan adherence percentage
Work permits and isolation (LOTO)	Safe work with permits and verified isolation	Permit management; work clearance; LOTO checklists; mobile permit steps	LOTO compliance percentage; incidents (zero target)
Execution and dismantling	Controlled dismantling and remediation	Maintenance execution; confirmations; mobile execution; quality checks	Ontime completion percentage; rework rate
Hazardous materials and waste handling	Classification, tracking, compliant disposal	EHS risk assessments; hazardous substance and waste management	Hazardous waste compliance percentage; violations
Material disposition (reuse/resale/recycle/scrap)	Value recovery and full traceability	Inventory and batch/serial management; logistics execution; resale via billing	Percentage reused/recycled; recovery value vs. plan

Table 2.7: Offboard-to-Decommission Process Map: SAP Capabilities, AI Enhancements, and Controls

SAP Cloud ERP Capability	What It Enables	Key Cloud Features	Example KPIs
Financial asset retirement (with/without revenue)	Accurate postings and derecognition	Asset accounting retirements: depreciation stop; gain/loss calculation	Retirement accuracy percentage; gain/loss vs. net book value
Impairment and derecognition (where needed)	Preretirement valuation adjustments	Impairment testing; revaluation; special postings	Impairment lead time; exception count
Real estate and land rights termination	Lease/rights close and property/unit retirement	Real estate: contract termination; unit/property retirement	Termination cycle time: penalties avoided
Records and evidence management	Controlled storage of closure artifacts	Document/records management; retention rules; legal hold	Documentation completeness percentage; retrieval time
Sourcing and contractor services	Procuring decommissioning and waste services	Source pay: RFQ/contract; purchase order; service entry; supplier qualification	On budget percentage; qualified supplier coverage
Sustainability and circularity	Waste and recycling outcomes, auditability	Footprint and circularity reporting; waste stream tracking	Waste diversion percentage; recycling rate percentage
Portfolio, risk, and analytics	Prioritization, progress tracking, value reporting	Dashboards and what-ifs; portfolio attributes and risk scoring	Portfolio risk reduction; value recovered

Table 2.7: Offboard-to-Decommission Process Map: SAP Capabilities, AI Enhancements, and Controls (Cont.)

The following optional AI capabilities can be layered on top of the standard decommissioning process to improve accuracy, reduce manual effort, and accelerate each stage of the offboard cycle:

- Generative workpacks from asset metadata and historical jobs (planning and approvals)
- Computer vision to verify dismantling progress and personal protective equipment (PPE)/hazard compliance (execution)

- Disposition recommender balancing resale, scrap, and recycle against financial and sustainability targets (material disposition)
- Retirement simulation for gain/loss timing and tax impact (finance)
- Autoclassification and linking of permits, manifests, and certificates (records)

The following master data, documents, and controls should be verified before initiating the offboard-to-decommission process to ensure a complete and auditable retirement:

- **Master data**
  Equipment, functional locations, linear assets, serials, fixed assets and assets under construction (AuCs), work breakdown structure (WBS)/internal orders, asset classes, depreciation areas, properties/units, contracts, land rights, vendor qualifications
- **Documents**
  Decommissioning plan, permits, work clearances, method statements, disposal manifests, certificates, inspection records, closure reports
- **Controls**
  Separation of duties (SoD) across plan/execute/post; permit and LOTO verification gates; chain of custody for hazardous materials; approval workflow for retirements; retention and legal hold policies

With this capability map and the six-movement storyline, your offboard-to-decommission section reads cleanly in one to two pages while staying concrete about what SAP Cloud ERP actually does, how value is measured, and where to add AI for speed, safety, and better financial outcomes.

2.6.5 Managing Assets

Let's look more closely at our foundational topic in more detail: managing asset master data. In the maintenance management functionality of SAP S/4HANA Cloud, as a master data specialist, you can maintain asset master data in terms of functional locations and equipment to represent your physical assets in a structured way by covering details such as location details, technical characteristics, documents, and business-relevant data such as organizational structure. In the process, you can lay the foundation to run all related business processes on top of the functionality in an efficient and effective way. By going a little further and paying attention to the planning and execution processes, asset master data also covers failure data definitions, predefined task lists, structured lists of spare parts (maintenance BOMs), and so on.

By having up-to-date, complete, and trustworthy asset master data and related data available, you can define best practices and become more efficient in subsequent processes, such as maintenance planning. This set of master data is also relevant for extending end-to-end processes with an asset performance management capability. With an aligned data model across the disciplines in asset management, SAP Asset Performance Management lets you overcome the siloed approaches from the past and integrate, extend, and make use of this foundational asset master data on both sides: maintenance execution and strategy management.

As maintenance management transforms into a more collaborative business, asset master data creation already starts outside the operator's four walls. The manufacturer of an asset has all the details belonging to its products in terms of technical specifications, data sheets, and recommended service and maintenance documentation. By using SAP Business Network for Asset Management with the maintenance management functionality of SAP S/4HANA Cloud, you can benefit from collaborative processes around asset master data management.

To enable a full transition into an intelligent asset management approach, the more ERP-led definition of an asset master dataset must be enhanced from the technology side. Connecting to the operation technology (OT) layer enables monitoring of physical assets. With SAP Asset Performance Management, you can directly connect to the OT layer and access accurate, real-time information from your assets and interpret them frequently. Thus, the digital twin of a physical asset moves closer to reality and helps you optimize and innovate processes in asset management. In addition, SAP Asset Performance Management helps you calculate standardized health scores for your connected assets to drive your processes using ML algorithms.

2.6.6 Public Cloud Versus Private Cloud

Asset management and maintenance capabilities share significant common ground across both editions. Corrective maintenance (BH1), preventive maintenance (BJ2), emergency maintenance (BH2), asset accounting (J62), asset management, order-based maintenance, service monitoring and analytics, and nonconformance management are all confirmed in both editions. Integration with SAP Asset Performance Management and SAP Business Network for Asset Management is supported in both deployment models.

The functional differences identified are summarized in Table 2.8.

Capability	Public Cloud Edition	Private Cloud Edition
Maintenance resource scheduling (43R) Scheduling of maintenance technicians and resources against work orders, integrated with the maintenance planning board.	Available as a dedicated scope item (43R and variant O3R).	Not confirmed as a scope item version. Resource scheduling relies on standard work center capacity planning or the SAP S/4HANA Asset Management for resource scheduling add-on.

Table 2.8: Functional Differences in Acquire-to-Decommission, Public Cloud Edition Versus Private Cloud Edition

Capability	Public Cloud Edition	Private Cloud Edition
Service order management and monitoring (3D2) Management and monitoring of service orders for in-house maintenance with analytics-driven oversight.	Available as a dedicated scope item.	Not confirmed as a scope item version. Service order management is available through standard maintenance order processing (IW31, IW32).
Asset retirement obligation (56E, 5IT, 6X6) Accounting for legal obligations associated with the retirement of long-lived assets, including provision calculation and unwinding.	Available as dedicated scope items across multiple variants.	Not confirmed as scope item versions. Asset retirement obligation accounting is handled through manual provision postings or custom configuration.
Compliance management for EHS (4C8) Management of EHS compliance obligations, incident tracking, and regulatory reporting.	Available as a dedicated scope item.	Not confirmed as a scope item version. Environment, health, and safety compliance is typically addressed through SAP Environment, Health, and Safety Management as a separate solution or custom development.
SAP Procurement Intelligence (add-on) Advanced procurement and asset spend analytics.	Not available as a separate add-on.	Available as a dedicated add-on product: SAP S/4HANA Cloud Private Edition, Procurement Intelligence.
User interface and custom maintenance logic	SAP Fiori exclusively. Extensions through released APIs and key user extensibility only.	SAP Fiori plus SAP GUI (Transactions IW21, IW31, IW38, IP10, etc.). Full ABAP available for custom maintenance strategies, condition-based maintenance rules, and integration with IoT platforms and SCADA systems.

Table 2.8: Functional Differences in Acquire-to-Decommission, Public Cloud Edition Versus Private Cloud Edition (Cont.)

The asset management differences cluster around two themes. First, SAP S/4HANA Cloud Public Edition delivers maintenance resource scheduling, asset retirement obligations, and EHS compliance as dedicated, ready-to-activate scope items, areas that in SAP S/4HANA Cloud Private Edition require either add-on products, manual configuration, or custom development. Second, SAP S/4HANA Cloud Private Edition provides the ABAP environment needed for organizations with highly specialized maintenance requirements, such as custom condition-based maintenance algorithms, proprietary IoT integration, or industry-specific regulatory compliance logic.

2.6.7 Outcome

David's challenge at Korindal is not just an IT architecture problem. It shows up in asset management as well: maintenance work orders that require custom integrations to reach the right technician, asset data spread across disconnected systems, and unplanned downtime that consumes the production capacity that planning assumed would be available. The asset management capabilities in this section address the operational side of that problem. Predictive maintenance converts unplanned failures into planned interventions. For James, every hour of unplanned downtime on a critical production line translates directly into delayed customer orders. The transition from calendar-based to condition-based maintenance reduces unnecessary preventive interventions while identifying incipient failures early enough to schedule repairs within planned production windows. The network dimension matters too. SAP Business Network for Asset Management enables asset data sharing with third-party maintainers and equipment manufacturers, so the service provider arrives with the right parts and the right context rather than starting from scratch. Over time, the maintenance cost decreases while asset availability increases. The 30 percent reduction in unplanned downtime reported by organizations at this maturity level does not just reduce costs. It also protects the OTIF commitment that James is measured on.

2.7 Summary

This chapter walked through the core business capabilities of SAP Cloud ERP, not as a feature catalog but as a demonstration of what happens when standardized processes, integrated data, and embedded intelligence operate together inside a single digital core. The goal was to show how each functional domain creates the conditions for autonomous operations by producing clean, contextually rich data that the intelligence layer can act on reliably.

Finance establishes the foundation through the Universal Journal, a single ledger structure that eliminates the reconciliation burden between entity and group accounting, between statutory and management views, and across currencies and reporting standards. On this foundation, capabilities like intelligent cash application, the collections worklist, accruals management, and group reporting transform the financial close from a labor-intensive monthly sprint into a more continuous, exception-driven process. Event-based revenue recognition captures financial impact as work completes rather than waiting for period-end processing. SAP Advanced Financial Closing orchestrates the entire close as a system-driven workflow rather than a spreadsheet-tracked exercise. These capabilities deliver their full value when they are organized into a continuous operating rhythm.

Sales translates customer demand into revenue through coordinated order capture, pricing governance, fulfillment orchestration, and billing across multiple business models. The emphasis is on low-touch execution for routine orders, with the human focus reserved for complex, relationship-driven scenarios. Centralized pricing eliminates inconsistency across channels. Automated credit and compliance checks let routine orders flow without manual intervention. Contract management handles long-running agreements with scheduled billing and renewal tracking. The outcome is fewer late deliveries, fewer pricing disputes, and operations that intervene before commitments are missed rather than after.

Sourcing and procurement connects sourcing decisions to their downstream payment obligations on a single operational spine. Source-to-contract puts sourcing events, the contract lifecycle, and supplier performance in one place, ensuring that negotiated terms actually reach daily buying. Central procurement provides global control without forcing a disruptive ERP consolidation. Invoice-to-pay digitizes capture, automates verification, and embeds approval workflows inside the business process rather than routing them through email. The compound effect is higher contract coverage, lower maverick spend, faster cycle times, and fewer payment inquiries.

Plan-to-make covers the full manufacturing lifecycle from demand planning through production execution and quality assurance. Planning capabilities balance supply against demand on multiple horizons, from strategic capacity decisions down to daily sequencing. Manufacturing supports discrete, process, repetitive, and outsourced production modes with shop floor integration that closes the loop between planning and execution. Quality management embeds inspection and control directly into procurement, production, and delivery rather than treating quality as a downstream gate.

Logistics provides the physical execution layer, moving materials through receiving, storage, picking, packing, shipping, and transportation. Integrated warehouse management handles stock placement and retrieval with wave-based processing and mobile execution. ATP confirms delivery dates against real inventory and production capacity. Transportation management optimizes carrier selection, route planning, and freight cost allocation. These capabilities ensure that what the planning system promises, the logistics system delivers.

Acquire-to-decommission manages physical assets across their full lifecycle, from initial capitalization through maintenance operations to eventual retirement. Intelligent asset management combines traditional maintenance execution with predictive and prescriptive approaches driven by sensor data and reliability engineering. Plan-to-optimize defines maintenance strategies based on risk and criticality assessment. Operate-to-maintain executes those strategies through structured work demand, planning, scheduling, and mobile-enabled field execution. Finally, offboard-to-decommission handles the controlled end-of-life process with full financial accuracy and compliance traceability.

Across every domain, two patterns repeat. Processes are designed for exception-driven operation: Routine transactions flow automatically while human attention is focused on genuine deviations. Each domain produces the structured, contextually rich data that the intelligence layer needs to reason and act. This is the practical meaning of the flywheel described in Chapter 1. Applications that run well generate better data, better data enables sharper intelligence, and sharper intelligence produces better execution; thus, the cycle compounds. SAP Cloud ERP is where that cycle begins.

The next chapter examines two-tier ERP landscapes, exploring how organizations manage the coexistence of multiple SAP systems across global operations and how the integration capabilities described here enable a coherent enterprise even when no single system handles everything. It examines how organizations manage the coexistence of global and subsidiary ERP systems, exploring when a two-tier model makes sense, how SAP S/4HANA Cloud integrates with subsidiary environments, and what this means for the clean core and integration strategies described throughout this chapter.

Chapter 3
Two-Tier ERP Landscapes

Now that you understand SAP Cloud ERP's business capabilities, in this chapter, we'll explore how organizations can implement a two-tier ERP landscape setup to optimize their processes.

Enterprises expand unevenly: into new markets, through acquisitions, into adjacent business models, and often under pressure to move faster than their core systems were designed to allow. Over time, what once felt like a well-structured enterprise resource planning (ERP) landscape becomes a constraint. The core system remains stable and essential, but everything around it begins to stretch: Subsidiaries operate differently, local requirements diverge, and new business units need capabilities that cannot wait for centralized change cycles.

The two-tier ERP landscape emerges as a response to this reality. It reflects a simple but powerful shift: instead of forcing the entire organization into a single system, the enterprise accepts that different parts of the business require different speeds, levels of control, and degrees of flexibility.

Global enterprises face an enduring tension: They must maintain centralized financial control and compliance while granting subsidiary operations the autonomy and agility they need to compete locally. A two-tier ERP landscape resolves this tension by connecting a powerful central system at headquarters with tailored cloud ERP systems at subsidiaries, enabling both standardization and flexibility.

This chapter reveals how leading enterprises, from multinational manufacturers to global consulting firms, use two-tier ERP to accelerate financial close, improve working capital, and unlock procurement savings. More importantly, we show you how to design, implement, and operate these landscapes to create competitive advantage.

3.1 What Is Two-Tier ERP?

Imagine a global manufacturing company operating across continents. Manufacturing plants in Germany and Ireland produce under tightly controlled conditions. Sales organizations in Brazil and Singapore engage local markets with different pricing expectations, regulatory requirements, and customer behaviors. At the headquarters in Switzerland, finance consolidates results, the treasury manages liquidity, and supply chain teams coordinate production plans.

The question is not whether these activities can be supported by a single system. In theory, they can. The question is whether they can be supported without slowing the business down.

This is where the two-tier ERP landscape emerges, not as an architectural preference but as a practical response to scale and diversity.

In this model, the enterprise operates with two coordinated layers. At the center sits the Tier 1 system, typically SAP S/4HANA at the headquarters. It holds the structures that must remain consistent: financial consolidation logic, master data definitions, procurement standards, and planning frameworks. Around it, subsidiaries operate on cloud ERP systems—often SAP S/4HANA Cloud Public Edition—designed for speed, standardization, and local adaptability. The distinction between these tiers is not technical but functional.

At the headquarters, processes are designed for control and coherence across the enterprise. At the subsidiary level, processes are designed for execution under local conditions. Now let's understand the key responsibilities of the command center (headquarters) and local operations (subsidiary):

- **Tier 1: The command center**
  The system at the headquarters, or Tier 1, acts as the organization's command center. Its primary responsibilities include the following:
  - *Group financial consolidation*: Consolidating results from all subsidiaries into unified financial statements compliant with Generally Accepted Accounting Principles (GAAP), International Financial Reporting Standards (IFRS), and local regulations.
  - *Master data governance*: Maintaining the central database of customers, suppliers, products, and finance dimensions that all subsidiaries reference.
  - *Strategic procurement*: Negotiating global supplier contracts and defining enterprise-wide purchasing standards, allowing subsidiaries to procure at negotiated rates.
  - *Supply chain planning*: Creating global demand forecasts and optimizing inventory across all locations to minimize working capital.
  - *Treasury and cash management*: Coordinating group-wide cash positions, optimizing payment timing, and managing liquidity across the organization.
  - *Compliance and risk*: Ensuring adherence to corporate policies, regulatory requirements, and audit standards across all entities.
- **Tier 2: Local operations**
  Each subsidiary's system, in Tier 2, is optimized for operational agility. Tier 2 systems handle the day-to-day transactions that drive revenue and execute the organization's strategy at the local level:
  - *Local sales management*: Processing customer orders, managing deliveries, invoicing customers in local currency and language, and handling collections.

- *Subsidiary procurement*: Ordering materials and services from suppliers, including those contracted centrally and occasionally local suppliers for urgent or region-specific needs.
- *Inventory and warehouse management*: Maintaining local warehouse operations, picking and packing orders, and managing inventory levels.
- *Employee and payroll management*: Managing local hiring, compensation, and payroll processing according to regional labor laws and tax requirements.
- *Operational reporting*: Tracking local performance metrics and reporting results to headquarters.

Organizations that attempt to run a single, monolithic ERP system globally face a painful trade-off. They can customize the system extensively to accommodate local requirements, a path that leads to bloated implementations, high maintenance costs, and inflexible processes that take months to change. Alternatively, they can enforce standardized processes globally, which improves maintainability but frustrates local teams who cannot adapt to their markets.

Two-tier ERP eliminates this false choice. By allowing headquarters to maintain tight control over financial processes, master data, and strategic supply chain decisions while empowering subsidiaries to adapt their operations to local market realities, organizations achieve both standardization and agility. The result: faster innovation, lower total cost of ownership, and competitive advantage.

3.2 Deployment Models and Use Cases

Two-tier ERP manifests in different forms depending on organizational structure. Let's examine four common deployment models from actual customer implementations:

- **Model 1: Headquarters with regional subsidiaries**
  The headquarters manages the global supply chain strategy, procurement, and financial consolidation. Each region operates independently for local operations and customer management.

  The headquarters creates global demand forecasts shared with subsidiaries for production scheduling. All intercompany transfers consolidate at headquarters for group-level cost analysis and supply chain optimization. This model is typical for multinational manufacturers for which global coordination of production and inventory is critical to profitability, but regional sales teams need autonomy to engage customers locally. Figure 3.1 shows more key use cases for the procure-to-pay, plan-to-product, finance, and order-to-cash processes. We'll dive into these use cases in the next sections.

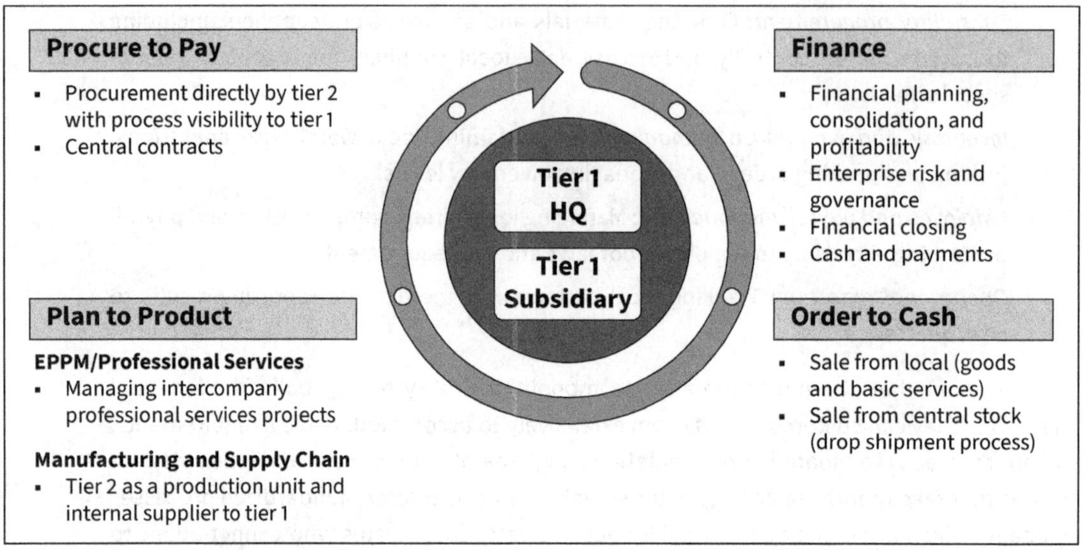

Figure 3.1: Headquarters with Regional Subsidiaries

- **Model 2: Subsidiary as sales office (sales from local stock)**
  A subsidiary operates as a regional sales office managing customer relationships and local inventory. The sales team builds relationships with local customers, understands their preferences and payment behaviors, and adapts offerings to regional market conditions. Headquarters maintains the central customer master with credit limits, payment terms, and pricing agreements. When a regional sales representative receives a customer purchase order, the sales order is created in the Tier 2 system. The system immediately checks the customer credit limit against the Tier 1 master; if the customer is already at their limit or has credit issues at another subsidiary, the order is flagged. Once approved, delivery is created and goods are picked from the subsidiary's local stock, as shown in Figure 3.2.

  The bill document is created in the local currency and local language, and payment is collected locally. Once the payment clears, the financial update flows back to Tier 1 for group consolidation. Tier 1 gains immediate insights into subsidiary sales activities, enabling faster business decisions. Regional sales teams move fast on local opportunities. Tier 1 maintains customer credit control and consolidated revenue visibility. Working capital improves through local inventory optimization.

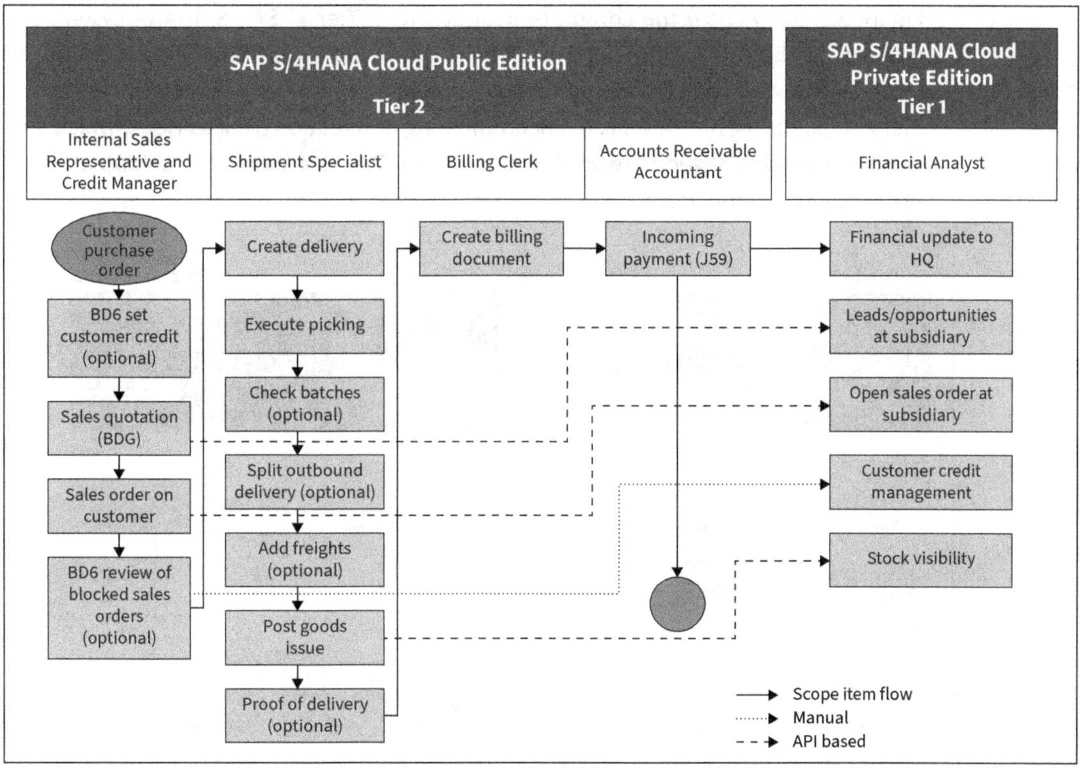

Figure 3.2: Subsidiary as Sales Office

- **Model 3: Subsidiary as production unit and internal supplier**
  A subsidiary operates as a manufacturing center supplying headquarters or other subsidiaries. Headquarters sends a demand forecast based on customer orders and strategic inventory plans. The subsidiary receives this forecast and schedules production accordingly. Once components are produced, they are shipped to headquarters or to other subsidiaries at a predefined transfer price. At month end, all intercompany transfers and manufacturing costs consolidate at headquarters, where the group supply chain team analyzes costs, efficiency, and resource utilization.

  This model reduces manual activities at Tier 2 because the subsidiary does not need to maintain complex pricing or billing logic; the transfer price is fixed, and the transaction is straightforward. Headquarters gains immediate insights about subsidiary activities, enabling faster business decisions and optimal resource utilization. The headquarters also improves its planning accuracy because it has visibility into the component stocks at subsidiaries, which facilitates inventory control and meeting customer demand promptly. Tier 2 receives accurate demand from Tier 1, eliminating guesswork. Master data maintenance becomes simpler because headquarters maintains one master definition that all subsidiaries reference. Good visibility into Tier 2 production data helps headquarters make better supply chain decisions.

Figure 3.3 shows how the corporate headquarters (Tier 1) SAP S/4HANA system integrates with the subsidiary (Tier 2) SAP S/4HANA Cloud Public Edition system. The key integrations work as follows: Tier 1 purchase orders become sales orders in Tier 2. When Tier 2 ships goods, it sends an advanced shipping notification back to Tier 1, which automatically creates a goods receipt. Finally, when Tier 2 bills for the shipment, Tier 1 receives a supplier invoice.

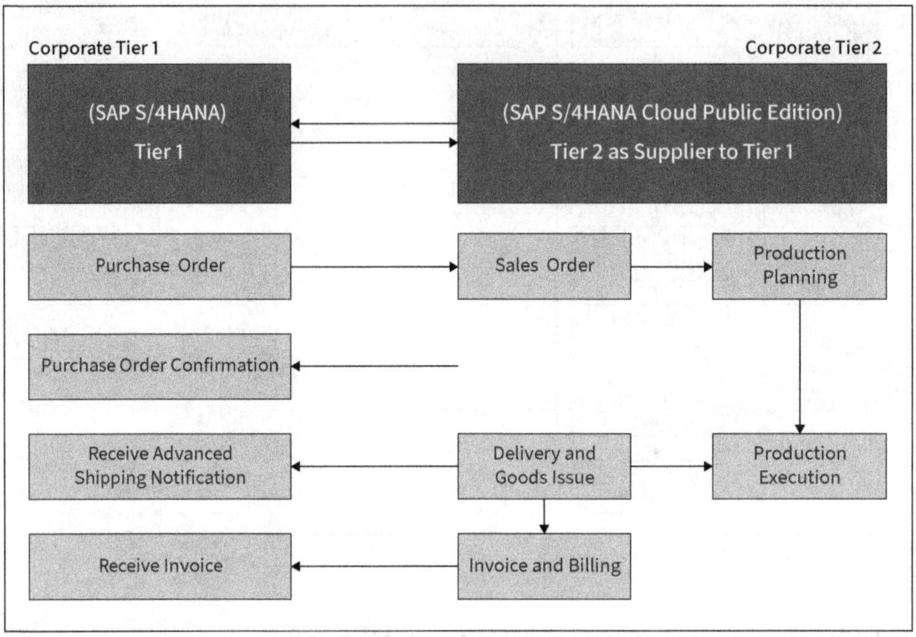

Figure 3.3: Subsidiary as Production Unit and Internal Supplier

- **Model 4: Shared services center**
  The organization centralizes finance shared services or central procurement in a Tier 2 instance, which processes transactions for all other subsidiaries. Rather than each subsidiary maintaining its own accounts payable team, for example, a centralized finance shared services center in one location handles invoicing, collections, and accounts payable processing for all regions. This team becomes expert in invoice matching, three-way matching, and payment processing. Results consolidate back to Tier 1 for group reporting. This model drives standardization and cost efficiency by centralizing transaction processing across all subsidiaries, reducing duplicative work and enabling economies of scale.

3.3 Finance Domain: From Consolidation to Strategic Planning

Finance is the backbone of two-tier ERP. All subsidiary transactions flow to headquarters for consolidation. This section covers master data governance, financial planning, group reporting, and consolidation.

Let's walk through the key areas in the finance domain that are impacted by a two-tier ERP landscape:

- **Master data governance and distribution**
  Master data objects—profit centers, cost centers, general ledger charts of accounts, asset masters, suppliers, customers, products—must be clean, consistent, and centrally governed. Tier 1 is the system of record. Tier 2 systems maintain local copies that replicate automatically from Tier 1. When headquarters needs to add a new profit center, update the chart of accounts, or define a new product, the change is made once in Tier 1. That change automatically propagates to all Tier 2 instances. Local teams never manually enter or maintain these structures. The result is consistency: all subsidiaries work from the same definitions, enabling reliable consolidation.
- **Financial planning and budgeting**
  The planning cycle begins at headquarters, where finance develops the group budget in SAP Analytics Cloud using the prior-year actuals and business strategy. The budget includes revenue projections, cost of goods sold, operating expenses, and headcount by cost center and profit center. Business leaders debate trade-offs, adjust assumptions, and reach consensus. Once the CFO approves the budget, it is shared with subsidiaries via an application programming interface (API). Each subsidiary imports the budget into their local general ledger.

  Throughout the year, each subsidiary tracks actuals against the imported budget. At month end, the subsidiary finance team reviews variances: If actual spending is more than 10 percent higher than budget, they investigate causes and discuss the issue with headquarters. Headquarters receives actual data from all subsidiaries and runs a consolidated plan versus actual report. The CFO can drill down to any subsidiary, any cost center, or any line item. This consistency in planning methodology enables organizations to have one conversation about performance rather than separate conversations in each region. Finance teams have real-time visibility into performance against plan, enabling faster course corrections when markets shift or opportunities emerge.

  Figure 3.4 shows the specific data flows that make this possible. At the subsidiary level (Tier 2), processes enabled by SAP S/4HANA Cloud Public Edition's J58 scope item generate financial statements in the local currency, the trial balance, tax information by country, and deviating valuation reports. The 1HB scope item handles plan data import from headquarters and produces plan versus actual reports for month-end reviews. At headquarters (Tier 1), business planning preparation creates the cost center, profit center, balance sheet, and foreign exchange (FX) rate planning data, which is sent to the subsidiary. SAP Analytics Cloud consolidates data from both Tier 1 and Tier 2, giving the CFO and finance leadership real-time visibility into performance across the entire organization. This two-tier architecture ensures that each subsidiary maintains local control over regional accounting while the headquarters maintains visibility into and governance over the consolidated enterprise.

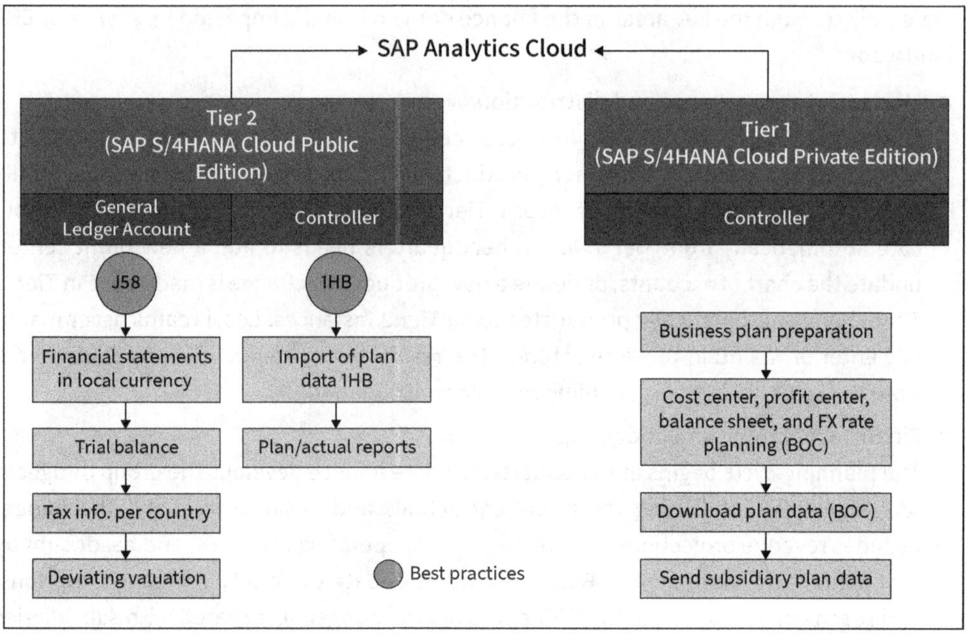

Figure 3.4: Financial Planning with Two-Tier ERP

- **Integrated planning: Connecting finance and operations**
  SAP Analytics Cloud consolidates actual data from all tiers, enabling collaborative planning across functions. Tier 1 performs group-level planning and integrated financial planning for all Tier 2 entities. Finance teams from both Tier 1 and Tier 2 participate in the same planning process. Once planning results are approved by leadership, they are shared back to Tier 2. This creates alignment: Finance plans in concert with operations, sales, and supply chain. Rather than finance creating a plan in isolation and throwing it over the wall, the entire organization plans together.

- **Liquidity management: Real-time cash visibility**
  Each subsidiary sends actual and forecasted cash positions to SAP Analytics Cloud on a regular cadence: daily, weekly, or monthly depending on the organization's requirements. The group cash manager models group-level cash needs, considering payment schedules, receivables forecasts, and strategic initiatives. By consolidating subsidiary forecasts, headquarters can optimize payment timing, accelerating collections in some regions, deferring payments in others, and moving cash between subsidiaries to minimize borrowing costs and maximize returns, as highlighted in Figure 3.5.

  The group cash manager can also negotiate better bank terms because banks see consolidated cash positions and volume. Rather than negotiating separately with each regional bank, the group has one voice with banks globally. Real-time cash visibility enables an optimized liquidity position across the group, resulting in the right amount of cash in the right place at the right time.

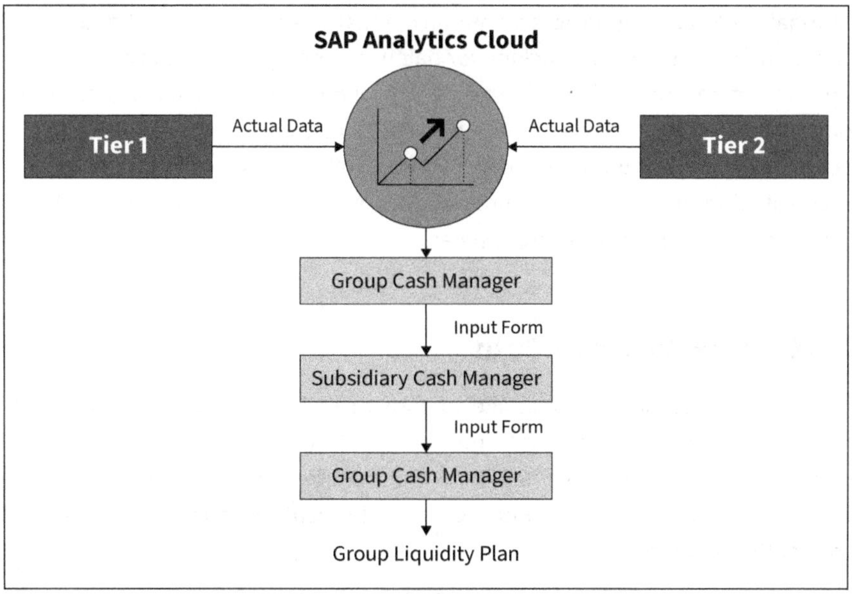

Figure 3.5: Liquidity Planning

- **Financial close: From manual spreadsheets to automated workflows**
  At month end, Tier 2 subsidiaries prepare local financial data: general ledger balances, trial balance, intercompany activities, and so on. A consolidation specialist in Tier 1 executes group reporting in SAP S/4HANA Cloud. The system consolidates from all Tier 2 instances, eliminates intercompany transactions automatically, and produces unified financial statements, as shown in Figure 3.6.

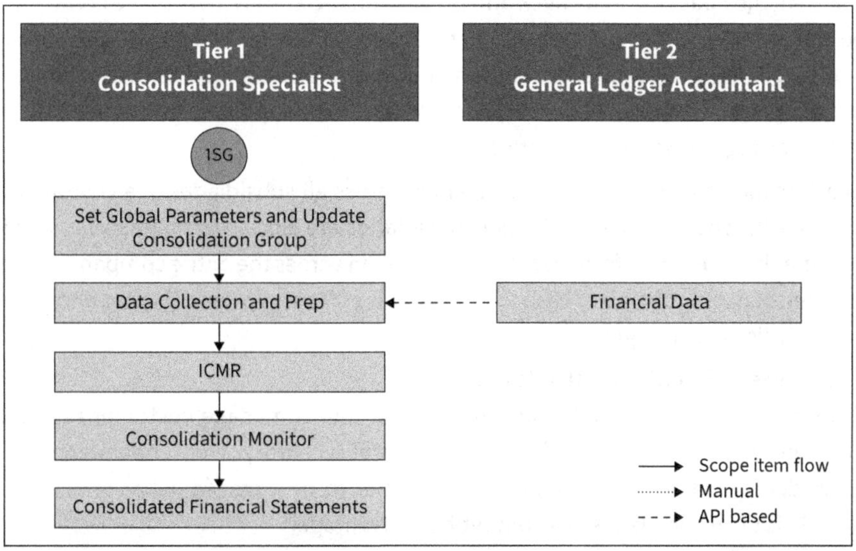

Figure 3.6: Group Reporting: Financial Consolidation

In a traditional landscape, this process required manual spreadsheet consolidation, pulling data from multiple systems, checking for balance, identifying and eliminating intercompany transactions manually, and building consolidated statements in Excel. Errors were common. The close took weeks. In a two-tier landscape with proper automation, the process takes days. Automated consolidation eliminates manual, spreadsheet-based reconciliation. Intercompany eliminations are performed systematically by the system, not by a person reading a spreadsheet.

3.4 Sales Domain: Order-to-Cash

Two-tier ERP enables subsidiaries to manage customer relationships with local autonomy while granting headquarters visibility into revenue, cash flow, and customer credit risk across the organization. This balance is critical: Subsidiaries must move fast to capture opportunities in their markets, yet headquarters must prevent revenue leakage and manage credit risk at the group level.

Let's discuss the key two-tier ERP use cases for the sales domain:

- **Tier 1: Customer master and credit control**
  Headquarters maintains the *central customer master*, the authoritative database of all customers the organization serves. This master includes credit limits, payment terms, pricing agreements, and customer classification. Large enterprise customers may have negotiated group prices that all subsidiaries honor. This master replicates to all Tier 2 systems so that subsidiaries can work from the same customer data.

  Headquarters also defines group-wide credit approval rules: "No customer can purchase on credit until they pass a credit check and provide financial statements. The credit limit is three times the monthly average purchase volume." These rules apply consistently across all subsidiaries. Revenue recognition policies are also defined centrally: When is revenue recognized? At invoice, delivery, or payment? These policies ensure consistent financial reporting across all subsidiaries.

  Headquarters monitors outstanding receivables across all subsidiaries. If a customer is slow-paying with one subsidiary, all other subsidiaries are alerted to tighten terms. This prevents one slow customer from becoming a problem across the entire company. Headquarters consolidates subsidiary cash forecasts to predict group-level inflows and plans treasury activities accordingly.

- **Tier 2: Sales execution with local autonomy**
  Subsidiaries execute sales locally. Subsidiary sales teams manage customer relationships, take orders, and manage fulfillment from local inventory. They understand their customers deeply: what they value, how they prefer to be engaged, what drives their decisions. This local knowledge is a competitive advantage.

Once goods are delivered, the subsidiary invoices the customer in the local currency and language, with payment terms defined in the customer master maintained by headquarters. The subsidiary collects payment from customers. Payments are deposited locally and then reported to headquarters for group cash management.

When a customer returns goods, the subsidiary processes the return and issues a credit memo. Subsidiary sales and revenue data flow automatically to headquarters for group consolidation and analysis. The sales representative does not need to worry about consolidation; it happens automatically in the background.

When a regional sales representative receives a customer purchase order, they enter the sales order into the Tier 2 system. The system immediately checks the customer credit limit against the Tier 1 master. If the customer is at their limit or has credit issues elsewhere, the order is flagged for the credit team to review. Once approved, delivery is created. Warehouse staff pick the goods from local stock and record the goods issue. The bill document is created in local currency and language. (Refer to Figure 3.2, which showcased the complete process.)

Payment is collected locally. Once the payment clears, the financial update flows back to Tier 1 for group consolidation. Tier 1 gains immediate insights into subsidiary sales activities, enabling faster business decisions about which regions are performing, which customer segments are most valuable, and where cash inflows come from. Regional sales teams move fast on local opportunities without waiting for corporate approval on every transaction. Tier 1 maintains customer credit control and consolidated revenue visibility, preventing revenue leakage and managing group-level credit risk. Working capital improves through local inventory optimization: Each subsidiary keeps what it needs locally, rather than maintaining excess inventory for corporate oversight.

There's an additional impact on intercompany sales and consolidation logic. When Tier 1 sells to Tier 2 at a transfer price, Tier 1 records a sale and accounts receivable, while Tier 2 records a purchase and accounts payable. These are real transactions in the accounting records. At consolidation, the system automatically eliminates the intercompany transaction, removing the sale from Tier 1 and removing the purchase from Tier 2, so that the group revenue reflects only external sales. Transfer pricing allocates profits fairly across jurisdictions, ensuring tax compliance.

3.5 Procurement Domain: Procure-to-Pay

Procurement is one of the most significant sources of opportunity in a two-tier landscape. Headquarters can consolidate purchasing power, negotiate better rates, and enforce quality standards globally. Yet subsidiaries must retain flexibility to source locally when market conditions, quality requirements, or supply chain timing demand it. If poorly executed, two-tier procurement becomes a bureaucratic bottleneck. If executed well, it drives significant savings while maintaining operational agility.

Let's now discuss the key two-tier ERP use cases for the procure-to-pay processes:

- **Tier 1: Strategic procurement and supplier management**
  Headquarters' procurement organization handles strategic sourcing. They maintain the central database of approved suppliers with negotiated terms, payment terms, and quality standards. This master replicates to all Tier 2 systems so that subsidiaries procure from known, vetted sources.

  Headquarters negotiates global framework agreements with major suppliers. For example: "ABC Corp will supply all capacitors at $10 per unit, volume discount to 25 percent at 1M units annually, delivery within 10 days." These contracts define the terms under which all subsidiaries source. Rather than each subsidiary negotiating separately, the company negotiates once and benefits globally.

  Headquarters creates global demand forecasts and shares them with suppliers. This visibility secures capacity reservations and ensures reliable supply. Suppliers know that they will receive consistent volume throughout the year, so they allocate capacity and negotiate favorable terms. Headquarters defines quality specifications and inspection procedures. All suppliers must comply with these standards regardless of which subsidiary they serve. This consistency ensures that a capacitor from ABC Corp meets the same quality standard whether it is used in Germany or Brazil.

- **Tier 2: Operational procurement with guardrails**
  Subsidiaries execute the daily procurement function. When a subsidiary team identifies a need, they create a purchase requisition against headquarters' approved supplier list. The system validates whether the supplier is approved and whether the request aligns with corporate contracts. A requisition for capacitors automatically validates against the ABC Corp contract: If the price and terms are standard, the requisition converts to a purchase order automatically. If the requisition asks for nonstandard terms or an unapproved supplier, it requires supervisor approval. See Figure 3.7 for an illustration of the end-to-end process for a two-tier procurement scenario.

  Once approved, the purchase order is sent to the supplier. When goods arrive, the subsidiary team receives them, inspects them against quality standards, and records receipt in the system. The supplier submits an invoice. The subsidiary system matches the invoice to the purchase order and goods receipt (three-way match). If everything aligns—the purchase order quantity matches the invoice quantity, the purchase order price matches the invoice price, the received quantity is correct—then payment is approved automatically. If discrepancies exist, the subsidiary investigates before paying.

  All Tier 2 purchase orders are visible to Tier 1, enabling headquarters to monitor spend, identify off-contract purchases, and consolidate supplier performance data, as highlighted in Figure 3.7. If a subsidiary is consistently paying premium prices or working with unapproved suppliers, then headquarters identifies the pattern and works with the subsidiary to align with corporate contracts.

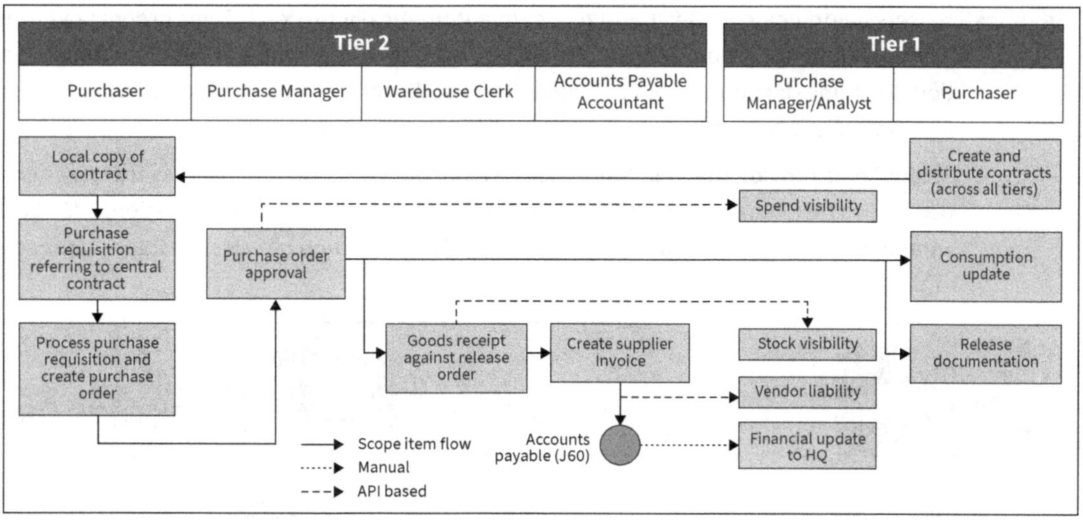

Figure 3.7: Central Contract with Purchasing for Two-Tier ERP

The procurement cycle is driven by demand. Each quarter, headquarters creates a global demand forecast based on customer orders, sales forecasts, and strategic inventory plans. This forecast is shared with suppliers, who reserve capacity and commit delivery dates. Subsidiaries receive the forecast and use it to plan local inventory levels.

Throughout the month, as local demand emerges, subsidiaries create purchase requisitions against the forecast. Most requisitions are fulfilled by suppliers already committed to the forecast, so delivery is reliable and pricing is fixed. If a subsidiary needs something urgently or off-forecast, they can source locally at local prices. At month end, all purchase orders consolidate at headquarters for spend analysis. Headquarters identifies trends: which regions are spending more, which suppliers are overutilized, where quality issues are emerging. Supplier rebates are calculated at the group level based on consolidated volume, then distributed back to subsidiaries.

3.6 Integration and Setup

Two-tier ERP requires robust integration between Tier 1 and Tier 2 as well as between subsidiaries. The integration architecture determines the data flow, synchronization frequency, and system health.

A typical two-tier system (see Figure 3.8) leverages SAP Business Technology Platform (SAP BTP) as the integration middleware. Tier 1 runs SAP S/4HANA Cloud Private Edition or on-premise SAP S/4HANA. Tier 2 systems run SAP S/4HANA Cloud Public Edition or SAP Cloud ERP. SAP BTP sits in the middle, orchestrating data flows between the tiers.

Master data flows from Tier 1 to Tier 2 using Data Replication Framework (DRF) or APIs. Product catalogs, customer lists, supplier lists, charts of accounts, profit centers, and cost centers all originate in Tier 1 and replicate automatically to Tier 2. Sales orders flow from Tier 2 to Tier 1 for group visibility. Financial data flows from Tier 2 to Tier 1 for consolidation. Planning data flows from SAP Analytics Cloud back to Tier 2 instances. Cash forecasts flow from Tier 2 to Tier 1.

Scope items provide the foundation: 1SG (Financial Consolidation) handles the consolidation logic, 1RO (Master Data Integration) orchestrates master data replication, J78 (Cash Operations) enables treasury workflows, and 2EJ (Integration of Sales) enables sales order visibility.

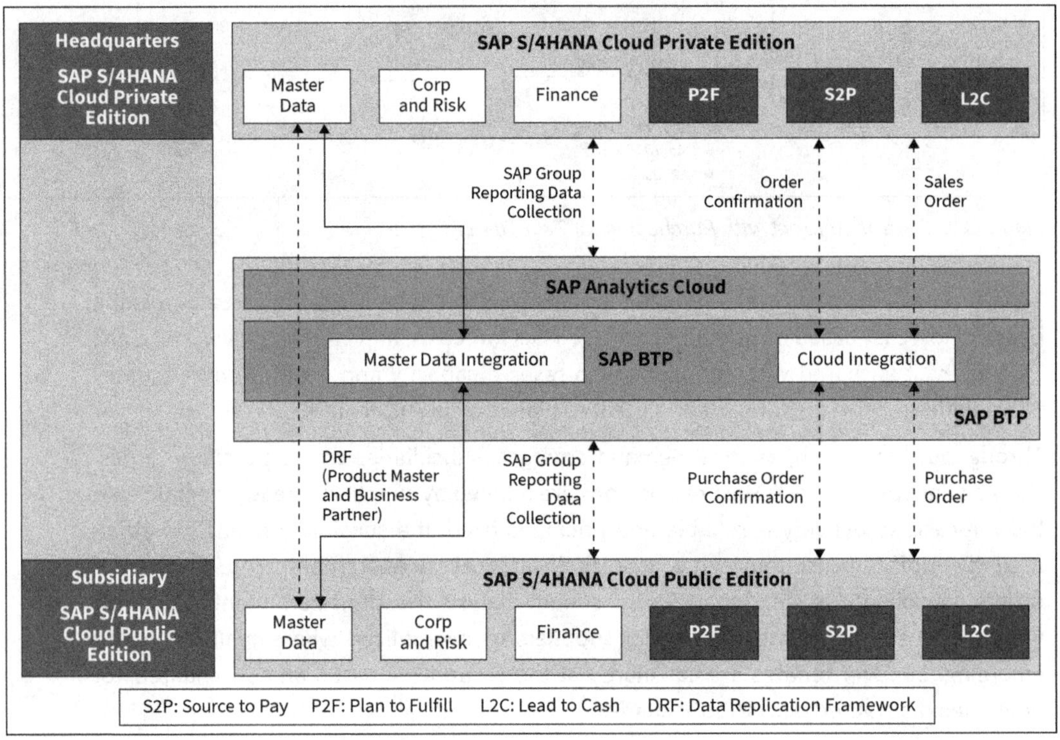

Figure 3.8: Reference Architecture for Two-Tier ERP

Master data, the reference information that underpins all transactions, must be clean, consistent, and governed centrally. When two-tier ERP setups fail, it's often because master data is misaligned between tiers. Fortunately, modern cloud ERP platforms and middleware solutions make master data synchronization straightforward if you follow certain principles.

Tier 1 is the system of record for all master data. Profit centers, cost centers, general ledger charts of accounts, asset masters, bank masters, customer lists, supplier lists, and product catalogs originate in Tier 1. Tier 2 systems maintain local copies of this master data, which they reference for all transactions. When headquarters needs to add a new profit center, update a chart of accounts, or define a new product, the change is made once in Tier 1 and automatically replicates to all Tier 2 systems.

SAP provides predelivered *integration patterns* specifically for two-tier landscapes, designed to simplify implementations and reduce integration costs. Master data—suppliers, product BOMs, customers, pricing—replicates from Tier 1 to Tier 2 via DRF. Sales transactions—customer orders, confirmations—flow between tiers so that Tier 1 has visibility into all subsidiary sales. Financial plan data, actual data for planning, and finance postings flow between tiers. Cash flow data and cash visibility flow from Tier 2 to Tier 1. Procurement transactions such as purchase orders and confirmation flow between tiers.

Figure 3.9 shows the integration patterns that enable different use cases in a two-tier landscape. It highlights how specific integrations support different functional areas. For consolidation scenarios, integrations connect Central Finance (Tier 1) with Tier 2 systems for financial accounting posting, country-specific payments, and financial consolidation reporting. For order-to-cash processes, integrations flow sales orders, customer confirmations, and customer invoices between tiers so that Tier 1 has full visibility into subsidiary sales and revenue. For procurement, purchase orders and supplier invoices flow between tiers, enabling centralized supplier management. For project management and finance, customer project status and project work breakdown structure data flow from Tier 2 to Tier 1 for consolidated project visibility and financial tracking. In addition, Tier 2 (SAP S/4HANA Cloud Public Edition) can integrate with Central Finance for external customer payment processing.

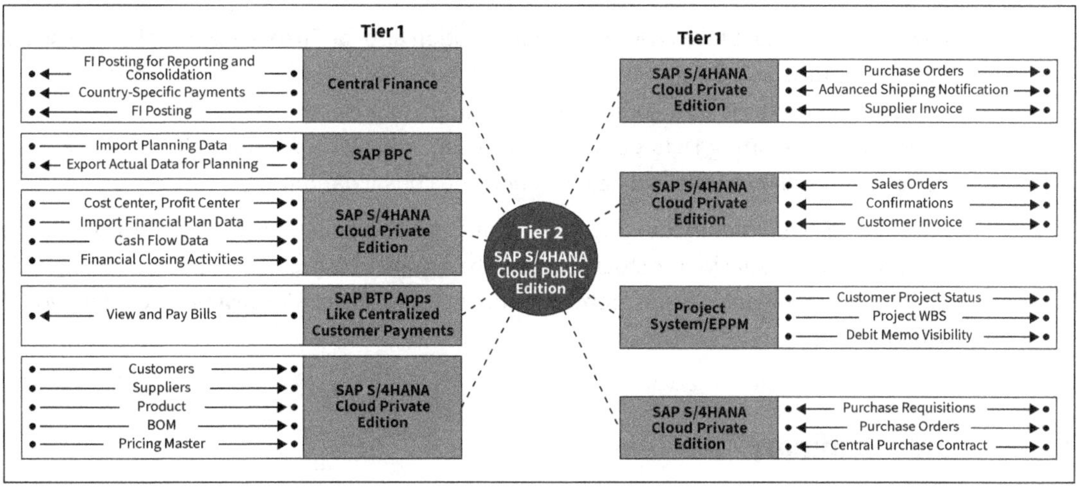

Figure 3.9: Integration Scenarios

Organizations choose their integration technology based on landscape topology: direct system-to-system integration for simple two-tier landscapes, integration via SAP BTP for cloud-centric landscapes, Central Finance for consolidation scenarios, or SAP Business Planning and Consolidation (SAP BPC) for advanced planning needs. The key is that these are pretested, certified patterns. Organizations do not build integration from scratch; they select from among proven templates.

3.7 Accelerators

SAP provides prebuilt accelerators and scope items designed to reduce implementation time and cost for two-tier ERP deployments. These are templates, workflows, and data models validated across many customer implementations.

Accelerators are preconfigured business process templates, data models, and integration patterns specific to two-tier ERP. Rather than designing each process from scratch, organizations can leverage SAP Best Practices that have been tested across dozens of customer implementations. A master data integration accelerator includes prebuilt data validation rules, reconciliation workflows, and error-handling procedures. A financial consolidation accelerator includes prebuilt general ledger posting logic, intercompany elimination procedures, and currency conversion frameworks. Organizations adopt these templates, customize them for their specific requirements, and deploy them.

Master Data Integration (scope item 1RO) includes preconfigured DRF replication for suppliers, customers, and products. This accelerator reduces data integration effort significantly by providing templates for data validation, transformation, and error handling. Financial Consolidation (1SG) includes prebuilt consolidation workflows with general ledger posting, intercompany elimination, and currency conversion built in. Rather than custom-coding consolidation logic, organizations use the prebuilt templates. This substantially accelerates consolidation processes.

Advanced Financial Close (4HG) is an orchestration app for financial closing tasks. The app automates task scheduling, monitoring, and audit trails. Rather than a finance manager managing close tasks in a spreadsheet, the app ensures consistency and provides audit trails. Group Reporting Data Collection (1SG, 287, 4R7) includes prebuilt forms and workflows for subsidiary financial data collection and consolidation.

Subsidiary Sales Order Processing (BD9) includes a preconfigured sales order workflow for local sales from subsidiary stock. Rather than configuring sales order processing from scratch, organizations use the prebuilt template that reflects SAP best practices for two-tier sales processes.

3.8 Summary

Two-tier ERP is not a temporary compromise or a band-aid solution. For large, geographically distributed enterprises, it is the strategic operating model that enables both standardization and agility. The headquarters maintains financial control, compliance, and strategic direction, while subsidiaries operate with the autonomy and agility they need to compete in local markets.

The financial benefits are substantial: faster financial close, procurement savings, and working capital improvement. But perhaps more important are the strategic benefits: the ability to scale globally without growing bureaucracy, to move fast without losing control, and to support innovation while maintaining compliance.

As enterprises navigate increasingly complex regulatory environments, volatile supply chains, and accelerating digital transformation, the two-tier model's flexibility and resilience become more valuable. Organizations that implement it well—with clean data, strong governance, and executive commitment—gain a competitive advantage.

The journey to two-tier ERP is substantial, requiring investment, discipline, and executive commitment. Yet organizations across fast-moving consumer goods, consulting, retail, and manufacturing industries have proven that the payoff is worth it. If your organization is struggling with the tension between global control and local autonomy, two-tier ERP may be your solution.

SAP S/4HANA Cloud provides the foundation, but Autonomous Enterprise requires specialized applications. The next chapter shows how line-of-business applications extend the core and how integration makes them work together as one operation.

Chapter 4
LoB Apps and Extended Capabilities

Line-of-business (LoB) applications extend the value of SAP Cloud ERP into specialized domains that require capabilities beyond what the digital core was designed to provide, which we covered in Chapter 2. Certain functions demand a depth that a core ERP system cannot deliver without becoming unwieldy, necessitating the applications that we'll discuss in this chapter.

SAP Cloud ERP provides the operational heartbeat, and the transactional core handles what most organizations need most of the time. However, certain functions demand depth that no core system can deliver without becoming something else entirely: planning a supply network across dozens of plants and thousands of SKUs simultaneously, executing production at the machine level with real-time equipment integration, orchestrating a high-volume distribution center with automated material handling, or predicting asset failures before they stop a production line. This chapter covers the LoB applications that extend SAP S/4HANA Cloud into that territory. These are not bolt-on tools; they are the extended operational layer of SAP Autonomous Suite, connected by shared context, governed by a single architecture, and designed to deepen the flywheel rather than sit alongside it.

LoB applications complement the ERP backbone rather than compete with it. Each is purpose-built for its domain, maintained on its own release cycle, and connected to SAP S/4HANA Cloud through native integration that shares context rather than merely exchanging data. When an SAP Integrated Business Planning for Supply Chain (SAP IBP) demand plan adjusts a forecast, SAP S/4HANA Cloud's material requirements planning (MRP) runs against the updated signal the same day. When a quality event triggers on the shop floor in SAP Digital Manufacturing, SAP S/4HANA Cloud's production order reflects the hold immediately. The LoB applications extend the ERP system's reach into territory where the transactional core alone cannot operate with sufficient precision.

This chapter covers the four major LoB areas: supply chain, finance and spend, customer experience, and human capital management (HCM). We begin with supply chain because it is where the extension model is the most visible and the operational stakes the highest.

4.1 Supply Chain

In Chapter 1, we described the Autonomous Supply Chain Management (SCM) as one of five domains in which agents, applications, and data operate as one. In Chapter 2, we covered what the ERP core provides: MRP, production order management, embedded warehouse management, and basic transportation planning. Those capabilities handle steady-state operations effectively, but the modern supply chain does not operate in steady state. Demand is volatile. Supplier networks are global and fragile. Customer expectations for delivery speed and visibility have compressed from weeks to days. The planning horizon that matters is not the next MRP run but the next quarter, modeled across 15 plants, 400 suppliers, and 12,000 SKUs simultaneously.

SAP addresses this gap through five LoB applications that extend the ERP core into advanced supply chain territory: SAP IBP for cross-functional planning and sales and operations planning (S&OP), SAP Digital Manufacturing for shop floor execution and real-time production intelligence, SAP Logistics Management for advanced warehouse and transportation operations at scale, SAP Asset Performance Management for predictive maintenance and asset strategy optimization, and SAP Supply Chain Orchestration for end-to-end visibility and exception management across the extended supply chain.

Each application shares the same business data foundation as SAP S/4HANA Cloud and connects through native integration maintained by SAP. The result is a supply chain architecture in which strategic planning, operational execution, asset reliability, and real-time monitoring operate on the same data without the latency, reconciliation effort, and context loss that characterize bolted-on, third-party solutions.

We'll describe each of these LoB applications in more detail in the following sections, but first, let's take a closer look at the underlying strategic challenge of the supply chain.

4.1.1 The Strategic Challenge

Consider James at Korindal, the chief supply chain officer we introduced as part of Case Study 2 in Chapter 1, Section 1.8.2. His on-time, in-full (OTIF) delivery sits at 86 percent, generating $30 million per year in penalties and eroding customer confidence. He runs MRP faithfully. His planners work long hours. The ERP core is not broken. What is broken is the gap between operational execution and strategic visibility. By the time his team discovers a demand spike in the Rotterdam region, the production schedule has already locked, the components are in transit from the wrong supplier, and the logistics team is scrambling to find capacity on lanes that were allocated weeks ago. The problem is not that any single function fails but that no single system can see the whole picture fast enough to act.

Supply chain management in SAP Cloud ERP is not a single value stream. It is the connective tissue that runs through almost every process the enterprise operates. Figure 4.1 illustrates this architecture, in which supply chain sits at the center of five integrated value streams:

- *Idea-to-market* brings new products from concept through engineering into production readiness. Source-to-pay secures the materials and services that manufacturing consumes.

- *Plan-to-fulfill* orchestrates production planning, manufacturing execution, quality, warehousing, and transportation.
- *Lead-to-cash* fulfills customer commitments through order management and service delivery.
- *Acquire-to-decommission* manages the physical assets (production lines, warehouse automation, fleet) that the supply chain depends on.
- *Finance* wraps around all the others, with financial planning and analysis at the top and record-to-report at the bottom, ensuring that every physical movement has a financial mirror.

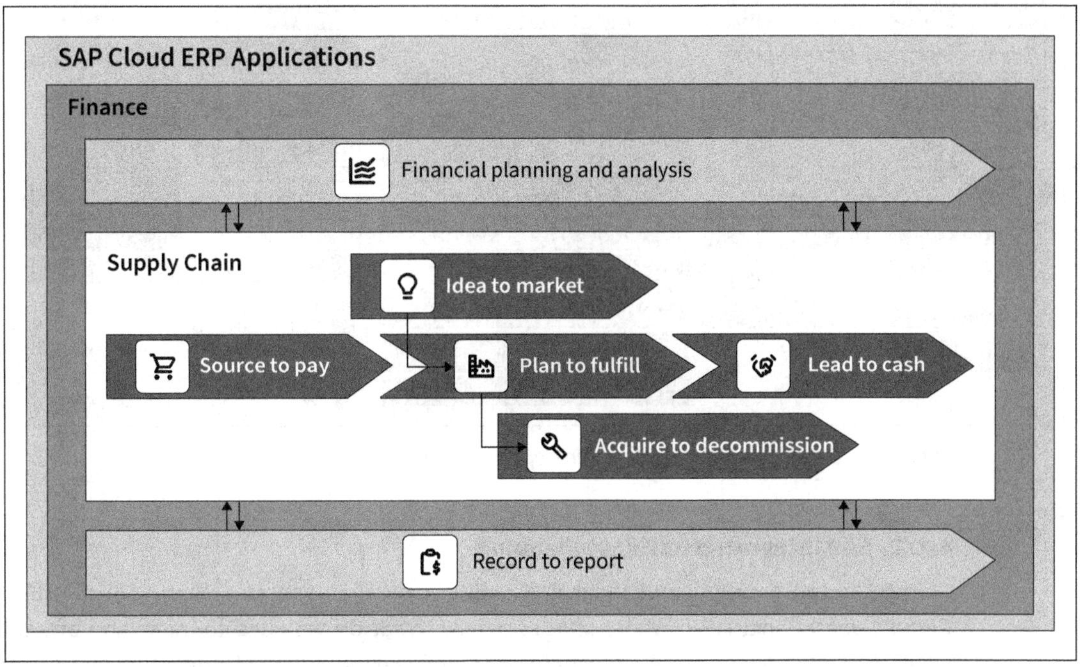

Figure 4.1: Supply Chain Management with SAP Cloud ERP

This cross-process position is precisely why supply chain requires the deepest set of LoB extensions. The ERP core provides the transactional backbone for each value stream, but the specialized depth needed at scale (multiechelon demand planning across 12,000 SKUs, real-time shop floor execution integrated with Internet of Things (IoT) sensors, autonomous warehouse operations processing 50,000 picks per shift, or end-to-end visibility across 400 suppliers) requires purpose-built applications that extend beyond what any single ERP module can deliver.

Figure 4.2 shows how SAP packages these capabilities. The base layer (SAP Finance and Supply Chain operations) provides the core transactional capabilities we covered in Chapter 2: MRP, production order management, basic warehouse and transportation management, quality management, and inventory control. SAP Supply Chain Base and SAP Supply

Chain Premium add progressively deeper functionality, including advanced available-to-promise (ATP), manufacturing operations management, and logistics material identification. Beyond these tiers, dedicated LoB applications (shown as add-ons) deliver the specialized depth that distinguishes operational adequacy from competitive advantage: SAP IBP for demand and supply planning, SAP Digital Manufacturing for shop floor execution, and SAP Asset Performance Management for predictive asset strategies.

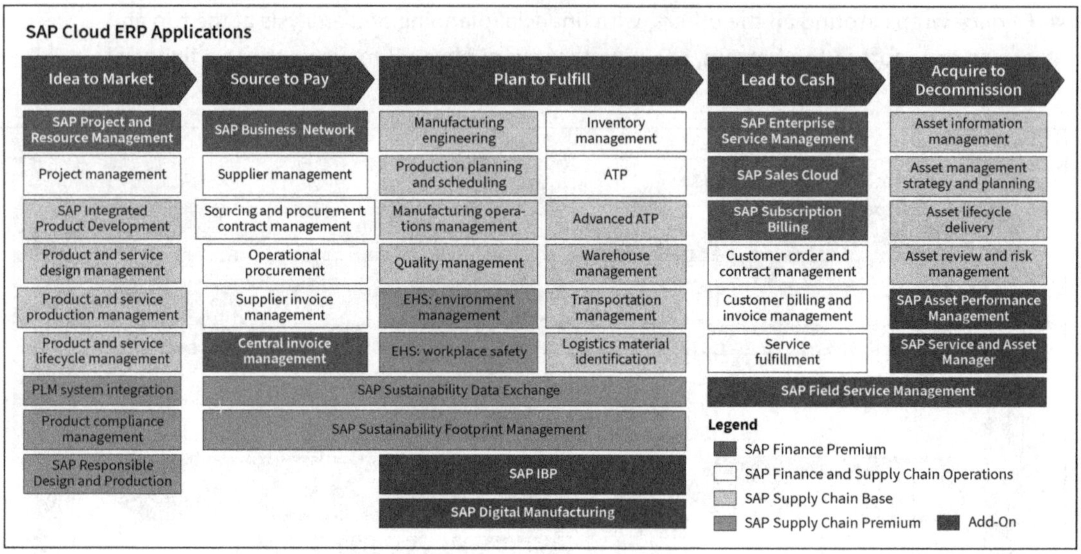

Figure 4.2: SAP Cloud ERP Supply Chain Capability Map

4.1.2 SAP Integrated Business Planning

SAP IBP is the strategic and tactical planning layer that operates above SAP S/4HANA Cloud's operational MRP. Where MRP answers, "What do we need to make and buy this week?," SAP IBP answers, "What should our supply network look like over the next 6 to 18 months, and how do we align demand, supply, inventory, and financial plans into a single executable consensus?"

The fundamental problem SAP IBP solves is fragmentation. In most organizations, demand planning lives in one tool, supply planning in another, inventory optimization in a spreadsheet, and S&OP in a PowerPoint deck reviewed monthly. The plans do not reconcile. When the demand planner adjusts the forecast, the supply planner does not see it until the next cycle. When the finance team publishes revenue targets, nobody verifies whether the supply chain can physically deliver the volume implied by those targets. SAP IBP eliminates this fragmentation by running all planning processes on a single data model with a unified time series.

Figure 4.3 shows how SAP IBP's planning processes connect to SAP S/4HANA Cloud's operational execution layer, with master data flowing outbound to keep planning grounded in current reality and planned orders and purchase requisitions flowing back for execution.

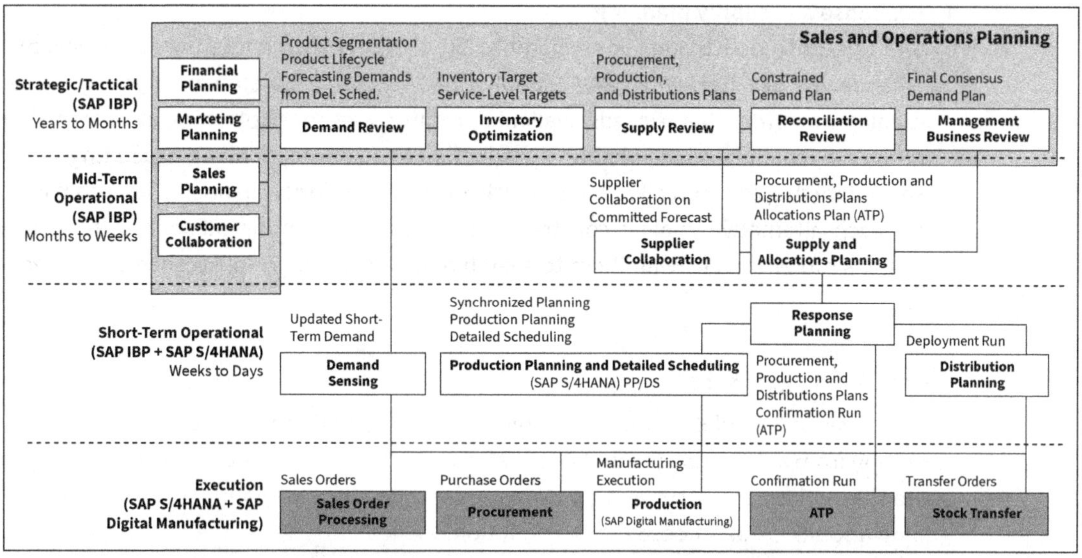

Figure 4.3: Supply Chain Planning Integration into SAP S/4HANA

SAP IBP consists of a subset of solutions that offer various functionalities, as follows:

- **Demand planning and sensing**
  SAP IBP for demand provides statistical forecasting, consensus demand planning, and short-term demand sensing. Statistical algorithms generate baselines from historical shipments. Demand planners layer in market intelligence, promotional uplift, and customer commitments. ML-based demand sensing incorporates external signals (weather, economic indicators, point-of-sale data, social sentiment) to improve near-term accuracy by 20 to 30 percent over pure statistical methods. The demand plan flows directly into supply planning without export, import, or manual reconciliation.
- **Supply planning and inventory optimization**
  SAP IBP for supply balances demand against constrained supply across the full network: multiple plants, distribution centers, contract manufacturers, and suppliers. The optimizer evaluates trade-offs between service-level targets, production capacity, transportation costs, and inventory carrying costs simultaneously, producing a feasible supply plan that respects real constraints rather than assuming infinite capacity.

  SAP IBP for inventory optimization operates across multiple echelons simultaneously, determining optimal safety stock levels at each node in the network given service-level targets, demand variability, and supply lead time variability. This is mathematically distinct from the single-location safety calculations in SAP S/4HANA Cloud's MRP. Multiechelon optimization typically reduces total network inventory by 10 to 20 percent while maintaining or improving service levels because it accounts for the statistical pooling effects across locations that single-location methods miss.

- **Response and supply planning**
  SAP IBP for response and supply handles the tactical execution of the consensus plan when reality deviates. When a key supplier declares force majeure or when a customer escalates an order that exceeds available inventory, response planning evaluates alternatives in real time, such as shifting production to an alternate plant, expediting from a secondary supplier, or reallocating stock from a lower-priority channel. The system models each alternative against constraints and costs, presenting planners with ranked options rather than leaving them to search manually through spreadsheets and phone calls.

Case Study Example

For James at Korindal, response planning is the capability that converts a supply disruption from a three-day fire drill into a 30-minute decision. The system already knows the alternatives, the costs, and the downstream impacts. The planner's job shifts from "find a solution" to "choose the best solution."

- **Sales and operations planning**
  SAP IBP for sales and operations planning provides a cross-functional collaboration platform via which demand, supply, finance, and executive leadership align in a single operating plan. Scenario comparison enables decision-makers to evaluate alternatives: "If we accept this large customer order, what is the capacity impact, and does the margin justify the disruption?" Financial integration translates operational scenarios into P&L and working capital impacts, enabling true demand/supply/financial alignment rather than sequential handoffs between functions.
  The S&OP process in SAP IBP typically runs on a monthly cadence with weekly refinement cycles. Each cycle produces an updated consensus plan that propagates down to operational execution. The discipline is not new; what has changed is the technology behind it. Previous S&OP tools forced organizations to choose between granularity and speed, but SAP IBP provides both: You can plan at the SKU-location level across thousands of combinations, run scenarios in minutes rather than days, and publish the results directly to the execution layer without manual translation.

The integration between SAP IBP and SAP S/4HANA Cloud operates bidirectionally. Master data (materials, plants, BOMs, routings, suppliers) flows from SAP S/4HANA Cloud into SAP IBP to ensure planning operates against current reality. Planned orders and purchase requisitions generated by SAP IBP flow back into SAP S/4HANA Cloud for operational execution. The SAP S/4HANA Cloud Supply Chain Integration add-on manages this exchange, synchronizing planning-relevant data on configurable schedules while supporting real-time, event-based updates for critical signals.

This means, for example, that a demand-sensing adjustment in SAP IBP at 8:00 a.m. can trigger an adjusted MRP run in SAP S/4HANA Cloud by 8:15 a.m. The planning decision and the operational execution happen in the same morning rather than the same week. For organizations managing volatile demand in short product lifecycle industries, this responsiveness is the difference between meeting customer commitments and generating exceptions that cascade through the network.

4.1.3 SAP Digital Manufacturing

SAP Digital Manufacturing extends SAP S/4HANA Cloud's production order management to the shop floor level, providing the execution layer where planning meets physical reality. Where SAP S/4HANA Cloud manages production orders, bills of materials, routings, and confirmations at the enterprise level, SAP Digital Manufacturing operates at the machine level: individual operations, operator instructions, real-time production data capture, and integration with manufacturing equipment through IoT connectivity.

The gap it addresses is specific. For example, say that SAP S/4HANA Cloud knows that Production Order 48721 is scheduled for Line 3 on Thursday. SAP Digital Manufacturing knows that Line 3's CNC machine is currently running at 94 percent of the target cycle time, that the operator on Shift B has completed training on the new assembly sequence, that the quality check at Operation 30 passed with values within two sigma of the specification center, and that tool wear on the milling head will require replacement in approximately 180 cycles. This shop-floor-level granularity is what enables the AI-driven manufacturing intelligence that we'll describe in Chapter 5.

Figure 4.4 shows the SAP Digital Manufacturing architecture: the Production Operator Dashboard, the machine integration layer that ingests real-time equipment data, and the bidirectional connection to SAP S/4HANA Cloud production orders.

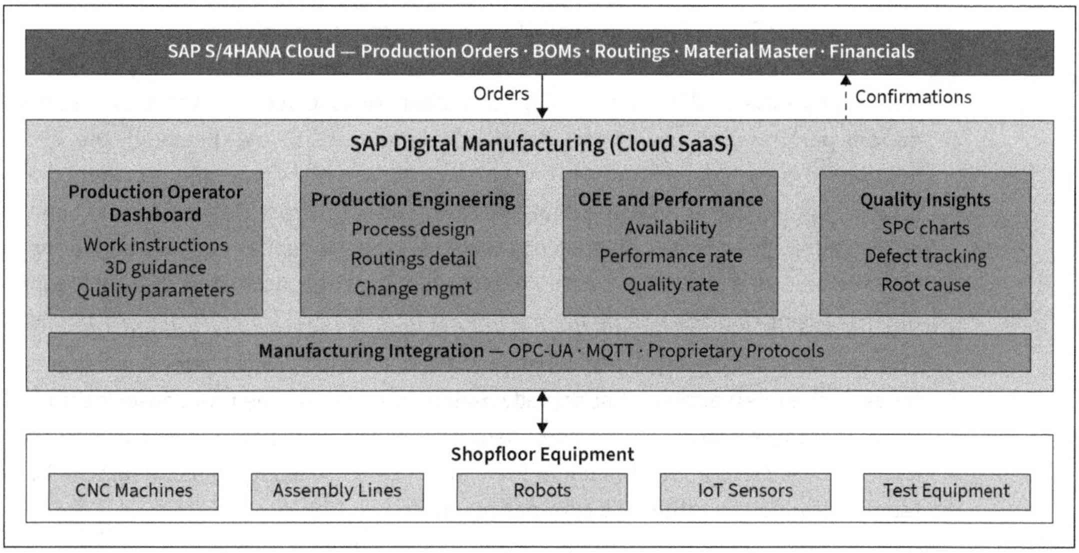

Figure 4.4: SAP Digital Manufacturing Architecture

Let's take a closer look at these key features:

- **Production execution and operator guidance**
  The production operator dashboard provides a configurable interface for shop floor operators, displaying work instructions, quality parameters, and real-time process data in context. The 3D work instructions guide operators through complex assembly sequences, reducing errors and accelerating onboarding for new personnel. SAP Digital Manufacturing supports paperless manufacturing with electronic records that satisfy good practice (GxP) requirements in regulated industries like life sciences and aerospace.
- **Machine integration and data collection**
  Manufacturing integration connects production equipment to the digital layer through standard protocols (OPC-UA, MQTT) and proprietary interfaces. Real-time machine data (cycle times, temperatures, pressures, vibration, tool wear) flows into the system continuously, creating the operational data foundation that predictive quality, predictive maintenance, and overall equipment effectiveness (OEE) optimization require. Without this data pipeline, AI-driven manufacturing intelligence has nothing to reason over.
- **Production engineering and process design**
  SAP Digital Manufacturing includes capabilities for managing production processes, routings, and work instructions at a level of detail beyond what SAP S/4HANA Cloud's routing master supports. Process engineers define detailed operation sequences with embedded quality specifications, setup parameters, and tooling requirements. Changes propagate to the shop floor in real time, eliminating the version control problems that plague paper-based or disconnected manufacturing execution system (MES) environments.
- **Overall equipment effectiveness (OEE) and performance analytics**
  OEE monitoring provides real-time visibility into equipment availability, performance, and quality rates at the machine, line, and plant levels. Operators and supervisors see current performance against targets. When a line drops below threshold, the system identifies the loss category (downtime, speed loss, or quality loss) and provides context for investigation. Over time, the analytics reveal patterns: specific products that consistently cause speed losses on specific machines, shift-dependent quality variations, or maintenance intervals that are either too frequent (wasting productive time) or too infrequent (causing unplanned stops).

  This operational intelligence is what enables continuous improvement programs to focus on the interventions that actually matter, discovering the root causes behind the losses, not the averages that obscure them. A team that knows Line 3 loses 4.2 percent of its available time to a specific changeover sequence on Tuesday nights can act on that. A team reviewing a monthly OEE summary cannot.

SAP Digital Manufacturing receives production orders from SAP S/4HANA Cloud and returns confirmations, goods receipts, and quality results. The integration is event-based: When a production order is released in SAP S/4HANA Cloud, it appears on the shop floor immediately. When an operator confirms an operation in SAP Digital Manufacturing, SAP S/4HANA Cloud updates the order status, posts material consumption, and triggers downstream logistics processes without batch delay. This tight coupling ensures that the enterprise view and the shop floor view remain synchronized in real time.

For organizations running multiple plants, SAP Digital Manufacturing provides a consistent execution layer across sites regardless of local equipment differences. A new product introduction designed and validated at one plant can be deployed to others through the same digital process definitions, reducing the time and risk associated with multisite rollouts.

4.1.4 SAP Logistics Management

SAP Logistics Management is a cloud-native software-as-a-service (SaaS) solution that serves as the advanced logistics execution platform for organizations whose warehouse and transportation requirements exceed what SAP S/4HANA Cloud's embedded capabilities are designed to handle. SAP S/4HANA Cloud includes integrated warehouse management and basic transportation management sufficient for most standard operations. Organizations running high-volume distribution centers processing tens of thousands of picks per shift, multisite warehouse networks with centralized orchestration, or complex global transportation operations across dozens of carriers need the depth that a dedicated logistics platform provides.

Figure 4.5 shows the SAP Logistics Management architecture, illustrating how wave management, labor optimization, and automation integration connect to SAP S/4HANA Cloud for inventory control and order management.

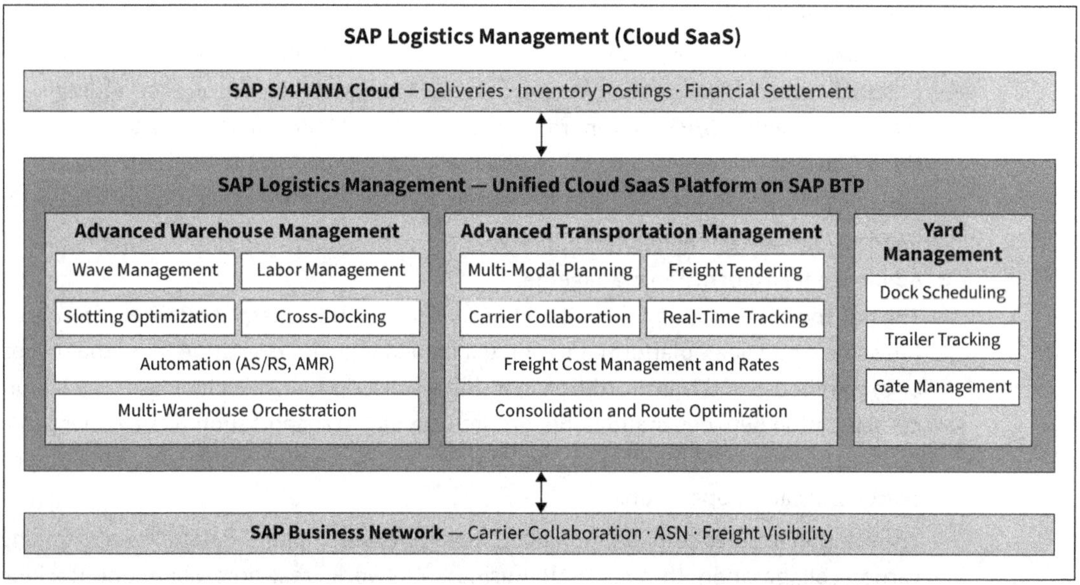

Figure 4.5: SAP Logistics Management Architecture

The SaaS delivery model is an architectural choice, not a packaging option. Unlike the legacy on-premises SAP Extended Warehouse Management (SAP EWM) offering, which required dedicated hardware per site and significant basis administration, this cloud-native platform delivers continuous innovation through quarterly updates; scales elastically during peak periods like Black Friday, seasonal spikes, and promotional events; and aligns cost with actual warehouse throughput rather than provisioned capacity. For multisite operations, a single tenant administers multiple warehouses from a unified layer, with site-specific configurations for processes, layouts, and automation profiles. The version fragmentation that characterized on-premise SAP EWM deployments, with each warehouse on a different patch level and each site carrying its own custom code, is eliminated by design. SAP manages the infrastructure, security patching, and feature delivery pipeline, while the organization manages the processes.

Let's take a closer look at the core capabilities provided by SAP Logistics Management:

- **Advanced warehouse management**
  SAP Logistics Management provides capabilities for complex warehouse operations: wave management for coordinating large-scale picking campaigns, labor management for workforce productivity tracking and optimization, slotting optimization for dynamic product placement, yard management for dock scheduling and trailer tracking, and cross-docking for flow-through operations that bypass storage entirely.

  The system supports multiwarehouse architectures in which a central planning layer orchestrates fulfillment across multiple physical sites based on inventory availability, proximity to customer, and transportation cost. For e-commerce and omnichannel operations, this enables order-level routing decisions: The system can ship from the warehouse closest to the customer or consolidate from two locations if no single site has complete inventory.

  Automation integration connects the warehouse management system to automated material handling equipment: conveyors, sortation systems, automated storage and retrieval systems (AS/RS), autonomous mobile robots (AMRs), and goods-to-person stations. The system orchestrates these physical assets as part of the fulfillment workflow, assigning work to human operators or automated systems based on the task characteristics and current workload distribution.

- **Advanced transportation management**
  The transportation capabilities within SAP Logistics Management are delivered as part of the same cloud SaaS platform, providing a unified logistics execution layer that covers both warehouse and transportation in a single subscription. This eliminates the historical separation between standalone, on-premise SAP Transportation Management (SAP TM) and SAP EWM, enabling seamless coordination between warehouse outbound processes and carrier operations.

  The solution handles complex global logistics scenarios: multimodal shipment planning, carrier collaboration through SAP Business Network, real-time shipment tracking, freight cost management with complex rate structures, and transportation analytics that identify optimization opportunities across the network.

Freight tendering automates the carrier selection process, distributing shipment requirements to contracted carriers based on lane assignments, rates, and service-level agreements. Carriers respond through the network, and the system evaluates responses against cost, transit time, and reliability criteria. For organizations managing thousands of shipments daily across dozens of carriers, this automation eliminates the manual coordination that consumes logistics team capacity.

Transportation planning optimization evaluates consolidation opportunities across shipments, balancing the trade-off between transportation cost (consolidation reduces cost per unit) and delivery speed (consolidation may delay individual shipments). The optimizer considers vehicle capacity, weight limits, delivery windows, and carrier availability simultaneously.

- **Yard management**
  SAP Logistics Management includes yard management capabilities that provide visibility and control over trailer movements, dock scheduling, and gate operations. Dock scheduling optimizes inbound and outbound appointment windows to reduce truck wait times and balance labor at receiving and shipping docks. Trailer tracking monitors the location and status of trailers across the yard, reducing detention charges and improving asset utilization. Gate management automates check-in and check-out processes, integrating with transportation planning to ensure timely loading and departure. Together, these capabilities connect the warehouse to the carrier network and extend the operational visibility of SAP Logistics Management from inside the four walls to the full logistics execution boundary.

SAP Logistics Management integrates with SAP S/4HANA Cloud through delivery documents, inventory postings, and financial settlement. Outbound deliveries created in SAP S/4HANA Cloud trigger warehouse execution in the logistics platform. Goods movements post back to SAP S/4HANA Cloud inventory. Freight charges accrue and settle through SAP S/4HANA Cloud's financial accounting. The integration ensures that the advanced logistics operations remain connected to the enterprise's financial and operational reality without requiring manual reconciliation between systems.

4.1.5 SAP Asset Performance Management

SAP Asset Performance Management extends the acquire-to-decommission value stream beyond the transactional asset lifecycle management provided in SAP S/4HANA Cloud. Where the ERP core handles asset master data, maintenance order processing, preventive maintenance scheduling, and financial depreciation, SAP Asset Performance Management focuses on the predictive and prescriptive strategies that maximize asset availability while minimizing total cost of ownership. The two systems are complementary by design: SAP S/4HANA Cloud executes the maintenance work, while SAP Asset Performance Management determines what maintenance work should be done and when.

Unplanned downtime on a production line costs orders of magnitude more than planned maintenance, and it cascades. A single critical equipment failure stops production; customer orders miss delivery windows; penalties accumulate. The logistics team scrambles to reschedule shipments that were built around a production run that never completed.

Case Study Example

For James at Korindal, every hour of unplanned downtime on his Rotterdam assembly line costs approximately $180,000 in lost throughput and downstream disruption. SAP Asset Performance Management exists to prevent those hours from occurring.

SAP Asset Performance Management consists of two main capabilities:

- **Predictive maintenance and condition monitoring**
  SAP Asset Performance Management ingests sensor data from production equipment (vibration, temperature, pressure, current draw, acoustic signatures) and applies machine learning models to predict failure before it occurs. The system establishes baseline operating profiles for each asset and detects deviations that indicate developing faults. For example, a bearing that shows increasing vibration amplitude at a specific frequency is likely three to six weeks from failure. SAP Asset Performance Management identifies this pattern and recommends maintenance intervention during the next planned window rather than waiting for the catastrophic failure that would halt production.

  The predictive models improve as they accumulate operational data specific to each asset in its actual context. A pump running in a tropical climate under heavy load degrades differently than the same model running in a temperate environment at moderate utilization. SAP Asset Performance Management accounts for these contextual differences rather than relying solely on manufacturer-recommended intervals built for average conditions. The longer the system runs on a given asset, the more precisely it can predict that specific asset's failure modes.
- **Asset strategy and risk management**
  Beyond individual asset prediction, SAP Asset Performance Management supports a portfolio-level asset strategy. Organizations define criticality classifications, risk tolerance levels, and maintenance investment priorities across their entire asset base. The system evaluates the health of each asset against its criticality and recommends where to allocate maintenance resources for maximum reliability improvement per dollar spent. Low-criticality assets with redundancy may run to failure by design. High-criticality, single-point-of-failure assets receive intensive monitoring and proactive intervention.

SAP Asset Performance Management integrates bidirectionally with SAP S/4HANA Cloud's plant maintenance functionality. Predicted failures generate maintenance notifications in SAP S/4HANA Cloud, which planners convert to maintenance orders scheduled within production windows. Completed maintenance activities in SAP S/4HANA Cloud feed back into SAP Asset Performance Management's models, confirming or refining the prediction accuracy. Integration with SAP Digital Manufacturing provides the real-time machine data that SAP Asset Performance Management's models consume, creating a closed loop: SAP Digital Manufacturing captures the operating data, SAP Asset Performance Management analyzes it for degradation signals, SAP S/4HANA Cloud executes the maintenance work, and the cycle repeats continuously.

4.1.6 SAP Supply Chain Orchestration

SAP Supply Chain Orchestration is a cloud-native SaaS solution that provides end-to-end visibility and exception management across the extended supply chain, spanning internal operations and external partners. Where SAP IBP plans the network, SAP Digital Manufacturing executes on the shop floor, and SAP Logistics Management handles physical movement, SAP Supply Chain Orchestration monitors the entire chain in motion and intervenes when reality deviates from the plan. It is a system that sees everything simultaneously, which no single application can do independently.

The platform's cloud-native design matters for the visibility use case specifically. Supply chain monitoring requires ingesting data from heterogeneous sources—including both SAP systems and third-party ERP instances, logistics providers, IoT devices, and external data feeds—and processing millions of supply chain signals daily across global operations. Delivered as a SaaS offering on SAP BTP, SAP Supply Chain Orchestration receives continuous updates, scales without infrastructure provisioning, and connects new data sources through a growing library of integration content. Organizations subscribe based on the scope of their monitored supply chain—number of nodes, lanes, and trading partners—rather than licensing capacity they may not use consistently.

Figure 4.6 shows the SAP Supply Chain Orchestration capability, illustrating how end-to-end visibility from the supplier through production to customer delivery is combined with exception alerts and downstream impact assessment.

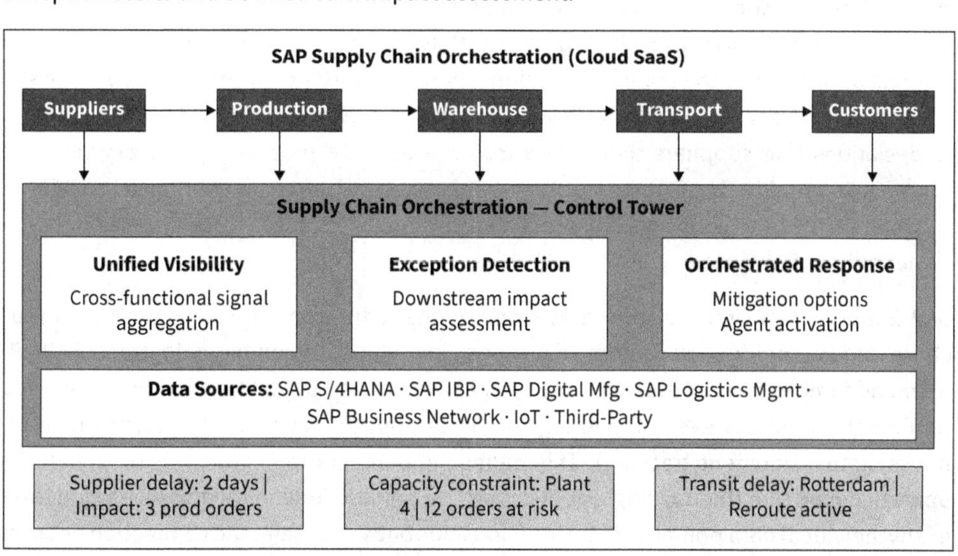

Figure 4.6: SAP Supply Chain Orchestration

Let's take a closer look at the key capabilities offered by SAP Supply Chain Orchestration:

- **End-to-end visibility**
  The platform aggregates signals from across the supply chain into a unified visibility layer: purchase order status, supplier shipments in transit, production progress, warehouse throughput, and outbound delivery tracking. This cross-functional visibility enables detection of problems that are invisible within any single system. A supplier shipment delayed by two days affects procurement, but it also affects the production schedule that depends on those materials, the customer orders that depend on that production, and the financial commitments made against those customer orders. SAP Supply Chain Orchestration connects these dependencies.
- **Exception management and orchestrated response**
  When deviations are detected, the system assesses the downstream impact across the chain and recommends response actions. A supplier delay triggers an impact analysis: which production orders are affected, which customer commitments are at risk, what alternative sources exist, and what each mitigation option costs. This moves exception management from *reactive investigation*—that is, discovering the problem after it has propagated—to *proactive intervention*—that is, detecting the signal and acting before the downstream impact materializes. The difference is not incremental. It is the difference between managing a crisis and preventing one.

  For the Autonomous SCM domain, SAP Supply Chain Orchestration provides the situational awareness that agents need to coordinate cross-functional responses. When the system detects a supply constraint, it can activate sourcing agents to identify alternatives, planning agents to adjust production schedules, and logistics agents to reroute shipments, all within the orchestrated response that Chapter 1 described as the operating model of the Autonomous Enterprise.
- **Supplier collaboration and network integration**
  Through integration with SAP Business Network, SAP Supply Chain Orchestration extends visibility into supplier operations. Purchase order confirmations, advance ship notices, and quality certificates flow through the network, providing early warning of deviations that suppliers themselves may not escalate proactively. For organizations managing hundreds of suppliers across global networks, this network-level visibility is what enables the continuous sensing that distinguishes autonomous operations from periodic review cycles.

SAP Supply Chain Orchestration draws operational data from SAP S/4HANA Cloud (purchase orders, production orders, deliveries, inventory), planning data from SAP IBP (demand forecasts, supply plans), execution data from SAP Digital Manufacturing (production progress, quality events), and logistics data from SAP Logistics Management (warehouse status, shipment tracking). This multisource integration creates a comprehensive operational picture that no single application can provide independently. The value is not in any individual data point but in the relationships between them: the connection between a supplier delay, a production schedule impact, and a customer commitment at risk.

4.1.7 The Integration Model

Figure 4.7 shows how all five supply chain LoB applications (SAP IBP, SAP Digital Manufacturing, SAP Logistics Management, SAP Asset Performance Management, and SAP Supply Chain Orchestration) connect to SAP S/4HANA Cloud and to each other through shared data and event-based integration.

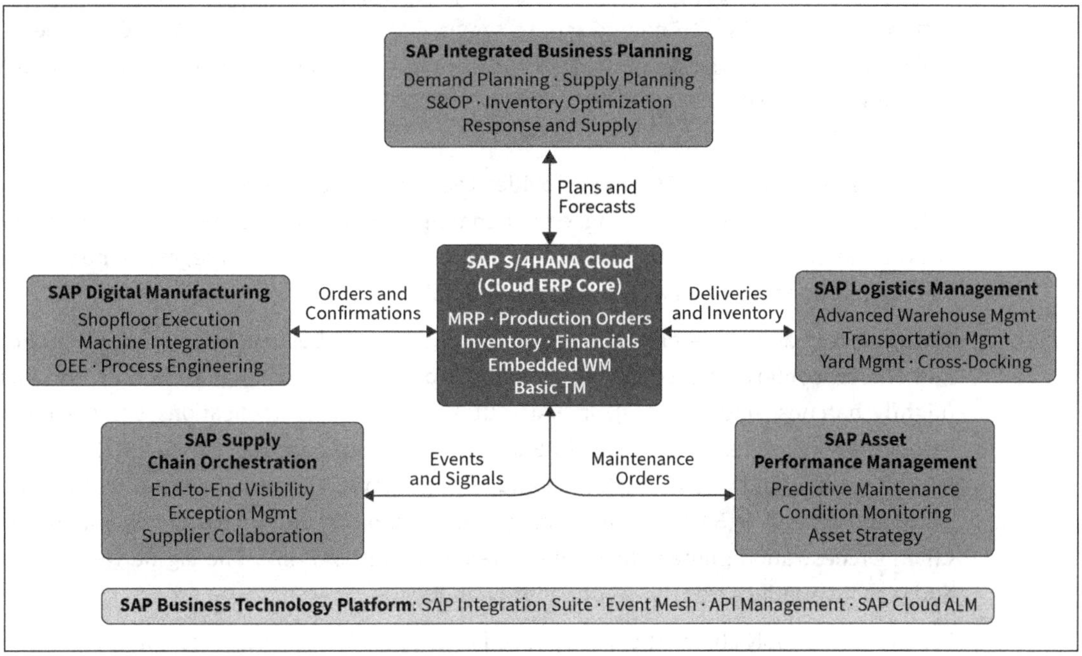

Figure 4.7: Supply Chain LoB Integration Summary

Across all five supply chain LoB applications, a consistent integration architecture applies. SAP maintains the integration content: prebuilt integration flows, data mappings, and event definitions that connect each application to SAP S/4HANA Cloud and to each other. Organizations activate these integrations rather than building them. When SAP updates an application, the integration adapts alongside it, thus eliminating the maintenance burden that defines custom-built connections, in which every upgrade breaks something that has to be rebuilt.

Case Study Example

For David at Korindal, this prebuilt integration model is the answer to the question his CEO asked: Why does every change feel harder than it should? With SAP-maintained integration content, his team activates connections rather than building them, and upgrades to any application no longer break the connections that depend on it.

Case Study Example

Likewise, the connection runs in both directions. For Maria at Korindal, the supply chain is not a separate domain. Every shipment James's team misses generates a dispute or a credit that lands in her receivables ledger. Every week of premium freight spend reduces the margin she is trying to report. When the supply chain LoB applications keep James's OTIF commitments, they are also protecting Maria's DSO number. The integration model described in this section is what makes that connection visible in time to act on it.

SAP Integration Suite on SAP BTP provides the runtime for these connections, handling data transformation, event routing, error handling, and monitoring. API Management governs access controls and rate limits, while SAP Cloud ALM monitors integration health and alerts operations teams when flows degrade or fail.

This is what Chapter 1 described as the practical difference between scheduled integration and shared context. In a conventional setup, applications exchange data on a cadence (nightly batches, hourly syncs). In SAP Autonomous Suite, applications share context through events: A demand change in SAP IBP immediately signals SAP S/4HANA Cloud's MRP; a quality hold in SAP Digital Manufacturing immediately blocks the corresponding outbound delivery in SAP Logistics Management; a supplier delay detected by SAP Supply Chain Orchestration immediately triggers replanning in SAP IBP. The lag between signal and response shrinks from hours or days to minutes.

For the chief supply chain officer introduced in Chapter 1, this integrated LoB architecture is what makes the OTIF improvement possible. No single application solves the 14 percent gap. The improvement comes from applications working together: SAP IBP catching the demand shift early, SAP S/4HANA Cloud adjusting the production schedule, SAP Digital Manufacturing executing the revised plan on the shop floor, SAP Asset Performance Management ensuring the equipment stays available to run the revised plan, SAP Logistics Management expediting the right shipments, and SAP Supply Chain Orchestration monitoring the entire chain to ensure the response is working. This is the flywheel in motion.

4.2 Finance and Spend

Finance and spend management in a modern enterprise are no longer limited to accounting and reconciliation. They operate as a connected system spanning financial control, procurement, employee spending, external workforce engagement, governance, and working capital management. SAP S/4HANA Cloud provides the financial foundation by organizing accounting, reporting, and controlling processes around a unified model, while LoB applications extend this model to manage how spend originates across the enterprise.

From this foundation, procurement governs how goods and services are sourced and acquired, travel and expense management captures decentralized employee spending, and external workforce solutions manage contingent labor and service-based delivery. Additional capabilities within the office of the CFO extend this into treasury, compliance, and working capital optimization. Figure 4.8 highlights these components and their relationships.

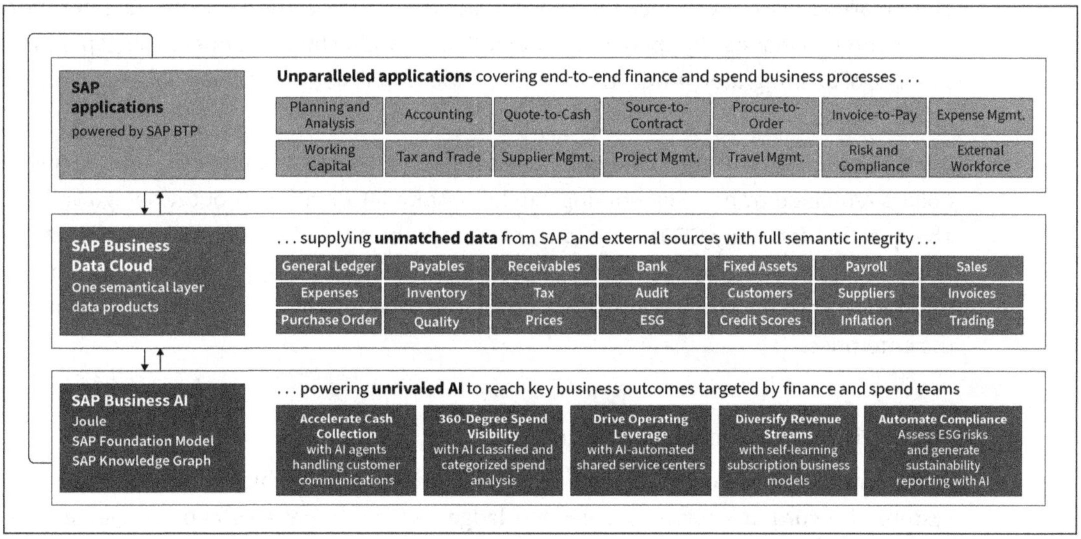

Figure 4.8: Key Components of SAP Finance and Spend

What makes this model effective is not the individual systems, but how they operate together. Spend originating from procurement, employee activity, and external workforce engagement flows into financial systems with its context preserved. The result is a unified financial view that reflects what was spent, but also how, where, and why.

Now let's discuss some of the typical challenges organizations face in the finance and spend areas.

4.2.1 The Strategic Challenge

ERP systems provide strong control over financial outcomes. They ensure that transactions are recorded accurately, financial statements are compliant, and reporting structures are consistent. These capabilities are essential, but they operate within a limited frame. They describe financial results after they occur rather than influencing how those results are created.

The challenge arises because enterprise spend originates outside the ERP core. Procurement decisions are made through sourcing events and supplier collaboration. Employee expenses are incurred through travel and operational activity. External workforce engagements are governed through contracts and service delivery. Each of these processes operates on its own timeline, often with limited connection to financial systems.

A finance system can report total spend, but it cannot determine whether procurement strategies were effective, whether employee expenses complied with policy, or whether external workforce costs contributed to intended outcomes. By the time transactions reach financial accounting, their operational context has already been lost.

This creates a disconnect between financial accuracy and decision-making. Organizations can report on what has happened, but not influence what should happen. The gap is in the absence of an integrated model that connects spend decisions with financial execution, not in accounting.

For Maria in Chapter 1, the challenge was not a lack of financial data. Four hundred million dollars remained trapped in working capital despite information about it being available. The question for many CFOs is no longer whether the core ERP can process transactions. They want to know how core finance capabilities and extended applications work together to improve cash flow, manage spend, reduce risk, and accelerate decision-making across the enterprise.

4.2.2 Financial Core and Office of the CFO

At the center of the finance architecture is SAP S/4HANA Cloud, which serves as the financial system of record. It governs the general ledger, accounts payable and receivable, asset accounting, and controlling processes. These capabilities ensure that all transactions are classified, validated, and reported consistently.

The office of the CFO portfolio extends this foundation beyond accounting into broader financial management. Financial close and consolidation processes ensure that organizations can produce accurate and timely financial results across entities. Governance, risk, and compliance capabilities embed controls into processes, ensuring adherence to regulatory and policy requirements.

Treasury and working capital management extend finance into liquidity planning and cash optimization. Organizations must balance payables, receivables, and financing strategies while maintaining operational flexibility. SAP Taulia complements this by enabling dynamic discounting, supply chain financing, and working capital optimization. It connects buyers and suppliers through financial transactions that improve liquidity while optimizing cost structures.

In this context, finance also evolves toward an autonomous model in which policy enforcement, anomaly detection, and decision logic are embedded directly within financial and spend processes, allowing validation, exception handling, and optimization to operate continuously alongside procurement, workforce, and expense activity rather than being applied after the fact.

Together, these capabilities ensure that finance operates not only as a reporting function but as a coordinating layer that aligns operational activity with financial performance.

Figure 4.9 highlights the key components of SAP's Autonomous Finance portfolio. Core finance capabilities are extended through specialized applications and assistants. At the top, Joule Assistants support key finance processes such as receivables, treasury, tax, planning, close, and governance. These experiences are built on finance capability areas including revenue management, treasury and working capital, tax and trade, accounting and close, financial planning and analysis, and risk and compliance. Underneath, a broad set of finance applications and business processes operate on the SAP Business AI Platform, providing the foundation for contextualized decisions, governance, and automation across the finance function.

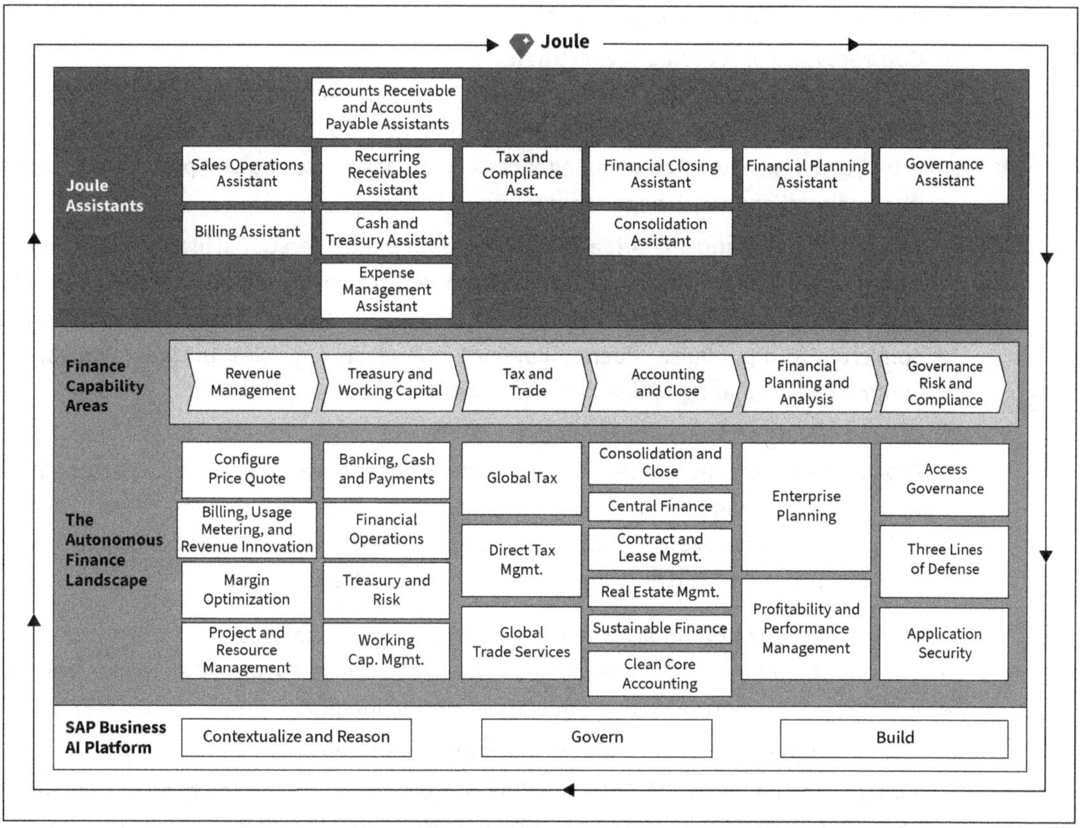

Figure 4.9: Autonomous Finance Portfolio

4.2.3 SAP Ariba

Procurement represents the largest structured category of enterprise spend. It operates as a lifecycle that begins with strategic planning and continues through sourcing, purchasing, delivery, and financial settlement. Let's walk through the key features offered within SAP Ariba:

- **Spend analysis and strategic sourcing**
  The procurement lifecycle begins with understanding spend. Spend analysis provides visibility into historical and projected expenditure, allowing organizations to identify opportunities for consolidation and cost optimization. This informs category management strategies, in which procurement approaches are defined based on market and organizational conditions.
  The SAP Ariba Strategic Sourcing suite supports this process by enabling supplier qualification, sourcing events, contract negotiation, and supplier risk management. Sourcing activities are executed through structured workflows, ensuring consistency and transparency across supplier engagement.
- **Contracts and procurement execution**
  Once suppliers are selected, procurement execution is governed through contracts that define pricing, delivery conditions, and compliance requirements. SAP Ariba Contracts provides a framework for managing these agreements, ensuring that procurement activity remains aligned with negotiated terms.
  SAP Ariba Buying and SAP Ariba Invoicing supports the operational lifecycle, from requisitioning to payment. Users create purchase requisitions based on approved suppliers and catalog content; these requisitions are then routed through approval workflows and converted into purchase orders. Suppliers interact through SAP Business Network to confirm orders, manage deliveries, and submit invoices.
- **Guided buying and user experience**
  Procurement execution extends beyond procurement teams to the broader organization. Guided buying provides a structured experience that directs users toward compliant purchasing decisions. It aligns user behavior with procurement policies, reducing uncontrolled spend and ensuring consistency across transactions.

SAP Ariba supports both indirect and direct procurement scenarios. *Indirect procurement* focuses on operational goods and services, where catalog management and guided buying ensure compliance and efficiency. *Direct procurement* integrates closely with SAP S/4HANA, linking sourcing decisions to production and planning requirements.

The integration between SAP Ariba and SAP S/4HANA ensures continuity across the lifecycle. Sourcing decisions flow into operational procurement, and procurement transactions are reflected in financial systems without loss of context.

In this context, procurement also moves toward an Autonomous Spend model (see Figure 4.10) in which policy, supplier intelligence, and exception detection guide decisions continuously during execution rather than after the transaction is complete.

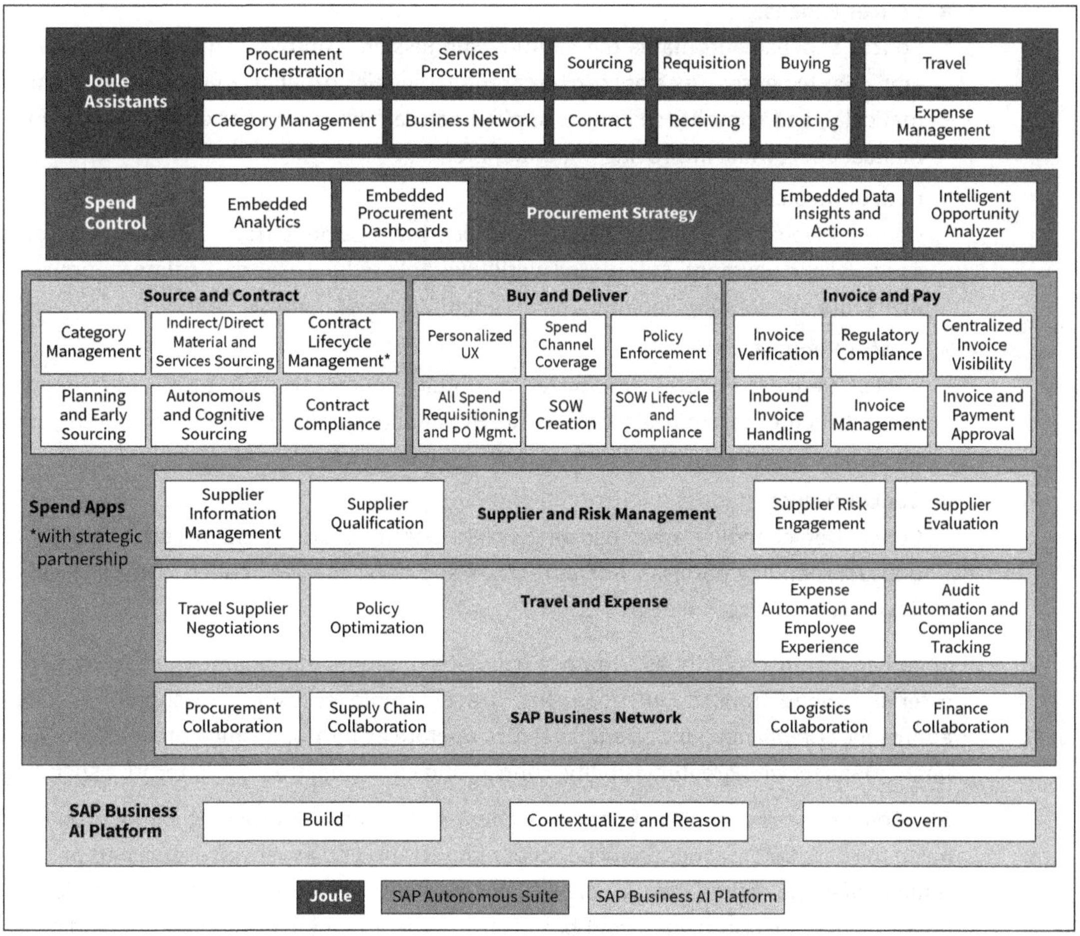

Figure 4.10: Autonomous Spend

4.2.4 SAP Concur

As organizations grow, spend management becomes more complex. Employees book travel through multiple channels, expenses arrive from dozens of sources, and finance teams often struggle to maintain visibility and control while preserving a positive employee experience. SAP Concur addresses this challenge by connecting the travel, expense, and invoice processes in a single, integrated environment.

The SAP Concur portfolio includes several key capabilities:

- **Concur Travel**
  Concur Travel helps employees plan, book, and manage business travel while enforcing corporate travel policies and preferred supplier agreements. Travel bookings, itinerary changes, and approvals are managed through a common platform, providing visibility into travel spend before costs are incurred.

- **Concur Expense**
  Concur Expense automates the capture, submission, approval, and reimbursement of employee expenses. Receipts can be captured digitally, expense reports generated automatically, and compliance checks applied before reimbursement occurs. This reduces manual effort while improving policy adherence.
- **Concur Invoice**
  Concur Invoice streamlines supplier invoice management from submission through approval and payment. Automated workflows help reduce processing time, improve visibility into outstanding liabilities, and strengthen spend controls.
- **Concur Request**
  Concur Request enables employees to obtain preapproval for travel and spending requests before commitments are made. By validating policy compliance up front, organizations can better manage budgets and reduce out-of-policy spending.
- **Concur Detect**
  Concur Detect applies advanced analytics to identify duplicate claims, policy violations, unusual spending patterns, and potential fraud risks, helping finance teams strengthen governance and compliance.

Together, these capabilities create a connected spend management ecosystem that extends beyond transaction processing. Travel, employee expenses, supplier invoices, approvals, compliance, and spend analytics operate as part of a single process, providing finance leaders with greater visibility, control, and consistency across enterprise spending.

SAP Concur does not replace financial accounting; rather, it improves the quality and timing of the data that reaches financial accounting. Travel and expense become part of a continuous lifecycle rather than a delayed administrative process. That shift is particularly important in project-driven or field-based environments, where discretionary spending can materially affect the margin, budget adherence, and compliance performance.

4.2.5 SAP Fieldglass

External workforce engagement represents a significant and growing category of enterprise spend. Organizations increasingly rely on contingent labor and service providers to meet operational and strategic needs. SAP Fieldglass extends the spend management landscape beyond suppliers and employees to the external workforce. It provides organizations with visibility, governance, and control over contingent labor, contractors, service providers, and statement-of-work (SOW) engagements, helping manage a growing portion of enterprise spend that often sits outside traditional procurement processes.

The key SAP Fieldglass capabilities include the following:

- **Contingent workforce management**
  Contingent workers are engaged to fill specific roles within the organization. They operate as an extension of internal workforce capacity and require structured management to ensure cost, quality, and compliance.

SAP Fieldglass provides the framework for managing contingent labor, including supplier engagement, rate management, and work tracking.

- **Services procurement and SOW management**
  Services procurement differs from contingent labor in that it focuses on outcomes rather than roles. Engagements are governed through SOWs that define deliverables, timelines, and pricing structures. Supplier organizations manage execution, while the buying organization validates outcomes.
- **Vendor management system**
  SAP Fieldglass operates as a vendor management system, providing a centralized platform for managing suppliers, workforce activity, and compliance. It ensures visibility into external workforce cost and performance. Integration with SAP S/4HANA ensures that external workforce activity is reflected in financial systems, linking service delivery with financial outcomes.

SAP Fieldglass is not simply a repository for supplier labor. It provides the operational structure that allows contingent labor and service-based work to be managed with the same financial discipline applied to procurement and internal labor. Work is validated against agreed-upon terms, exceptions are surfaced before they become cost overruns, and the financial consequences of external workforce decisions remain visible throughout the project lifecycle.

4.2.6 The Integration Model

The finance and spend architecture is based on a clear separation of responsibilities, combined with a shift toward continuous orchestration across domains. SAP S/4HANA Cloud governs financial execution and accounting, while SAP Ariba, SAP Concur, and SAP Fieldglass manage how spend originates across procurement, employee activity, and external workforce engagement. Integration connects these domains into a unified system in which transactions flow with their business context preserved.

This integration model extends beyond structured data exchange into coordinated execution. Procurement events, expense transactions, and workforce activities generate signals that are reflected immediately across systems. A purchase requisition created in SAP Ariba is validated against budget and policy in real time. An expense submitted in SAP Concur is checked against travel policy and project allocation before approval. A service milestone recorded in SAP Fieldglass triggers invoice validation and payment readiness without requiring manual reconciliation.

At the center of this model is an intent-driven interaction layer in which business users define outcomes rather than execute individual steps. For example, say that a manager requests external engineers for a project, submits a travel expense, or initiates procurement activity. Instead of navigating multiple systems, the request is interpreted and executed across sourcing, contracting, invoicing, and financial posting processes.

This execution is enabled by the coordinated interaction of assistants and agents, as highlighted in Figure 4.11.

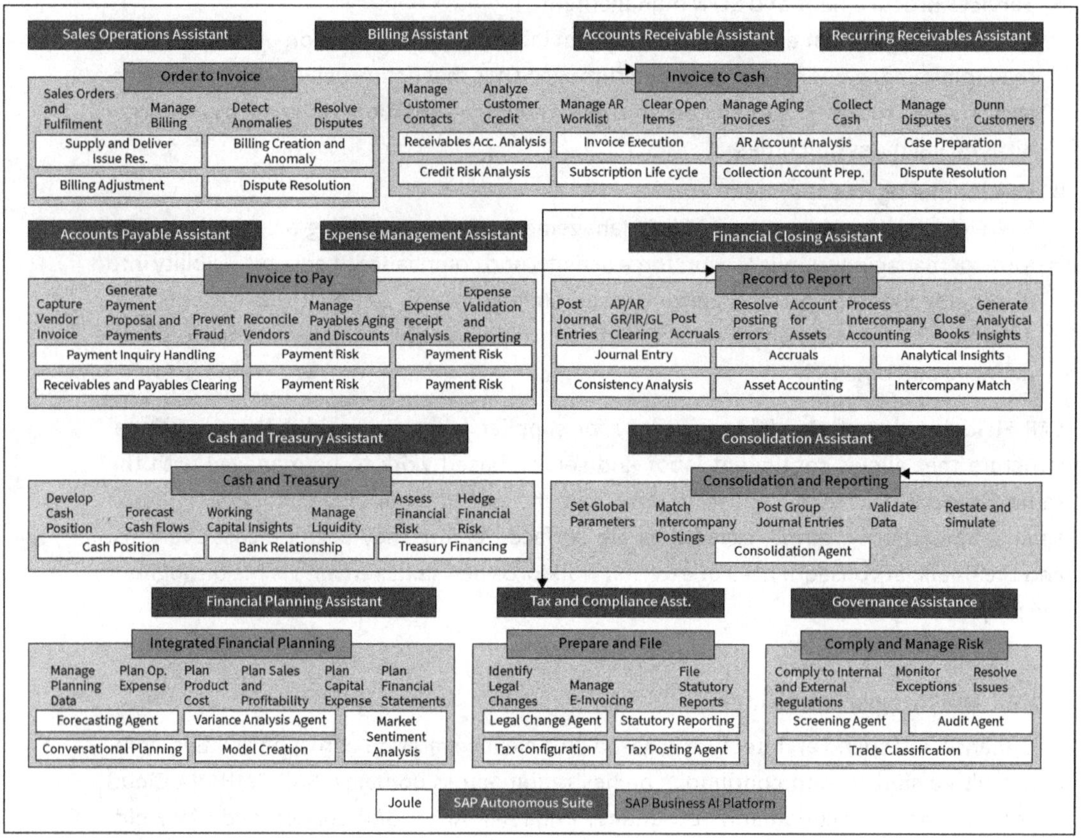

Figure 4.11: Autonomous Finance: Key AI Assistants

Assistants provide contextual guidance at decision points. For example, during procurement, an assistant can recommend preferred suppliers based on contract terms, historical pricing, and performance metrics. During expense submission, it can identify missing receipts, flag out-of-policy items, and suggest corrections before submission. In external workforce scenarios, assistants can highlight supplier performance trends, contract utilization, and rate benchmarks, as shown in Figure 4.12.

Agents operate in the execution layer and perform actions across systems. A sourcing agent can analyze supplier responses and highlight cost deviations. An expense validation agent can automatically detect duplicate or noncompliant claims and route them for review. A workforce agent can monitor timesheet entries and validate service delivery against contractual commitments. These actions occur continuously and are triggered by transactional events rather than manual intervention.

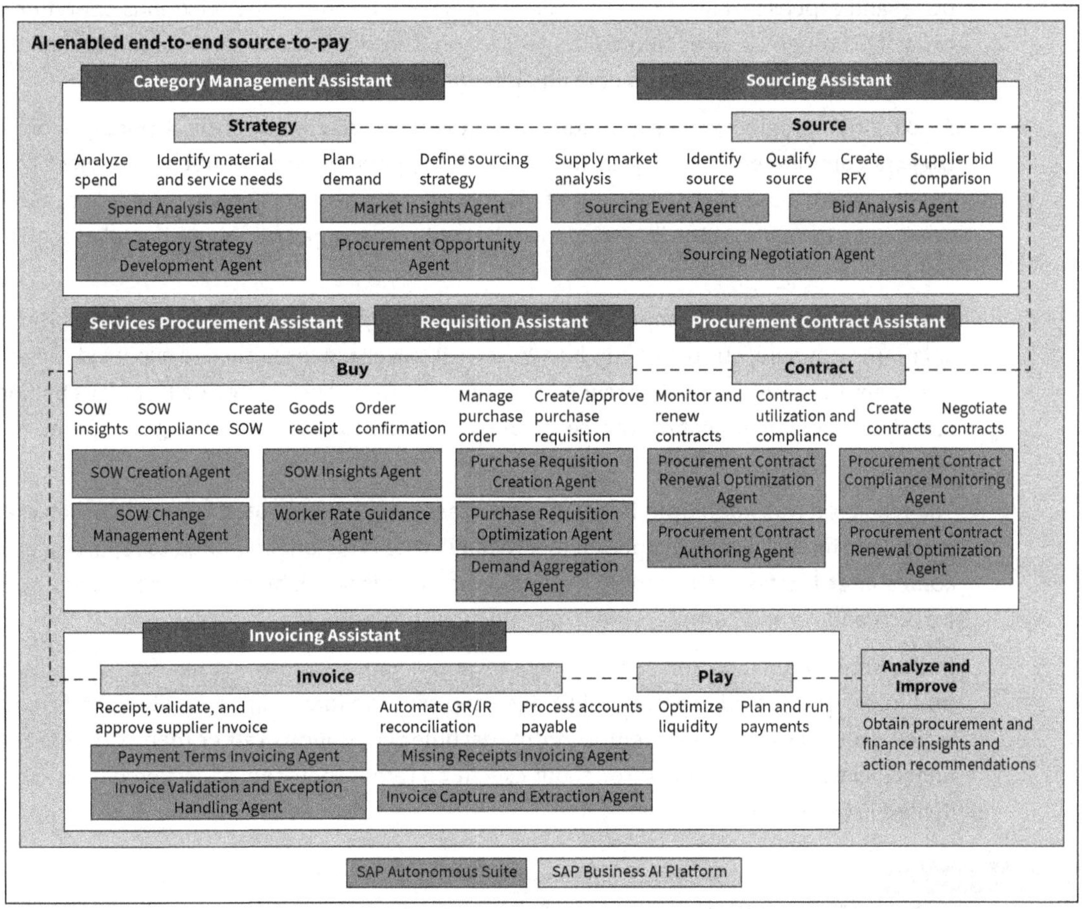

Figure 4.12: SAP Autonomous Spend: Key AI Assistants and Agents

The interaction between assistants and agents creates a continuous execution model. When a requisition is created, agents validate budget availability and pricing thresholds. If conditions are met, approval can be expedited. If deviations occur, the transaction is routed with context for review. Similarly, when an expense is submitted, validation occurs automatically before reimbursement. In external workforce scenarios, service delivery and invoice processing are aligned through event-based validation.

A key characteristic of this model is embedded governance. Policies are enforced directly within processes rather than through periodic audits. Procurement transactions are checked against contracts and approval thresholds. Expense claims are validated against policy and historical patterns. Workforce engagements are monitored for compliance with agreed-upon rates and deliverables. This ensures that financial control operates continuously rather than retrospectively.

Finance and spend processes therefore operate as a connected lifecycle. Procurement establishes supplier relationships and purchasing activity, employee actions generate travel and expense transactions, and external workforce engagement extends operational capacity through services and contingent labor. These processes execute in parallel but remain aligned through shared data and integrated workflows.

As execution progresses, the monetary impact is reflected continuously. A purchase order triggers commitment tracking in financials. An approved expense updates project cost structures. A validated service invoice flows directly into accounts payable. Because validation occurs throughout the lifecycle, financial posting becomes a continuation of execution rather than a separate reconciliation phase.

This model also enables continuous optimization. If supplier pricing deviates from historical norms, sourcing strategies can be adjusted. If travel expenses exceed expected ranges, policy enforcement can be tightened or routing updated. If workforce delivery falls behind schedule, supplier allocation can be modified. These actions occur during execution, not after financial close.

The result is a transition from fragmented processes to a coordinated system in which execution and financial control operate together. Finance is no longer limited to reporting outcomes after the fact. It becomes a continuous layer that aligns procurement, employee activity, and workforce engagement with financial performance.

In this context, finance and spend evolve toward an autonomous model. Assistants interpret intent and provide context, while agents execute validation, coordination, and transaction processing across systems. Control becomes continuous rather than periodic, and financial insight is available throughout execution rather than only at settlement, as highlighted in Figure 4.13.

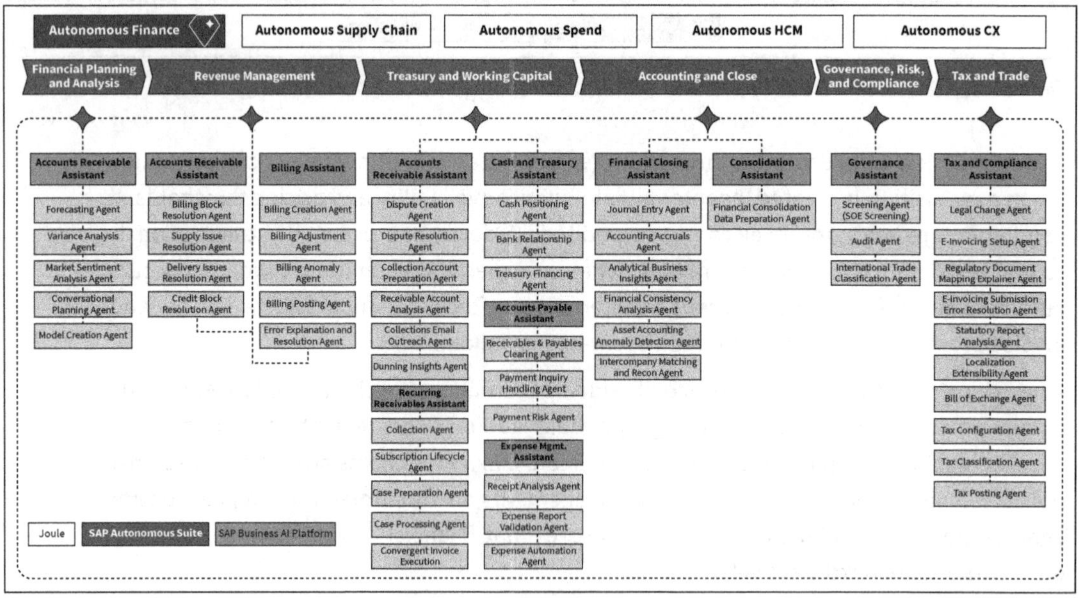

Figure 4.13: Autonomous Finance: Key AI Assistants and Agents

This interaction redefines the role of finance within the enterprise. Instead of acting primarily as a retrospective reporting function, finance becomes an active participant in operational decisions. Procurement, workforce, and expense processes remain aligned with financial objectives at every stage, enabling organizations to manage spend proactively rather than reactively.

4.3 Customer Experience

Customer-facing processes extend beyond sales transactions. Organizations must manage customer relationships, marketing interactions, commerce activities, service requests, and customer data across multiple channels and touchpoints. As businesses grow, these activities often become distributed across different teams and systems, making it difficult to maintain a consistent view of the customer throughout the relationship lifecycle.

The SAP Customer Experience (SAP CX) portfolio provides a set of integrated solutions designed to support customer engagement from initial awareness and acquisition through sales, fulfillment, service, and ongoing loyalty. Rather than focusing on a single function, SAP CX brings together capabilities across customer data management, marketing, commerce, sales, and service, helping organizations connect customer-facing activities with the operational processes running in the enterprise.

The following sections examine the major components of the SAP CX portfolio, including SAP Customer Data Cloud, SAP Emarsys Customer Engagement, SAP Commerce Cloud, SAP Sales Cloud, and SAP Service Cloud, and the role each plays in supporting end-to-end customer processes.

4.3.1 The Strategic Challenge

Customer experience has become one of the most complex operational domains within the enterprise. Organizations engage with customers across marketing, commerce, sales, and service processes, each supported by different systems and data structures. While individual functions may be optimized locally, the overall experience often remains fragmented and inconsistent.

The root of this fragmentation lies in the separation between customer-facing processes and core business operations. Customer data is typically distributed across multiple systems, with each function maintaining its own representation of the customer. Marketing platforms capture behavioral interactions, commerce systems record transactions, sales systems store pipeline data, and service applications manage issues and requests. These datasets are rarely aligned in real time, and integration often relies on batch synchronization or custom interfaces.

At the same time, customer interactions are frequently disconnected from operational realities. Campaigns may promote products that are unavailable, pricing may not reflect current conditions, and service agents may lack visibility into orders or delivery status. This misalignment creates a gap between what is communicated to the customer and what the organization can actually deliver.

Attempts to close this gap through middleware and point integrations have introduced additional complexity rather than solving the underlying problem. These approaches increase the cost of maintenance, slow down change, and still fail to provide the real-time consistency required for modern customer interactions. As a result, organizations are left with fragmented processes that require manual coordination across systems and teams.

This structural fragmentation is illustrated in Figure 4.14, where customer data, operational processes, and engagement systems are shown as separate layers with limited integration.

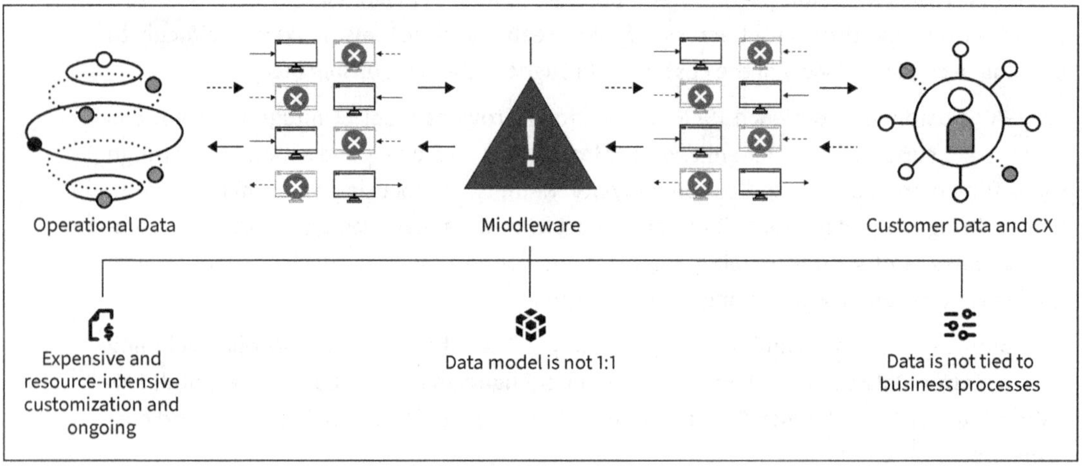

Figure 4.14: Key Challenges in Customer Experience

The consequence is not only a degraded customer experience but also operational inefficiency. Teams spend significant effort reconciling data and coordinating execution instead of focusing on creating value.

To address these challenges, customer experience must be restructured as an integrated operational domain rather than a collection of disconnected applications. The core principle of this architecture is the direct connection between front-office processes and core business systems.

In this model, marketing, commerce, sales, and service processes are no longer executed independently. Instead, they operate on shared data structures and synchronized workflows. Customer interactions are grounded in real-time business data, including inventory levels, pricing conditions, and fulfillment constraints.

A key enabler of this approach is the alignment of data models between customer experience systems and ERP. Core business objects such as customer, product, and order are defined consistently across the landscape, eliminating the need for reconciliation between systems. This allows all functions to operate on a common representation of the business.

This architectural model is illustrated in Figure 4.15, which shows how customer data, business processes, and engagement systems are connected across the enterprise.

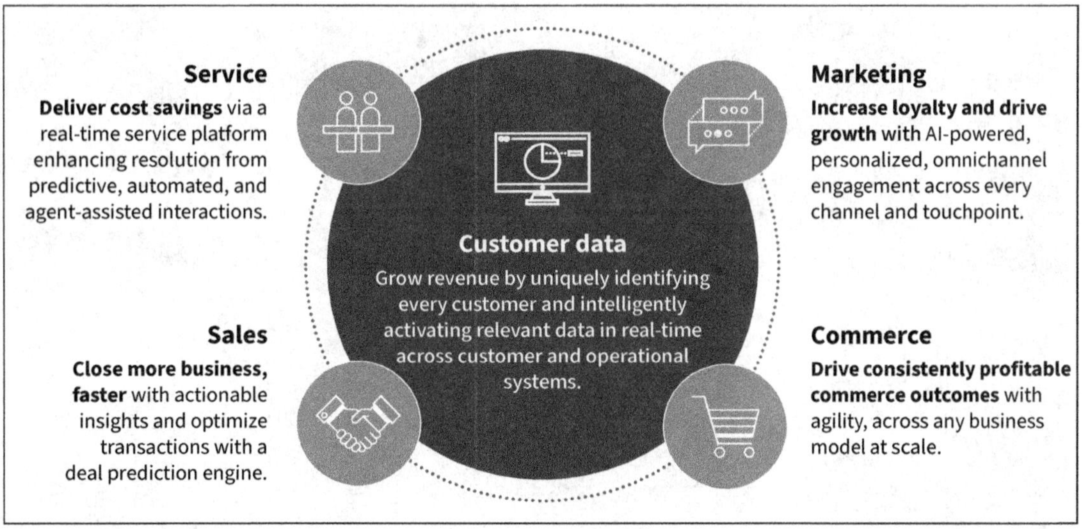

Figure 4.15: SAP Engagement Cloud

The impact of this integration is best understood through the flow of customer interaction. A marketing campaign no longer operates in isolation; it is informed by operational data such as inventory and pricing. When a customer engages and initiates a purchase, the commerce system validates availability and pricing in real time. The resulting order flows directly into fulfillment and financial processes without manual intervention. Service interactions are then executed with full visibility into the customer's history.

Customer experience is therefore transformed from a series of disconnected touchpoints into a continuous lifecycle that reflects the actual state of the business.

Now, let's discuss various components of the SAP CX portfolio.

4.3.2 SAP Customer Data Cloud

The foundation of the SAP CX portfolio is SAP Customer Data Cloud. Customer interactions increasingly occur across websites, mobile applications, commerce platforms, marketing systems, sales applications, and service channels. Without a common customer identity, organizations struggle to maintain a consistent understanding of customer relationships.

SAP Customer Data Cloud provides identity, consent, and profile management capabilities that unify customer information across multiple systems. It enables organizations to maintain a consistent customer profile while managing permissions, consent preferences, and privacy requirements. Customer information can then be made available to other customer-facing applications, ensuring that sales, marketing, commerce, and service teams are working from the same customer record.

Figure 4.16 illustrates this data layer, showing how customer data is unified, enriched, and activated across the enterprise landscape.

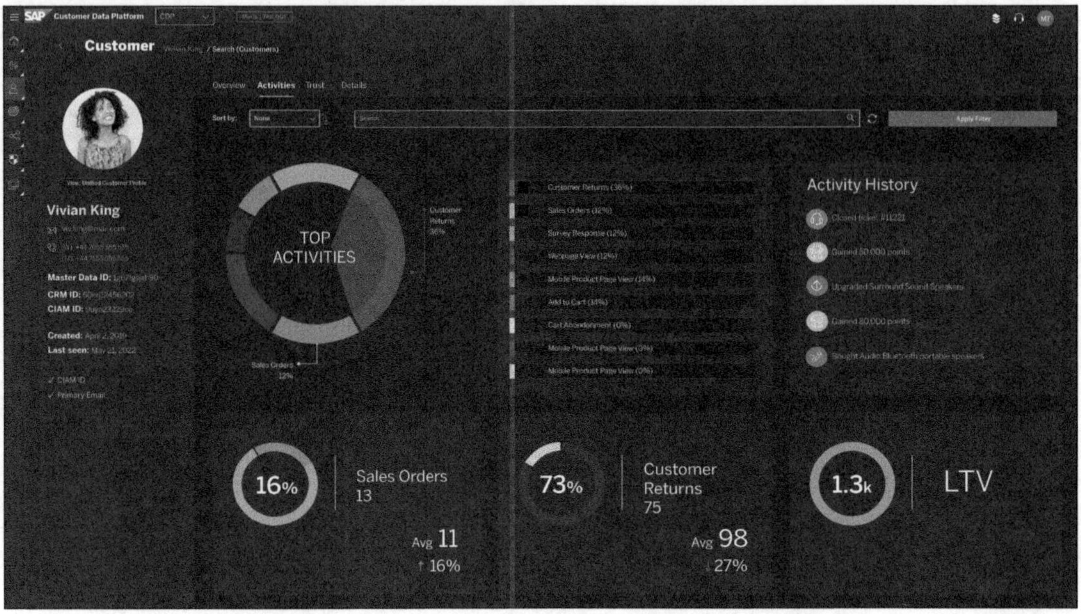

Figure 4.16: SAP Customer Data Cloud Unifies Customer Profiles

The unified profile is not static. It is continuously updated as new interactions occur, whether through marketing engagement, commerce transactions, sales activities, or service interactions. This allows every process to operate with the most current understanding of the customer.

Equally important is the management of consent and data privacy. Customer data must be collected and used in accordance with regulatory requirements, and customers must be given transparency and control over their information. These requirements are embedded directly within the data layer, ensuring that compliance is maintained as part of normal operations.

With this foundation in place, customer data becomes a shared asset across the enterprise, enabling consistent and context-aware interactions.

4.3.3 SAP Marketing Cloud

SAP Marketing Cloud supports customer engagement and campaign execution across digital and traditional channels. Marketing organizations use the platform to define audiences, execute campaigns, manage customer journeys, and measure engagement across multiple channels.

The platform combines customer profile information with behavioral, transactional, and engagement data to support segmentation and campaign planning. Marketing teams can coordinate communications across email, web, mobile, digital advertising, and other channels while maintaining a consistent customer experience.

Figure 4.17 illustrates the multichannel engagement model, showing how campaigns are executed across channels while maintaining alignment with underlying data and processes.

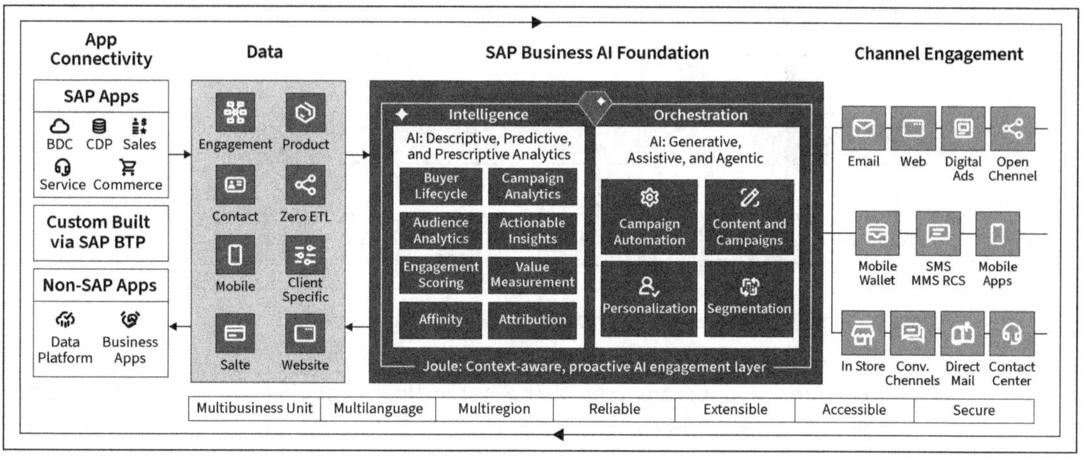

Figure 4.17: Multichannel Engagement Model

This integration enables a shift from static campaigns to continuous engagement. Marketing interactions are adjusted dynamically based on customer behavior and operational signals. For example, if inventory levels change, campaigns can adapt to promote available products. If customer preferences evolve, segmentation is updated in real time.

Marketing therefore becomes an ongoing process that feeds into commerce and sales activities, rather than a series of isolated campaigns. Each interaction contributes to the progression of the customer lifecycle and is informed by the same data foundation.

4.3.4 SAP Commerce Cloud

SAP Commerce Cloud supports digital commerce processes, including product discovery, catalog management, pricing, shopping cart management, ordering, and customer self-service. For many organizations, it serves as the primary digital transaction platform through which customers interact with products and services.

A key capability of SAP Commerce Cloud is its integration with ERP and supply chain systems. Product information, pricing, inventory availability, fulfillment status, and order history are shared between commerce and operational systems, ensuring that customer interactions accurately reflect business realities.

This integrated model is illustrated in Figure 4.18, which shows how commerce processes connect to supply chain and operational systems.

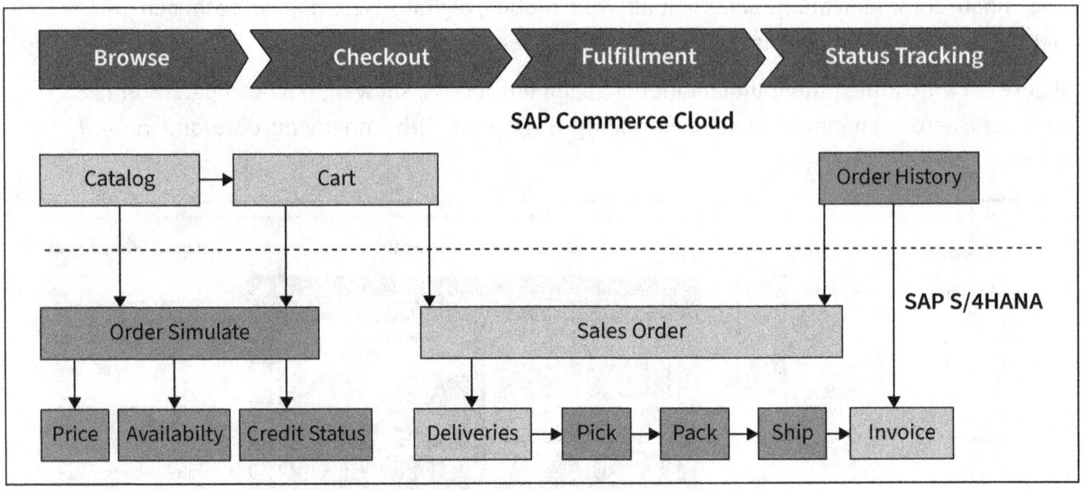

Figure 4.18: SAP Commerce Cloud and SAP S/4HANA Integration

When a customer initiates a purchase, the system verifies pricing conditions and inventory availability before confirming the order. The order then triggers downstream processes such as allocation, shipment, and financial posting. These processes are executed automatically and remain synchronized across systems.

Commerce therefore acts as a bridge between customer engagement and operational execution, ensuring that transactions are both accurate and fulfillable.

4.3.5 SAP Sales Cloud

SAP Sales Cloud supports lead management, opportunity management, forecasting, account management, quoting, and sales execution processes. It provides sales organizations with a consolidated view of customer relationships, sales opportunities, and pipeline performance.

By connecting customer information, historical transactions, engagement history, and operational data, SAP Sales Cloud enables organizations to manage opportunities throughout the sales lifecycle. The platform supports account planning, forecasting, deal management, and quote generation while maintaining alignment with downstream fulfillment and revenue processes.

Figure 4.19 illustrates this flow, showing how lead management, quoting, and order creation are integrated. When a lead is generated, it is evaluated and assigned based on predefined criteria. As the opportunity progresses, pricing and product information are retrieved directly from operational systems. Quotes reflect current conditions, and once a deal is closed, the order is created and passed directly into fulfillment.

This integration eliminates manual handoffs and ensures that sales execution is aligned with operational capabilities.

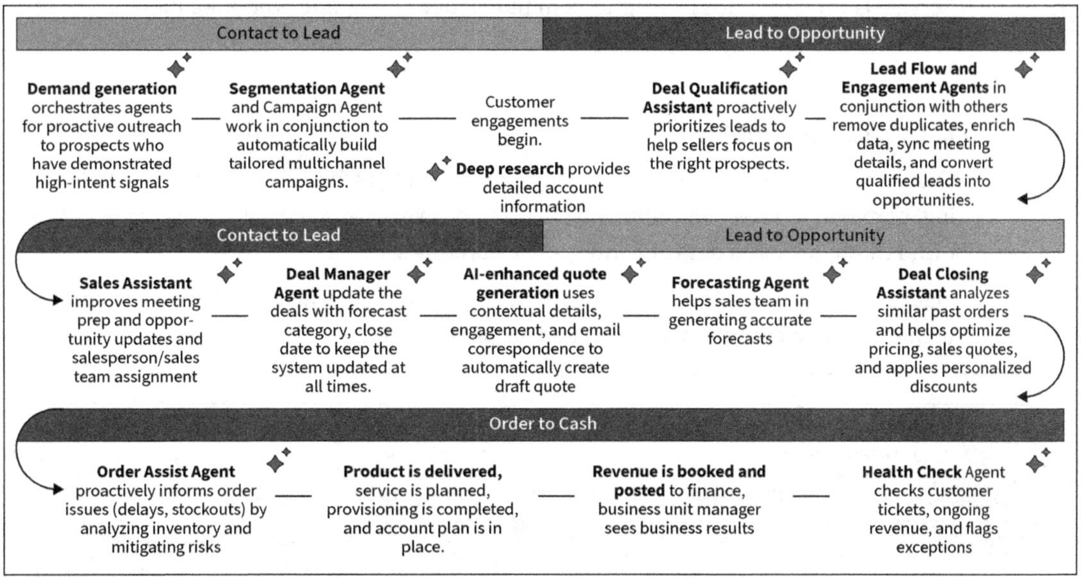

Figure 4.19: Integrated Contact-to-Cash Process

4.3.6 SAP Service Cloud

SAP Service Cloud supports customer service and issue resolution processes following a sale or service interaction. Service organizations use the platform to manage customer inquiries, cases, requests, complaints, and ongoing support activities.

The platform provides service agents with access to customer profiles, order history, product information, service agreements, and previous interactions. This allows service teams to diagnose issues more effectively and coordinate corrective actions across multiple business functions when necessary.

Figure 4.20 illustrates this integrated service model. When a customer raises an issue, the system automatically links the request to relevant data such as orders, shipments, and invoices. This allows the agent to diagnose and resolve the issue quickly. If necessary, actions are triggered in downstream systems such as logistics or billing.

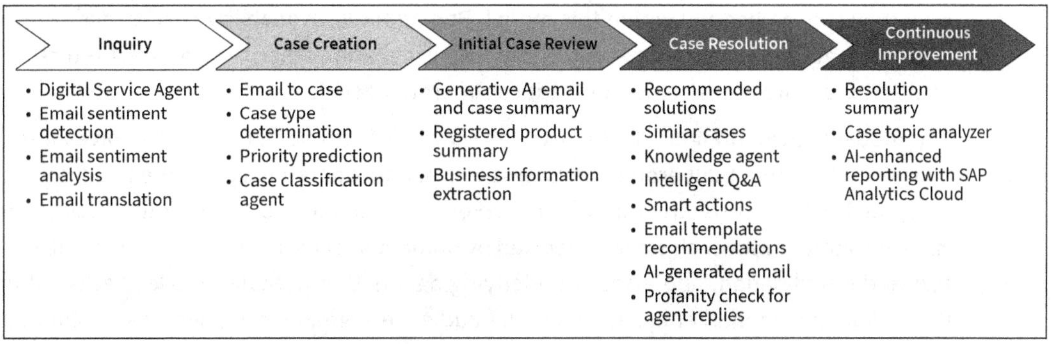

Figure 4.20: Integrated Service Management process

Service is therefore not limited to communication; it includes the execution of corrective actions across the organization.

4.3.7 The Integration Model

Customer experience operates as a continuous system rather than a linear process. Marketing, commerce, sales, and service activities are executed in parallel and remain synchronized through shared data and integrated workflows.

Figure 4.21 illustrates this integrated domain, showing how all components are connected within a unified execution model called Autonomous CX.

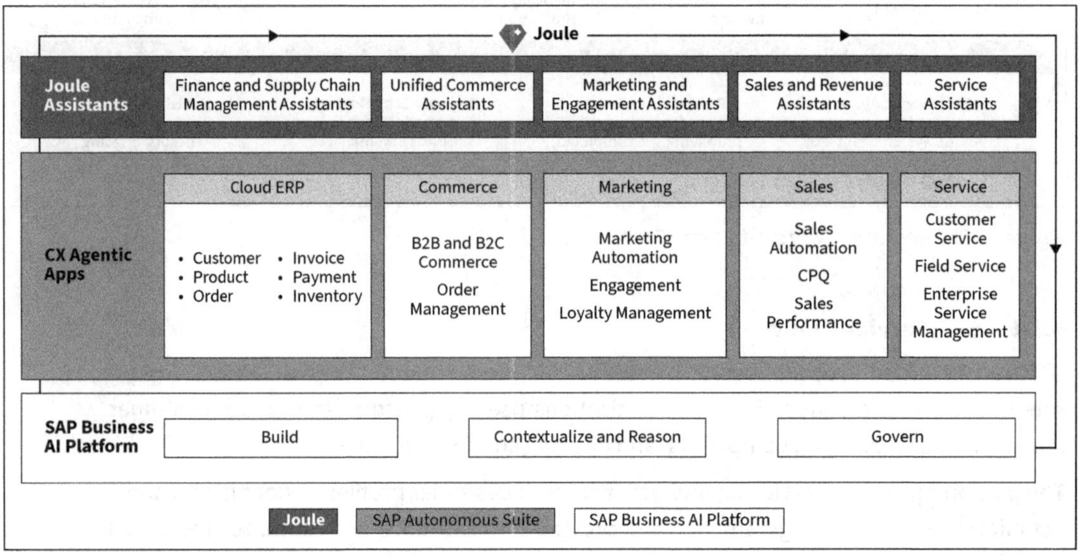

Figure 4.21: Autonomous CX Portfolio

Each customer interaction generates signals that are shared across the system. These signals update customer data, trigger processes, and inform decision-making. As conditions change, processes adapt automatically, ensuring that execution remains aligned with business reality.

Governance is embedded within this model. Pricing rules, availability checks, and service commitments are enforced as part of process execution rather than applied retrospectively. This ensures consistency and reduces operational risk.

The model shown in Figure 4.22 illustrates how the customer experience evolves into a coordinated system built around assistants and agents operating across the full customer lifecycle. The lifecycle is structured from market engagement through revenue execution to issue resolution, with each stage supported by domain-specific assistants that provide context and coordination, and a set of underlying agents that execute detailed tasks within those domains. In marketing, agents refine audience segmentation, personalize content, and evaluate campaign performance. In sales, they support lead qualification, pipeline analysis, pricing, and deal execution. In commerce, they manage product recommendations, assortment decisions, order processing, and fulfillment alignment. In service, they handle case preparation, knowledge retrieval, resolution workflows, and ongoing customer interactions. What changes in this model is not the existence of these activities, but how they are coordinated. Instead of users manually navigating systems and triggering each step, assistants help provide a continuous interaction layer that connects intent to execution, while agents operate across systems to carry out validation, analysis, and transaction processing. This creates a new way of working in which customer processes are executed as connected flows rather than isolated steps, allowing marketing, commerce, sales, and service to operate with shared context and continuous alignment to the underlying business operations.

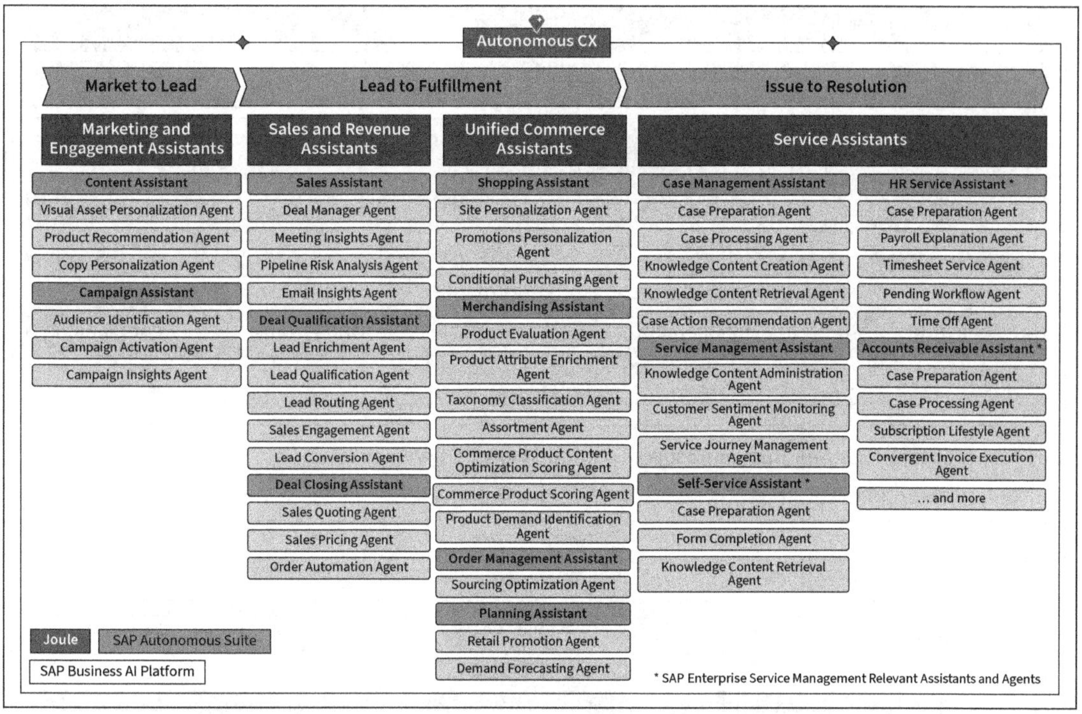

Figure 4.22: Autonomous CX: Humans Decide, Assistants Orchestrate, and Agents Execute

Customer experience, in this model, becomes more than a front-office capability. It is an integrated operational discipline in which every interaction contributes directly to execution, performance, and business outcome.

4.4 Human Resources

HCM in a modern enterprise is no longer a set of isolated HR processes. It operates as a connected system spanning core workforce data, operations, performance, talent, and development. SAP SuccessFactors provides this structure by organizing these capabilities around an integrated employee record, covering identity, role, compensation, and organizational context; all workforce processes depend on this.

From this foundation, workforce planning and people operations address day-to-day questions of capacity and execution, while performance and compensation processes translate organizational goals into measurable outcomes at the individual level. Talent acquisition and development extend this further, ensuring that the organization can attract, deploy, and continuously build the skills required to meet evolving business demands. Figure 4.23 highlights the various components of the SAP SuccessFactors suite of solutions and Figure 4.24 highlights the key elements of the Autonomous HCM portfolio.

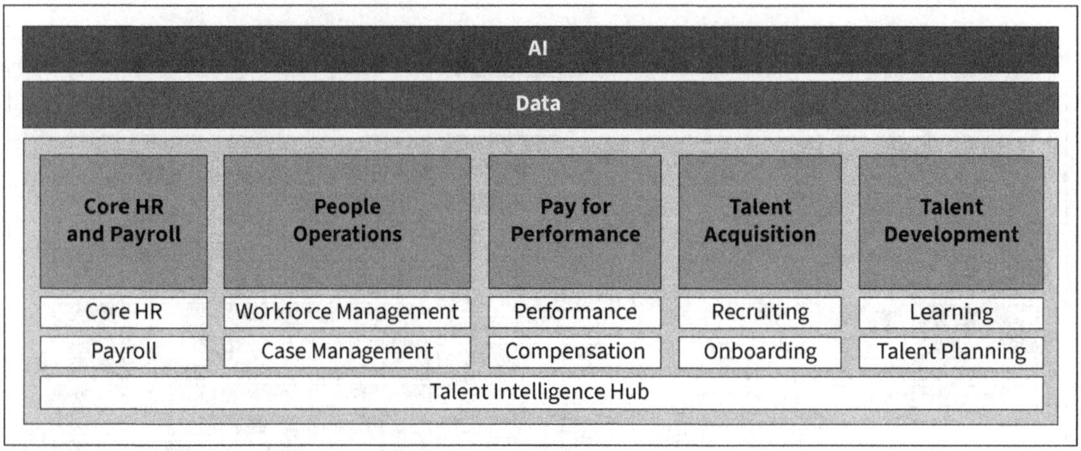

Figure 4.23: SAP Human Resource Management Components

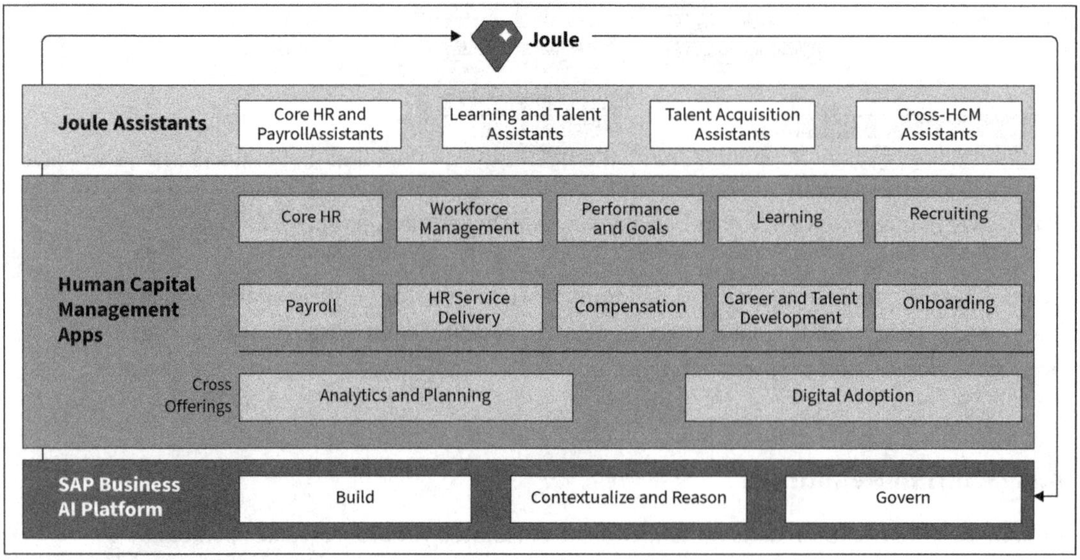

Figure 4.24: Autonomous HCM Portfolio

What makes this model effective is not the individual components, but how they operate together. Through shared data and consistent integration, workforce information can be shared across processes in a coordinated way. The result is a shift from fragmented HR functions to a more unified view of the workforce—one that allows organizations to understand not just how many people they employ but whether they have the right capabilities, organized effectively, to support business outcomes.

We'll dive deep into the key solutions of SAP's HR management portfolio. However, before we do that, let's discuss some key challenges companies face today in the area of human resource management.

4.4.1 The Strategic Challenge

ERP systems handle workforce administration with high levels of consistency and control. They maintain payroll accuracy, ensure proper cost allocation, and enforce organizational structures. These functions are essential, but they operate within a limited frame; they describe the current state of the workforce rather than its ability to meet future demand.

The challenge emerges when organizations attempt to answer questions that cut across time horizons and domains. A finance system can report labor costs at a given moment but cannot determine whether that cost structure is aligned with future revenue. A project system can show demand for resources but cannot determine whether those resources exist or what skills they possess. A hiring system can track open positions but cannot determine whether those positions are aligned with the organization's strategic direction.

This creates a disconnect between operational reliability and strategic decision-making. Workforce plans, hiring processes, resource allocation, and execution operate on different timelines and often in different systems. As a result, organizations either overcommit without capacity or maintain capacity that is not aligned to demand. The gap is created not by poor execution but by the absence of an integrated model that connects these elements into a continuous flow.

4.4.2 SAP SuccessFactors

SAP SuccessFactors provides a cloud-based HCM suite that supports the full employee lifecycle. It combines core HR processes with talent management, workforce planning, learning, performance, and employee experience capabilities within a unified platform. Rather than treating human resources as a collection of administrative functions, SAP SuccessFactors provides a common system for managing workforce information, employee development, organizational structures, and talent processes across the enterprise.

The platform is organized into a set of integrated applications that address different stages of the workforce lifecycle. SAP SuccessFactors Employee Central serves as the system of record for workforce and organizational data, while applications such as SAP SuccessFactors Recruiting, SAP SuccessFactors Onboarding, SAP SuccessFactors Learning, SAP

SuccessFactors Performance & Goals, SAP SuccessFactors Compensation, SAP SuccessFactors Succession & Development, and SAP SuccessFactors Workforce Analytics support specialized HR and talent management processes. Together, these capabilities provide organizations with a connected view of their workforce and help ensure that people-related decisions are aligned with operational and strategic objectives.

Now let's discuss the key applications of the SAP SuccessFactors suite.

Core HR and Payroll

At the center of the workforce architecture is the employee record. Core HR maintains this record as the authoritative source of workforce data. It defines who is employed, where they sit within the organization, and how they are compensated. This includes structural elements such as cost centers, reporting relationships, and role definitions.

The importance of this layer lies in its role as the single source of truth. Every other workforce process depends on it: Workforce planning requires accurate headcount and role definitions; talent acquisition depends on consistent job structures; performance and compensation depend on aligned reporting relationships. Without this consistency, downstream processes lose coherence.

Integration between SAP SuccessFactors and SAP S/4HANA Cloud helps align workforce data with financial processes, enabling organizations to better understand the cost and impact of their workforce decisions. Core HR data such as employee information, organizational structure, and compensation is typically maintained in SAP SuccessFactors Employee Central and shared with SAP S/4HANA to support downstream processes like payroll, cost accounting, and financial reporting.

Updates to employee data, such as role changes or compensation adjustments, can be transferred to SAP S/4HANA through configured integration processes, helping ensure that payroll and financial systems operate on up-to-date information.

This integration reduces data silos between HR and finance and supports more consistent reporting while allowing organizations to define the timing, controls, and validation steps appropriate for their business and compliance needs.

SAP SuccessFactors supports time capture and workforce data management that can be integrated with SAP S/4HANA to enable project-based cost tracking and analysis. Time entered and approved in SAP SuccessFactors (e.g., via SAP SuccessFactors Time Tracking) can be transferred to downstream systems, where it is combined with payroll results and financial configuration in SAP S/4HANA to determine labor costs and allocate them to projects or cost objects. Depending on the integration design, organizations can transfer time and cost-relevant data to SAP S/4HANA on a scheduled or near-real-time basis, supporting more frequent visibility into project effort and cost trends.

This approach helps reduce manual data transfers and improves alignment between HR, payroll, and project accounting processes while allowing organizations to define the level of detail, timing, and validation controls required for accurate financial reporting.

Figure 4.25 shows this relationship with the employee master data serving as the anchor for payroll processing and financial reporting within ERP. The top screenshot shows the **Org Chart** view. This provides a visual representation of the reporting hierarchy, allowing managers and employees to navigate organizational structures, reporting relationships, team composition, and position hierarchies. Users can move between managers, peers, and direct reports while viewing role information, organizational assignments, and employee details. The profile panel on the bottom provides additional information about the selected employee, including their position, location, and contact information.

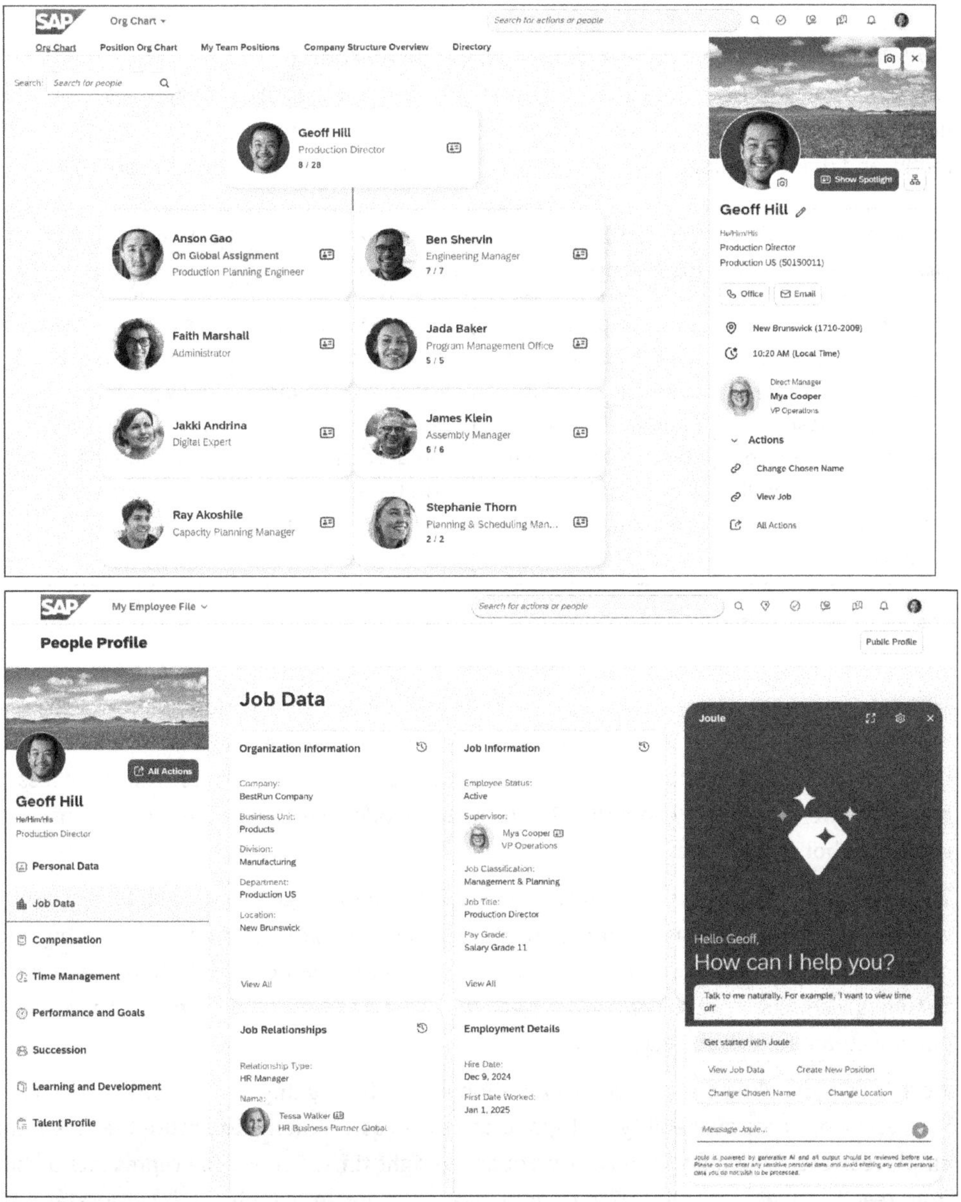

Figure 4.25: SAP SuccessFactors Employee Central

The lower-right screenshot shows the **Job Data** view in the people profile within SAP SuccessFactors Employee Central. This serves as the employee's core HR record and displays organizational information, job details, reporting relationships, and employment data. Information such as business unit, department, location, manager, position, pay grade, employment dates, and HR contacts is maintained here as part of the employee master record. The screen also shows how users can navigate to related areas such as compensation, performance, learning, succession, and talent management from a single employee profile.

Workforce Management: Time, Absence, and Scheduling

Workforce management translates the workforce structure into operational execution. This helps ensure that employees are assigned to work appropriately, that time is captured correctly, and that compliance requirements are consistently met, as shown in Table 4.1.

Absence Management	Time and Attendance	Workforce Scheduling	Activity Management
Streamline employee leave and absence processes by automating requests, accruals, and approvals, ensuring compliance with local regulations.	Enhance productivity with accurate time tracking, ensuring operational efficiency, compliance with labor laws, and streamlined workforce management.	Optimize workforce efficiency by forecasting, scheduling, and managing employee deployment based on availability, skills, and business needs.	Use activity-based time recording to allow people to record what they worked on in addition to when they worked.

Table 4.1: SAP SuccessFactors Workforce Management

Time management and absence processes form the basis of this domain. They provide mechanisms to record when employees work and when they are unavailable, while ensuring compliance with regulatory requirements and organizational policies. These processes have become increasingly complex due to global workforces, hybrid work patterns, and evolving labor regulations.

Time tracking extends this capability by capturing work at a granular level, linking hours worked to specific activities or assignments. This enables organizations to move beyond measuring time presence toward understanding actual work contribution. Time data becomes a critical input for payroll, project costing, and financial reporting, and this supports accurate reflection of labor costs.

Workforce scheduling completes the operational layer by aligning demand with capacity. It connects workforce availability, skills, and business requirements to ensure that the right employees are assigned to the right work at the right time. Scheduling represents a shift from reactive workforce allocation to proactive, data-driven planning. It incorporates visibility into skills, availability, compliance constraints, and operational demand, allowing organizations to manage workforce deployment with greater flexibility. This integration of absence management, time tracking, and scheduling creates a continuous loop in which workforce availability, assignment, and execution are closely connected. As a result, organizations improve their visibility into and control over workforce utilization, cost, and compliance simultaneously.

Once the workforce is defined, the next challenge is aligning it with demand. People's operations connect a workforce structure to its execution by supporting how resources are allocated, how capacity is utilized, and how work activities are captured. This domain operates across multiple time horizons. Workforce planning looks ahead to estimate future capacity needs, while staffing and assignment decisions address near-term requirements by aligning available resources to projects or operational tasks. Time tracking then captures work performed, providing input into cost, utilization, and performance analysis.

The connection between these layers is important. Planning without execution insight can lead to unrealistic assumptions, while execution without planning can result in reactive decisions. By linking these processes through integration and shared data, organizations can better align demand with capacity.

When demand is identified, organizations can evaluate available capacity and make informed decisions about allocation. Where gaps exist, this insight may inform hiring or development planning. As work is performed, time data can be transferred to downstream systems, such as ERP systems, based on the configured integration approach, thus supporting visibility into cost and performance outcomes over time.

Figure 4.26 illustrates this flow, showing how demand, capacity evaluation, staffing, and time capture connect in a continuous cycle.

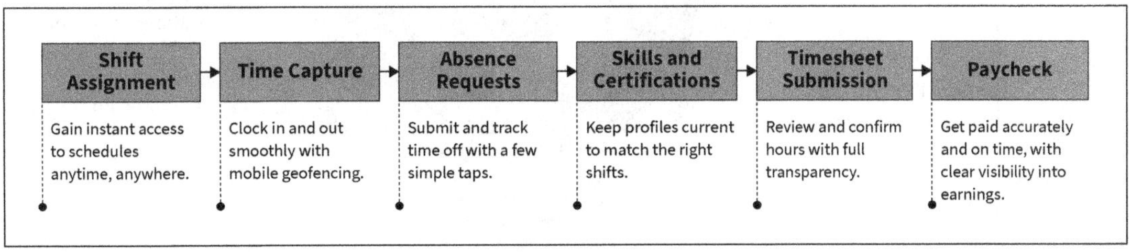

Figure 4.26: Workforce Management: From Shift to Paycheck

Payroll and Total Compensation

Payroll represents the point at which workforce data is processed into compensation outcomes. It supports accurate payment of employees and provides input to financial systems so that labor costs can be reflected through defined processes and cycles. Integration between payroll and core HR helps ensure that changes in employee status or compensation are incorporated into payroll calculations based on configured timing and controls.

Modern payroll capabilities extend beyond execution by providing visibility into payroll processes, supporting compliance monitoring, and enabling more streamlined operations.

Automated calculations, reporting, and workflow integration can help reduce manual effort and support accuracy, depending on system configuration.

Total compensation management expands this domain by incorporating a broader view of rewards, including base pay, incentives, and benefits. It helps organizations manage compensation planning in alignment with market data, internal policies, and workforce strategies while improving visibility across reward components.

By connecting compensation processes within the broader HCM landscape, organizations can work toward a more consistent view of employee rewards. This supports alignment with performance outcomes and business priorities while helping improve transparency and employee understanding of total compensation.

Figure 4.27 shows how performance evaluation flows into compensation decisions and ultimately into ERP financial structures.

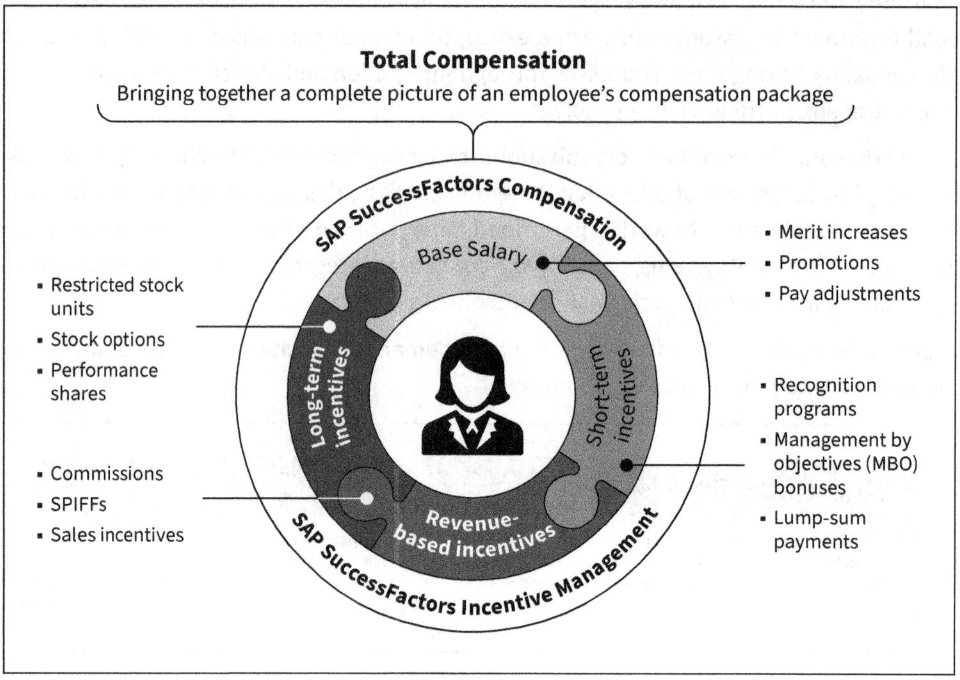

Figure 4.27: SAP SuccessFactors Compensation Management

Talent Acquisition

Talent acquisition connects workforce demand with external talent markets by supporting how organizations identify, attract, and hire works with the skills needed to meet business priorities.

Workforce requirements are translated into hiring needs, with increasing emphasis on skills and capabilities rather than purely role-based hiring. Modern recruitment approaches incorporate market conditions, talent availability, and organizational priorities to guide hiring strategies.

Recruitment processes can be integrated with the broader HCM landscape, allowing candidate information—such as skills and experience—to be captured and, where appropriate, incorporated into employee records. This supports continuity between hiring, onboarding, and subsequent talent processes.

While recruitment remains a distinct process, integration with workforce planning, development, and performance management enables organizations to better align hiring decisions with longer-term workforce strategies.

As a result, talent acquisition contributes to building workforce capability over time by linking external hiring with internal talent management and evolving business needs.

Figure 4.28 highlights how the Recruiting Assistant support the hiring process by helping recruiters and hiring managers manage activities more efficiently across the recruitment lifecycle. They can assist with tasks such as identifying relevant candidates based on job requirements, supporting candidate engagement, and helping coordinate interview scheduling through integrated workflows. These capabilities build on recruitment data, including job requisitions and candidate profiles, to provide recommendations and streamline process steps.

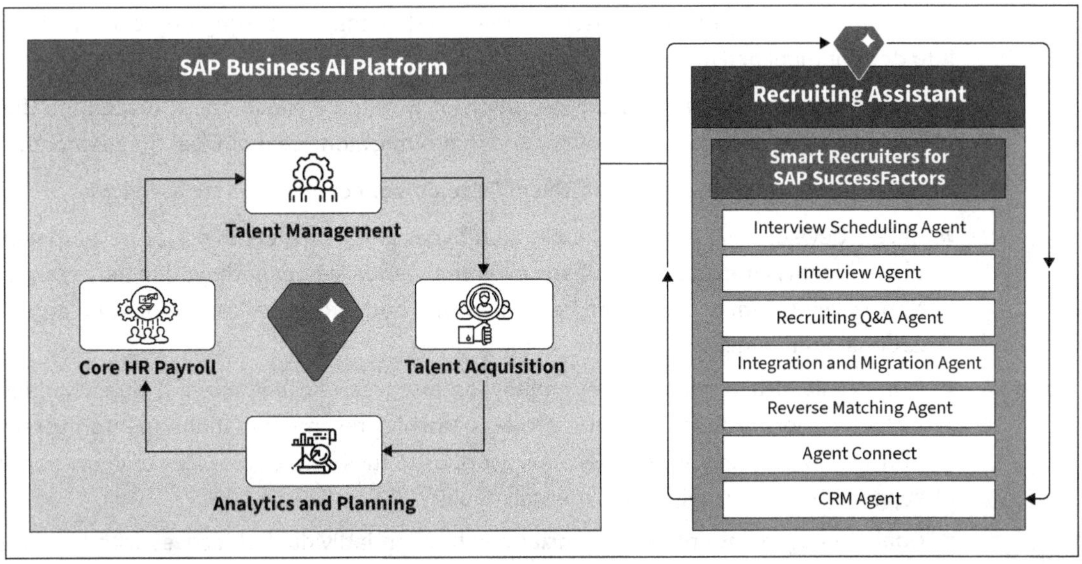

Figure 4.28: Recruiting Assistant

Rather than fully automating hiring decisions, recruiting assistants augment the recruitment process by improving responsiveness and consistency. For example, they can help accelerate candidate communication, surface qualified candidates earlier in the process, and provide visibility into pipeline status.

The impact of these capabilities depends on system configuration and organizational processes. When effectively implemented, they can support a more efficient and responsive recruitment experience for both candidates and hiring teams.

Talent and Career Development

Talent management focuses on developing workforce capability over time by supporting performance management, career development, and skill growth. These processes help organizations align employee development with evolving business needs.

Career and talent development processes provide visibility into skills, strengths, and aspirations. They support identifying opportunities for employees, including learning paths and career progression options, enabling a more responsive approach to workforce development.

A key concept in this domain is the use of skills as structured data. This allows organizations to better identify skill gaps, guide development activities, and track progress over time, depending on the maturity of their skills framework and data quality.

Integration between career development and workforce planning can help align capability development with anticipated business demand, providing input into hiring, reskilling, and workforce strategies.

Learning solutions extend these capabilities by providing structured training and content. Through integration with internal or external content providers, organizations can deliver targeted learning experiences based on identified needs, supporting improved visibility into development activities and outcomes.

Talent development addresses the evolution of workforce capability. It recognizes that demand for skills changes continuously and that hiring alone cannot meet that demand.

Key SAP SuccessFactors Career and Talent Development capabilities are as follows:

- Career Development Planning allows employees and managers to create development plans, identify competency gaps, and explore potential career paths within the organization. Development activities can be aligned with both individual aspirations and organizational requirements.
- Opportunity Marketplace helps employees discover relevant internal opportunities, including short-term assignments, projects, learning recommendations, mentoring relationships, and open positions. By making opportunities visible across the organization, it supports workforce agility and internal mobility.
- Goals Management enables organizations to align individual objectives with business priorities. Employees can establish personal and professional goals, track progress throughout the year, and connect development activities to broader organizational outcomes.
- The Mentoring and Coaching capabilities support knowledge transfer and employee development by helping organizations establish formal mentoring programs. Employees can be matched with mentors based on skills, experience, career interests, and development objectives.
- The Assignments and Internal Mobility capabilities help organizations identify employees for temporary assignments, project-based work, rotational programs, and new positions.

These processes enable employees to gain experience while helping organizations address evolving workforce requirements.

- Succession and Development Planning support the identification and development of future leaders and key talent pools. Organizations can evaluate readiness, identify successors for critical roles, assess talent risk, and establish targeted development plans to strengthen leadership pipelines.

Together, these capabilities create a structured approach to talent development, helping organizations align workforce skills, career growth, succession planning, and business needs within a single talent management framework.

Learning and Skill Development

Learning is increasingly positioned as a continuous process rather than a periodic activity. It plays a critical role in maintaining workforce relevance in an environment characterized by rapid technological change and evolving skill requirements.

The integration of learning with workforce data ensures that training is aligned with actual business needs. Organizations can identify skill gaps, assign learning interventions, and track outcomes in a structured manner. This approach transforms learning from a generic activity into a targeted mechanism for capability building, as shown in Figure 4.29. It highlights the SAP SuccessFactors Learning experience across both desktop and mobile devices. The platform provides employees with a centralized environment for managing assigned training, professional development, certifications, compliance requirements, and learning activities.

In the desktop view, the employee landing page is organized around learning priorities. The upper section highlights organizational learning initiatives and announcements, while the **Required Assignments** area displays mandatory training courses, certifications, and compliance programs along with due dates and completion status. Below this, the system presents recommended learning opportunities, enabling employees to discover courses and learning paths relevant to their roles and career objectives.

The mobile view demonstrates how the same learning experience is available through a mobile interface. Employees can review required training, access assigned courses, search the learning catalog, and complete learning activities directly from their mobile devices. This allows training and certification requirements to be managed consistently regardless of location or device.

The figure also highlights one of the key strengths of SAP SuccessFactors Learning: the ability to combine compliance-driven training requirements with employee-driven development. Mandatory courses, certifications, learning recommendations, and career-related learning opportunities are presented within a single environment, creating a unified approach to workforce development.

The shift toward skills-based organizations reinforces the importance of learning. Skills become the central element around which workforce processes are organized, influencing recruitment, development, performance, and compensation. This creates a continuous cycle in which skills are identified, developed, validated, and applied.

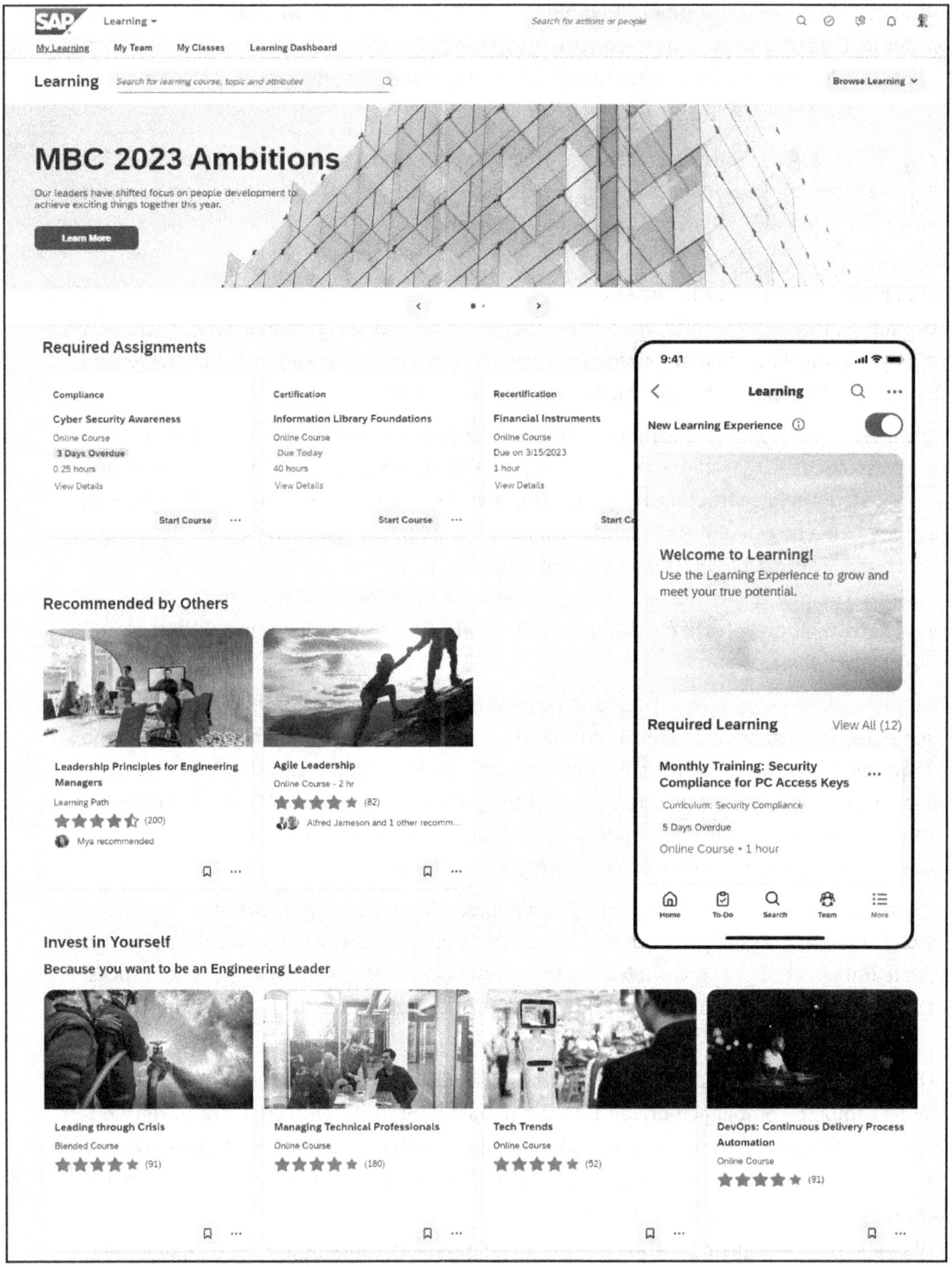

Figure 4.29: Integrated Learning Experience with SAP SuccessFactors

4.4.3 The Integration Model

The architecture supporting SAP SuccessFactors and SAP S/4HANA is based on a clear separation of responsibilities. SAP SuccessFactors manages workforce data and HR processes, while SAP S/4HANA governs financial execution and accounting. Integration connects these domains to support end-to-end business processes. The key scenarios are as follows:

- **Employee data alignment (SAP SuccessFactors → SAP S/4HANA)**
  Employee and organizational data from SAP SuccessFactors can be shared with SAP S/4HANA to support downstream processes such as payroll, cost allocation, and reporting. This reduces manual data entry and helps maintain consistency across systems.
- **Organizational structure alignment (SAP SuccessFactors → SAP S/4HANA)**
  Organizational structures maintained in SAP SuccessFactors can be aligned with financial structures in SAP S/4HANA through integration. This supports consistent reporting relationships and cost assignment across HR and finance.
- **Cost structure integration (SAP S/4HANA → SAP SuccessFactors)**
  Cost-related information, such as cost centers and financial attributes, can be shared from SAP S/4HANA to SAP SuccessFactors. This helps align workforce assignments with financial structures and supports workforce planning and analysis.
- **Payroll and compensation processing (SAP SuccessFactors ↔ SAP S/4HANA)**
  Workforce data from SAP SuccessFactors is used as input for payroll processing and financial activities. Compensation and payroll results can then contribute to financial reporting and planning processes, based on configured integration and processing cycles.
- **Time and cost tracking integration**
  Time data captured in SAP SuccessFactors can be transferred to downstream systems, where it contributes to payroll, project costing, or financial reporting. The timing and processing logic depend on the integration design and financial configuration.
- **End-to-end workforce and financial alignment**
  By connecting HR, workforce management, and financial systems, organizations can align workforce activities with financial outcomes. Integration supports visibility across hiring, development, compensation, and cost management processes, while allowing each system to perform its core function.

4.5 Summary

This chapter examined how SAP's LoB applications extend the capabilities of the ERP core to support end-to-end business processes across finance, spend, supply chain, human resources, and customer experience. While SAP S/4HANA provides the system of record for enterprise execution, applications such as SAP Concur, SAP Fieldglass, SAP Ariba, SAP SuccessFactors, SAP IBP, and the SAP CX suite of solutions address specialized business requirements that go beyond transactional processing.

We explored how these applications work together to create autonomous business domains. By sharing process context, business data, and common workflows, they help organizations move from isolated functional activities to connected, end-to-end operations. Much like the challenges faced by Maria, James, and David in the Korindal case study, value is created not by adding more systems but by connecting processes, decisions, and actions across the enterprise.

However, applications alone do not create autonomy. The next step is understanding how intelligence becomes embedded directly within business processes. In the next chapter, we examine how AI is integrated into enterprise workflows, enabling systems to assist, orchestrate, and increasingly execute work as part of everyday operations.

Chapter 5

Embedded AI in Business Processes

Autonomy is not a destination for the enterprise, but the fundamental shift that transforms a passive system of record into an active system of execution. Its true value is realized when the technology itself recedes and the focus shifts to outcomes, allowing the business to sense what is happening, decide what should happen next, and act as one connected system so that people can focus on the judgment, strategy, and decisions that no machine should make alone.

The preceding chapters of this book have traced a clear trajectory. Enterprise resource planning (ERP) modernization establishes the transactional backbone; cloud migration unlocks elastic scalability; embedded analytics surfaces real-time insight. Each wave of capability has elevated the enterprise, yet each has also revealed the same uncomfortable truth: Intelligence without autonomy still depends on humans to interpret, decide, and act. Systems that once recorded what happened can now help execute what is next. The evolutionary step SAP is building toward with its entire portfolio is the Autonomous Enterprise.

The Autonomous Enterprise is SAP's vision of a business in which artificial intelligence (AI) assistants and agents manage and execute end-to-end processes while people focus on judgment, strategy, and value creation. It is not a single product announcement or a feature toggle. It is an architectural philosophy that reorganizes SAP's entire technology investment around a single premise: When conditions change, the business responds as one connected system so that leaders can turn intent into action in real time. Autonomy does not mean human removal. It is what SAP calls the *new human advantage*, freeing people from disjointed systems, manual handoffs, and time-consuming reconciliation so they can lean into the strategy, creativity, relationships, and decisions that move the business forward.

Three value dimensions define the impact of this shift. First, the business runs in real time: Agents coordinate across the enterprise continuously, and there is no lag between signal and action. A supply disruption detected in Southeast Asia triggers replanning, customer notification, and financial reforecasting within minutes rather than days. Second, work moves up the value chain: People stop chasing, checking, and reconciling. Agents handle routine transactions from end to end so that employees can focus on the decisions that genuinely require human judgment. Third, the enterprise builds and operates with total control: Every AI action is governed, auditable, and traceable. Governance is not added as an afterthought; it is built in from the start, ensuring that speed and control are never a tradeoff.

This chapter applies these concepts to the end-to-end value streams introduced earlier in the book. We'll begin with finance, showing how Autonomous Finance, the most mature realization of the Autonomous Enterprise, transforms the plan to optimize financials, invoice-to-cash, invoice-to-pay, and treasury and working capital through 13 Joule Assistants coordinating more than 60 specialized agents. We then turn to lead-to-cash, tracing autonomous capabilities from opportunity to quote through order-to-fulfill, invoice-to-cash, and manage customers and channels. Next we focus on source-to-pay, where Autonomous Spend enhances source-to-contract, procure-to-receipt, and invoice-to-pay with document understanding, sourcing agents, and proactive supplier collaboration. We then cover plan-to-fulfill, following Autonomous Supply Chain Management (SCM) through planning, manufacturing, quality, and logistics with examples such as AI-assisted planning, optimizer explainability, and shop floor agents. Finally, we'll examine acquire-to-decommission, showing how intelligent technologies support the plan to optimize assets, acquire-to-onboard, operate-to-maintain, and offboard-to-decommission through the Asset and Service Assistant, Predictive Maintenance Agent, and Maintenance Planner Agent. Taken together, these sections demonstrate how the Autonomous Enterprise is not a distant aspiration but an operational reality being delivered today, turning connected applications and governed data into intelligent, adaptive processes at the enterprise scale.

First, let's establish the technologies that underpin SAP's ecosystem for the Autonomous Enterprise.

5.1 How SAP Makes the Autonomous Enterprise Real

Four architectural elements come together to deliver this vision. Understanding their interplay is essential to appreciating how the capabilities described throughout this chapter operate in practice.

Joule serves as the engagement layer, the single interface where people set intent and the system brings together the right insights, workflows, and agents to deliver the outcome. Chapter 7 covers Joule and Joule Work in more detail. Joule Work is the dynamic workspace built around intent, enabling users to interact with the full SAP application landscape through a single conversational interface.

Joule supports multiple interaction patterns: informational queries grounded in live business data, navigational prompts that jump directly to the right application screen, transactional instructions that create or change business objects, and analytical queries that surface live KPIs and trends. Because Joule operates within SAP's authorization model and the organization's live operational data, it carries enterprise-grade context and security that generic AI tools cannot replicate. The experience comes to life through Joule Work, a dynamic workspace where users express what they want to accomplish and the system surfaces the right insights, coordinates agents, and executes across systems without requiring navigation through application menus.

SAP Autonomous Suite forms the operational core of the business. It combines SAP Cloud ERP applications, a unified data foundation, and embedded AI into a single integrated system organized into five autonomous business functions: Autonomous Finance, Autonomous Supply Chain Management, Autonomous Spend, Autonomous HCM, and Autonomous CX. Each function describes how SAP's applications, data, and agents work together to run a business domain from end to end. The existing application brands, SAP S/4HANA Cloud, SAP SuccessFactors, SAP Ariba, SAP Concur, and others, remain. What changes is how they operate: Agents can now act on the data, process logic, and controls those applications already contain, turning them from passive systems of record into active participants in business execution. Figure 5.1 shows SAP Autonomous Suite, illustrating how SAP Cloud ERP applications, a unified data layer, and embedded AI combine into a cohesive operational core.

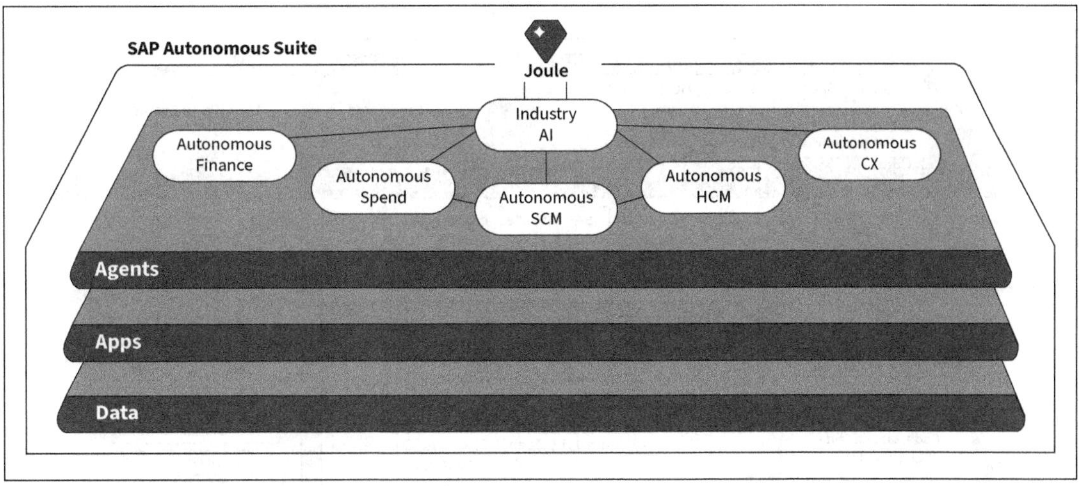

Figure 5.1: SAP Autonomous Suite

Inside each function, the operating model follows a clear hierarchy: People direct, assistants orchestrate, and agents execute. People set the strategic direction and define the guardrails within which automation operates. Joule Assistants are human-facing interfaces grounded in a specific role or process that interpret user intent and coordinate specialized agents. Joule Agents are goal-driven systems that can plan, reason, and act autonomously across applications, processes, and business data to complete multistep tasks. In practice, assistants and agents work together seamlessly; the assistant interprets what the user wants to accomplish and orchestrates one or more agents to carry out the underlying actions, while people retain oversight for governance, exceptions, and strategic decisions.

Consider a concrete example. An Accounts Receivable Assistant understands the context of a collections specialist's role. When the specialist asks about high-risk overdue accounts, the assistant coordinates multiple agents: a Receivable Account Analysis Agent to score customer risk, a Collection Account Preparation Agent to compile relevant documentation, and a Collections Email Outreach Agent to draft customer communications, all working in concert to deliver an actionable recommendation. The specialist reviews, adjusts, and approves, and the agents execute. The cycle time from signal to action compresses from days to minutes.

SAP Business AI Platform provides the governed foundation that makes everything above it trustworthy at enterprise scale. It is the enterprise AI foundation that combines deep process context, unified business data, and purpose-built models with enterprise governance so that agents and applications understand the business, not just the prompts. Figure 5.2 illustrates the SAP Business AI Platform structure, detailing the build, contextualize and reason, and govern pillars that provide the trusted foundation for enterprise AI.

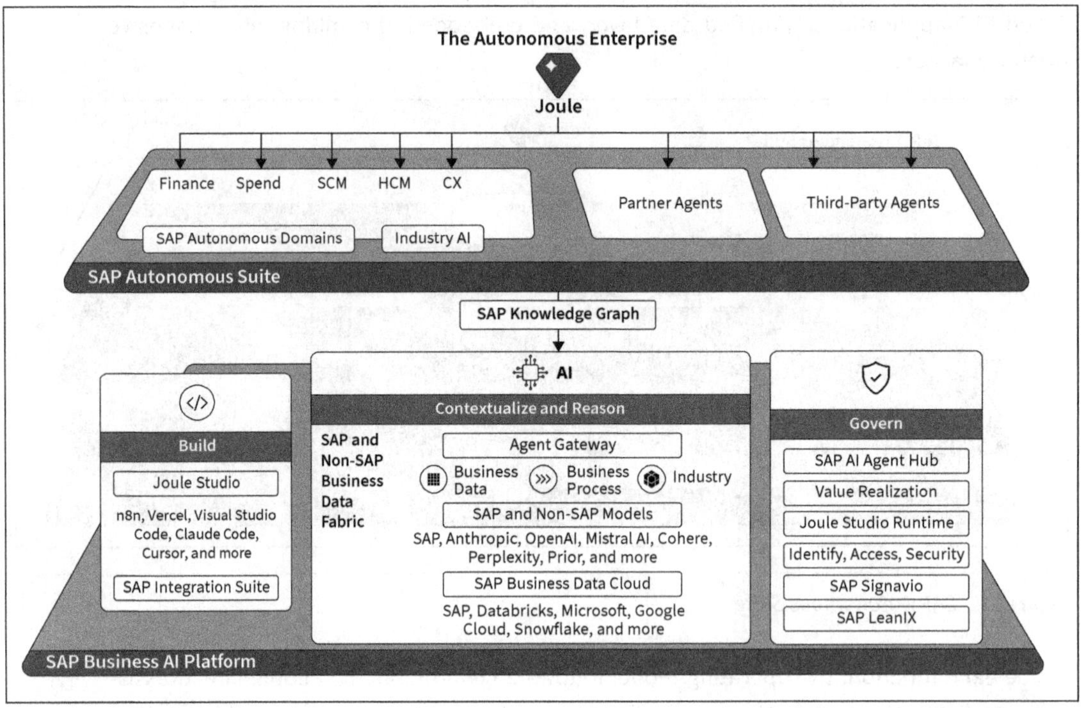

Figure 5.2: SAP Business AI Platform

The platform is organized around three pillars, which we'll discuss in more detail in Chapter 7:

- The *build* pillar provides a unified environment to create AI agents, applications, and workflows grounded in business context, deploy them without operational overhead, and integrate them across the enterprise landscape. Joule Studio, SAP's AI-first development experience, enables both professional developers and business users to go from intent to deployed agent in minutes rather than days. SAP Integration Suite connects agents across SAP and third-party landscapes using Model Context Protocol (MCP) and application programming interface (API) management, ensuring that agentic capabilities span the entire enterprise rather than stopping at SAP system boundaries.

- The *contextualize and reason* pillar delivers AI that reasons with deep, universal business context. At its core sits SAP Business Data Cloud, a business data fabric that serves as the trusted knowledge core for every enterprise application and agent. It unifies SAP and third-party data so that every AI initiative operates from the same governed foundation. SAP Knowledge Graph, with over 7.3 million data fields encoding 50 years of business process expertise, provides the semantic relationships between customers, orders, materials, suppliers, and business rules that allow agents to reason far beyond simple pattern matching. Purpose-built SAP models such as SAP-RPT-1 for operational predictions and SAP-ABAP-1 for development productivity arrive prebuilt in this context, while governed access to leading third-party models ensures that organizations always have the right model for the job.
- The *govern* pillar provides the command center for enterprise-wide data and AI agent governance. SAP AI Agent Hub enables organizations to manage agents across their full lifecycle, from planning and building through discovery, provisioning, runtime execution, observability, optimization, and decommissioning. Identity, access, and security controls ensure that agents operate within organizational policies. SAP Signavio and SAP LeanIX provide the architecture and process intelligence to continuously track where AI is creating value and identify where it could create more. This is governance not as an afterthought but as a first-class architectural capability that allows enterprises to scale AI with the same rigor they apply to financial controls.

Industry AI extends each autonomous function with vertical depth, embedding sector-specific process knowledge, data models, and regulatory logic directly into industry workflows. Where autonomous functions organize AI around a single business domain, Industry AI spans multiple domains along an industry's value chain, such as asset management in energy, regulated manufacturing in life sciences, and subscription monetization in high tech. These are problems that cut across functions and require the vertical depth that comes from 50 years of industry coinnovation across more than 25 industries. Figure 5.3 shows the Industry AI domains, illustrating how sector-specific process knowledge and agentic capabilities are embedded vertically within each autonomous function.

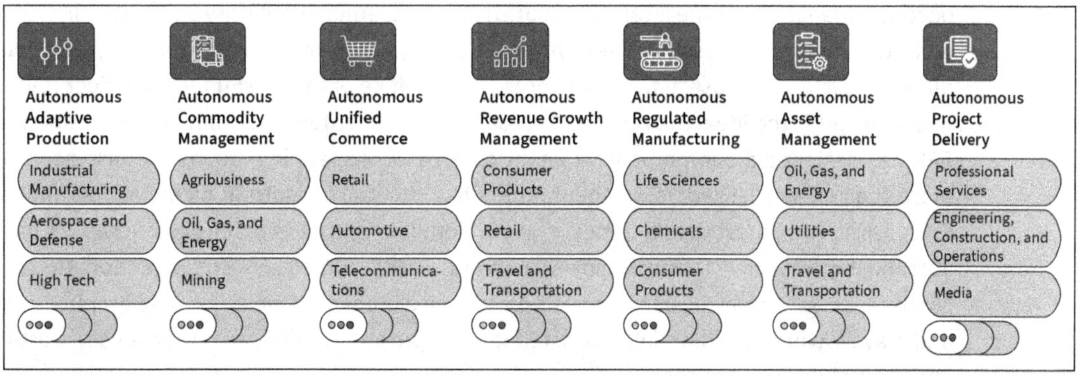

Figure 5.3: Industry AI Domains

From a process perspective, the Autonomous Enterprise manifests in three recurring patterns that you will encounter throughout the sections that follow:

1. First, the system becomes *proactive*. Situation handling means that users no longer search for problems. Instead, the system surfaces exceptions: an expiring contract, an anomalous journal entry, a delayed shipment, a liquidity shortfall with context and recommended actions. Agents detect conditions in real time and initiate responses within preapproved guardrails, escalating to humans only when the situation exceeds their authority or requires judgment that cannot be codified.
2. Second, work becomes *conversational*. Instead of navigating through transaction codes and complex application menus, users describe what they want in plain language and let Joule orchestrate the right combination of assistants, agents, analytics, and transactions. The interface disappears, and what remains is the outcome.
3. Third, the system enables *agentic execution*. Specialized agents pursue goals over multiple steps, consult other agents, access external services, and execute transactions all while leaving a transparent trail of decisions for audit and governance. An Accounts Receivable Assistant does not merely surface overdue invoices. It analyzes risk, prioritizes accounts, drafts customer communications, coordinates with dispute resolution agents, executes collection actions, and reports outcomes autonomously, continuously, and with full auditability.

> **Note**
>
> The autonomous assistants and agents discussed in this book are based on available and roadmap features at the time of writing (summer 2026). Details may change as AI capabilities evolve and business scenarios emerge across the enterprise.

5.2 Finance

Autonomous Finance is the most mature realization of SAP's Autonomous Enterprise vision because finance sits at the intersection of strategic steering, regulatory accountability, and daily transaction processing. Every revenue event, supplier payment, hedging decision, planning cycle, and close activity converges in the office of the chief financial officer (CFO), which makes it the ideal proving ground for the agentic enterprise. To meet this mandate, SAP has designed a comprehensive agentic framework spanning 13 Joule Assistants and more than 60 specialized Joule Agents that cover the full breadth of financial operations. Each assistant orchestrates a critical finance domain—accounts receivable, accounts payable, billing, financial closing, financial planning, cash and treasury, consolidation, tax and compliance, governance, sales operations, expense management, enterprise architecture, and travel—while coordinating a constellation of purpose-built agents that sense, reason, and act on live financial data.

We'll provide an overview of the Autonomous Finance capabilities next. The sections that follow take each of the four critical value streams in turn: plan to optimize financials, invoice-to-cash, invoice-to-pay, and treasury and working capital, demonstrating how SAP embedded intelligence, Joule Assistants, and Autonomous Finance agents execute these processes with quantifiable impact.

5.2.1 Overview of Autonomous Finance

The architecture follows the three-layer model that defines the Autonomous Enterprise. Joule serves as the engagement layer, where finance professionals' express intent and Joule Assistants surface the right insights, detect risks, simulate scenarios, and coordinate agents across systems. The SAP Autonomous Suite provides the operational core where SAP Cloud ERP applications execute financial transactions from end to end, for accounting, planning, treasury, tax, billing, and close. SAP Business AI Platform supplies the governed foundation that combines deep process context, unified business data through SAP Knowledge Graph, and enterprise-grade governance so that every agent action is auditable, compliant, and traceable. Figure 5.4 shows the Autonomous Finance overview, presenting the three-layer architecture of Joule, assistant orchestration, and agent execution across all finance subdomains.

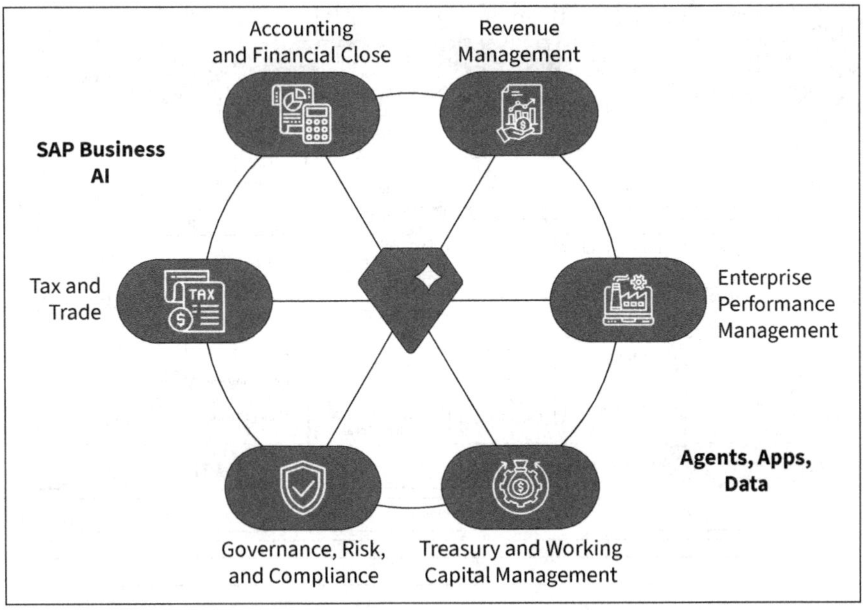

Figure 5.4: Autonomous Finance Overview

Joule Work, the new dynamic workspace introduced at SAP Sapphire 2026, fundamentally changes how finance teams interact with this architecture. Rather than navigating between applications, finance professionals work within a single, intent-driven environment in which they can get answers to complex questions, find information across any connected system, put agents to work on multistep tasks, and build custom tracking applications or monitoring agents, all personalized to their role: A new accountant asks Joule how to run month-end reconciliation and is guided through the process live; a controller requests live finance KPIs, risk indicators, and alerts in one personalized view; a CFO asks what drove Q3 variance in APAC operating expense and receives an instant narrative with detailed drill-down. Beyond navigational assistance, this is a fundamentally new way of executing financial work.

The operating model is precise: People direct, assistants orchestrate, and agents execute. Inside each domain, the cycle runs continuously. People set the direction, applications generate signals, data provides context, agents act, and the value compounds. This hierarchy separates the Autonomous Enterprise from isolated point solutions. Finance professionals retain full authority over strategy, judgment, and exception handling while agents handle the routine coordination, analysis, and execution that previously consumed 60 to 80 percent of their productive time. Figure 5.5 shows the Autonomous Finance portfolio overview, mapping all assistants and agents across accounts receivable, accounts payable, treasury, financial close, and compliance.

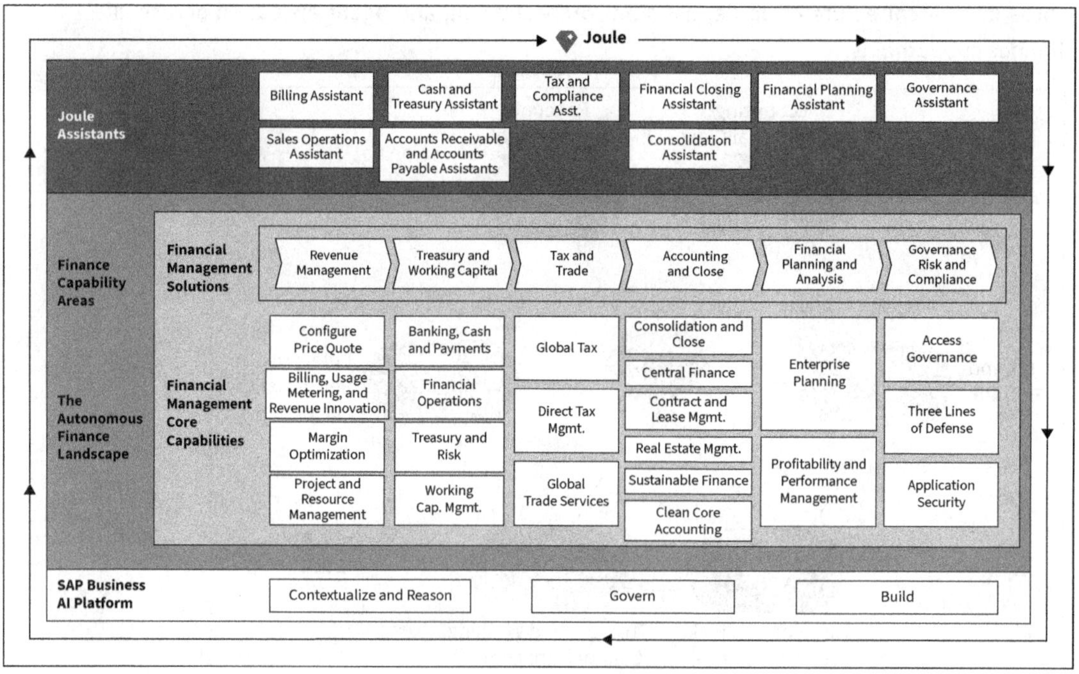

Figure 5.5: Autonomous Finance Portfolio Overview

What sets this approach apart is cross-functional orchestration. Assistants collaborate across boundaries to resolve exceptions and optimize outcomes in real time. When the Accounts Receivable Assistant detects a high-value overdue invoice tied to an open dispute, it invokes the Dispute Resolution Agent to accelerate settlement. When the Cash and Treasury Assistant identify a looming liquidity shortfall, it synthesizes signals from both accounts receivable and accounts payable to pinpoint root causes and recommend countermeasures. When the Financial Closing Assistant encounters an intercompany mismatch, it coordinates with the Consolidation Assistant to resolve the discrepancy before it delays group reporting. This choreography, mediated through Joule, separates Autonomous Finance from disconnected automation. Agents do more than automate tasks; they reason over shared data, anticipate risks, and act in concert.

The impact is material and measurable. SAP's value analysis, developed with KPMG and SAP Value Engineering and modeled on a €1 billion manufacturing company, quantifies the gains across every finance domain (see Figure 5.6):

- Revenue management contributes a 60 to 80 percent billing productivity uplift and a 10 to 20 percent reduction in order-to-cash cycle time.
- Accounts receivable delivers a 10 to 20 percent reduction in days sales outstanding (DSO) through accelerated collections and faster dispute resolution.
- Accounts payable generates a 50 to 70 percent cash cycle time reduction with 70 percent less cash management effort.
- Accounting and close drives a 20 to 30 percent productivity increase in general ledger and closing activities with a 10 to 20 percent time reduction in the closing cycle.
- Financial planning frees 40 to 60 percent more time for analysis and delivers a 60 percent reduction in time spent on reporting and forecasts.
- Treasury and working capital achieve a 50 to 70 percent cash cycle time reduction with a 5 to 20 percent improvement in key quality metrics.
- Tax and trade realize a 50 percent decrease in trade classification effort with automated e-invoicing setup.
- Governance drives a 50 to 60 percent productivity increase in risk management and auditing.

Autonomous Enterprise

Explore the latest Autonomous Enterprise assistants, agents, and their value analysis at *https://www.sap.com/products/artificial-intelligence/ai-assistant.html*.

Figure 5.6 shows the value creation model for Autonomous Finance, illustrating how a unified application, data, and AI architecture delivers measurable and compounding financial impact.

Note

All of the value analysis numbers in this chapter are indicative only and are sourced from benchmarking studies from SAP Value Advisory.

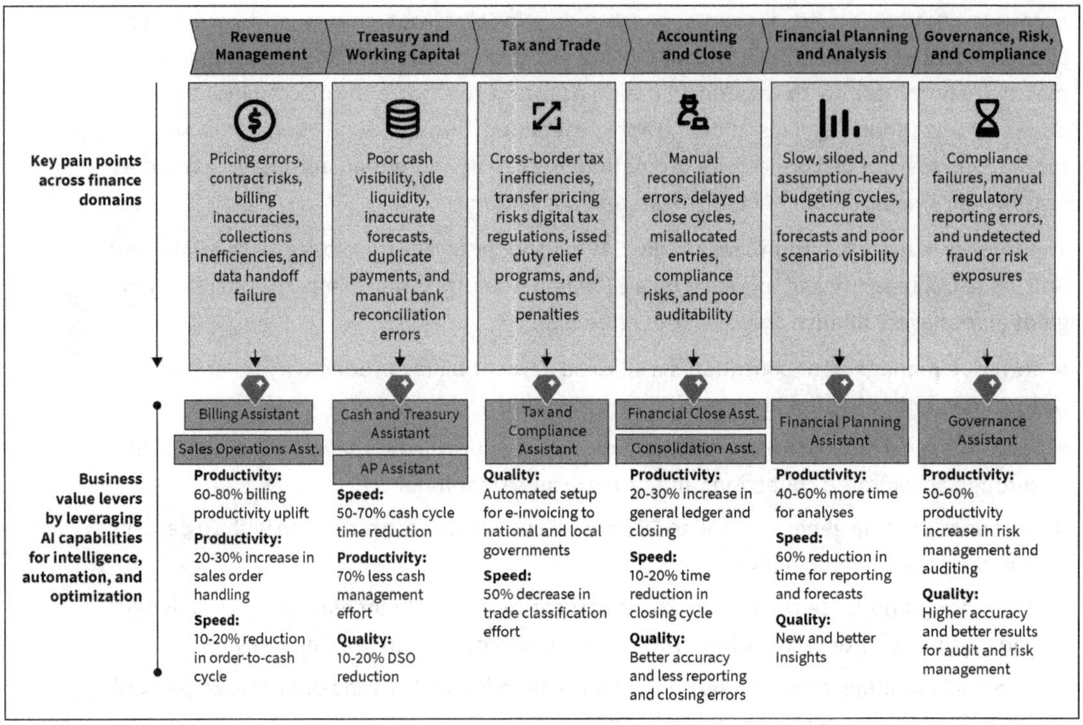

Figure 5.6: Value Creation Model for Autonomous Finance

This value comes from a unified architecture in which applications, data, and AI work together as a cohesive management fabric. SAP Cloud ERP provides the transactional backbone across accounting, planning, treasury, tax, billing, and revenue processes. SAP Business Data Cloud, now the data fabric within the contextualize and reason pillar of SAP Business AI Platform, supplies a curated semantic layer of finance data products, ensuring consistent definitions, lineage, and governance. SAP Business AI completes the stack with embedded machine learning (ML), generative capabilities, and agent systems that deeply understand posting logic, master data, regulatory requirements, and cross-process relationships. In practice, finance teams work side by side with Joule and with specialized assistants and agents that continuously monitor enterprise data, suggest next-best actions, and, where appropriate, execute autonomously within defined guardrails.

SAP differentiates Autonomous Finance through three structural advantages that competitors building from the outside in cannot replicate. First, deep process and industry knowledge: 50 years of business process and industry intelligence encoded directly in SAP applications and SAP Knowledge Graph, producing 3.5 times higher prediction quality than generic large language models (LLMs) and reducing hallucination risk. Second, semantically rich business data: 7.3 million data fields in SAP Knowledge Graph where customers link to contracts, materials tie to orders, and suppliers score for risk, not with raw tables but with genuine business context. Third, enterprise-grade governance: All agents follow team-defined guardrails that enforce corporate policies and regulatory requirements, with lifecycle management, security, and policy enforcement built in from planning through production. SAP offers the only enterprise AI platform with ISO/IEC 42001 certification for AI governance, aligned to the EU AI Act and NIST Tier 3 standards.

The timing is urgent. CFOs are moving from traditional scorekeepers to strategic value architects in a world of constrained growth, higher capital costs, and escalating regulatory complexity. Many remain constrained by fragmented legacy systems, manual reconciliations, and limited predictive analytics. Autonomous Finance answers this challenge by replacing fragmentation with continuous, context-aware intelligence and action. It compresses close cycles, accelerates cash, improves forecast accuracy, strengthens compliance, and unlocks trapped working capital, not as a patchwork of tools but as a coordinated system of Joule Assistants and Agents operating on a shared foundation.

The path to Autonomous Finance is not a rip-and-replace transformation. It is a deliberate progression through four stages:

1. First, modernize and harmonize data by establishing clean, governed financial master data and a unified chart of accounts.
2. Second, automate the core by deploying SAP Cloud ERP with embedded intelligence for touchless processing, automated matching, and continuous close.
3. Third, add agentic intelligence by activating Joule Assistants and Joule Agents that reason across processes, predict outcomes, and execute autonomously within guardrails.
4. Fourth, defend and reduce risk by embedding continuous controls, real-time compliance monitoring, and AI-powered audit capabilities.

Each stage compounds the value of the prior one, and each is available today or arriving within the current roadmap.

5.2.2 Plan to Optimize Financials

Financial planning, budgeting, and forecasting represent the strategic nerve center of the finance function. Every capital allocation decision, every headcount authorization, every market expansion commitment ultimately traces back to a planning process that translates business ambition into financial commitments. Yet in most organizations, this process remains stubbornly manual. Planning cycles stretch four to six weeks as teams consolidate data from dozens of cost centers, business units, and geographies into monolithic spreadsheets. Forecast accuracy hovers at plus or minus 15 to 20 percent because models rely on static assumptions that cannot adapt to changing market conditions in real time. Variance analysis is retrospective; controllers spend days after period close manually investigating deviations and writing commentary that arrives too late to influence the decisions that caused them. The result is a planning function that consumes enormous effort while delivering insights that are perpetually stale.

At Korindal, the financial planning and analysis (FP&A) team produces quarterly rolling forecasts, annual budgets, and monthly variance reports across twelve business units spanning North America, Europe, and Asia Pacific. The cycle is familiar to every finance organization. Three weeks before a board meeting, the head of FP&A sends templates to business unit controllers, who manually populate them with local estimates, often derived from conversations rather than data-driven models. The central team then spends the next two weeks consolidating, reconciling discrepancies, and preparing presentation-ready commentary. By the time the forecast reaches the CFO, the underlying assumptions are already outdated. When the board asks "What happens if we delay the European expansion by one quarter?," the answer takes days to model because each scenario requires manual rework of the entire consolidation chain.

In the following sections, we'll see how embedded AI capabilities optimize the FP&A process.

SAP Analytics Cloud and Joule

SAP Analytics Cloud for planning fundamentally restructures this process by embedding ML directly into the planning workflow. Predictive planning uses automated ML to turn historical data into accurate forecasts, automatically selecting the optimal algorithm, exponential smoothing, linear regression, or advanced time series models for each data series. Planners generate predictive forecasts directly within time series or line charts, projecting future values with a single click rather than building manual forecast models. The system continuously updates forecasts as new actuals flow in, eliminating the static snapshot problem that plagues traditional planning cycles. SAP quantifies the impact as a 60 percent reduction in time spent on reporting and forecasts, compressing what previously took four to six weeks into a process that runs continuously.

The intelligence extends beyond raw forecasting into analytical reasoning. Smart Insights uses deviation analysis to automatically identify the drivers behind significant variances, presenting textual explanations combined with supporting visualizations. The AI-assisted analytical business insights capability in the Review Booklet applications transforms complex financial data into clear, actionable narratives through a generative AI side panel that offers quick actions for common reporting tasks, summarizing cost center performance, explaining budget-to-actual deviations, and highlighting trends that warrant management attention. For group-level reporting, the same approach extends to the Group Financial Statement application, enabling consolidated variance analysis across entities.

Natural language query democratizes access to financial data by allowing any user to ask questions in plain language, such as "What were the top three cost centers by budget overrun last quarter?" or "Show me revenue variance by region compared to the prior year," and receive instant, data-grounded answers. This capability, powered by the SAP HANA Cloud vector engine for enhanced response accuracy, eliminates the traditional bottleneck in which business users must wait for analysts to build custom reports. Joule with SAP Analytics Cloud takes this further by embedding a full conversational AI layer into the planning workflow, enabling users to create planning scenarios, modify forecast values, generate data actions, and build analytical stories through natural language interaction.

For cost accounting and management reporting, Joule provides conversational access to budget consumption, allocation results, and cost postings. Cost center owners can ask Joule for current available budget information to react quickly to critical consumption situations. Business analysts can view amounts allocated across cost objects, profitability segments, or profit centers through natural language queries, eliminating the need to navigate complex transaction hierarchies. When production cost postings generate errors during period close, the AI-assisted error-resolution capability helps cost accountants identify root causes and resolve them promptly, ensuring all costs are settled before deadlines with minimal manual effort.

The Financial Planning Assistant

The Financial Planning Assistant, part of the Autonomous Finance framework, orchestrates five specialized agents that collectively transform financial planning from a periodic, labor-intensive exercise into a continuous, always-on capability. SAP quantifies the headline impact: 40 to 60 percent more time freed for analysis and a 60 percent reduction in time spent on reporting and forecasts, with measurably higher forecast accuracy and reporting insights.

Let's examine each agent offered via the Financial Planning Assistant:

- **Financial Forecasting Agent**
  The Financial Forecasting Agent applies AI-driven forecasting with driver-based modeling and external data enrichment to automate rolling forecasts. Rather than producing a single-point estimate, the agent continuously updates projections as new data arrives, sensing changes in revenue patterns, cost behaviors, and market conditions and thus enabling adaptive planning that responds to shifting business dynamics without requiring manual forecast rework.
- **Variance Analysis Agent**
  The Variance Analysis Agent performs multidimensional variance decomposition automatically, identifying not just that a variance exists but precisely which combination of drivers, volume, price, mix, foreign exchange (FX), and timing caused it. It generates automated narrative explanations and provides drilldown automation, delivering instant variance insights that previously required days of manual investigation.
- **Market Sentiment Analysis Agent**
  The Market Sentiment Analysis Agent extends the planning horizon beyond internal data by monitoring external signals, commodity prices, currency movements, industry indicators, and competitive developments, and quantifying their potential impact on financial plans. This transforms planning from an internally focused exercise into an externally aware discipline that can anticipate disruptions before they hit the income statement.
- **Model Creation Agent**
  The Model Creation Agent addresses the structural challenge of building and maintaining planning models. Using AI-assisted model generation, it accelerates the creation of

driver-based models, allocation logic, and forecasting structures that traditionally required weeks of configuration by specialized planning consultants. Figure 5.7 shows the Financial Planning Assistant, illustrating how five specialized agents transform planning from a periodic exercise into a continuous, always-on capability.

- **Conversational Planning Agent**
  The Conversational Planning Agent enables finance professionals to interact with the entire planning process through natural language. Planners describe what they want, such as "Create a revenue scenario assuming 8 percent growth in EMEA and flat performance in Americas" or "Show me the P&L impact if raw material costs increase 12 percent in Q3," and the agent translates intent into executable planning actions, scenario configurations, and analytical outputs.

Figure 5.7 shows the Financial Planning Assistant, illustrating how the five specialized agents work together to transform planning from a periodic, labor-intensive exercise into a continuous, always-on capability.

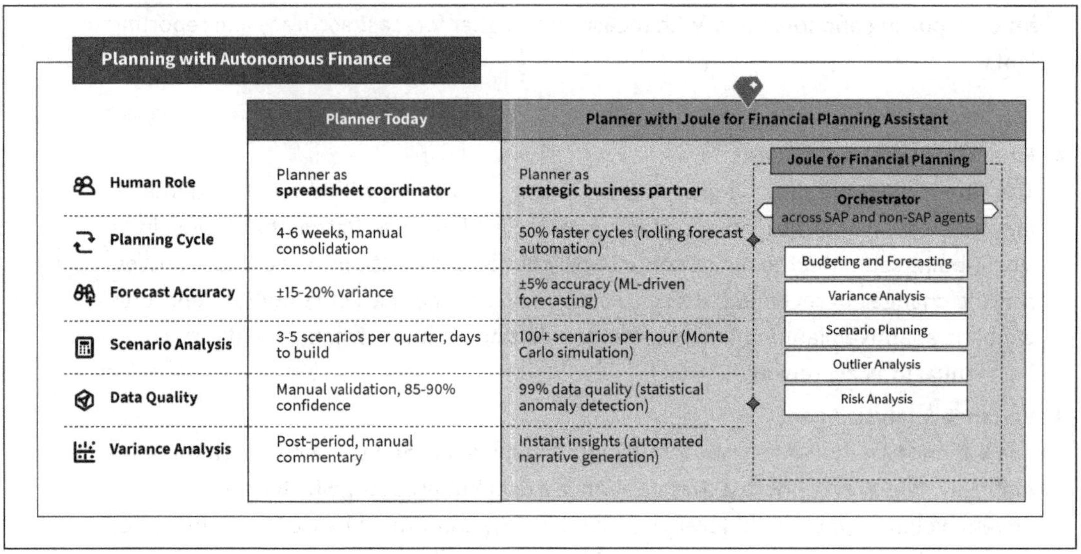

Figure 5.7: Financial Planning Assistant

Case Study: Planning Transformation

At Korindal, the planning transformation delivers immediate, measurable impact. The quarterly forecast cycle compresses from three weeks to four days as SAP Analytics Cloud's predictive planning generates baseline forecasts automatically and the Financial Forecasting Agent enriches them with driver-based adjustments. Forecast accuracy improves from plus or minus 18 percent to plus or minus 6 percent within three planning cycles as ML models learn Korindal's revenue patterns, cost behaviors, and seasonal dynamics.

The most dramatic change occurs during the next board meeting. When the CEO asks about the financial impact of accelerating the Asia-Pacific expansion while simultaneously managing a raw material cost increase, the CFO does not promise to get back to them next week. Instead, the Conversational Planning Agent generates twelve scenarios in real time, varying expansion timing, cost assumptions, and FX impacts. The Variance Analysis Agent explains the key sensitivity drivers, and the Market Sentiment Analysis Agent quantifies the external risk factors for each scenario. The board makes an informed decision in the meeting rather than a week later, a transformation that changes how planning works and how strategic decisions are made.

5.2.3 Invoice-to-Cash

The invoice-to-cash process is the arena in which companies ultimately win or lose the working capital battle. Revenue recognition may technically occur at the point of product delivery or invoicing, but the true test of commercial success is actual cash collection. In most global organizations, that chain is long, exceptionally fragile, and rife with manual interventions. Payments arrive with incomplete remittance data; customers dispute charges weeks after initial invoicing; collection priorities are dictated by human intuition or static spreadsheet formulas rather than data-driven risk models; billing errors create downstream exceptions that propagate through receivables. Consequently, finance teams spend the weeks following month end analyzing why DSO drifted instead of managing it in real time.

SAP intelligent technologies across the invoice-to-cash value stream are designed to completely flip that script. By deploying ML and autonomous agents through three coordinated assistants, the Billing Assistant, the Accounts Receivable Assistant, and the Sales Operations Assistant, SAP turns the entire order-to-cash cycle from a reactive reporting function into a proactive control system that manages cash conversion day by day, customer by customer, and invoice by invoice.

At Korindal, the CFO reviews the identical set of metrics every single month. Revenue is precisely on plan and margins remain stable despite global market volatility. The foundational ERP systems have been successfully upgraded across all operating regions. Yet, despite these operational victories, one critical number continues to drift in the wrong direction: free cash flow. And there is one primary culprit: DSO.

DSO is the ultimate measure of the ability of a company to translate revenue into actual cash in the bank. On paper, the cash has already been earned. In practice, it remains locked in unresolved customer disputes, delayed collections outreach, billing errors that block payment, and fragmented follow-ups spread across regional teams. The finance teams spend the first two weeks of every new month explaining why the metric moved, rather than acting proactively to prevent the movement. The frustration for Korindal's CFO is not a lack of data; comprehensive reports and executive dashboards exist. What is fundamentally missing is operational control, from the moment a billing document is created through final cash receipt.

Let's take a closer look at the embedded AI capabilities that enhance the invoice-to-cash process and enable the employees in our case study to address their business problems.

The Billing Assistant

Cash collection begins with accurate billing. If invoices contain errors, incorrect quantities, wrong prices, missing references, or misapplied tax rates, they inevitably trigger disputes, payment delays, and write-offs downstream. The Billing Assistant brings agentic intelligence to revenue operations by orchestrating four specialized agents that automate billing creation, detect anomalies before invoices reach customers, and resolve disputes at the source. It includes the following agents:

- **Billing Creation Agent**
  The Billing Creation Agent automates the generation of billing documents by analyzing sales orders, delivery confirmations, and contractual terms to produce accurate invoices without manual intervention. It prioritizes billing queues based on revenue impact, payment terms, and customer credit status, ensuring that high-value documents process first.
- **Billing Anomaly Agent**
  The Billing Anomaly Agent applies pattern recognition across historical billing data to detect outliers before documents are released: unusual quantities, pricing deviations from contractual terms, tax calculation errors, or inconsistencies between order and delivery data. By catching errors at the point of creation rather than after customer complaint, the agent eliminates the single largest source of downstream disputes.
- **Billing Adjustment Agent**
  The Billing Adjustment Agent handles correction workflows when discrepancies are identified. Rather than routing adjustment requests through manual approval chains, the agent evaluates the adjustment against policy rules, proposes the appropriate correction mechanism—credit memo, debit memo, or invoice cancellation and reissuance—and executes the approved action.
- **Billing Posting Agent**
  The Billing Posting Agent ensures that billing documents post correctly to the general ledger, accounts receivable subledger, and revenue recognition schedules. It validates account assignments, tax codes, and period allocations before posting, eliminating the journal entry corrections that traditionally consume close-cycle effort.

Figure 5.8 shows the Billing Assistant, illustrating how specialized agents handle billing creation, anomaly detection, adjustment workflows, and general ledger posting.

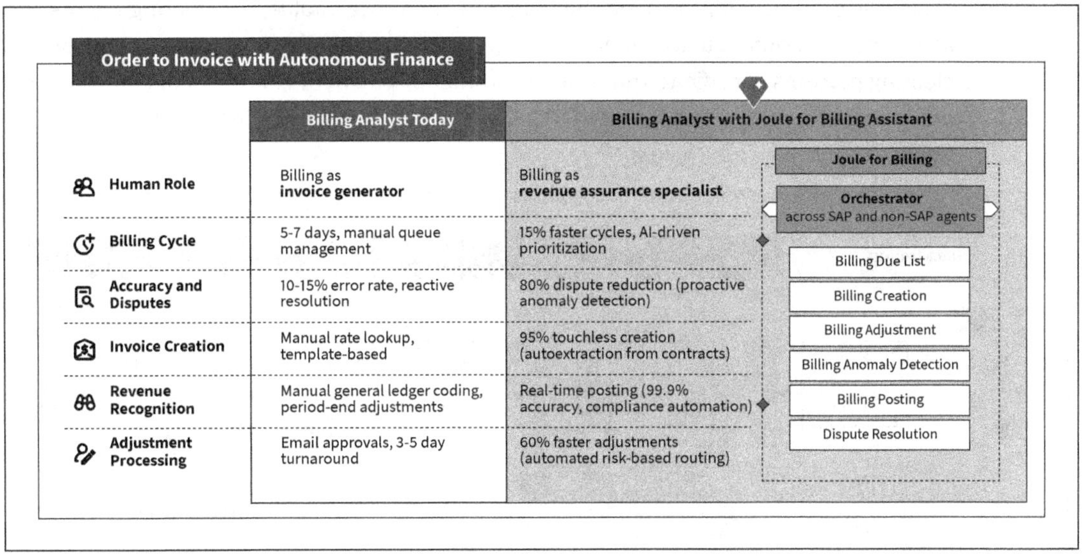

Figure 5.8: Billing Assistant

SAP quantifies the combined impact: 70 percent less effort creating prioritized billing documents, 40 percent less time correcting disputed invoices, and 95 percent less time resolving posting issues. For the broader order-to-cash cycle, the Billing Assistant contributes to a 60 to 80 percent billing productivity uplift and a 10 to 20 percent reduction in end-to-end order-to-cash cycle time. These are not aspirational targets. They reflect the elimination of manual handoffs and error-correction loops that currently consume the majority of billing team effort.

SAP Cash Application

SAP Cash Application, delivered as an embedded AI service in SAP S/4HANA Cloud and SAP Cloud ERP Private, fundamentally changes the economics of payment processing. It utilizes advanced ML to match incoming payments to open invoices with accuracy rates that approach or exceed human performance. Unlike rigid rule-based matching, which requires explicit configuration for every conceivable payment scenario, SAP Cash Application continuously learns from historical behavior. It analyzes exactly how accountants have matched payments in the past, identifies hidden patterns in remittance formats and customer-specific quirks, and builds predictive models that improve continuously as more enterprise data accumulates. Figure 5.9 shows receivables automation with SAP Cash Application, illustrating how bank statement data is matched against open items using machine learning.

The technical workflow is straightforward but immensely powerful. Bank statements flow into SAP S/4HANA through standard interfaces. Payment line items that fail standard rule-based clearing are immediately routed to SAP Cash Application, which runs securely on SAP AI Core within SAP Business Technology Platform (SAP BTP). The ML model rigorously evaluates each payment against the ledger of open receivables, considering factors such as payment amount, customer account, reference text, memo lines, dates, and historical clearing patterns specific to that exact customer or payment channel. For each payment, the model generates one or more matching proposals, each assigned a mathematical confidence score.

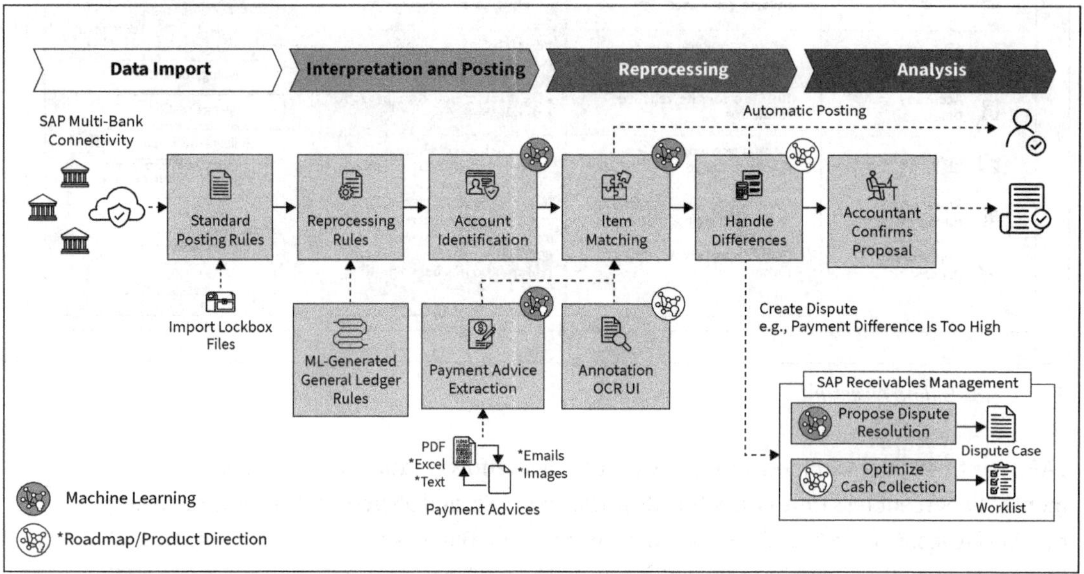

Figure 5.9: Receivables Automation with SAP Cash Application

Proposals that score above a predefined threshold configured by the finance team, typically 85 percent or higher, are automatically posted, seamlessly clearing both the bank item and the receivable without any human review. Lower-confidence proposals automatically appear in the Manage Incoming Payments application, ranked by confidence, so that accountants can efficiently review, accept, reject, or modify the proposed matches. Critically, every single decision the accountant makes feeds directly back into the continuous learning loop. If a proposal is accepted, the model reinforces that pattern; if rejected or modified, it adjusts its future behavior accordingly. Over time, the model learns intricate, customer-specific behaviors, such as from a customer who always rounds payments to the nearest hundred or one who references multiple invoices in a single free text field; in this way, accuracy steadily improves.

SAP's published value studies report dramatic operational improvements. Organizations utilizing SAP Cash Application have increased automatic matching rates from 30 percent to 90 percent, achieving a corresponding 71 percent reduction in manual reconciliation effort. For mid-sized companies processing moderate payment volumes, this yields annual benefits in the range of $59,000 to $89,000. For high-volume businesses that process tens of thousands of micropayments monthly, the savings are even more pronounced as the model effortlessly handles complex payment advice extraction and line-item matching at massive scale without manual setup.

Note

SAP Cash Application has been upgraded from a traditional ML model to a generative AI tabular data model using SAP-RPT-1.5, SAP's proprietary foundation model purpose-built for structured business data prediction. SAP-RPT-1.5 is pretrained on millions of SAP financial transactions, which means it arrives with deep contextual understanding of payment-matching patterns across industries and does not require organizations to build training datasets from scratch. The result is faster time to value, higher matching accuracy, and continuous model improvement as it learns from each organization's specific remittance behavior.

Explore the SAP Cash Application functionality and its benefits on SAP Discovery Center at *http://s-prs.co/v633503*.

The broader strategic benefit for the CFO is absolute visibility and control. When 90 percent of payments clear automatically within hours of receipt, unapplied cash balances shrink rapidly, receivables aging becomes accurate in near real time, and treasury teams can forecast cash positions with absolute confidence. Cash application is no longer an operational bottleneck; it becomes a distinct competitive advantage.

The benefits of automated matching are well illustrated by organizations that have deployed SAP Cash Application in production, for example, Winterhalter (*http://s-prs.co/v633505*). For their accounting teams, managing a high volume of payments from a diverse global customer base created a significant manual workload that caused particular strain during peak periods or staff absences. After implementing SAP Cash Application, Winterhalter utilized ML models that learned from their historical data to automatically match payments to open items. The results were immediate and transformative. Within less than two months of deployment, the organization successfully reduced the administrative burden on its accounting staff, allowing them to shift their focus from repetitive data entry to more complex, value-added financial tasks. This successful implementation demonstrates that even for mid-sized enterprises with sophisticated manufacturing operations, intelligent automation provides a scalable foundation to handle increased workloads without expanding headcount.

The Accounts Receivable Assistant

Where SAP Cash Application addresses the high-volume transactional matching problem, the Accounts Receivable Assistant brings full agentic intelligence to collections management, dispute resolution, and receivable risk control. This is one of the most agent-rich assistants in all of Autonomous Finance, orchestrating 11 specialized agents that collectively transform receivables from a reactive reporting function into a proactive, continuously optimizing cash-recovery engine.

The architecture spans the full receivables lifecycle. The Receivable Account Analysis Agent continuously examines customer balances, dunning history, disputed items, historical payment behavior, credit exposure, and external signals to identify financial risk and prioritize action. The Collection Account Preparation Agent assembles the complete context for each collection case, including open items, payment promises, communication history, and dispute linkages, enabling collectors to engage customers with full information rather than starting each conversation from scratch. The Collection Insights Agent applies predictive models to score every open item for late-payment risk, updating scores daily as new data accumulates. Unlike traditional aging buckets that treat all 60-day-overdue invoices identically, this agent analyzes customer payment patterns, dispute frequency, and credit trends to predict which specific invoices will age beyond 90 days.

The Collections Email Outreach Agent automates customer communications by generating personalized, contextually appropriate collection messages based on customer relationship status, overdue amount, payment history, and configured escalation rules. It determines the optimal tone, timing, and channel for each outreach: a gentle reminder for reliable payers with minor delays, a more assertive follow-up for accounts showing deteriorating patterns. The Dunning Insights Agent provides analytics on dunning effectiveness, such as which dunning levels, intervals, and communication styles produce the highest response rates for different customer segments, enabling continuous optimization of the dunning strategy.

The Case Preparation Agent and Case Processing Agent handle complex collection scenarios that require structured investigation. When an account requires escalation beyond standard dunning, these agents assemble the complete case file, identify the appropriate resolution path, and execute approved actions, from payment plan proposals to legal collection initiation.

The Dispute Creation Agent and Dispute Resolution Agent address the coordination nightmare of customer disputes. When a dispute is received, through email, customer portal, or direct entry, the Dispute Resolution Agent immediately activates and begins a structured investigation. It validates the dispute, retrieves the relevant documents (invoice, sales order, delivery records, contract, prior credit memos, correspondence history), reasons through the data to identify the root cause, and proposes resolution options. For example, it might detect a quantity discrepancy where the invoice shows 100 units but the delivery document confirms only 95 shipped, or a pricing error where the invoice price diverges from the contractual rate for orders above a volume threshold. Once the root cause is identified, the agent proposes a resolution—a corrected invoice, goodwill adjustment, or escalation—drafts the customer communication, and executes the resolution in the system upon approval.

The Invoice Execution Agent and Subscription Lifecycle Agent extend the Accounts Receivable Assistant's reach into billing-adjacent territory. The Invoice Execution Agent handles convergent invoicing scenarios in which multiple revenue streams, including subscription fees, usage charges, and one-time purchases, must be consolidated into unified billing documents. The Subscription Lifecycle Agent manages the unique receivables challenges of recurring revenue models, contract renewals, usage true-ups, mid-cycle modifications, and churn-related credit handling.

Figure 5.10 shows the Accounts Receivable Assistant, mapping the specialized agents that handle collections outreach, dispute management, and invoice execution.

SAP quantifies the combined impact of the Accounts Receivable Assistant as a 30 to 40 percent productivity increase in cash collection and a 10 to 20 percent reduction in DSO. For a billion-dollar revenue company with 60 days DSO, a 10 percent improvement translates to approximately 16 million dollars in released working capital, cash that was already earned but trapped in receivables.

Importantly, the Accounts Receivable Assistant is deeply grounded in SAP Knowledge Graph and SAP Business Data Cloud. This provides the semantic context and cross-process relationships that allow agents to reason beyond simple aging buckets. The agent understands that a customer with a disputed invoice should not receive aggressive dunning on that specific item, or that a customer with a flawless, on-time payment history across thousands of prior invoices deserves fundamentally different treatment than a new customer with no track record. This contextual intelligence separates agent-driven accounts receivable from traditional rules-based dunning.

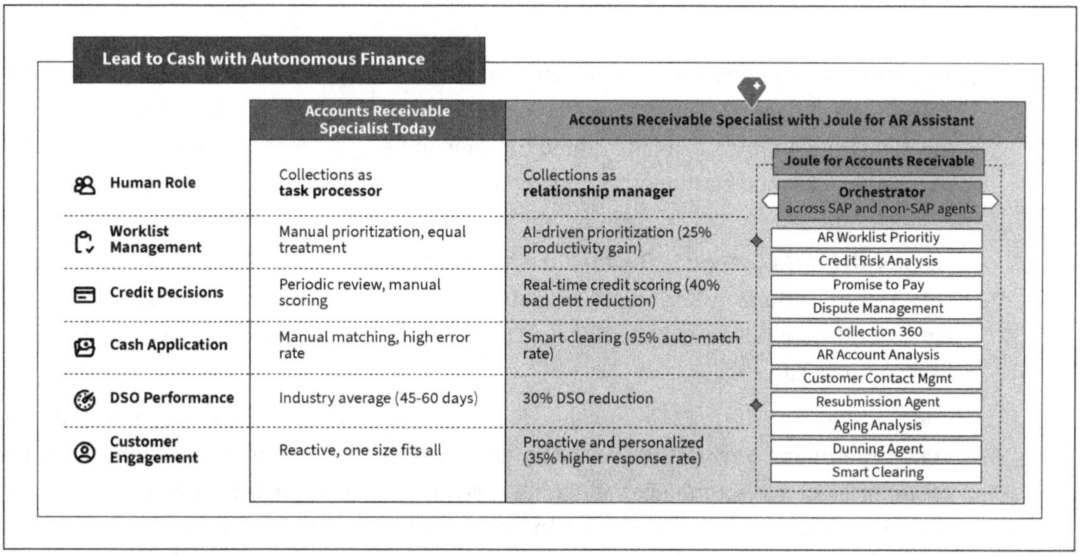

Figure 5.10: Accounts Receivable Assistant

Custom AI for Invoice-to-Cash Differentiation

Although SAP's standard AI scenarios and assistants address the core 80 percent of invoice-to-cash workflows, the remaining 20 percent often contains the highest-value opportunities for differentiation. Organizations can build custom AI models using the build pillar of SAP Business AI Platform to address unique business patterns.

Customer Payment Behavior Prediction for Dynamic Dunning trains an ML model on historical invoice, payment, and dunning data to predict which customers respond to early email reminders, which require immediate phone follow-up, and which will pay on schedule regardless of dunning intensity. The model integrates into collections management, dynamically adjusting dunning actions at runtime: gentle reminders for reliable payers, aggressive escalation for high-risk accounts, and no dunning for customers whose behavior shows they will honor terms without intervention.

Intelligent Situation Handling for Dispute Escalation combines SAP S/4HANA's situation-handling framework with SAP Build Process Automation to detect dispute conditions automatically and orchestrate cross-functional responses. Finance teams configure rules such as "dispute cases open longer than 14 days with no update" or "dispute amounts greater than 10,000 dollars and reason code equals pricing error," which trigger workflows the moment conditions are met. Organizations report 20 to 30 percent reductions in dispute-handling time, cutting resolution cycles from 18 days to under 12 days.

Note

Explore the dispute resolution benefits on SAP Discovery Center at *http://s-prs.co/v633506.*

The result is a composable AI architecture in which standard assistants and agents handle common patterns, and custom AI scenarios address the nuances that differentiate one business from another, all orchestrated through unified governance that ensures auditability, compliance, and continuous improvement.

Korindal: Maria Converts DSO from a Metric to a Lever

Let's return to Maria at Korindal. A breakthrough happens when the working capital problem is fundamentally reframed. DSO is no longer viewed as a reporting issue but recognized as a control issue.

The transformation unfolds across the full invoice-to-cash chain. The Billing Assistant eliminates the billing errors that previously generated 23 percent of all customer disputes, catching pricing discrepancies, quantity mismatches, and tax-calculation errors before invoices reach customers. The Billing Anomaly Agent alone prevents approximately 840 disputed invoices per quarter that previously took an average of 14 days to resolve.

SAP Cash Application raises automatic matching rates from Korindal's baseline of 35 percent to 88 percent within three months, eliminating the two-person team previously dedicated to full-time payment reconciliation. The Accounts Receivable Assistant's Collection Insights Agent identifies that 31 percent of Korindal's overdue balance is concentrated in 47 accounts that share a common pattern: They pay reliably, but consistently 12 to 18 days late due to internal approval processes. Rather than wasting collection effort on these predictable late payers, the agent recommends adjusted payment terms for this segment and redirects collection resources to the 89 accounts that exhibit genuinely deteriorating payment behavior.

The Dispute Resolution Agent transforms dispute handling from an 18-day average resolution cycle to nine days. In the first full quarter, it autonomously resolves 62 percent of disputes without human intervention: cases where the root cause is clear, the policy allows automated correction, and the amounts fall within preapproved thresholds. The remaining 38 percent reach human specialists with a full investigation context already assembled, reducing the per-case handling time from 22 minutes to eight minutes.

The narrative for the CFO changes completely. Before Autonomous Finance, the story was this: "We upgraded our ERP system." After deploying intelligent billing, predictive collections, and autonomous dispute resolution, the story becomes this: "We released the cash that was already earned." DSO improves by seven days in the first two quarters, representing approximately 19 million dollars in released working capital for Korindal's revenue base. Free cash flow returns to target, and the CFO delivers a different message to the board: "DSO is no longer a metric we report. It is a lever we manage."

5.2.4 Invoice-to-Pay

The invoice-to-pay process is where operational efficiency and working capital optimization collide head-on. Every supplier invoice that enters an organization triggers a chain of events—data extraction, validation, matching, approval, payment scheduling, and reconciliation—that collectively determine how quickly and accurately cash flows outward. When this process runs poorly, the consequences compound in every direction. Late payments damage supplier relationships and forfeit early payment discounts. Duplicate payments drain cash that may take months to recover. Manual three-way matching creates bottlenecks that leave invoices in queues for weeks. For the CFO, accounts payable is simultaneously the least glamorous and most consequential operational function in finance.

A distinguishing feature of Autonomous Finance in the invoice-to-pay domain is that its intelligence spans both SAP S/4HANA Cloud and SAP Ariba, working in concert. Invoice capture, matching, and payment execution operate within SAP S/4HANA's financial core, while procurement collaboration, supplier network integration, and central invoice management leverage SAP Ariba and SAP Business Network. The Accounts Payable Assistant and its agents orchestrate across both platforms, ensuring that intelligence flows seamlessly from requisition through payment regardless of where the underlying transaction originates. This dual foundation means that organizations benefit from ERP-grade accounting controls and network-grade supplier collaboration in a single coordinated workflow.

At Korindal, the accounts payable team processes over 40 thousand supplier invoices per month across 15 countries. Invoices arrive in every conceivable format: PDF attachments in email, scanned paper documents, EDI transmissions, and supplier portal submissions. Data entry errors cascade downstream, triggering mismatches that require manual investigation. The three-way matching process catches legitimate discrepancies but also generates enormous volumes of false exceptions. Payment runs execute on rigid weekly schedules with no optimization for cash positioning or discount capture. The team knows that millions of dollars in early payment discounts expire uncaptured every quarter, but they lack the bandwidth to systematically identify and prioritize them.

The Accounts Payable Assistant, part of SAP's Autonomous Finance framework, brings the full power of agentic AI to the invoice-to-pay process. This assistant orchestrates eight specialized Joule Agents, each addressing a distinct dimension of payables management. Together, they transform accounts payable from a labor-intensive data-processing function into an autonomous, exception-driven operation. We'll discuss the capabilities of the Accounts Payable Assistant in more detail in the following sections.

Invoice Capture

The *Invoice Capture Agent* serves as the intelligent entry point for the accounts payable workflow. It detects invoice emails and attachments; routes them to the appropriate capture engine; extracts metadata including the vendor, invoice number, date, currency, amounts, tax breakdowns, and purchase order references; and validates completeness before forwarding to the validation stage. Using AI and ML document extraction with multiformat recognition and continuous learning, the agent achieves 90 percent touchless capture rates with 98 percent field accuracy, creating an exception-only workflow in which human intervention is required only for genuinely ambiguous documents.

SAP Document AI fundamentally changes the economics of invoice processing. Delivered as an AI-powered document-processing service on SAP BTP, SAP Document AI extracts relevant information from supplier invoices regardless of format, layout, or language. The solution provides multilingual support for over 100 languages out of the box and learns instantly from user feedback to continuously improve extraction accuracy. Unlike traditional template-based optical character recognition (OCR) that requires manual configuration for each supplier's invoice format, SAP Document AI uses machine learning models that generalize across document layouts, achieving field-level extraction accuracy of 98 percent or higher.

The technical architecture is straightforward but powerful. Invoices arrive through any channel: email, supplier portal, EDI, or scan. SAP Document AI processes the document image or PDF; it then identifies and extracts key fields including the supplier name, invoice number, date, currency, line-item details, tax breakdowns, and purchase order references. The extracted data populates draft invoice records in SAP S/4HANA or SAP Ariba Invoicing, where it enters the validation workflow. The AI-assisted multimodel data-extraction capability further enhances accuracy by leveraging multiple AI engines, both ML and generative AI models, and selecting the highest-confidence value for each field across all engines. Figure 5.11 shows Intelligent Invoice Recognition, illustrating how SAP Document AI captures and extracts invoice data from any channel to eliminate manual keying.

For organizations using SAP Ariba Invoicing, the AI-augmented Invoice Capture capability adds a second intelligence layer. Beyond raw field extraction, it validates data and applies tax and accounting data based on rules and Data Attribute Recommendation services. The AI-assisted general ledger line item determination automatically enriches draft invoices by predicting the correct general ledger account, cost center, and work breakdown structure (WBS) element based on patterns learned from historical invoice data. This eliminates one of the most error-prone manual steps in the payables process, account coding, and thus dramatically reduces the volume of invoices requiring human review.

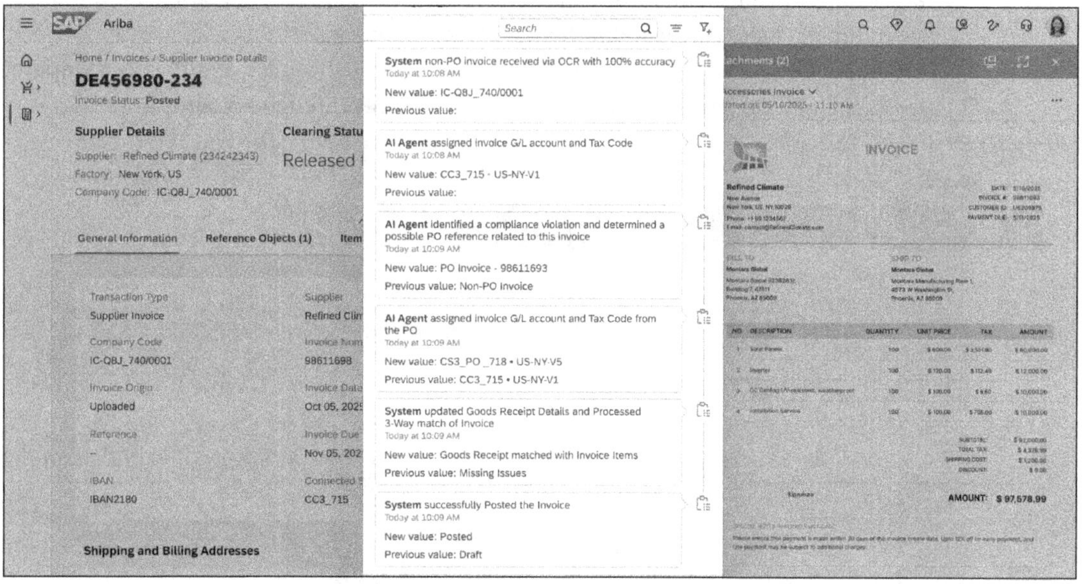

Figure 5.11: Intelligent Invoice Recognition

From this point, three additional agents within the Accounts Payable Assistant can further optimize the invoice capture process:

- **Invoice Validation Agent**
  The Invoice Validation Agent applies smart three-way matching, price variance analysis, and contract compliance validation. It automatically approves low-risk invoices while intelligently routing exceptions to the appropriate handlers based on variance type and dollar impact. The agent achieves 85 percent automatic validation rates, meaning that only 15 percent of invoices require any human review whatsoever.
- **Duplicate Detection Agent**
  The Duplicate Detection Agent employs fuzzy matching algorithms and cross-system identification to prevent duplicate payments before they occur. Unlike simple exact match checks, the agent evaluates combinations of vendor, invoice number, amount, currency, tax, purchase order number, date windows, bank details, and document hash features to identify likely duplicates even when individual fields differ slightly. The result is 99.9 percent duplicate prevention, protecting organizations from one of the most common and preventable sources of cash leakage.

- **Payment Proposal Agent**
  The Payment Proposal Agent acts as a final validation gate between invoice approval and payment execution. It harmonizes financial data between cloud and backend systems, ensuring that payment terms, methods, blocks, and banking details are fully compliant with both SAP S/4HANA vendor master data and the original purchase order before a payment request is finalized. Critically, the agent applies cash optimization algorithms and discount return on investment (ROI) analysis to determine optimal payment timing, balancing working capital preservation against discount capture opportunities. Figure 5.12 shows the Invoice Upload Agent, illustrating the automated workflow for ingesting, validating, and routing supplier invoices through accounts payable.

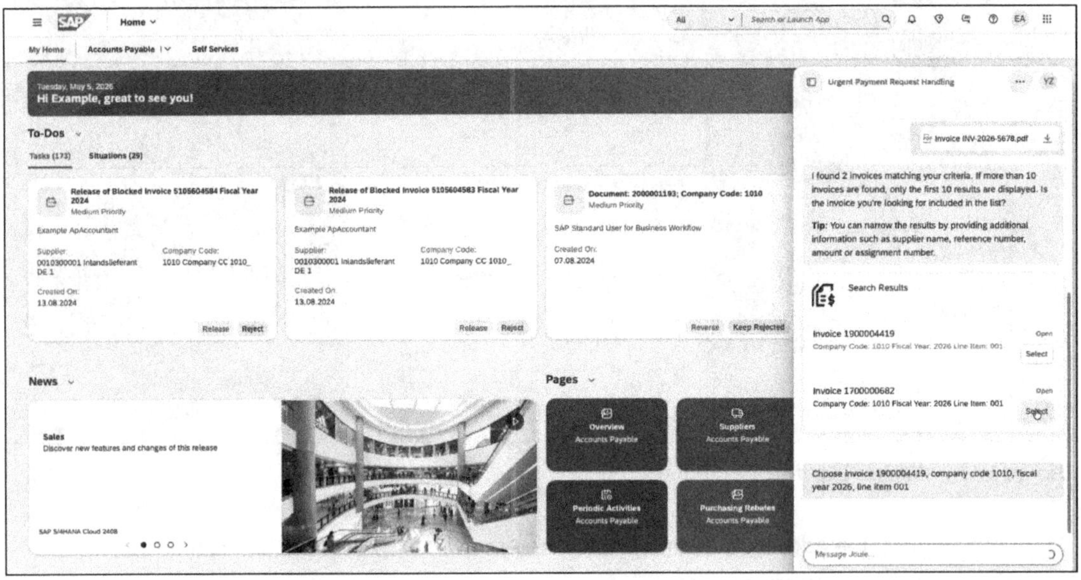

Figure 5.12: Invoice Upload Agent

Intelligent Three-Way Matching and Validation

Once invoice data is captured, the critical validation step begins. SAP's intelligent three-way matching compares each invoice line item against the corresponding purchase order and goods receipt to verify that the price charged matches the price agreed upon, the quantity invoiced matches the quantity received, and the delivery terms align with contractual commitments. This is not simple field comparison. The matching engine applies configurable tolerance thresholds, handles partial deliveries and partial invoicing, and accounts for currency conversions, tax calculations, and unit-of-measure differences.

The intelligence layer transforms three-way matching from a pass-fail gate into a risk-based routing engine. Invoices that match perfectly across all dimensions are automatically approved and posted without human intervention for true touchless processing. Invoices with minor variances within defined tolerance bands are posted with documented exceptions. Invoices with material discrepancies are intelligently routed to the appropriate investigator based on exception type, supplier history, and dollar impact. This risk-based approach means that accounts payable specialists spend their time exclusively on genuine exceptions rather than reviewing invoices that could have been approved automatically.

SAP's AI-assisted intelligent invoice error classification further accelerates exception handling. When invoices submitted through SAP Business Network are rejected, the system automatically categorizes them into appropriate rejection categories using machine learning, enabling systematic reprocessing rather than ad hoc investigation. Combined with AI-assisted supplier invoice verification through Joule, which allows accounts payable accountants to query invoice status, payment information, and approval chains using natural language, the entire validation workflow becomes conversational and transparent.

Several agents optimize the intelligent three-way matching process further:

- **Payment Execution Agent**
  The Payment Execution Agent optimizes the physical act of payment through bank selection optimization, fraud detection, and compliance validation. It automates multibank routing, format selection, and payment tracking, targeting zero-payment errors across all payment methods and geographies.
- **Vendor Reconciliation Agent**
  The Vendor Reconciliation Agent shifts vendor statement reconciliation from a periodic, manual exercise to a continuous, automated process. Using statement matching ML and variance root cause analysis, the agent achieves 80 percent automatic reconciliation and proactively creates dispute cases for identified discrepancies rather than waiting for month-end discovery.
- **Accounts Payable Aging Agent**
  The Accounts Payable Aging Agent applies cash flow impact modeling and supplier risk analysis to optimize days payable outstanding (DPO) strategically. Rather than paying all invoices on the same schedule, the agent recommends DPO optimization strategies and working capital recommendations tailored to each supplier's risk profile, payment terms, and strategic importance. SAP quantifies the impact as a 10-day improvement in DPO.
- **Discount Optimization Agent**
  The Discount Optimization Agent analyzes discount economics, performing cash versus discount trade-off calculations to ensure that every available early payment discount is evaluated and captured when the return exceeds the cost of accelerated payment. Through dynamic discounting and integration with early payment programs, the agent targets 2 to 3 percent savings on addressable spend, a figure that translates to millions of dollars annually for large enterprises.

Figure 5.13 shows the Accounts Payable Assistant, mapping the specialized agents that handle duplicate detection, payment proposals, vendor reconciliation, and discount capture.

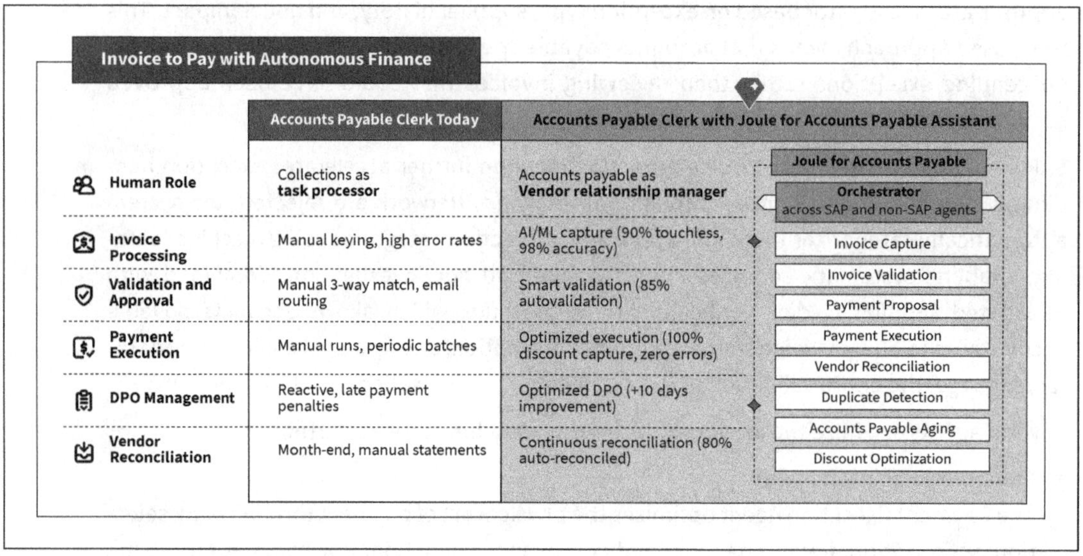

Figure 5.13: Accounts Payable Assistants

SAP quantifies the combined impact of the Accounts Payable Assistants at three to eight million dollars in annual value for a billion-dollar enterprise. The headline metrics are compelling: 85 percent touchless invoice processing end to end, a 10-day improvement in DPO, and 100 percent discount capture on eligible invoices. These numbers reflect the transformation from a manual, error-prone process to an autonomous operation in which the accounts payable team's role shifts fundamentally from data entry specialists to vendor relationship managers, focused on strategic supplier partnerships, payment term negotiations, and working capital optimization rather than chasing paper invoices and resolving data entry errors.

Case Study: The Accounts Payable Transformation

At Korindal, the accounts payable transformation unfolds in stages. The first breakthrough comes when SAP Document AI eliminates the data entry bottleneck entirely. Invoice capture accuracy jumps from 80 percent under the legacy OCR system to 98 percent, and the percentage of invoices requiring manual data correction drops from 40 percent to under 5 percent. The second wave arrives when intelligent three-way matching begins, routing only genuine exceptions for human review. The false positive rate on matching exceptions drops by 70 percent, freeing investigators to focus exclusively on real discrepancies.

The most significant financial impact, however, comes from the Payment Proposal and Discount Optimization Agents. For the first time, Korindal's treasury and accounts payable teams have a unified view of which invoices should be paid early to capture discounts, which should be held to optimize DPO, and which suppliers qualify for supply chain finance programs. In the first full quarter of operation, Korindal captures an additional 1.8 million dollars in early payment discounts that had previously expired unclaimed, while simultaneously improving DPO by eight days, freeing approximately 12 million dollars in working capital.

The accounts payable team's daily experience changes fundamentally. Instead of opening stacks of invoices and manually keying data, accounts payable specialists start their day by asking Joule: "Show me all invoices with payment blocks" or "Which supplier reconciliations have open variances above $5,000?" The Accounts Payable Assistant surfaces the exceptions that genuinely require human judgment, provides full context including historical patterns and recommended actions, and executes approved transactions, all through natural language interaction. What was once a 40-person data-processing operation becomes a 12-person strategic function managing supplier financial relationships and optimizing working capital.

5.2.5 Treasury and Working Capital

Treasury and working capital management is the domain where enterprise finance becomes most genuinely strategic—and most dangerously exposed to real-time market dynamics. Unlike accounts receivable and accounts payable, where underlying transactions are discrete and data is structured, treasury operates in a world of continuous cash flows, fluctuating interest rates, volatile FX markets, and counterparty risks that can materialize without warning. The treasurer's fundamental mandate is deceptively simple: Ensure the organization always has enough cash in the right currencies, in the right accounts, at the right time, while maximizing the return on idle balances and minimizing the cost of external financing.

At Korindal, the treasury team consists of six professionals managing cash positions across 42 bank accounts in 14 currencies. Their weekly rhythm has not fundamentally changed in a decade despite the ERP modernization. Every Monday begins with a manual aggregation of bank statement balances downloaded from multiple banking portals, loaded into spreadsheets, and reconciled against the SAP general ledger. Cash flow forecasts are constructed by pulling open receivables, scheduled payables, and known treasury transactions, then applying manual judgment to estimate timing. The resulting forecast, which typically takes two full days to compile, has an accuracy range of plus or minus 20 percent, a margin of error that forces the team to maintain excessive precautionary cash balances, forgoing interest income on idle funds. FX exposures are reviewed monthly, hedging decisions are made reactively after significant rate movements, and working capital metrics are analyzed quarterly in isolation rather than managed holistically as components of the cash conversion cycle.

The strategic cost of this manual, periodic approach is substantial but largely invisible. Maria, Korindal's CFO, knows that the company maintains approximately 15 million dollars more in non-interest-bearing operating accounts than necessary because the treasury team cannot position cash quickly enough to invest surpluses or cover shortfalls in real time. Early payment discount capture is suboptimal because the treasury lacks the cash visibility to approve accelerated payments with confidence. FX losses from reactive hedging exceed the benchmark by approximately 200 basis points annually.

Now, let's see how embedded AI capabilities can change and enhance the course of business for treasury.

The Cash and Treasury Assistant

The Cash and Treasury Assistant orchestrates three specialized agents that transform treasury from a periodic monitoring function into a continuous, AI-driven optimization engine. SAP quantifies the combined impact: 50 to 70 percent cash cycle time reduction, 5 to 20 percent improvement in key quality metrics, and cost reductions ranging from 3 to 8 million euros for a billion-euro enterprise.

Let's walk through the agents within the Cash and Treasury Assistant that make these benefits possible:

- **Cash Positioning Agent**
  The Cash Positioning Agent provides the intelligence foundation. It ensures bank statements are processed accurately upon receipt, aggregates opening balances across all accounts, and accounts for expected cash flows from open receivables, scheduled payables, and known treasury transactions. The result is an up-to-date, consolidated cash position that would traditionally require hours of manual spreadsheet work, delivered in seconds through Joule. Beyond position reporting, the agent proposes transfers to optimize operating cash by accurately identifying shortages and surpluses across accounts, currencies, and entities. It projects forward-looking cash positions by combining confirmed cash flows with ML forecasts, enabling proactive liquidity management rather than reactive firefighting.
- **Treasury Financing Agent**
  The Treasury Financing Agent addresses the investment and borrowing decisions that directly determine interest income and expense. Using AI-driven analysis of cash surplus patterns, interest rate environments, and counterparty limits, the agent recommends optimal investment horizons for surplus cash, identifies the most cost-effective funding sources for anticipated shortfalls, and proposes financing structures that balance liquidity, return, and risk. It enables treasury specialists to create and process financial transactions, interest rate instruments, money market fund activities, and FX trades through conversational interaction, regardless of the format in which transaction data arrives from financial service providers.
- **Bank Relationship Agent**
  The Bank Relationship Agent enables cash managers to instantly prepare comprehensive summaries of banking relationships, consolidated positions, fee analyses, service utilization, and negotiation preparation, all through natural language queries. This capability transforms bank relationship management from a quarterly reporting exercise into an

on-demand strategic capability. When the treasurer asks, "What is our total fee exposure across all banking partners, and which banks have the highest cost-to-service ratio?," the agent assembles the analysis in seconds rather than the days traditionally required.

The Joule integration with SAP Multi-Bank Connectivity extends the assistant's reach by providing conversational access to bank communication and payment status across the entire banking landscape, creating a unified treasury command center.

The true strategic power of the Cash and Treasury Assistant emerges when it collaborates with the Accounts Receivable and Accounts Payable Assistants to optimize working capital holistically. Rather than managing DSO, DPO, and days inventory outstanding (DIO) as isolated departmental metrics, the system models their interplay as components of the cash conversion cycle.

This integrated view enables trade-off analysis that no single function can perform alone. Accelerating collections by two days might cost customer goodwill. Extending payment terms by five days might strain supplier relationships; reducing inventory investment might risk stockouts. The Cash and Treasury Assistant evaluates these competing objectives simultaneously, recommending the combination of collection acceleration, payment timing, and inventory investment that maximizes overall working capital efficiency within the CFO's defined constraints Figure 5.14 shows the Cash and Treasury Assistant, illustrating how the Cash Positioning, Financing, Bank Relationship, and Foreign Exchange Agents transform treasury into a continuous optimization function.

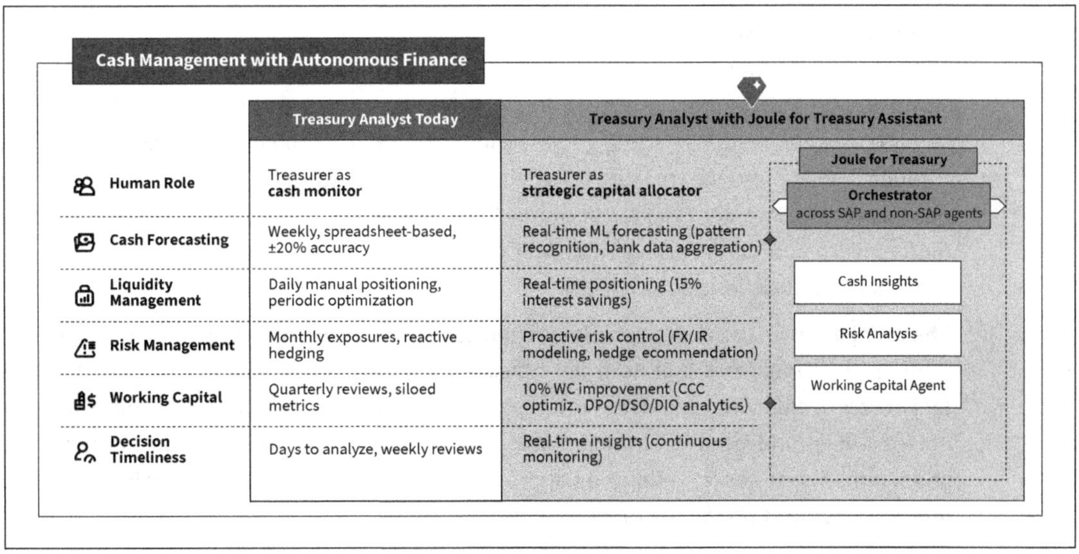

Figure 5.14: Treasury Assistant

SAP targets a 10 to 20 percent improvement in overall DSO performance and significant improvements in cash cycle efficiency. The working capital gains are not theoretical; they represent cash that is freed from the operating cycle and available for investment, debt reduction, or strategic deployment.

Integration with SAP Taulia

The treasury and working capital story extend beyond SAP S/4HANA through integration with SAP Taulia, SAP's working capital management platform. The Spend Data Analytics capability allows treasurers to interact with their spend data conversationally, understanding how much spend is available for early payment programs, how much has been offered and accepted by suppliers, and how these patterns vary by currency and region. The Working Capital Optimization capability provides deeper strategic guidance, highlighting how much working capital can be unlocked through supply chain finance programs, predicting the volume of early payment that can be accelerated through various scenarios, and helping configure optimal payment term extensions.

This integration creates a powerful closed loop. The Cash and Treasury Assistant identifies surplus cash that could be deployed for early payment programs. SAP Taulia's platform executes those programs with suppliers, capturing dynamic discounts that flow directly to the bottom line. The Cash Positioning Agent monitors the impact on overall liquidity, and the working capital optimization logic recalibrates recommendations based on actual program performance. The result is a continuously self-improving working capital engine that adapts to changing business conditions.

Case Study: Treasury Transformation

At Korindal, the transformation begins with the Cash Positioning Agent eliminating the Monday morning aggregation ritual. Bank statements are processed automatically upon receipt, opening balances are aggregated in real time, and the cash position is available through Joule before the treasury team's first coffee. Cash forecast accuracy improves from plus or minus 20 percent to plus or minus 5 percent within two quarters as the agent's ML models learn Korindal's receivable collection patterns, payable disbursement rhythms, and seasonal cash flow variations.

The immediate financial impact comes from cash positioning optimization. The Cash Positioning Agent identifies that Korindal consistently maintains an average of 12 million dollars in excess balances across non-interest-bearing operating accounts. By proposing automated sweep and pooling arrangements, the agent frees this capital for overnight and short-term investments, generating approximately 1.8 million dollars in additional annual interest income at prevailing rates.

The Treasury Financing Agent transforms Korindal's approach to short-term funding. Previously, when cash shortfalls arose in specific entities or currencies, the treasury team relied on expensive overdraft facilities because they lacked the real-time visibility to arrange cost-effective intercompany loans or draw on committed credit lines optimally. The agent now identifies funding needs three to five days in advance, recommends the lowest-cost source, and enables execution through conversational interaction. Annual interest expense on short-term borrowing declines by approximately $400,000.

The Bank Relationship Agent surfaces an insight that no manual analysis had previously identified: Korindal's banking fees have crept upward by 18 percent over three years as services were added incrementally without competitive benchmarking. Armed with the agent's comprehensive fee analysis and service utilization data, the treasurer renegotiates banking arrangements in the next annual review cycle, achieving a 12 percent fee reduction across the primary banking panel.

The most strategically significant transformation comes from the holistic working capital view. By modeling DSO, DPO, and DIO as interconnected components of the cash conversion cycle rather than isolated departmental metrics, the system reveals that Korindal's accounts payable team is paying invoices an average of six days earlier than necessary, not as a deliberate strategy but because the lack of integrated cash visibility makes it safer to pay promptly than risk late-payment penalties. Simultaneously, the account receivable team's collection patterns show systematic delays in following up on invoices in the 15-to-30-day aging bucket. The coordinated recommendation: Extend payment timing on low-risk suppliers by five days while accelerating collection efforts on medium-aging receivables. The net impact is a seven-day improvement in the cash conversion cycle, freeing approximately 19 million dollars in working capital without any deterioration in supplier or customer relationships.

FX management transforms from reactive to proactive. Previously, the treasury team reviewed FX exposures monthly and placed hedging trades reactively. The system now monitors exposures continuously, identifies optimal hedging windows based on market conditions and Korindal's risk tolerance parameters, and recommends specific trades with full value-at-risk context. In the first year, FX losses decline by 40 percent, not because market conditions improve but because hedging decisions are made proactively rather than after adverse movements.

The treasurer's daily workflow transforms completely. Instead of spending two days compiling cash forecasts and preparing board-ready liquidity reports, the treasurer starts each day by asking Joule: "What is our consolidated cash position, and are there any accounts with surplus balances above two million dollars?" The Cash Positioning Agent responds with a real-time position, highlights specific accounts with investable surplus, and proposes transfer and investment actions. The treasurer reviews, adjusts if necessary, and approves, all within a single conversational session that takes minutes rather than days.

The narrative for Korindal's board changes fundamentally. Before the Cash and Treasury Assistant, the quarterly treasury update was a retrospective report: "Here is what happened to our cash position last quarter." After deployment, the update becomes forward-looking and actionable: "Our cash conversion cycle has compressed by seven days. We are capturing 100 percent of eligible early payment discounts. Our FX hedging program is outperforming the benchmark by 180 basis points. And our working capital position has improved by

19 million dollars, providing cash that is now deployed in short-term instruments that generate returns rather than sitting idle in operating accounts."

The transformation is complete. Autonomous Finance has turned the office of the CFO from retrospective reporting into continuous financial intelligence. The CFO operates not as a financial steward but as an enterprise value architect, enabled by 13 assistants and more than 60 agents that sense, predict, and execute with speed, delivering high efficiency gains and greater ROI. The result: a business that runs leaner, moves faster, and bends without breaking.

5.2.6 Accounting and Close

The period-end close and financial consolidation cycle represents one of the most time-pressured, high-stakes processes in all of finance. Every organization faces the same paradox: The close must be fast to deliver timely results, yet it must be accurate to satisfy auditors, regulators, and executive stakeholders. Traditional close processes resolve this tension through brute-force labor, with accountants working extended hours to reconcile accounts, post accruals, clear intercompany items, and investigate exceptions under extreme deadline pressure. For organizations with complex, multientity structures, group consolidation adds yet another layer of effort, with currency translations, elimination entries, minority interest calculations, and intercompany matching consuming entire teams for days after the legal entity close completes.

At Korindal, the close cycle has historically consumed ten working days each month. The shared services team of 12 accountants spends the first five days posting accruals, reconciling accounts, and clearing open items across 14 legal entities. The consolidation team then takes an additional five days to collect submissions, resolve data quality issues, process eliminations, and produce the consolidated group financials. By the time the CFO receives results, the information is already two weeks old, limiting its value for operational decision-making.

Now, let's walk through the embedded AI capabilities that accelerate period-end close and improve group-level intelligence and see how they impact our case study.

The Financial Closing Assistant

The Financial Closing Assistant orchestrates six specialized agents that accelerate and automate close activities. Figure 5.15 shows the Financial Closing Assistant, illustrating how six agents coordinate accruals, anomaly detection, intercompany reconciliation, and consistency analysis.

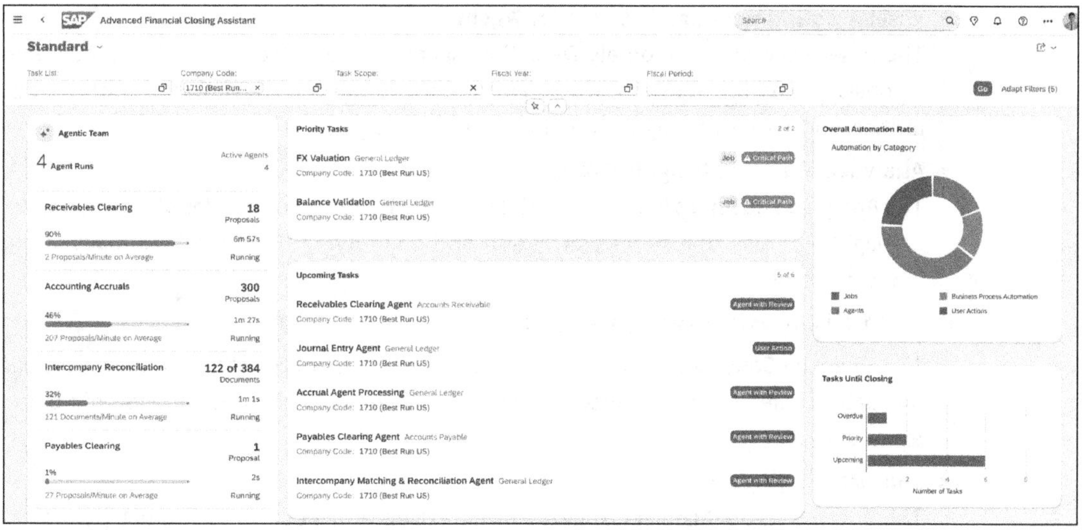

Figure 5.15: Financial Closing Assistant

Let's examine each agent in turn:

- **Accounting Accruals Agent**

  The Accounting Accruals Agent automates the identification, calculation, and posting of period-end accruals by analyzing open commitments, recurring patterns, and contractual obligations.

  Figure 5.16 shows the Accounting Accruals Agent, demonstrating how it automates identification, calculation, and posting of period-end accruals.

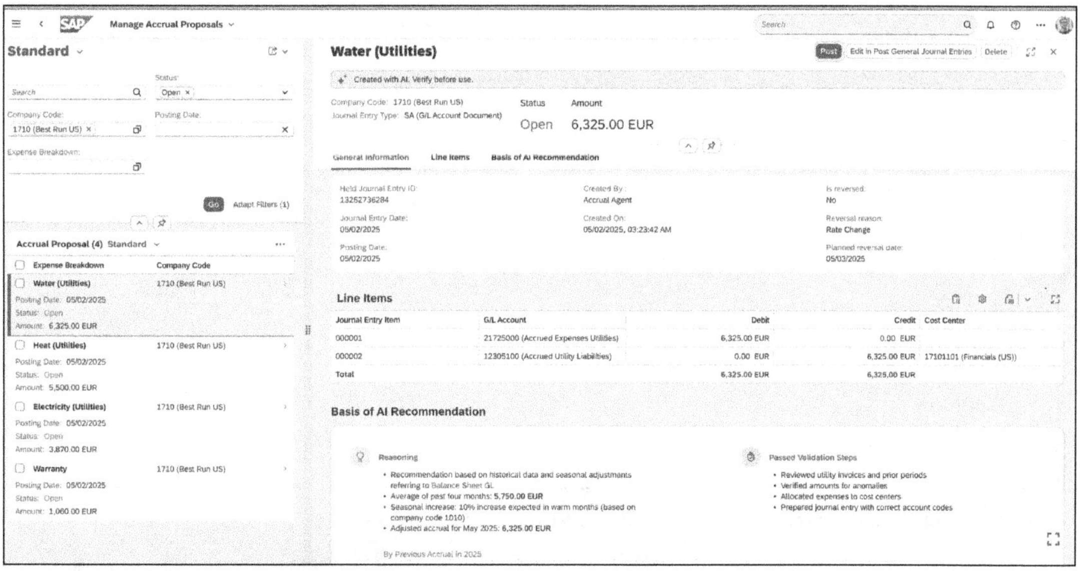

Figure 5.16: Accounting Accruals Agent

- **Asset Accounting Anomaly Detection Agent**
  The Asset Accounting Anomaly Detection Agent monitors fixed asset transactions for inconsistencies, unusual depreciation patterns, capitalization errors, or disposal irregularities before they propagate into financial statements.
- **Analytical Business Insights Agent**
  The Analytical Business Insights Agent generates narrative summaries of close progress, highlighting areas that require attention and quantifying the financial impact of outstanding items.
- **Financial Consistency Analysis Agent**
  The Financial Consistency Analysis Agent validates cross-account relationships and reconciliation rules, detecting inconsistencies that would otherwise surface only during external audit.
- **Journal Entry Agent**
  The Journal Entry Agent automates the creation and posting of routine period-end entries based on predefined rules and historical patterns.
- **Intercompany Matching and Reconciliation Agent**
  The Intercompany Matching and Reconciliation Agent addresses one of the most labor-intensive close activities by automatically matching intercompany transactions across legal entities and flagging discrepancies for resolution.

  Figure 5.17 shows the Intercompany Matching and Reconciliation Agent, illustrating how it automates resolution of intercompany discrepancies to accelerate the group close.

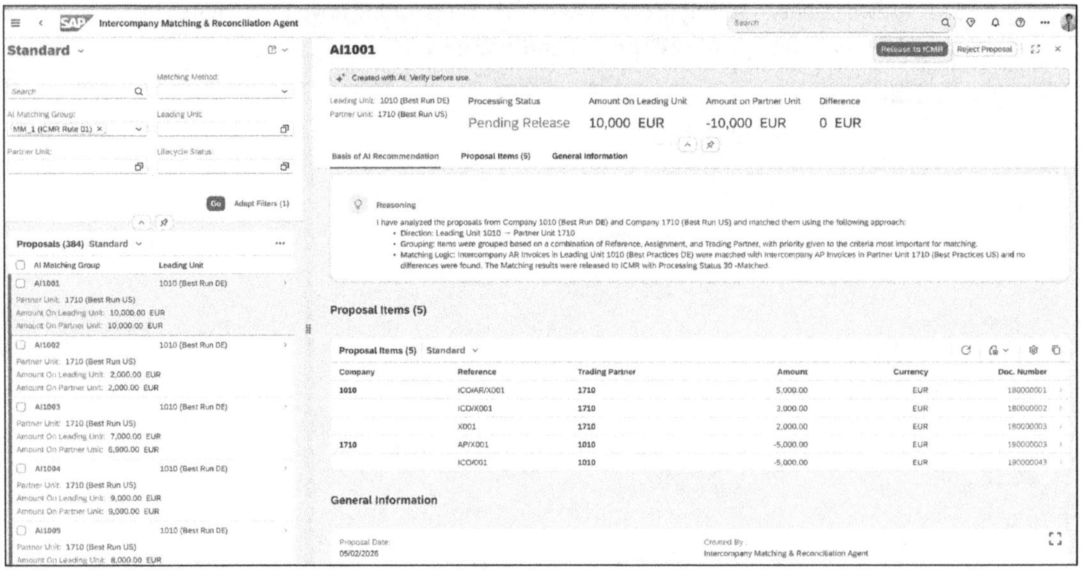

Figure 5.17: Intercompany Matching and Reconciliation Agent

Figure 5.18 shows the Financial Closing Assistant value story, quantifying the productivity gains and cycle time reductions delivered across the close process.

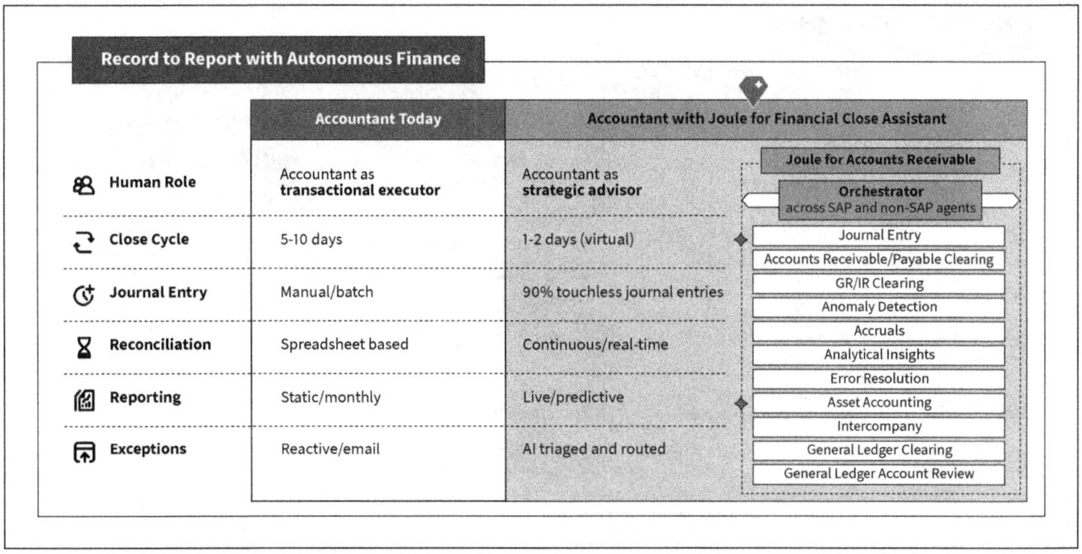

Figure 5.18: Financial Closing Assistant Value Story

SAP quantifies the combined impact: 20 to 30 percent increase in general ledger and closing productivity, 10 to 20 percent reduction in the closing cycle, and higher reporting accuracy with fewer errors. For organizations pursuing continuous close, with the aspiration to compress month end from 22 hours to two hours, these agents provide the automation foundation that makes that ambition achievable.

The Consolidation Assistant

For organizations with complex legal entity structures, financial consolidation adds another layer of close complexity. The Consolidation Assistant automates preconsolidation error resolution by detecting data quality issues across reporting entities, analyzing patterns in submission errors, applying standardized fixes, and engaging finance users only when human judgment is required. Its Consolidation Error Resolution Agent identifies and resolves common consolidation failures—currency translation mismatches, elimination errors, minority interest calculation discrepancies—that traditionally require hours of specialist investigation. Figure 5.19 shows the Reporting Assistant, illustrating how AI enables finance professionals to generate and distribute financial reports through natural language.

SAP targets a 20 to 30 percent increase in group closing productivity and a 10 to 20 percent reduction in the group closing cycle.

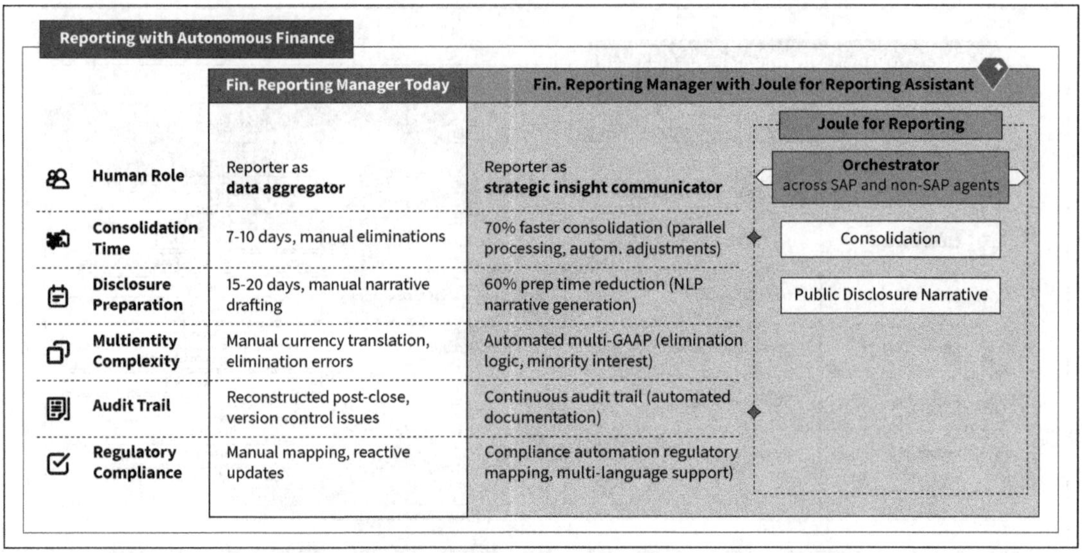

Figure 5.19: Reporting Assistant

Case Study: Accounting and Close Transformation

The Financial Closing Assistant transforms Korindal's month-end close from a 10-day marathon to a streamlined, four-day process. What previously required accountants working evenings to post accruals, reconcile intercompany balances, and resolve exceptions compresses to six days in the first quarter and four days by the third quarter. The Intercompany Matching and Reconciliation Agent eliminates the two full days that Korindal's shared services team previously spent manually matching transactions across 14 legal entities. The Asset Accounting Anomaly Detection Agent catches a systematic capitalization error in the third close cycle that would have required a restatement if discovered during external audit.

The Consolidation Assistant reduces Korindal's group close from five days to two. Previously, the consolidation team spent the first day chasing late submissions and correcting data-quality errors from reporting entities. The Consolidation Error Resolution Agent now identifies and resolves common submission errors automatically, eliminating the back-and-forth communication cycle. Currency translation mismatches that previously required hours of specialist investigation are detected and resolved in minutes. The CFO receives consolidated group results by day six of the close rather than day 10, gaining four additional days of decision-making relevance from each period's financials.

Figure 5.20 shows financial business insights, illustrating how AI synthesizes operational data into forward-looking narratives and anomaly highlights.

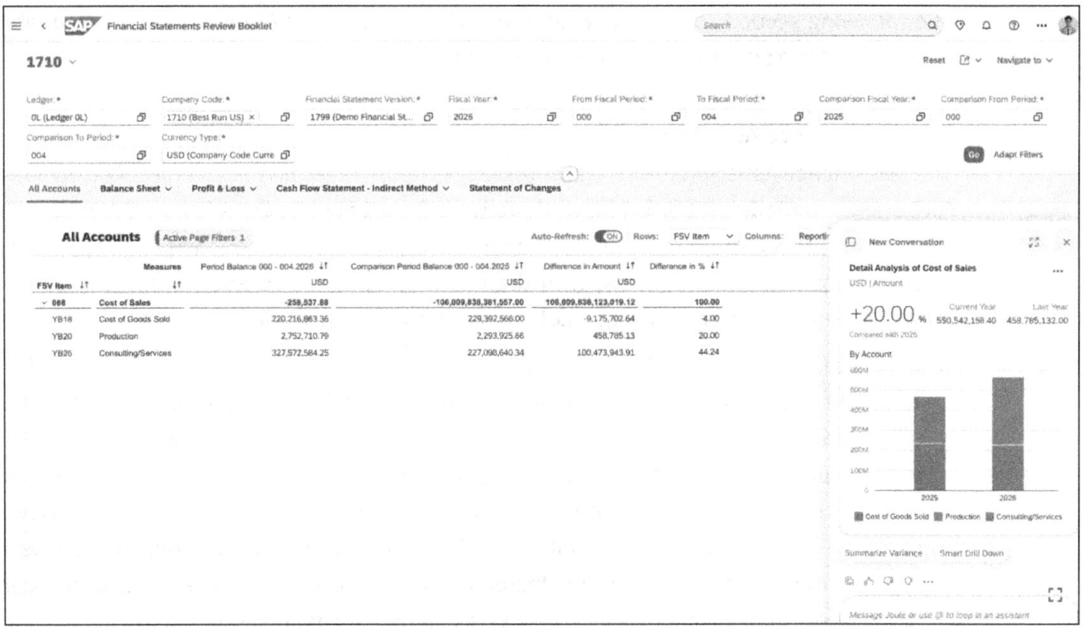

Figure 5.20: Financial Business Insights

5.2.7 Governance, Risk, and Compliance

Every financial process described in the preceding sections operates within a framework of controls, regulations, and risk boundaries. Governance, risk, and compliance (GRC) is not an afterthought bolted onto the end of a close cycle or triggered reactively by an audit notification. In Autonomous Finance, GRC is a continuous, cross-cutting discipline embedded into transactional flows from the moment a journal entry posts, an invoice clears, or a payment executes. The shift from periodic compliance checks to always-on assurance represents one of the most consequential transformations in the CFO's operating model because it eliminates the trade-off between speed and control. Organizations no longer have to choose between fast closes and rigorous governance; agentic intelligence delivers both simultaneously.

The regulatory landscape reinforces this urgency. Multinational organizations must comply with an expanding web of requirements: IFRS sustainability disclosures, EU Corporate Sustainability Reporting Directive (CSRD), sanctions screening obligations, export control regulations, and evolving AI governance standards such as ISO/IEC 42001. Manual compliance programs cannot scale to match this complexity. They depend on sampling rather than full-population testing, quarterly attestations rather than continuous monitoring, and reactive remediation rather than preventive detection. The result is audit findings that arrive months after root causes occurred, regulatory penalties driven by process gaps rather than intentional misconduct, and GRC teams perpetually fighting the last war rather than anticipating the next risk.

The following sections will expand on the various embedded AI capabilities that enable continuous controls and intelligent compliance.

The Governance Assistant

The Governance Assistant orchestrates three specialized agents that collectively transform GRC from a periodic, labor-intensive function into a continuous, AI-powered assurance capability. SAP quantifies the headline impact: 50 to 60 percent productivity increase in risk management and auditing tasks, with dramatically compressed audit preparation timelines and broader control coverage. Let's walk through each available agent:

- **The Audit Agent**
  The Audit Agent automates continuous controls monitoring across the entire transaction population rather than relying on statistical sampling. It identifies anomalies, segregation-of-duties violations, and control exceptions in real time, generating risk-scored findings with root cause analysis and remediation recommendations. When the Audit Agent detects a pattern of manual journal entries bypassing automated approval workflows, it does not merely flag the exception. Instead, it traces the entries to their originators, correlates them with access privileges, assesses materiality against defined thresholds, and proposes specific control enhancements to prevent recurrence. The shift from periodic sample-based auditing to continuous full-population monitoring eliminates the assurance gap that traditional audit cycles create.
- **International Trade Classification Agent**
  The International Trade Classification Agent applies AI-powered product classification against harmonized tariff schedules and export control lists, ensuring that every cross-border transaction carries the correct commodity codes, license determinations, and duty calculations. For organizations managing thousands of SKUs across dozens of trade jurisdictions, manual classification is error-prone and unsustainable. The agent maintains classification accuracy as products evolve, regulations change, and new trade agreements take effect, reducing classification errors that can trigger customs penalties, shipment delays, and compliance findings.
- **Screening Agent**
  The Screening Agent performs real-time sanctions, denied-party, and politically exposed persons (PEP) screening against continuously updated global watchlists. Every business partner, beneficiary, and counterparty is screened at the point of transaction, not in a batch process that runs overnight. When a potential match occurs, the Screening Agent applies fuzzy logic and contextual analysis to reduce false positives, escalates genuine hits through defined workflows, and maintains a complete audit trail of screening decisions for regulatory examination.

Figure 5.21 shows the Governance Assistant, illustrating how the Audit, Trade Compliance, and Screening Agents provide continuous controls monitoring and regulatory compliance.

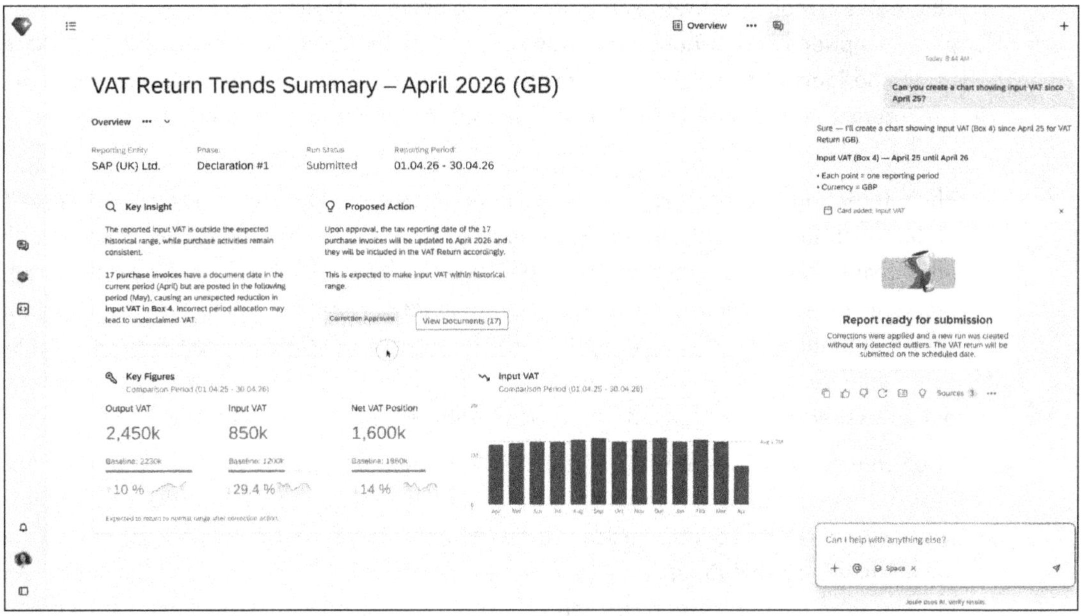

Figure 5.21: Governance Assistant

SAP GRC for SAP HANA

Underpinning the Governance Assistant is SAP GRC for SAP HANA 2026, the next-generation governance platform reaching general availability in Q3 2026 with support extending to 2040. This platform unifies access control, process control, risk management, and audit management in a single SAP HANA foundation, eliminating the fragmented GRC tool landscapes that many organizations have accumulated over decades. The architecture enables real-time risk analytics, continuous controls monitoring at scale, and native integration with SAP S/4HANA Cloud processes, meaning that controls execute within the same transaction pipeline rather than as external bolt-on checks.

The platform's compliance posture extends to the AI capabilities themselves. SAP Business AI Platform maintains certifications including ISO/IEC 42001 for AI management systems, ISO 27001 for information security, SOC 2 for service organization controls, and alignment with General Data Protection Regulation (GDPR) requirements. These certifications ensure that the AI agents operating within GRC workflows meet the same governance standards they are designed to enforce, addressing the emerging regulatory requirement that organizations demonstrate responsible AI governance alongside traditional financial controls.

Case Study: GRC Transformation

At Korindal, the GRC transformation begins with continuous controls monitoring replacing the quarterly internal audit cycle. Previously, the internal audit team of six spent three weeks preparing for each quarterly review: pulling transaction samples, testing control effectiveness, documenting findings, and preparing management reports. The Audit Agent eliminates sampling entirely by monitoring 100 percent of transactions continuously, flagging exceptions within hours of occurrence rather than discovering them months later during periodic testing. Audit preparation compresses from three weeks to four days, and the internal audit function shifts from retrospective testing to forward-looking risk advisory.

The International Trade Classification Agent addresses a persistent pain point in Korindal's global operations. With 8,400 active SKUs shipped across 23 countries, the trade compliance team previously spent 40 hours per week manually classifying new products and reclassifying existing items when regulations changed. The agent automates initial classification and monitors regulatory updates, reducing manual classification effort by 70 percent while improving accuracy from 91 percent to 98.5 percent, eliminating the customs penalties that previously cost Korindal approximately 2.3 million dollars annually.

The Screening Agent transforms Korindal's third-party risk management. Previously, business partner screening ran as a nightly batch process, creating a window during each business day where new counterparties could be onboarded without completed screening. The agent moves to real-time, transaction-level screening, ensures zero-gap coverage, and reduces false positive rates by 65 percent through contextual analysis, freeing the compliance team from investigating thousands of spurious alerts each quarter.

The cumulative impact on Korindal's risk posture is transformative. The CFO presents to the audit committee not a retrospective summary of control testing results but a real-time risk dashboard showing continuous assurance metrics: percentage of transactions monitored, mean time to exception detection, control effectiveness scores by process area, and predictive risk indicators for emerging exposure areas. The conversation shifts from "Were we compliant last quarter?" to "Where are the emerging risks we need to address proactively?," fulfilling the promise of governance as a continuous capability rather than a periodic obligation.

The transformation is complete. Autonomous Finance has turned the office of the CFO from retrospective reporting into continuous financial intelligence. The CFO operates not as a financial steward but as an enterprise value architect, enabled by 13 assistants and more than 60 agents that sense, predict, and execute with speed, delivering high-efficiency gains and greater return on investments. The result: a business that runs leaner, moves faster, and bends without breaking.

5.3 Lead-to-Cash

The lead-to-cash value stream represents the commercial heartbeat of the enterprise, the end-to-end sequence of activities that begins with identifying a potential customer, progresses through opportunity qualification and quoting, executes through order fulfillment and delivery, and concludes with invoicing and cash collection. For the chief revenue officer, chief sales officer, and chief financial officer alike, lead-to-cash is where revenue ambition meets operational reality. Every inefficiency in this chain—a misqualified lead, a delayed quote, a fulfillment exception, an invoicing error—directly reduces the velocity at which the organization converts market demand into recognized revenue and collected cash.

SAP's autonomous capabilities across the lead-to-cash value stream target every stage of this commercial lifecycle. In the opportunity-to-quote phase, ML tools score leads, predict deal outcomes, and automate quote generation. In order-to-fulfill, AI-assisted order creation, intelligent availability checking, and autonomous fulfillment monitoring eliminate the manual bottlenecks that delay delivery. In invoice-to-cash, the capabilities explored in Section 5.2 for finance, like SAP Cash Application, the Accounts Receivable Assistant, and the Dispute Resolution Agent, close the final mile from delivery to deposited cash. And across the entire chain, intelligent master data management ensures that the customer, product, and pricing foundations are accurate and continuously maintained.

After providing an overview of the Autonomous CX and Autonomous Finance Order Management functions ahead, we will walk through the embedded AI capabilities for key areas of lead-to-cash: order-to-fulfill, opportunity-to-quote, invoice-to-cash, and customer management.

5.3.1 Overview of Autonomous CX

Autonomous CX, one of the five autonomous functions within the SAP Autonomous Suite, organizes these capabilities into a comprehensive agentic architecture. A Deal Qualification Assistant automates lead enrichment and qualification, converting sales-ready prospects into opportunities faster while improving pipeline quality. A Sales Assistant orchestrates agents for deal management, meeting preparation, risk analysis, and outreach to boost productivity and help teams close more deals at scale. A Deal Closing Assistant guides sellers through quotes, pricing, and orders with agentic precision, reducing friction and shortening sales cycles. A Case Management Assistant empowers service teams to triage, manage, and resolve cases faster by surfacing real-time insights grounded in full business context. A Self-Service Assistant provides instant answers around the clock, resolving policy questions and invoice disputes without agent involvement. Together, these assistants and their constituent agents transform the lead-to-cash process from a sequence of disconnected manual handoffs into an orchestrated, AI-driven commercial engine.

SAP quantifies the impact across key metrics, validated in a 2026 BCG study (*http://s-prs.co/v633507*) on agentic AI for customer engagement: a 22 percent increase in expansion revenue rate, a 15 percent increase in average deal value, a 29 percent decrease in seller time spent on administration versus selling, and a 34 percent increase in first-contact resolution rate for service scenarios. These are not aspirational targets. They reflect the measured impact of coordinated assistants and agents operating on shared commercial data.

At Korindal, the commercial organization mirrors the complexity that most global enterprises face. There are 1,400 sales representatives who operate across four channels: direct enterprise sales, inside sales, partner-assisted sales, and digital commerce. They manage

42,000 active customer accounts, process an average of 8,000 sales orders per week, and operate in markets that demand rapid order confirmation, precise delivery commitments, and flawless invoicing. The sections that follow examine how SAP Business AI transforms each phase of this commercial lifecycle.

5.3.2 Order-to-Fulfill

Order-to-fulfill is the operational crucible of the lead-to-cash process, the phase where customer commitments become physical actions. A customer has been qualified, an opportunity has been won, a quote has been accepted, and a sales order has been created. From this point forward, the enterprise must confirm product availability, schedule production or procurement, pick and pack goods, arrange transportation, deliver to the customer, and confirm that everything arrived as promised. Every step in this chain is a potential failure point: a stockout that delays shipment, a credit hold that freezes the order, a pricing discrepancy that blocks billing, a delivery exception that triggers a customer complaint. For the internal sales representative, fulfillment monitoring is a daily exercise in navigating these exceptions, and it consumes an enormous share of their productive time.

The challenge intensifies at scale. Organizations processing thousands of orders daily cannot afford to have sales representatives manually checking each order's status across production, warehouse, transportation, and billing systems, yet this is precisely what happens in most enterprises: A customer calls asking when their order will ship. The sales representative opens multiple transactions, checking the sales order, the production order, the outbound delivery, the shipment, trying to piece together a coherent status picture. When an exception exists, the representative must diagnose whether it is a credit hold, a material shortage, a warehouse bottleneck, or a billing configuration issue, then coordinate with the appropriate team to resolve it. This process can take 30 minutes or more per inquiry, and in organizations with high order volumes, it represents millions of dollars in unproductive labor annually.

The following sections cover the embedded AI capabilities that address this challenge in order to optimize the order-to-fulfill process flow.

Low-Touch Sales Order Management

SAP S/4HANA Cloud addresses the order entry challenge through a comprehensive low-touch sales order management architecture designed to minimize human intervention from the moment an order enters the system. The framework spans multiple entry channels and automation technologies, creating a multilayered approach to touchless order processing.

At the foundation, SAP S/4HANA Cloud provides optimized and automated order entry from multiple sources. Sales orders can be created through simplified user experiences designed for fast entry, through uploading Microsoft Excel or Google Sheets for batch processing, through standard APIs for system-to-system integration, or through robotic process automation bots that import orders from external formats. The system performs a comprehensive suite of automated checks at creation: order completeness validation, pricing calculation, credit check, product availability check and confirmation, foreign trade compliance check, product compliance check, and payment card authorization. This ensures that orders enter the fulfillment pipeline fully validated.

The *AI-assisted sales order creation* capability takes this further by enabling internal sales representatives to create sales orders directly from customer purchase order files in PDF or image formats. Using SAP Document AI, the system automatically extracts information from unstructured documents (customer identification, product numbers, quantities, requested delivery dates, and pricing references) and proposes values for the corresponding sales order fields. The sold-to party is determined by matching extracted data against the business partner master, and material numbers are resolved through intelligent matching against the product catalog. The sales representative reviews the proposed order, corrects any exceptions, and converts the request into a confirmed sales order, reducing order entry time from minutes of manual data entry to seconds of review and approval.

The *AI-assisted sales order automatic completion* capability addresses a related but distinct challenge: orders that enter the system through any channel but arrive with missing or incomplete fields. Rather than routing these orders to a manual incompletion worklist where they may sit for hours awaiting attention, the system uses ML to predict and populate missing values based on historical order patterns, customer preferences, and product defaults. This eliminates the fulfillment delays caused by incomplete data and reduces the error rate from manual field population.

For high-volume environments, the *Joule Sales Mass Change* capability enables sales representatives to execute bulk updates across multiple orders through natural language. A representative can instruct Joule to "remove the delivery block for all sales orders between 24817 and 24830," for example, or "set the requested delivery date to next Friday for all open orders from customer ABC," and the system executes the mass change automatically. This capability is particularly valuable during supply disruptions or logistics delays, where hundreds of orders may need simultaneous date adjustments.

AI-Assisted Backorder Processing with Advanced Available-to-Promise

Product availability confirmation is the critical moment in order-to-fulfill, the point where the enterprise commits to delivering specific quantities by specific dates. When standard available-to-promise (ATP) logic cannot confirm the requested delivery date because insufficient stock or production capacity exists, orders enter the backorder processing queue. Traditional backorder processing is one of the most labor-intensive and expertise-dependent activities in sales operations. Planners must evaluate hundreds or thousands of unconfirmed order items, determine optimal rescheduling sequences, balance customer priority against material availability, and execute complex reallocation decisions, a process that can take days to complete manually.

SAP's *AI-assisted backorder processing* reimagines this process by leveraging prompt-driven AI to simplify scheduling and significantly enhance fulfillment outcomes. The solution offers two operating modes that represent different points on the autonomy spectrum:

- **Fast Track mode**
  The Fast Track mode provides a one-step experience in which users provide only the material and plant details. The AI autonomously executes the entire backorder processing run in the background, evaluating all unconfirmed orders for the specified materials; determining optimal confirmation sequences based on customer priority, delivery urgency, and available supply; and executing the rescheduling. The system delivers optimized fulfillment results without requiring the user to configure complex scheduling parameters, sort sequences, or selection criteria. This dramatically reduces the time required for backorder processing from hours to minutes.
- **Guided Intelligence mode**
  The Guided Intelligence mode provides more control for experienced planners. Users describe their scheduling requirements in natural language—for example, "Reschedule all unconfirmed orders for material XYZ-100 in Plant 1000, prioritizing orders with delivery dates in the next two weeks"—and the AI parses this input, extracts the relevant backorder processing parameters (material selections, plant assignments, priority rules, date ranges), generates the appropriate backorder processing configuration, and executes the run. This conversational approach eliminates the need to navigate complex configuration screens while preserving the planner's ability to apply business judgment to the scheduling logic.

For SAP S/4HANA Cloud Private Edition customers running advanced ATP, the Fast Track Backorder Processing capability brings the same Joule integration to the more sophisticated advanced ATP scheduling engine. The capability accepts user-provided text input; parses it to extract BOP requirements and preferences; transforms the input into structured BOP-relevant data including selection conditions, sort rules, and execution parameters; and triggers the advanced ATP backorder processing run. This integration ensures that organizations using advanced ATP rules, including product allocation, backorder processing with substitution, and multilevel availability checking, can access the same conversational AI experience as standard ATP users.

The Sales Operations Assistant

The Sales Operations Assistant, part of the Autonomous Finance framework, represents a fundamental shift from human-initiated monitoring to autonomous, agent-driven exception resolution. This assistant orchestrates four specialized agents, each targeting a distinct category of fulfillment exception. SAP quantifies the combined impact as a 20 to 30 percent increase in sales order handling efficiency and a 10 to 20 percent reduction in order-to-cash cycle time.

Let's take a closer look at each capability:

- **AI-Assisted Sales Order Fulfillment Monitoring**
  Once an order enters the fulfillment pipeline, the most operationally impactful AI capability in the order-to-fulfill domain is sales order fulfillment monitoring. SAP has invested through three successive releases of this capability, each expanding the scope of intelligence and automation, to create a comprehensive, Joule-powered monitoring and resolution experience.

  The capability enables sales departments to leverage AI to monitor and resolve order fulfillment issues in real time. Sales orders are summarized at both the header and item level in natural language, fulfillment issues are identified and explained, resolution actions are proposed and executed, and document flows are made transparent. Rather than navigating multiple transaction screens to piece together an order's status, the sales representative asks Joule a natural language question, such as "What is the fulfillment status of sales order 24817?" or "Show me all open orders for customer Meridian Corp that have delivery issues."

  Joule responds with a structured, contextual explanation. It identifies whether the order is on track or has issues, and for each issue, it provides the specific blocking factor (a credit hold, a missing production confirmation, an incomplete delivery, a billing block) along with the business context needed to resolve it. For each identified issue, Joule either proposes a resolution action that the representative can approve or executes navigational guidance to the appropriate application for manual resolution.

  The Track Sales Order capability, evolved through version 3, provides a complete overview of fulfillment status by examining the entire document chain: the sales order itself, production orders, purchase requisitions, outbound deliveries, shipments, and billing documents. The system evaluates each document's status against predefined business rules and offers guidance or direct actions based on the specific exception type. This transforms fulfillment monitoring from a reactive, inquiry-driven process into a proactive, exception-driven workflow in which the system surfaces issues before customers call to complain. Figure 5.22 shows the Sales Operations Assistant, illustrating how it monitors open sales orders and provides structured resolution guidance to sales teams.

- **Delivery Issue Resolution Agent**
  The Delivery Issue Resolution Agent continuously monitors open sales orders for delivery issues, found in orders where the total delivery status shows partial delivery or no delivery based on exceeded or approaching delivery dates. When an issue is detected, the agent evaluates predefined business rules, provides suggestions and actions, displays delivery and shipping status along with related customer details, and proposes creating outbound deliveries for selected items. It can suggest splitting deliveries, rescheduling shipments, or escalating to logistics management, all autonomously, without waiting for a customer complaint to trigger investigation. Figure 5.23 shows the Delivery Issue Resolution Agent, illustrating how it continuously monitors sales orders and orchestrates resolution across logistics and procurement.

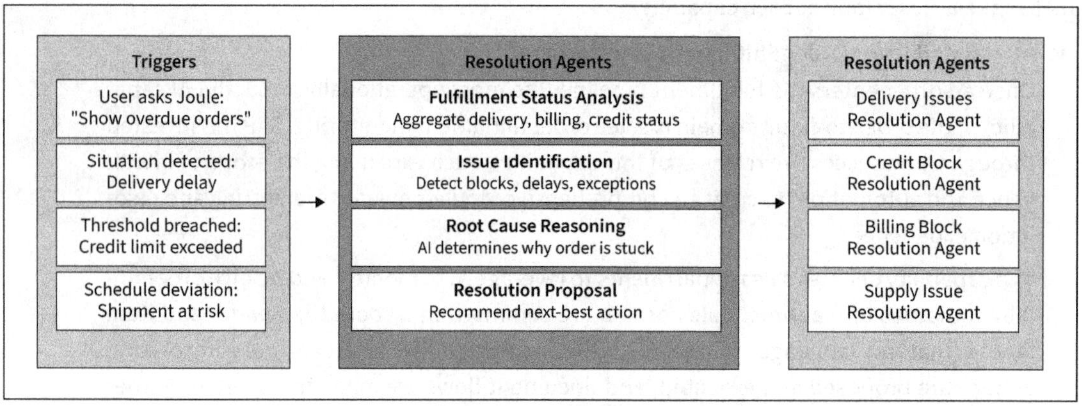

Figure 5.22: AI-Assisted Sales Order Fulfillment Monitoring with Sales Operations Assistant

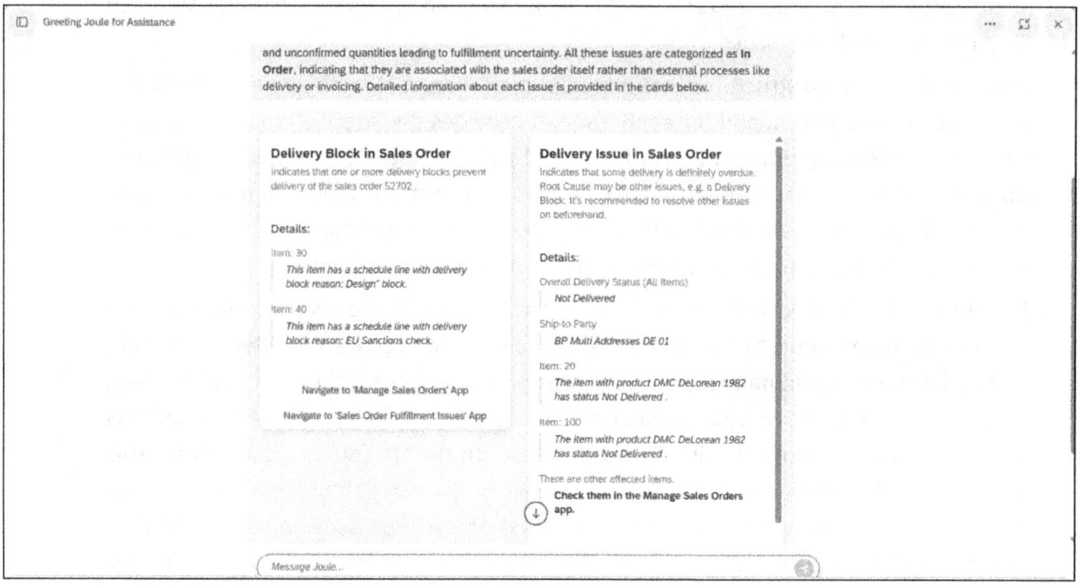

Figure 5.23: Delivery Issue Resolution Agent

- **Credit Block Issue Resolution Agent**
  The Credit Block Issue Resolution Agent addresses one of the most common fulfillment bottlenecks: sales orders that are blocked because the customer's credit exposure exceeds predefined limits or fails specific credit checks. The agent analyzes the credit situation (current exposure, credit limit, payment history, and outstanding receivables) and provides contextual information to help the sales representative or credit analyst make an informed release decision. It can suggest releasing the block with justification, escalating to the credit manager, or recommending a payment-on-delivery arrangement.

- **Billing Block Issue Resolution Agent**
  The Billing Block Issue Resolution Agent handles orders that are blocked due to billing configuration issues: missing pricing calculations, changed payment terms, incomplete billing-relevant master data, or contractual discrepancies. The agent displays header information and the list of blocked items, provides detailed analysis of each block reason, and suggests specific remediation actions such as removing the block, updating pricing conditions, or routing to financial accounting for review.
- **Supply Issue Resolution Agent**
  The Supply Issue Resolution Agent detects purchasing and manufacturing issues within sales orders. It monitors whether purchase requisitions have been converted to purchase orders, whether purchase order confirmations and goods receipts are on track, and whether production orders are progressing on schedule. When supply chain disruptions affect order fulfillment, the agent suggests creating purchase orders with reference to outstanding requisitions, expediting production orders, or arranging alternative sourcing, providing the sales representative with actionable supply chain visibility without requiring them to navigate procurement or production planning systems.

Together, these four agents transform the sales representative's role from reactive exception investigator to proactive relationship manager. Instead of spending hours diagnosing fulfillment issues across multiple systems, the representative receives autonomous alerts with preanalyzed root causes and recommended resolutions. The representative's judgment is still essential (approving credit releases, confirming delivery splits, authorizing expedited shipments), but the analytical and investigative work that previously consumed 30 to 40 percent of their day is now handled by agents.

Pricing Intelligence

Pricing is the silent revenue lever in order-to-fulfill and one of the most complex configuration challenges in any ERP environment. SAP S/4HANA's condition technique supports thousands of pricing condition records across materials, customer groups, distribution channels, and validity periods. Managing this complexity manually—particularly, ensuring that expiring prices are renewed, that discount schedules reflect current commercial agreements, and that pricing anomalies are detected before they impact revenue—requires constant vigilance.

The AI-assisted expiring price handling capability enables pricing specialists to interact with Joule conversationally to identify prices, discounts, or surcharges approaching expiration. A specialist can ask "Show me all pricing conditions expiring in the next thirty days for distribution channel 10" and receive an immediate inventory of at-risk conditions. They can then create new conditions through the same conversational interface, setting validity periods and amounts without navigating the traditional condition maintenance transactions. SAP quantifies the impact: Mass price change updates that previously required several weeks can now be completed in three to four days.

The Manage Sales Pricing Condition Records capability extends Joule's pricing intelligence further, enabling full create, read, update, and delete (CRUD) operations on pricing business objects through natural language. Pricing analysts can query condition records, understand the pricing waterfall for specific customer-material combinations, identify inconsistencies across pricing hierarchies, and make corrections directly through Joule. This conversational access to pricing data is particularly valuable for sales managers reviewing deal profitability, who can ask, "What is the net price for material ABC-200 for customer Meridian Corp, including all applicable discounts and surcharges?" and receive an instant, fully calculated response.

The AI-assisted sales order summary capability provides sales managers with a complementary analytical view, generating natural language insights about pricing changes from original prices (whether prices have increased or decreased and by what percentage), on-time delivery performance, and fulfillment status across their order portfolio. This transforms the sales management review from a spreadsheet exercise into a conversational intelligence briefing.

Delivery Prediction and Logistics Intelligence

The final dimension of order-to-fulfill intelligence addresses the physical movement of goods. SAP S/4HANA Cloud embeds ML for delivery-delay prediction, enabling the system to predict potential delivery delays before they occur by analyzing historical patterns of order-processing times, warehouse throughput, transportation lead times, and seasonal demand fluctuations. When a delay is predicted, situation handling triggers proactive notifications to the relevant stakeholders (the sales representative, the logistics coordinator, the customer), enabling corrective action before the customer's expected delivery date passes.

For shipping specialists, Joule provides conversational access to outbound delivery and sales order information. Specialists can query the delivery status, identify shipments requiring attention, and take corrective actions through natural language interaction rather than by navigating transactional screens. The embedded SAP Analytics Cloud dashboards provide real-time visualization of order confirmation rates, delivery performance, and customer 360 perspectives, giving operations managers continuous visibility into fulfillment health.

Case Study: Order-to-Fulfill Transformation

For James at Korindal, the order-to-fulfill transformation produces the most immediately visible operational impact of any AI deployment in the commercial organization. Before the transformation, the inside sales team of 300 representatives spent an estimated 35 percent of their time on order status inquiries and fulfillment exception investigation. Each inquiry averaged 25 minutes: opening multiple transactions, piecing together document flows, diagnosing exception root causes, and coordinating with logistics, credit, and billing teams. The team processed approximately 1,200 fulfillment inquiries per day, consuming over 500 person-hours daily.

The first phase of transformation begins with AI-assisted sales order creation from unstructured data. Korindal's largest customers, representing 40 percent of its order volume, still submit purchase orders as PDF email attachments. Historically, an inside sales representative manually keyed each order, a process averaging eight minutes per order with a 6 percent error rate. SAP Document AI reduces the average order entry time to under 90 seconds of review and approval, and the error rate drops below 1 percent. Across 1,200 PDF-originated orders per week, Korindal saves approximately 130 person-hours in order entry labor alone.

The second phase arrives with Joule-powered fulfillment monitoring. Sales representatives no longer open six different transactions to understand an order's status. They ask Joule: "Show me all orders for Meridian Corp with fulfillment issues." Joule returns a prioritized list of exceptions (two orders with credit holds, one with a delivery delay prediction, and three with partial shipment status), each with a natural language explanation of the root cause and recommended resolution. The representative resolves the credit holds by reviewing Joule's analysis of Meridian's credit position and approving the release, confirms the delivery delay notification has been sent to the customer, and approves the partial shipment completions. What previously took 2.5 hours of investigation takes 12 minutes.

The backorder processing transformation is equally dramatic. Korindal's planning team previously spent two full days per week running manual backorder processing cycles, configuring parameters, executing runs, reviewing results, adjusting sequences, and rerunning. The Fast Track mode reduces this to a 45-minute process in which the planner specifies materials and plants, and the AI handles the optimization. The Guided Intelligence mode allows planners to apply nuanced business logic ("Prioritize all backorders for the automotive segment, then industrial, then distribution") through conversational prompts that would previously have required complex configuration screen navigation.

These fulfillment capabilities directly address the challenge introduced in Chapter 1, in which Korindal's chief supply chain officer (CSCO) faced an on-time, in-full (OTIF) delivery metric stuck at 86 percent. The root cause was not poor planning; it was late decision-making. By the time a fulfillment miss became visible, it was already too late to prevent it. The CEO's question, "Why do we only find out after we miss?," captured the fundamental problem. With Joule-powered fulfillment monitoring surfacing at-risk orders before shipment deadlines pass; the Sales Operations Assistant autonomously resolving delivery, credit, billing, and supply exceptions; and AI-assisted backorder processing enabling deliberate backorder prioritization rather than first-come, first-served allocation, Korindal shifts from reactive firefighting to proactive commitment management. Supply shortfalls now trigger prioritization rather than panic. Backorders are processed deliberately rather than automatically. Customer commitments stabilize because the organization acts early enough to change the outcome. The CSCO's story changes from "We hit the plan" to "We keep our promises," and OTIF moves from a lagging indicator reviewed on Monday mornings to a managed metric governed in real time.

The cumulative impact is quantifiable: Korindal's inside sales team recovers approximately 35 percent of its productive capacity from fulfillment monitoring and order entry automation, equivalent to reassigning 105 representatives to revenue-generating activities (prospecting, relationship development, and upselling) rather than operational firefighting. Order processing capacity increases sevenfold, and the average time from order receipt to confirmed shipment decreases by 40 percent. The customer experience improves in parallel: Proactive delay notifications replace reactive complaint handling, and order accuracy approaches 99 percent.

5.3.3 Opportunity-to-Quote

The opportunity-to-quote phase is where commercial strategy meets execution, the critical window between identifying a potential sale and committing the enterprise to deliver specific products at specific prices on specific terms. For sales organizations, this phase determines win rates, deal velocity, and margin quality. A poorly qualified lead wastes expensive sales capacity, a delayed quote loses to a faster competitor, and an inaccurate quote erodes margin or creates fulfillment problems downstream. Figure 5.24 shows the Autonomous CX portfolio overview, mapping all assistants and agents across opportunity-to-quote, order-to-cash, and manage customer and channels.

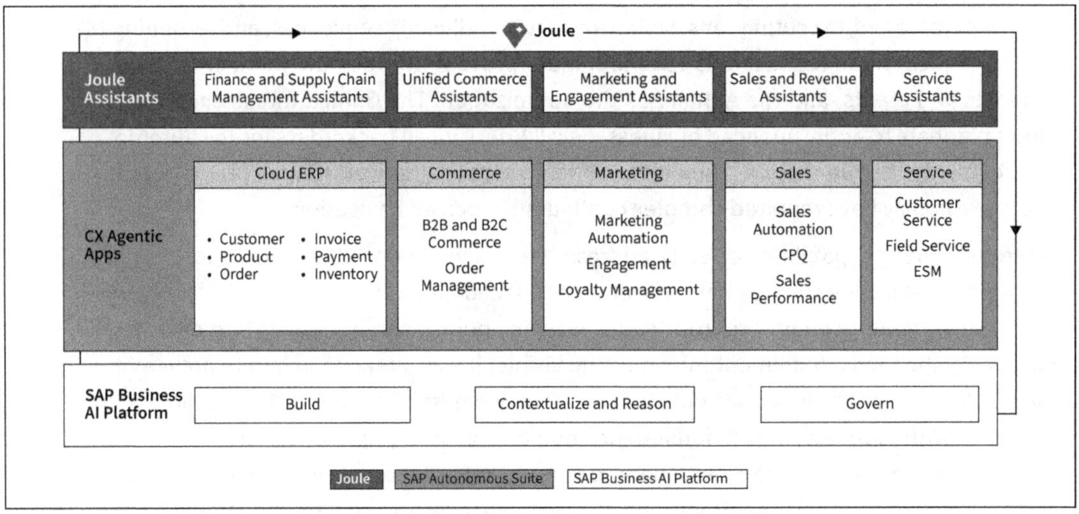

Figure 5.24: Autonomous CX Portfolio Overview

A defining characteristic of this phase is that its intelligence spans both SAP Sales Cloud (the Autonomous CX platform) and SAP S/4HANA, working in concert. The Deal Qualification Assistant, Sales Assistant, and Deal Closing Assistant operate within SAP Sales Cloud, where they orchestrate lead scoring, opportunity management, pipeline analytics, and guided selling through native customer relationship management (CRM) agents. Simultaneously, SAP S/4HANA contributes configure, price, quote (CPQ) capabilities, real-time pricing condition management, credit and availability checks, and the commercial data backbone that ensures quotes reflect accurate inventory, pricing, and delivery commitments. This dual foundation means that front-office selling intelligence and back-office commercial precision operate as a single coordinated system rather than requiring manual handoffs between CRM and ERP.

We'll walk through these assistants in detail in the following sections and see how they impact our case study. Figure 5.25 shows the Opportunity to Quote Assistant, illustrating how AI supports deal qualification, configuration, pricing, and proposal generation.

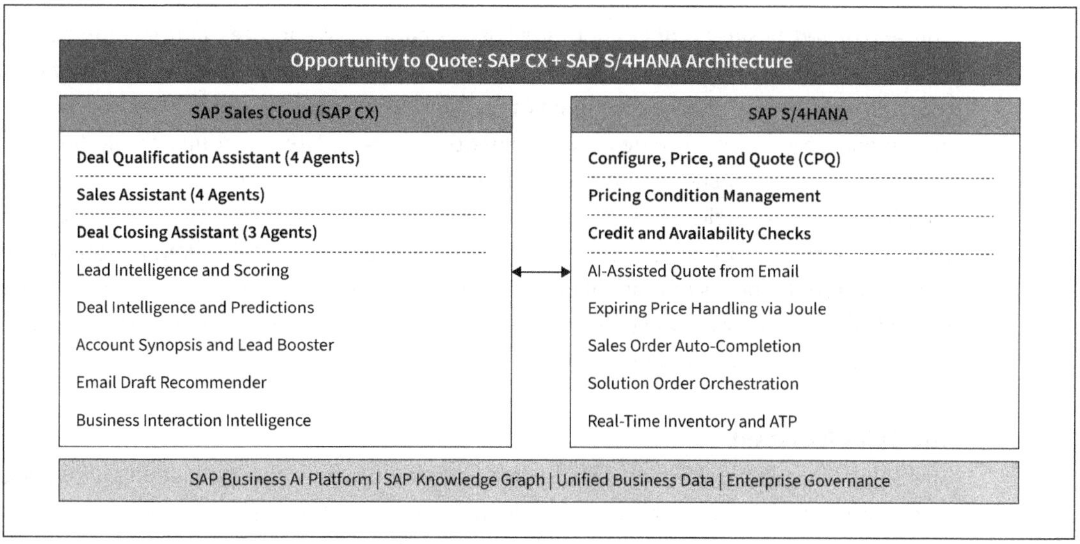

Figure 5.25: Opportunity to Quote Assistant

The Deal Qualification Assistant

The Deal Qualification Assistant automates lead enrichment and qualification, converting sales-ready prospects into opportunities faster while improving pipeline quality and scalability without sales representative involvement. The assistant coordinates four specialized agents that collectively ensure that only qualified, high-potential leads consume expensive selling resources. SAP quantifies the impact as increased lead conversion rates, increased pipeline value and volume, and faster lead conversion and deal cycles.

The foundation of effective selling is knowing which opportunities to pursue and how to win them. SAP's *AI-assisted lead intelligence* applies machine learning to historical lead data, analyzing which leads converted and which declined, what characteristics predicted success, and what patterns differentiate high-value opportunities from time-wasting pursuits. The resulting predictive model provides each new lead with a conversion score: the higher the score, the better the chance of conversion. Sales managers use these scores to prioritize team capacity, ensuring that the highest-potential leads receive immediate attention while lower-probability leads are nurtured through automated sequences.

AI-assisted deal intelligence extends this predictive capability to active opportunities. The model analyzes deal characteristics (product mix, customer segment, competitive context, deal stage duration, engagement patterns) to determine which deals can close, which will not, and which could go either way with the right intervention. Sales representatives see deal scores directly in their opportunity views, enabling them to allocate their most intensive selling efforts to the deals for which intervention will have the greatest impact. For sales managers, this creates a data-driven pipeline review process that replaces gut-feel forecasting with statistically grounded probability assessments.

The *AI-assisted business interaction intelligence* capability captures and synthesizes the intelligence generated through customer interactions. Before meetings, sales teams receive AI-generated premeeting briefs that synthesize relevant account history, recent interactions, open opportunities, and competitive context. After meetings, the system generates postmeeting insights that extract action items, commitments, and next steps from meeting notes. For field sales representatives conducting customer visits, previsit briefs consolidate all relevant account information into a single, actionable summary that would otherwise require hours of manual preparation. This capability transforms meeting preparation from a discretionary activity that busy representatives often skip into an automatic, zero-effort intelligence briefing.

The Sales Assistant

The Sales Assistant orchestrates four agents that automate updates, meeting assistance, risk analysis, and outreach to boost productivity, sharpen forecasts, deepen engagement, and help sales teams close more deals faster and at scale. SAP quantifies the impact as a 22 percent increase in expansion revenue rate and a 29 percent improvement in lead response time. The assistant continuously monitors the opportunity pipeline, identifying deals that are stalling, approaching critical decision dates, or showing engagement patterns that predict either positive or negative outcomes.

It generates proactive recommendations that keep deals moving through the pipeline. It also automates the administrative burden of CRM updates, interaction logging, and forecast reporting, freeing sales representatives to spend their time on actual selling rather than system maintenance. The Opportunity Close Date Prediction capability uses ML to predict the expected close date for open opportunities, providing sales managers with a statistically grounded view of when revenue will materialize rather than relying on the often-optimistic estimates that sales representatives enter manually.

The AI-assisted email draft recommender accelerates the communication workflow that surrounds every opportunity. The system generates email drafts based on the conversation context (responding to customer inquiries, following up after meetings, sending proposals, or requesting next steps) that the sales representative can review, personalize, and send. This capability is particularly valuable in high-velocity, inside sales environments, where representatives manage large numbers of concurrent opportunities and cannot afford to craft each email from scratch.

The Deal Closing Assistant

The Deal Closing Assistant guides sellers through quotes, pricing, and orders with agentic precision, reducing friction, ensuring accuracy, shortening sales cycles, and closing complex deals faster with confidence. SAP quantifies the impact as a 15 percent increase in average deal value and a 29 percent decrease in seller time spent on administration versus selling. The assistant coordinates three specialized agents that collectively address the most time-consuming and error-prone aspects of the quoting and closing process.

The quoting process in complex sales environments often represents the single largest bottleneck in deal velocity. Configuring products, applying the correct pricing, validating commercial terms, generating the quote document, and routing it for approval can take days in organizations selling configurable products or solution bundles. SAP addresses this challenge through multiple AI-powered capabilities that accelerate each step.

The AI-assisted sales quote creation from emails streamlines the most common quoting scenario, in which a customer requests a quote via email. The system intelligently scans email contents for crucial details (product names, quantities, prices, and contact information) and automatically creates a quote in SAP Sales Cloud. What previously required a sales representative to read the email, open the quoting application, manually enter each line item, look up pricing, and save the quote is reduced to a review-and-confirm workflow.

For organizations using SAP S/4HANA's CPQ capabilities, the quoting process supports complex solution selling for which a single quote may contain physical goods, services, projects, and subscriptions. The CPQ process enables detailed configuration and pricing negotiation, producing a solution order that orchestrates fulfillment across all offering types. The system supports complex pricing models, captures configuration requirements directly in the sales quote, and maintains a holistic order progress view for operational transparency.

Account Intelligence and Relationship Management

The AI-assisted account synopsis generates comprehensive account summaries that enhance the sales team's understanding of each customer's needs, business context, and competitive landscape. Using generative AI, the system synthesizes information from CRM records, interaction history, order patterns, and external data sources into a narrative account overview, the kind of institutional knowledge that traditionally exists only in the heads of veteran account managers. When an account manager leaves or a new representative is assigned, this AI-generated account intelligence provides immediate context rather than a cold-start learning curve.

The AI-assisted lead booster takes account intelligence in a proactive direction, generating tailored information that identifies upsell and cross-sell opportunities within existing accounts. By analyzing purchase patterns, product adoption gaps, and industry benchmarks, the system surfaces specific recommendations for expanding the customer relationship ("This customer uses Product A but has not adopted the complementary Product B, which 78 percent of similar accounts in this industry have purchased"). These data-driven expansion recommendations transform account management from relationship maintenance into strategic growth.

Joule with SAP Sales Cloud Version 2 provides the conversational layer that makes all this intelligence instantly accessible. Sales teams can interact with account data, opportunity pipelines, and customer histories through natural language, receiving intelligent recommendations to help win deals faster and close them more effectively. Rather than navigating multiple screens and reports, a representative asks: "Give me a summary of the Meridian Corp account, including all open opportunities and recent interactions," and they receive an instant, synthesized response. Figure 5.26 shows the Autonomous CX assistants, illustrating how account intelligence, lead booster, and conversational Joule capabilities enhance sales team effectiveness.

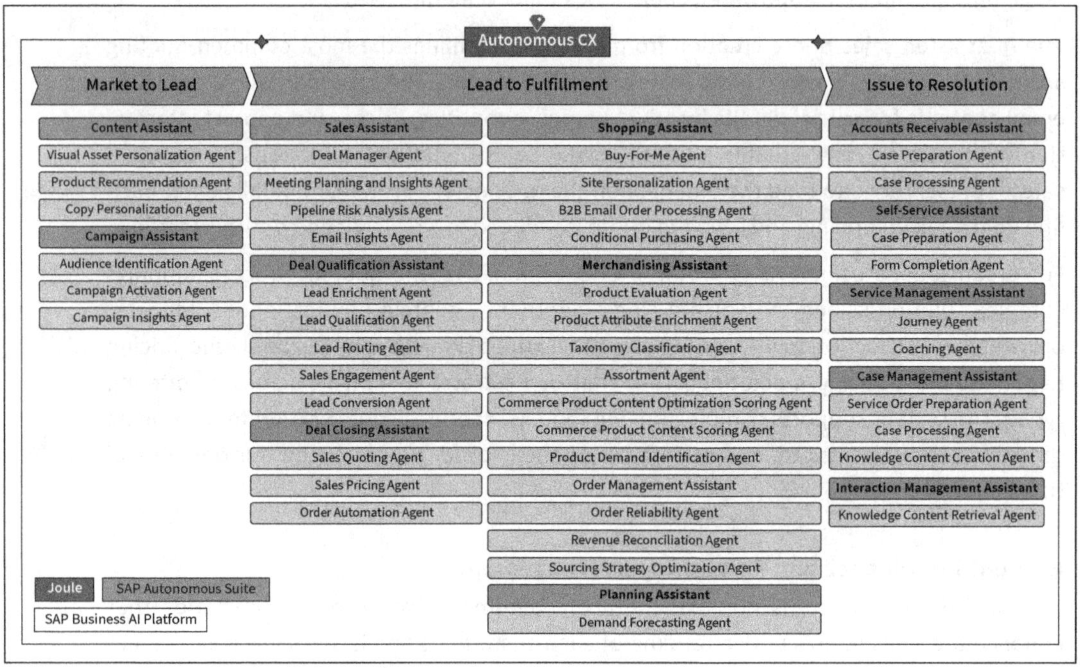

Figure 5.26: Autonomous CX Assistants

Case Study: Opportunity-to-Quote Transformation

At Korindal, the opportunity-to-quote transformation begins with lead intelligence fundamentally changing how the sales team prioritizes its pipeline. Before AI scoring, all leads received roughly equal treatment: the same outreach cadence, the same qualification process, the same time investment. Lead intelligence reveals that 22 percent of Korindal's inbound leads generate 78 percent of eventual revenue, and these high-potential leads share identifiable characteristics that the model captures. By routing high-score leads to senior representatives and automating nurture sequences for lower-score leads, Korindal increases lead-to-opportunity conversion rates by 18 percent while reducing the total time invested in lead qualification.

Deal intelligence transforms the weekly pipeline review. Previously, the sales VP reviewed opportunities in a three-hour meeting in which each regional manager presented a subjective assessment of their top deals. Now, the meeting begins with an AI-generated pipeline analysis that identifies the 15 deals most at risk, the 10 deals most likely to accelerate, and the five deals for which competitive displacement signals have been detected. The discussion shifts from "tell me about your pipeline" to "here are the specific interventions needed this week," cutting the review time in half while dramatically improving the quality of pipeline management decisions.

The quoting acceleration has the most direct revenue impact. Korindal's average time from customer quote request to delivered proposal drops from 4.2 days to 1.1 days. For a business in which 60 percent of competitive situations are won by the first vendor to submit a technically compliant quote, this acceleration translates directly into an 11 percent improvement in the win rate for competitive deals. The AI-assisted email quote creation capability alone eliminates approximately 400 hours per month of manual quoting labor across the inside sales team.

5.3.4 Invoice-to-Cash

We covered the core invoice-to-cash SAP Autonomous Suite capabilities in Section 5.2.3. Apart from the SAP S/4HANA Cloud ERP capabilities, the assistants that can help on the CX side are the Case Management Assistant and Self-Service Assistant: Post-Sale Service Excellence.

The commercial relationship does not end at cash collection. Postsale service quality directly influences customer retention, expansion revenue, and brand reputation. The *Case Management Assistant* empowers service teams to triage, manage, and resolve cases faster by surfacing real-time insights grounded in full business context and automating workflows from end to end. SAP quantifies the impact as an 11 percent reduction in operational support cost, a 34 percent increase in first-contact resolution rate, and a 75 percent increase in ticket resolution without service representative involvement.

The *Self-Service Assistant* complements this by providing customers with instant answers around the clock without waiting. It resolves policy questions, invoice disputes, order status inquiries, and routine service requests, reducing costs while freeing service teams for complex support. SAP quantifies the impact as a 33 percent increase in deflection rate, an 80 percent reduction in case creation and categorization time, and a 60 percent reduction in cost per contract.

Together, these assistants ensure that the customer experience extends beyond the transaction, creating the satisfaction and trust that drives repeat purchases and expansion revenue. In a commercial model in which acquiring a new customer costs five to seven times more than retaining an existing one, postsale service intelligence is not a cost center but a revenue-protection mechanism.

5.3.5 Manage Customer and Channels

The master data foundation is the invisible infrastructure upon which every lead-to-cash process depends. Customer records, business partner relationships, pricing agreements, distribution channel configurations, and product assignments collectively form the data foundation that determines whether orders process smoothly or fail at validation. When master data is accurate and complete, orders flow through creation, pricing, credit checking, availability confirmation, and billing without interruption. When it is not—and in most enterprises, master data quality is a persistent challenge—every downstream process suffers. Pricing errors, delivery misroutes, credit check failures, billing rejections, and customer duplicate records are overwhelmingly master data problems masquerading as process failures.

The challenge of master data quality in the lead-to-cash context is structural. Customer and business partner records are created and modified by multiple departments, like sales, finance, logistics, and compliance, each with different data requirements and different quality standards. Address formats vary by country. Tax classification codes change with regulatory updates. Bank details must be validated against external registries. Credit limits require periodic review. Across all of this, duplicate records proliferate as different departments create entries for the same legal entity without recognizing existing records. The cost of poor master data is rarely measured directly, but it manifests in every fulfillment exception, every pricing error, and every billing rejection that requires manual investigation and correction. Figure 5.27 shows the master data lifecycle, illustrating how AI-assisted business partner maintenance and data quality management underpin the lead-to-cash processes.

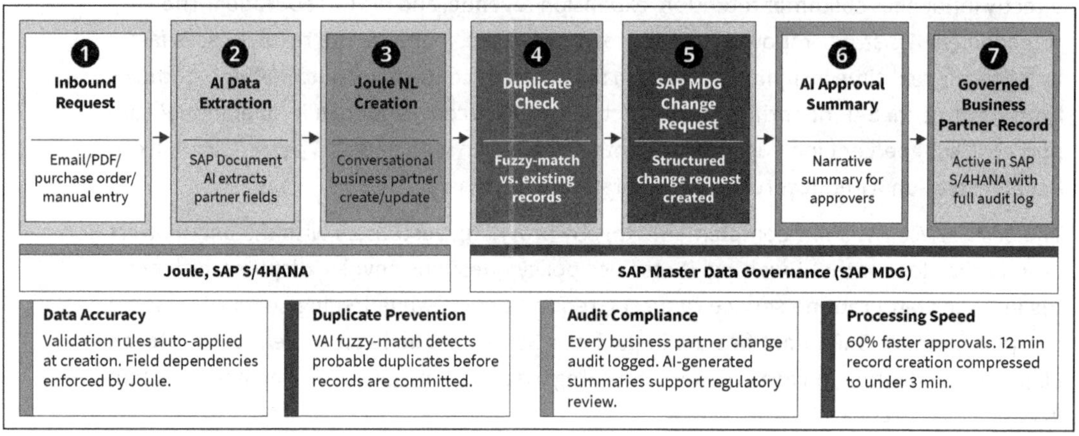

Figure 5.27: Master Data Lifecycle

The following sections will discuss the embedded AI capabilities that simplify customer data management and enable master data transformation for organizations.

Simplified Business Partner Maintenance with Joule

The *business partner* is the central master data object in SAP S/4HANA for managing customer, supplier, and general partner information. Historically, creating or modifying a business partner record required navigating complex transaction screens with dozens of fields organized across multiple tabs for the general data, company code data, sales area data, purchasing data, bank details, tax information, and more. The expertise required to correctly populate these fields, understand which are mandatory for which business scenarios, know which validation rules apply, and determine which dependent fields must be consistent represents a significant training investment and a persistent source of data-entry errors.

Joule integration simplifies and accelerates the lifecycle management of business partners by enabling the creation and change of business partner data through natural language interaction. A sales operations specialist can tell Joule: "Create a new customer business partner for Meridian Corporation, headquartered at 1400 Industrial Boulevard, Chicago, Illinois 60607, with sales organization 1000, distribution channel 10, and payment terms net 30," and Joule will create the record with all mandatory fields populated, validation rules applied, and dependent configurations set automatically. For changes, the specialist can say: "Update the shipping address for business partner 10004721 to 2800 Commerce Drive, Suite 200, Atlanta, Georgia 30339," and the system will execute the change with appropriate audit logging.

The AI-assisted easy-fill capability extends this principle beyond business partners to any business object in SAP S/4HANA Cloud. Users can populate fields using natural language input, describing what they need in conversational terms rather than navigating entry screens field by field. This streamlines data entry across the entire lead-to-cash process: Creating sales orders, updating pricing conditions, modifying delivery schedules, and adjusting billing configurations can all be performed through natural language that the system interprets and maps to the appropriate fields.

Master Data Governance with AI

For organizations requiring formal governance over master data changes, particularly those operating in regulated industries or managing complex global hierarchies, SAP Master Data Governance (SAP MDG), SAP's dedicated master data governance solution, provides a structured framework for change management, quality assurance, and approval workflows. The AI-assisted master data governance capability enhances this framework by enabling users to leverage natural language to create and update master data change requests and then generate summaries of all updates contained in a change request.

This capability is particularly valuable in organizations where master data changes require multilevel approval. Rather than reviewing dozens of individual field changes to understand the business impact of a change request, approvers receive an AI-generated summary: "This change request modifies the credit limit for customer Meridian Corp from $500,000 to $750,000, adds a new ship-to address in Atlanta, and updates the payment terms from net 30 to net 45. The credit limit increase is supported by the customer's improved payment history over the last four quarters." This narrative summarization transforms the approval process from a tedious field-by-field review into an efficient business decision.

The Intelligent Business Partner Creation from Documents capability takes automation further by enabling the creation of business partner records directly from source documents: purchase orders, quotations, contracts, or customer correspondence received as PDFs or emails. The system extracts partner information from these documents, matches it against existing records to prevent duplicate creation, and proposes a complete business partner record for review and approval. This is particularly impactful for organizations onboarding large numbers of new customers or suppliers, where manual data entry from paper documents has traditionally been a major bottleneck.

Customer Data Quality and Duplicate Management

The AI-assisted duplicate detection capability addresses one of the most pervasive and costly master data problems: duplicate customer records. The system performs intelligent checks during account creation, comparing configured fields against the list of existing accounts to identify potential matches. Unlike simple exact-match checks, the AI-powered detection uses fuzzy matching and field correlation to identify probable duplicates even when names are spelled differently, addresses use different formats, or phone numbers include different country codes. Sales agents see potential matches ranked by probability score, enabling them to either merge with an existing record or confirm that the new entry is genuinely distinct.

For ongoing data quality, the display capabilities through Joule enable business partner, customer, and supplier master data to be queried and reviewed conversationally. Rather than navigating transaction screens to look up a customer's credit terms, delivery preferences, or tax classification, sales representatives and operations specialists can simply ask Joule "Show me the credit limit and payment terms for customer 10004721" or "What are the active ship-to addresses for Meridian Corp?" and receive instant, accurate responses. This conversational access to master data reduces both the time required for data lookups and the likelihood of errors caused by misreading transaction screen fields.

Korindal: David's Team Cuts Data Quality Issues and Customer Record Errors

At Korindal, the master data challenge is quantifiable. An internal audit reveals that 8 percent of customer records have at least one critical data quality issue: incorrect tax classifications, outdated addresses, duplicate entries, or misaligned credit terms. These issues generate an estimated 2,400 fulfillment exceptions per month that require manual resolution, each averaging 15 minutes of investigative and corrective effort. The total monthly cost of poor master data, measured purely in exception-handling labor, exceeds $150,000, not counting the revenue impact of delayed orders and dissatisfied customers.

The transformation begins with Joule-enabled business partner maintenance, reducing the average time to create a new customer record from 12 minutes of complex screen navigation to under three minutes of conversational interaction. More importantly, the error rate on new records drops from 7 percent to under 1 percent because Joule applies validation rules automatically rather than relying on the data entry specialist's knowledge of field dependencies. The duplicate detection capability prevents approximately 180 duplicate customer records per month from entering the system, duplicates that would have generated pricing inconsistencies, credit limit confusion, and reporting inaccuracies downstream.

The master data governance transformation has an equally significant but less visible impact. Change request processing time decreases by 60 percent as AI-generated summaries enable approvers to make informed decisions in a few minutes rather than spending 30 minutes reviewing individual field changes. The quality of approvals improves in parallel: Approvers now understand the business context of each change rather than simply rubber-stamping field-level modifications they do not have time to fully analyze.

5.4 Source-to-Pay

The source-to-pay value stream governs how the enterprise acquires everything it needs to operate, from strategic sourcing and supplier negotiation through procurement execution, goods receipt, invoice processing, and payment. For the chief procurement officer, this value stream determines whether the organization captures competitive pricing, maintains supply continuity, enforces contractual compliance, and optimizes the cash outflow that represents the single largest controllable cost in most enterprises. Every inefficiency in this chain—an unsourced category renewed on autopilot, a maverick purchase that bypasses contracted suppliers, a late delivery that disrupts production, a duplicate payment that leaks cash—directly erodes the margin advantage that procurement is expected to protect.

The challenge is one of scale and fragmentation. Procurement organizations manage thousands of supplier relationships, tens of thousands of material codes, and hundreds of thousands of transactions annually, yet the processes connecting these elements remain heavily manual. Sourcing cycles stretch four to six weeks. Contract compliance goes unmonitored between periodic audits. Purchase requisitions are miscategorized because employees lack commodity expertise. Supplier delivery predictions rely on confirmed dates that bear little relationship to actual arrival. And invoice processing consumes entire teams in data extraction, three-way matching, and exception resolution.

These challenges bring us to the introduction of Autonomous Spend, which we'll discuss in the following sections, diving into the relevant key processes: source-to-contract and procure-to-receipt.

5.4.1 Overview of Autonomous Spend

SAP's autonomous capabilities address these gaps across the full source-to-pay lifecycle. AI-assisted spend categorization and supplier matching create the visibility foundation. Agentic sourcing automates event construction, supplier discovery, and negotiation. Contract intelligence ensures that negotiated terms are enforced and continuously optimized. Intelligent requisitioning and guided buying route demand through compliant channels. ML predicts supplier delivery dates before delays disrupt operations. Finally, SAP Document AI transforms invoice processing into touchless, exception-only workflows.

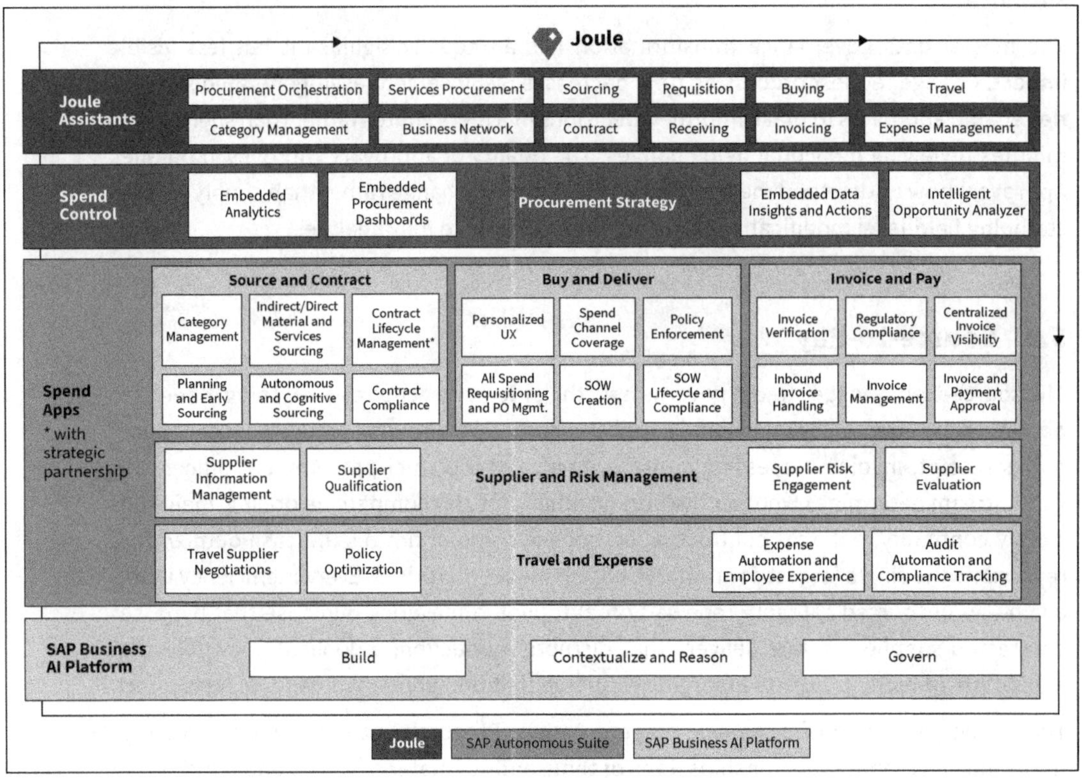

Figure 5.28: Autonomous Spend Portfolio

The Autonomous Spend domain within the Autonomous Enterprise framework orchestrates these capabilities through specialized assistants and agents: a Sourcing Assistant coordinating supplier discovery, request for *x* (RFx) orchestration, and negotiation agents with 5 to 15 percent total cost of ownership (TCO) reductions; a Requisition Assistant routing demand through optimal purchasing paths with 10 to 20 percent full-time equivalent (FTE) productivity uplift; a Procurement Contract Assistant managing the full contract lifecycle with 40 to 50 percent productivity gains; a Procurement Orchestration Agent enabling end-to-end autonomous execution for routine transactions; and an Accounts Payable Assistant orchestrating eight specialized agents across the invoice-to-payment cycle. At Korindal, the procurement transformation spans 1.2 billion dollars in addressable spend across 3,400 suppliers and 140 spend categories, where the gap between strategic intent and operational execution has historically cost the organization tens of millions of dollars annually in uncaptured savings, contract leakage, and maverick spend. Figure 5.28 showed the Autonomous Spend Portfolio, presenting the full range of AI capabilities across source-to-pay, from spend visibility through invoice automation.

Figure 5.29 shows the Autonomous Spend assistants, mapping the specialized agents that address spend analysis, sourcing, contract management, and invoice processing.

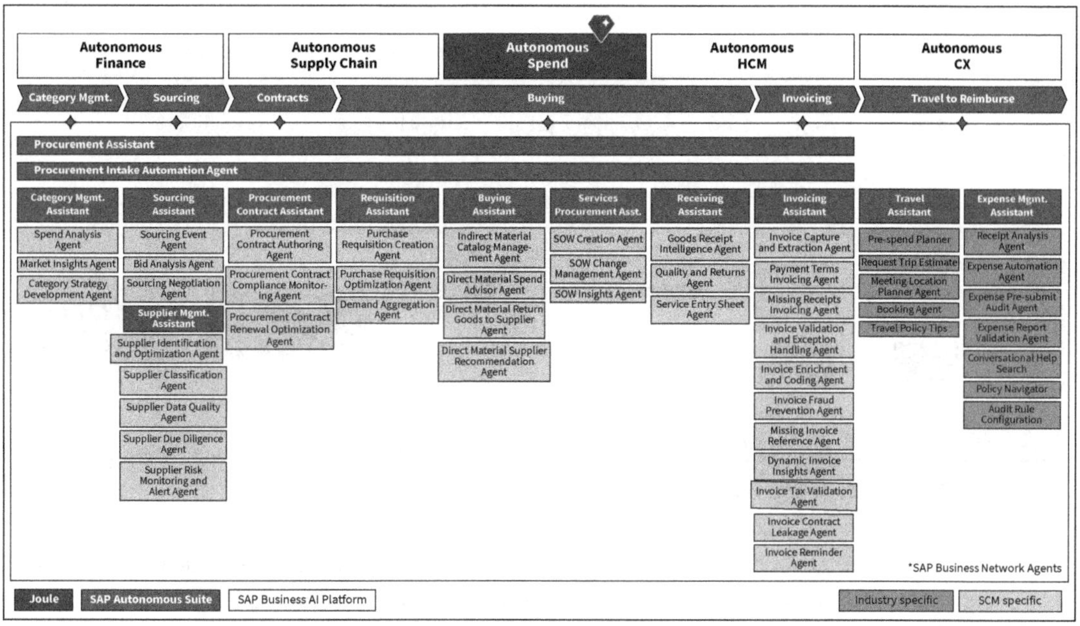

Figure 5.29: Autonomous Spend Assistants

5.4.2 Source-to-Contract

The source-to-contract process is where procurement creates strategic value, transforming raw demand signals into optimized supplier relationships, competitively sourced materials, and enforceable contractual commitments. For the chief procurement officer, this phase determines whether the organization pays market-competitive prices, maintains supply resilience, and enforces compliance with negotiated terms.

The challenge is structural. The typical sourcing cycle takes four to six weeks, from market research through supplier evaluation, negotiation, and contract documentation. In organizations managing thousands of contracts across hundreds of categories, the volume of work far exceeds the capacity of even well-staffed procurement teams. The result is predictable: 30 to 40 percent of organizational spend flows through noncontracted channels, paying spot prices that exceed what competitive sourcing would achieve.

Let's explore the embedded AI capabilities that enable AI-driven insights, negotiation, and management.

Working Capital Insights Dashboard

Spend visibility is not limited to what the organization buys and from whom. It extends to how payment behavior across the supplier base effects working capital, and this is where the Working Capital Insights dashboard delivers strategic value within the source-to-contract function. Figure 5.30 shows the Spend Visibility dashboard, providing real-time insight into spend by category, supplier, and cost center alongside DPO and DSO analysis.

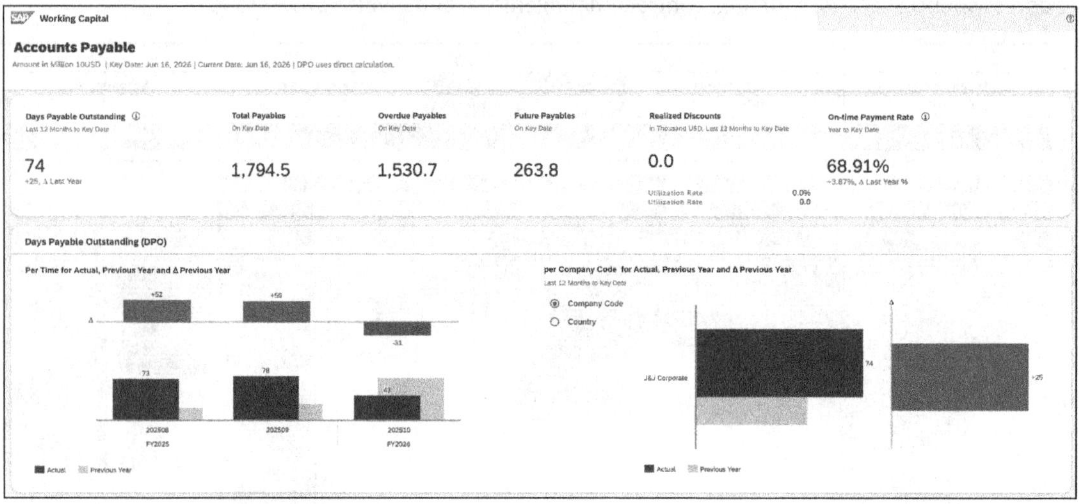

Figure 5.30: Spend Visibility Dashboard

The DPO analysis capability enables procurement and finance to examine every dimension of payables behavior across the supplier portfolio. The dashboard surfaces which supplier segments, spend categories, payment terms, and business units are driving DPO compression or extension, and it allows drill-through to the underlying transaction data to identify root causes. A supplier segment consistently paid ahead of contracted terms represents an unrecognized early payment cost. A category in which payment terms have drifted below organizational policy represents a negotiation opportunity. The dashboard makes these patterns visible and actionable rather than buried in general ledger transaction data.

The DSO analysis counterpart provides the receivables perspective, revealing how customer payment behavior and applied credit terms affect the cash conversion cycle from the revenue side. Together, DPO and DSO analytics give the CFO and chief product officer (CPO) a shared working capital language: The organization's ability to extend DPO through payment term negotiation with suppliers, capture early payment discounts where the arbitrage favors acceleration, and tighten DSO through collections management creates a compounding working capital improvement that reduces dependency on external financing.

The Joule-powered Working Capital Optimization capability (J390), delivered through SAP Taulia, extends this dashboard intelligence into action. It highlights the quantum of working capital that can be unlocked through dynamic discounting and supply chain finance programs, predicts the volume of early payments that can be accelerated through various working capital scenarios, and it assists treasury teams in configuring payment term extensions that balance supplier relationships against cash-positioning objectives. For procurement, this closes the loop between sourcing decisions and treasury outcomes: A negotiated 30-day improvement in supplier payment terms on a 400-million-dollar payables base generates approximately 33 million dollars in additional working capital, cash that can be redeployed into growth initiatives or used to reduce short-term borrowing.

The Autonomous Sourcing Assistant

The Autonomous Sourcing Assistant within the Autonomous Spend framework orchestrates three specialized agents that together transform sourcing from a labor-intensive administrative process into an AI-driven strategic function. We'll discuss these agents in the following sections. The long-term vision is an autonomous sourcing agent that ties these three agents together for holistic end-to-end automation, from event creation through supplier negotiation to bid analysis and award, enabling fully autonomous sourcing for routine categories while escalating complex events to human sourcing managers.

Supplier Discovery

The Supplier Discovery Agent leverages AI-powered matching across SAP Business Network's five-million-plus trading partners, combining capability scoring, financial risk assessment, and environmental, social, and governance (ESG) profiling to expand the qualified supplier pool by a factor of three compared to manual research methods. The RFx Orchestration Agent automates sourcing event construction using natural language processing to analyze requirements, validates bids through price anomaly detection and should-cost benchmarking, and models award scenarios using TCO analysis, reducing sourcing cycle times by 40 percent. The Sourcing Negotiation Agent performs BATNA and ZOPA analysis with should-cost decomposition based on real commodity price indices, regional labor benchmarks, and logistics rates, predicting outcomes and recommending concession strategies. SAP quantifies the impact as 8 to 15 percent in negotiated savings.

Sourcing Event Intelligence

The Sourcing Event Agent represents a fundamental shift in how sourcing events are created and managed. Rather than manually constructing requests for quotation from scratch, selecting suppliers, drafting requirements, configuring evaluation criteria, and setting timelines, the sourcing manager describes the sourcing need, and the agent analyzes historical sourcing data and past events, shortlists qualified suppliers based on performance history and capability profiles, and suggests evaluation questions tailored to the specific category. The agent automates event configuration, recommends optimal lot strategies for maximizing competition, and provides TCO-based supplier rankings that go beyond simple unit price comparisons.

SAP S/4HANA Cloud integrates this intelligence with operational sourcing execution. Purchase requisitions that cannot automatically find a source of supply are routed to the sourcing workflow, either going through SAP S/4HANA's built-in request for quotation process or escalated to SAP Ariba Sourcing for complex events. The integration with SAP Business Network provides access to more than five million suppliers, enabling supplier discovery that dramatically expands the competitive field beyond the organization's existing base. The AI-assisted supplier prediction capability, powered by SAP Ariba's machine learning capabilities, recommends the best suppliers for guided sourcing events based on historical data from similar events, analyzing item names, event types, departments, commodities, regions, and past supplier performance to generate ranked supplier recommendations.

Contract Intelligence and Lifecycle Management

Once sourcing events conclude, the results must be captured in enforceable contracts that protect negotiated terms over their validity period. SAP S/4HANA Cloud supports the creation, renewal, and monitoring of procurement contracts with real-time data augmented by ML-enabled models and visualization tailored to the needs of procurement officers. The AI-assisted contract creation from prompts capability enables contract managers to create contracts through natural language, while the AI-assisted contract creation from documents enables extraction of metadata, header details, and line items from existing contract PDFs to rapidly onboard legacy agreements. The AI-assisted contract field recommendations capability further accelerates new contract creation by predicting appropriate values for the supplier, company code, payment terms, and other dependent fields based on organizational patterns. Purchase contracts are enabled for the situation handling framework, with predelivered templates for purchase contract expiration that alert procurement officers before critical renewal deadlines pass.

The Contract IQ Agent is a generative AI-powered conversational assistant that enables procurement and legal professionals to interact with contracts using natural language. It performs semantic searches, generates clause-level summaries, and extracts obligations. The Contract Authoring Agent automates draft generation from awarded sourcing events with smart clause assembly. The Contract Renewal Optimization Agent evaluates contracts nearing expiration and recommends actions, such as renew, renegotiate, or re-source, based on supplier performance and current market pricing, targeting 5 to 15 percent renewal savings. The Compliance Monitoring Agent provides continuous contract enforcement through real-time deviation detection: When purchases are routed outside contracted channels or at prices above contracted rates, the agent detects the deviation, quantifies the impact, and triggers remediation workflows, preventing an estimated loss of two million dollars or more in annual contract value leakage.

SAP S/4HANA Cloud's supplier evaluation capabilities analyze supplier performance regarding accuracy of delivery, price competitiveness, and quality. The Digital Assistant for Supplier Management using Joule makes evaluation data conversationally accessible, enabling procurement managers to query performance and identify underperformers that require corrective action.

Korindal: Procurement Reduces Maverick Spend and Sourcing Cycle Time

At Korindal, the procurement organization manages approximately 2,800 active contracts across 140 spend categories, with an annual addressable spend of 1.2 billion dollars. Before the transformation, 34 percent of this spend flowed through noncontracted channels at spot prices averaging 12 to 18 percent above competitively sourced rates. The sourcing team of 18 processes 220 sourcing events per year, but the four- to six-week cycle time per event means that lower-priority categories are renewed automatically without market testing.

The Sourcing Event Agent reduces the average cycle from 32 to 14 days, with supplier shortlists drawn from SAP Business Network, introducing genuine competition into categories that had been single-sourced for years. The Contract IQ Agent enables conversational portfolio queries, such as "Show me all contracts expiring in the next 90 days above $100,00," while the Compliance Monitoring Agent detects approximately 340 price deviations per month, preventing an estimated 1.8 million dollars in annual leakage. The cumulative savings total 7 to 10 percent of addressable spend, translating to 85 to 120 million dollars in annual procurement value creation.

5.4.3 Procure-to-Receipt

The procure-to-receipt process is the operational engine of source-to-pay, the transactional sequence in which planned requirements become purchase requisitions, requisitions become purchase orders, orders are confirmed by suppliers, goods are received, and the procurement transaction is ready for invoicing. For the operational purchaser, this is where the volume challenge is the most acute: Organizations processing thousands of purchase orders daily cannot afford manual intervention at every step, yet the traditional procurement process is riddled with manual touchpoints that consume purchaser capacity without adding value.

SAP S/4HANA Cloud addresses this through a comprehensive automation architecture targeting 60 percent automation of purchasing processes. The system automates purchase requisition creation through MRP runs and intelligent bots, streamlines approvals through AI-powered confidence scoring, and enables end-to-end digitized supplier collaboration through SAP Business Network integration.

Let's dig deeper into the embedded AI capabilities that bring intelligence to your requisitioning and goods receipt processes.

Intelligent Requisitioning and Guided Buying

The procurement process begins with demand capture, identifying what is needed and routing the request through the appropriate purchasing channel. SAP Ariba Buying provides a guided purchasing experience integrated with SAP S/4HANA Cloud, converting purchasing requests to purchase orders and tracking the process from end to end. The Digital Assistant for Guided Buying using Joule enables employees to search catalogs, check statuses, and view approvers through conversational interactions. Figure 5.31 shows the Autonomous Spend Buying Agent, illustrating how AI automates requisitioning, guided buying, and purchase order processing to achieve touchless procurement.

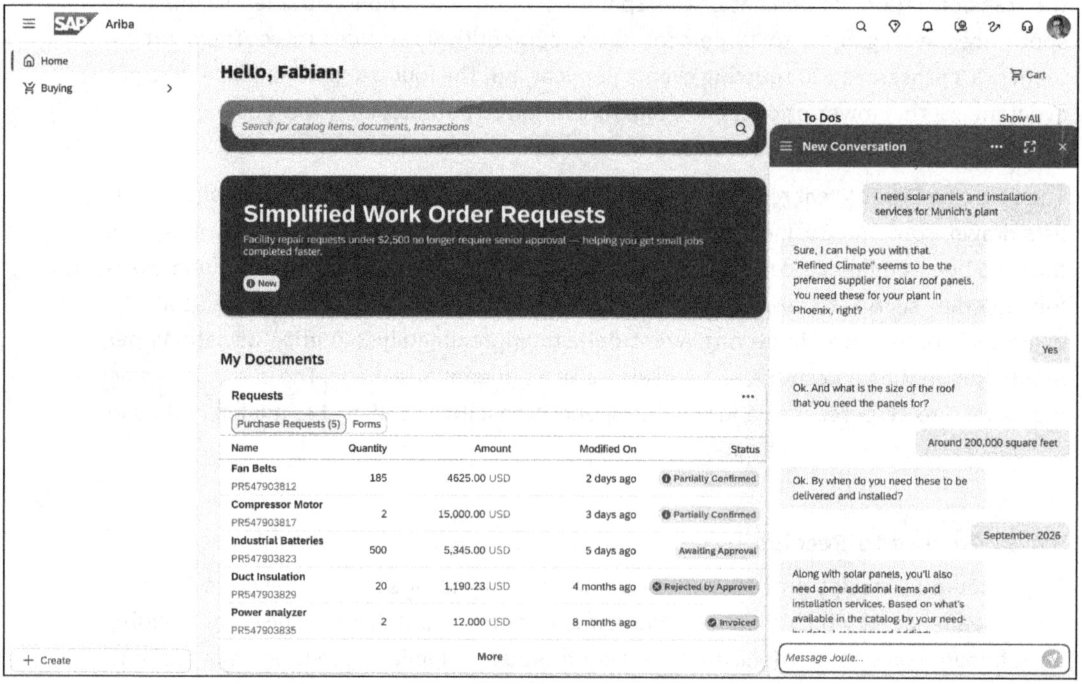

Figure 5.31: Autonomous Spend Buying Agent

The AI-assisted recommended material groups capability addresses a persistent challenge: Users selecting incorrect codes leads to miscategorized requests and procurement delays. The system suggests the most suitable material group based on the product's name and description. The AI-assisted catalog item recommendations feature suggests products based on past purchasing patterns, guiding users toward contracted items to reduce maverick spend. Joule's purchase requisition creation capability simplifies purchasing further. For example, a maintenance technician can tell Joule, "I need to order 10 units of bearing assembly 4721 for Plant Chicago," and the system creates a properly categorized requisition with the correct material number, supplier assignment, and delivery address.

The Requisition Intelligence Agent

The Requisition Intelligence Agent represents the agentic evolution of purchasing. This agent routes users through optimal paths, like catalog, punchout, or sourcing, using natural language processing (NLP) for autocategorization, compliance prediction, budget checking, and duplicate detection. SAP targets 90 percent autocategorization with a 30 percent compliance improvement. The Demand Capture and Routing Agent complements this with a conversational interface that guides users toward compliant purchasing paths before the transaction occurs rather than identifying violations afterward.

Purchase Order Management and Supplier Collaboration

Once requisitions are converted to purchase orders, SAP S/4HANA Cloud enables automated creation, modification, and monitoring with full SAP Business Network integration for digitized supplier collaboration, order confirmation, shipping notices, and delivery status. The Operational Purchaser Joule integration enables purchasers to manage orders and requisitions conversationally.

The AI-assisted supplier delivery date prediction capability is among the most impactful features in procure-to-receipt. Using ML tools trained on historical goods receipt data, the system predicts whether a purchase order will be delivered on time by analyzing patterns across material, supplier, plant, and seasonal factors. Purchasers see predicted dates alongside confirmed dates, enabling proactive action before late deliveries disrupt downstream operations. The AI-assisted supplier delivery date mass update extends this, enabling bulk date updates through Joule's natural language interface. The intelligent approval workflow analyzes purchase requisitions and calculates confidence levels for each item; high-confidence items are autoapproved within policy parameters, while low-confidence items are routed with full context, transforming approvals from a blanket review process to an exception-based workflow.

Goods Receipt Intelligence

The final step in procure-to-receipt, posting the goods receipt, is traditionally a manual, error-prone process. SAP's AI-assisted goods receipt processing automates this by extracting information from delivery notes through SAP Document AI, matching against open purchase orders, and proposing goods receipt postings. For SAP S/4HANA Cloud Public Edition, warehouse clerks can take a photo of a delivery note in the SAP Fiori application, and OCR then extracts details and matches them to the relevant purchase order. The Goods Receipt Acceleration Agent handles the scenario in which suppliers cannot send advance shipping notifications, the inbound clerk scans paper delivery notes, and the agent proposes the goods receipt posting, transforming a 20- to 30-minute manual process into one completed in just a few minutes.

We will cover logistics features in Section 5.5.

At the highest level, the Procurement Orchestration Agent enables autonomous orchestration of the entire procurement process. This agent analyzes each transaction's context, material category, supplier relationship, contract coverage, and risk profile, and then it selects the appropriate combination of capabilities to execute with minimal human intervention. Routine transactions flow from demand to receipt without manual touchpoints, while complex or high-value transactions are escalated with full context to procurement professionals.

Korindal: Purchasing Team Achieves Touchless PO Processing

At Korindal, the procurement team processes approximately 4,200 purchase orders per week across 12,000 active materials and 3,400 suppliers. Before the transformation, 62 percent of requisitions require manual source of supply determination, and the average time from requisition to purchase order creation is 3.8 days.

The guided buying transformation with Joule reduces the requisition creation time from 11 minutes to under four minutes. AI-assisted material group recommendations eliminate approximately 1,400 misclassified requisitions per month. Catalog item recommendations increase catalog utilization from 58 to 79 percent, reducing maverick spend by 2.1 million dollars annually.

The supplier delivery date prediction delivers the most visible operational improvement. The predictive model flags 73 percent of delayed deliveries three to five days before the confirmed date, giving purchasers time to expedite or arrange alternatives. Production schedule disruptions from late receipts decrease by 38 percent, a direct contribution to the OTIF improvement in Chapter 1. The goods receipt automation eliminates the most labor-intensive step: Clerks who spent 22 minutes per delivery manually matching documents now process through AI-assisted scanning in under five minutes, saving approximately 50 person-hours daily across 180 deliveries. The cumulative impact: Korindal processes 30 percent more purchase orders with the same team size, while improving on-time delivery rates and freeing experienced purchasers to focus on supplier relationship management rather than transactional processing.

5.5 Plan-to-Fulfill

The plan-to-fulfill value stream is where physical operations meet market demand. The end-to-end sequence of activities begins with translating demand signals into production schedules, progresses through manufacturing execution and quality assurance, and concludes with the storage and movement of finished goods to the customer. For the CSCO, this is where commercial promises are either kept or broken. Every planning lag, production exception, quality hold, and logistics delay accumulates into the OTIF performance that customers measure and competitors exploit.

The challenge is structural. Most enterprises manage planning, manufacturing, quality, and logistics through separate tools, separate teams, and separate decision cycles, creating handoff gaps where exceptions propagate unchecked and response speed is determined by the slowest link in the chain. When a supplier delivers late, the impact on the production schedule remains invisible until the next batch run. When a quality issue emerges on the line, its resolution depends on whoever happens to have encountered a similar problem before. When a shipment is delayed, the customer finds out before the logistics manager does.

To address these challenges, SAP introduced Autonomous SCM. We'll provide an overview and dive into the key process areas for plan-to-fulfill in the following sections.

5.5.1 Overview of Autonomous Supply Chain Management

SAP's autonomous capabilities address each of these gaps systematically. Predictive MRP and real-time capacity planning eliminate the planning lag. AI-driven manufacturing agents detect and resolve production exceptions before they cascade. ML-powered quality intelligence transforms defect response from reactive correction to proactive prevention. An autonomous logistics layer, spanning warehouse, transportation, and inventory, converts physical execution into a precision fulfillment engine.

The Autonomous Supply Chain Management (SCM) domain within SAP's Autonomous Enterprise framework organizes these capabilities across Planning, Manufacturing, Logistics, and Asset and Service assistants. SAP quantifies the combined impact at 20 to 30 percent faster planning cycles, 10 to 15 percent overall equipment effectiveness (OEE) improvement in manufacturing, up to 50 percent faster goods receipt processing, and up to 12 percent reduction in transportation costs. At Korindal, the plan-to-fulfill transformation directly addresses the OTIF gap the CSCO raised in Chapter 1, not through heroic effort but through intelligence embedded at every operational step. Figure 5.32 shows the Autonomous SCM portfolio, mapping AI capabilities across plan-to-fulfill, including planning, manufacturing, quality, logistics, and asset management.

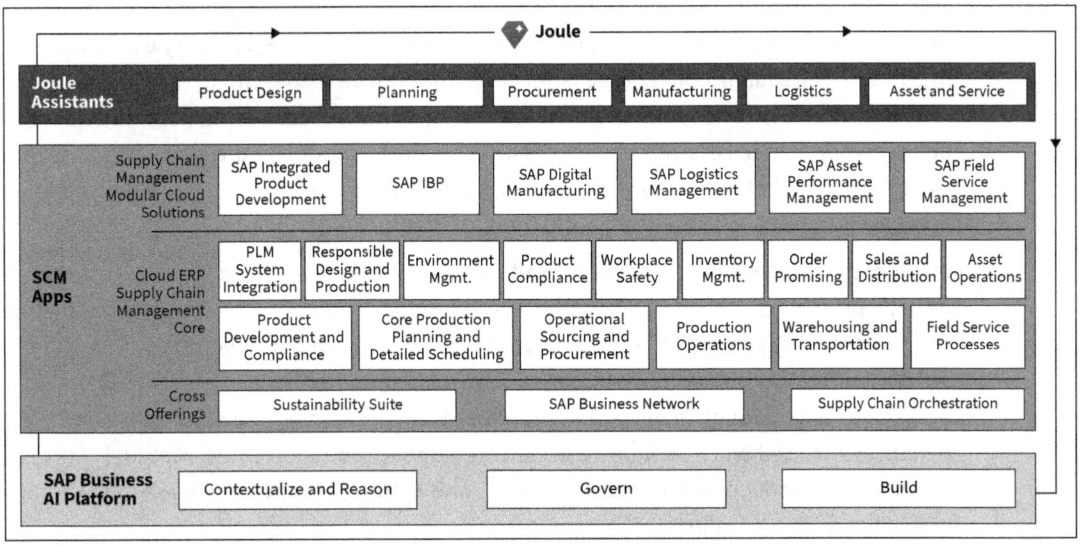

Figure 5.32: Autonomous SCM Portfolio

Figure 5.33 shows the Autonomous SCM Assistants, illustrating how the Planning Assistant, Asset and Service Assistant, and supporting agents orchestrate end-to-end supply chain execution.

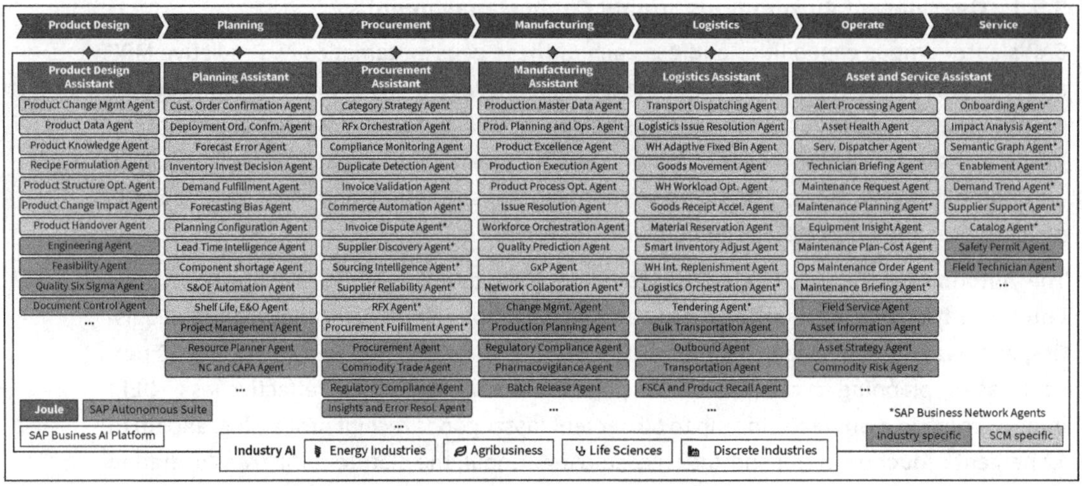

Figure 5.33: Autonomous SCM Assistants

5.5.2 Plan-to-Schedule

The plan-to-schedule process translates demand signals into executable production and procurement plans. This is the orchestration phase in which MRP, capacity evaluation, and production scheduling converge to determine what will be made, when, where, and in what sequence. For the production planner, balancing customer commitments, material availability, work center capacity, and supplier lead times produces the daily decisions that determine whether the enterprise meets its delivery promises or generates fulfillment exceptions.

The fundamental challenge is speed. Traditional MRP runs operate on batch cycles, daily or weekly, producing planned orders that may already be outdated by the time the planner reviews them. Capacity constraints, material shortages, and supplier lead time changes arrive continuously, but the planning system responds periodically, forcing planners to bridge the gap through manual intervention and spreadsheet-based workarounds. The result: Overloaded work centers go undetected until production stalls, material shortages surface after production orders are released, and delivery commitments are made against plans that no longer reflect reality.

In the following sections, we'll discuss the embedded AI capabilities that address this challenge and enable an automated planning and scheduling workflow.

Predictive and Demand-Driven MRP

SAP S/4HANA Cloud transforms MRP through predictive MRP (pMRP), a simulation-based approach that provides planners with a holistic view of demand, supply, and capacity before committing to plan execution. Three integrated simulation modes enable planners to evaluate the complete impact of changes across materials, capacity, and delivery commitments before committing to action: demand plan simulation, evaluating how changes cascade through the material structure; multilevel material simulation, tracing requirements across the entire bill of materials hierarchy; and capacity plan simulation, projecting work center loads.

When a simulation identifies capacity issues, the system recommends preproduction to build inventory ahead of demand peaks, suggests alternative sources of supply, and enables rescheduling or reallocation of operations across work centers. Graphical visualization of work center capacities provides real-time identification of critical overloads through KPIs that direct planners to specific resolution actions, transforming capacity management from post hoc discovery into proactive, simulation-driven optimization.

The interactive real-time MRP capability addresses the batch processing limitation directly. Rather than waiting for overnight runs, planners trigger MRP instantly through an intuitive SAP Fiori cockpit, visualize bottleneck situations as they emerge, and simulate resolution options based on system-guided proposals. The graphical simulation approach enables planners to resolve bottlenecks with system guidance, the system proposes solutions, the planner evaluates them visually, and the selected resolution executes immediately. MRP change requests integrate with SAP Business Network, enabling direct collaboration with suppliers on requirement changes, closing the gap between planning decisions and supplier awareness.

For organizations where traditional forecast-driven MRP produces excessive nervousness, SAP S/4HANA Cloud supports demand-driven MRP (DDMRP) following the standard of the Demand Driven Institute. This buffer-based approach decouples supply from demand variability by positioning strategic inventory buffers at key points in the supply chain. The process includes buffer positioning and sizing based on demand variability, supply lead times, and service-level targets, followed by replenishment planning that generates orders to maintain buffer levels. ML-based buffer-level optimization further refines sizing based on observed demand patterns, reducing both excess inventory and stockout risk.

Work Center Visualization

Once material requirements are determined, capacity planning and production scheduling translate planned orders into an executable production sequence. SAP S/4HANA Cloud provides a graphical visualization of work center capacities in real time, with KPIs identifying overloads and directing resolution options. The interactive capacity scheduling board with predefined scheduling strategies enables planners to optimize production sequences visually using an interactive Gantt chart, preventing overloads and ensuring that commitments align with available capacity.

Scheduling via pacemaker work centers enables prioritization based on customer delivery dates, material availability, and efficiency considerations. The handover of the production schedule to SAP Digital Manufacturing provides seamless integration between planning and shop floor execution, with role-specific SAP Fiori dashboards and a configurable Production Operator dashboard orchestrating the shop floor. Static and time-dependent safety stock methods adapt replenishment to business requirements, while planned independent requirements (PIRs) enable planners to enter or import demand forecasts. Bots automate the upload and maintenance of PIRs, eliminating manual data entry that consumes hours of planner time each cycle.

The Planning Assistant: AI-Driven Planning Intelligence

The Autonomous SCM framework organizes planning intelligence through a Planning Assistant that orchestrates multiple specialized agents, targeting 20 to 30 percent faster planning cycles with improved demand fulfillment, fewer stockouts, and reduced working capital.

Let's walk through the available agents within the Planning Assistant:

- **Production Planning and Release Agent**
  The Production Planning and Release Agent validates material and capacity availability for production orders, automates prerequisite checks for release, and recommends workarounds for material and scheduling conflicts, evaluating substitution materials, alternative routings, and schedule adjustments when orders cannot proceed. SAP quantifies this as accelerating order-to-delivery by saving planners time spent on manual release checks.
- **Lead Time Agent**
  The Lead Time Agent performs daily monitoring by observing ERP execution history and extracting historical lead time data. It compares observed execution values with planning values, triggers updates when discrepancies are detected, and feeds lead-time intelligence into inventory optimization, replacing periodic manual reviews with continuous monitoring that ensures planning parameters reflect current reality.
- **Exception and Scenario Management Agent**
  The Exception and Scenario Management Agent evaluates intelligent alerts, generates mitigation options, and enables scenario-based decision-making when disruptions require rapid replanning, generating scenarios with projected impacts on delivery commitments, inventory levels, and costs.

- **Inventory Investment Allocation Agent**
  The Inventory Investment Allocation Agent addresses one of the most consequential planning decisions: where to invest in inventory and where to reduce it, while maintaining target service levels. The agent operates through a six-step autonomous workflow:
  - First, a business trigger initiates the process when service-inventory trade-offs require decisions—for example, at the start of a planning cycle or when service-level deviations are detected.
  - Second, the agent explores possible service-inventory scenarios, modeling different stocking strategies across material clusters and locations.
  - Third, the agent generates trade-off insights by identifying efficiency frontiers and diminishing returns, showing precisely where additional inventory investment yields meaningful service improvement and where it does not.
  - Fourth, the agent proposes concrete decision options: Invest in specific materials to close service gaps or reduce inventory in areas where current stock far exceeds what is needed to maintain the target service level.
  - Fifth, a planner reviews the proposals in an optional human-in-the-loop step, approving, modifying, or requesting reanalysis through Joule.
  - Sixth, approved decisions are applied directly to planning parameters, and the agent tracks projected service and inventory outcomes against the baseline.

  The agent delivers four categories of key insight. The Best Next Move identifies the highest service-level-per-cost improvement available. Cheap Wins surfaces low-cost improvement opportunities where small inventory additions yield disproportionate service gains. The Efficiency Gap quantifies how much inventory value can be saved while maintaining current service levels. Finally, Diminishing Returns identifies the service-level threshold beyond which additional inventory investment yields negligible improvement, helping planners avoid overinvestment. Each recommendation carries a confidence score based on scenario run stability, and planners can approve, ask Joule for further explanation, or trigger reanalysis directly from the recommendation interface. For organizations managing thousands of SKUs across multiple locations, this agent transforms inventory investment from a periodic spreadsheet exercise into a continuous, AI-driven capital allocation discipline. Figure 5.34 shows the Inventory Investment Analysis Assistant, highlighting the four insight categories that guide inventory reallocation to improve service levels.

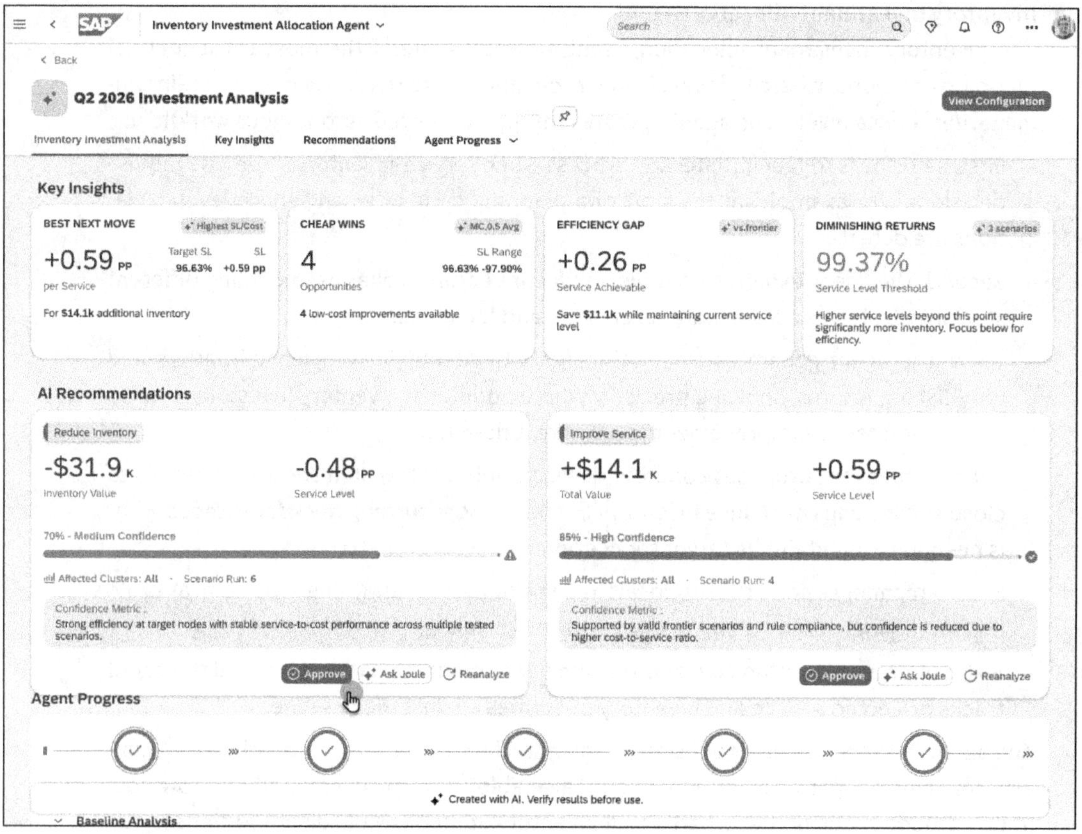

Figure 5.34: Inventory Investment Analysis Assistant

- **Demand Fulfillment Agent**

  The Demand Fulfillment Agent optimizes revenue at risk by balancing constrained supply allocation to demand. In supply-constrained environments where available materials cannot satisfy all demand simultaneously, planners must identify which products face shortages, understand why, and decide how to allocate limited supply across competing requirements. This agent automates that entire analytical workflow, transforming what was previously a multiday investigative exercise into a continuous, AI-driven process.

  The agent operates through a three-phase cycle:

  - In the insights and summary phase, it generates AI-powered summaries that explain gaps between demand and supply, compares fulfillment metrics across successive planning runs to detect deterioration, analyzes demand fulfillment percentage and quantity changes relative to the previous supply planning run, performs trend analysis across planning horizons, identifies the top N gating materials driving the largest finished goods shortages, and produces detailed stakeholder-specific demand fulfillment analysis with both top-down and bottom-up views, including root-cause identification and gating factor decomposition.

- In the recommended actions phase, the agent proposes concrete mitigation steps for the planner, such as reviewing supplier escalation scenarios, prioritizing supplier negotiation before demand reallocation, or enabling component substitution.
- In the what-if scenario planning phase, the agent creates simulation scenarios and the planner executes changes within a sandbox environment, evaluating the impact of proposed mitigations before committing them to the live plan.

Figure 5.35 shows the Demand Fulfillment Agent, illustrating how it optimizes supply allocation to demand in constrained scenarios to protect revenue.

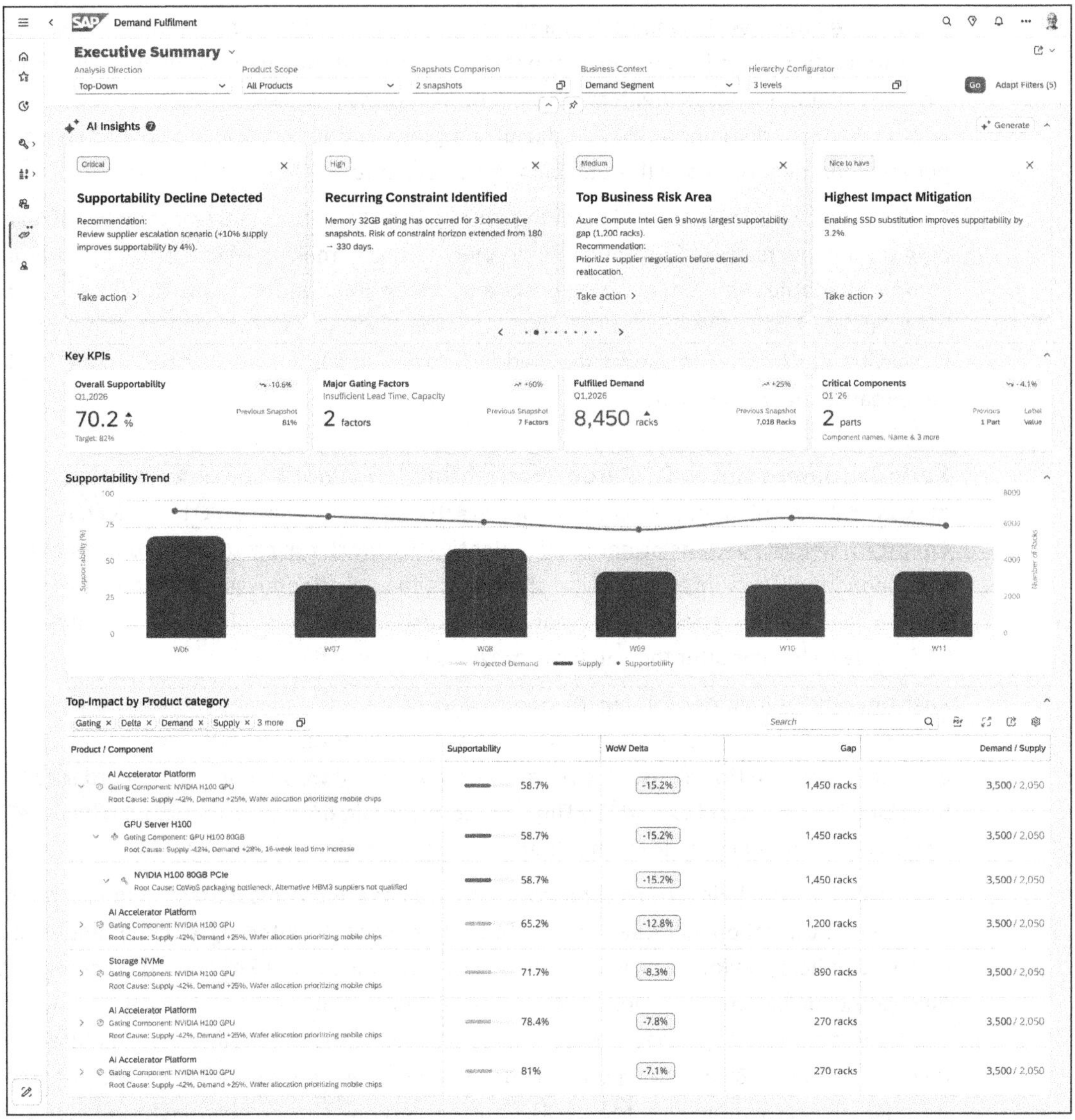

Figure 5.35: Demand Fulfillment Agent

The Executive Summary dashboard provides planners with immediate visibility into four key KPIs: overall supportability percentage against target, the number of major gating factors (with week-over-week comparison), fulfilled demand in units with trend direction, and critical components requiring attention. AI Insights cards are prioritized by severity (critical, high, medium, or nice-to-have) and include actionable recommendations, such as "Supportability Decline Detected: review supplier escalation scenario with plus 10% supply to improve supportability by 4 percent" or "Recurring Constraint Identified: Memory 32 GB gating has occurred for three consecutive snapshots, risk of constraint horizon extended from 180 to 330 days." A supportability trend chart visualizes projected demand against supply week over week, while a Top Impact by Product Category table ranks products by supportability gap with root-cause annotations, enabling planners to focus intervention where revenue impact is greatest. The agent is currently in development, co-innovating with a customer, demonstrating its value in complex multitier supply networks where component constraints cascade through bills of material into finished goods availability.

SAP's AI-assisted conversational planning capability enables planners to interact with planning data conversationally through Joule with SAP IBP, making efficient decisions amid complex variables while minimizing costs and environmental footprint. The Excel add-in for AI-assisted planning enables planners to generate SAP IBP formulas and formatting rules without extensive technical knowledge, bringing generative AI directly into the familiar spreadsheet environment.

Korindal: James Moves OTIF from Firefighting to Forward-Looking Planning

At Korindal, the planning transformation directly supports James's OTIF objective from Chapter 1. Before the transformation, the nightly MRP batch run produces results that planners spend mornings interpreting, scrolling through exception messages, manually cross-referencing capacity reports, and updating spreadsheets. The planning cycle operates on a 24-hour lag; changes after the batch run are invisible until the next day.

Real-time MRP eliminates this lag. Planners trigger MRP runs interactively when significant changes occur and see cascading impacts immediately. Predictive MRP simulation allows evaluation of resolution alternatives before committing: "If we shift production order 48721 from Line 3 to Line 5 and extend the Thursday second shift, does the capacity plan balance?" The answer comes in seconds rather than requiring overnight reruns.

The capacity scheduling transformation produces the most measurable improvement. Work center utilization increases from 74 to 82 percent as the interactive scheduling board reveals conflicts previously invisible until production stalled. The Production Planning and Release Agent automates prerequisite checking that previously consumed an estimated two hours per planner per day. The planning cycle compresses from five days to under two days, achieving the 80 percent reduction in planning cycle time that SAP quantifies for plan-to-fulfill transformations. For the CSCO, shorter planning cycles mean faster response to change and fewer OTIF misses caused by stale plans.

5.5.3 Manufacturing

The manufacturing process transforms planned production into finished goods, where raw materials and components are converted into products through discrete, repetitive, process, Kanban, just-in-time (JIT), or outsourced manufacturing operations. SAP S/4HANA Cloud supports all manufacturing types within a unified framework, enabling make-to-stock, make-to-order, and engineer-to-order processes with full integration across production planning, warehouse management, quality management, and cost accounting.

For the production supervisor, the daily challenge is visibility and response speed. Production orders must be released, components staged, operations confirmed, quality inspections triggered, and finished goods received, all while managing exceptions: machine breakdowns, component shortages, quality holds, and schedule changes. Each exception requires manual investigation and cross-system coordination that consumes supervisory capacity and slows production flow.

Let's take a closer look at the embedded AI capabilities for autonomous production that support a complete manufacturing transformation for organizations.

Production Execution Across Manufacturing Types

SAP S/4HANA Cloud provides real-time management and release of production orders with continuous monitoring. The complete lifecycle is covered, from conversion of planned orders through confirmation, goods receipt, quality inspection, and cost settlement.

For *discrete manufacturing*, production order management includes manual or automatic creation from planned orders, flexible order splitting, component staging from integrated warehouse management, and confirmation tracking with scrap and variance capture. For repetitive manufacturing, run schedules assigned to production lines determine dates and quantities, with pull lists controlling in-house material flow and a lean confirmation approach including backflush of components via dedicated cost collectors.

For *Kanban manufacturing*, automated procedures optimize material flow based on physical stock, with classic, event-driven, and one-card procedures, graphical Kanban Board monitoring, and automated calculation of optimal container quantities and fill levels. For process manufacturing, process orders manage production through defined phases, picking ingredients, processing, confirming, and receiving product batches, with dedicated SAP Fiori apps for monitoring.

For *JIT manufacturing*, the system supports JIT or just-in-sequence (JIS) supply-to-customer based on sales scheduling agreements, with API-based receipt of JIT calls, management of summarized or sequenced calls, and logistics execution including packing and outbound delivery processing. For outsourced manufacturing, subcontracting capabilities manage multiple subcontractors with the subcontracting cockpit monitoring provisioning and subcontracting MRP areas tracking inventory at subcontractor sites.

Integration with SAP Digital Manufacturing

Integration with SAP Digital Manufacturing extends execution to the shop floor level, with role-specific SAP Fiori dashboards, a configurable Production Operator Dashboard, and 3D work instructions that guide operators through complex assembly sequences. Joule with SAP Digital Manufacturing provides conversational access to shop floor information, production status, performance data, and exception details through natural language.

The AI-assisted production engineering capability helps engineers analyze error logs with AI to identify root causes and generate resolution instructions. The system also generates 3D work instructions that reduce operator errors and accelerate onboarding. AI-assisted production-routing proposals recommend existing routings that can be copied when creating new routings, leveraging historical data to accelerate manufacturing engineering.

The Manufacturing Assistant

Autonomous SCM's Manufacturing Assistant orchestrates agents that transform production management from reactive supervision to proactive, AI-driven operations, targeting a 10 to 15 percent OEE improvement and a 10 to 15 percent reduction in defect rate with end-to-end operational visibility.

Let's walk through the available agents:

- **Shop Floor Supervisor Agent**
  The Shop Floor Supervisor Agent continuously monitors production performance, generates structured shift summaries, and recommends next-best actions. When disruptions occur, the agent evaluates the situation, identifies affected orders, and proposes schedule adjustments to minimize delivery impact, transforming the supervisory role from investigation to decision-making.
- **Production Scheduling and Operations Agent**
  The Production Scheduling and Operations Agent monitors disruptions in real time and orchestrates corrective actions across the production network, ensuring schedule changes propagate correctly through material requirements and capacity plans.
- **Quality Prediction and Prevention Agent**
  The Quality Prediction and Prevention Agent detects cycle-time deviations, scrap trends, and machine anomalies in real time, recommending corrective actions before defects propagate. The Root Cause Analysis Agent complements this by detecting recurring defects, correlating production, quality, sensor, and maintenance data, and recommending systemic corrective actions.
- **Operator Assist Agent**
  The Operator Assist Agent supports production operators by analyzing production logs and generating step-by-step resolution instructions for unfamiliar error conditions, capturing institutional knowledge that would otherwise exist only in veteran operators' heads.

- **Machine Connectivity Agent**
  The Machine Connectivity Agent automates mapping of programmable logic controller (PLC) and machine tags, creates digital twins, and continuously validates mappings as equipment changes.
- **Continuous Improvement Agent**
  The Continuous Improvement Agent simplifies the creation and enhancement of process definitions, generating standardized documentation for ongoing improvement initiatives.

Korindal: Shop Floor Supervisors Shift from Exception Investigation to Production Leadership

At Korindal, discrete manufacturing operations across three plants produce approximately 2,200 production orders per week. Before the transformation, production supervisors spend an estimated 30 percent of their time on exception investigation, diagnosing stalled orders, coordinating component availability, and manually rescheduling around machine downtime.

The Shop Floor Supervisor Agent transforms this to exception resolution rather than investigation, presenting preanalyzed disruptions with recommended actions. When a CNC machine on Line 2 develops a tool wear issue, increasing cycle times by 12 percent, the Quality Prediction and Prevention Agent detects the deviation before scrap rates increase, the Shop Floor Supervisor Agent proposes rescheduling to Line 4, and the Production Scheduling and Operations Agent propagates the change through material staging and delivery commitments. What previously required 90 minutes of cross-functional coordination happens in a few minutes.

AI-assisted production engineering enables conversational analysis of error logs, identifying recurring failure patterns that drive root cause elimination. The Production Planning and Operations Agent automates prerequisite checks, eliminating manual verification loops that added a half-day lag between planning and production start. OEE improves by 8 percentage points, delivering the 20 percent improvement in production efficiency and real-time serial number traceability that SAP quantifies for plan-to-fulfill implementations.

5.5.4 Quality Management

Quality management is the discipline that ensures products meet specifications at every stage of the manufacturing and logistics process, from incoming material inspection through in-process quality checks to final product certification. SAP S/4HANA Cloud provides a holistic quality management framework integrated directly into production, procurement, and logistics processes, covering quality planning, inspection, and improvement in a unified system. For the quality manager, the challenge is twofold: maintaining rigorous inspection standards while operating at production speeds, and converting defect data into systematic prevention rather than repetitive correction.

The following sections will unpack the core features and embedded AI capabilities that enable optimized quality management.

Quality Planning and Master Data

The quality management process begins with master data, inspection methods, sample procedures, and inspection plans that define material-specific inspection procedures for each quality-relevant transaction. Inspection lots are created automatically during goods receipt, production confirmation, or outbound delivery, initiating inspection without manual intervention. The system supports result recording with inspection points, building the foundation for usage decisions that determine whether materials are accepted, rejected, or conditionally released, with automation logic enabling automatic acceptance for materials consistently meeting specifications.

Quality Improvement and Nonconformance Management

Dynamic modification of the inspection scope adapts the inspection depth based on previous results: Materials from suppliers with consistently high quality receive reduced inspection, while deteriorating quality triggers intensified checking. This risk-based approach concentrates inspection resources where they create the most value.

Nonconformance management provides holistic defect processing from automated recording through structured resolution. Quality notifications record and process complaints from customers and toward suppliers, creating a closed feedback loop. The 8D methodology is embedded as a predefined workflow for systematic root cause analysis, with integration into SAP Digital Manufacturing for issue resolution for structured collaboration with suppliers and customers. Failure mode and effect analysis (FMEA) enables early identification of potential failures with a dedicated toolset. The AI-assisted description enhancement capability helps quality managers maintain detailed problem, issue, task, root cause, and closing note descriptions, generating clear, structured text that improves consistency and searchability.

Planning and management of quality certificates enables certification of goods quality in PDF form. Analytical overviews for quality technicians and engineers, along with quality analytics tracking inspection lots, results, and defects, provide dashboards for optimized quality decisions.

AI-Powered Quality Intelligence

The AI-assisted processing of incoming quality certificates with SAP Document AI automates a traditionally manual step. After uploading a certificate, SAP Document AI extracts data and identifies corresponding inspection lots, inspection plans, and material specifications. Extracted data is matched against inspection requirements with discrepancies flagged for review, eliminating manual comparison that consumed significant inspection labor and introduced transcription errors.

The AI-assisted issue resolution guidance helps quality managers resolve recurring issues faster by automatically detecting and connecting similar problem-solving cases. The system scans historical event data, compares error descriptions, and evaluates resolution effectiveness to suggest the most effective resolution path. This capability operates across both standard SAP Digital Manufacturing and SAP Digital Manufacturing for issue resolution, transforming quality resolution from an expertise-dependent activity into a systematically guided process that captures institutional knowledge.

The AI-assisted visual inspection assists in monitoring conditions to identify critical quality issues such as wear, surface defects, or assembly errors, with ML models running on the device via the Production Operator Dashboard to identify nonconformances, representing a pathway toward automated inline quality inspection at production line speeds.

The AI-assisted safety observation reporting capability helps production operators and industrial hygienists report safety observations using natural language. The system processes the input, detects missing critical details, and guides users with clarifying questions, reducing the barrier to safety reporting and increasing the volume of observations that feed into the prevention system.

Korindal: Quality Team Reduces Manual Inspection and Accelerates Issue Resolution

At Korindal, the quality management transformation addresses two specific operational pain points that directly affect production throughput and customer satisfaction.

First, incoming material inspection for the suppliers delivering components to the manufacturing plants generates approximately 1,800 inspection lots per month. Processing the accompanying quality certificates, matching supplier-provided data against inspection plans, recording results, and making usage decisions previously required 4.2 FTEs of quality technician capacity. Each certificate required manual comparison of supplier test results against Korindal's specification limits, transcription of values into the system, and a documented usage decision. Error rates in manual transcription averaged 3.5 percent, and each error triggered a rework loop that consumed an additional 20 minutes of technician and supervisor time. SAP Document AI reduces the certificate processing effort to 1.5 FTEs by automating extraction and matching, while the dynamic modification of inspection scope reduces the number of lots requiring full inspection by 40 percent for suppliers with demonstrated quality performance. Quality technicians are freed to focus on physical inspection activities, root cause analysis, and supplier development rather than administrative processing.

Second, the AI-assisted issue-resolution guidance transforms how Korindal addresses recurring quality problems. Before the transformation, resolution effectiveness depended entirely on individual expertise: A veteran quality engineer recognizes a defect pattern and applies a proven corrective action, while a newer team member starts the investigation from scratch, potentially trying approaches that have already proven ineffective for similar issues. The AI-driven guidance ensures that historical resolution knowledge is systematically applied regardless of who handles the notification, reducing the average time to resolution for recurring issues by 40 percent and improving first-time-fix rates from 62 to 81 percent. For the CSCO, better quality resolution means fewer production delays caused by quality holds, directly supporting the OTIF improvement objective.

5.5.5 Logistics

The logistics process encompasses the physical movement and storage of goods across the supply chain, warehouse management, transportation management, inventory management, and the cross-cutting logistics capabilities that connect production, procurement, and customer fulfillment. For the logistics manager, this is where operational efficiency directly determines cost, service level, and working capital performance. Every minute of warehouse processing time, every mile of transportation, and every unit of inventory represents a cost that must be optimized against the competing demands of customer service and operational resilience.

Now, let's explore the automated capabilities for managing warehouses, transportation, and inventory processes to enable logistics transformation.

Warehouse Management

SAP S/4HANA Cloud provides embedded warehouse management with full stock transparency down to the storage bin level, mobile applications for efficient operations, and tight integration into the production, procurement, and quality management processes. Handling unit management enables package-based logistics, serial number management provides component-level traceability throughout the lifecycle, and batch management groups and tracks materials based on configurable characteristics.

The inbound process creates deliveries to trigger warehousing for incoming purchase orders and production orders, with handling unit packing during receipt, goods receipt posting into inventory, and putaway execution using mobile devices. Customizable putaway strategies determine the best bin assignment during warehouse task creation, with ABC analysis categorizing product importance to optimize putaway decisions. The outbound process mirrors this flow in reverse: Outbound deliveries trigger warehouse task creation, picking bins are calculated for optimal efficiency, picking is executed with mobile devices, and handling units are packed for shipment before goods issue posting.

Internal warehouse processes include replenishment—planned, automatic, and order-based—to ensure picking areas maintain adequate stock levels. The warehouse monitoring dashboard provides drill-through to operational details across all warehouse processes, with embedded SAP Analytics Cloud content for daily operations monitoring and KPI tracking including stock value by stock type and warehouse throughput history.

Mobile warehousing provides out-of-the-box radio frequency (RF) framework capabilities with more than 50 predelivered transactions and a screen persona engine for personalization. The SAP Warehouse Operator native Apple iOS application brings powerful warehouse features to iPhones with fast and reliable barcode scanning; picking for serialized, nonserialized, and batch-managed products; and handling unit putaway.

The AI-assisted product slotting capability uses machine learning to determine optimal product placement within the warehouse. By analyzing historical picking patterns, product velocity, physical dimensions, and storage constraints, the system recommends bin assignments that minimize travel distance and maximize picking efficiency, a deceptively simple optimization that can improve warehouse throughput by 10 to 15 percent. The enhanced slotting leverages machine learning to determine product master attributes, providing recommendations that would require extensive manual analysis to develop.

The AI-assisted predictive labor demand planning capability predicts planned durations for picking and packing processes based on historical workload data, enhancing the efficiency of resource and workload planning. By utilizing historical workload patterns, the system forecasts current and future warehouse workloads in inbound and outbound processes, enabling logistics managers to forecast and allocate warehouse resources proactively rather than reacting to bottlenecks after they form. The evolved Warehouse Workload Optimization Agent extends this into agentic territory, helping customers optimally allocate resources across warehouse outbound areas and prevent potential resource shortages based on predicted workload.

Joule with SAP Logistics Management enables logistic clerks to query and create storage bins, manage freight tendering, create and query pickup schedules, and track pickup and delivery documents through conversational interaction. The Joule skills integrated into warehouse outbound processes enable warehouse supervisors to get the status of outbound orders and make necessary changes on the delivery item and header level, triggering follow-up actions to accelerate process completion through natural language rather than navigating transactional screens.

Transportation Management

SAP S/4HANA Cloud provides end-to-end transportation management, from order-based planning through carrier selection, execution monitoring, and freight cost settlement. Freight units are automatically created from purchase orders, sales orders, or stock transport orders as granular elements for transportation planning. The transportation cockpit provides a proven interface for manual planning by assigning freight units to resources to create freight orders, with carrier selection for road freight order execution, freight order monitoring, and delivery creation based on the freight order. The system calculates freight charges, posts accruals, and settles freight costs after successful transportation and invoice receipt.

Integration with SAP Business Network enables carrier collaboration and freight tendering, providing seamless logistics execution and transport preparation through integration into cloud warehouse management. Advanced shipping and receiving capabilities provide a dedicated cockpit for process monitoring across the transportation lifecycle.

The Logistics Assistant

Autonomous SCM's Logistics Assistant orchestrates agents across warehouse and transportation operations, targeting up to a 30 percent reduction in tender processing effort, up to 50 percent faster goods receipt processing, and up to a 12 percent reduction in transportation costs.

Let's explore the available agents for warehouse management first:

- **Warehouse Internal Replenishment Agent**
  The Warehouse Internal Replenishment Agent ensures that all picking areas and production feeding zones are continuously supplied with necessary stock. Triggered by a warehouse manager, warehouse clerk, or autonomously by the system, this agent assigns fixed bins, checks open orders, triggers replenishment dynamically, and monitors stock levels in active picking locations, initiating replenishment from reserve storage before stockouts disrupt outbound operations or production supply.
- **Cross Dock Agent**
  The Cross Dock Agent matches inbound deliveries with outbound requirements, reducing walking time and warehouse storage requirements by routing goods directly from receiving to shipping.
- **Parcel Optimization Agent**
  The Parcel Optimization Agent optimizes parcel space utilization, reducing the number of packages and improving transportation cost efficiency.
- **Shipping Agent**
  The Shipping Agent ensures timely creation of outbound deliveries with coordinated picking and goods issue status updates.
- **Goods Movement Agent**
  The Goods Movement Agent handles goods receipts, posting changes, stock information updates, and ad hoc goods issues, automating the transactional overhead of inventory movements.
- **First Aid Agent**
  The First Aid Agent serves as an operational health monitor for the warehouse system itself, watching for issues, detecting problems in system integration, analyzing status conditions, and applying countermeasures. When self-healing procedures are not successful, it notifies human operators with diagnostic context.
- **Process Consultant Agent**
  The Process Consultant Agent checks process performance, compares against company and SAP-wide benchmarks, and identifies constellations of deviations that may indicate systemic improvement opportunities.

Next, let's walk through the available agents for transportation management:

- **Freight Visibility Agent**
  The Freight Visibility Agent provides end-to-end shipment transparency with early risk

detection for delays, disruptions, and service-level agreement breaches, enabling proactive exception handling to stabilize delivery performance.

- **Freight Audit Agent**
  The Freight Audit Agent validates carrier invoices against agreed contract rates, detects deviations, and proposes dispute actions.
- **Dispatcher Agent**
  The Dispatcher Agent optimizes carrier and routing selection, identifies consolidation potential across shipments, simulates impact before execution, and replans in case of disruptions.
- **Traceability Agent**
  The Traceability Agent traces material movements, batch updates, serial number history, and handling unit changes across the entire logistics chain, providing the end-to-end visibility that regulatory compliance, customer inquiries, and quality investigations require.
- **Packing Regulatory Agent**
  The Packing Regulatory Agent provides context-based packing guidance, automated compliance retrieval, and step-by-step packing reminders to ensure regulatory requirements are met.

Inventory Intelligence

Between warehouse operations and transportation, inventory management provides the analytical foundation for stock optimization. Machine learning predicts the potential duration of stock transfers, while ML-based identification of slow-moving or nonmoving materials enables strategic inventory management. Dead stock analysis supports time-dependent investigation of obsolete inventory. Situation handling streamlines physical inventory counting by proactively approaching users with tasks, with support for periodic, continuous, and cycle counting procedures.

The AI-assisted conversational planning capability enables transportation planners to execute planning operations directly within the transportation cockpit using natural language. Rather than navigating complex screen sequences to assign freight units, evaluate carrier options, and balance capacity, planners issue multistep instructions that the system interprets and executes as coordinated action chains. The underlying LLM proposes transportation plans based on available capacities and planning requirements, displays projected outcomes, and applies the selected plan upon confirmation. This reduces the cognitive load of managing multiple variables, including cost, capacity, carrier availability, and environmental impact, allowing planners to optimize route and resource decisions conversationally rather than transactionally. Figure 5.36 shows the conversational planning capability for SAP Transportation Management, illustrating how natural language enables transportation planning operations.

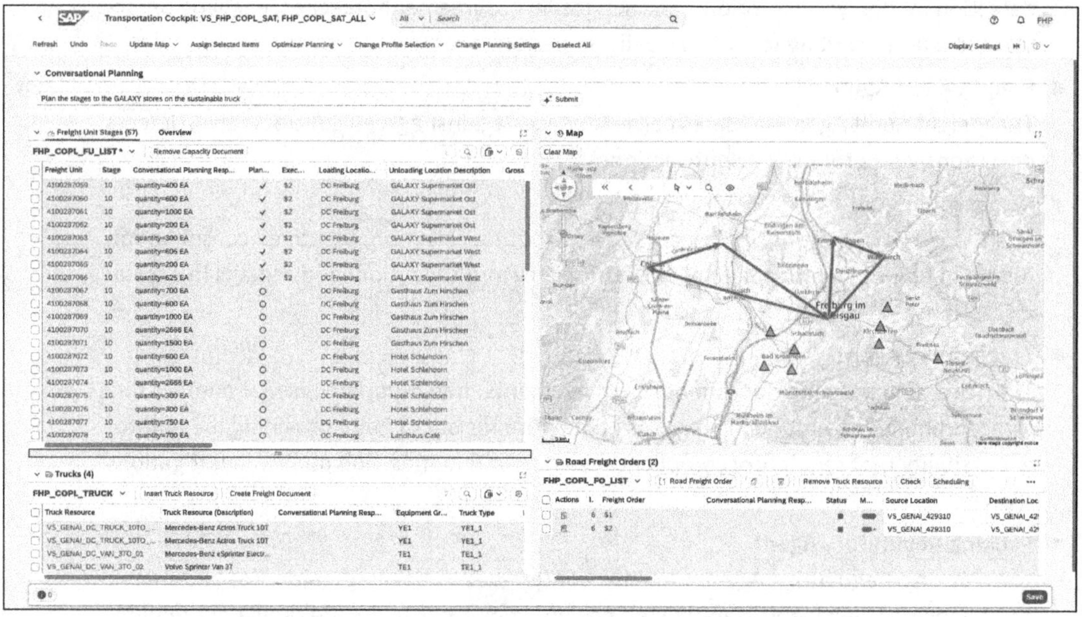

Figure 5.36: Conversational Planning in SAP Transportation Management

Korindal: Logistics Team Improves Delivery Performance and Recovers Freight Overcharges

At Korindal, the logistics transformation spans three distribution centers processing an average of 1,400 outbound deliveries per day. Before the transformation, warehouse resource allocation is based on the previous day's volume, a reactive approach that consistently under-resources peak periods and over-resources quiet periods.

The AI-assisted predictive labor demand planning capability transforms this by predicting next-day workload by area and time window, while the Warehouse Workload Optimization Agent recommends specific resource allocation adjustments across outbound areas. Resource utilization improves from 68 to 83 percent, and the average order-to-ship cycle time decreases by 22 percent. The AI-assisted product slotting optimization repositions the top 500 high-velocity SKUs to optimized zones, decreasing average pick path distance by 18 percent. The Replenishment Agent ensures optimized picking zones remain stocked through dynamic triggering based on open order demand.

The transportation transformation addresses both cost and service. The Freight Visibility Agent eliminates the inability to provide accurate delivery status after goods leave the distribution center; customers now receive proactive delay notifications, and the service team queries shipment status conversationally through Joule. The Dispatcher Agent consolidates shipments that were previously tendered individually, reducing average LTL shipments by 14 percent and freight cost by approximately $420,000 annually.

The Freight Audit Agent automates verification of approximately 4,200 carrier invoices per month, detecting an average of 340 billing discrepancies and recovering approximately $180,000 annually in freight overcharges. The Traceability Agent provides end-to-end material tracking for Korindal's regulated product lines, replacing manual batch-tracking spreadsheets with system-managed traceability that reconstructs material history in seconds.

The cumulative logistics impact directly supports the CSCO's OTIF improvement: Better warehouse resource allocation means faster order processing, optimized slotting means faster picking, proactive transportation visibility means fewer delivery surprises, and automated freight audit means lower logistics costs. The logistics operation transforms from a cost center managing daily crises into a precision-execution engine, closing the final mile between promise and performance that the CSCO identified as the OTIF gap in Chapter 1.

5.6 Acquire-to-Decommission

The acquire-to-decommission value stream encompasses the entire lifecycle of physical assets, from defining asset strategy and investment planning through acquisition, onboarding, operational maintenance, and eventual retirement. For the VP of operations, plant manager, and CFO alike, this value stream determines whether capital investments generate their expected returns over their full useful lives. Every hour of unplanned downtime, every reactive maintenance call, every premature asset replacement represents a failure to extract the value that the original investment was designed to deliver.

The challenge is one of visibility and foresight. Most enterprises manage asset data, maintenance planning, field execution, and financial depreciation through disconnected systems and manual processes, creating blind spots where equipment degrades undetected, maintenance schedules drift from actual asset condition, and decommissioning decisions are made on accounting timelines rather than engineering reality. When a critical asset fails unexpectedly, the cost extends far beyond the repair: Production stops, customer orders are delayed, and the maintenance team is pulled from planned work into emergency response.

SAP's autonomous capabilities within Autonomous SCM (which we introduced in Section 5.5.1) address these gaps across the asset lifecycle. AI-assisted anomaly detection identifies degradation before failures occur; predictive and condition-based maintenance strategies replace calendar-based schedules with data-driven interventions; agentic field service management optimizes technician dispatch, skill matching, and job execution; and AI-powered asset master data management ensures that the information foundation supporting all these capabilities remains accurate and complete.

The Asset and Service Assistant within SAP's Autonomous SCM domain orchestrates agents targeting a 5 to 10 percent first-time-fix-rate improvement, 3 to 7 percent workforce productivity increase, and 3 to 10 percent reduction in incident rate. At Korindal, the asset management transformation spans approximately 4,200 pieces of production equipment across three manufacturing plants and nine distribution centers, assets whose reliability directly determines the OTIF performance the CSCO committed to improving. Figure 5.37 shows the Asset Intelligence Assistants, illustrating how AI capabilities span anomaly detection, maintenance planning, visual inspection, and field service.

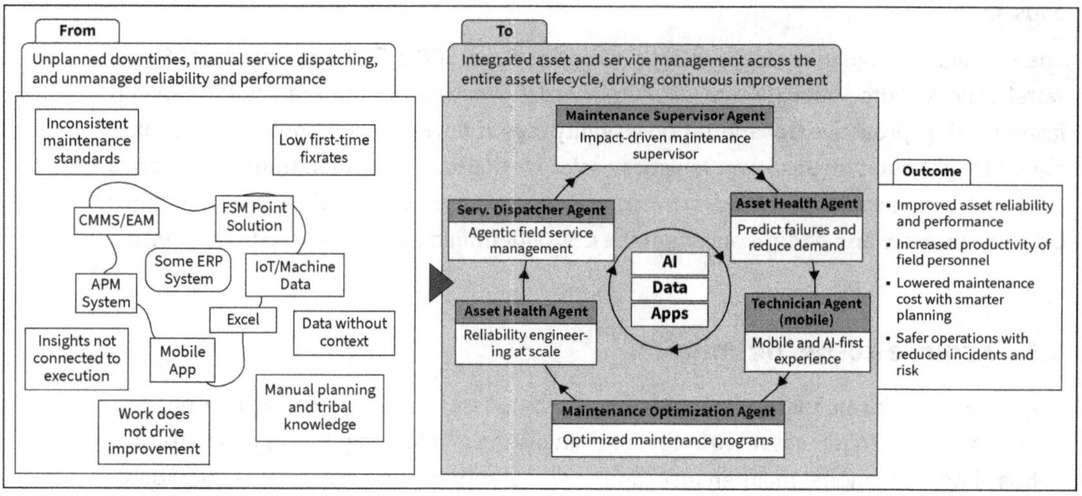

Figure 5.37: Asset Intelligence Assistants

The following sections will walk through the key capabilities for automating the acquire-to-decommission process, including plan-to-optimize, acquire-to-onboard, operate-to-maintain, operate-to-decommission, and managing assets.

5.6.1 Plan-to-Optimize Assets

The plan-to-optimize assets process is where asset strategy meets operational intelligence, the phase where reliability engineers analyze asset performance data, define maintenance strategies, and establish the plans and rules that govern how assets are maintained throughout their operational lives. For most organizations, the gap between asset strategy and operational execution is enormous: Reliability engineers define maintenance programs based on manufacturer recommendations and historical experience, but these programs rarely adapt to the actual condition and performance trajectory of individual assets. The result is a portfolio of maintenance plans that is simultaneously too conservative for low-risk assets, generating unnecessary interventions that consume technician capacity, and too relaxed for high-risk assets, for which degradation patterns go undetected until failure occurs.

Let's walk through the embedded AI capabilities to optimize asset planning and strategy.

Asset Performance Intelligence

SAP Asset Performance Management provides the strategic planning foundation. The *Maintenance Strategy capability* enables reliability engineers to segment assets based on risk and criticality, focusing maintenance investment on assets where failure consequences are most severe. Engineers analyze potential failures and their operational impact to develop the right maintenance and service strategy for each asset class, then manage the resulting maintenance and service recommendations through structured workflows. This risk-based segmentation ensures that a critical production line motor receives a fundamentally different maintenance approach than a noncritical office HVAC unit, a distinction that calendar-based programs cannot make.

The *Asset Health capability* extends strategy into real-time operations by establishing asset connectivity to receive sensor data for continuous health monitoring. Rules automate the creation of maintenance demand; when sensor readings cross defined thresholds, the system generates alerts that transfer into maintenance notifications or work orders without requiring human monitoring. AI-assisted anomaly detection enables reliability engineers to identify unusual patterns in sensor data streams, automatically recognizing inconsistent behavior that would require extensive manual review to detect. The system trains anomaly-detection models using historical sensor data and business context, manages the model lifecycle including threshold optimization for alerting, and provides a health check for equipment based on detected anomalies. The visual inspection capability further extends monitoring to assets for which sensor instrumentation is impractical, using ML models trained on image data to identify wear, corrosion, or structural deterioration from photographs captured during routine inspections.

The Asset and Service Assistant

The Alert Processing Agent qualifies alerts with a final recommendation for how to proceed, filtering the signal from the noise so that maintenance planners receive actionable intelligence rather than raw sensor data. In organizations with thousands of monitored assets, the volume of sensor alerts can overwhelm human reviewers; the agent evaluates each alert against the asset's performance baseline, maintenance history, and criticality classification to determine whether the alert warrants immediate intervention, planned investigation, or acknowledgment without action.

The Maintenance Strategy Optimization Agent recommends appropriate maintenance programs and creates or changes maintenance plans based on those recommendations, continuously optimizing the maintenance program using performance data from SAP Asset Performance Management. When an asset class consistently demonstrates longer-than-expected mean time between failures, the agent recommends extending maintenance intervals; when failure patterns emerge earlier than predicted, it tightens them, ensuring that maintenance programs evolve with operational reality rather than remaining static.

Korindal: James Reduces Unplanned Downtime Through Condition-Based Maintenance

At Korindal, the transition from calendar-based to condition-based maintenance planning reduces the frequency of unnecessary preventive interventions by approximately 20 percent on critical production equipment, while the anomaly detection capability identifies 85 percent of incipient failures more than 48 hours before they would have caused unplanned downtime, providing sufficient lead time for planned intervention rather than emergency response. The Maintenance Strategy Optimization Agent reviews Korindal's 340 active maintenance plans quarterly, recommending interval adjustments based on accumulated performance data, a continuous optimization process that previously required a dedicated reliability engineering initiative conducted once per year at most.

5.6.2 Acquire-to-Onboard

The acquire-to-onboard process covers the commissioning of new assets and the handover of asset data from manufacturing or construction into maintenance operations. SAP S/4HANA Cloud manages asset master data throughout the entire lifecycle of technical assets, including equipment hierarchies, functional location structures, technical documentation, maintenance task lists, maintenance plans, and measuring points. Partners, risks, and warranty data are maintained as additional asset attributes.

Let's explore the AI capabilities and see their impact on our case study.

Asset Onboarding and Master Data

SAP Business Network for Asset Management standardizes equipment management through asset master data and spare parts information available for update from the network. Organizations can achieve collaborative, end-to-end maintenance and service processes for complex installation and maintenance procedures, streamline maintenance activities through access to the latest maintenance instructions and service bulletins, and install a digital twin in their enterprise asset management system to represent each physical machine.

The AI-assisted creation of fixed asset master data streamlines the creation process through a conversational interface, enabling asset accountants to establish new asset records through Joule rather than navigating complex master data transactions. The AI-assisted explanation of fixed asset key figures empowers users to understand asset values in natural language, explaining depreciation calculations, book values, and key figures throughout the asset's entire lifecycle. The Maintenance Request Agent enables any employee to report damage to assets and infrastructure in seconds without prior SAP knowledge; the AI analyzes the input, extracts all relevant information, and creates a structured maintenance notification automatically.

Korindal: David's Team Cuts Equipment Onboarding from Two Weeks to Three Days

At Korindal, the onboarding process for new production equipment, previously a two-week effort involving manual data entry across multiple systems, compresses to three days as SAP Business Network provides manufacturer-supplied master data and digital twin configurations that flow directly into the asset management system.

5.6.3 Operate-to-Maintain

The operate-to-maintain process is the operational heart of asset management, the ongoing cycle of monitoring, planning, scheduling, and executing maintenance that determines asset reliability and uptime. SAP S/4HANA Cloud supports all types of maintenance processes within a standardized phase model: reactive maintenance for unplanned failures, proactive maintenance driven by maintenance plans and schedules, improvement maintenance for capital projects that enhance asset capability, and operational and overhead maintenance for routine activities such as greasing and calibration.

The following sections will dive into the embedded AI features and capabilities to automate and optimize the maintenance planning and execution processes.

Maintenance Planning and Scheduling

The reactive maintenance process begins with maintenance request creation featuring an embedded risk and impact evaluation, a customizable questionnaire that determines request priority based on a risk-impact matrix. Requests are screened and prioritized through a dedicated app, with potential duplicates flagged automatically. Maintenance orders are created and routed through automatic approval workflows before entering the planning and scheduling pipeline.

For proactive maintenance, maintenance strategies derived from SAP Asset Performance Management feed into maintenance plans, including multiple-counter, single-cycle, and strategy-based, which trigger maintenance operations automatically according to plan. The system supports condition-based maintenance, where sensor data triggers interventions based on the actual asset condition rather than fixed calendar intervals and on ML-based predictive and prescriptive maintenance to anticipate failures before they occur.

Resource scheduling provides full transparency into work centers through dedicated monitoring of key figures, such as due maintenance orders, work center utilization, and capacity availability. Time-bucket-based planning and monitoring of maintenance orders ensures that all prerequisites are met, including material and lean service availability, before work is dispatched. The interactive maintenance scheduling board offers drag-and-drop features for adjusting orders and operations, with simulation of different schedules to identify optimal work center utilization before dispatching. Material availability checks ensure operations run smoothly after dispatch. Figure 5.38 shows the Maintenance Order Planning

Agent, illustrating how AI assists maintenance planners in creating, scheduling, and resourcing work orders.

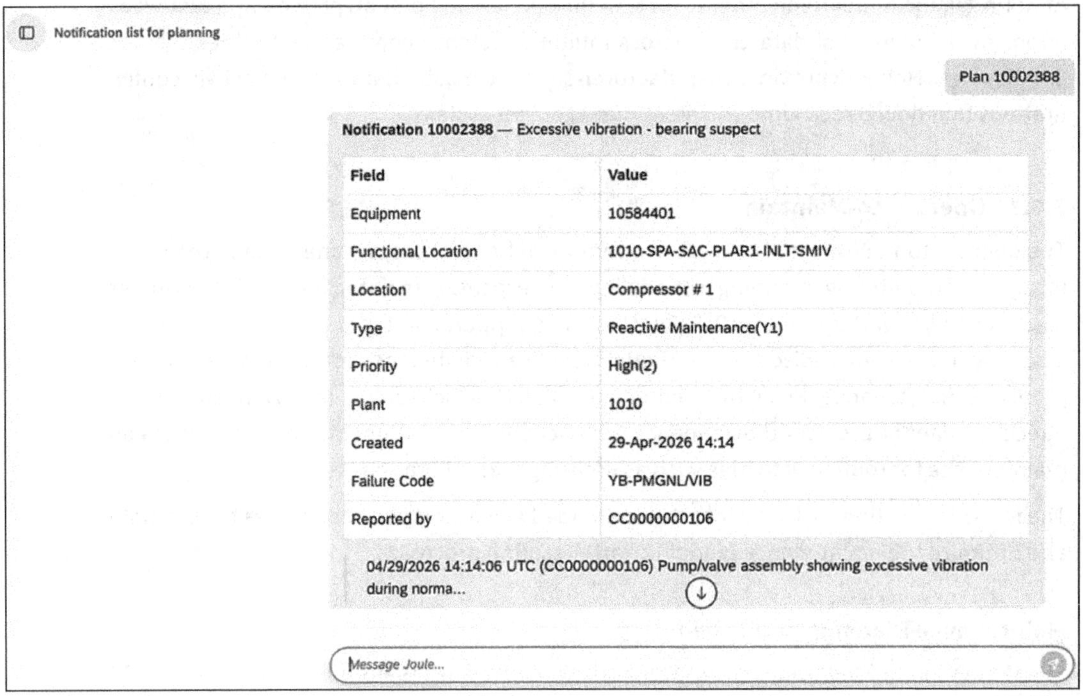

Figure 5.38: Maintenance Order Planning Agent

Maintenance Execution and Field Service

Maintenance execution leverages responsive SAP Fiori apps for mobile devices and the SAP Maintenance Assistant as a native mobile alternative, enabling technicians to capture execution data, record time, track material consumption through direct goods issue postings, and perform postcompletion activities. AI-assisted maintenance order recommendations let AI suggest maintenance orders that solved similar incidents from the maintenance history, enabling technicians to copy proven approaches rather than starting investigations from scratch. The AI-assisted mobile execution capability enables technicians to use voice input to create structured text entries, describing observations, actions taken, and parts consumed through natural speech that the system converts into standardized maintenance records. This eliminates the friction of typing detailed technical notes on mobile devices while wearing work gloves on a plant floor. AI-assisted spare parts identification further accelerates execution by helping technicians locate the correct replacement components from catalogs containing thousands of items.

Agentic Maintenance Planning and Dispatch

The Maintenance Event Planning Agent, part of the Asset and Service Assistant, helps maintenance planners efficiently manage and prioritize a growing list of maintenance tasks while ensuring equipment reliability and minimizing downtime. The Maintenance Supervisor Screening Agent provides impact-driven notification summaries with risk-based prioritization and autoscreening. The Operational Maintenance Planning Agent serves as a strategic command center, enabling planners to analyze maintenance workloads across time horizons, identify scheduling conflicts before they cascade into delays, and optimize resource allocation across multiple work centers simultaneously. The Long-Term Spare Parts Forecasting Agent fetches mid-to-long-term demands, identifies critical spares, and projects availability based on forecasted demand, preventing the stockouts that turn planned maintenance into delayed maintenance. Figure 5.39 shows the Maintenance Event Planning Agent, illustrating how it helps planners manage and prioritize maintenance tasks to minimize equipment downtime.

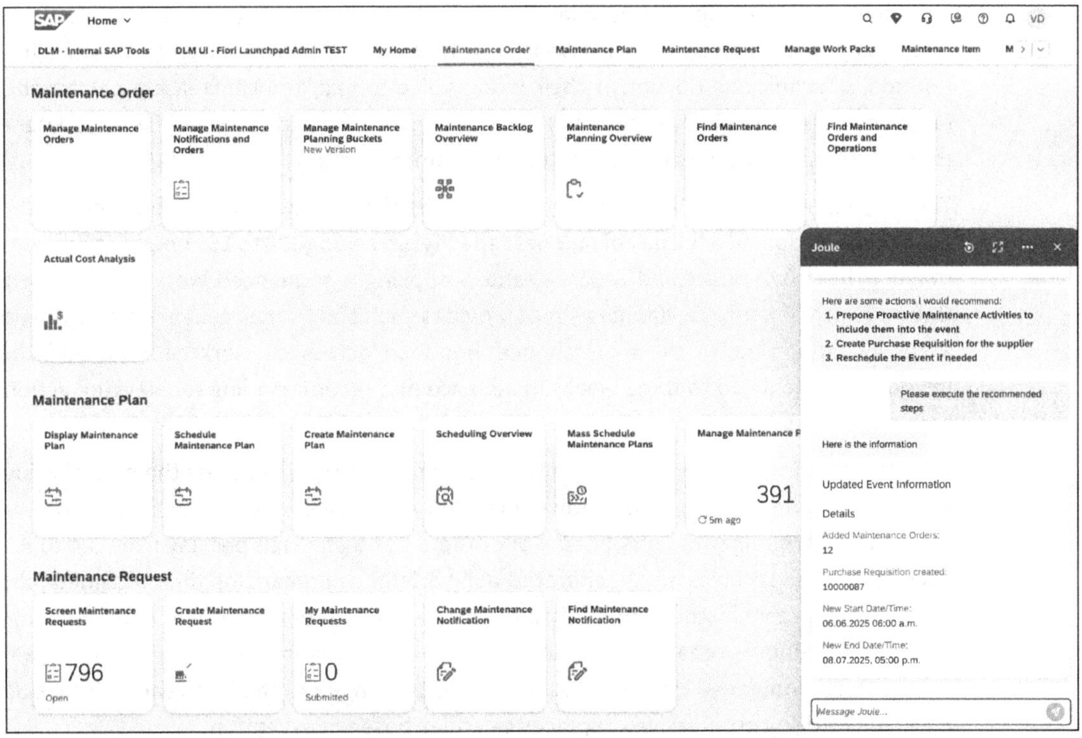

Figure 5.39: Maintenance Event Planning Agent

The Field Service Dispatcher Agent enables dispatchers to efficiently plan and optimize service orders by leveraging real-time data, technician availability, and intelligent recommendations, proposing the right technician for the right job at the optimal time. AI-assisted automated scheduling analytics provides dispatchers with performance insights into scheduling effectiveness, while predictive routing optimizes technician travel paths based on geographic proximity and traffic patterns. Joule with SAP Field Service Management provides conversational access to dispatching capabilities, while AI-assisted equipment insights provide a centralized repository for detailed service history records. The Technician Agent provides mobile technicians with question answering based on historical activities, quick access to tasks through daily briefings, AI-generated job summaries, and access to a field service knowledge base built from historical work orders, ensuring that institutional knowledge is preserved and accessible even when experienced technicians retire or transfer.

Korindal: James's Maintenance Team Moves from Calendar-Based to AI-Driven Operations

At Korindal, the maintenance transformation produces the most directly quantifiable impact in the asset management domain. The AI-assisted maintenance order recommendation capability reduces the average time to resolution for reactive maintenance by 35 percent; when a conveyor motor fails, the system identifies three similar historical failures and their resolution approaches, enabling the technician to apply a proven fix rather than diagnosing from scratch. The AI-assisted mobile execution capability transforms how Korindal's technicians document their work: Voice-to-text input increases maintenance record completeness from 60 percent to over 90 percent, building a richer historical dataset that in turn improves the quality of future AI recommendations.

The Maintenance Event Planning Agent automates the weekly prioritization exercise that previously consumed a full day of planner capacity, screening 120 to 150 open notifications, ranking them by impact and urgency, and proposing a sequenced work schedule. The Operational Maintenance Planning Agent provides each plant's maintenance manager with a consolidated view of the maintenance workload across all work centers, surfacing resource conflicts two to three weeks in advance and recommending rebalancing actions before backlogs develop.

The Field Service Dispatcher Agent optimizes technician routing across the nine distribution centers, reducing average travel time between assignments by 22 percent and increasing the number of completed work orders per technician per day from 3.4 to 4.1. Each technician receives an AI-generated daily briefing summarizing their assigned jobs, relevant equipment history, required materials, and safety considerations, eliminating the 30-minute morning preparation routine that previously required manual review of multiple systems. Unplanned downtime decreases by 18 percent in the first year as condition-based monitoring progressively replaces calendar-based intervention.

5.6.4 Offboard-to-Decommission

The offboard-to-decommission process addresses the end-of-life phase of the asset lifecycle, defining exit strategies, executing asset sales or trades, and decommissioning productive assets from operational service. Although often overlooked in digital transformation initiatives, this phase has significant financial and operational implications. Assets maintained beyond their economically useful life consume maintenance budgets disproportionate to their productive contribution, while assets decommissioned prematurely forfeit residual value.

SAP S/4HANA Cloud supports the financial dimensions of decommissioning through integrated fixed asset accounting, retirement postings, scrapping transactions, and sale or trade processing that automatically update the asset's financial records, close associated maintenance plans, and remove the asset from operational scheduling. The AI-assisted asset scrapping capability enables asset accountants to fully scrap assets through Joule's conversational interface across all accounting principles, simplifying a process that traditionally required navigation of multiple financial transactions.

At Korindal, the disciplined decommissioning process, informed by asset health data from SAP Asset Performance Management and TCO analysis from maintenance history, enables the CFO to replace the legacy annual depreciation review with a continuous, data-driven retirement assessment that identifies assets for which ongoing maintenance costs exceed replacement economics, recovering approximately $2.4 million annually in avoided maintenance spend on assets that should have been retired.

5.6.5 Manage Assets

The manage assets process provides the cross-cutting analytical and governance capabilities that span the entire asset lifecycle. SAP S/4HANA Cloud delivers embedded analytics for asset performance, maintenance cost tracking, and equipment reliability, enabling asset managers to move from periodic reporting to continuous, data-driven portfolio management.

We'll explore the asset service intelligence capabilities in more detail in the following sections.

Asset Service Intelligence

Joule with SAP Asset Performance Management provides conversational access to asset health and performance data, enabling reliability engineers to query asset conditions, review anomaly histories, and assess fleet-wide trends through natural language rather than navigating analytical dashboards. The AI-assisted retrieval of equipment information in service management enables service managers to quickly access comprehensive details for equipment installed at customer sites, including warranty status, active or completed service transactions, and an AI-generated summary of services performed with actionable recommendations. This capability transforms equipment inquiries from multiscreen investigations into single conversational interactions. The AI-assisted reliability assessments capability supports failure mode and effects analysis by identifying potential failure modes, evaluating their impact, and recommending mitigation strategies, bringing structured engineering methodology into the digital maintenance environment.

Visual inspection augments quality control with AI that detects defects, anomalies, and assembly issues directly from images and video. By training models on labeled examples and integrating them into production lines, manufacturers can standardize inspections and reduce human variability. The result is faster detection, fewer false rejects, and continuous learning from feedback loops, driving higher yield and safer operations. This scenario scales from edge devices on the shop floor to cloud-driven analytics for fleet-wide visibility. Figure 5.40 shows visual inspection in SAP Asset Performance Management, illustrating how AI-powered image analysis detects defects and anomalies to augment quality control.

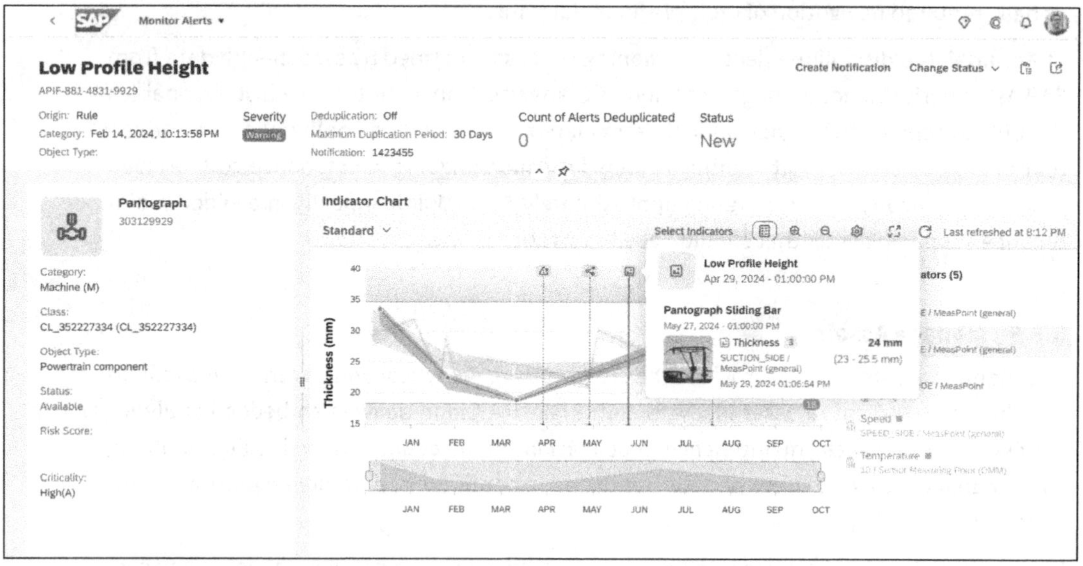

Figure 5.40: Visual Inspection in Asset Performance Management

Korindal: James Gains Real-Time Asset Intelligence Across Three Plants

At Korindal, reliability engineers previously queried asset conditions by navigating multiple SAP Asset Performance Management screens and cross-referencing maintenance history spreadsheets. A typical asset health review consumed two to three hours of a senior engineer's time per plant per week. With Joule integrated into SAP Asset Performance Management, the same review takes minutes. Engineers ask conversational questions: Which assets show anomaly trends this week? Which have upcoming maintenance windows? Which are running outside specification? They receive grounded answers drawn from live sensor data and maintenance history across all three plants simultaneously.

Visual inspection compounds the benefit. Components arriving from suppliers are screened automatically against defect models, and the system flags anomalies before they reach the production line. The combination of conversational asset intelligence and AI-driven visual inspection shifts the reliability team's work from reactive investigation to proactive intervention, directly supporting James's goal of reducing the unplanned downtime that costs Korindal approximately $180,000 per hour in lost throughput.

5.6.6 Industry AI for Asset Management

The Autonomous Enterprise framework organizes AI capabilities around five horizontal business domains: finance, SCM, spend, HCM, and CX. Each domain addresses the processes that define its functional area. However, many of the most complex and valuable business challenges that enterprises face are not purely horizontal. They are industry-specific, cutting across multiple domains along a vertical value chain. Asset management in energy, regulated manufacturing in life sciences, adaptive production in automotive, project delivery in engineering and construction: These problems require the vertical depth that comes from 50 years of industry coinnovation across more than 25 industries. This is the purpose of Industry AI.

Industry AI embeds sector-specific process knowledge, data models, and agentic capabilities directly into industry workflows. Where autonomous domains organize AI around a single business function, Industry AI spans multiple domains along an industry's value chain. Industry AI autonomous scenarios are groups of highly connected, agent-driven business capabilities that span end-to-end industry processes. These scenarios are realized through autonomous Industry AI solutions that are validated through forward-deployed engineering in close coinnovation with customers; orchestrate agentic workflows across applications, data, and processes; and are refined from customer engagements into standardized solutions for wider commercialization. Industry AI builds on and extends SAP's existing industry solutions, embedding autonomous decision-making and execution into them. The solutions are delivered on SAP Business AI Platform, which provides the shared business context, unified data, purpose-built models, and enterprise-grade governance required for reliable autonomous execution.

The *Autonomous Asset Management* scenario represents one of the flagship Industry AI scenarios unveiled at SAP Sapphire 2026. It targets industries in which physical assets form the central value chain: energy and utilities, oil and gas, mining, transportation, and heavy manufacturing. In these industries, asset reliability is not merely an operational concern but a safety imperative, a regulatory obligation, and the primary determinant of enterprise value creation.

The scenario combines Autonomous SCM domain agents from the Asset and Service Assistant with industry-specific agents that bring sector knowledge into the decision loop. SAP quantifies the combined impact as 30 percent reduction in unplanned downtime, 25 percent total maintenance cost reduction, and 40 percent first-time-fix-rate improvement. Figure 5.41 shows the Asset and Service Assistant, illustrating how industry-specific AI extends standard asset management with sector-specific routing and field service intelligence.

The following sections will explore the embedded AI capabilities introduced by Industry AI in more detail.

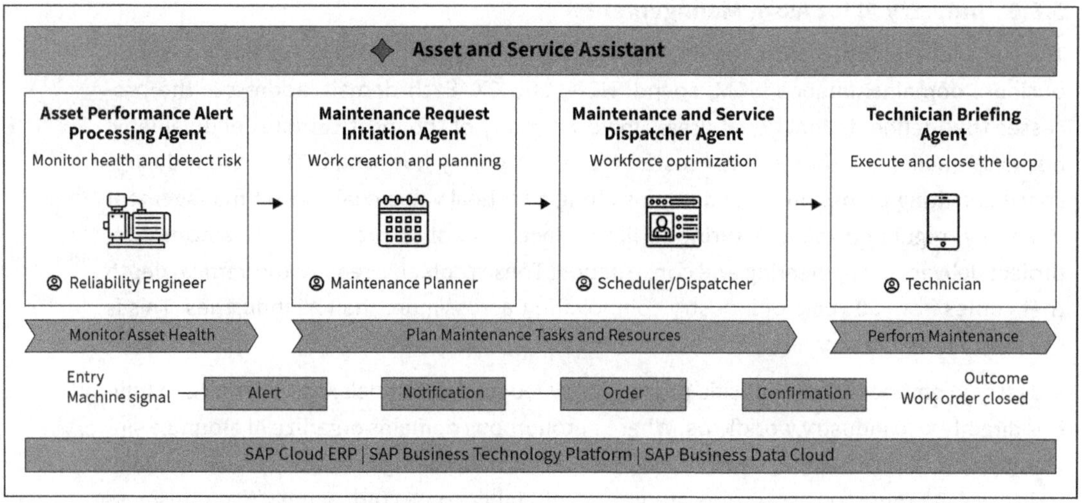

Figure 5.41: Asset and Service Assistant

Asset and Service Assistant Extension

The Asset and Service Assistant within the Autonomous Supply Chain domain provides the operational foundation: alert processing, asset health monitoring, maintenance planning, maintenance request management, equipment insight, maintenance plan cost simulation, technician briefing, operational maintenance order management, service dispatching, and maintenance briefing. These agents handle the universal asset management processes that apply regardless of industry.

Industry AI extends this foundation with agents that carry sector-specific intelligence:

- **Asset Strategy Agent**
  The Asset Strategy Agent embeds industry-specific lifecycle models that reflect how assets degrade under sector-specific operating conditions. A gas turbine in a combined-cycle power plant degrades differently from a centrifugal pump in a refinery or a conveyor system in a mining operation. The Asset Strategy Agent incorporates industry failure mode libraries, regulatory inspection schedules, and sector-specific reliability models to recommend maintenance strategies that reflect actual operating conditions rather than generic manufacturer guidelines.

- **Safety Permit Agent**
  The Safety Permit Agent automates the permit-to-work process that is mandatory in hazardous environments such as oil refineries, chemical plants, offshore platforms, and power generation facilities. Before any maintenance intervention can begin on critical equipment, a safety permit must be issued, confirming that isolation procedures are complete, atmospheric monitoring is in place, and the work area meets regulatory safety standards. The Safety Permit Agent integrates with the maintenance planning workflow to ensure that permit requirements are identified during planning, prerequisite safety checks are scheduled in the correct sequence, and work cannot proceed until all safety

conditions are verified. In industries subject to OSHA Process Safety Management or equivalent regulations, this automation replaces manual paper-based permit systems that historically create bottlenecks and compliance risk.

- **Field Service Agent**
  The Field Service Agent extends standard service dispatching with industry-specific routing logic. In utilities, field service must account for geographic distribution across thousands of asset locations with varying access restrictions and safety classifications. In oil and gas, field service may involve helicopter transport to offshore platforms or coordination with vessel scheduling. The Field Service Agent incorporates these industry constraints into dispatch optimization, ensuring that the technician assignment considers not only skill matching and proximity but also site access authorization, safety certification status, and industry-specific regulatory requirements.
- **Field Technician Agent**
  The Field Technician Agent provides mobile technicians with industry-specific guidance during maintenance execution. Rather than generic work instructions, the agent delivers context-aware procedures that account for the specific asset type, its operating history, regulatory classification, and the industry-standard resolution approach. In energy, this includes compliance documentation for grid-connected assets. In mining, it includes environmental monitoring requirements during maintenance of processing equipment. The agent also captures execution data in the format required by industry-specific regulatory reporting.
- **Asset Information Agent**
  The Asset Information Agent serves as the single point of access for comprehensive asset intelligence across the installed base. In asset-intensive industries, a single piece of equipment may have engineering specifications, manufacturer documentation, operational history, maintenance records, inspection reports, regulatory compliance certificates, and sensor data streams distributed across multiple systems. The Asset Information Agent synthesizes this information on demand, providing maintenance planners, reliability engineers, and field technicians with the complete picture required for informed decision-making.
- **Resource Planner Agent**
  The Resource Planner Agent optimizes workforce allocation across asset-intensive operations for which specialized skills, certifications, and security clearances constrain the available labor pool. In nuclear energy, only technicians with specific Nuclear Regulatory Commission (NRC) certifications can perform work on safety-classified systems. In aviation maintenance, repair, and overhaul (MRO), only mechanics with type-specific ratings can release aircraft to service. The Resource Planner Agent accounts for these industry-specific constraints in long-term resource planning, identifying certification gaps that would otherwise surface as execution blockers during maintenance campaigns. Figure 5.42 shows the Maintenance Event Notification Agent, illustrating how AI automates the creation and routing of maintenance notifications from sensor signals and field observations.

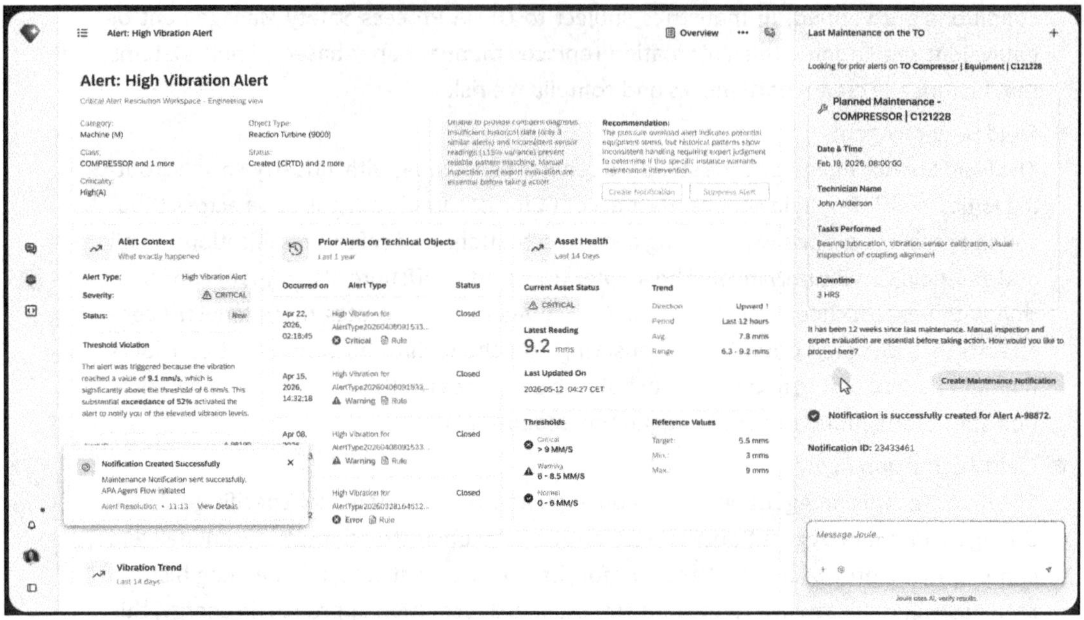

Figure 5.42: Maintenance Event Notification Agent

Agent-to-Agent Orchestration Across Domains

The distinguishing characteristic of Industry AI scenarios is agent-to-agent orchestration that crosses domain boundaries. In the Autonomous Asset Management scenario, Industry AI agents collaborate with agents from the Autonomous SCM domain and the Autonomous Finance domain simultaneously.

Consider a scenario in which the Asset Health Agent detects an anomaly pattern in a critical turbine. The Alert Processing Agent qualifies the alert and triggers a maintenance demand. The Maintenance Planning Agent schedules the intervention, while the Safety Permit Agent initiates the permit-to-work process in parallel. The Resource Planner Agent confirms that a certified technician is available and assigns the work order. The Field Service Agent optimizes routing and scheduling. During execution, the Field Technician Agent provides step-by-step guidance and captures regulatory documentation. The Asset Information Agent supplies historical context. The Maintenance Plan-Cost Agent evaluates the financial impact and updates the asset's total cost of ownership projections. Finally, the Asset Strategy Agent evaluates whether the anomaly pattern suggests a shift in degradation trajectory that warrants adjusting the long-term maintenance strategy.

This end-to-end orchestration spans asset monitoring, maintenance planning, safety compliance, workforce scheduling, field execution, and financial analysis in a single connected workflow. No individual domain agent possesses the cross-functional scope to manage this sequence alone. Industry AI provides the orchestration layer that connects them, grounded in sector-specific process knowledge.

Delivery and Availability

Industry AI solutions are delivered via SAP Business AI Platform and are validated through forward-deployed engineering in close coinnovation with pilot customers. The Autonomous Asset Management scenario targets industries including energy, utilities, oil and gas, mining, and industrial manufacturing. Specific Industry AI assistants already in the roadmap include the IS Oil and Gas Assistant (with Tax Compliance Agent and Returned Items Processing Agent planned for Q3 2026 and Smart Ticketing Agent planned for Q4 2026) and the IS Utilities Assistant (with Outsorted Utilities Billing Document Resolution Agent planned for Q3 2026). Additional industry-specific agents for the Autonomous Asset Management Scenario continue to be developed through coinnovation partnerships and will be refined into standardized solutions for broader commercialization.

SAP does not commercialize Industry AI as a separate subscription. Industry AI solutions follow value-based pricing, linking commercial models directly to the outcomes they deliver. The underlying Joule Agents remain activated through AI units within the premium AI capabilities packages, consistent with the commercialization model for all Joule Agents across the Autonomous Enterprise.

Korindal: James Achieves 30 Percent Downtime Reduction with Industry AI

At Korindal, the asset portfolio includes several categories of equipment that benefit from Industry AI depth. The three manufacturing plants include process equipment subject to pressure vessel regulations, rotating machinery requiring vibration-based condition monitoring, and environmental compliance systems for which maintenance failures trigger regulatory reporting obligations.

For the rotating equipment fleet of approximately 340 critical motors, compressors, and pump assemblies, the Asset Strategy Agent applies industry-specific failure mode libraries for rotating machinery, incorporating bearing wear curves, lubrication degradation models, and vibration signature analysis to recommend maintenance interventions at the optimal point in the degradation cycle. The Safety Permit Agent ensures that lock-out/tag-out procedures are completed and documented before any invasive maintenance begins on equipment with stored energy hazards.

The Resource Planner Agent addresses Korindal's challenge with specialized maintenance skills. Approximately 15 percent of maintenance work orders require technicians with specific certifications (high-voltage electrical, pressure vessel inspection, or confined space entry). Without Industry AI resource planning, these constraints surface late in the scheduling process, causing delays that extend the maintenance window. The Resource Planner Agent identifies certification requirements during the planning phase and schedules certified technicians proactively, reducing maintenance schedule slippage by an estimated 20 percent for regulated equipment classes.

The cumulative impact at Korindal is measurable: The 30 percent reduction in unplanned downtime directly supports James's OTIF improvement objective as every hour of unplanned production equipment downtime translates directly into delayed customer orders. The 25 percent maintenance cost reduction enables the CFO to reallocate maintenance budgets toward capital improvements that further extend asset productive life. And the 40 percent first-time-fix-rate improvement means that maintenance interventions are resolved correctly on the first visit, eliminating the repeat dispatches that consume technician capacity and extend equipment outage durations.

5.7 Summary

This chapter has traced how SAP Business AI transforms every end-to-end value stream in SAP Autonomous Suite from a transactional system of record into an intelligent, adaptive system of work. The transformation follows a consistent architectural pattern across all five domains: embedded machine learning delivers predictive and analytical intelligence within standard applications, Joule provides conversational access that makes that intelligence actionable for every user regardless of technical expertise, and specialized Joule Agents orchestrate multistep workflows autonomously while keeping humans in the loop for governance and strategic judgment.

In finance, the Autonomous Finance framework deploys eight assistants and 48 specialized agents across billing, accounts receivable, accounts payable, close, planning, treasury, reporting, and governance. At Korindal, the invoice-to-cash transformation shifts DSO from a lagging metric reviewed after month end into a controllable operational lever managed daily, with SAP Cash Application achieving 90 percent automatic matching rates, the Accounts Receivable Agent reducing manual analysis effort by 75 percent, and the Dispute Resolution Agent compressing resolution cycles by 30 percent. The invoice-to-pay transformation achieves 85 percent touchless invoice processing, while the Treasury Assistant compresses the cash conversion cycle by seven days, freeing 19 million dollars in working capital.

In lead-to-cash, autonomous capabilities span the full commercial lifecycle. AI-assisted sales order creation from unstructured documents, Joule-powered fulfillment monitoring, and the Sales Operations Assistant with its four specialized resolution agents transform inside sales from reactive exception investigation into proactive commitment management. Deal intelligence, lead scoring, and conversational quoting accelerate the opportunity-to-quote cycle, while master data governance with AI eliminates the data quality issues that cascade into fulfillment failures.

In source-to-pay, the Sourcing Event Agent reduces sourcing cycle times by over 50 percent; the Autonomous Sourcing Assistant orchestrates supplier discovery, RFx management, and AI-driven negotiation; and the Procurement Orchestration Agent enables end-to-end autonomous procurement for routine transactions. Contract intelligence, compliance monitoring, and goods receipt automation close the operational gaps that historically allowed spend leakage and supplier delivery failures.

In plan-to-fulfill, real-time interactive MRP, predictive simulation, and the Manufacturing and Logistics Assistants transform planning, production, quality, and warehouse operations. The Planning Assistant accelerates planning cycles by 20 to 30 percent. The Manufacturing Assistant drives a 10 to 15 percent OEE improvement. The Logistics Assistant achieves up to 50 percent faster goods receipt processing and 12 percent lower transportation costs, collectively closing the gap between plan and execution that drives OTIF performance.

In acquire-to-decommission, SAP Asset Performance Management, the Asset and Service Assistant, and field service intelligence transform asset lifecycle management from calendar-based maintenance into condition-driven, AI-optimized operations, targeting a 5 to 10 percent first-time-fix-rate improvement, 3 to 7 percent workforce productivity increase, and 3 to 10 percent reduction in incident rate.

Across all five value streams, the pattern is consistent: Intelligence senses what is happening, optimization determines what should happen next, and automation executes those decisions at enterprise scale. The Autonomous Enterprise is not a future vision. It is the operational reality that SAP Autonomous Suite delivers today, one value stream at a time.

The next chapter, on industry scenarios, applies the embedded AI and autonomous capabilities described throughout this chapter to specific industry contexts, showing how SAP's vertical depth in sectors such as manufacturing, retail, and utilities extends these capabilities with industry-specific process knowledge, data models, and agentic scenarios.

Chapter 6
Industry-Specific Autonomous Scenarios

This chapter examines how integrated enterprise processes operate across industries. The focus is not on individual applications; instead, we'll look at how processes connect, from planning through execution to financial outcome. Each section highlights how operational fragmentation is reduced when systems, data, and workflows are aligned, and how measurable outcomes improve as a result.

Enterprise transformation is often described in terms of platforms, architectures, and capabilities. In practice, however, the effectiveness of those capabilities is revealed only in the context of specific industries. Each sector operates under its own constraints: regulatory frameworks, supply dynamics, product structures, and customer expectations. These constraints shape how decisions are made, how processes are executed, and where inefficiencies emerge.

We will explore how the Autonomous Enterprise translates universal principles into industry-specific solutions by examining five representative scenarios: regulatory submission orchestration in life sciences, demand-driven supply chain in consumer products, supply chain resilience in high-tech manufacturing, supply constraints and production continuity in industrial manufacturing, and predictive asset maintenance in oil, gas, and energy. For each scenario, we illustrate the specific orchestration required, the data flows involved, and the coordination mechanisms that enable organizations to operate more responsively and with greater visibility.

6.1 Life Sciences: Traceability, Compliance, and Controlled Supply Chains

Life sciences organizations operate in an environment where product movement is inseparable from regulatory accountability. Drug development spans 10–15 years from discovery through regulatory approval. Manufacturing operates under Good Manufacturing Practice (GMP) standards. Supply chains are global and subject to import/export regulations. Patient safety and data security are non-negotiable requirements. The challenge is not simply making and distributing products, but proving, at any point in time, where those products are, how they moved, and whether they comply with regulatory requirements.

Let's begin by examining an industry challenge. Pharmaceutical companies must manage serialized products across global supply chains. Each product unit carries a unique identifier, and every transaction, from packaging to dispensing, must be recorded and shared across ecosystem partners. These partners often operate different systems and comply with different national regulations. When these processes are not aligned, even minor inconsistencies in data can halt distribution. Products are delayed not because of physical constraints, but because the underlying records cannot be reconciled in time.

SAP provides a connected traceability and compliance layer that integrates manufacturing, warehouse, and distribution processes with serialization and regulatory data exchange, as shown in Figure 6.1. This capability ensures that every physical movement is matched by a verified digital record. The upper portion of Figure 6.1 highlights the role of SAP Traceability Hub in supporting serialization compliance. It provides capabilities for regulatory reporting, trading partner onboarding, and the management of serialization events across the supply chain. These capabilities allow organizations to exchange compliance information with manufacturers, distributors, logistics providers, and regulatory agencies using a common platform.

The lower portion of the figure contrasts traditional exception handling with an integrated traceability process. In the traditional approach, data-transmission errors require manual investigation, often involving quality, logistics, and IT teams before affected products can move forward. The traceability platform centralizes exception information and provides a single view of affected products, transactions, and business partners, allowing organizations to identify, investigate, and resolve compliance issues more efficiently while maintaining an auditable record of events.

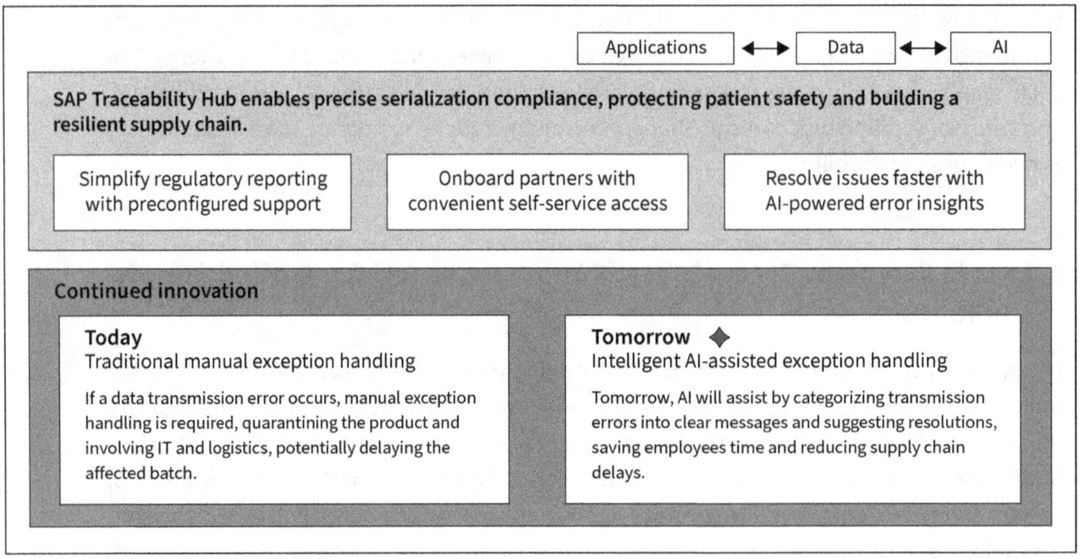

Figure 6.1: SAP Traceability Hub

Let's consider the results: A pharmaceutical manufacturer produces a batch of serialized products for distribution across multiple regions. As the products move through packaging, shipment, and distribution, each transaction is registered and validated automatically. If a discrepancy occurs, such as a mismatch in serial numbers or delayed reporting, it is detected immediately and surfaced within the operational workflow. Instead of halting the entire distribution, only affected units are isolated.

The impact is measurable. Serialization compliance approaches full coverage, delays caused by data inconsistencies are reduced, and regulatory reporting becomes a direct output of operations rather than a separate process.

Organizations typically see improved traceability completeness, reduced batch quarantine events, and faster resolution of compliance exceptions.

6.2 Consumer Products: Demand Volatility and Omnichannel Fulfillment

Consumer products organizations operate in an environment defined by variability. Demand shifts across channels, product portfolios expand, and fulfillment requirements become increasingly complex.

The primary challenge is coordinating demand signals across retail, e-commerce, and direct channels while ensuring that fulfillment decisions remain economically viable. Inventory may exist, but in the wrong location. Orders may be fulfilled, but at a higher cost than necessary.

Disconnected systems amplify the problem. Planning, fulfillment, and logistics operate independently, resulting in inefficient allocation and increased cost to serve.

Traditional systems separate demand planning (monthly cycles based on historical patterns), sales data (lagged one to two weeks), and supply planning (based on outdated demand forecasts). The result is a disconnect between actual demand signals and supply chain actions.

SAP provides unified demand and fulfillment orchestration that connects planning, inventory visibility, and order execution across channels, as shown in Figure 6.2. The lower section of the figure highlights the operational applications supporting consumer products organizations, including customer experience, supply chain, financial management, and industry-specific solutions such as SAP Omnichannel Promotion Pricing, SAP Order Management Foundation, SAP Revenue Growth Management, and SAP Billing and Revenue Innovation Management. Together, these applications support planning, pricing, inventory management, order fulfillment, and revenue execution across the product lifecycle.

The middle layer shows SAP Business Data Cloud providing a common semantic foundation across customer, supply chain, finance, spend, and industry data products. This shared data model enables demand, fulfillment, commercial, and operational functions to operate from a consistent view of customers, products, inventory, and orders.

The upper layer illustrates how domain-specific agents operate across customer experience, supply chain, finance, spend, analytics, and human resources. Rather than functioning within individual applications, these agents coordinate activities across end-to-end business processes spanning product innovation, manufacturing, supply chain operations, sales channels, and customer engagement.

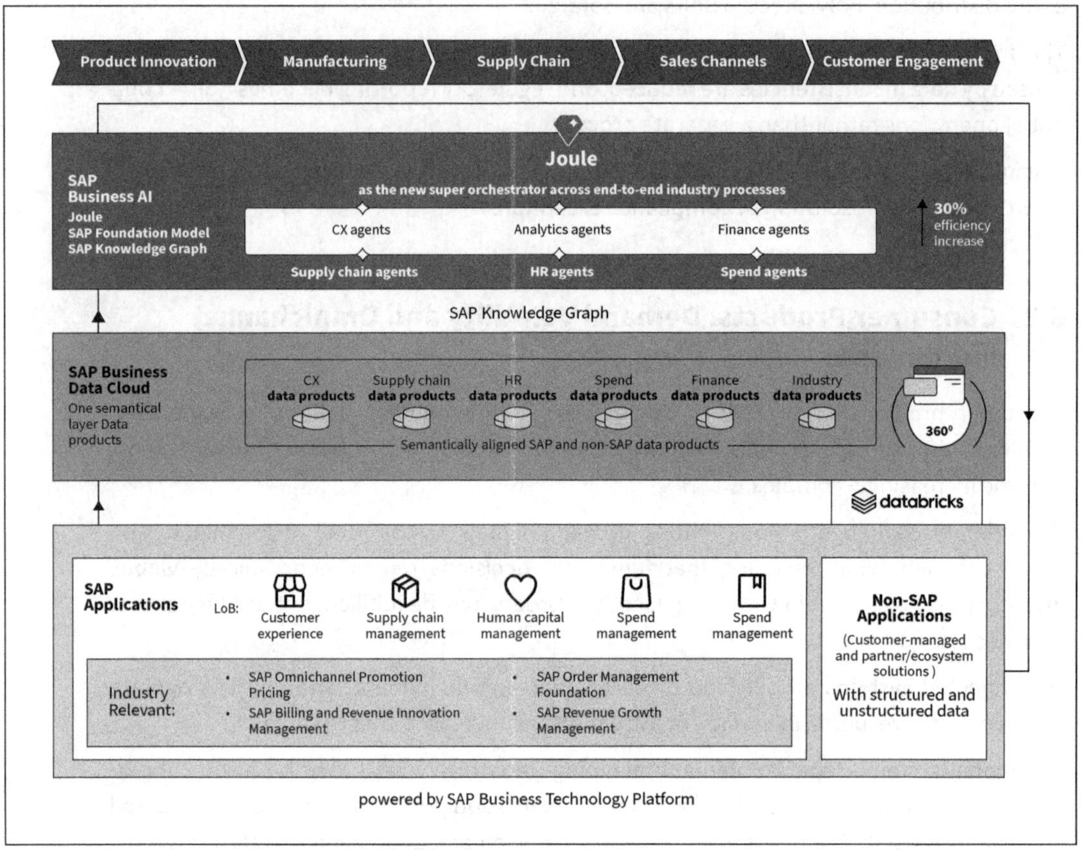

Figure 6.2: SAP Reference Architecture: Consumer Products

Let's consider the results: A consumer products company receives orders from retail and online channels simultaneously during a promotion. The system continuously evaluates inventory across distribution centers and stores, balancing fulfillment cost with delivery commitments. Instead of processing each order independently, fulfillment decisions are coordinated. Orders are routed to optimal locations based on availability, proximity, and cost.

Over time, the impact becomes evident. Order-fill rates improve, inventory turnover increases, and unnecessary stock transfers decrease. Most importantly, the organization reduces variability in fulfillment performance, even under fluctuating demand conditions.

6.3 High Tech: Configurable Products and Rapid Innovation Cycles

High-tech organizations face a different type of complexity. Rather than managing large volumes of identical products, they deal with continuous variation. Products are configured to customer specifications, and innovation cycles are short.

The challenge lies in translating customer configurations into executable production and fulfillment processes. Each configuration can affect pricing, bills of materials, and production timelines. Without tight alignment, delays and inconsistencies become common.

SAP connects product configuration, pricing, and manufacturing execution into a unified process, ensuring that configured products can be validated and produced without manual intervention. SAP provides an integrated lead-to-cash architecture that connects sales, pricing, ordering, billing, and receivables processes across the enterprise, as shown in Figure 6.3. The figure is organized around the major business process stages of lead-to-quote, quote-to-order, order-to-bill, and bill-to-cash, creating a continuous flow from customer engagement through revenue collection.

The lower section shows the applications supporting each stage of the process. SAP Sales Cloud, SAP Commerce Cloud, and SAP CPQ support sales and quoting activities, while SAP Cloud ERP, SAP Subscription Billing, SAP Billing and Revenue Innovation Management, and related solutions manage order execution, billing, revenue management, and cash collection. The upper layer illustrates how assistants operate across these process steps, helping coordinate activities that span sales, finance, and project execution.

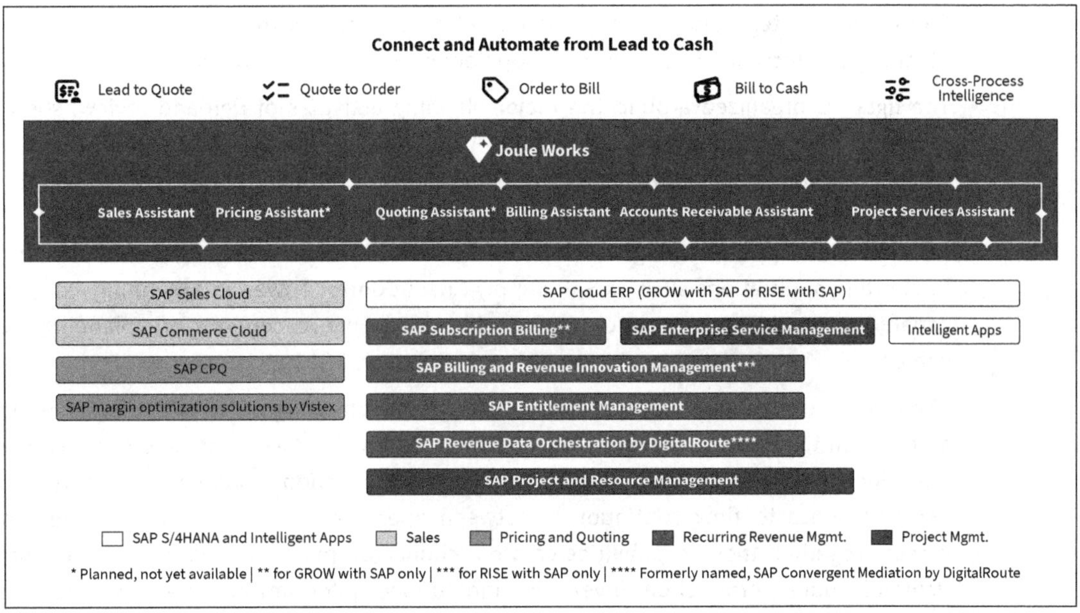

Figure 6.3: High-Tech Industry: Integrated Lead-to-Cash Process

Let's consider the results: A customer configures a product with specific features and submits an order. The configuration is validated immediately against available components and production capacity. Pricing is generated based on current conditions, and delivery commitments are confirmed in real time.

Once confirmed, the configuration flows directly into manufacturing planning. There is no need for reentry or manual validation across systems, as highlighted in Figure 6.3.

The effect is a reduction in order confirmation time, improved accuracy in delivery commitments, and better alignment between demand and supply. High-tech firms often see reduced cycle times for custom orders and higher customer satisfaction due to reliable commitments.

6.4 Industrial Manufacturing: Supply Constraints and Production Continuity

Industrial manufacturing operates under constant pressure to maintain production continuity despite supply disruptions. Material shortages, supplier variability, and volatile demand all impact production planning.

The challenge is to maintain production commitments when material availability is uncertain. Traditional planning systems operate in cycles, creating delays between detection of issues and response.

SAP provides integrated planning and execution capabilities that connect supply chain planning, production scheduling, and shop floor execution, as shown in Figure 6.4.

The figure is organized around the major planning activities of demand review, supply review, network collaboration, and management review, reflecting the end-to-end sales and operations planning (S&OP) process. At the center of the architecture is SAP Integrated Business Planning for Supply Chain (SAP IBP), which provides capabilities for S&OP, response and supply planning, network collaboration, and variant configuration planning. Supporting applications such as SAP Supply Chain Control Tower and planning analytics capabilities provide visibility into supply chain performance, exceptions, and planning outcomes.

The lower section shows the operational foundation consisting of SAP S/4HANA, external systems, and SAP Business Technology Platform (SAP BTP). Master data and transactional data, including sales orders, materials, suppliers, production resources, inventory, and demand forecasts, flow continuously between operational systems and planning processes. Together, these capabilities create a connected planning environment in which demand, supply, production, inventory, and supplier information remain synchronized across the supply chain.

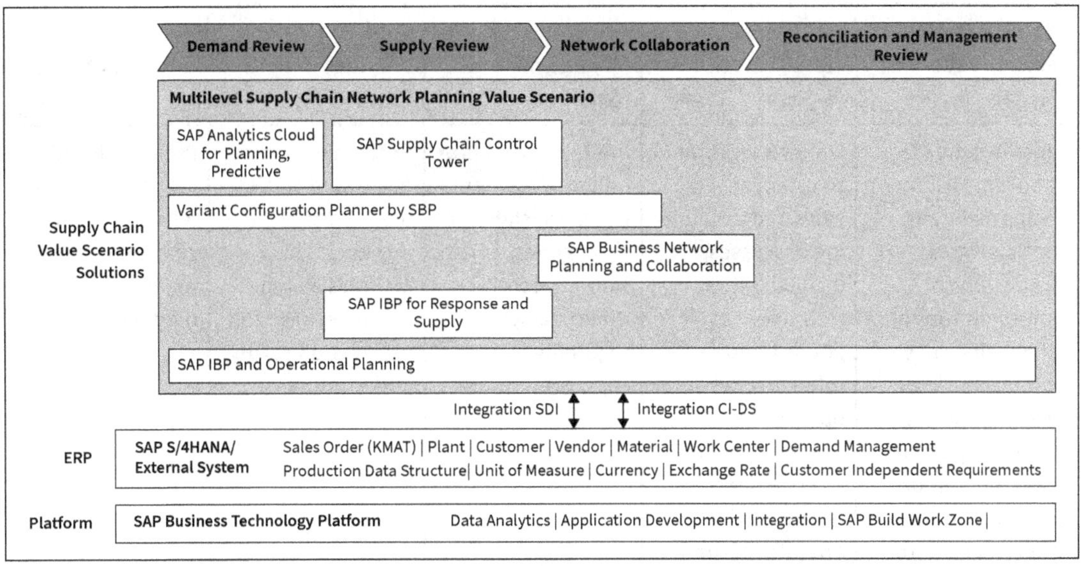

Figure 6.4: Reference Architecture: Industrial Manufacturing

Let's consider the results: A manufacturer experiences a shortage of critical components. Instead of halting production planning, the system evaluates available inventory, alternative suppliers, and order priorities simultaneously. Production is adjusted dynamically; high-priority orders are fulfilled first, and alternative sourcing options are incorporated into the plan.

The outcome is not the elimination of disruption, but a reduction in its impact. Production schedules remain aligned with actual conditions, and customer commitments are maintained more consistently. Manufacturers typically see improved schedule adherence and reduced backlog.

6.5 Oil, Gas, and Energy: Distributed Operations and Asset Coordination

Energy companies operate across wide geographic networks, where coordination between supply, distribution, and retail operations is critical.

This brings us to an industry challenge. Retail fuel networks must balance supply availability, pricing strategies, and operational efficiency across hundreds or thousands of locations. Disconnected systems lead to stock imbalances, inefficient replenishment, and delayed financial reconciliation.

SAP provides integrated operational and financial management across fuel distribution and retail operations. Table 6.1 highlights the key features of SAP S/4HANA Oil & Gas for retail fuel network operations.

Service Station Master Data Management	Wet Goods Management (Sales and Stock)	Dry Goods Management (Sales)	Card Settlement (Payment/Fleet Card)	Point-of-Sale Integration
Just-in-time station replenishment helps maintain optimal stock levels, which improves supply chain efficiency while reducing waste. Seamless fleet and payment card integration simplifies transactions for customers, enhancing their experience and fostering loyalty.	Makes it possible to establish all sales and accounting processes, including the movement of goods for fuels. retail fuel network operations (RFNO) accounts for contractual specifics regarding the operator of the service stations.	Enables all types of billing processes to be established for shop goods, considering contractual specifics regarding the process owner.	Settlement against banks and settlement partners of any kind. This includes all types of banks, fleet and association cards, and third-party cards.	Provides RFNO-standard IDocs for inbound point-of-sale data, meter readings, and dip readings.

Table 6.1: SAP S/4HANA for Retail Fuel Network Operations

Let's consider the results: A fuel retailer manages replenishment across multiple service stations. Sales data flows directly into planning systems, triggering replenishment based on real consumption patterns. Inventory levels stabilize across the network. Stations are less likely to experience stockouts, and excess inventory is reduced.

Financial processes are aligned with operations. Sales transactions translate directly into financial records, reducing reconciliation effort and improving reporting accuracy.

6.6 Summary

Across industries, the specifics of operations vary, but the underlying pattern remains consistent. Performance is determined not by individual capabilities, but by how effectively those capabilities are connected. Where systems remain fragmented, organizations operate with delayed visibility and increased inefficiency. Where processes are integrated, those inefficiencies are reduced, and decisions flow more naturally across the enterprise.

Although each industry presents distinct challenges, a common architectural pattern emerges. Operational processes are connected across planning, execution, and financial layers. Data flows continuously between systems, and decisions are evaluated within the context of the entire enterprise rather than individual functions.

This architecture ensures that decisions made in one area—such as demand planning or customer ordering—are immediately reflected across production, logistics, and financial processes.

The examples in this chapter illustrate how this integration translates into measurable outcomes: improved compliance in life sciences, better fulfillment efficiency in consumer products, faster response in industrial manufacturing, more reliable execution in high tech, and coordinated operations in energy.

The transition is not from manual to automated operations; it is from isolated processes to connected systems. It is this shift that enables organizations to align execution with intent and to manage complexity without being overwhelmed by it.

These industry solutions deliver business value at the process level, but they depend on a common platform foundation. The next chapter examines SAP Business AI Platform and the capabilities it provides for grounding, orchestration, development, and governance across the enterprise.

PART II

SAP Business AI Platform

Chapter 7
The New Platform

This chapter explores how SAP has assembled its artificial intelligence capabilities into a unified platform that underpins every application, every process, and every agent across the SAP Autonomous Suite.

Artificial intelligence (AI) is no longer a capability that sits alongside your enterprise systems waiting to be called upon. It is becoming the operating layer through which work happens. The vision SAP articulated at SAP Sapphire 2026 is precise on this point: The autonomous enterprise is not one that has added AI features to existing applications but one in which AI agents actively perform work, make decisions, and orchestrate processes across the full breadth of the business, in finance, supply chain, procurement, human resources, and customer operations, with humans in the loop where judgment and accountability require it. Reaching that state of autonomous operation is not a matter of picking the right large language model (LLM) or deploying a single clever automation. It requires a platform: a coherent, integrated stack of AI infrastructure, data foundations, development tooling, governance mechanisms, and user experience that works together as a system. That platform is what SAP has built.

This chapter examines SAP Business Data Cloud and SAP Knowledge Graph as the semantic data foundation that gives AI agents the business context they need to reason accurately. It explores the generative AI hub as the governed access layer for frontier AI models, and SAP AI Core and SAP AI Launchpad as the operational infrastructure for deploying and managing those models in production. It describes Joule Studio and SAP AI Agent Hub as the development and governance environment for the agents that operate across your SAP landscape. It covers SAP Signavio and Business Transformation Management as the process intelligence layer that tells AI where to act. It also examines how Joule Work, SAP's vision for the autonomous enterprise, brings all these capabilities together into a single, unified experience that fundamentally changes the way work gets done. By the end of this chapter, you will have a clear understanding of how SAP Business AI Platform is designed, how its layers connect, and how your organization can use it to move from AI experimentation toward genuine enterprise-wide automation.

Note

At the time of writing (summer 2026), SAP Business AI has just been introduced by SAP. As such, there may be some enhancements released between the time of writing and the release of this book.

7.1 The Platform Architecture

SAP organizes its AI platform around three strategic pillars:

- The first pillar is about building AI-powered experiences, providing the development tools, model access, and agent frameworks that allow developers, business analysts, and partners to create custom AI agents and skills that are grounded in the SAP ecosystem.
- The second pillar is about contextualizing AI reasoning, ensuring that the models operating on your behalf have access to the right enterprise data in the right semantic form so that AI decisions are based on your actual business context rather than generic knowledge.
- The third pillar is about optimizing and governing AI, providing the operational discipline, monitoring capabilities, process intelligence, and compliance controls that allow AI to be deployed responsibly at enterprise scale.

None of these pillars work effectively in isolation. The most sophisticated agent development tooling produces unreliable results without a strong data context layer beneath it, the richest data foundation is wasted if it cannot be accessed by well-governed agents, and governance frameworks without development velocity create compliance theater rather than operational AI. It is the combination that makes the platform.

Before examining each pillar in detail, it is worth understanding the overall architectural model SAP has published for SAP Business AI. SAP describes its *North Star architecture* for the Autonomous Enterprise as a four-layer system. At the top is the user experience layer, where Joule, SAP's AI copilot, serves as the primary interaction point for end users, whether they are navigating standard SAP Fiori applications, operating within SAP Build Work Zone, or using the new Joule Work interface that SAP introduced at SAP Sapphire 2026. Beneath the user experience is the process layer, where SAP S/4HANA, SAP SuccessFactors, SAP Ariba, SAP Customer Experience (SAP CX), and the rest of the SAP Autonomous Suite function as capability providers. These applications expose their data and operations through the APIs, OData services, and event streams that you have explored in previous chapters, and they are increasingly augmented by domain-specific Joule Agents that automate work within their process boundaries. Below the process layer is the foundation layer, which contains the reasoning services, SAP Business Data Cloud with its SAP Knowledge Graph, the trained and third-party models accessible through the generative AI hub, and the long-term memory infrastructure that agents use to accumulate operational knowledge over time. Finally, at the base is the platform layer, comprising the runtime infrastructure, governance services, and security frameworks that run the whole system reliably.

What makes this architecture distinctive relative to generic AI platforms is the second layer from the bottom—that is, the foundation layer—and specifically the role of SAP Business Data Cloud. Most enterprise AI deployments fail not because the AI model is inadequate but because the model lacks the business context it needs to reason correctly. A general-purpose language model asked to assess a customer's payment risk knows nothing about the customer's twelve-month payment history, their current dispute count, the aging breakdown of their open invoices, or the credit risk score that your collections team has been tracking. SAP Business Data Cloud solves this by providing a governed, semantically harmonized layer of enterprise data that carries business meaning: Beyond raw numbers and codes, there are entities with understood relationships, processes with understood semantics, and data with the governance provenance that makes it trustworthy as AI input. SAP Knowledge Graph, which lives within SAP Business Data Cloud, is the mechanism through which this meaning is encoded and made available to the AI agents that need to reason about it.

7.2 Build with SAP BTP and Joule Studio

The build pillar of SAP Business AI Platform consists of SAP Business Technology Platform (SAP BTP) and Joule Studio. Within SAP BTP, the key tool to discuss here is SAP Integration Suite. We'll discuss these solutions in more detail in the following sections.

7.2.1 Joule Studio

If the generative AI hub and SAP Business Data Cloud form the data and model foundation of SAP Business AI Platform, then Joule Studio is the construction layer, the development environment in which AI agents are built, tested, and deployed. Joule Studio was announced at SAP Connect in October 2025 and became generally available in December 2025. It is the first comprehensive, enterprise-grade agentic development environment that is purpose-built for the SAP ecosystem, and it represents a significant departure from the way custom AI capabilities have been built on SAP BTP historically.

Joule Studio is available within SAP Build, which means it shares the same platform environment as SAP Build Code, SAP Build Process Automation, and SAP Build Apps. This colocation is architecturally intentional. When you build a Joule Agent in Joule Studio, the skills that agent executes can call SAP Build Process Automation workflows, consume SAP Cloud Application Programming Model services built in SAP Build Code, and trigger low-code automations built in SAP Build Apps all within the same SAP BTP subaccount, using the same destination and authentication infrastructure. The agent does not need to build its own integration layer to reach these capabilities; it reaches them through the platform services that SAP Build provides as a unified environment.

Joule Studio supports both no-code and pro-code development paths, which reflects the reality that different types of AI agents require different levels of technical sophistication to build. A business analyst who wants to create a simple retrieval agent that answers questions about company policy documents by querying a document store connected to the SAP HANA Cloud vector engine can build that agent in Joule Studio using a visual, no-code interface to describe the agent's purpose in natural language, select the data sources it should query, and configure the response format, all without writing any code. A development team building a complex, multiagent system that orchestrates actions across SAP S/4HANA, SAP Build Process Automation, and SAP CX, like the Joule Collections Agent, works in Joule Studio's pro-code environment, which supports LangChain, LangGraph, CrewAI, and other leading agentic frameworks. The same studio environment hosts both development modes, and agents built through either path are deployed and governed through the same runtime and monitoring infrastructure.

For development teams that prefer to work in their own IDEs, Joule Studio provides a VS Code extension and a command-line interface (CLI) that integrate Joule Studio's capabilities, like project scaffolding, runtime connectivity, and agent testing, directly into Visual Studio Code, Cursor, or Windsurf. The CLI supports continuous integration/continuous delivery (CI/CD), allowing agent configurations and skill definitions to be version-controlled and deployed through the same automated pipelines that manage your other SAP BTP development artifacts in SAP Cloud Transport Management. As of SAP Sapphire 2026, design-time access to Joule Studio is available at no additional cost through the end of 2026 for companies running SAP and SAP partners, lowering the barrier to beginning agentic development on the platform.

7.2.2 SAP Integration Suite

The role of SAP Integration Suite in SAP Business AI Platform extends well beyond its traditional function as a process integration platform. As the enterprise's central hub for API management and event-driven integration, SAP Integration Suite provides the connectivity layer that allows AI agents to act upon backend systems safely, at scale, and in accordance with your organization's integration governance standards. In the context of the AI platform, there are three specific SAP Integration Suite capabilities that deserve attention.

The first is the Model Context Protocol (MCP) Gateway, which SAP announced at SAP TechEd 2025. MCP is the emerging open standard for how AI agents discover and interact with external tools and data sources. The MCP Gateway in SAP Integration Suite exposes your enterprise APIs, both SAP standard APIs and custom integration flows, as MCP tools that any compatible AI agent can discover and invoke. This means that an AI agent built in Joule Studio, or indeed any MCP-compatible AI agent from a third-party vendor, can access the full catalog of APIs managed in your SAP Integration Suite API Management layer as tool definitions, without requiring custom skill code for each API. The governance policies, access controls, and usage monitoring that you have configured in API Management automatically apply to every agent invocation of those APIs through the MCP Gateway, giving you centralized governance of AI-initiated API calls without requiring a separate governance layer for the AI use case.

The second SAP Integration Suite capability relevant to the AI platform is the AI-powered anomaly detection and self-healing capability announced at SAP TechEd 2025. SAP Integration Suite's monitoring infrastructure, which was already capable of tracking message flow health across your entire integration landscape, has been augmented with intelligent pattern recognition that identifies anomalous conditions like rising error rates, unusual message volumes, and unexpected latency patterns and can, in certain scenarios, automatically apply remediation steps without requiring human intervention. For organizations operating AI agents that make high volumes of API calls through SAP Integration Suite, this automated monitoring and healing capability ensures that the integration infrastructure supporting the agents remains healthy without requiring your integration operations team to manually respond to every transient issue.

The third capability is the support for agent-to-agent communication patterns within SAP Integration Suite's event mesh infrastructure. As your AI landscape matures and you begin to deploy multiple agents that need to coordinate with each other—for example, a procurement agent that needs to notify a collections agent when a vendor payment has been made against a disputed invoice—the event mesh in SAP Integration Suite provides the reliable, asynchronous communication backbone for that coordination. Agents publish events to named topics in the event mesh, subscribe to the events they need to consume, and the event mesh handles the delivery, buffering, and replay guarantees that make the coordination reliable even when individual agents are temporarily unavailable. This event-driven approach to agent coordination is the same publish-and-subscribe pattern that connects SAP Autonomous Suite applications, which means your AI agent architecture can be built on the same integration foundations that your application landscape already depends on.

For more information on SAP Integration Suite, see Chapter 11.

7.3 Knowledge Core with SAP Business Data Cloud

SAP Business Data Cloud is the data intelligence platform that serves as the contextual foundation for AI across SAP Autonomous Suite. It represents SAP's answer to one of the most persistent challenges in enterprise AI deployment: The data that AI needs to perform business operations is scattered across dozens of applications, formatted differently in each, carrying technical identifiers that have no meaning outside their source system, and governed by access controls that make it genuinely difficult to bring the right data to the right place at the right time. SAP Business Data Cloud is SAP's systematic solution to all these problems simultaneously.

The architecture of SAP Business Data Cloud is built around the concept of data products. A *data product* in SAP's framework is not simply a query result or a data extract; it is a reusable, semantically annotated, lifecycle-managed dataset that carries its business meaning with it. SAP publishes its own standard data products covering the core business domains of SAP Autonomous Suite: finance and controlling, procurement and sourcing, supply chain and logistics, human resources, customer experience, and sustainability. These standard data products are built on the same business semantics as SAP One Domain Model, the canonical conceptual model that SAP has developed over decades to represent how enterprise business objects relate to each other across application boundaries. When you consume the finance standard data product in SAP Business Data Cloud, you are not working with raw financial accounting document tables. You are working with a semantically structured dataset that expresses relationships between customers, invoices, payments, clearing documents, and cost objects in business language that AI agents can reason about directly.

Beyond the standard data products that SAP delivers, SAP Business Data Cloud provides the Data Product Studio, a visual tooling environment in which your organization's data teams can create and publish their own data products. A custom data product might combine data from SAP S/4HANA's standard finance data product with data from a proprietary credit risk model, an external credit bureau feed, and the custom business objects your key user team has created via the extensibility tools described in Chapter 10. Once published, that custom data product becomes a governed, reusable asset that any AI agent in your landscape can discover and consume, rather than a bespoke integration that a single application team maintains in isolation.

SAP Knowledge Graph is the semantic backbone of SAP Business Data Cloud, and it is the capability that elevates SAP Business Data Cloud from a data-distribution platform into an AI reasoning foundation. A *knowledge graph* represents information not as isolated records but as a network of entities and the named relationships between them. In the SAP context, the entities are business objects (customers, products, vendors, orders, invoices, employees, cost centers, projects) and the relationships are the real-world connections between them: A customer places orders, an order generates invoices, an invoice references products, a product is supplied by a vendor, a vendor is associated with purchasing agreements, and so on through the full web of enterprise business operations.

What SAP Knowledge Graph adds to these relationships is *semantics*: the encoded understanding of what the relationships mean in a business context. SAP has 50 years of enterprise software experience encoded in SAP Knowledge Graph: knowledge about what constitutes a healthy order-to-cash cycle, what patterns in accounts receivable indicate payment risk, what sequences of events in a supply chain indicate a potential disruption, and what combinations of HR metrics suggest workforce planning interventions. When a Joule Agent queries SAP Knowledge Graph as part of an agentic reasoning step, it is not simply retrieving data. It is accessing the accumulated business intelligence of SAP One Domain Model, which gives the agent's responses a level of business accuracy that a generic model reasoning over raw database records cannot match.

As of early 2026, SAP HANA Cloud can automatically generate knowledge graphs from the metadata of connected data sources in a matter of minutes. This is a significant practical improvement from what was previously a weeks-long manual process. An enterprise connecting SAP Business Data Cloud to a new source system can expect SAP Knowledge Graph to incorporate the relationships between that source and the existing enterprise data landscape almost immediately, without requiring a specialized knowledge engineering engagement. SAP HANA Cloud is the underlying database engine that powers SAP Knowledge Graph, and it is the same engine that provides the vector store capabilities that agents use for retrieval-augmented generation (RAG), the in-memory analytical capabilities that underpin SAP Business Data Cloud's domain content, and the agentic memory store where Joule agents accumulate long-term operational context about your business.

SAP's standard *data products* are the prebuilt semantic datasets that SAP delivers as part of SAP Business Data Cloud to give your AI agents immediate access to the key data domains of SAP Autonomous Suite without requiring your data team to build from scratch. SAP committed at SAP Sapphire 2025 to delivering hundreds of standard data products across all SAP Autonomous Suite domains throughout 2025, and by early 2026 that commitment spans finance and controlling, procurement, supply chain, human capital management, customer experience, and sustainability.

Each standard data product carries its business semantics with it. The finance standard data product does more than expose raw posting documents; it expresses the relationships between posting items and the customers, cost objects, and payment terms they reference in a form that allows an AI agent to ask a question like "Which customers in the Western Europe region have invoices more than 60 past due that do not have an active collections case?" and receive a meaningful answer, because the data product has encoded the relationships between the invoicing domain, the collections domain, and the geographic organizational structure in a way that makes that query semantically coherent.

Standard data products are also the input for SAP's domain content, a growing catalog of analytical and operational applications built on SAP Business Data Cloud that provide AI-powered insights across Finance Intelligence, Spend Intelligence, Supply Chain Intelligence, Revenue Intelligence, People Intelligence, and Customer Intelligence. Domain content is effectively prebuilt AI use cases that your organization can activate against your own data products, giving you a starting point for AI value realization that is significantly faster than building analytical AI from scratch.

We'll discuss SAP Business Data Cloud, SAP Knowledge Graph, and its data products in more detail in Chapter 9.

7.4 SAP AI Core and SAP AI Launchpad

SAP AI Core is the managed execution environment within which the generative AI hub runs and within which your organization's custom AI workloads' model training, batch inference, and long-running analytical jobs execute. SAP AI Core is a Kubernetes-based multitenant runtime that abstracts the underlying compute infrastructure from the AI workloads running on it. You interact with SAP AI Core through the AI API, a stable, versioned REST interface that covers all operational concerns: creating resource groups to isolate workloads, submitting training executions, registering model artefacts, creating inference deployments, and reading execution logs and metrics. The AI API is the same interface that Joule Studio, the generative AI hub, and external tooling all use to interact with SAP AI Core, which means that any capability accessible through the SAP AI Launchpad UI is also accessible programmatically through the same API surface.

SAP AI Launchpad is the operational control plane for SAP AI Core, the interface from which your AI platform team manages the full lifecycle of AI workloads in production. Through SAP AI Launchpad, your MLOps engineers can monitor the health of running inference deployments, review the metrics of completed training executions, compare the performance of multiple trained model versions side by side, and promote a selected model version from a staging deployment to the production endpoint with a tracked, auditable promotion action. For organizations running custom AI models on SAP AI Core, SAP AI Launchpad is the operational dashboard from which the AI engineering team manages the model's lifecycle, monitors for performance drift, and triggers retraining when the model's accuracy begins to degrade.

SAP AI Core achieved ISO 42001 AI governance certification in the third quarter of 2025, making it one of the first enterprise AI platforms to achieve formal third-party certification against the international standard for AI management systems. For your organization's risk and compliance teams, this certification provides independent assurance that the AI governance controls embedded in SAP AI Core—access management, data handling, model documentation, and operational monitoring—meet the requirements of the emerging international regulatory framework for enterprise AI.

7.5 The Generative AI Hub

The generative AI hub is the model access layer of SAP Business AI Platform. It is the service hosted within SAP AI Core on SAP BTP, through which every Joule agent, every custom AI application, and every AI-assisted feature in SAP Autonomous Suite accesses the LLMs that provide generative and reasoning capabilities. The generative AI hub's central value proposition is not the breadth of its model catalog, though that catalog has expanded rapidly, but the governance and operational controls it wraps around every model call. When a Joule Agent calls a frontier LLM through the generative AI hub, the call is authenticated against your SAP BTP subaccount, the input is passed through a content safety and data masking filter before it reaches the model provider, and the interaction is logged in SAP BTP's audit service for compliance. Your enterprise data never trains the underlying foundation models, and the model provider never receives data that your governance policies classify as confidential. This boundary, which is technically enforced, not just contractually promised, is what makes it possible for your organization to use the most capable frontier AI models for business-critical scenarios without creating unacceptable data privacy exposure.

As of the first quarter of 2026, the generative AI hub supports an extensive and rapidly evolving catalog of foundation models from multiple providers. On the Anthropic side, Claude Sonnet 4.6 and Claude Opus 4.6 are available, representing the current generation of Anthropic's most capable reasoning models. OpenAI's GPT-5, GPT-5 Mini, and GPT-5 Nano are accessible alongside the reasoning-optimized models in the o-series family. Google's Gemini 3.0 Pro and Gemini 2.5 Flash are available for organizations that prefer Google's model family. Amazon's Bedrock Nova Premier model is accessible for organizations with

AWS-aligned workloads. The hub also supports models from Cohere and Mistral for specific enterprise use cases, and Perplexity's Deep Research capability for scenarios that require synthesizing information across internal enterprise data and public external sources.

In addition to the third-party frontier models, SAP has introduced two proprietary models that are unique to the generative AI hub:

- **SAP-RPT-1**
  SAP-RPT-1 is a relational foundation model specifically designed for reasoning over the structured, tabular business data that makes up most of SAP's data estate. Unlike general-purpose language models that are optimized for text generation, SAP-RPT-1 is trained on the structural patterns of business data across SAP's application domains, making it significantly more capable at tasks like anomaly detection in financial posting patterns, demand forecasting from structured sales history, and classification of business documents. It is available in three sizes, with a smaller, faster variant for high-volume operational tasks and a larger, more accurate variant for complex analytical scenarios, and SAP has released the open-source weights, allowing your data science teams to fine-tune it on your organization's specific data patterns.
- **SAP-ABAP-1**
  SAP-ABAP-1 is a code-generation model trained on more than 250 million lines of ABAP code, providing development assistance for ABAP developers that is significantly more accurate than general-purpose coding models for SAP-specific patterns, APIs, and idioms.

The most practically important capability within the generative AI hub is the *orchestration service*, which allows developers and agent builders to compose multistep AI workflows without writing the coordination logic from scratch. The orchestration service provides a declarative pipeline that chains prompt construction, model invocation, content safety filtering, data masking, and RAG grounding into a single callable endpoint. Instead of building the orchestration logic themselves—calling the vector store, assembling the retrieved context into the prompt, handling model errors, filtering the response for sensitive content—developers declare the pipeline configuration and the orchestration service manages the execution.

This is the capability that makes RAG practical at enterprise scale. When a Joule Agent needs to answer a question that requires access to current enterprise data, the orchestration service accepts the user's query, retrieves the relevant context from the SAP HANA Cloud vector store or SAP Knowledge Graph, assembles the retrieved context into a prompt along with the user's question and any relevant conversation history, calls the selected foundation model, applies the content safety and data masking postprocessing, and returns a structured response. The agent developer configures this pipeline once; the orchestration service executes it reliably for every invocation without the agent needing to manage the complexity of model interaction directly.

The orchestration service also provides a prompt optimizer capability that automatically tests and refines prompt formulations across multiple models to identify the combination of model and prompt structure that produces the best results for a specific use case. For business AI scenarios in which response quality has direct operational consequences, such as a collections agent giving an inaccurate customer risk summary or a procurement agent misidentifying the correct contract terms, this systematic prompt optimization replaces the ad hoc prompt engineering that is often the weakest link in enterprise AI deployments.

7.6 Joule Work

Joule is SAP's AI copilot, the natural language interface through which users across SAP Autonomous Suite interact with the AI capabilities embedded in their applications, and through which Joule Agents perform work on their behalf. Since its introduction, Joule has grown from a feature within individual SAP applications into the unifying user experience layer of the entire SAP Business AI Platform. As of the first quarter of 2026, Joule operates across 35 SAP solutions, with more than 2,500 skills available across more than 30 specialized agents. SAP has delivered more than 400 AI use cases embedded across SAP Autonomous Suite, representing a transformation in the scope of what AI does within SAP applications compared to just two years earlier.

The breadth of Joule's deployment across SAP Autonomous Suite is worth appreciating concretely. In finance, Joule skills cover accounting accruals, cash management, dispute resolution, international trade classification, fixed asset calculations, and e-invoicing error handling. In supply chain, Joule provides a Production Planning and Operations Agent, a Shop Floor Supervisor Agent, and AI-assisted materials requirements analysis. In human resources, Joule assists with performance preparation, skill identification from candidate resumes, and succession planning. In procurement, Joule supports tender analysis, statement of work automation, and trade promotion management. In customer experience, Joule powers digital service agents, shopping assistants, and catalog optimization capabilities. Every one of these is a domain in which repetitive, time-consuming cognitive work that previously required human attention is now being automated, accelerated, or augmented by the AI capabilities embedded in the Joule layer.

Joule Work is SAP's vision for the next evolution of the user interface. It was announced at SAP Sapphire 2026, with early adopter access available at announcement and full desktop availability planned for the second half of 2026. Where today's Joule experience is primarily a chat panel embedded within existing SAP Fiori applications, Joule Work is a purpose-built workspace that inverts the relationship between the user and their applications. In Joule Work, the conversation is the workspace. Instead of navigating to an application and then optionally invoking Joule within it, the user interacts with Joule Work as their primary interface, delegating tasks to AI agents using natural language; the agents then navigate to the appropriate applications, retrieve the relevant data, execute the required actions, and return structured results, all within the conversational workspace, without the user needing to open multiple applications and manually correlate the results.

Joule Work is designed for desktop, mobile, and voice interaction, which means that the same delegated task experience is available to a treasury manager at their workstation, a field sales representative on a mobile device, and an operations supervisor who needs hands-free access in a plant environment. *SAP Domain Model*, the structured schemas that tell Joule how to interpret and respond to domain-specific requests in finance, supply chain, procurement, and other areas, are targeted for general availability in the third quarter of 2026, which will extend Joule Work's ability to handle complex, domain-specific tasks with the accuracy that business-critical operations require.

7.7 SAP AI Agent Hub

SAP AI Agent Hub, managed as a capability within the SAP LeanIX product family, is the governance and discovery layer for AI agents operating across your SAP landscape. Think of it as SAP Business Accelerator Hub for agents: a central catalog in which agents, skills, and the LLMs and MCP servers they depend on are registered, documented, and governed throughout their lifecycle. As of May 2026, SAP AI Agent Hub manages over 100,000 agents across more than 150 companies globally, giving the enterprise an unprecedented view of the scale at which autonomous AI work is now operating.

SAP AI Agent Hub functions as a system of record for all AI actors in your landscape. Every custom Joule Agent built in Joule Studio, every prebuilt Joule Agent that SAP delivers with its SAP Autonomous Suite applications, and every third-party agent from SAP's partner ecosystem is registered in the hub with its documentation, its skill inventory, the data sources it accesses, the APIs it calls, and the SAP BTP authentication scope it requires. This registration makes agent governance tractable at scale: An enterprise running hundreds of agents across dozens of teams cannot govern those agents through manual review processes. SAP AI Agent Hub gives governance teams the visibility to see what agents exist, what they do, what data they access, and whether they are operating within expected parameters without requiring them to inspect each agent's code individually.

SAP AI Agent Hub also provides the *discovery capability* that accelerates agent development. Before a development team builds a new agent for a business capability, they can search the hub to determine whether a suitable agent or a set of reusable skills already exists in the organization's agent catalog. This prevents redundant development and creates the foundation for a composable agent architecture in which new agents are assembled from existing, tested skill components rather than built entirely from scratch. The hub surfaces adoption analytics: which agents are being used most frequently, which skills have high success rates, and which agents are generating user feedback that indicates they need improvement. This gives agent development teams the operational intelligence they need to prioritize their improvement efforts based on real usage data.

SAP's commitment to open agent interoperability is reflected in the hub's support for the Agent2Agent (A2A) protocol, which SAP has aligned with, alongside AWS, Google, Microsoft, and ServiceNow. This means that agents registered in SAP AI Agent Hub can be discovered and invoked by orchestration layers running outside the SAP ecosystem, and external agents registered in compatible hubs can be discovered and called by Joule Agents operating within the SAP landscape. For organizations running multicloud, multivendor AI architectures, this interoperability is what prevents the SAP AI ecosystem from becoming an isolated island; instead, it positions it as a participant in the broader enterprise AI fabric.

7.8 Business Transformation Management

An autonomous enterprise requires more than AI agents that can perform individual tasks. It requires an ongoing understanding of how work actually flows across the organization, where the process inefficiencies and friction points are, and which areas represent the highest-value opportunities for AI-driven improvement. This process intelligence capability is what SAP Signavio and SAP LeanIX provide within SAP Business AI Platform, organized under what SAP calls *Business Transformation Management*.

Business Transformation Management is the umbrella framework that brings together three products: SAP Signavio, which handles process transformation and process mining; SAP LeanIX, which handles enterprise architecture management and SAP AI Agent Hub; and WalkMe, which provides digital adoption tooling that guides users through new processes and capabilities as they are deployed. Together, these three capabilities give your organization the intelligence to understand your current process landscape, identify where AI can create the most value, deploy AI agents to address those opportunities, and then measure whether the agents are delivering the expected outcomes, completing the loop between process design, AI deployment, and continuous improvement.

We'll take our first look at SAP Signavio and SAP LeanIX in the following sections. For detailed guidance, see Chapter 13.

7.8.1 SAP Signavio

SAP Signavio is the process intelligence and transformation platform that serves as the process awareness layer of SAP Business AI Platform. Its core capabilities—process mining, process modeling, process simulation, and process benchmarking—allow your organization to build a data-driven picture of how your business processes actually operate, as opposed to how they are documented to operate. Process mining in SAP Signavio reads the event logs of your SAP and non-SAP systems and constructs an empirical model of how process steps are sequenced, how long each step takes, where rework and exceptions occur, and how the actual process variants in use across your organization compare to the intended process design.

In the context of SAP Business AI Platform, this process intelligence capability takes on a new significance. Before deploying AI agents to automate a business process, you need to understand the process well enough to know what the agent should do, in what order, and how it should handle exceptions. SAP Signavio's process-mining output provides precisely this foundation: a data-driven map of the process as it actually operates, with the variant frequency, cycle time, and exception rate that allow an AI agent designer to understand which process paths account for the majority of cases and which represent the edge conditions that require special handling. SAP Signavio's AI-enabled Transformation Advisor, which became available at SAP TechEd 2025, allows process analysts to query this process intelligence using natural language, asking questions like "Where in our procure-to-pay process are we losing the most time compared to SAP's best-practice benchmark?" and receiving a structured analysis with recommended next steps, including recommendations for specific Joule Agents that could address identified inefficiencies.

As of the first quarter of 2026, five dedicated Joule agents are generally available within the SAP Signavio product: the Screen Guide Agent, the Value Case Creation Agent, the Dashboard Analyzer Agent, the Process Content Recommender Agent, and the Workspace Administration Agent. These agents bring the Joule conversational interface to the process intelligence work itself. A process analyst can ask the Dashboard Analyzer Agent to summarize the key findings from a process analysis without manually reviewing every metric, or ask the Process Content Recommender Agent to identify which SAP standard process content is most relevant to the process variant they are modeling. Joule became generally available within SAP Signavio in February 2026, representing the completion of the integration between SAP's process intelligence platform and its AI copilot.

The Agent Mining capability within SAP Signavio represents a particularly forward-looking element of the platform. Agent Mining analyzes how AI agents are contributing to process performance as they operate in production tracking: not just whether agents are executing correctly but whether their execution is actually improving the process outcomes that the original business case projected. For a collections agent, Agent Mining might track the correlation between agent invocations and collections case resolution rates, showing whether the accounts that the Joule Collections Agent works on are being resolved faster than those managed without agent assistance. This closes the feedback loop between AI deployment and business outcome measurement that many enterprise AI programs struggle to establish.

7.8.2 SAP LeanIX and Enterprise Architecture Management

SAP LeanIX provides the enterprise architecture dimension of Business Transformation Management. Where SAP Signavio gives you a process-level view of your operations, SAP LeanIX gives you the architectural view: a map of the applications, capabilities, and technology assets that support your business, and the relationships between them. In the context of the AI platform, SAP LeanIX's role extends naturally into the architecture of your AI landscape. SAP AI Agent Hub, described earlier in this chapter, is an SAP LeanIX capability;

it is the enterprise architecture catalog applied to AI agents rather than application systems. Just as SAP LeanIX helps your enterprise architects understand which applications support which business capabilities and what the dependencies between them are, SAP AI Agent Hub helps your AI governance team understand which agents support which business processes, what data and APIs they depend on, and what the risk profile of each agent is.

SAP LeanIX launched its own MCP server in November 2025, connecting AI models and agents to the enterprise architecture data that SAP LeanIX manages. This means that a Joule Agent or a custom AI application can query the SAP LeanIX application portfolio to understand which SAP and non-SAP systems are involved in a particular business process, what the integration dependencies between those systems are, and whether any of the systems in the landscape have been flagged as technically outdated or scheduled for decommissioning. For IT planning and transformation scenarios, this capability gives AI the architectural awareness it needs to make recommendations that are consistent with your organization's technology roadmap rather than simply optimizing for immediate process efficiency without regard for underlying technical constraints.

7.9 Governing AI at Enterprise Scale

The three pillars of SAP Business AI Platform, building AI experiences, contextualizing AI reasoning, and governing AI, are genuinely coequal. Organizations that invest heavily in the first two pillars but treat governance as an afterthought will eventually encounter the consequences: an agent that acts on inaccurate data, a model whose performance has drifted undetected, a skill that executes actions users did not intend, or an audit finding that the organization cannot demonstrate what its AI agents are doing and why. SAP has made governance a platform-level concern rather than leaving it to individual application teams, which means the governance capabilities are available from the first agent you deploy rather than something you need to build after a governance failure has occurred.

The governance layer of SAP Business AI Platform operates at several levels. At the model level, the generative AI hub's content safety filtering, data masking, and model-level access controls ensure that every model call is governed regardless of which application or agent initiates it. At the agent level, SAP AI Agent Hub provides the system of record, the access control enforcement, and the adoption analytics that give your AI governance team continuous visibility into what agents are operating and how. At the process level, SAP Signavio's agent-mining capability tracks whether deployed agents are delivering the business outcomes they were designed to produce. Finally, at the compliance level, SAP AI Core's ISO 42001 certification and SAP BTP's audit log provide the independent assurance and audit trail that your risk and compliance functions require.

The human-in-the-loop principle is also embedded throughout the platform's design. SAP AI is explicit that the Autonomous Suite is not designed to remove humans from consequential decisions. Instead, it is designed to remove humans from the routine, repetitive execution steps that consume time without requiring human judgment, while preserving human oversight for the decisions that carry business, legal, or ethical weight. The Joule Collections Agent's TriggerDunningAction skill validates prerequisites and logs every action, but the collections manager retains the ability to review and override any action the agent takes. The Autonomous Close Assistant automates routine reconciliation steps but escalates exceptions to the finance team for review. This design principle, *automate the mechanics, but preserve the judgment*, is what allows AI to take on increasing responsibility over time as your organization builds confidence in its reliability, rather than requiring a binary choice between full automation and no automation.

For a deep dive into AI governance and responsible AI, see Chapter 14.

7.10 Summary

SAP Business AI represents SAP's answer to the fundamental challenge of enterprise AI: not how to access a capable AI model, but how to deploy AI that is grounded in the right business context, governed by the right controls, and connected to the right processes to produce outcomes that actually improve how your organization operates. The platform's three strategic pillars, building AI experiences, contextualizing reasoning with enterprise data, and optimizing and governing AI at scale, are designed to work together as a system, and it is the integration between them that creates value that none of them could deliver independently.

SAP Business Data Cloud and SAP Knowledge Graph provide the semantic data foundation that gives every AI agent operating in your landscape access to business-contextual, semantically harmonized enterprise data, the raw material that makes the difference between a generic AI response and an accurate, actionable business recommendation. Standard data products accelerate time to value by delivering prebuilt, semantically rich datasets across the core SAP Autonomous Suite domains, while the Data Product Studio allows your data teams to create and publish custom data products that extend the platform to your organization's specific data landscape.

The generative AI hub, running on SAP AI Core, provides governed access to the broadest catalog of frontier AI models in the SAP ecosystem, including SAP's own purpose-built models for tabular business data and ABAP code generation. The orchestration service within the hub makes RAG and multistep AI pipelines practical at enterprise scale by managing the complexity of model interaction behind a declarative configuration interface. SAP AI Launchpad provides the operational visibility and control that MLOps teams need to run custom AI workloads reliably in production.

Joule Studio gives development teams a comprehensive, framework-agnostic environment for building the Joule Skills and Joule Agents that bring AI capabilities to life in your business processes, while SAP AI Agent Hub provides the governance and discovery layer that makes a growing catalog of agents manageable at enterprise scale. Joule Work introduces a new mode of enterprise work in which the user delegates tasks to AI agents using natural language and the agents handle the application navigation, data assembly, and action execution, compressing the time and effort of routine business operations dramatically.

SAP Signavio and the Business Transformation Management suite provide the process intelligence layer that makes AI deployment strategic rather than ad hoc, identifying where in your process landscape AI can create the most value, measuring whether deployed agents are delivering the expected outcomes, and governing the growing population of agents through the SAP AI Agent Hub. SAP Integration Suite, through the MCP Gateway and the event mesh infrastructure, provides the connectivity backbone that allows AI agents to act on backend systems safely, at governed scale, using the same API management and integration governance standards that your organization has established for all application-to-application communication.

Together, these capabilities constitute a platform designed not just for the AI use cases of today but for the Autonomous Enterprise that SAP is building toward, an enterprise in which AI agents are active operational participants, humans focus on the judgment and creativity that genuinely requires human intelligence, and the organization as a whole operates with a speed, accuracy, and consistency that was not achievable through human effort alone.

In the next chapter, we'll dive deeper into SAP BTP and walk through the various capabilities in SAP Build.

Chapter 8
SAP Business Technology Platform

You've taken your first look at SAP Business AI Platform. Now, let's dig deeper into SAP Business Technology Platform, the foundation for the Autonomous Enterprise.

SAP Business Technology Platform (SAP BTP) is a platform-as-a-service (PaaS) offering from SAP that acts as the glue for all SAP Autonomous Suite applications. SAP BTP provides integration services, extension services, artificial intelligence (AI) services, and data storage services to safely customize and integrate your back-office applications. SAP BTP plays a key role in SAP's clean core strategy, providing a safe, decoupled environment that can access SAP Autonomous Suite applications through standard APIs and events. This provides an environment in which you can customize applications, workflows, and processes without building technical debt inside a system that would delay upgrades or add refactoring work later.

Customizations built on SAP BTP can be removed easily if features built on SAP BTP end up overlapping with new functionality. SAP releases allow you to take an easier fit-to-standard approach with your applications and maximize the value of your SAP Autonomous Suite investment. SAP BTP also provides integration tooling that contains a large repository of prepackaged integrations between SAP Autonomous Suite applications and non-SAP applications, allowing you to easily and seamlessly integrate your applications. We'll discuss extensibility and integration in more detail in Chapter 10 and Chapter 11, respectively.

SAP BTP is also the environment in which SAP provides custom AI services and tooling to embed AI in your applications and processes. In SAP BTP, you also can extend the functionality of the standard Joule agents released by SAP and create agent-to-agent scenarios so that Joule can communicate with other, third-party agents. Finally, SAP BTP offers SAP Joule for Developers, an embedded AI assistant to help developers with creating customizations and integrations for their SAP Autonomous Suite and third-party applications.

This chapter introduces SAP BTP as the foundation for SAP Autonomous Suite, enabling a clean core architecture and seamless extensibility. It explains how SAP BTP underpins application integration, data management, and AI enablement across the suite. You'll learn about SAP Build, runtime choices for flexibility, and extensibility patterns that preserve system integrity while accelerating innovation. We'll also explore platform capabilities, including integration services, event-driven architecture, and lifecycle management, and highlight ABAP AI updates that bring intelligence into custom development. By the end, you'll understand why SAP BTP is the strategic anchor for agility, scalability, and continuous innovation in the intelligent enterprise.

8.1 Clean Core

SAP's clean core strategy is a multipillared framework to guide companies toward a modern governance framework that allows them to realize the full value of their ERP system. SAP provides the clean core framework to help companies maximize their investment in their SAP Autonomous Suite systems, adopt the latest features and enhancements, and reduce the level of effort required during a system upgrade.

The core principles of clean core focus on building resilient business processes, upgrade-safe extensions, clean and trusted data, and seamless integration, leading to more efficient operations. A major emphasis of clean core is to adopt standard processes and extensions just for business, differentiating processes and applications. SAP's clean core strategy does not mean a company cannot create extensions in its SAP Autonomous Suite systems. However, the strategy means that when an extension is created, it is done in a framework that does not cause additional overhead work during an update or exclude companies adopting newly released functionality.

SAP defines five clean core pillars to help you achieve its clean core standard, acting as an easy guide to help you better realize the value from your SAP systems while staying agile and innovative:

- **Processes**
  Measures current process alignment to SAP standard and performs fit-gap analysis on custom process extensions. The goal is to adopt SAP Best Practice processes out of the box and reserve customization only for genuinely differentiating business processes, avoiding bespoke logic that adds upgrade overhead without competitive value.
- **Extensibility**
  Tracks custom code quality via the A–D grading scale, ABAP Test Cockpit (ATC) findings, percentage of unused objects, and the business impact of modifications. New extensions must use the ABAP RESTful application programming model, key-user extensibility, or SAP BTP side-by-side patterns to remain in Tier A, the highest level of clean core adherence.
- **Data**
  Ensures data quality, completeness, and governance across SAP Autonomous Suite. Clean, well-governed master and transactional data is a prerequisite for accurate AI outcomes, reliable integration, and confident business decision-making. Poor data quality is one of the leading causes of failed ERP modernization programs.
- **Integration**
  Assesses upgrade stability of integration technology, end-to-end monitoring coverage, interface documentation, and adoption of modern APIs (OData, events). Legacy remote function call (RFC) and IDoc interfaces should be wrapped or replaced with released APIs to remove integration as a source of upgrade risk and operational fragility.

- **Operations**
  Covers the operational health of the SAP Autonomous Suite landscape: system monitoring, lifecycle management, transport hygiene, and the adoption of SAP Cloud ALM for change and operations management. Clean operations reduce the manual effort of running the system and create the stable platform on which the Autonomous Enterprise is built.

We'll dive into the clean core in more detail in the following sections, including its grading system and how it applies to SAP's extensibility frameworks.

8.1.1 Grading System

SAP has delivered a methodology for measuring clean core maturity to help companies adopt a governance model for the highest clean core level. The governance and maturity approach determines a current clean core score and a target score based on the company's goals and ambitions with SAP Autonomous Suite. Maturity assessments are done across the pillars of clean core.

SAP BTP provides tooling to help companies execute their clean core strategy. It helps primarily across the process, extensibility, integration, and operations pillars. In the process pillar, clean core looks at current process alignment to SAP standard and performs a fit-gap analysis to see how custom process extensions are built and where they deviate from standard. Integration looks at upgrade stability with the usage of clean integration technology for data integration and process integration, end-to-end integration monitoring coverage, interface documentation, and adoption of modern integration technologies. Extensibility stretches across many of the clean core verticals, with its metrics coming from the code line supporting SAP Autonomous Suite. This area looks at the level-grading of the custom code inside of SAP Autonomous Suite, a weighted score of priority findings surfaced from ATC, the percentage of unused custom development objects, and the impact of business modification.

Table 8.1 shows SAP's clean core grading system, with Tiers A and B being where your organization should strive to move its development toward.

Tier	What?	Why?	Upgrade Risk	Goal
A	**Highest clean core adherence** ABAP RESTful application programming model, key-user extensibility tools, and SAP BTP side-by-side extensions that communicate with SAP S/4HANA exclusively through released OData APIs and standard events.	These extensions do not touch internal SAP objects directly. A SAP BTP side-by-side extension lives entirely outside the SAP S/4HANA system boundary and can be removed or replaced with zero impact on the core.	*None:* Upgrade-safe by design.	All new development should target Tier A.

Table 8.1: SAP's Clean Core Tier-Grading System

Tier	What?	Why?	Upgrade Risk	Goal
B	**Documented upgrade-safe APIs** Classic SAP APIs and extension points that predate the ABAP RESTful application programming model framework but are explicitly documented by SAP as upgrade-safe. Examples include business add-ins (BAdIs) and enhancement spots implemented via the classic ABAP Workbench.	While not aligned with modern ABAP RESTful application programming model conventions, these extension points are part of SAP's published API contract and will be maintained across system updates. Risk is low, but higher than that of Tier A.	*Low:* SAP-documented as stable.	Acceptable for legacy extensions not yet refactored to the ABAP RESTful application programming model.
C	**Legacy frameworks: Upgrade risk** Older SAP development frameworks that are not aligned with clean core standards and are not guaranteed to survive a system upgrade without manual remediation. Classic module pool programs, older function module exits, and nonreleased enhancements fall into this category.	These objects rely on SAP internal structures that can change between releases. Each system upgrade demands regression testing and potential rework, adding cost and slowing down the adoption of new SAP functionality.	*Medium-high:* Requires testing on every upgrade.	Existing Tier C objects should be systematically refactored toward Tier A or B.
D	**Core modifications: Avoid** Hard modifications to SAP core code, direct writes to SAP database tables bypassing the application layer, and the use of unsupported or undocumented extension points. These are changes made inside SAP's own namespace.	When SAP delivers an update, Tier D modifications conflict directly with the changed code or data structures. Every upgrade becomes a high-risk, high-effort remediation project. These objects are the primary source of the technical debt that makes long-running SAP landscapes expensive to maintain.	*Critical:* Will cause high-impact issues on upgrade.	Tier D objects must be eliminated. No new Tier D development should ever be created.

Table 8.1: SAP's Clean Core Tier-Grading System (Cont.)

To better understand the clean core metrics, it's important to understand the custom code grading that feeds the extension grading metrics. SAP provides an A through D grading of custom objects within the system:

- *Tier A* represents the highest level of adherence to clean core. It uses the modern ABAP RESTful application programing model and key-user extensibility to make enhancements in an SAP S/4HANA system, and it also uses SAP BTP to safely extend the core ERP platform. Companies on SAP S/4HANA Cloud Public Edition are locked in to only leveraging Tier A development.
- *Tier B* is the second-highest level for clean core. These are development artifacts that use classic SAP APIs and extension points that do not align with the modern ABAP RESTful application programming model framework but are deemed upgrade-safe based on SAP S/4HANA's roadmap.
- *Tier C* has conditionally clean objects and represents priority objects for refactoring. This tier uses legacy SAP development frameworks that will introduce risk during a system upgrade.
- *Tier D* represents high-risk objects that should be prioritized as they will cause high impact during a system upgrade. These customizations use outdated technology, unsupported extension points, hard modifications in the core code, and direct table modifications that are unstable.

SAP provides the clean core framework to help companies embrace the latest technology from SAP and realize the full value of their investment in SAP Autonomous Suite. The pillars of clean core are in place to help guide companies to a modern governance model for transformation with SAP Autonomous Suite. SAP Business Technology platform provides a safe extension and integration layer to add flexibility to extend applications and processes without creating technical debt in their core systems.

8.1.2 In App Versus Side by Side

Many companies running an SAP S/4HANA system will still need to customize the content inside their system beyond what is available from the standard delivered business content. When a company identifies a need for customization, it will need to decide how that customization is built in adherence to SAP's clean core principles and where to deploy that customization: either on stack in SAP S/4HANA or side by side in SAP BTP.

Figure 8.1 shows the extensibility model in alignment with the clean core grading system. You can also see the in-app extensibility tools: key-user extensibility and developer extensibility. Sitting side by side with SAP S/4HANA is SAP BTP, offering tooling to help you extend SAP S/4HANA safely while meeting the highest standard of clean core.

For on-stack customizations, SAP provides two frameworks for handling extensions: key-user extensibility and developer extensibility. Key-user extensibility takes a low-code configuration approach to creating modifications of an SAP S/4HANA system.

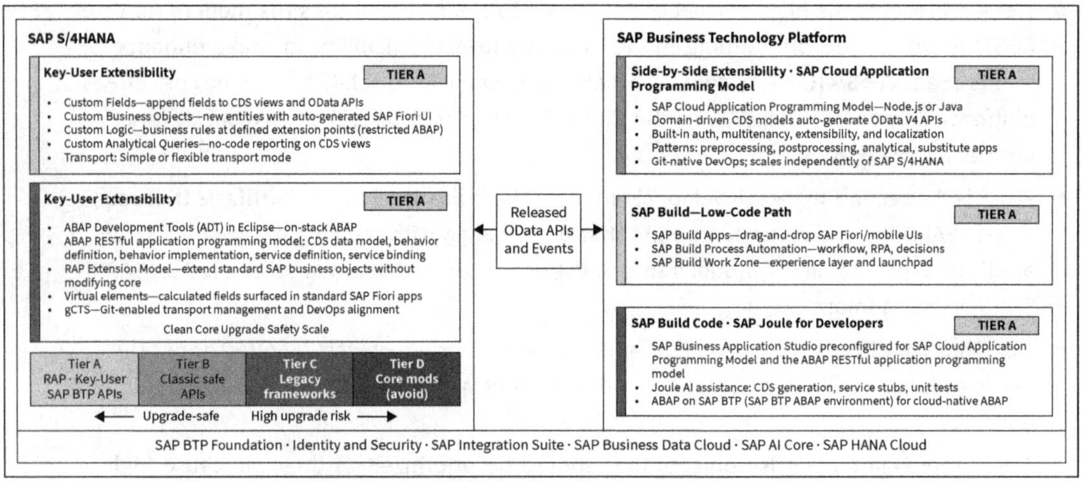

Figure 8.1: Extensibility Model for SAP S/4HANA

Key-user extensibility provides a catalog of extensibility applications to help customize applications and their user interfaces, reports, email templates, and form templates. Using key-user extensibility, you can also create database tables for segmentation and design queries. Key-user extensibility provides a framework for creating simple extensions more quickly than with developer extensibility tooling. Assets created via key-user extensibility adhere to Tier A, the highest level of SAP's clean core standard. Common scenarios include adding custom fields, creating custom business objects, creating custom objects, and implementing custom logic in defined business extension points. Key-user extensibility is granted to users via a role collection, and a dedicated namespace is created for objects, code, and fields made via this functionality. For users to leverage key-user extensibility on a standard object in SAP S/4HANA, the object must first be activated; then it can be safely extended via the released local API point within the SAP S/4HANA system.

With the custom fields feature, key users can create, edit, and delete custom fields, which then can be appended to core data services (CDS) views and their respective backend APIs. The fields can be used in reports, forms, processes, and search and appended to the user interface of standard SAP S/4HANA applications. Custom business objects can be created simply, and the respective tables and CDS views will be generated for the user. From a custom business object, you can create SAP Fiori user interfaces for reports and data entry. Custom business logic can be implemented for custom fields and objects or inserted into standard BAdIs. Custom code written via the key user custom business logic application must adhere to the Tier A standard for custom code and will not be able to be published if it uses classic ABAP instead of ABAP RESTful application programming model–supported Tier A framework.

Key-user extensions can be transported between systems via either simple transport mode or flexible transport mode. *Simple* transport mode offers basic functionality for transporting by supporting only one business configuration transport; transport types are configured automatically, but there is the option for individual instant transports as well. *Flexible* transport mode offers fully flexible transports, with the possibility to create multiple business configuration transports and the addition of being able to create transport automations. ATC checks are available for artifacts created via key-user extensibility, and extensibility templates can be created for key-user extensibility artifacts.

Figure 8.2 outlines the steps for transporting key-user extensibility artifacts between environments in SAP S/4HANA Cloud, and it shows the options for simple transport mode and flexible transport mode to migrate artifacts safely from development to production. For more information on key-user extensibility, see Chapter 10, Section 10.1.

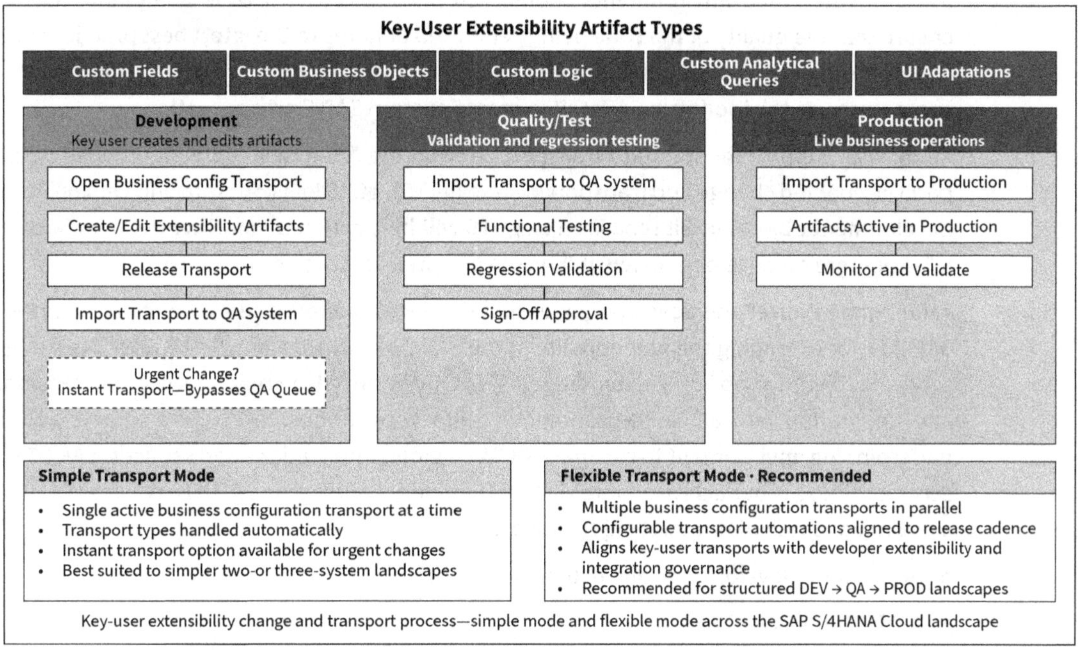

Figure 8.2: SAP S/4HANA Key User Extensibility: Change and Transport Process

When the need for an extension goes beyond what is possible in key-user extensibility, SAP offers *developer extensibility* for on-stack development of the SAP S/4HANA system. SAP's environment for developer extensibility is delivered via the ABAP development tools inside Eclipse. Developer extensibility in the ABAP development tools allows a developer to display and edit multiple ABAP objects in parallel, and it includes features for syntax checks, code completion, pretty printing, forward navigation, searches, and implementation of quick fixes.

Developer extensibility in the ABAP development tools should always adhere to the ABAP RESTful application programming model development framework. The ABAP RESTful application programming model framework was created to help developers efficiently create SAP Fiori applications and their respective backend data models, business logic, and APIs. Companies running the latest editions of SAP S/4HANA Cloud Private Edition, SAP

S/4HANA Cloud Public Edition, or SAP BTP's ABAP environment get access to SAP Joule for Developers to help them in creating and testing development objects in the ABAP development tools environment. ABAP RESTful application programming model and SAP Joule for Developers will be discussed in further detail in later sections of this chapter.

In SAP S/4HANA Cloud Private Edition environments, companies can create both classic and ABAP RESTful application programming model development objects in ABAP. To reduce technical debt and to safeguard their systems against refactoring work during future upgrades, it is important that companies adhere to the modern cloud development guidelines when using developer extensibility in the ABAP development tools. Companies using private cloud environments should run periodic custom code analysis checks to ensure that the quality of the code being generated adheres to the latest best practice standards from SAP. For more robust SAP Fiori application development, we recommend that companies use the tooling in SAP BTP to create custom SAP Fiori applications.

Along with support for standard transport control, the ABAP development tools also support Git-enabled change and transport system (gCTS). gCTS lets you store and version ABAP development objects in Git repositories. gCTS will be discussed in further detail in Section 8.5. For more on developer extensibility, see Chapter 10, Section 10.2.

Aside from key user and developer extensibility, SAP also offers *side-by-side extensibility* via SAP BTP for extending the functionality of the SAP S/4HANA system. SAP BTP offers a large service catalog for not only extending SAP S/4HANA but also creating system extensions and integration into all SAP Autonomous Suite applications. SAP BTP also acts as the underpinning environment to run many of SAP's cloud applications and services. SAP BTP's service catalog is publicly available via SAP Discovery Center. The SAP BTP services fall into a few major categories, including the following:

- Application development and extension
- Systems integration
- Identity and authentication
- Artificial intelligence
- Data storage and analytics

SAP BTP runs side-by-side SAP applications in hyperscaler data centers, allowing companies to choose where they want to deploy their services on SAP BTP. SAP has partnerships with Amazon Web Services, Microsoft Azure, Google Cloud, and Alibaba Cloud, so companies can choose a data center near where their business suite applications are deployed in order to closely geolocate their SAP BTP services and their core applications.

SAP BTP services communicate with SAP Autonomous Suite solutions through dedicated API endpoints, so custom logic lives outside the core system, side by side on SAP BTP. Building outside of core SAP S/4HANA allows companies to gain flexibility in runtime choices for building their extension and scalability, with options to scale up or down the resources needed to run a side-by-side extension on SAP BTP. SAP BTP also makes it easy to have multiple call outs to SAP or non-SAP systems to retrieve or send data based on the requirement of the extension. Loosely coupled extensions make it easier to remove customizations that overlap with standard functionality in future system releases and to embrace a fit-to-standard approach, thus maximizing the investment value in SAP Autonomous Suite. Asynchronous extension scenarios are often the best candidates to be developed side by side on SAP BTP. Scenarios include the following:

- Proxy applications
- Preprocessing/postprocessing applications
- Substitute applications
- Analytical applications

For more on side-by-side extensibility, see Chapter 10, Section 10.3.

8.2 Accounts and Runtime Environments

SAP BTP is organized via *global accounts* and *subaccounts* for service deployment to allow companies to easily monitor and govern activity within their SAP BTP environment. Traditionally, companies running SAP BTP will have at least one global account assigned to their organization.

Within your global account, you can create directories and subaccounts. *Directories* provide a semantic layer above subaccounts for you to group subaccounts into logical buckets. Subaccounts are always tied to a hyperscaler data center and are the environments in which SAP BTP services are provisioned and run. For example, imagine that Acme Global has multiple divisions running SAP Autonomous Suite and is using SAP BTP. It would have one global account for Acme Global, then could have multiple directories for divisions like Acme Medical, Acme Research, and so on. Within each directory, Acme Medical and Acme Research could each have three subaccounts; Acme Medical could have a development subaccount, test subaccount, and production subaccount, while Acme Research could have development, test, preproduction, and production subaccounts.

Figure 8.3 shows an example SAP BTP account structure for Acme Global. In its environment, there are two dedicated directories for Acme Medical and Acme Research so that the two organizational entities can have access to a single governed SAP BTP environment with their own dedicated spaces. Below their environments are the shared SAP BTP resources that both teams leverage.

You will always have at least one global account assigned to your organization, but can have as many subaccounts and directories as best fits your organization. Directories can be layered to support subdirectories as well. Users can be added to an SAP BTP environment at the global account, directory, or subaccount level. Users at a global account level can view or manage directories and subaccounts based on their roles, and they can also manage the entitlements that a directory or subaccount is granted. *Entitlements* set the services and service quantities that can be consumed within a directory and its subaccounts. Users at a directory level can be granted access to subaccounts and the entitlements each subaccount has, while users at a subaccount level can view, deploy, and develop in their respective subaccounts based on their roles and the service entitlements granted at the global account or subaccount level. Business users can be granted access to applications that are deployed on SAP BTP; for example, end users of a custom application built on SAP BTP would be added the role to access the application running on SAP BTP.

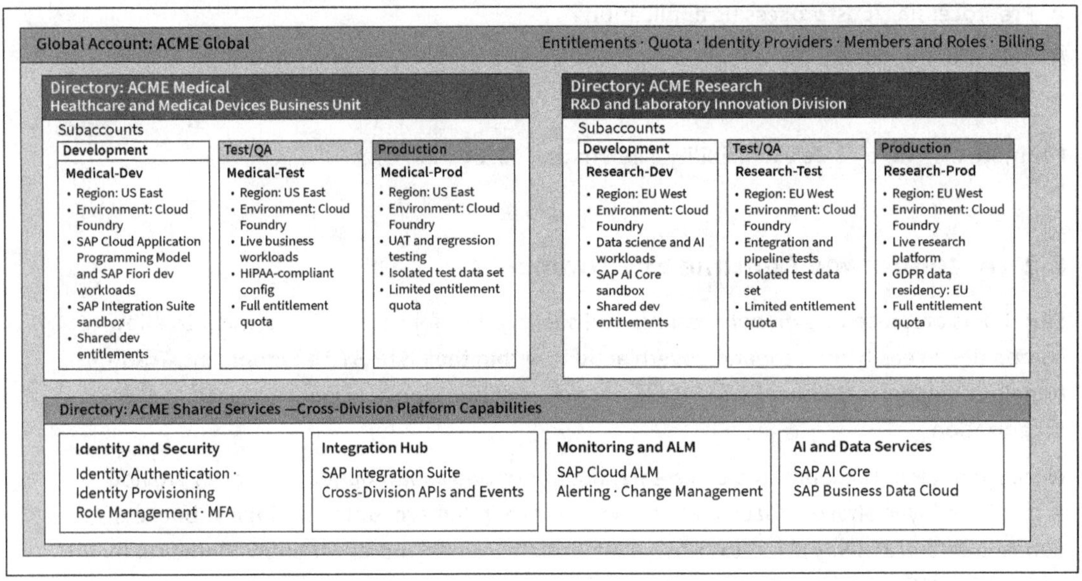

Figure 8.3: SAP BTP Account Structure Example

Figure 8.4 shows an example of how SAP BTP subaccounts can be nested within directories. Users can choose the environment inside SAP BTP that best fits their organization's needs and processes for development.

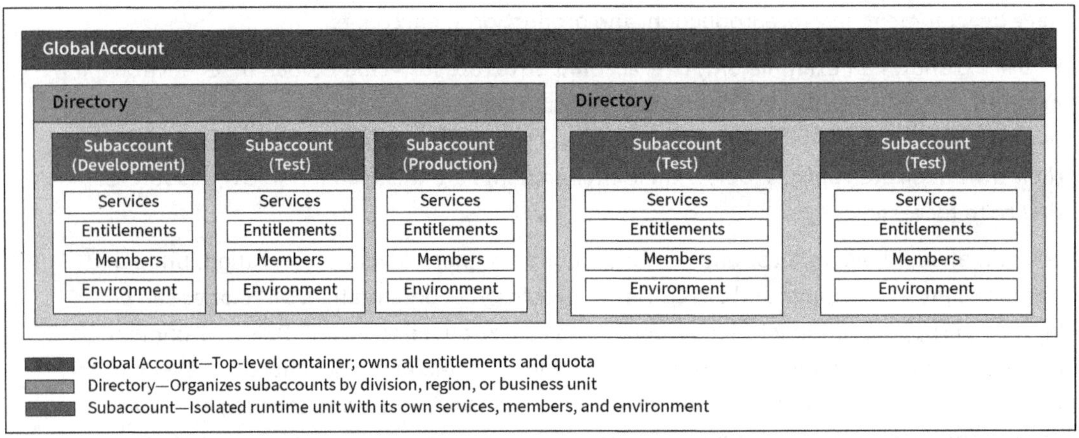

Figure 8.4: SAP BTP Account Hierarchy

SAP BTP offers three major runtime environments to choose from:

- **Cloud Foundry**
  The Cloud Foundry environment is the largest ecosystem for services on SAP BTP. Cloud Foundry is used as the backing runtime for many SAP BTP services, such as SAP Integration Suite, SAP Build, SAP HANA Cloud, and others. SAP's Cloud Foundry environment is based on the open-source Cloud Foundry Foundation application platform. If you use the service marketplace in the SAP BTP cockpit, you're leveraging Cloud Foundry behind the scenes when you create services or deploy applications. You can also use the Cloud Foundry command-line interface (CLI) to manage services within the Cloud Foundry runtime on SAP BTP.
- **ABAP**
  The ABAP environment provides a cloud ABAP runtime for customers to create extensions for ABAP-based products such as SAP S/4HANA Cloud. The ABAP environment can be created through SAP Build, though the developer environment is still delivered via ABAP development tools in Eclipse. A remote connection is created via ABAP development tools to link to the ABAP environment in SAP BTP. The ABAP environment comes bundled with a dedicated SAP HANA database to provide a layer of persistence, and the database is linked one-to-one with the ABAP runtime. The ABAP environment also includes a standard SAP Fiori launchpad with applications for user management, custom code checks, system management, and monitoring.
- **Kyma**
  Kyma is a fully managed, cloud-native, Kubernetes-based runtime inside SAP BTP. Kyma permits native consumption of SAP and non-SAP applications as well as supporting frameworks for cloud-developed SAP applications. Kyma implements a dedicated application runtime to deploy highly scalable containerized microservices. A Kyma environment runs on one or more Kubernetes clusters created in a specific subaccount. Kyma services can be bound to other services, so applications such as an SAP Application can call a Kyma function or microservice. Kyma also provides its own CLI to help developers streamline their activities so they can easily push source code or docker files to Kyma's in-cluster registry, eliminating manual setup processes.

8.3 SAP Build

SAP Build on SAP BTP provides an end-to-end environment to easily extend SAP Autonomous Suite solutions. SAP Build comes with three major styles of tooling: low-code tooling, pro-code tooling, and runtime services. You can use SAP Build to create custom extensions, applications, automated workflows, Joule Skills, and business sites. The tool is designed to cater to a spectrum of developers, ranging from citizen developers to seasoned application developers. This allows for better collaboration between IT development teams and the business units they build applications for, fostering a creative development environment to solve unique business problems for an organization. Users can onboard SAP Build within the subaccounts of their SAP BTP environment; SAP Build developer users get access to SAP Build Lobby, which provides templates and prebuilt scenarios to speed up development time. SAP Build Lobby also allows developers to use a *develop once, use multiple times* approach to reduce rework on future projects. SAP Build is also the home of SAP Joule for Developers (Section 8.3.3), an embedded AI assistant to help create development artifacts inside SAP Build.

SAP Build Lobby is the central hub of SAP Build. One of the key features of SAP Build Lobby is to create and manage projects in SAP Build. Projects can be created from templates, cloned from Git repositories, created from scratch, or imported from other SAP Build environments. From the lobby, existing projects can be managed, project users can be added or removed, and projects can be renamed, transported, or deleted. In the lobby, you can also manage connections to external systems. These connections in SAP Build are referred to as *actions*, and they represent system APIs or events. Once an action is created, it can be reused across multiple SAP Build projects.

SAP Build Lobby also provides access to SAP Build Store. The store offers prebuilt content packages, built by SAP and authorized SAP partners. Inside the store, you can filter content based on industry, line of business, or solution and find a catalog of ready-to-use content and templates that can be directly imported as projects into SAP Build already configured and extended. This reduces the development time of building a project from scratch. In the store, there is also a wizard to help you get the updates pushed to packages you have imported into your SAP Build environment.

In the following sections, we'll explore key features of SAP Build for low-code and pro-code tooling. We'll also discuss the related Joule capabilities for app development.

8.3.1 Low-Code Tooling in SAP Build

Low-code tooling in SAP Build is delivered in the key areas of creating custom business sites, creating custom workflow automations, creating custom applications, and creating custom Joule Skills and Agents. The low-code tooling in SAP Build allows for citizen developers to better engage with traditional SAP development teams in order to create innovative solutions and provide feedback on development activities earlier, thus reducing rework later. The drag-and-drop development features of the low-code capabilities in SAP Build allow for faster time to value when creating a custom extension or application for SAP Autonomous Suite.

Let's take a closer look at the SAP Build solutions for low-code tooling: SAP Build Work Zone, SAP Build Process Automation, and SAP Build Apps.

SAP Build Work Zone

SAP Build Work Zone offers the capability to create low-code business sites. SAP Build Work Zone comes in two different varieties: standard edition and advanced edition. SAP Build Work Zone, standard edition provides the ability to create custom SAP Fiori launchpads that can act as a landing page for access to standard and custom SAP Fiori applications. With SAP Build Work Zone, standard edition, you can create a launchpad to host standard applications with custom applications built on SAP BTP in a single location. Both versions of SAP Build Work Zone use SAP BTP's Identity Management service to authenticate users. This makes it easy to extend existing user roles to custom-built SAP Work Zone tenants so that end users have one singular access point for all their applications.

With SAP Build Work Zone, advanced edition, you can go beyond the standard SAP Fiori user experience and create a consumer-grade site for business partner portals, such as a customer or supplier portal via which external users can log in and safely view content from SAP Autonomous Suite, SAP BTP applications, and non-SAP systems relevant to them. SAP Work Zone also can be integrated with SAP Task Center so that users can see a centralized inbox with their open task items and so that it can be a landing page to host SAP Joule. This way, business users can interact with Joule to assist with their day-to-day tasks. Content developed on SAP Build can be used across multiple SAP Work Zone sites easily, making it easy to reuse content needed for multiple business units.

In SAP Build Code, you can create custom SAPUI5 cards that act as lightweight widgets to display KPIs and summary information linked to an application. SAP Build Work Zone creates an easy landing page for users to have a single view to access all the applications relevant to their day-to-day operations. Users can collaborate with one another on projects; upload and store documents; see their open tasks; access SAP, non-SAP, and SAP BTP applications; and interact with Joule Agents.

SAP Build Process Automation

SAP Process Automation is the low-code capability of SAP Build used for creating custom automation workflows. SAP Build Process Automation allows citizen developers to adapt, improve, and automate business processes with a low-code workflow and a robotic process automation designer.

Figure 8.5 shows the low-code designer for SAP Build Process Automation. Here you can see an example of an invoice-processing custom automation, including custom forms, business rules, and automation steps.

SAP Build Process Automation allows you to easily extend or automate business processes across SAP Autonomous Suite. When you need to customize a business process such as sales order preprocessing, you can use SAP Build Process automation to rapidly adapt the process to fit your requirements. SAP Build Process Automation includes low-code tooling capabilities for modeling custom business processes, business rules, user forms, workflow logic, and bot logic, as well as other AI features such as document information extraction, which can extract information from semistructured data such as PDFs or image files. SAP Build Process Automation workflows can be triggered from SAP Autonomous Suite via a standard system event or API call, by a bot running on a local machine looking at an email outbox or file directory, by a low-code user form built in the workflow designer, or by a Joule Agent.

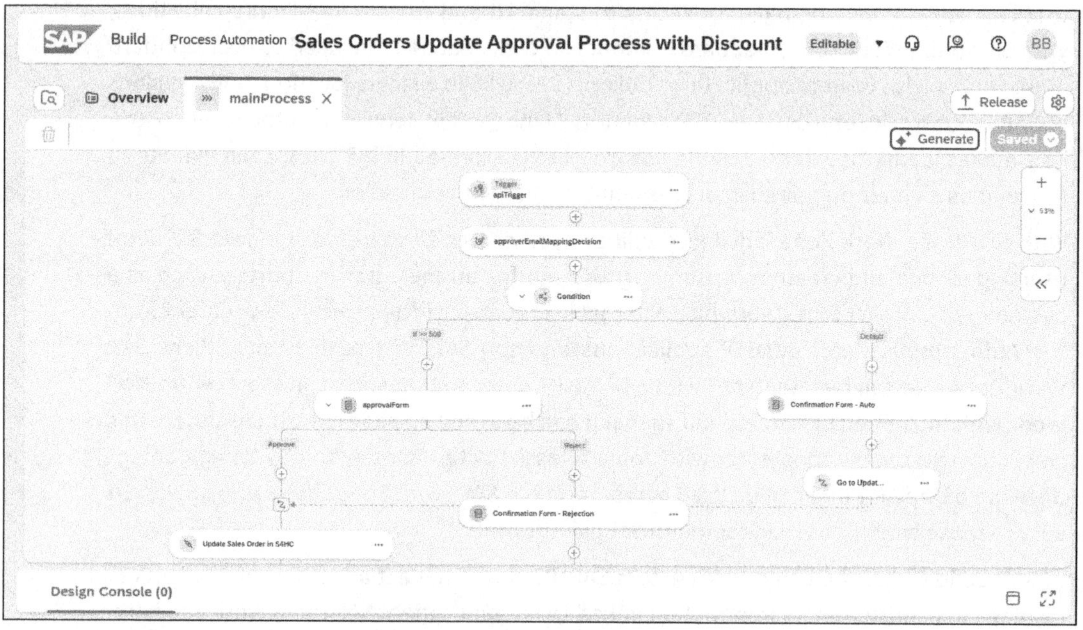

Figure 8.5: Low-Code Designer for SAP Build Process Automation

Figure 8.6 shows an example of how SAP Build Process Automation can be used to extend a standard workflow for credit management in SAP S/4HANA. SAP Build Process Automation picks up a standard event and then passes the data through a custom workflow running side by side with the ERP system to give flexibility to the process while staying within the lines of SAPs clean core framework.

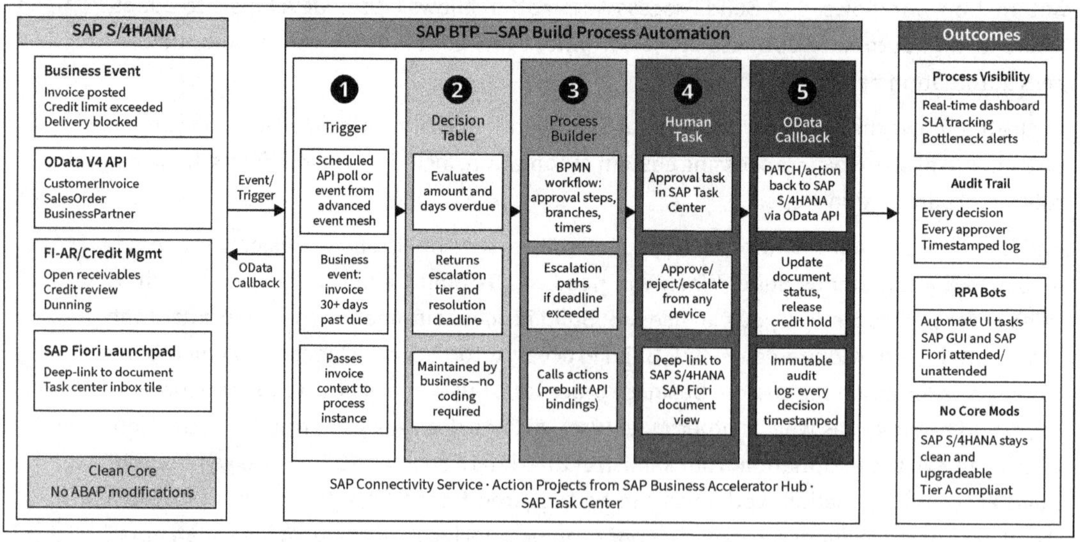

Figure 8.6: Extending SAP S/4HANA Workflows with SAP Build Process Automation

Like SAP Build Work Zone, SAP Build Process Automation can use existing user roles to send tasks to a user's centralized inbox for approval or rejection. If a scenario requires a custom form to initiate a custom process delivered via SAP Build Process Automation, that form can be easily integrated into an SAP Build Work Zone site for a user.

With SAP Build Process Automation's low-code features, citizen developers can assist IT teams by creating low-code workflows, making use of their strong process knowledge about how a process variant should execute and the business rules associated with that process variant. Tasks in SAP Build Process Automation can either run fully automated or semiautomated with a human in the loop where needed based on business rules or when an error is received in the data given to trigger the workflow process.

SAP Build Process Automation also includes the capabilities of process visibility dashboards. These dashboards allow a process analysis to view the health of how a custom workflow built in SAP Build Process Automation is running and assign KPIs to the workflow process. The process visibility dashboards provide a window into areas where a process can be automated or refactored for further efficiency.

SAP Build Process Automation can execute workflows across multiple systems due to its architecture running on SAP BTP. For example, a process automation might first create a user in SAP Success Factors, then create the user in SAP S/4HANA, then in SAP Concur for traveling for work training, then as a vendor in SAP S/4HANA for travel reimbursement, then add the user to SAP Ariba for equipment procurement for their job, and finally send the user back into SAP SuccessFactors for assignment of an onboarding mentor. In this scenario, you have to orchestrate a workflow across four different systems. It needs to be monitored and may need to follow different business rules depending on your company's industry or culture. SAP Build Process Automation gives you the flexibility to create custom business processes that extend beyond the boundaries of one system so that you can adapt agilely to fill process gaps and better monitor the health of your operations.

SAP Build Apps

SAP Build Apps is the capability of SAP Build for creating full-stack applications via low-code tooling. SAP Build Apps can create mobile native applications or web applications that can be accessed independently or through SAP Build Work Zone.

SAP Build Apps contains a catalog of content for creating the frontend design of an application as well as its processing logic. The low-code features of SAP Build Apps allow users to confidently build enterprise-grade applications without ever touching a single line of code. SAP Build Apps recommends following the *don't repeat yourself* (DRY) principle; that is, a custom application built on SAP Build Apps should not repeat the logic of a standard SAP application but should instead have a narrowly defined scope meant to fill a functional gap within SAP Autonomous Suite.

SAP Build Apps can serve the purpose of consolidating multiple applications into one, easy-to-consume interface for end users. This use case is strong when blending data from SAP and non-SAP systems. SAP Build Apps can be used as the trigger or step in a workflow built on SAP Build Process Automation when the requirement for a user interface goes beyond what is capable out of the box in the form designer of SAP Build Process Automation.

SAP Build Apps also includes a marketplace of tooling to handle more complex user interface or processing logic than what is provided out of the box by the standard features in SAP Build Apps. SAP Build Apps includes software development kits (SDKs) to deploy built applications to iOS or Android devices so that field technicians or service workers can access data and submit transactions to SAP systems while they are in areas of limited service.

Like the rest of the SAP Build services, SAP Build Apps can extend existing user roles defined in SAP S/4HANA or other SAP systems so that user access can be federated rather than being maintained in multiple locations.

8.3.2 Pro-Code Tooling

When the need for development extends beyond what is delivered out of the box in the low-code tooling of SAP Build or when a developer prefers to code rather than use a low-code solution, SAP offers pro-code tooling on SAP Build for creating custom applications and extensions. The developer environment for the pro-code environment of SAP Build depends on the framework the developer wants to use. For developers leveraging the SAP Cloud Application Programming Model delivered via Node.js or Java, SAP Build Code is a turnkey environment in which a developer can create an application. For custom SAP Fiori user interfaces, you will also use SAP Build Code. For ABAP RESTful application programming model development in the SAP BTP ABAP environment, a developer sets up the environment via SAP Build but continues their development in ABAP development tools in Eclipse as they would for developer extensibility in an SAP S/4HANA system.

SAP Build Code is a turnkey environment that creates a developer IDE environment in which you can select plugins for your development based on the tools you would need, such as an iOS SDK or a connection to an SAP HANA Cloud database. These plugins help accelerate the time to value when creating a custom application on SAP Build.

In the next section, we'll explore how to use SAP Build Code to create custom SAP Cloud Application Programming Model applications. We'll then explore two other avenues for pro-code development: the ABAP RESTful application programming model and SAPUI5.

SAP Cloud Application Programing Model

SAP Cloud Application Programming Model is a framework of languages, libraries, and tools for building cloud applications. It incorporates proven best practices for creating custom applications and extensions to create a golden path for developers to follow. SAP Cloud Application Programming Model focuses on domains, leveraging the concepts of CDS from SAP S/4HANA but bringing those concepts to developers who have a background in Node.js and Java. This model has been called ABAP for the non-ABAP world. It is designed for developers to grow as they learn or gather new requirements for their use cases, making it easy to add in new features such as adapted SAP Fiori user interfaces, message queues, or different databases.

The SAP Cloud Application Programming Model framework helps bring enterprise structure to development by providing out-of-the-box features for localization of data, data federation, verticalization, and extensibility by default. SAP Cloud Application Programming Model promotes the use of plugins to easily add SAP BTP services that can augment the capabilities of a nSAP Cloud Application Programming Model application. The model is open but opinionated; its framework is designed to give guidance, but to do so in a manner that does not sacrifice openness or flexibility.

Figure 8.7 shows an example of a customer loyalty application built in the SAP Cloud Application Programming Model framework to extend SAP S/4HANA. The application calls the SAP S/4HANA system through standard APIs and maintains all of its custom logic inside of SAP BTP.

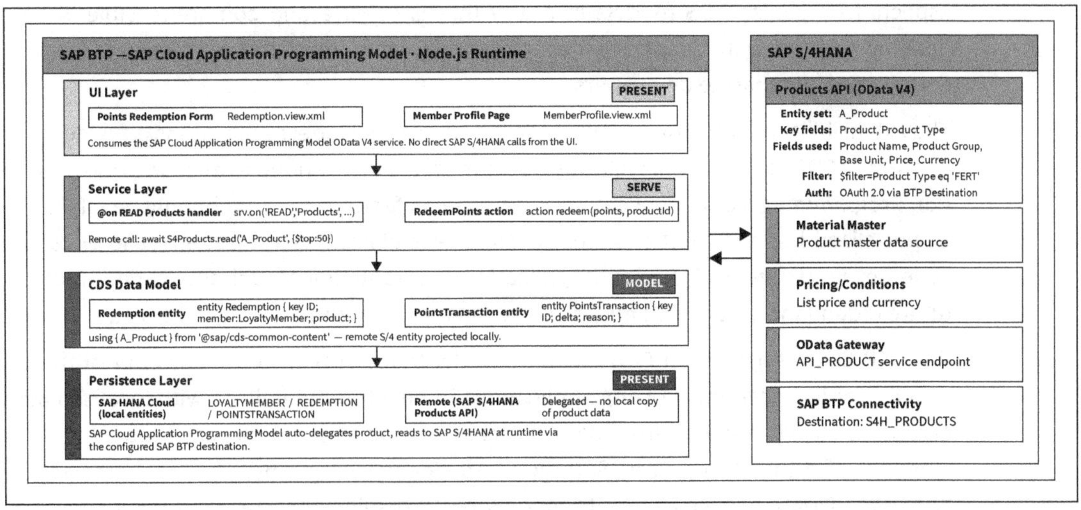

Figure 8.7: SAP Cloud Application Programming Model Framework: Customer Loyalty Application

SAP Cloud Application Programming Model places its primary focus on domains, and CDS is its backbone. It uses a conceptual definition of language to make a human-readable schema of a business object. This allows developers to better collaborate with business experts to validate that their domain is correct for the entities created for an SAP Cloud Application Programming Model application. CDS focuses on conceptual modeling, aiming to capture intent, which means that it aims to capture the *what* of an entity, not the *how*, when it comes to implementation. SAP Cloud Application Programming Model focuses on entity relationship modeling, taking its models and defining the relationship between them. For example, a dispute application built with SAP Cloud Application Programming Model would have relationships to a customer model, an order model, and an employee model, representing the things an extension would need to handle a custom dispute resolution process.

CDS models can contain managed associations, one-to-many associations, and many-to-many associations. The CDS models define what data we are interacting with and what we do with that data, including create, read, update, and delete (CRUD) operations.

SAP Cloud Application Programming Model supports out-of-the-box authentication and authorization by annotating services and entities. Like SAP S/4HANA, SAP Cloud Application Programming Model breaks down security into the key areas of authentication and authorization. *Authentication* refers to validating who the end user is and if they are assigned the necessary roles to access an SAP Cloud Application Programming Model application. *Authorizations* define what actions a user can perform with the data inside an application—that is, what restrictions they have on viewing, creating, updating, and deleting data in a SAP Cloud Application Programming Model application.

Annotations can also be used to provide context to a frontend SAP Fiori application to make UI development easier. SAP Cloud Application Programming Model uses the concept of aspects to organize CDS models in a way that is concise and comprehensible. *Aspects* decompose models into separate files so that they can potentially have separate lifecycles contributed by separate developers.

The extensibility of SAP Cloud Application Programming Model applications is fueled by the concept of CDS aspects, which allow for custom fields to be added to an SAP Cloud Application Programming Model application and toggle on features that might only be relevant to a specialized business unit or industry. This flexibility allows for better reuse of custom code developed in an SAP Cloud Application Programming Model application, reducing future redundant rework that might be needed for a new enhancement request.

The SAP Cloud Application Programming Model framework also provides out-of-the-box support for localization, allowing applications to be internationalized with localized models and localized data. Localized data allows for an application to be easily adapted to the language of a target end user. SAP Cloud Application Programming Model creates fully functional OData services from its models by leveraging the created CDS models. The OData APIs make it easy to expose the functionality of an SAP Cloud Application Programming Model service to an SAP Autonomous Suite system or to rapidly create a standalone SAP Fiori application from the SAP Cloud Application Programming Model service.

SAP Cloud Application Programming Model can also integrate easily with remote services from any application, SAP or non-SAP. SAP Cloud Application Programming Model provides a platform-agnostic way to consume remote services by representing them as SAP Cloud Application Programming Model services so that they can be consumed as if they were local to the SAP Cloud Application Programming Model application. The model manages the communication details under the hood and can mock remote services automatically for testing. This flexibility allows developers to create custom applications that combine data from multiple SAP or non-SAP systems and build an application that an end user can interact with without having to hop between multiple systems. SAP Cloud Application Programming Model features a mix of both SAP and open-source technologies built on Express.js and Spring. SAP Cloud Application Programming Model applications have native support for SAP HANA, SQLite, and PostgreSQL.

ABAP RESTful Application Programming Model

The ABAP RESTful application programming model is a powerful framework for SAP ABAP to build cloud-ready applications. This model was created by using lessons learned from classic ABAP development to enhance the ABAP programing model for SAP Fiori applications. The ABAP RESTful application programming model is designed to support integration into SAP HANA and create OData APIs to feed SAP Fiori applications. The core components of the ABAP RESTful application programming model are as follows:

- Business objects
- Behavior definitions
- Behavior implementations
- Service definitions
- Service bindings

Like SAP Cloud Application Programming Model, the ABAP RESTful application programming model uses CDS as its data-modeling layer for business objects. CDS is used for semantically rich data models, entities, associations, and projections. The CDS layer acts as the bridge between SAP HANA and the ABAP runtime to perform optimized, pushed-down queries at runtime so that developers can focus on business logic inside their customization.

A *business object* represents a real-world artifact, such as a product, sales order, purchase order, or so on. Business objects have hierarchical nodes tied to CDS views; for example, a sales order will have a sales order heard with a one-to-many relationship to sales order items, and items will have product information such as the SKU number, name, price, unit, and so on. Business objects will have standard functionality shared across different implementations of types of sales order; this information is captured for reuse in the behavior definition. Sales orders can be of many different types for an organization, so there will be a root sales order object and then implementations of the sales order, and the implementation of object will include defined behaviors, validations, and determinations.

SAPUI5

SAPUI5 is the user interface technology that underpins the SAP Fiori UX. SAPUI5 is provided with all the main SAP platforms, but it can be used outside of SAP systems as well. SAPUI5 provides an out-of-the-box frontend framework to build enterprise applications, and an open-source version has been developed via OpenUI5. SAPUI5 is a client user interface framework using HTML5, CSS, and JavaScript to run applications on any device via a web browser, built with a mobile-first design pattern. SAPUI5 applications scale to fit desktop, mobile, and tablet browsers. SAPUI5 is the default framework for SAP S/4HANA and for the templates for building custom SAP Fiori applications via SAP Business Application Studio.

SAPUI5 offers a consistent user experience across SAP Autonomous Suite applications, allowing for easier adoption of new applications by users coming from legacy SAP systems or non-SAP systems to SAP Autonomous Suite. It features rich UI controls for handling complex user interactions and provides predefined templates for common application use cases. SAPUI5 applications have predefined extension points so that you can easily adapt applications to your specific business requirements.

SAPUI5 leverages the model-view-controller (MVC) concept for structuring applications, which allows for easier maintenance and extension. MVC breaks applications into three major components to create clear distinctions between the processing control, data model structure, and user interface.

The *model* is an application of a business object and represents a layer of data administration for the application. It responds to information requests about its status from the view and handles statements of change from the controller. The model should be used only to process data internally within the SAPUI5 application.

The *view* handles graphical and textual output to the user interface. The view contains elements for input and output in the form of push buttons, menus, dialog boxes, and other UI elements. The view handles the visualization activities of an application, displaying data and status changes to a user. SAPUI5 supports two view types: an XML view and a typed view. Views can be loaded asynchronously, allowing for multiple elements of a view to be loaded based on the complexity of the content that needs to be rendered to the user.

The *controller* acts as the bridge between the model and the view. It monitors and interprets data that is input from the user via the view, and it uses model methods to change the internal status and then return updates to the view. Because the model does not recognize either views or controllers, internal data processing can be detached entirely from the user interface.

This MVC framework provides better readability, maintainability, and extensibility, allowing you to change the view without ever touching the underlying business logic.

Figure 8.8 shows the reference architecture for separation of concerns and communication between elements of SAPUI5 applications in adherence with the MVC guidelines.

SAP provides SAP Fiori elements to give you the benefit of faster development from a template and a metadata-driven approach. SAP Fiori elements provide design patterns and predefined floor plans for common application use cases. You can leverage the SAP Fiori elements catalog to create applications based on OData services from SAP Autonomous Suite applications. SAP Fiori elements help increase developer efficiency by allowing you to focus on business logic and minimize the amount of UI code that needs to be written. This helps create a consistent user experience that is optimized for and supports enterprise-class features such as translation support, search, application navigation, and accessibility.

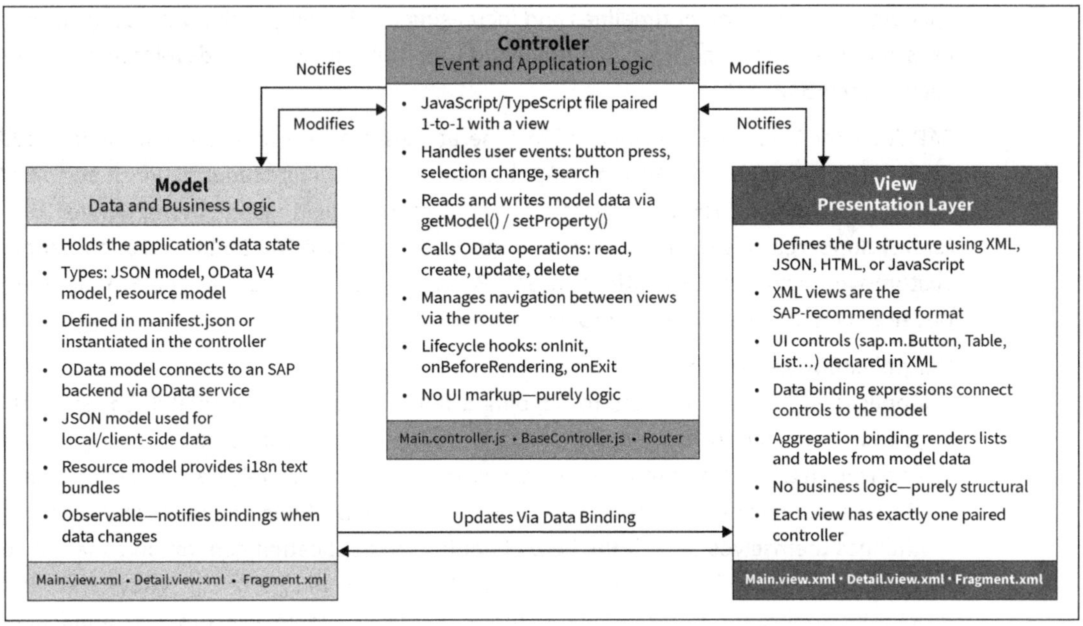

Figure 8.8: MVC with SAPUI5

Like key user and developer extensibility in SAP S/4HANA, SAPUI5 supports extensions of applications via key user and developer adaptation. Key-user adaptation allows users to make changes to the user interface at runtime: adding or removing elements, renaming or rearranging elements, and embedding content in an SAPUI5 application. Developer adaptation allows users in SAP Build to perform the same tasks as key user adaptation, but it goes a layer deeper by allowing the user to go into the supporting code to make the needed changes to an SAPUI5 application.

SAP Build provides out-of-the-box templates, frameworks, and SAP Fiori elements for extending the features of standard SAP Autonomous Suite applications. You can link SAP Autonomous Suite environments to SAP Build and make customizations to standard SAP Fiori applications easily, extending the features and floorplan of an application without having to develop an application fully from scratch. This allows you to safely customize a standard application to fit your business requirements while still staying aligned with SAP's clean core framework.

8.3.3 SAP Joule for Developers

SAP Autonomous Suite and SAP BTP users get access to a generative AI assistant for creating custom extensions and applications via SAP Joule for Developers. SAP Joule for Developers is embedded in all features of SAP Build, from SAP Build Lobby to its low-code and pro-code tools. SAP Joule for Developers understands SAP development intricacies, is trained on SAP development best practices, and can anticipate a developer's needs and offer intelligent suggestions to reduce manual tasks associated with development, thus speeding up development timelines and increasing productivity. Behind the scenes, Joule uses purpose-built large language models (LLMs) tailored to these development needs, such as the ABAP LLM.

SAP Joule for Developers can be used to generate custom code for extensions built on SAP BTP. Joule can code for SAP Cloud Application Programming Model, SAPUI5, and ABAP RESTful application programming model based on natural language prompts. SAP Joule for Developers can also be used to create low-code artifacts in SAP Build such as content for custom workflows or automations in SAP Build Process Automation. The SAP Joule for Developers functionality can be broken down into three key areas:

- **Application generation**
  Application generation refers to building a full application from scratch based on the prompt delivered to Joule. In this scenario, Joule will create data models, sample data, application logic, user interface, unit tests, and more. It will apply best practices at each step; developers can then review the code generated by Joule and make any necessary changes themselves or with the help of Joule. The application generation capability of Joule is available for both the SAP Cloud Application Programming Model and ABAP RESTful application programming model frameworks. Joule also supports application generation when it comes to building native mobile applications in the pro-code tooling in SAP Build. In this scenario, Joule creates an application using the mobile development kit in SAP Build to build native iOS or Android applications. The application generation capability is also available in the low-code environments of SAP Build. In SAP Build Process Automation, you can give Joule a prompt describing your process requirements, and it will generate the process flows, automations, forms, and business decisions based on your given requirements.
- **Code optimization**
  Code optimization in SAP Joule for Developers is delivered via automatic code completion and through commands directed to Joule in SAP Build. SAP Joule for Developers offers code explanation features that can provide detailed explanations for custom code or CDS views, allowing developers to easily understand the work of others. It can also provide context-aware insights that consider factors like ABAP language version and development object relationships. Expanding this feature, SAP Joule for Developers offers predictive code completion in the form of ghost text suggestions that a user can accept, speeding up the time it takes a developer to create custom CDS view elements or business logic. SAP Joule for Developers can also create unit tests for customizations built, reducing the time and effort of testing. These capabilities can reduce development time and cost and enhance code quality by providing developers with code generated in adherence with SAP Best Practices.
- **Code refactoring**
  Code refactoring is an exciting capability if you are looking to move from legacy SAP technology to SAP Autonomous Suite or are looking at refactoring legacy customizations to align with SAP's clean core standards for extensibility. Custom code developed in a

legacy SAP system or release of SAP S/4HANA must be assessed, analyzed, and adapted during a system conversion or upgrade. This is often a highly complex and time-intensive task. SAP Joule for Developers reduces this burden by providing AI-based custom code analysis and support in the Custom Code Migration application of the SAP BTP ABAP environment in SAP Build. You can run an ATC readiness check with the Custom Code Migration application, and when the readiness check is complete, you can perform an analysis of the findings inside the application and receive help from Joule. Joule can create a code explanation of the applications surfaced in the results and can then use ATC to explain features, give a summary of the ATC check findings, and provide a step-by-step guide for how to resolve the issues in the legacy custom code. Developers can then access Joule in Eclipse with ABAP development tools and ask for further details of how to resolve issues with their legacy code. Finally, Joule can generate code proposals for developers to add into their applications to resolve the issues surfaced during the ATC checks. Code proposals can be added from the Joule chat directly into the project. This reduces the time and effort needed to refactor legacy code that cannot be fixed by deterministic quick fixes.

- **Artifact generation**
  Artifact generation is a targeted way to use SAP Joule for Developers to create content for custom extension application. Artifact generation has numerous capabilities today. In the pro-code environment of SAP Build, SAP Joule for Developers can be used to generate application templates, custom CDS views, application logic, service implementation, and SAPUI5 components. SAP Joule for Developers can create complex CDS views with associations, compositions, and business constraints. It can add schema definitions, annotations, and validation rules. It can also be used to create custom code snippets or add documentation and comments to existing code in an application. In the low-code environment, SAP Joule for Developers can create processes, forms, business rules, and automation steps inside SAP Build Process Automation. SAP Joule for Developers can also be used in SAP Build Apps to generate a user interface based on the connected data sources in the SAP Build Apps project.

SAP Joule for Developers is an exciting area of growth in SAP BTP. It provides a turnkey agent trained on SAP Best Practices to reduce the manual work and time needed to create a customization on SAP BTP. This allows developers to focus on refining the code or content generated by Joule instead of creating a project from scratch, with the assurance of knowing the content developed by Joule adheres to SAP Best Practices and SAP's clean core standards.

8.3.4 Joule Studio

Joule Studio, which we introduced in Chapter 7, lets you extend the features of standard Joule agents delivered by SAP to create custom Joule Agents and Skills tailored to your organization's needs. Joule Studio is now embedded as a capability of SAP Build. It serves as the centralized workplace for building, deploying, customizing, and managing Joule Agents and Skills across SAP and non-SAP systems. Joule Studio has two key components: Joule Skills and Joule Agents.

A *Joule Skill* is a deterministic, rule-based task that follows a predefined sequence to perform a specific repetitive business operation. Joule Skills developed in Joule Studio represent custom tasks that are tailored to your specific business process. They are activated based on triggers. A *trigger* is a natural language prompt that describes what the skill intends to do; a trigger can also receive data from other Joule Skills or APIs, and this data is defined via parameters that also have natural language descriptions, semantics about the types of data it is expected to receive, and if that data is required for the skill to be called. A Joule Skill also can call *actions*, which are API calls to remote systems, automations, and processes. Joule Skills can also have conditional branches to add logic based on information gathered from the trigger or other artifacts in the Joule Skill. Finally, a Joule Skill will have a *response* that can be sent to other Joule Skills or to an end user via the Joule Agent. Joule Skills should have a single purpose; that is, each skill should be self-contained and perform one specific task to avoid confusion and improve performance. It is important that Joule Skills have clear descriptions so that a Joule Agent can determine when to trigger a Joule Skill.

Joule Agents are comprised of Joule Skills. They are autonomous, multitooled systems that can be used to solve complex problems by selecting and executing the appropriate tools dynamically with human-like reasoning, reflection, and adaptability. A Joule Agent is given the context of and instructions for the problems and tasks it is asked to perform, and it then can perform multistep reasoning to handle multistep workflows with contextual awareness, dynamically adjusting based on changing business conditions. Joule Agents come in three distinct forms:

- **Conversational agents**
  Engage in natural language interactions with the user to understand their needs and provide answers based on the context of the conversation
- **Autonomous agents**
  Independently execute workflows and make decisions without human intervention
- **Specialized business agents**
  Specialize in business domains like procurement, finance, compliance, and others with deep expertise in these topic areas

Joule Agents can be connected to SAP systems and non-SAP systems via API or event integration into SAP Build. This enables them to retrieve or send data to these systems to execute tasks and receive information. Joule Agents also can be connected to existing MCP servers connected to SAP BTP as a remote destination and document sources such as Microsoft SharePoint or S3 buckets to reference internal knowledge bases inside your organization.

Joule Studio is an exciting area of growth inside of SAP BTP. As Joule becomes the engagement layer for users to interact with functionality inside of SAP Autonomous Suite, Joule Studio provides the flexibility of extending Joule to meet the needs of your specific business processes.

8.4 Platform Integration Capabilities

Integration is a key pillar of SAP BTP. Because an SAP BTP environment sits decoupled from your SAP Autonomous Suite application, this environment is positioned well to handle all the necessary activities needed to safely send and receive data between SAP and non-SAP applications. In this section, we will look at the integration services provided in SAP BTP and how your organization can use these services to reduce manual data entry in multiple systems, errors from incorrect data, and issues due to data being out of sync between business systems.

We'll take our first look at integration services and event-driven integration in the following sections. For a deeper discussion, see Chapter 11.

8.4.1 Integration Services

SAP BTP offers a few key services for integration. Foundationally, SAP BTP provides the SAP Destination and Connectivity services via the SAP BTP cockpit in each subaccount you have created.

The *SAP Destination service* connects your SAP BTP subaccount to a service or system. This approach is similar to creating a remote destination in Transaction SM59, but in the SAP BTP environment with better flexibility. Destinations allow you to define the remote service, the protocol the service will be called over, and the authentication mechanism you will use to connect to that remote service. This allows you to create a connection for development or integration that can be reused in additional scenarios. Destinations make transporting artifacts easier by parametrizing connections, so they do not require editing when moved from development to production.

The *SAP Connectivity service* of SAP BTP allows you to connect to your on-premises environment via the *cloud connector*, a tool installed within your on-premises or private cloud environment to initiate communication to SAP BTP via a secure communication tunnel without having to open your firewall and put your systems at risk. The cloud connector can only connect your systems to an SAP BTP environment; once that connection is established, a proxy of your systems will be viewable via the SAP Connectivity service in SAP BTP. The cloud connector and the SAP Connectivity service support OData, HTTPS, LDAP, MAIL, and SFTP communication protocols to make it easy to extend and integrate your environment with SAP BTP. The destination and connectivity services of SAP BTP are foundational to every SAP BTP subaccount and are provided by default.

Expanding on the foundational integration services provided in SAP BTP, there is also SAP Integration Suite, SAP Integration Suite, advanced event mesh, and SAP Master Data Integration. SAP Integration Suite, which we'll discuss in greater detail in Chapter 11, is SAP's integration-platform-as-a-service (iPaaS) offering. This is a fully managed process integration platform that handles SAP to SAP, SAP to non-SAP, and SAP to third-party integration scenarios. SAP Integration Suite has tools for process integration, API management, business-to-business integration, business-to-government integration, a localized event broker, and migration from legacy integration content (from SAP Process Orchestration or SAP Process Integration) to SAP Integration Suite. SAP Integration Suite has over 3,400 prebuilt integration artifacts and 180 adapters to third-party systems. SAP Integration Suite also comes with the Edge Integration Cell, a localized runtime of SAP Integration Suite's process integration engine that allows you to process integrations in a managed, on-premise or private cloud environment. SAP Integration Suite, advanced event mesh expands upon the capabilities of SAP Integration Suite's event broker and allows you to deploy a network of connected brokers to create a true mesh architecture for processing system events. We'll talk about event-driven integration in greater detail in the next section.

SAP Master Data Integration is a service focused on integrating master data. It allows SAP applications to perform initial loads of master data and stay in sync with the master data database of the service. SAP Master Data Integration adheres to SAP One Domain Model and supports its various elements, such as equipment, functional location, business partner, profit center, and so on.

SAP BTP provides many services for integration. In Chapter 11, we will explore the various features in these SAP BTP services and how they can accelerate integration between your applications.

8.4.2 Event-Driven Integration

Event-driven integration is a new pattern of integration between applications that can help reduce the time it takes to build new integrations or connect existing applications. Traditionally, applications were connected via point-to-point integrations, in which one application would call either a remote application or an integration platform either via an API call or by periodically dropping a batch data file. This required integration teams to develop new integrations every time a new system or functionality was onboarded in an organization, creating high lead times to connect systems together and creating more technical debt for integration teams to maintain. Event-driven integration supports a *publish-and-subscribe* pattern that allows for a system to hand data off to an event broker that will in turn deliver the data to all subscribing applications. Events represent a change in state for an application object, such as a business partner being created, updated, or deleted. Event-driven integration is an asynchronous pattern of integration that supports a multitude of integration scenarios, such as the following:

- **One-to-many distribution**
  Cascading updates when an element is changed for multiple subscribers

- **Shock absorber**
  Queueing high-volume transactions to distribute them to a subscriber at the rate they can consume it without disrupting operations
- **High availability**
  Storing transactions while a system is experiencing an unexpected outage or going through a maintenance window

SAP Autonomous Suite applications have standard events that can be enabled via configurations like APIs, as well as the capability to create custom events. Custom events can be created via SAP Cloud Application Programming Model or the ABAP RESTful application programming model; they can be created from existing intermediate documents via the SAP Application Integration Framework add-on or via the Event Enablement add-on (a plugin for SAP S/4HANA). Events in SAP come in two types: *notification events*, which provide summary details of an event with limited information, like the type of change and identifying information like the ID or sales order ID; or *data events*, which contain the full details of an event, like an API payload.

Often the best scenarios for building extensions on SAP BTP are asynchronous data exchange scenarios, as such events provide a strong framework for passing data to extensions running on SAP BTP. Workflows and automations running on SAP BTP can be triggered by an event; SAP Cloud Application Programming Model and ABAP RESTful application programming model applications can send events to and from SAP Autonomous Suite applications; and custom Joule agents can call events in Joule Skills to execute an action in a SAP Autonomous Suite system. By using event mesh inside SAP Integration Suite or SAP Integration Suite, advanced event mesh, you can create resiliency with a circuit breaker design pattern to safely exchange data between multiple applications.

8.5 Lifecycle Management

Building an application on SAP BTP is only the beginning of the story. Once a custom extension, integration flow, or AI agent is deployed into a productive subaccount, it becomes a living part of your business operations. It must be monitored for health and performance, updated as business requirements evolve, transported reliably across development and production landscapes, and eventually decommissioned without disrupting the processes that depend on it. SAP BTP provides a set of integrated lifecycle management capabilities that address every stage of this journey, organized primarily through SAP Cloud ALM.

The central role that SAP Cloud ALM plays in an SAP BTP–based landscape is analogous to the role that SAP Solution Manager traditionally played in on-premise landscapes, but it is redesigned from the ground up for the cloud era. Where SAP Solution Manager was a heavyweight, on-premise application that required its own system administrators and its own upgrade cycle, SAP Cloud ALM is a cloud-native SaaS service that SAP provisions and maintains on behalf of every customer with an eligible SAP support contract. Your team connects your SAP BTP subaccounts, SAP S/4HANA systems, and SAP Integration Suite tenants to SAP Cloud ALM through a lightweight configuration step, and the platform immediately begins collecting operational signals, change records, and health telemetry from across your landscape without any on-premise infrastructure investment.

For applications built with the SAP Cloud Application Programming Model framework or deployed as Cloud Foundry applications, the deployment pipeline is managed through SAP Cloud Transport Management. This service provides a central transport inbox that mirrors the concept of the ABAP transport management system, but it operates at the level of multitarget application (MTA) archives, the packaging format used by SAP BTP applications. A developer finishes a feature in the development subaccount, exports the MTA archive, and uploads it to the SAP Cloud Transport Management inbox, where it waits for a release manager to approve and schedule its forward transport into the test and production nodes. This process ensures that no change reaches production without a deliberate approval step and that every transport is recorded in the change log with its author, timestamp, and target subaccount.

For integration artifacts deployed in SAP Integration Suite, such as integration flows, API policies, and event mesh configurations, the transport mechanism is provided natively by the SAP Integration Suite's content transport capabilities, which can be connected to the same SAP Cloud Transport Management pipeline so that integration changes follow the same governed promotion path as application changes. Organizations that adopt this unified transport approach gain a single audit trail for all SAP BTP changes regardless of artifact type, which significantly simplifies the evidence collection required during an external audit or an SAP upgrade validation.

Once an SAP BTP application is running in production, SAP Cloud ALM's operations capabilities provide the continuous monitoring that keeps it healthy. The platform collects health check data from Cloud Foundry applications, integration flows, and event mesh topics, and presents this data in a unified operations dashboard that gives the SAP BTP operations team a single view across the entire landscape. When a health check fails, an integration flow stops processing messages, an application crashes, or an event mesh queue depth exceeds a threshold, SAP Cloud ALM raises an alert and, depending on your configuration, can trigger an automatic notification to the on-call engineer via email or through an integration with an external IT service management (ITSM) tool such as SAP Solution Manager or ServiceNow.

For integration flows running in SAP Integration Suite, SAP Cloud ALM provides end-to-end message monitoring that tracks every inbound message through its complete processing path. An operations engineer can search by message ID, source system, target system, or time window and drill into the processing log of any individual message to identify exactly which step in the integration flow failed and what the error payload contained. This capability is a significant improvement over the native SAP Integration Suite monitoring tools for multihop scenarios, because it correlates the message trace across systems showing not just that a message failed in the integration layer, but how it was received from SAP S/4HANA, what transformation was applied, and whether the downstream target system rejected it. This level of operational insight is captured in the integration pillar of the clean core framework's operations metrics, which measures end-to-end monitoring coverage as one of the indicators of a healthy, well-governed integration landscape.

Modern SAP BTP development is not a waterfall process in which a project team builds an application, deploys it to production, and hands it over to operations. The platform and the business requirements it supports evolve continuously, and the lifecycle management capabilities on SAP BTP are designed to support a DevOps delivery model in which development, testing, deployment, and monitoring form a continuous feedback loop rather than a sequential handoff.

SAP BTP integrates with standard DevOps toolchains through the SAP Continuous Integration and Delivery service, a cloud-native continuous integration/continuous delivery (CI/CD) pipeline service that connects to Git repositories hosted on GitHub, GitLab, or SAP's own source code management service. When a developer commits a code change to the repository, the CI/CD pipeline can automatically execute unit tests, perform a static code analysis using the SAP BTP SAPUI5 linter and the ATC for on-stack objects, then package the application as an MTA archive ready for transport. This automation eliminates the manual steps between writing code and having a validated, deployable artifact, and it ensures that quality gates are applied consistently on every commit rather than depending on an individual developer's discipline.

For ABAP-based developer extensibility on SAP S/4HANA, gCTS brings the same DevOps principles to on-stack development. ABAP objects are versioned in a Git repository alongside SAP BTP application code, enabling development teams to apply branching strategies, pull request reviews, and automated testing to their ABAP development with the same tooling and governance they use for their SAP Cloud Application Programming Model and SAPUI5 work. The combination of gCTS for on-stack changes and SAP Cloud Transport Management for SAP BTP changes gives organizations a unified DevOps lifecycle that spans both sides of the clean core boundary.

Lifecycle management on SAP BTP also encompasses the commercial and governance dimension of running a platform at scale. As an organization grows its SAP BTP footprint, adding subaccounts for new divisions, provisioning additional service instances, and consuming more platform resources, the risk of uncontrolled cost growth and service sprawl increases. SAP BTP addresses this through the entitlement management capabilities of the global account, which allow a central SAP BTP administrator to define precisely which services and how much of the quota each directory and subaccount is permitted to consume.

The SAP BTP cockpit provides cost estimation and consumption tracking that shows which subaccounts are consuming the most resources and which services are approaching their quota limits. For organizations with a large SAP BTP footprint, this visibility is essential for budget forecasting and for identifying subaccounts where services have been provisioned but which are no longer actively used, a common source of unnecessary cost in cloud platform environments. A disciplined entitlement governance process, supported by the SAP BTP cockpit's usage reporting, ensures that the organization's investment in SAP BTP is concentrated in the subaccounts and services that deliver measurable business value and that unused resources are decommissioned before they accumulate significant cost.

Taken together, the lifecycle management capabilities of SAP BTP change transport, operational monitoring, CI/CD integration, and entitlement governance form the operational backbone that allows your organization to run a growing portfolio of custom extensions and integrations with the same discipline and predictability that you bring to the management of your SAP S/4HANA core. The platform was designed with the assumption that the applications and integrations it hosts will evolve continuously, and its lifecycle management tooling reflects that reality.

8.6 Summary

SAP BTP offers an environment for developers to extend and integrate their SAP Autonomous Suite solutions to maximize the value of their investments. SAP BTP provides a layer for faster development via low-code tooling and pro-code tooling to meet unique business challenges. SAP embeds AI inits SAP BTP tools to speed up time to value so that developers can rapidly create applications and test them with a Joule Agent trained on SAP Best Practices.

In the next chapter, we'll dive into another key element of SAP Business AI Platform: SAP Business Data Cloud.

Chapter 9
SAP Business Data Cloud

You've seen a glimpse of SAP Business Data Cloud in previous chapters. Now, it's time to dive into the solution and understand its critical role in the Autonomous Enterprise.

This chapter introduces SAP Business Data Cloud, the knowledge core within SAP Business AI Platform of the Autonomous Enterprise, and how it tackles the challenges in handling the data that generates the Autonomous Enterprise. You'll see how that data can be used to derive superior insights and create AI-driven outcomes, powering the intelligent, modern, autonomous enterprise.

We'll talk about different components of SAP Business Data Cloud as the trusted knowledge core and the capabilities that it brings to bear to support various data and analytics needs of organizations. We'll also cover integration of line-of-business (LoB) applications' data, the corresponding data products, and the intelligent content aligned to some business processes. The chapter will cover how partnerships with other leading data engineering platforms work. We'll conclude with some use case patterns and the value that can be derived from those patterns.

Let's get started with the architecture and vision of SAP Business Data Cloud to provide business semantics and context.

9.1 Vision and Architecture

SAP Business Data Cloud as the trusted knowledge core is a fully managed SaaS solution designed to deliver unified, governed, and curated SAP data with context and rich semantics while seamlessly integrating with third-party sources such as non-SAP systems, data lakes, and spreadsheets. By bringing all relevant enterprise data into a single, harmonized environment, SAP Business Data Cloud empowers LoB leaders to make high-impact decisions rooted in deep business context and semantic consistency.

In this section, we will discuss the vision of SAP Business Data Cloud and the architecture of the solution. We'll look at the different components and the role of each of those components in achieving the vision.

9.1.1 Positioning and Value Drivers

Positioned as a modern *business data fabric*, SAP Business Data Cloud is the strategic pathway forward for organizations struggling with scattered, inconsistent, and siloed enterprise data. It enables AI-driven insights by providing access to the most comprehensive and mission-critical business data enriched with original SAP semantics. SAP Business Data Cloud is the one-stop solution for all SAP data. In practice, SAP Business Data Cloud serves as the central data lake for all SAP data, acting as a crucial node within an organization's broader data mesh architecture.

SAP Business Data Cloud was specifically developed to address the following evolving requirements from organizations:

- Easier access to their SAP data without the need for complex data-preparation steps
- Easier access to all SAP data from all LoB applications, such as SAP S/4HANA, SAP SuccessFactors, SAP Concur, and SAP Ariba without the need for complex extraction, transformation, or preparation steps
- A unified business view of their data regardless of which SAP application it originated from
- A secure, trusted data foundation suitable for building AI-powered applications and analytical solutions
- Ready-to-use intelligent content, enabling business users to quickly derive insights and improve decision-making
- Reduced landscape complexity, achieved by consolidating tools and minimizing the number of disparate technologies in use

The key value drivers of SAP Business Data Cloud include the following:

- Trusted and semantically rich data products, harmonized and governed according to SAP One Domain Model
- AI-driven intelligent content and intelligent packages built with context-rich SAP data
- Seamless interoperability with best-in-class data engineering platforms through zero-copy data sharing

The Autonomous Enterprise (via RISE with SAP or GROW with SAP, and SAP's LoB SaaS applications) publishes a rich set of standard data products out of the box in SAP Business Data Cloud. These data products are fully harmonized according to SAP One Domain Model, ensuring consistent semantics across LoBs. They are governed, curated, and maintained to provide a trusted and semantically rich foundation for downstream use cases.

Once available in the SAP Business Data Cloud object store, these data products are used within SAP Datasphere to build advanced analytical models and semantic views. These analytical views serve as the basis for SAP Analytics Cloud content, enabling the creation of dashboards, reports, and interactive stories that deliver actionable insights to business users.

Beyond analytics, the same data products are also consumed directly by cloud applications built using SAP Build. These applications are further enhanced with SAP Business AI capabilities, allowing organizations to take advantage of intelligent workflows, predictive insights, and contextual automation, all powered by harmonized, governed SAP data.

Figure 9.1 illustrates how SAP Business Data Cloud evolved to become a business data fabric solution delivering data products out of the box, following a harmonized data model that supports use cases and analytical and domain content leveraging SAP Business AI.

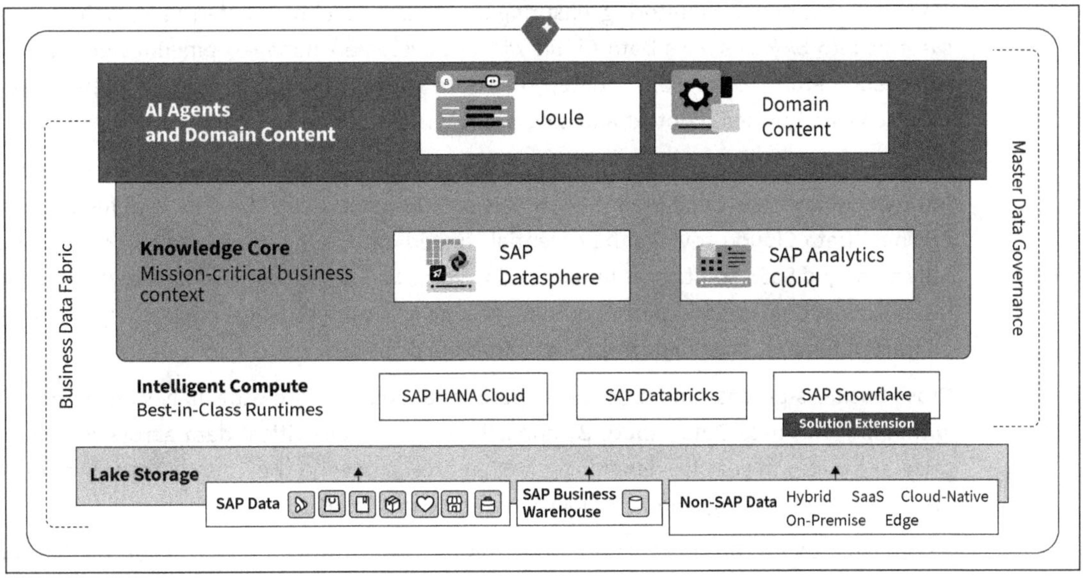

Figure 9.1: SAP Business Data Cloud

9.1.2 Applications, Data, and Business AI

The Autonomous Enterprise is not a collection of isolated tools. The tools function as an integrated ecosystem that connects business processes from end to end. This deep integration ensures that every transaction, event, and action captured within the suite produces trusted, business-critical data at its point of origin, preserving accuracy, context, and semantic integrity.

The analytics strategy for the Autonomous Enterprise spans multiple dimensions to support varying organizational needs. At the foundational level, task-oriented operational analytics provide users with real-time insights directly within their daily workflows. Beyond individual tasks, cross-business-process and cross LoB operational analytics offer a unified view of activities spanning multiple domains, enabling better coordination and decision-making across the enterprise. At the most advanced layer, organizations can leverage AI- and ML-driven analytics, unlocking predictive insights and intelligent scenarios that anticipate outcomes and recommend actions.

Embedded analytics, delivered as an integral part of the Autonomous Enterprise, play a central role in supporting task-oriented operational reporting. These insights are rendered through SAP Fiori applications and SAP Analytics Cloud content, providing business users with intuitive, real-time visualizations that guide operational decisions.

The data generated by the Autonomous Enterprise is centralized, harmonized, curated, and enriched within SAP Business Data Cloud as the trusted knowledge core through a fully managed push-based replication mechanism. SAP oversees all aspects of this replication from orchestration to monitoring, ensuring that data is reliably transferred from the source systems into SAP Business Data Cloud without customer-managed pipelines or complex extraction setups. Once consolidated, the data is unified into a single source of truth, enriched with business context and consistent semantics.

Reliable, contextualized data of this kind becomes foundational for advanced analytics and AI-driven innovation. In an era in which AI is only as powerful as the data that fuels it, SAP Business Data Cloud provides the essential trusted data foundation for all SAP data. As a fully managed SaaS solution, it unifies and governs SAP data from end to end while seamlessly connecting to third-party sources. This gives business leaders the comprehensive context required to make informed and impactful decisions.

Cross-business-process and cross-LoB operational analytics are especially well supported within SAP Business Data Cloud. By consolidating mission-critical data across the enterprise, the platform enables richer insights that extend beyond the boundaries of individual applications, delivering a holistic view of organizational performance.

A core pillar of SAP Business Data Cloud is its role as the SAP *data lake*, a central, harmonized repository that can be shared with external platforms through a zero-copy approach using the Delta Sharing protocol (or other protocols in the future). This allows organizations to securely exchange SAP data with the following leading data and AI platforms, all without physically moving or duplicating data:

- SAP Databricks
- SAP Snowflake
- Google's BigQuery
- Microsoft Fabric
- Amazon Athena
- Palantir

SAP Business Data Cloud includes an original equipment manufacturer (OEM) version of SAP Databricks, enabling organizations to run sophisticated AI and ML workloads directly within the environment. This tightly integrated AI/ML capability ensures that advanced scenarios like predictive analytics, automated decisioning, and generative AI use cases can be executed close to the governed SAP data that powers them.

Through this unified architecture, SAP Business Data Cloud becomes the strategic data foundation for modern analytics, cross-process insights, and domain content across the SAP ecosystem.

As SAP applications execute business processes, they continuously generate data. This data is enriched with semantic meaning and business context, plus associations that make it significantly more valuable once it is centralized and governed within SAP Business Data Cloud. From there, this context-aware, high-quality data becomes the foundation for generating meaningful insights. These insights feed directly into SAP Business AI, where intelligent tools and agents interpret the information and identify opportunities for improvement. SAP Business AI then invokes AI agents that optimize business processes: adjusting workflows, recommending next actions, and enabling better outcomes across the enterprise.

This creates a closed-loop intelligence cycle: processes in applications → generate data → governed data produces insights → insights drive AI → AI optimizes processes in applications.

Figure 9.2 demonstrates how applications, SAP Business Data Cloud, and SAP Business AI come together. The data generated from applications is managed and governed by SAP Business Data Cloud. Semantically rich data products and their relationships with each other drive SAP Business AI. SAP provides unrivaled AI tools that use context-rich data and advanced applications to drive automation, innovate, and adapt quickly.

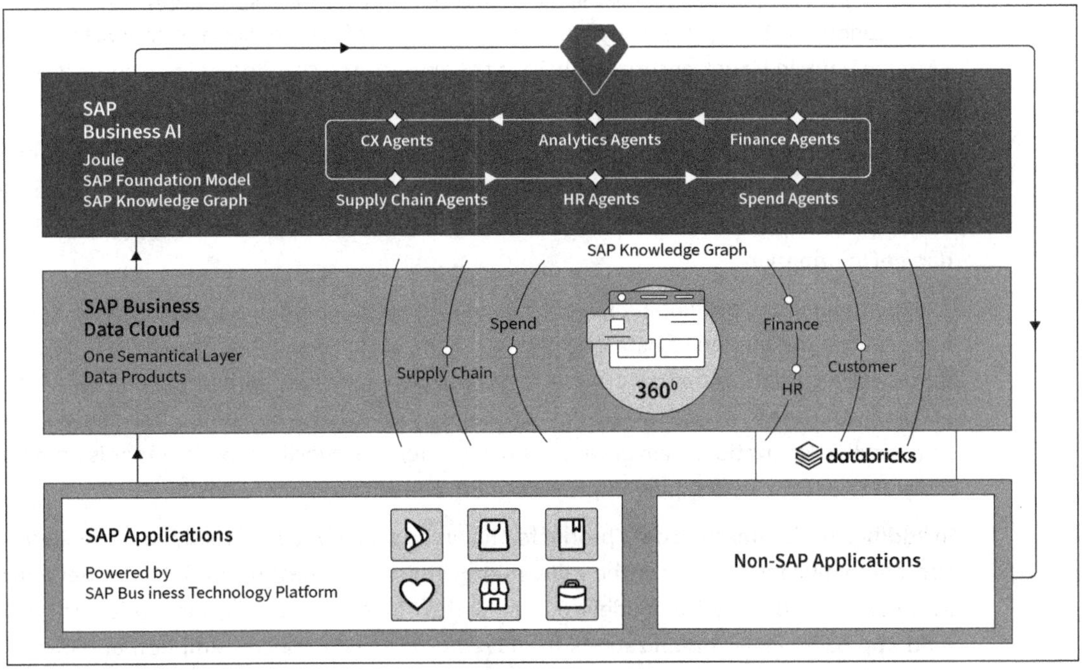

Figure 9.2: Application, Data, and AI

9.1.3 Current Data Challenges

For many organizations, unlocking the true value of enterprise data, especially the data embedded in the Autonomous Enterprise, remains one of the most persistent challenges in their digital transformation journey. Historically, the difficulty has not been the lack of data but rather the inability to access governed, curated, and context-rich data in a reliable and scalable manner. As the role of AI expands, this challenge has only grown more urgent. In essence, AI is limited by the quality of the data that fuels it, making structured, well-governed, semantically consistent data indispensable.

The imbalance in how organizations spend their time illustrates this problem starkly. Studies show that less than 20 percent of the average data professional's time is spent performing actual analysis (*http://s-prs.co/v633508*), the activity that generates insight and business value. The remaining 80 percent or more is consumed by the tedious and fragmented work of searching for data, preparing it, reconciling definitions, and validating its governance posture. Even after this investment, data quality and accuracy are not always guaranteed, creating friction for downstream processes and undermining trust among business users.

To address these long-standing issues, SAP introduced the concept of *data products* within SAP Business Data Cloud. These data products represent a significant evolution: They are standardized, ready-to-use business data models spanning key domains such as finance, human resources, procurement, and sales. Unlike traditional extracted datasets, data products preserve the original business semantics and contextual meaning embedded in the Autonomous Enterprise. They follow a common set of rules to achieve compliance with SAP One Domain Model, ensuring consistent semantics across different LoBs and across all use cases.

Each data product includes comprehensive technical and business metadata, describing the purpose, rules, and usage expectations of the data. Metadata serves as the semantic backbone of SAP Business Data Cloud, enabling consistent understanding and proper handling of information.

These products are developed, governed, curated, and maintained by SAP, relieving customers from the burden of data engineering at the basic level. SAP ensures that the standard-delivered data products adhere to established service-level agreements (SLAs) for accuracy, freshness, update frequency, and latency relative to the Autonomous Enterprise. As a result, organizations gain access to high-quality, trustworthy SAP data that is immediately usable for analytics, planning, integration, and AI-driven scenarios.

In addition to the standard data products, the broader ecosystem of ready-made analytical content significantly enhances the value proposition for business users. SAP provides a rich set of predefined analytical models, SAP Analytics Cloud dashboards, and stories. This layered approach allows organizations to move beyond raw data consumption and directly use actionable insights, predictive capabilities, and intuitive visualizations. As a result, users gain immediate access to business-ready analytics and intelligent workflows, reducing the need for custom development from scratch while accelerating time to value across the enterprise.

9.1.4 Key Components

SAP Business Data Cloud consists of multiple capabilities accessed through different components. Let's walk through them:

- **SAP Business Data Cloud cockpit**
  The SAP Business Data Cloud cockpit manages the system, browses/installs domain content and data packages, and shares data with SAP Databricks for AI/ML. It serves as the primary dashboard for managing the environment. The formation of different components and integrating different source systems are managed in the cockpit.
- **SAP Datasphere**
  At the heart of SAP Business Data Cloud is SAP Datasphere, which provides the foundational data models on top of the data products. The models are provided by SAP to support the predelivered domain content, but custom data models can be added. SAP Datasphere and its open data ecosystem are the foundations for a business data fabric. It allows you to accelerate time to value by automatically reusing the semantic definitions and associations from SAP applications to access authoritative data, harmonizing heterogeneous data into a diverse data landscape business semantic model so that all the data can be enriched and accessed across hybrid and cloud environments no matter where it resides.
- **Object store**
  The object store serves as the scalable, cloud-native storage layer within the SAP Business Data Cloud architecture. Designed to handle large volumes of unstructured data such as files, images, and binary large objects (BLOBs), it provides the flexible storage required for modern lakehouse patterns. The object store integrates seamlessly with major hyperscalers' storage services, including Amazon S3, Azure Blob Storage, and others, enabling organizations to securely stage and manage data within their preferred cloud environments. It also supports delta lake formats and facilitates zero-copy data sharing, significantly reducing the cost and complexity typically associated with extract, transform, and load (ETL)–heavy data movements. Within SAP Business Data Cloud, the object store layer hosts both SAP standard data products and customer-generated or -enriched data products. These data products become visible within SAP Datasphere, where they can be easily discovered and explored through the unified data catalog. Data replication from SAP source applications flows into the foundation services layer of SAP Business Data Cloud. Here, the raw data is harmonized with other relevant business information originating from different applications.
- **SAP Analytics Cloud**
  SAP Analytics Cloud brings together advanced analytics and enterprise planning capabilities to help organizations fully realize the value of their mission-critical business applications and their most important data assets. As a unified analytics and planning platform, SAP Analytics Cloud empowers users to explore data, model scenarios, and make

informed decisions, all within a single, integrated environment closely aligned with SAP's business processes. SAP Analytics Cloud also embraces the power of generative AI, enabling users to automate reporting, uncover hidden insights, and develop business plans more efficiently. Through Joule, SAP's AI copilot, business users can ask questions in natural language, generate visualizations automatically, and accelerate the planning and forecasting process. SAP Analytics Cloud further supports collaborative planning, unifying financial, supply chain, workforce, and operational planning within a single solution. Through this combination of analytics, planning, and generative AI, SAP Analytics Cloud becomes a critical component of the modern enterprise's decision-making landscape.

- **Domain content**
  Domain content extends and enhances traditional transactional systems such as enterprise resource planning (ERP), human capital management (HCM), and spend management by leveraging the underlying data models, ML capabilities, and analytical assets within SAP Business Data Cloud. By grounding intelligence in harmonized, context-rich SAP data, this content delivers personalized user experiences that help business users better understand the many dimensions of their operations. Domain content provides insights, recommendations, and guided actions designed to improve the outcomes that matter most to LoB users. Instead of requiring users to navigate complex datasets or interpret raw reports, through this approach, domain content offers ready-to-use intelligence embedded directly into daily workflows. It integrates multiple layers of SAP's data and analytics stack, including data products, SAP Datasphere models, SAP Analytics Cloud stories, and cloud applications, into a cohesive experience that empowers users to respond effectively to opportunities and risks. Ultimately, domain content bridges the gap between data and action, transforming curated SAP data into meaningful, contextualized insights that drive improved business performance.

- **SAP Business Warehouse**
  SAP Business Data Cloud is the path forward for SAP Business Warehouse (SAP BW) customers. SAP BW in the private cloud within SAP Business Data Cloud provides a fully hosted cloud environment that preserves existing SAP BW investments while enabling modern cloud consumption patterns. This environment allows organizations to continue operating their SAP BW data models and processes without disruption, ensuring full compatibility with their current SAP BW artifacts, queries, and data flows. Within this architecture, SAP BW data stored in the managed object store can be accessed seamlessly and exposed to downstream consumers. A key capability enabling this is the data products generator, a new functionality designed to transform SAP BW objects directly into governed data products without rebuilding models or recreating logic. Through this mechanism, SAP BW InfoProviders, composite providers, and other SAP BW modeling artifacts can be published as standardized data products in SAP Business Data Cloud,

making them easily consumable by analytics platforms, AI tools, and third-party data solutions.

- **SAP HANA Cloud**
  SAP HANA Cloud serves as the foundational in-memory database and high-performance processing engine within the SAP Business Data Cloud architecture. Because SAP HANA Cloud is natively integrated into the SAP Business Data Cloud stack as part of the SAP Business Data Cloud formation, cloud applications built using the SAP Build portfolio can leverage its capabilities seamlessly. These applications can take advantage of SAP HANA's graph engine, spatial processing, and advanced analytical functions to deliver rich, context-aware SAP Business AI features. Moreover, SAP HANA Cloud calculation views—core modeling artifacts—can be published directly as data products within SAP Business Data Cloud. This allows organizations to expose curated, modeled, and semantically enriched views as governed data products that can be consumed by analytics, AI applications, and external data platforms.
- **SAP Databricks**
  SAP Databricks is an OEM version of the Databricks platform, tightly integrated into the SAP ecosystem. When combined with SAP Datasphere, organizations gain a powerful and unified foundation for modern data management, analytics, and AI. SAP Datasphere contributes business context, semantic enrichment, and governed SAP data, while SAP Databricks adds advanced AI/ML capabilities. This synergy allows organizations to derive deeper, more meaningful insights by applying the analytical and ML capabilities of Databricks on top of richly contextualized SAP data.
- **SAP Business Data Cloud Connect**
  SAP Business Data Cloud Connect offers more than just bidirectional data sharing. While it utilizes data sharing, it also integrates numerous technologies to securely enroll partner platforms, enabling discovery and self-service requests to access data products (not just generic database-level data). SAP Business Data Cloud Connect also will eventually support other formats beyond delta tables for SAP Databricks, SAP Snowflake, Google, and other partners to be announced in the future. In short, SAP Business Data Cloud Connect introduces several innovations, with data sharing being just one component. By offering simplified zero-copy access to mission-critical SAP data, SAP Business Data Cloud Connect eliminates the need for complex and costly ETL operations and reduces the risk of data inconsistencies.
- **SAP Master Data Governance (SAP MDG)**
  SAP MDG serves as the core governance and cleansing layer within the SAP Business Data Cloud. It unifies and validates master data—creating golden records across customer, product, supplier, and other domains—so it can be securely consumed by AI and analytics applications.

Figure 9.3 shows the key components of SAP Business Data Cloud and how they come together.

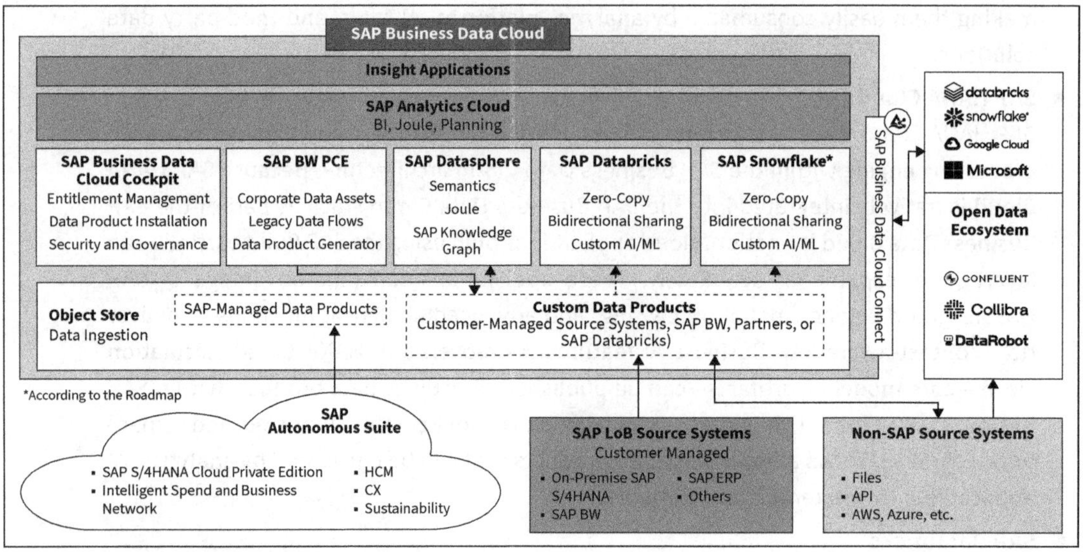

Figure 9.3: SAP Business Data Cloud Components

9.2 Setting Up SAP Business Data Cloud

Setting up SAP Business Data Cloud involves validating entitlements in SAP for Me, provisioning the tenants, and creating a *formation* in the SAP Business Data Cloud cockpit to link different source systems. Provisioning SAP Business Data Cloud requires a clear understanding of the entitlements, infrastructure, security, and data requirements.

There are various steps involved, which we'll walk through in the following sections:

- Performing technical and administrative steps
- Provisioning the SAP Business Data Cloud environment
- Creating the SAP Business Data Cloud formation
- Establishing user access and connecting data

9.2.1 Technical and Administrative Steps

Before configuring the system, ensure that all technical, administrative, and commercial prerequisites are met. Here's an overview of the requirements:

- **Verify S-user IDs**
  Confirm there is a valid S-user ID linked to the customer account associated with the SAP Business Data Cloud order. The system uses the S-user ID to verify entitlements. If the ID is not correctly linked, provisioning options will not appear in the SAP for Me portal.

- **Confirm required authorization**
  Ensure the S-user ID has the **Edit Cloud Data** authorization. This role is necessary to trigger and manage cloud system provisioning in SAP for Me.
- **Check the contract status**
  Verify that the SAP Business Data Cloud contract has officially started. Capacity units are only available for allocation after the contract start date.
- **Validate SAP Business Data Cloud entitlements**
  Log into SAP for Me to confirm that SAP Business Data Cloud entitlements are correctly assigned to the customer number.
- **Decide on landscape topology**
  Set the data center and region.
- **Identity provider**
  Choose an identity provider, such as SAP Cloud Identity Services, to handle authentication across SAP Business Data Cloud.

9.2.2 Provisioning the SAP Business Data Cloud Environment

The provisioning process follows a specific sequence. First, the SAP Business Data Cloud core (SAP Business Data Cloud cockpit) needs to be provisioned. Once the core is active, individual applications such as SAP Datasphere, SAP Analytics Cloud, and optionally SAP Databricks can be added. All provisioning tasks are managed through the SAP for Me portal.

Let's walk through the steps for provisioning the SAP Business Data Cloud cockpit:

1. Log into your SAP for Me account. Under the **Portfolio & Products** tab, you'll see the **SAP Business Data Cloud** tile, as shown in Figure 9.4.

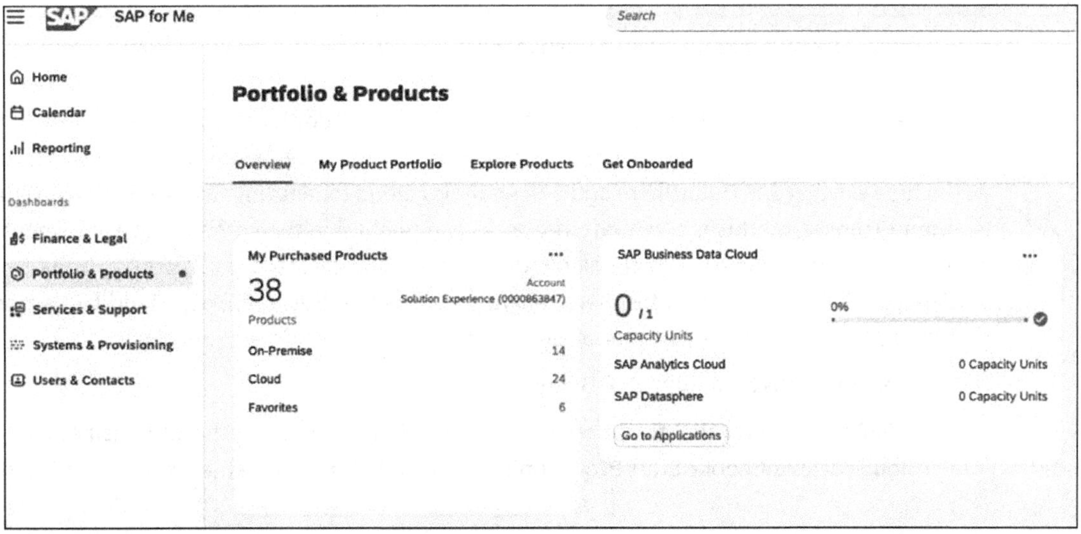

Figure 9.4: SAP for Me Portfolio and Products Tab

2. After you select the **SAP Business Data Cloud** tile, choose **Enable Now**.
3. In the screen that appears, enter a solution name and resource group.
4. The SAP Business Data Cloud tenant now is initiated for provisioning. The system will take you to the next screen, where you provide relevant input parameters: region, data center, owner email, owner name, and host name.
5. Review and confirm your entries.

Once the provisioning is completed, you can explore the solution and tenant resources. If you check the resource group and the tenant, you will see that the SAP Business Data Cloud core is enabled (see Figure 9.5).

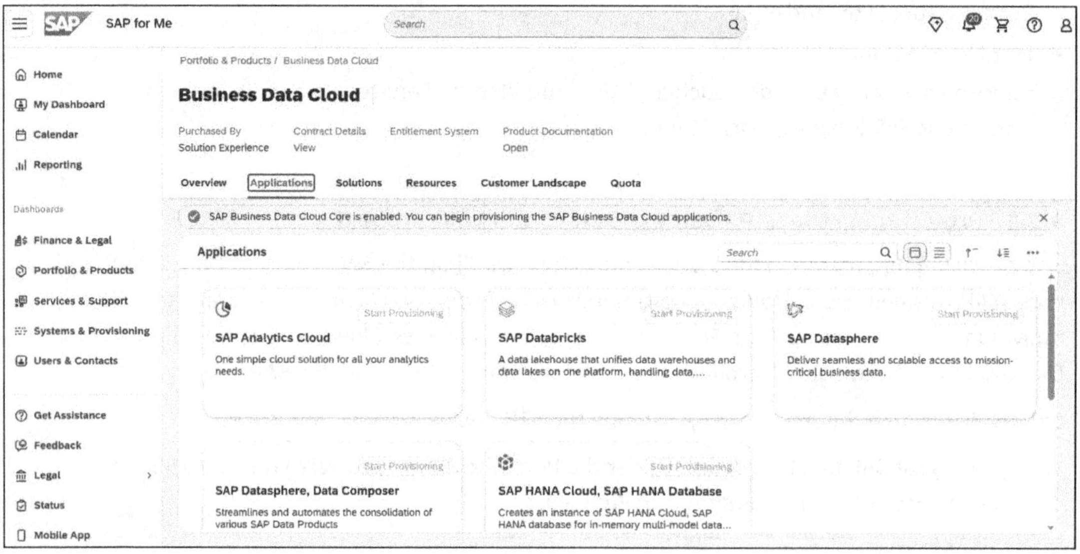

Figure 9.5: SAP Business Data Cloud Provisioned

Once SAP Business Data Cloud cockpit is provisioned, additional components like SAP Datasphere and SAP Analytics Cloud can be provisioned independently. It is possible to include existing SAP Datasphere and SAP Analytics Cloud tenants into an SAP Business Data Cloud landscape. This is often referred to as *rewiring*. Once the commercial adjustments have been made, there are minimal, nondisruptive, technical steps to carry out to bring SAP Analytics Cloud and SAP Datasphere tenants into the SAP Business Data Cloud landscape through rewiring. There is no migration required and no downtime necessary.

Here are the steps for adding SAP Datasphere:

1. In SAP for Me, the **SAP Datasphere** tile is visible after provisioning the SAP Business Data Cloud cockpit. Choose **Start Provisioning** in the **SAP Datasphere** tile, as shown in Figure 9.6.

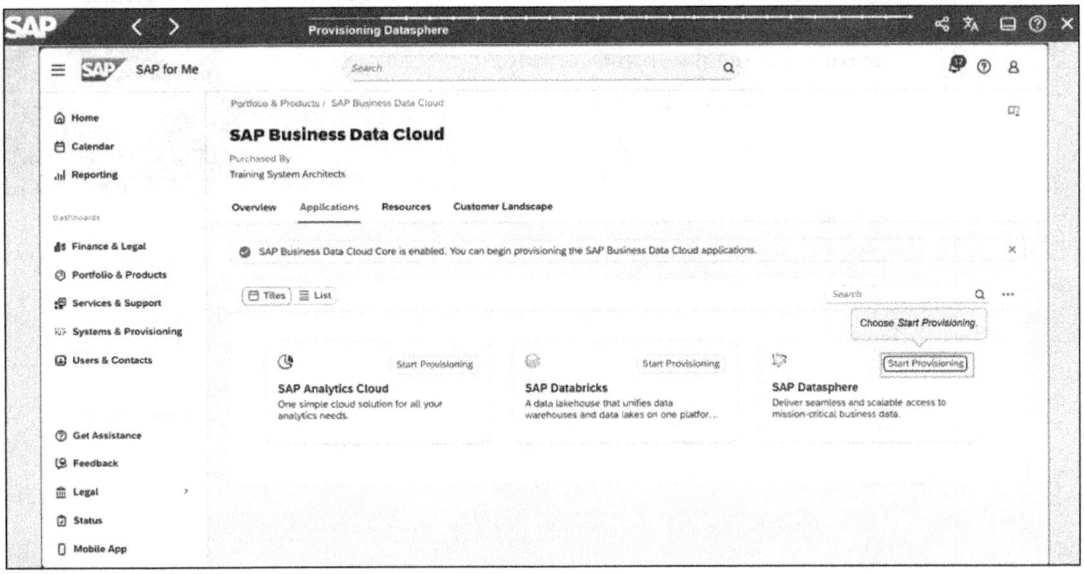

Figure 9.6: Provision SAP Datasphere

2. Enter the solution name, resource group, and business type. The solution name is the name of the SAP Datasphere tenant. Resource groups can be selected from the drop-down menu. Set **Business Type** to **Production**.
3. Once these details are entered, input the number of capacity units that are needed for the SAP Datasphere component. (A proper sizing exercise should be done to figure out the number of SAP Business Data Cloud capacity units needed for the SAP Datasphere tenant.)
4. Enter the parameters needed to set up the SAP Datasphere tenant. The parameters required are region, data center (same as the cockpit data center), owner's email, owner's name, and tenant host name for the tenant URL used to access the SAP Datasphere tenant.
5. Review the details and finish the process.

With a similar approach to provisioning SAP Datasphere, you can provision SAP Analytics Cloud and SAP Databricks. Figure 9.7 shows the provisioned SAP Business Data Cloud cockpit home screen.

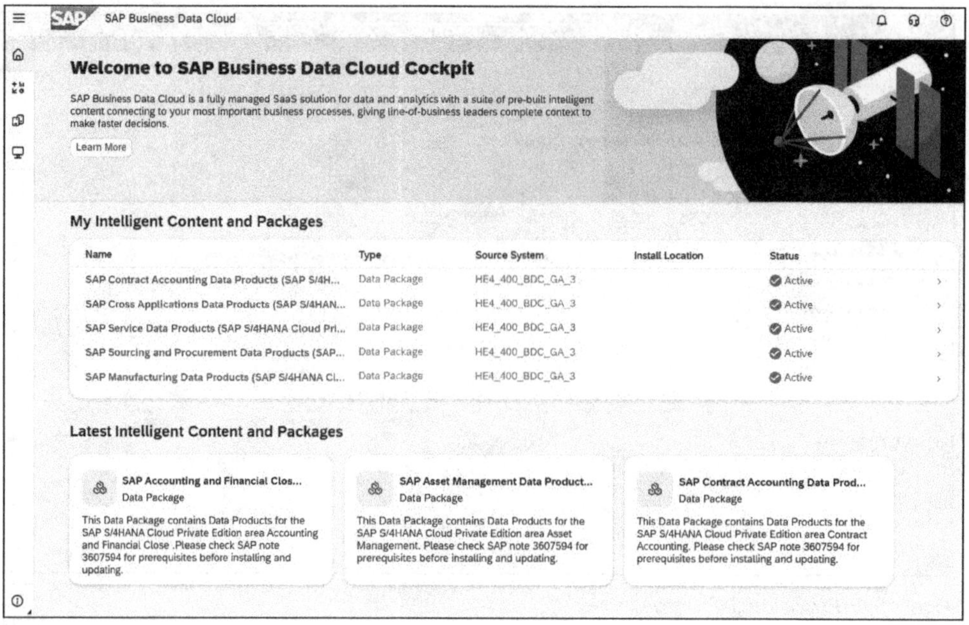

Figure 9.7: Finish Provisioning SAP Datasphere

9.2.3 Creating the SAP Business Data Cloud Formation

An *SAP Business Data Cloud formation* is a logical grouping of SAP components that enables automated integration between them. Formations act as structured configurations that define how SAP data products and applications are orchestrated across the landscape, simplifying connectivity and providing a unified view of all systems.

The formations can be created for different groups of users or for different regions, with separate formations for development, quality, and production environment.

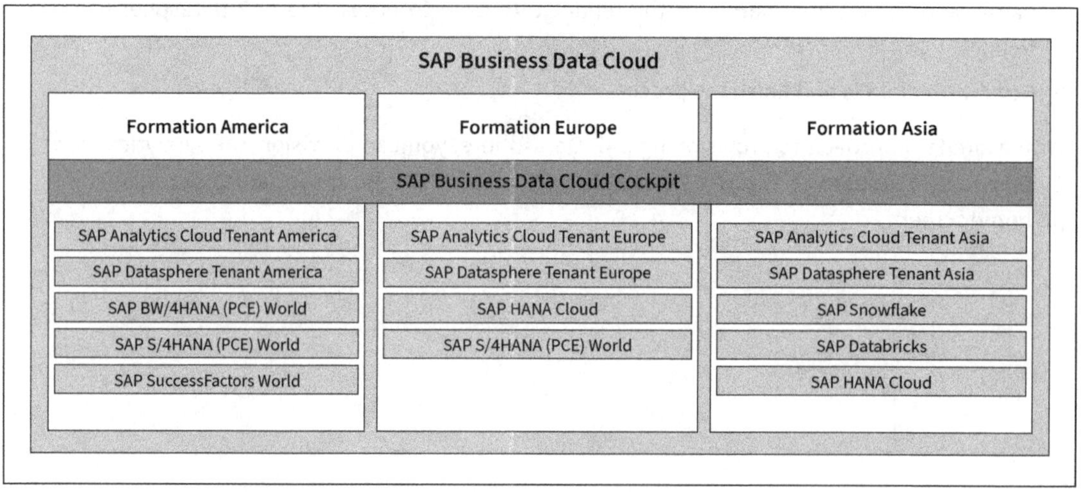

Figure 9.8: SAP Business Data Cloud Intelligent Content

Once a formation is defined, domain content is discovered using the SAP Business Data Cloud cockpit. The cockpit contains a list of the intelligent packages that are available to install, represented in Figure 9.8. The list includes all the intelligent packages that are provided by the source systems that are included in the SAP Business Data Cloud formation. Using the SAP Business Data Cloud cockpit, the domain content can be previewed before installing the content, and the list of data products that are consumed by the intelligent packages are listed. The installation of the intelligent packages is almost entirely an automated process. During installation, a check is performed to ensure it is at the minimum version required to support the intelligent package. If it is not, installation cannot proceed.

In the SAP Business Data Cloud cockpit home screen, you can see the list of all intelligent content available to install once the formation is completed (see Figure 9.9 for an example formation):

- A formation always includes SAP Business Data Cloud cockpit, so you do not need to add the cockpit to a formation.
- You can choose other SAP Business Data Cloud components to include in the formation, such as SAP Datasphere, SAP Analytics Cloud, and SAP Databricks.
- Formations must include all the source systems that provide data, such as SAP S/4HANA Cloud and SAP SuccessFactors.
- Some components can be reused across different formations. For example, the same SAP S/4HANA source could be included in all regional formations.
- Formations are managed and created using the SAP for Me portal.

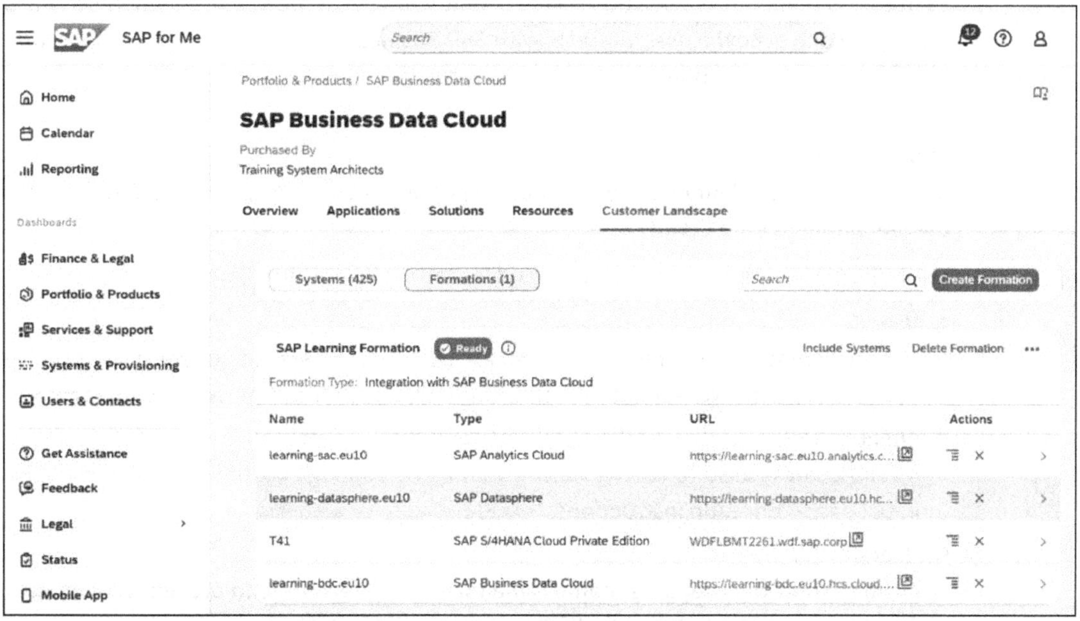

Figure 9.9: SAP Business Data Cloud Formation

9.2.4 Users and Connecting Data

Once the SAP Business Data Cloud landscape is set up, user access can be configured. We'll provide an overview of the different processes in the following sections.

Managing Roles

Roles determine the specific permissions and privileges granted to a user. Depending on their job responsibilities, a user can be assigned one or multiple roles. SAP provides some standard roles. Custom roles can be created to meet an organization's needs, either by starting with a blank role or by using a standard role template as a foundation.

SAP provides the following standard roles:

- **BD administrator**
  Grants full administrative access to the SAP Business Data Cloud cockpit.
- **BD viewer**
  Provides read-only access to the SAP Business Data Cloud cockpit.
- **Catalog administrator**
  Manages data governance tasks, such as connecting source systems to extract metadata, maintaining business glossaries, creating classification tags, and publishing assets.
- **Catalog user**
  Enables users to search for and discover data products or domain content. This role is ideal for modelers creating new content or viewers consuming governed assets.

Each role is made up of specific privileges that you choose from a predefined list. These privileges control how a user interacts with SAP Business Data Cloud. Some complex tasks may require a combination of different privileges.

Managing Users

Users can be created individually, or you can perform a bulk import using a CSV file within the SAP Business Data Cloud cockpit.

The following properties should be defined for setting up a user:

- **User ID**
  A unique identifier that must use uppercase letters, numbers, and underscores (20 characters maximum). Note that the user ID cannot be modified after the account is created.
- **Email address**
  A valid address used to send a welcome email containing account activation steps and initial password setup instructions.
- **User details**
  Although only the user's last name is mandatory, we recommend providing a first name and a display name to help identify users across the system.
- **Role assignments**
  These settings define the specific access levels and permissions for the user.

Connecting Data

During installation of intelligent packages, many artifacts are generated. The sequence is as follows:

1. Dependent data products are installed, and data is loaded to the object store of the SAP Business Data Cloud foundation services.
2. SAP-managed spaces are automatically created in SAP Datasphere to store intelligent packages, artifacts, and data.
3. Replication flows, local tables, views, and analytical models are generated in the SAP Datasphere spaces.
4. A protected SAP Analytics Cloud workspace is generated.
5. SAP Analytics Cloud stories are generated in the protected workspace.
6. Scoped roles are generated to enable the business users.

Once all the artifacts are generated, the replication flows in the ingestion space are executed with the initial and delta load type, to onboard the data from foundation services to the local tables in the ingestion space of SAP Datasphere.

Once the domain content has been installed, there are usually postinstallation tasks to complete. The tasks will be specific to the domain content.

9.3 Integration with SAP Cloud ERP and LoB Applications

Once SAP Cloud ERP systems and LoB applications are integrated into the SAP Business Data Cloud landscape, the installation of intelligent packages and the data hydration into the corresponding data products becomes a streamlined and efficient process.

However, the prerequisites and implementation steps vary across different applications, requiring careful evaluation and preparation for each system to ensure successful integration and data replication.

The following sections discuss the prerequisites and the steps for integrating SAP Business Data Cloud with SAP S/4HANA Cloud Private Edition, SAP S/4HANA Cloud Public Edition, and LoB applications.

9.3.1 Prerequisites

Before establishing connectivity between SAP S/4HANA Cloud Private Edition and SAP Business Data Cloud, ensure the following prerequisites are met:

- The ABAP data integration component—specifically, ABAP Push Replication Service (APRS)—within the SAP S/4HANA system enables the replication of data from released core data services (CDS) views to SAP Business Data Cloud.
- The set of CDS views available for replication is determined by the data products delivered with the specific SAP S/4HANA version in use.

- Certain capabilities of ABAP data integration may only be accessible in systems that are either fully patched or running on the latest SAP S/4HANA release, making it essential to verify the system's version and patch level before implementation.
- We recommend using the Cloud Integration Automation service's guided workflow to set up the integration of SAP S/4HANA Cloud Private Edition with SAP Business Data Cloud.
- Make sure the SAP cloud connector version installed is the minimum required or higher.

Cloud Integration Automation

The Cloud Integration Automation workflow for integrating SAP S/4HANA Cloud Private Edition with SAP Business Data Cloud streamlines and simplifies the overall integration process by providing step-by-step, role-based guidance along with built-in monitoring capabilities. It is a semiautomated workflow that leverages tenant-specific information, such as system target URLs and reusable configuration settings, to accelerate setup.

These capabilities not only reduce manual effort but also help ensure faster, more consistent, and reliable integration across systems.

9.3.2 SAP S/4HANA Cloud Private Edition

After setting up the prerequisites, you're ready to proceed with the integration steps. We'll walk through them in the following sections.

Step 1: Install the Relevant Components in SAP S/4HANA Cloud Private Edition

The ABAP data integration components (APRS) in the SAP S/4HANA system can be used to replicate data from released CDS views to SAP Business Data Cloud. Which CDS views are available depends on the available data products defined in the used SAP S/4HANA version. Some features of the ABAP data integration for SAP Business Data Cloud may only be available on fully patched systems or systems of the latest release.

If the `DHAFR_FOS_VERSION_GET` function module is absent, then the system is not yet enabled for integration.

Refer to SAP Note 3500131 (ABAP Data Integration for SAP Business Data Cloud) for more information.

Step 2: Set Up Inbound Communication for SAP S/4HANA Cloud Private Edition

To send a message to SAP S/4HANA Cloud Private Edition, SAP Business Data Cloud requires a technical user with the correct authorization.

Once the technical user is created in Transaction SU01, run the role maintenance transaction (Transaction PFCG). Choose **Role • Upload**, then upload the role file that is attached to SAP Note 3499606.

Generate the authorization profiles after completing the upload. Follow the steps to copy the SAP_BDC_REMOTE template role to a name from the customer namespace. See the following resource for more details: *http://s-prs.co/v633509*.

In addition, to make a secure call to SAP Business Data Cloud, SAP S/4HANA Cloud Private Edition needs a client certificate for authentication. Follow the steps detailed at *http://s-prs.co/v633510*.

Step 3: Create a Formation of Type Integration with SAP Business Data Cloud

We discussed creating a formation in Section 9.2.3.

Step 4: Set Up Outbound Communication for SAP S/4HANA Cloud Private Edition

In this step, you perform the client registration, creation of RFC connections, and registration of outbound connections. This establishes that data will be flowing from SAP S/4HANA Cloud Private Edition to SAP Business Data Cloud. The process is as follows:

1. **Upload the client certificate to SAP S/4HANA Cloud Private Edition**
   Manually upload the client certificate into Transaction STRUST (Trust Manager). This includes two important elements:
   - Client certificates (see *http://s-prs.co/v633511* for more information)
   - SAP Business Data Cloud connection properties (the configuration includes details of SAP Business Data Cloud connection properties, such as the host, port, and path prefix)
2. **Create an RFC destination in SAP S/4HANA Cloud Private Edition**
   Use the formation configuration details available in the SAP Business Data Cloud connection properties and manually create a destination in SAP S/4HANA Cloud Private Edition. Configure the destination using the formation configuration details. This destination enables SAP S/4HANA Cloud Private Edition to call SAP Business Data Cloud.

   Create RFC connections in Transaction SM59. Perform a quick SSL test to check whether connectivity between SAP S/4HANA Cloud Private Edition and SAP Business Data Cloud works by choosing the **Connection Test** button.
3. **Register the outbound connection in SAP S/4HANA Cloud Private Edition**
   Once RFC connection is created, register the outbound connection in the ABAP Integration Administration cockpit (Transaction DHADM) in SAP S/4HANA Cloud Private Edition. This is an administration tool used in different scenarios to extract data from an SAP system.
4. **Set up virtual connection**
   Use the RFC connection to set up a virtual connection that enables the SAP Business Data Cloud integration to access the cloud file storage system. Create this virtual connection in edit mode; the system will add it to the list of outbound connections.

5. **Add custom roles**
   To use the Monitoring and Support cockpit and perform expert functions, a custom role needs to be created from the `SAP_APRS_ADMINISTRATOR` template role.

Step 5: Configure the Cloud Connector

SAP S/4HANA Cloud Private Edition systems are installed in an isolated virtual private network provided and managed by SAP. SAP S/4HANA Cloud Private Edition systems are not exposed to the internet and are not directly accessible from SAP for Me. To access SAP S/4HANA Cloud Private Edition systems from SAP for Me, the cloud connector and SAP Connectivity service should be used.

Complete instructions for cloud connector configuration are beyond the scope of this book. For more information, see *http://s-prs.co/v633512*.

9.3.3 LoB Applications

Unlike SAP S/4HANA Cloud Private Edition integrations, there is no need to create users or certificates in SAP SuccessFactors manually. The certificate exchange happens automatically when the SAP SuccessFactors tenant is added to the formation in SAP for Me. Once the BD administrator role is set up in SAP Business Data Cloud, you can follow the steps we'll outline in this section.

Note

At the time of writing, integration with LoB apps like SAP Concur is not yet generally available. We'll focus on the currently available integration option, SAP SuccessFactors.

First, follow these initial steps in SAP Business Data Cloud:

1. Activate data packages one at a time (not in parallel).
2. Once all required data packages are activated, install the People Intelligence application from the SAP Business Data Cloud cockpit. This creates spaces and models in SAP Datasphere and imports stories into SAP Analytics Cloud.

Next, you can perform the following steps in SAP SuccessFactors:

1. Assign analytics admin authorization to the SAP SuccessFactors admin user.
2. In SAP SuccessFactors, go to **Admin Center • Manage Analytics Data Products** and configure the derived data product settings.

Finally, proceed with the following steps in SAP Business Data Cloud:

1. Maintain data access control by navigating to the data builder. Select the preparation space, then upload the permission table to the data access control table with the technical name `sap.sf.wrk.Permissions`.

2. With the BD administrator role, create SAP Business Data Cloud users and manage secure access through roles and privileges to make the domain content and the data products accessible to other employees, including admins, designers, and consumers of analytical dashboards in the organization.
3. Monitor the import in SAP Datasphere; review import logs for tables, replication flows, and model deployments.
4. Once all steps are complete, prebuilt SAP Analytics Cloud stories can be viewed. As a test, open the SAP Analytics Cloud story for workforce composition. If data appears correctly, then the installation is complete and verified!

9.3.4 SAP S/4HANA Cloud Public Edition

The SAP S/4HANA Cloud Public Edition integration to SAP Business Data Cloud is fully automated by SAP once it is added to a formation of type integration with SAP Business Data Cloud and the following steps have been completed:

1. Set up trust between SAP Cloud Identity Services (specifically, the Identity Authentication service) and both tenants.
2. Prepare SAP S/4HANA Cloud Public Edition by assigning communication management roles (e.g., `SAP_BR_ADMINISTRATOR` or admin equivalents) to create communication artifacts.
3. Whitelist CDS views. Ensure the views required for analytics are released/activated in the SAP S/4HANA Cloud tenant.
4. Configure the connection in SAP Business Data Cloud. Create an SAP S/4HANA Cloud connection with the target URL: S/4HC<service root> (fill the service root URL for your instance).
5. Once the formation is active, navigate to the **Data Products** section in the SAP Business Data Cloud cockpit. Select, activate, and install the specific SAP S/4HANA data products needed to make the extracted data available to domain content.

9.4 Intelligent Content and Data Products Economy

Customers today face significant challenges when accessing and utilizing data for key activities such as building new applications, integrating systems, developing analytical models, designing dashboards, and training AI solutions. One of the primary obstacles is the difficulty in discovering relevant data, largely due to inconsistencies and lack of clarity in the APIs that expose both data and metadata.

In many cases, available metadata is highly technical and does not provide sufficient business context, making it difficult for business users to interpret and use effectively. In addition, certain critical datasets are not exposed via APIs at all, further limiting usability and integration capabilities.

To address these issues, having a standardized and consistent metadata model established through data products is essential. By exposing landscape-specific data products directly within each customer environment, users can more easily discover, understand, and consume the data they need. This process is enabled through a centralized catalog, which ensures visibility and accessibility.

We'll discuss data products in more detail in the following sections. We'll also explore intelligent packages and domain content built on these data products.

9.4.1 Data Products

Data products are the fundamental building blocks of SAP Business Data Cloud. The domain content is built using these data products.

In SAP Business Data Cloud, a data product is a curated data asset that is exposed for consumption beyond the boundaries of the originating application through standardized APIs. It is enriched with high-quality, business-relevant metadata, which is made accessible via the data product catalog, enabling easy discovery and understanding.

Data products are purpose-built by LoB experts, ensuring that the data reflects accurate business context, meaning, and semantics—not just technical structures. This business-centric design makes them highly usable across a wide range of consumers, from analysts to application developers.

A key value of data products is that they address the traditionally complex challenge of unifying data across multiple SAP applications. They provide a single, consistent data layer while preserving the semantics and context of the data, ensuring that it can be interpreted correctly across different domains and use cases.

Data products are optimized for analytical workloads. They are designed for intensive, read-only usage, making them ideal for advanced analytics, dashboards and reporting, and AI and ML scenarios, but they are not intended for transactional processing. Data products are optimized for performance and scalability in analytical scenarios rather than operational or transactional workloads.

Any application supporting the Delta Sharing protocol (or other protocols in the future) can consume SAP Business Data Cloud data products, including partner tools or SAP HANA Cloud. The protocol can enable SAP HANA Cloud workloads such as CDS views, calculation views, and SAP Cloud Application Programming Model applications to use governed SAP Business Data Cloud data without replication.

The key characteristics of data products are as follows:

- **Business datasets**
  Comprise one or more business object entities, including related objects, analytical data (measures and dimensions), documents, graph data, spatial data, and more.

- **Consumable**
  Accessible via governed interfaces APIs or events. Supported interfaces include SQL (including SQL interface views), Delta Sharing, Event, REST, and OData, making them reusable across LoBs and systems without replication-heavy integrations.
- **Described**
  Accompanied by high-quality metadata exposed through Open Resource Discovery (ORD), following the standardized ORD schema for data products. This ensures that both the technical and the business context are clearly defined.
- **Discoverable**
  Indexed and searchable through the data product directory, a service of the unified customer landscape (UCL) in SAP Business Data Cloud. The directory aggregates metadata from all data products, enabling efficient discovery across the entire landscape.

The following sections will take a closer look at data products, including the technical aspects, types, and how they're organized into data packages.

Technical Aspects of Data Products

Foundation services facilitate the discovery, extraction, and consumption of SAP-managed data products. These services will onboard the different LoB applications in SAP:

- Data pipeline management
- Discovery via ORD protocol, enabling decentralized discoverability of data products
- Replication to object store
- SAP HANA Cloud data lake, storing data in data lake tables by following data sharing principles
- File-based storage

The core protocols that are supported are as follows (more options will be available in the future):

- **Open Resource Discovery (ORD)**
  Consumers query ORD aggregators to find data products and associated metadata. Source LoBs provide data product definitions, which are read and listed by the SAP Business Data Cloud catalog. These definitions are then used to generate data lake tables, making data products available for consumption.
- **Delta Sharing**
  Delta Sharing is an open protocol for sharing data stored in cloud-based data lake tables. It enables the following:
  - *Zero-copy data access:* No need to move or copy data, ensuring efficient data usage
  - *Central governance:* Controlled access and scalable sharing
  - *Multitool support:* Compatible with tools like Apache Spark and Python

Types of Data Products

There are two types of data products. The SAP standard data products are managed by SAP. Customers also can create and manage custom data products. There are two main approaches to implementing data products in SAP Business Data Cloud, as shown in Figure 9.10:

- **SAP-managed data products**
  SAP-managed data products are delivered and maintained by SAP. These are standardized and regularly updated as the source application evolves. They are fully governed and aligned with SAP standards and have the following traits:
  - Defined, owned, and managed by SAP (complete lifecycle).
  - SAP manages data extraction, transformation, load, and harmonization.
  - SAP manages metadata, business logic, semantics, associations, schemas, and APIs.

  SAP-managed data products can be found via the SAP Business Accelerator Hub (*https://api.sap.com/dataproducts*) or SAP Business Data Cloud catalog.

> **Note**
> SAP data products also support customer extensions within any SAP sources and locally defined extension fields to build data product customer extensions. A system's custom (Z) fields can be seamlessly appended to these data products from SAP S/4HANA if the underlying source CDS views are also extended. You can extend standard data products to reflect your customizations in SAP S/4HANA while adhering to clean core principles (see *http://s-prs.co/v633513*).

- **Customer-managed data products**
  Customer-managed data products are owned and managed by customers. Customers leverage custom CDS views from SAP S/4HANA or raw tables or extract information from legacy databases and third-party systems. Data is processed and stored locally within SAP Datasphere's SAP HANA data lake files (HDLF) layer. Replication flow is used to load data from the source entity into the SAP Datasphere object store. Customers handle their own transformations, versioning, and calculations, which have the following traits:
  - Owned and managed by customer.
  - Customer manages metadata, business logic, semantics, associations, schemas, and APIs.
  - Data extractions are performed using SAP Datasphere–native integrations and the SAP BW data product generator.

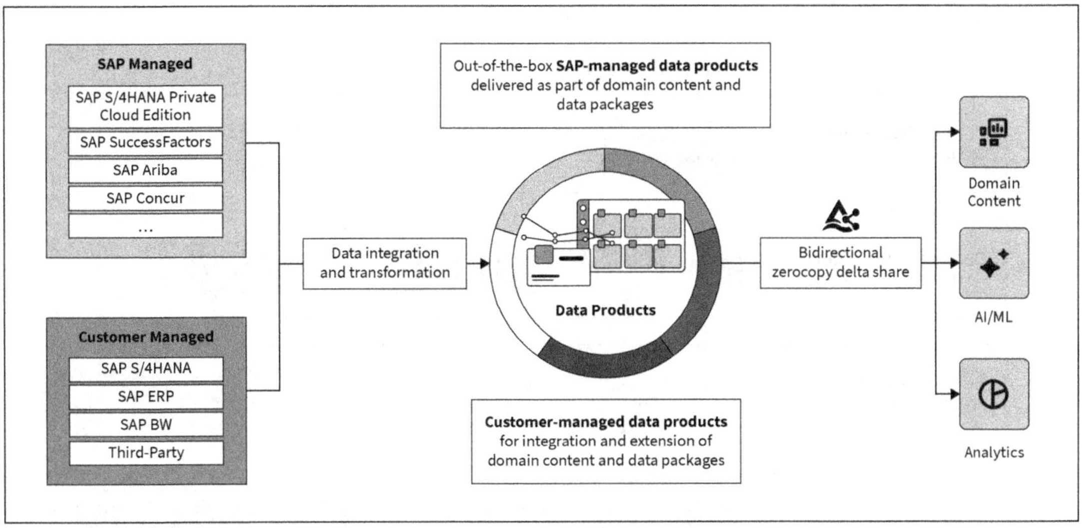

Figure 9.10: Data Products

There are two types of data products, as depicted in Figure 9.11:

- **Primary data product**
  A *primary* data product is directly generated and exposed by a source application, representing the original, authoritative data from that system. It is not derived from or dependent on other data products, ensuring that it remains a true representation of the underlying business data created and maintained in the source.
- **Derived data product**
  A *derived* data product is curated and delivered by SAP, built by combining and transforming data from one or more existing data sources. Derived data products generate value-added insights by enriching and integrating data from multiple sources, supporting business, analytical, and AI use cases. They offer a broader, unified view of business information and can be progressively refined through layered transformations to create highly reusable, consumption-ready data assets.

This layered approach enables the creation of a scalable and modular data architecture, where each level is built upon the previous one. The overarching objective is to establish a portfolio of highly reusable, semantically consistent data products that can efficiently support a wide range of use cases across the enterprise.

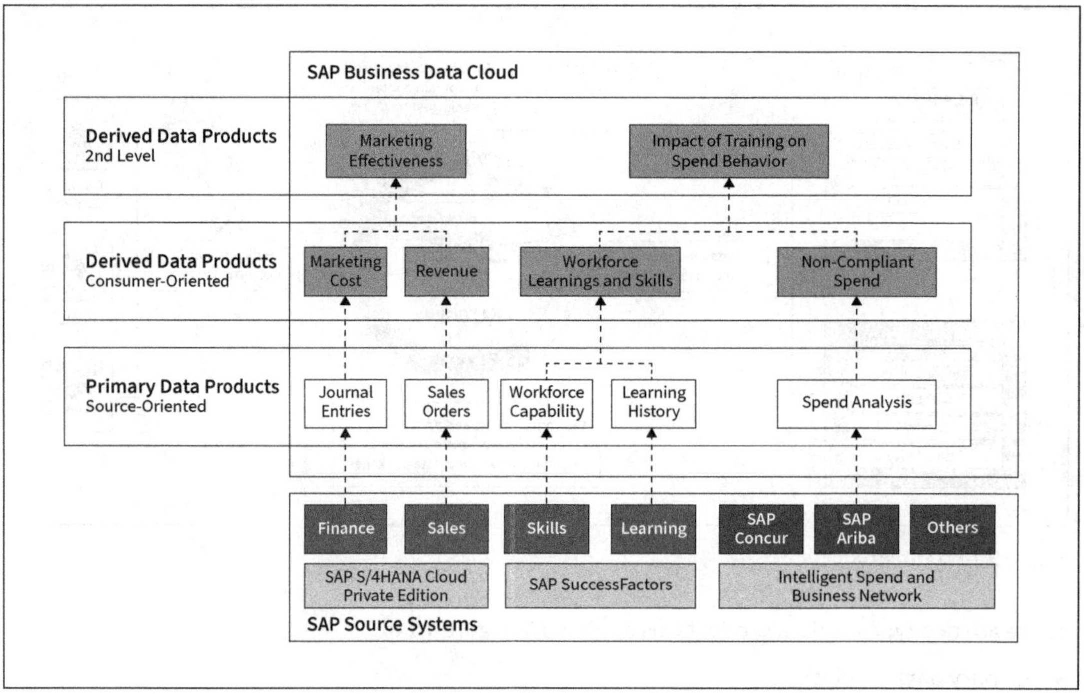

Figure 9.11: Primary and Derived Data Products

Data Package

To simplify discovery, deployment, and lifecycle management, data products are organized into data packages. A data package acts as a logical grouping of related data products, allowing users to work with them collectively rather than individually. Data products are grouped by data packages, making them easy to manage. For example, instead of activating a long list of data products one by one, you can simply activate the data package that contains all the related data products.

A data package may span multiple business domains. For example, the SAP Sales data products data package (SAP S/4HANA Cloud Private Edition) may contain multiple data products: billing document, customer return, sales order, sales quotation, and so on.

9.4.2 Intelligent Packages and Domain Content

SAP Business Data Cloud intelligent packages are the commercial offerings of adaptive and AI-powered applications, intelligent business content, and SAP-managed data products. These are fully SAP-managed analytical content and data products. Intelligent content is a suite of adaptive, AI-powered applications that learn from data, understand business context, and act on customers' behalf.

SAP Business Data Cloud provides the trusted data foundation that powers prebuilt intelligent content for every line of business. This intelligent content delivers data products, semantic models, analytical models, and dashboards aligned to specific use cases from each LoB (e.g., workforce planning for HCM, working capital for finance, etc.).

We'll explore the intelligent packages and content in more detail in the following sections.

Intelligent Package and Intelligent Content

An intelligent package for a given LoB/industry is a set of intelligent content along with all the requisite components. SAP delivers the following intelligent packages (see Figure 9.12):

- Cloud ERP Private Intelligence (released 2025)
- People Intelligence (released 2025)
- Finance Intelligence (controlled availability as of January 2026)
- Spend Intelligence

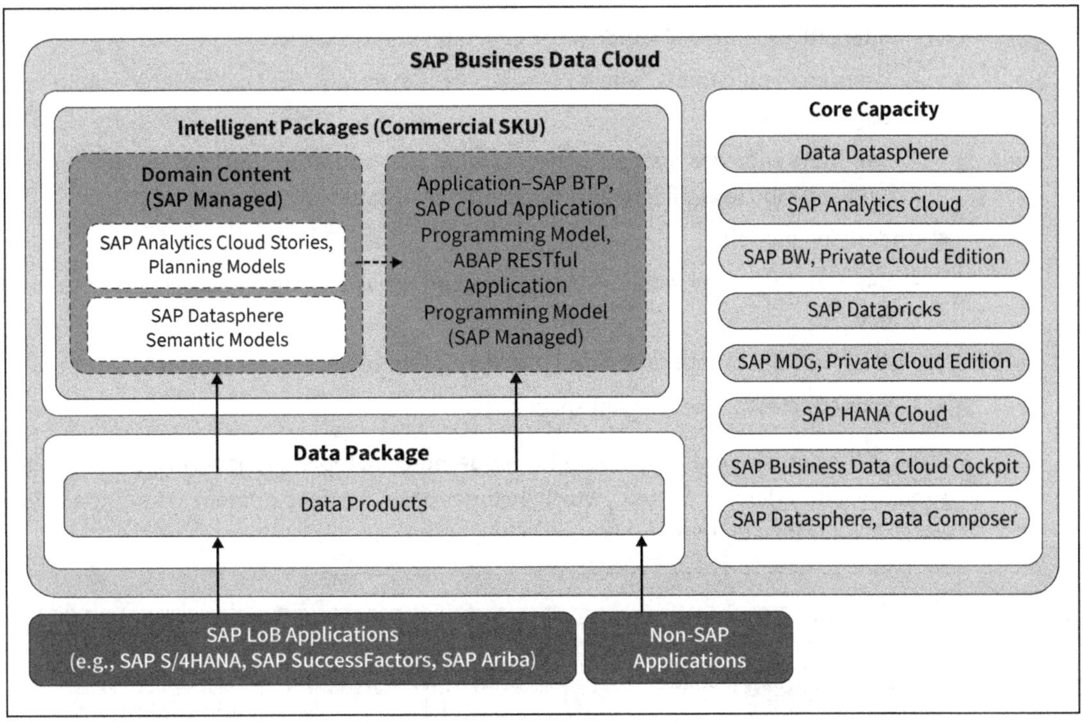

Figure 9.12: Intelligent Packages Leveraging Data Products

Domain content simplifies the visualization and analysis of data in SAP Business Data Cloud. They contain an SAP-managed collection of analytical assets and functionalities to provide value to customers through analytical, process, or domain insights. By leveraging

SAP Analytics Cloud for dashboards and SAP Datasphere for data preparation, domain content offers preconfigured, SAP-managed solutions that reduce complexity and enhance usability. The architecture integrates foundation services, data products, and analytic models, ensuring seamless deployment and scalability while allowing customization for advanced scenarios.

The new domain content released by SAP also includes SAP Cloud Application Programming Model and ABAP RESTful Application Programming Model, leveraging SAP BTP's SAP Build portfolio. When paired with SAP Business Data Cloud to leverage data products, these capabilities can be a powerful governed platform to build custom applications.

The high-level object structure of the domain content consists of the following components:

- **Visualization objects**
  - SAP Analytics Cloud stories serve as dashboards based on live data connections to SAP Datasphere.
  - Interactive elements include diagrams, tables, and charts.
  - Users can apply filters, select members or dimensions, and set variable values (e.g., target currency).
  - SAP Build provides a single platform with drag-and-drop tools for visually building web and mobile applications and generative AI capabilities.
- **Underlying models**
  - SAP Datasphere-based analytic models and views leveraging associated data products.
  - Automated data replication and transformation services.
  - Scoped roles for the relevant spaces.

Some of the domain content includes low-code/no-code tools, as shown in Figure 9.13. SAP is releasing new domain content, which includes pro-code development in addition to low-code/no-code development tools.

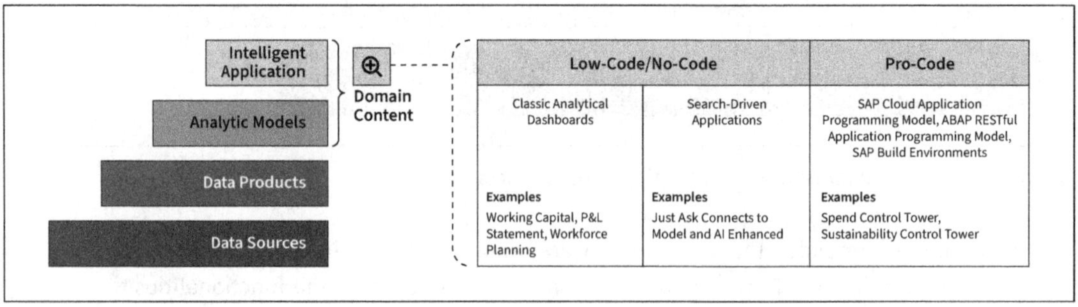

Figure 9.13: Domain Content with Low-Code/No-Code Plus Pro-Code Tools

SAP-managed domain content and its dependencies cannot be directly modified until extensibility capabilities are fully introduced in SAP Business Data Cloud. However, components can still be reused by copying and adapting them to meet specific business requirements, as shown in Figure 9.14.

For example, users can copy SAP Analytics Cloud stories to customize visualizations or tailor them to their reporting needs. For more advanced scenarios, users may also copy and modify the underlying analytic models. It is important to note that changes made to analytic models will impact both the original and any derived (copied) stories, requiring careful governance and version management.

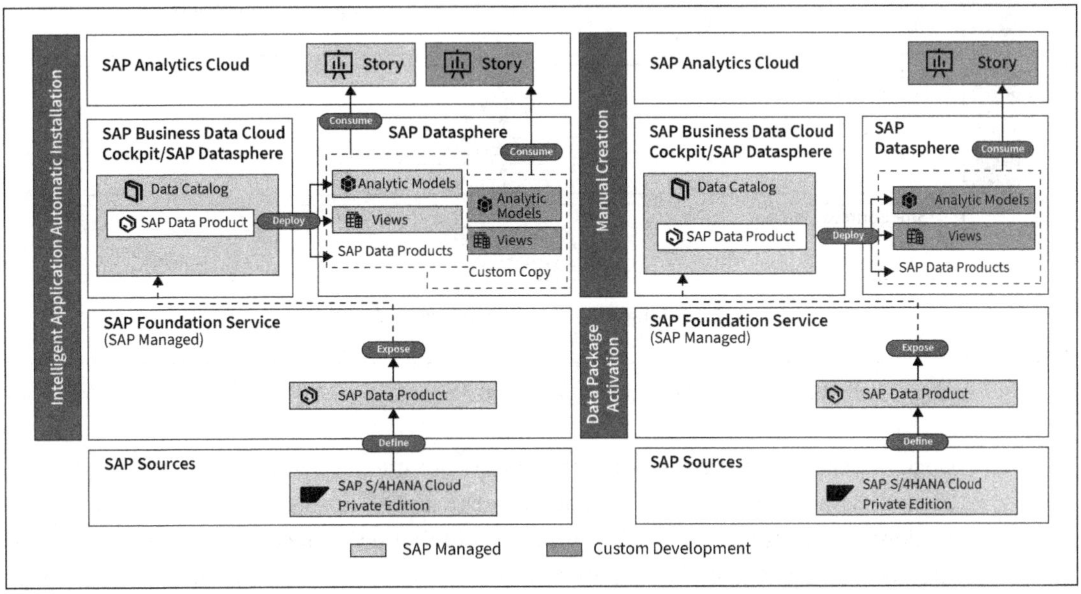

Figure 9.14: Custom Copy of Domain Content for Enhancements/Creation of Custom Domain Content

More sophisticated AI/ML use cases can be enabled by sharing data products with modern data engineering platforms such as SAP Databricks and SAP Snowflake, and soon, Microsoft Fabric and Amazon Athena. These platforms provide scalable environments for advanced analytics and ML model development.

Figure 9.15 depicts the concepts to leverage leading data engineering platforms like SAP Databricks or SAP Snowflake along with SAP Business Data Cloud to leverage AI/ML scenarios without moving data.

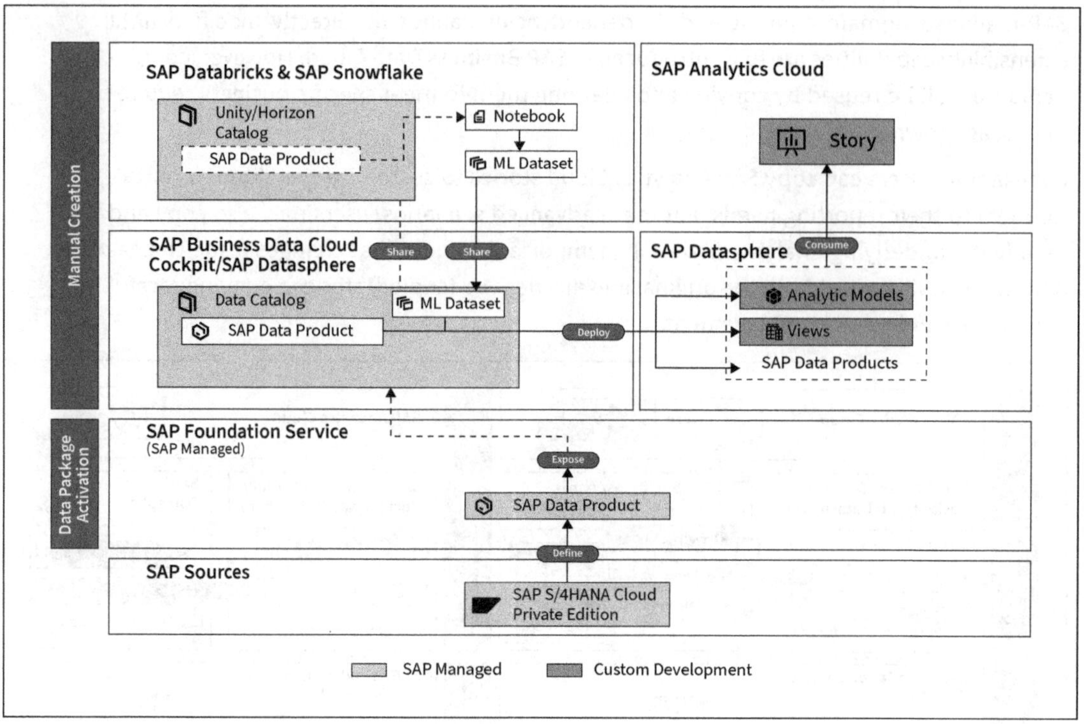

Figure 9.15: Advanced Analytics with AI/ML Content

Once enriched ML datasets or model outputs are created within these data engineering platforms, they can be seamlessly shared back into SAP Business Data Cloud through bidirectional data sharing using SAP Business Data Cloud Connect. The shared datasets can then be leveraged to build advanced, custom, domain-specific content, enabling deeper business insights, enhanced analytics, and more intelligent insights.

SAP-Released Domain Content

SAP currently provides the following out-of-the-box content packages to enable organizations to quickly gain value through ready-to-use analytics and insights in key business areas:

- Working Capital Insights
- People Intelligence Insights

SAP is continuously expanding its portfolio, and additional domain content packages are planned for release. These upcoming offerings are expected to cover more business domains, helping organizations further accelerate data-driven decision-making.

Working Capital Insights

Effective working capital management is critical for ensuring that a company maintains sufficient liquidity to meet its short-term obligations while optimizing cash flow to support growth. It involves the strategic management of key components such as accounts payable, accounts receivable, inventory, and overall cash flow.

By managing working capital efficiently, organizations can unlock cash that would otherwise remain tied up in the balance sheet. This improved liquidity reduces reliance on external financing and provides greater financial flexibility. As a result, companies are better positioned to expand operations, pursue mergers and acquisitions, invest in research and development, and drive long-term business growth.

Working Capital Insights contains the following pages:

- **Overview**
- **Liquidity & Cash**
- **Accounts Receivable**
- **Accounts Payable**
- **Inventory**

Each page is divided into a heading row of leading key performance indicators (KPIs) and charts related to the KPIs.

The Working Capital Insights package provides a comprehensive set of predefined KPIs that help organizations monitor liquidity, operational efficiency, and overall financial performance.

Users can interact with the widgets displayed on the story page by selecting the desired widget and clicking the **More Actions** button when it becomes visible.

The story has the following global filters: currency, company code, and country.

The key metrics in Working Capital Insights are as follows:

- **Key Financial & Liquidity Metrics**
  - Net Working Capital
  - Cash Conversion Cycle
  - Cash Asset Ratio
  - Quick Ratio
  - Liquidity
- **Profitability & Performance Metrics**
  - Return on Capital Employed (ROCE)
  - EBIT (Earnings Before Interest and Taxes)
  - Gross Margin
  - Net Income

- **Operational Efficiency Metrics**
  - Inventory Turnover Rate
  - Cost of Sales
- **Revenue & Sales Metrics**
  - Total Net Sales
- **Capital Structure & Financial Position**
  - Capital Employed
  - Total Assets
  - Current Liabilities
  - Marketable Securities
- **Expense & Tax Components**
  - Interest
  - Income Tax

Figure 9.16 shows the overview page of Working Capital Management Intelligent Insights. Filters and navigation options are in the left pane.

Figure 9.16: Working Capital Insights Overview Page

From the overview page, users can navigate to other pages, like **Accounts Receivable**, which is shown in Figure 9.17. The same filters selected in the overview page carry to other pages.

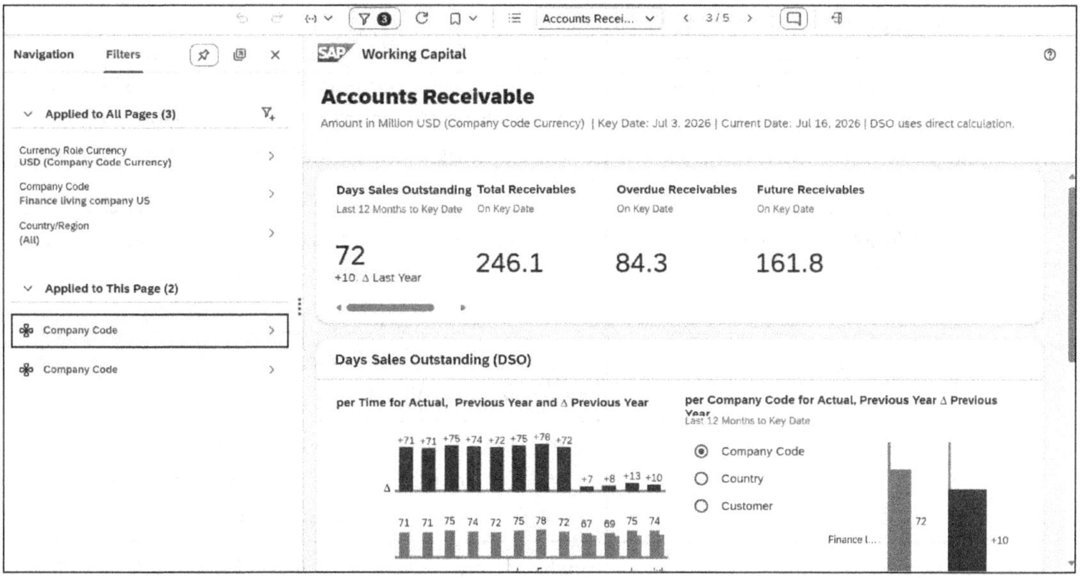

Figure 9.17: Working Capital Insights Accounts Receivables Page

People Intelligence Insights

The People Intelligence package in SAP Business Data Cloud is a people analytics solution that integrates HR, skills, and business data to deliver actionable, AI-powered insights. It enables organizations to make more informed decisions.

The People Intelligence insights are primarily built on HR data sourced from SAP Success-Factors, providing a strong foundation for comprehensive workforce analysis.

Sensitive personal data entered in source systems through uncontrolled methods such as generic data containers or custom configurations and extensions may inadvertently surface in SAP Business Data Cloud data products. Therefore, it is essential that organizations review these data products carefully to ensure compliance with applicable data protection standards.

Fields in the source systems that are marked for log read access are excluded from People Intelligence. As a result, these fields are not available in the corresponding data products or insights. This limitation is intentionally designed to prevent the exposure of sensitive personal data until People Intelligence fully supports read-access logging functionality.

The People Intelligence package offers the intelligent content shown in Table 9.1.

Name	Description
Workforce Composition Insights	Workforce Composition Insights provides visibility into your workforce trends. It empowers HR leaders to strengthen employee satisfaction, boost retention, and build trust.
Workforce Compensation Insights	Workforce Compensation Insights helps you analyze compensation data to uncover pay disparities and provide actionable recommendation for adjustment.
Skills Insights	Skills Insights helps increase organization-wide transparency into skill proficiency.
Learning Insights	Learning Insights is designed to provide actionable insights into employee training, development, and skills acquisition across the organization.
Goals Insights	Goals Insights is structured to provide powerful insights into goal-setting and progress within the organization by various segments.
Performance Insights	Performance Insights empowers organizations to maximize workforce performance, ensure equity, and align talent with business strategy, ultimately driving better business outcomes.
Succession Insights	Succession Insights empowers HR, talent management, and executives with actionable insights so that they can build strong succession pipelines, enhance leadership readiness, and ensure long-term organizational success.
Career Development and Planning Insights	Career Development and Planning Insights offers actionable insights into workforce development progress and areas of focus to support organizational growth and talent strategy.
Recruiting Insights	Recruiting Insights empowers HR and leadership teams to optimize hiring processes, improve resource allocation, accelerate time-to-hire, measure effectiveness, and ultimately support business growth through smarter talent-acquisition strategies.
Time Management Insights	Time Management Insights enables HR and management to monitor, analyze, and address patterns in employee absences by providing a detailed analysis of employee absences within an organization.

Table 9.1: KPIs and Definitions Included in People Insights

Name	Description
Onboarding Insights	Onboarding Insights provides a comprehensive view of the organization's onboarding process that enables HR teams to monitor, improve, and ensure a smooth and efficient onboarding experience.
Compensation Planning Insights	The Compensation Planning Insights dashboard tracks compensation metrics, comparing preplanning and post-planning cycle data. It displays KPIs including average annual salary, compensation ratio, merit increases, range penetration, and equity awards.
Rewards and Recognition Insights	Rewards and Recognition Insights provides a snapshot of reward and recognition activity, tracking awards sent and received and participation levels, and it highlights trends across recognition types, time, and locations to uncover engagement patterns.
Benefits Insights	Benefits Insights provides a comprehensive view of employee benefit utilization, enrollment trends, and associated costs across the organization.
Pay Transparency Insights	Pay Transparency Insights provides a comprehensive analysis of gender pay disparities across the organization, covering overall gaps, median differences, and breakdowns by pay components.
Employee Availability Insights	Employee Availability Insights provides a detailed view of workforce availability for a selected period, including employee identifiers, locations, and departmental distribution.

Table 9.1: KPIs and Definitions Included in People Insights (Cont.)

Figure 9.18 shows an example of the Workforce Composition Insights page.

The scope of information that a user can access within People Intelligence Insights can be managed using a data access control file. Data access control file *sap.sf.wrk.DAC* is associated with the data models for each insight and enables the definition of permissions for fields included in its **Criteria** section.

Within this file, operator-based and value-based data access controls are applied to protect specific criteria, thereby ensuring controlled and secure access to sensitive information.

Figure 9.19 shows an example of a Performance Insights page.

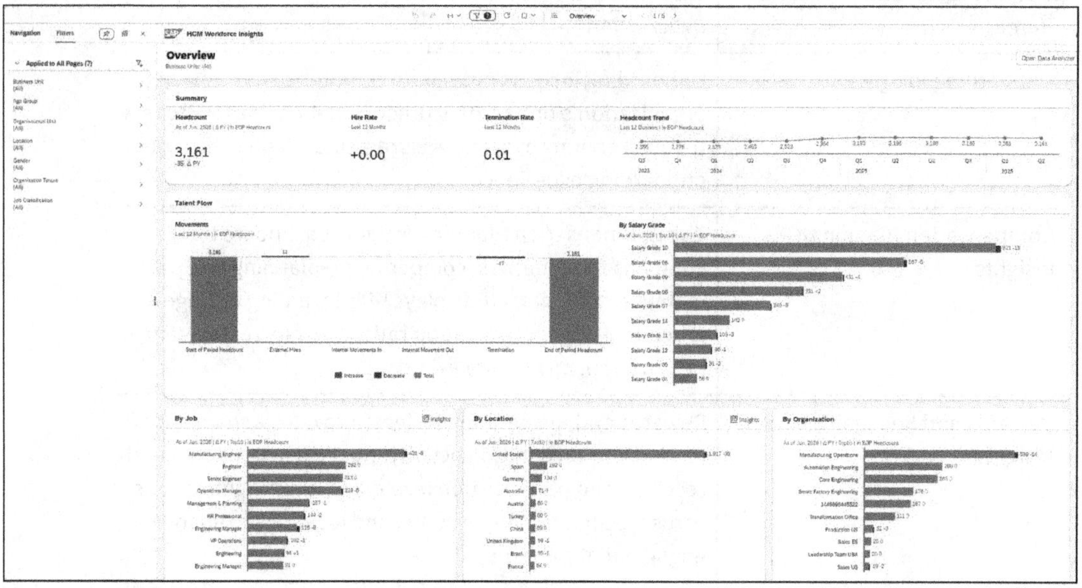

Figure 9.18: Workforce Composition Insights Page

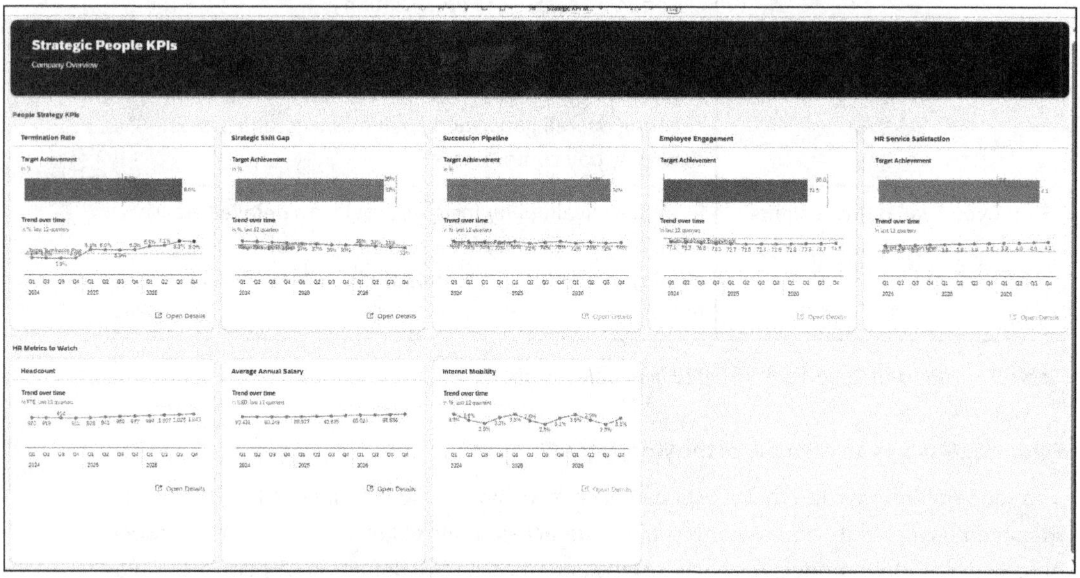

Figure 9.19: Performance Insights Page

The People Intelligence solution enables organizations to analyze compensation equity across different demographic groups and monitor trends in engagement, performance, and employee development. These capabilities help improve retention, promote fairness, and maximize the value of HR data across the enterprise.

The solution enhances operational efficiency by replacing manual and fragmented reporting processes with automated, self-service dashboards, standardized metrics, and an AI copilot (Joule). This approach reduces data-extraction costs and supports faster, more informed decision-making, ultimately improving business agility and optimization.

Note

SAP is planning to release more domain content, for Spend Intelligence, Finance Intelligence, Supply Chain Intelligence, Revenue Intelligence, and Consumer Products Intelligence.

9.5 Zero-Copy Data Sharing

Traditionally, extracting data from SAP applications required complex replication processes that often led to the loss of valuable semantics and business context. As a result, teams had to perform intricate reconciliation procedures in external systems to reconstruct the missing metadata. These approaches were labor-intensive, error-prone, and frequently resulted in data silos, ultimately undermining data trust and limiting innovation.

Zero-copy data sharing in SAP Business Data Cloud enables seamless, bidirectional data exchange with external platforms. External systems can securely access SAP data in real time without the need for data movement, replication, or complex ETL pipelines. By sharing schemas and semantic context directly, you can preserve business meaning while eliminating data duplication. Likewise, external platforms can share data back into SAP Business Data Cloud, enabling a truly collaborative and integrated data ecosystem.

SAP Business Data Cloud Connect enables secure and seamless, zero-copy sharing of semantically rich SAP data into your existing data platform environments. SAP Business Data Cloud Connect has the following unique capabilities:

- Uses industry standards, with a focus on data sharing for zero-copy data access
- Enables secure and seamless integration between SAP Business Data Cloud and external data platforms
- Facilitates bidirectional sharing and consumption of data products
- Dedicated to supporting customer choice and expanding the business data fabric with a zero-copy approach

SAP Business Data Cloud Connect is generally available with the following partners today: SAP Databricks, SAP Snowflake, and Google's BigQuery. It is planned for Microsoft Fabric and Amazon Athena.

Once SAP Business Data Cloud Connect is provisioned and connections are established with partners through SAP Business Data Cloud Connect, data products can be seamlessly shared with the partner system's catalog. This enables external platforms to discover, access, and consume shared data within their own ecosystem while preserving governance, security, and business context.

SAP Business Data Cloud includes SAP Databricks, with native data sharing capabilities that enable seamless data exchange within the SAP Business Data Cloud ecosystem.

Organizations can leverage SAP Business Data Cloud Connect for Databricks and Snowflake to perform zero-copy data sharing to their existing SAP Databricks/SAP Snowflake platforms. This allows enterprises to connect their SAP Databricks or SAP Snowflake environments directly to SAP Business Data Cloud, enabling real-time access to semantically rich SAP data products across all business processes without data duplication.

The SAP Business Data Cloud central catalog, accessible via the SAP Business Data Cloud cockpit, enables the discovery and sharing of data products from diverse source systems. It provides a self-service experience in which users can easily search, discover, and share data products with platforms such as SAP Databricks or SAP Snowflake, enabling secure data exchange and live, zero-copy access.

Figure 9.20 shows an example of connections to SAP Databricks and SAP Snowflake.

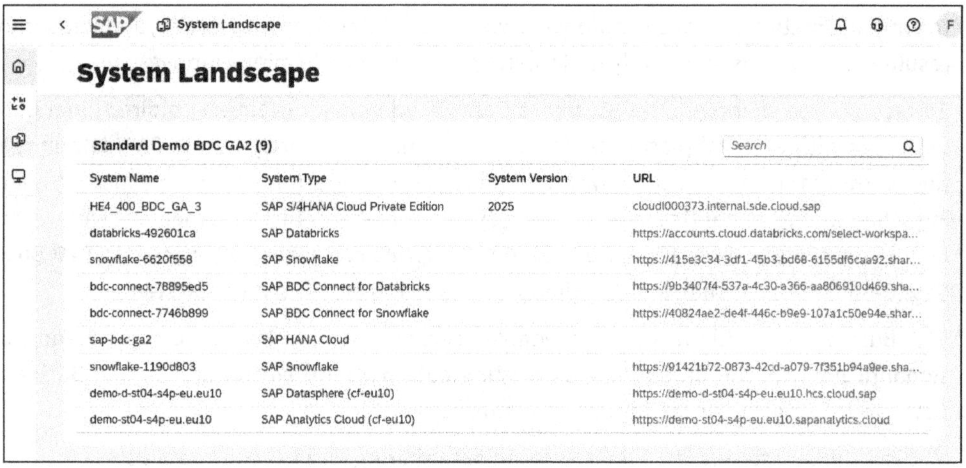

Figure 9.20: SAP Business Data Cloud Landscape with SAP Databricks and SAP Snowflake Connections

Once the data products are shared to SAP Databricks, they are available in the unity catalogue.

Figure 9.21 shows an example of the unity catalogue in SAP Databricks. Here, the data products shared from SAP Business Data Cloud are visible like native catalogue objects.

Data scientists can build advanced analytics and derived data products in their SAP Databricks or SAP Snowflake environments and seamlessly share them back into SAP Business Data Cloud (see Figure 9.22). Once published, these data products become available in the SAP Business Data Cloud Catalog within minutes, ensuring rapid availability across the enterprise.

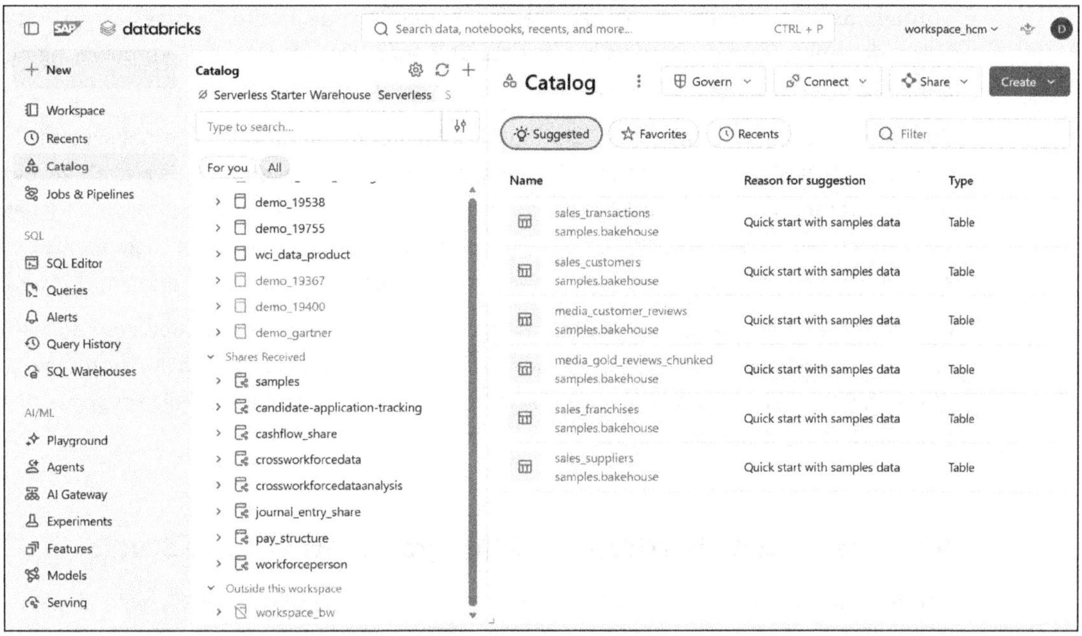

Figure 9.21: SAP Databricks Unity Catalog with Shared Data Products

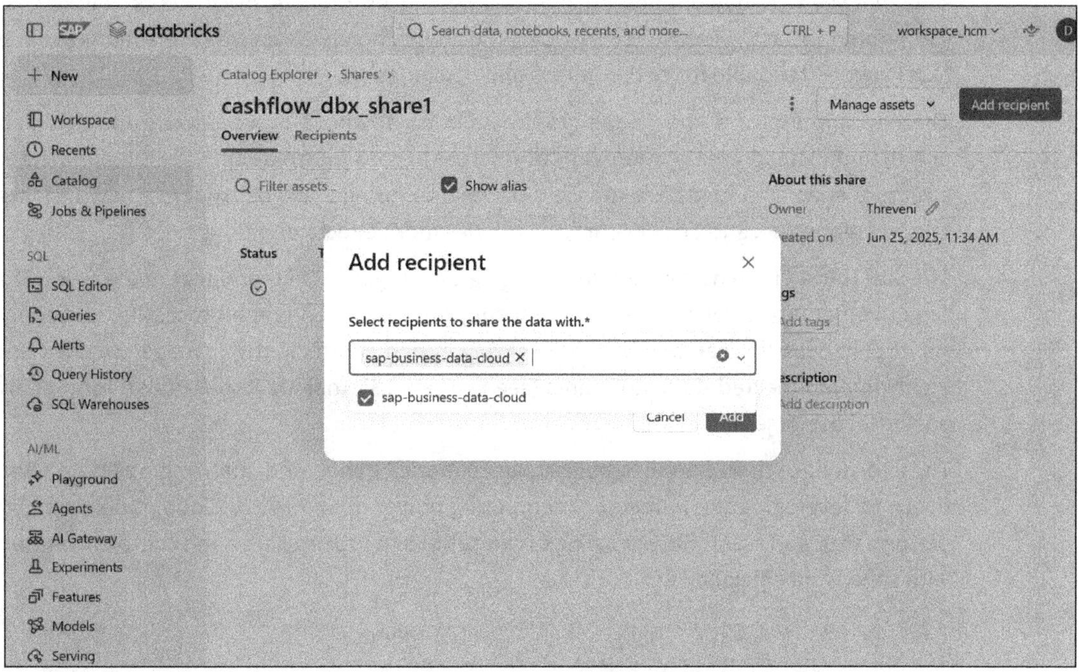

Figure 9.22: SAP Databricks Unity Catalogue Sharing Back to SAP Business Data Cloud

Through SAP Datasphere, users can then consume these shared data products to install them, develop analytical models, and create SAP Analytics Cloud stories, empowering rich data visualizations and actionable, insight-driven decision-making.

The integration of SAP Business Data Cloud Connect and supported external systems for zero-copy data sharing is more than just a technological advancement; it's a strategic evolution that redefines how businesses access and utilize data. By curbing unnecessary data movement and governance challenges, businesses can significantly lower their operational costs and total cost of ownership (TCO), allowing them to prioritize innovation and growth.

As enterprises advance into AI-driven analytics, these strategic partnerships exemplify efficiency and trust, facilitating rapid business transformation and enhanced decision-making. Through SAP Business Data Cloud Connect, companies can accelerate their advanced analytics and AI/ML initiatives.

9.6 SAP Knowledge Graph and Integration with Joule Studio

SAP Business Data Cloud offers two key embedded AI capabilities, which we'll discuss in this section: SAP Knowledge Graph and Joule Studio.

A *knowledge graph* is a structured method of organizing information that captures relationships between various entities, such as people, places, objects, or concepts. Rather than simply storing data, it focuses on mapping the connections between data elements in a way that is understandable for both humans and machines.

The key strength of a knowledge graph lies in its ability to move beyond isolated data points and instead create a network of interconnected information. This interconnected structure enables a more holistic view of data, allowing systems to derive new insights, infer relationships, and answer complex queries more effectively.

The SAP HANA Cloud database's knowledge graph engine is a component of the SAP HANA Cloud in-memory database platform that supports advanced graph processing and analysis. It enhances the capabilities of SAP HANA Cloud to efficiently manage complex and highly interconnected data structures (see Figure 9.23), making it well-suited for modern analytical use cases.

The knowledge graph engine enables users to model, query, and analyze graph data effectively. By leveraging the in-memory computing power of SAP HANA Cloud, it delivers high performance and scalability, allowing organizations to process large and complex datasets with speed and efficiency.

Note
SAP HANA Cloud can be added to the SAP Business Data Cloud formation.

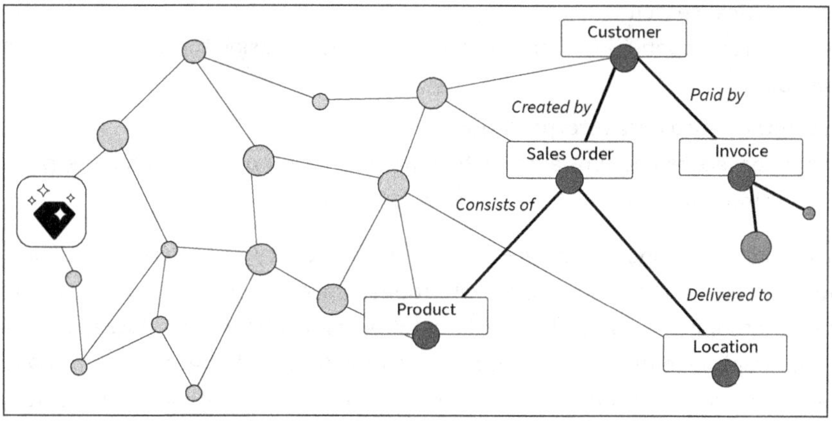

Figure 9.23: Knowledge Graph Engine

SAP Knowledge Graph in SAP Datasphere enhances the efficiency of data modelers by enabling several key capabilities:

- It allows for the rapid identification of relevant data models that meet specific conditions.
- It supports the derivation of where-used lists, providing visibility into how and where a model is utilized.
- It helps identify related and dependent entities, facilitating the creation of new and well-structured data models.

Although SAP Knowledge Graph plays a critical role in improving modeling efficiency, it is not directly visible or accessible to end users. Instead, it is utilized implicitly by Joule, which leverages the ontology to enable more effective and intelligent querying.

SAP's conversational AI, Joule, leverages SAP Knowledge Graph to query metadata of SAP Datasphere entities. This capability enables users to efficiently identify the appropriate data entities that contain specific fields, such as the customer number. In addition, it supports the derivation of where-used lists, providing visibility into how particular entities are utilized—for example, identifying that the customer dimension is used within the accounts payable and receivable analytical models.

Let's shift gears to SAP Analytics Cloud. The Repository Metadata Ontology in SAP Analytics Cloud enhances story generation and querying capabilities by leveraging rich metadata and relationships across analytical assets. It provides the following key benefits:

- **Improved story generation and querying**
  Enables more efficient and intelligent story creation by utilizing structured metadata
- **Enhanced narrative and analytical capabilities**
  Supports advanced analytics through story metadata, including pages, widgets, composites, linked analysis, and underlying data models

- **Improved analytical search**
  Strengthens search capabilities within content, making it easier to discover relevant insights and data
- **Support for intents and data interpretation**
  Enables intent-based interactions along with improved data explanation and interpretation, delivering more meaningful and context-aware insights

Joule Studio, as a native component of SAP Build, provides a built-in understanding of organizational automations and custom-built applications, while integrating seamlessly with the broader SAP Autonomous Suite. Through integration with SAP Business Data Cloud and SAP Knowledge Graph, agents gain semantic access to enterprise knowledge across both SAP and non-SAP applications, ensuring operations are performed with greater accuracy, contextual awareness, and reliability. These capabilities are unified within the Joule ecosystem, where custom agents are developed with consistent context, purpose, data, and models, enabling them to collaborate seamlessly with SAP-delivered agents to effectively address complex business scenarios.

The data products in SAP Business Data Cloud are used to hydrate the out-of-the-box knowledge graphs, enabling them to represent and connect enterprise data effectively. Looking ahead, SAP plans to introduce a framework that will allow organizations to customize and build their own knowledge graphs within SAP Business Data Cloud. As an alternative approach, organizations can use the knowledge graph engine in SAP HANA Cloud to design and implement custom knowledge graphs tailored to their specific business requirements.

Refer to Chapter 1, Figure 1.3, for an illustration of how SAP Business Data Cloud serves as the data foundation for AI agents leveraging Joule Studio.

SAP Knowledge Graph is being made available across major SAP applications, with each knowledge graph tailored to the specific context and requirements of each application. These application-specific knowledge graphs provide deep, domain-focused insights by modeling relationships within individual business processes.

In addition, SAP Knowledge Graph in SAP Business Data Cloud offers a broader, cross-domain perspective by integrating data from multiple SAP applications as well as non-SAP sources. This enables a more comprehensive view of enterprise data, supporting advanced analytics, contextual understanding, and intelligent decision-making across interconnected business landscapes.

Instead of merely searching text or running complex SQL, AI agents query this semantic graph to understand exactly how your specific business processes, supply chains, and finance operations interact. When prompted, Joule uses knowledge graphs to verify facts in real time, delivering logical, explainable, and trustworthy business answers rather than probabilistic guesses

In short, SAP Business Data Cloud is the knowledge core that handles the contextualization and reasoning for AI, and the data products are the foundation that support and hydrate SAP Knowledge Graph.

9.7 Embedded AI Capabilities

There are multiple AI-assisted capabilities available across SAP Business Data Cloud, with significant innovations delivered through its components, SAP Analytics Cloud and SAP Datasphere. These AI-driven features enhance productivity, streamline analytics, and improve decision-making processes.

The agentic experience in SAP Business Data Cloud delivers a robust library of agents designed for data products, analytics, and planning. These agents enable the development of the next generation of SAP's autonomous, agent-driven applications, empowering organizations to automate processes, enhance decision-making, and drive intelligent business outcomes. We'll discuss these in more detail in the following sections.

9.7.1 AI-Assisted Features in SAP Datasphere

Joule is embedded in SAP Datasphere, as well as many other embedded, AI-assisted capabilities:

- **Natural language search**
  This capability enables users to query data models, catalogs, and repositories using conversational language—for example, "show objects created last year containing sales"—thus eliminating the need for manual filter configuration.
- **AI-assisted catalog content generation**
  Generative AI is used to automatically create business-rich metadata, including asset summaries, detailed descriptions, and relevant tags, improving data discovery and understanding.
- **Semantic generation and modeling**
  This feature enhances datasets by automatically suggesting semantic usage, identifying attributes and measures, and assigning appropriate semantic types. It is particularly valuable for quickly establishing structure and meaning in unstructured or non-SAP data.

Figure 9.24 shows how you can enable natural language search queries directly within SAP Datasphere and automate the creation of business-rich metadata in the catalog.

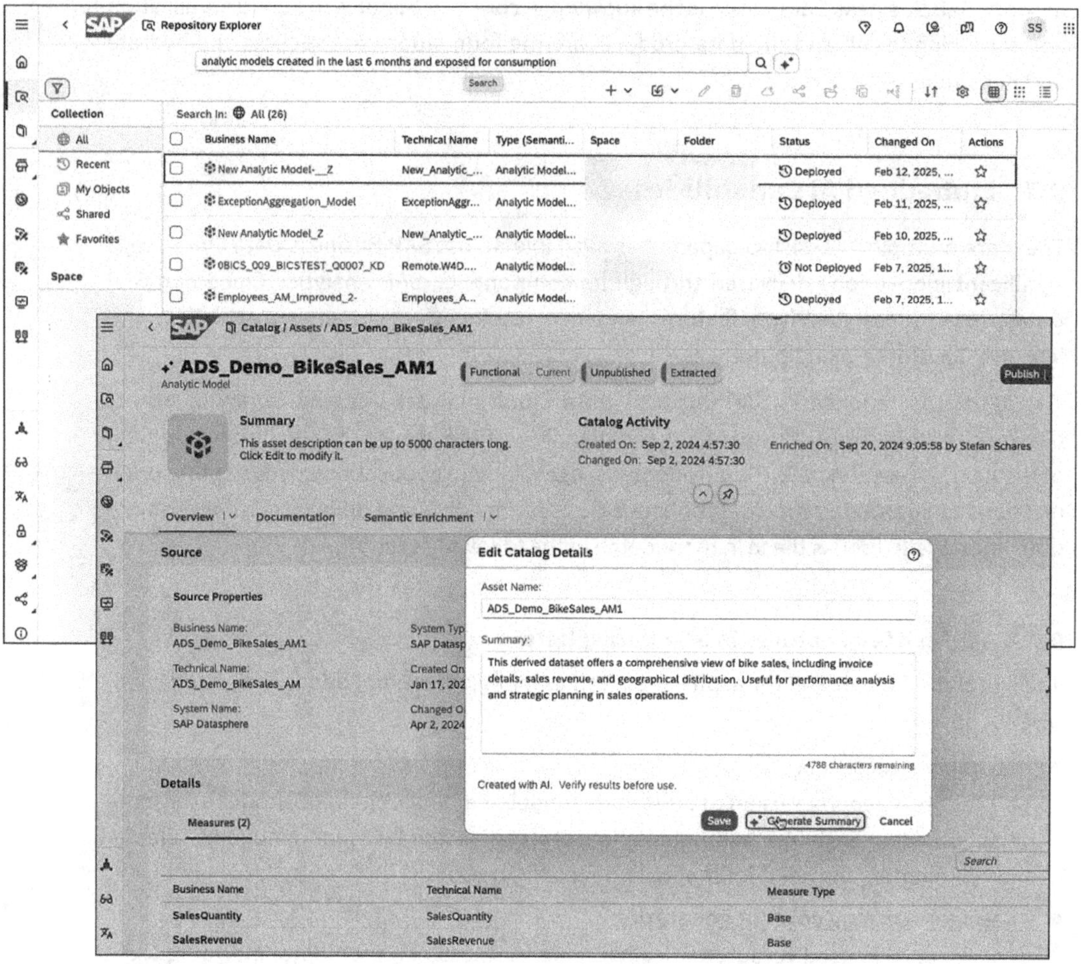

Figure 9.24: Creation of Semantic Rich Metadata in Catalog

9.7.2 AI-Assisted Features in SAP Analytics Cloud

Joule is embedded in SAP Analytics Cloud, as well as many other embedded, AI-assisted capabilities:

- **AI-assisted calculations**
  This capability enables users to generate complex formulas using natural language or to receive clear explanations of existing calculations and code, simplifying advanced analytical tasks. Figure 9.25 shows an example of AI-assisted calculations in SAP Analytics Cloud.

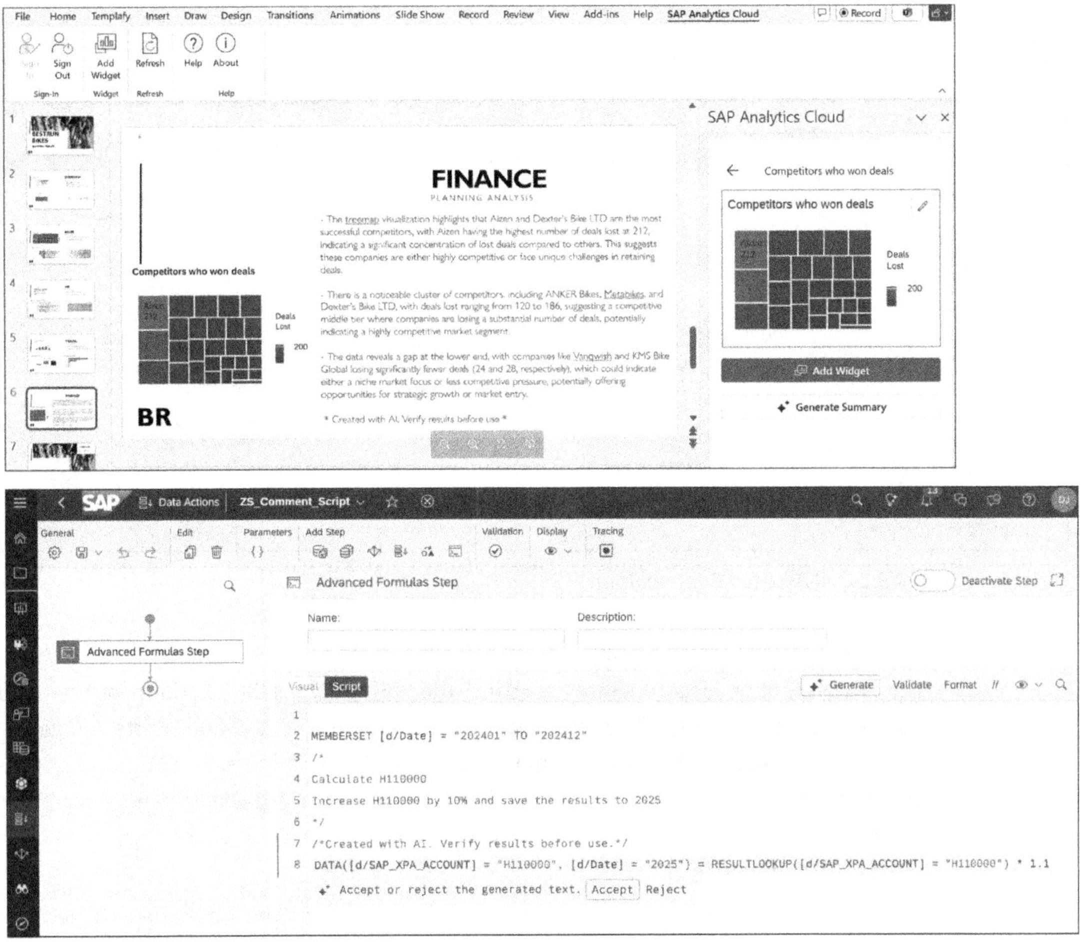

Figure 9.25: AI-Assisted Calculations

- **Compass**

  Compass is a powerful simulation engine that leverages Monte Carlo methods and driver-based relationships to model uncertainty and quantify risks. This supports more accurate planning and enhances confidence in decision-making.

- **AI-assisted commenting**

  This feature streamlines collaboration by aggregating, summarizing, translating, and rephrasing cell comments and extended discussions, helping to save time and improve clarity.

  Figure 9.26 shows an example of AI-assisted commenting functionalities in SAP Analytics Cloud.

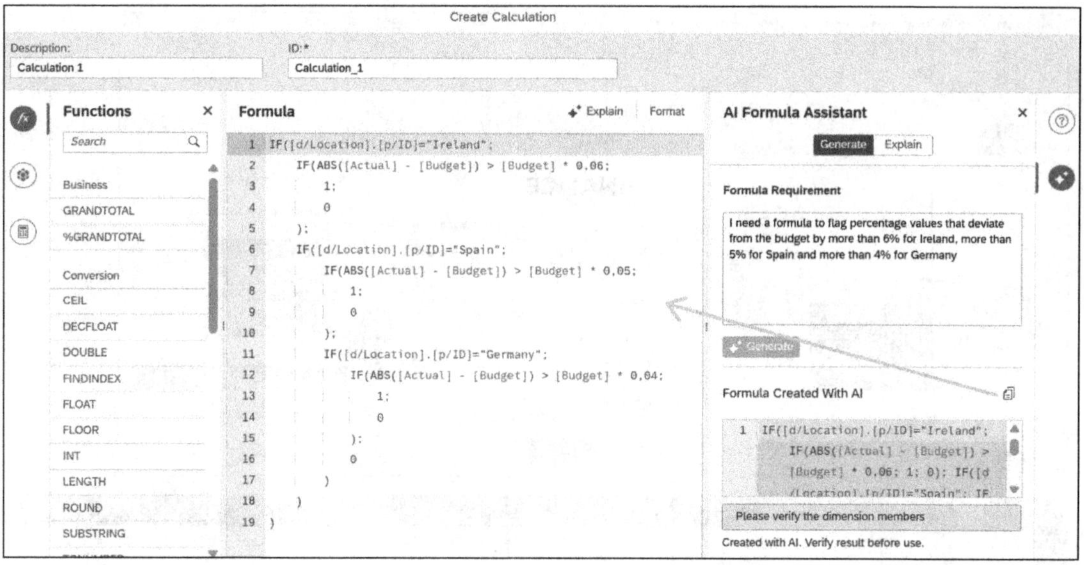

Figure 9.26: AI-Assisted Commenting

- **The AI-assisted chart summary**

  This is a generative AI feature integrated into the SAP Analytics Cloud add-in for Microsoft PowerPoint. It automatically analyzes embedded SAP Analytics Cloud charts and generates concise, natural-language summaries, typically in a three-bullet format, thus enabling presenters to quickly convey key insights. This capability helps save time, enhances storytelling, and allows users to communicate data-driven narratives with clarity and confidence.

  AI-assisted advanced formula generation in SAP Analytics Cloud uses generative AI to translate plain-language business inputs into precise technical scripting logic. This capability significantly accelerates the creation of complex planning scenarios while reducing the need for deep scripting expertise. When working in the script editor within an advanced formula step in SAP Analytics Cloud, users can simply enter a plain-language business instruction (e.g., "Copy prior year data to the current year and increase it by 10 percent"). The AI instantly interprets the request and generates the corresponding technical script logic. It presents the output with highlighted suggestions, allowing users to review, accept, or refine the generated code, thus streamlining development and improving productivity.

  Figure 9.27 shows an example of AI-assisted chart summary and AI-assisted advanced formula generation.

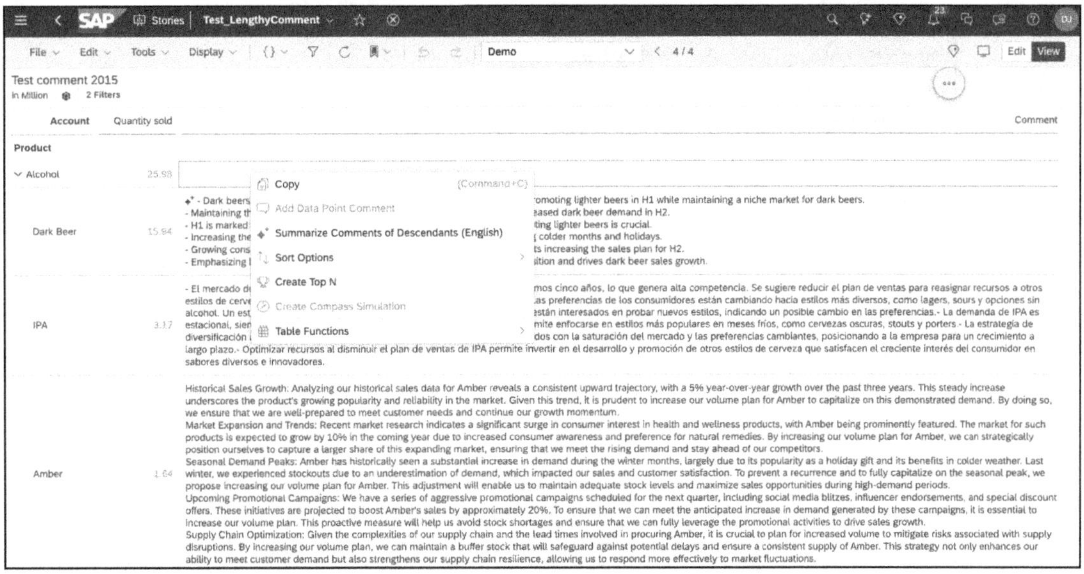

Figure 9.27: AI-Assisted Chart Summary and AI-Assisted Advanced Formula Generation

9.7.3 Agents

Multiple new agents are targeted to be released in SAP Business Data Cloud (check *https://roadmaps.sap.com/welcome* for the exact dates), including the following:

- **Insight Generation Agent**
  The Insight Generation Agent automatically converts data, business context, and user queries into structured, data-driven SAP Analytics Cloud stories and dashboards. By leveraging LLMs and SAP Knowledge Graph, this agent delivers clear, actionable insights in a user-friendly format.
- **Narrative Agent**
  The Narrative Agent generates real-time, context-aware executive summaries from data and analytics content using user prompts via Joule. It collaborates with specialized agents to deliver a unified, business-ready narrative with insights, analysis, and recommendations tailored for decision-makers.

  Figure 9.28 shows an example of the Analytics Narrative Agent.
- **Planning Agent**
  The Planning Agent in SAP Business Data Cloud leverages generative AI to automate the creation and execution of planning processes using natural language input. It enables users to define planning logic, adjust values, run calculations, and dynamically generate simulation versions—all with minimal manual effort. This capability streamlines planning workflows, enhances agility, and empowers users to perform sophisticated scenario analysis with greater speed and accuracy.

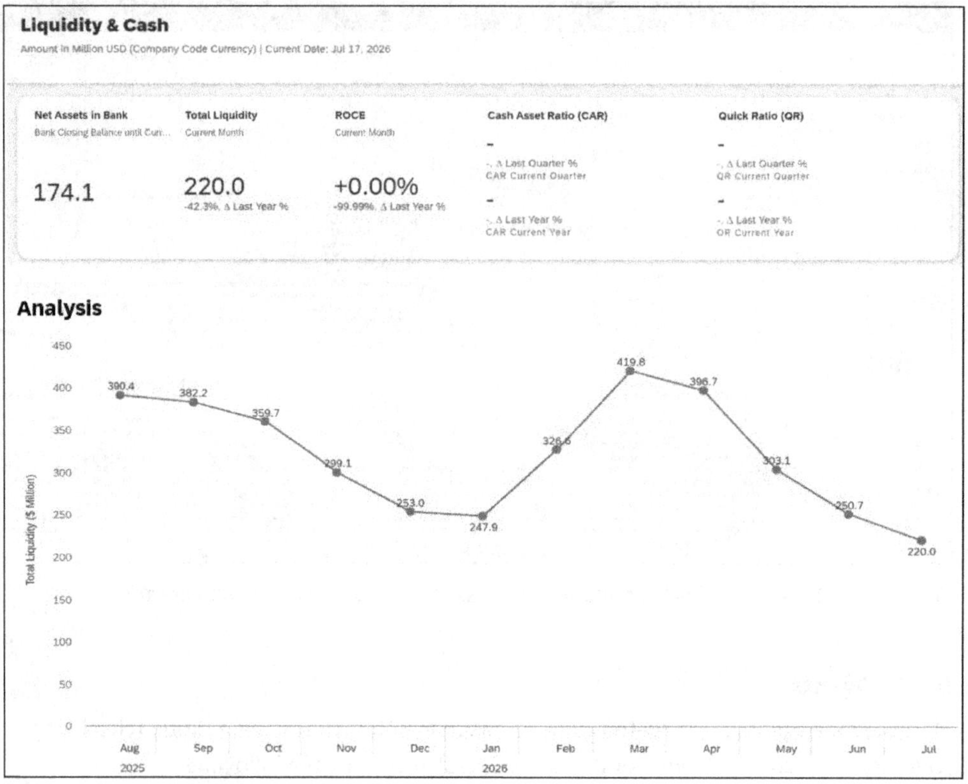

Figure 9.28: Analytics Narrative Agent

- **Data Product Creation Agent**
  The Data Product Creation Agent in SAP Business Data Cloud empowers users to both discover existing data products and create new ones using natural language. By leveraging SAP Knowledge Graph, it intelligently identifies relevant data sources, generates the necessary transformations, and produces reusable, high-quality data products. These data products are automatically registered within SAP Business Data Cloud, making them easily discoverable and reusable for future analytics and business scenarios.

  Figure 9.29 shows an example of the Data Product Creation Agent.
- **Model Creation Agent**
  The Model Creation Agent enables SAP Analytics Cloud users to create an analytics and/or planning model in SAP Business Data Cloud based on a natural language prompt, leveraging SAP Business Data Cloud data products from underlying source-system data.
- **SAP HANA Cloud Discovery Agent**
  The SAP HANA Cloud Discovery Agent helps users explore and understand database objects in SAP HANA Cloud by leveraging a knowledge graph built from system metadata. It identifies and connects relevant objects such as tables, views, and procedures, enabling efficient data discovery without manual schema navigation.

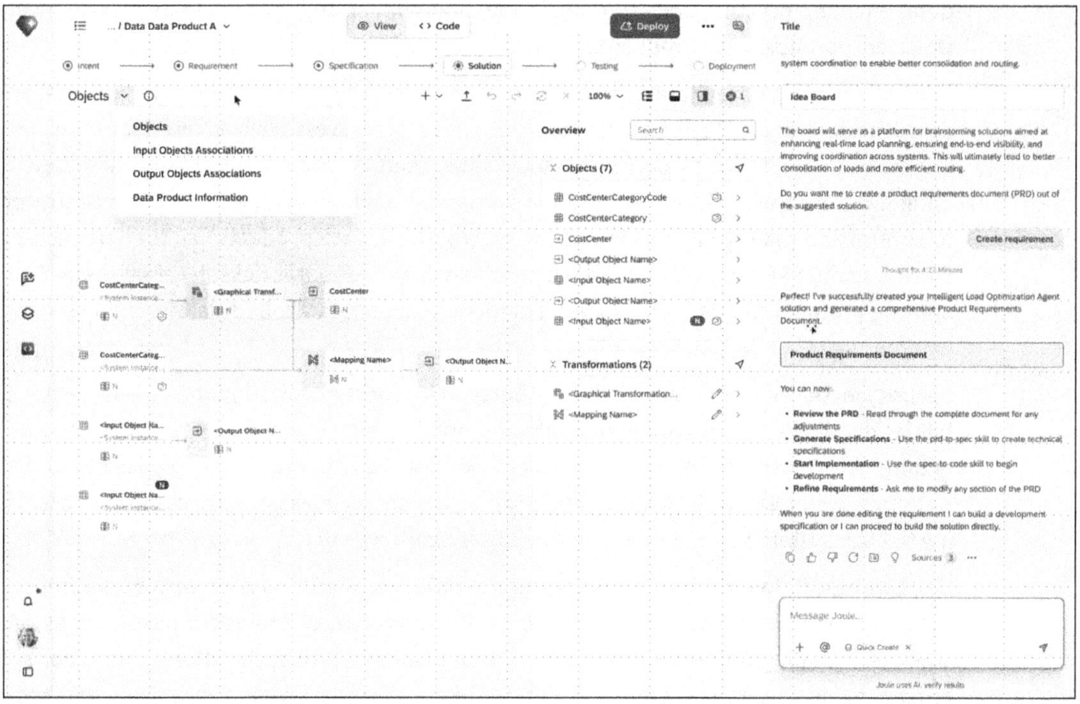

Figure 9.29: Data Product Creation Agent

- **SAP HANA Cloud Data Agent**
  The *SAP HANA Cloud Data Agent* enables users to query SAP HANA Cloud using natural language by automatically translating intent into executable SQL queries. It uses insights from the Discovery Agent to identify relevant objects and relationships, allowing seamless data retrieval, function execution, and procedure calls.

9.8 Governance

Governance in SAP Business Data Cloud plays a critical role in establishing a unified, trusted, and business-ready data foundation across the enterprise. It brings together master data, governance policies, and semantic context into a cohesive business data fabric, ensuring that all data consumers, from analytics applications to AI agents, operate on consistent and reliable information.

From a compliance and security perspective, governance in SAP Business Data Cloud enforces enterprise-grade controls, including data access restrictions, role-based permissions, and data privacy safeguards. These measures ensure that sensitive information is protected and that all data usage aligns with regulatory and organizational policies.

Ultimately, governance in SAP Business Data Cloud ensures that AI-driven applications, analytics, and business processes are powered by high-quality, trusted, and well-governed data, enabling organizations to drive innovation with confidence while maintaining control, transparency, and compliance.

The SAP Business Data Cloud catalog enables users to effectively discover, understand, and consume data products across the enterprise. It provides a centralized interface for exploring available data assets along with their associated business context and metadata. In addition, the catalog supports robust governance capabilities by allowing administrators to control and manage access to data products. These data products can then be securely consumed by SAP Datasphere and platforms such as SAP Databricks, leveraging the Delta Sharing protocol to ensure seamless, standardized, and controlled data exchange across environments.

Security in SAP Business Data Cloud is managed in a decentralized manner, with responsibilities distributed across different platforms. Administrative and viewer roles are governed within their respective environments, including SAP Datasphere, SAP Analytics Cloud, SAP Databricks, and SAP BW, ensuring that user access and permissions are controlled according to each system's specific roles and security requirements.

SAP Business Data Cloud supports self-service data discovery and promotes data democratization across the organization. It enables the extraction of metadata from various SAP sources, including SAP Analytics Cloud, SAP Datasphere, SAP applications, and other systems, providing users with greater visibility into available data assets. By making data more accessible and understandable, it empowers users across business functions to become more data-driven, fostering informed decision-making and reducing dependency on centralized data teams.

SAP Business Data Cloud ensures the proper use of data across the organization by establishing consistent governance practices. It facilitates the exchange of metadata with third-party catalog providers, enabling integration within broader data ecosystems. It also supports the monitoring of data activity, tracking of workflows, and enforcement of organizational policies, ensuring compliance, transparency, and effective data management throughout the enterprise.

User and role management in SAP Business Data Cloud is handled centrally through the SAP Business Data Cloud cockpit, ensuring streamlined administration and governance of access rights. The following are the key user roles in SAP Business Data Cloud:

- **SAP Business Data Cloud administrator**
  The SAP Business Data Cloud administrator role provides full privileges within SAP Business Data Cloud. Users assigned to this role have administrative control across integrated environments, including acting as an administrator in the SAP Datasphere tenant, with the ability to create users, define custom roles, and assign users to roles; and acting as an administrator in the SAP Analytics Cloud tenant, with similar capabilities to manage users, roles, and role assignments.

- **SAP Business Data Cloud viewer**
  The SAP Business Data Cloud viewer role provides read-only access to SAP Business Data Cloud. Users in this role can view data and insights but do not have permissions to make administrative or configuration changes.

SAP Business Data Cloud uses spaces as secure logical boundaries to control access to data assets, allowing only authorized users to access specific tables, views, and data models. This design ensures strong data isolation, security, and governance across teams and business domains.

Figure 9.30 shows how spaces are created and managed in SAP Business Data Cloud.

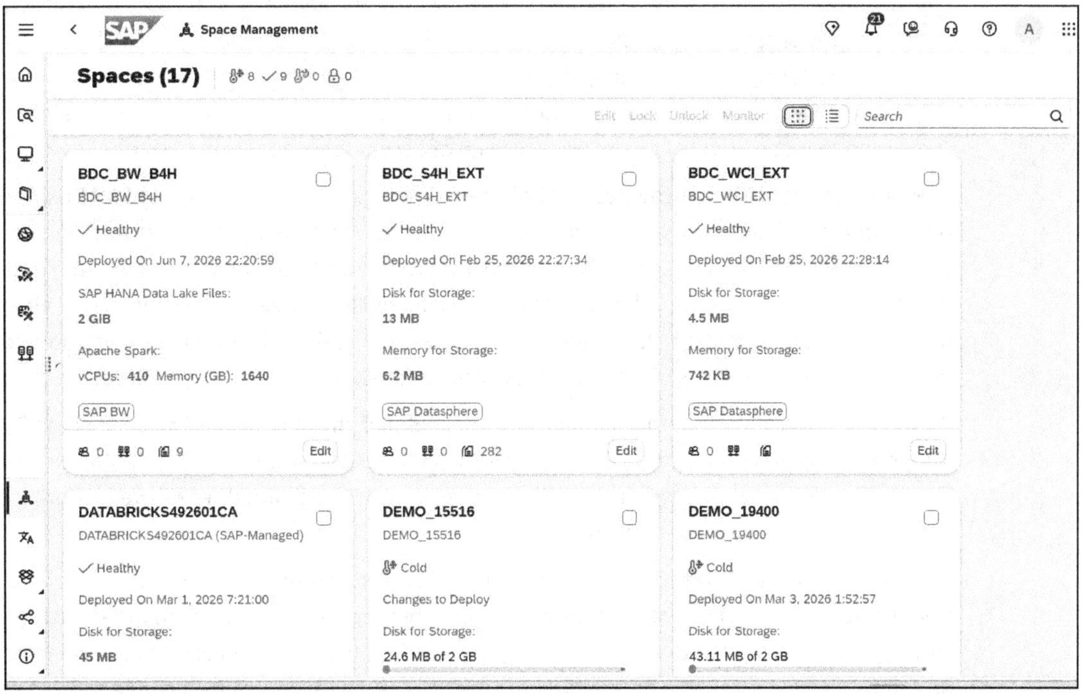

Figure 9.30: Space Management

In addition, SAP Business Data Cloud enforces segregation of duties through clearly defined roles and the separation of preparation and application spaces. Preparation spaces are dedicated to data engineering activities, while application spaces are used for data consumption. This separation prevents any single role from both creating and consuming sensitive datasets without proper governance oversight, thereby strengthening security, compliance, and accountability across the data lifecycle.

SAP Business Data Cloud maintains comprehensive audit logs that capture security-relevant activities, including role assignments, configuration changes, and catalog administration events. Access to these logs is restricted to authorized administrators with explicitly assigned privileges. Although the logs are not described as fully tamper-proof, they are safeguarded against unauthorized alteration or deletion. Log retention is managed in accordance with compliance and audit requirements, ensuring traceability, accountability, and effective monitoring across the platform.

Security is embedded throughout the entire SAP Business Data Cloud lifecycle, spanning secure development, configuration, and ongoing operations. Core security capabilities—including role-based access control, data access controls, encryption, and secure integrations—are natively integrated into the platform. This ensures that data protection is not treated as an afterthought but as a foundational element of the platform's architecture, enabling consistent, enterprise-grade security across all processes and data flows.

SAP Business Data Cloud manages user authorization through a structured framework of roles and privileges. Roles group together specific privileges that define both the actions users can perform, such as managing data packages or configuring governance, and the data they can access. Permissions, including create, read, update, delete, execute, maintain, share, and manage, determine the precise level of access granted within each privilege.

SAP Business Data Cloud provides a set of predefined standard roles, such as SAP Business Data Cloud administrator, SAP Business Data Cloud viewer, catalog administrator, and catalog user, which align with common responsibilities. In addition, administrators have the flexibility to create custom roles tailored to specific business needs, allowing fine-grained permission assignment at the object level. Roles are assigned during user creation or import, and user accounts can be activated, deactivated, or removed as required.

Within SAP Datasphere, authorization is further refined through scoped roles such as DW integrator, DW modeler, and DW consumer, which control access to designated spaces (e.g., ingestion, preparation, and application spaces). Fine-grained, row-level security is enforced using data access controls.

SAP Business Data Cloud further strengthens its security framework by supporting additional safeguards such as data sharing for governed, zero-copy data exchange, ensuring secure and controlled data access without duplication. It also provides secure connectors and data agents for reliable integration pipelines, along with centralized connection management for consistent control over data connections. IP allowlisting also is supported to restrict data exchange to trusted networks and systems, thus enhancing overall security and protecting sensitive enterprise data.

All SAP Business Data Cloud components encrypt customer data, metadata, logs, and backups at rest to protect information even if underlying storage is compromised. Encryption uses strong standards (typically AES-256) with keys managed via SAP or the underlying hyperscaler key management service. Let's take a closer look at encryption for SAP Datasphere and SAP Analytics Cloud:

- **SAP Datasphere encryption**
  Runtime data in SAP HANA and the data lake is encrypted with customer-specific encryption keys. Metadata and configuration artifacts are stored in a shared database, encrypted with keys managed by SAP.

- **SAP Analytics Cloud encryption**
  Data replicated into SAP Analytics Cloud's SAP HANA database is encrypted at rest in line with SAP's strong encryption-at-rest policies.

SAP Business Data Cloud encrypts all data in transit using strong, industry-standard protocols to protect against interception or tampering during transmission.

Within SAP Business Data Cloud and SAP services, all communication between SAP Business Data Cloud components (including SAP Datasphere and SAP Analytics Cloud) and connected SAP services is encrypted using TLS 1.2 or higher. Unencrypted connections (e.g., JDBC/ODBC to SAP HANA) are explicitly rejected. Web UI, command-line interface, and API access are also protected by TLS/SSL.

For customers requiring direct control over encryption keys, SAP Datasphere supports customer-controlled encryption keys (CCEKs) through SAP Key Management Service. With CCEK, customers provide their own keys, which SAP uses to encrypt the database. Once enabled, CCEK cannot be reverted to SAP-managed keys, and the key ID cannot be changed.

9.9 Use Case Patterns

The use case patterns for SAP Business Data Cloud can be generally classified into the following segments:

- Leverage SAP's intelligent packages to derive insights (domain content such as insight apps and SAP data products), accelerating time-to-value.
- Design and develop custom data products and analytical solutions directly within SAP Business Data Cloud. Tailor solutions to specific business requirements and advanced analytics use cases, such as the following:
  - SAP S/4HANA Cloud Private Edition, leveraging CDS views or raw tables
  - Third-party systems
- Modernize SAP BW in SAP Business Data Cloud with the data product generator.
- Use SAP Business Data Cloud Connect and leverage partners' capabilities to enhance the solution with AI/ML.

The following sections cover some of the key use cases for SAP Business Data Cloud.

9.9.1 Working Capital

SAP Business Data Cloud provides several high-value KPIs out of the box through the Working Capital Insights domain content and applications. Among these, days sales outstanding (DSO) and days payable outstanding (DPO) are some of the most widely used and critical metrics.

Customers often focus on a range of KPIs within supply chain management, such as on-time, in-full (OTIF) delivery, on-time delivery, and spend insights, to monitor operational efficiency and optimize performance.

Some of these KPIs are readily available as part of standard domain content in SAP Business Data Cloud, enabling quick access to prebuilt, business-ready insights. However, certain KPIs may require the integration of external or third-party data sources to provide a more comprehensive and accurate view, especially in scenarios involving supplier performance, logistics partners, or broader spend analysis.

DSO is one of the most commonly tracked KPIs by business users. It measures the average number of days it takes to collect payment after a sale has been made, making it a key indicator of accounts receivable efficiency.

It is calculated as follows:

DSO = (Total Accounts recievables x Numbre of days) ÷ Total sales during the period

DSO plays a crucial role in cash-on-hand and liquidity management scenarios, helping organizations understand how quickly they can convert receivables into cash. Within SAP Business Data Cloud, this KPI is available within the **Accounts Receivable** section of the Working Capital Management Insights application, providing ready-to-use, business-context-rich analytics.

Figure 9.31 illustrates the data models and analytical models that are part of the Working Capital Analytics Insight apps.

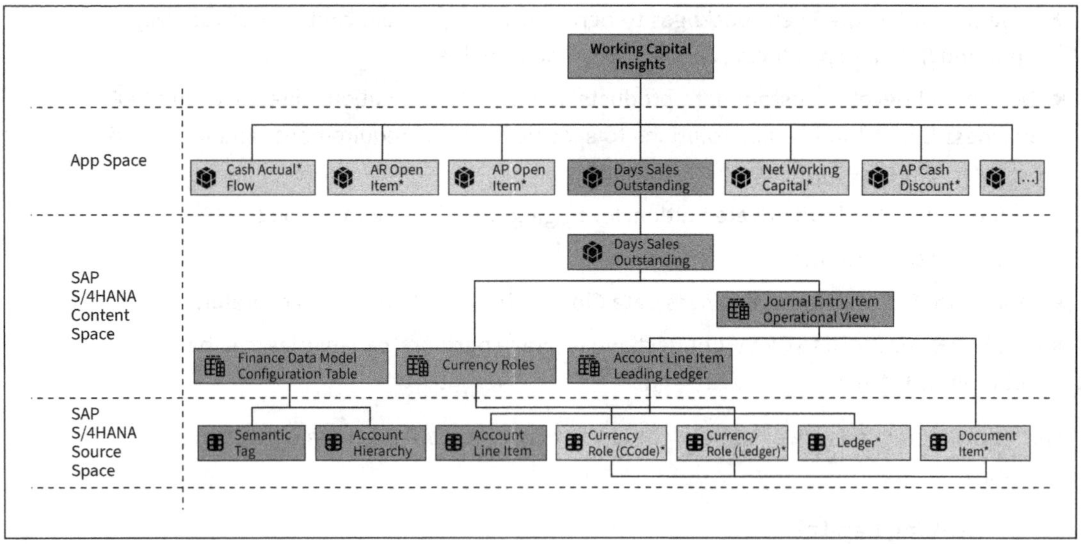

Figure 9.31: Working Capital Insights Models

Because DSO is an out-of-the-box KPI within the Working Capital Insights domain content, it enables a very fast time to value for organizations. The standard domain content is maintained by SAP, including the regular refresh of underlying data products sourced from SAP

S/4HANA. This ensures that users have access to up-to-date, consistent, and business-context-rich data without the need for additional modeling or complex data integration efforts.

Inefficiencies in the order-to-cash cycle, such as delayed invoice delivery or late payments from specific customer segments or regions, can significantly impact an organization's financial health. A rising DSO is a key indicator of liquidity risk and highlights challenges in receivables management.

This is a critical area in which AI can add significant value. By leveraging advanced analytics and machine learning, organizations can do the following:

- Predict overdue amounts and payment delays with greater accuracy
- Generate automated risk scores for customers and regions
- Proactively identify high-risk accounts and take corrective actions

Figure 9.32 shows an example of a DSO chart within the accounts receivable area in Working Capital Insights.

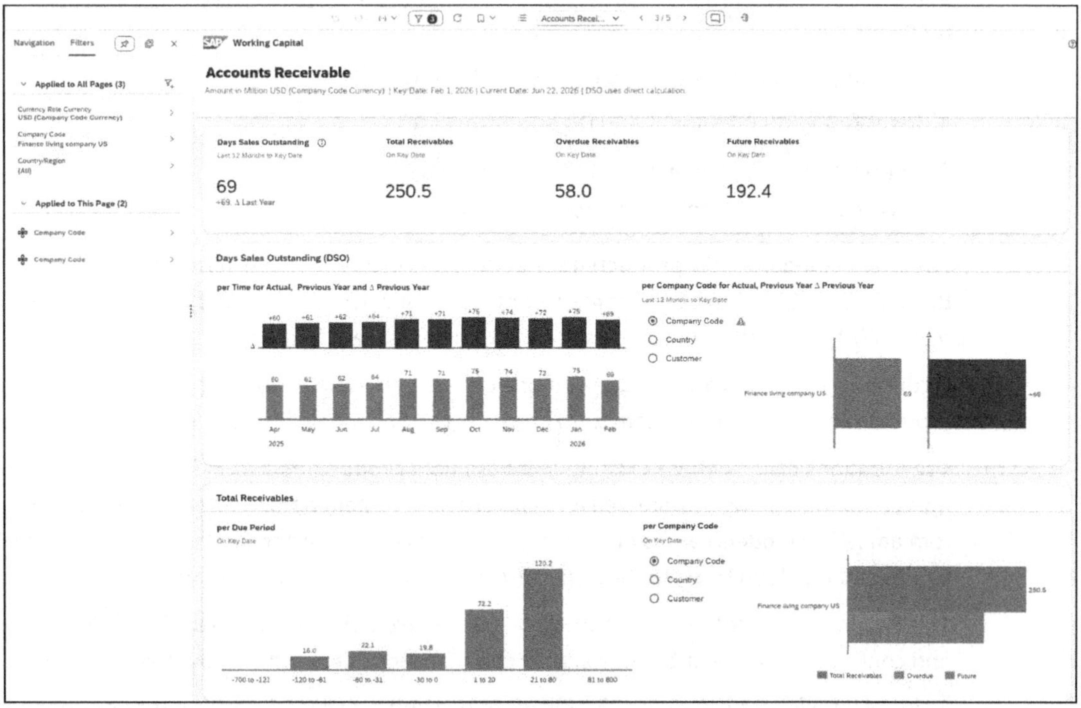

Figure 9.32: Daily Sales Outstanding as Part of Accounts Receivables in Working Capital Insights

By leveraging accounts receivable aging data products in combination with customer financial health signals and historical payment behavior, organizations can use SAP Databricks or SAP Snowflake to build predictive models that accurately identify invoices at risk of late or nonpayment.

These insights enable collections teams to prioritize high-risk, high-value accounts proactively, focusing efforts where they have the greatest impact.

As SAP continues to deliver an expanding portfolio of SAP-managed data products, an increasing number of customer-relevant use cases can be addressed out of the box, accelerating time to value. At the same time, organizations can create custom data products to complement these SAP-provided assets, enabling deeper insights and tailored KPI development aligned with specific business needs.

9.9.2 On Time, In Full (OTIF)

In the supply chain domain, metrics such as OTIF and delivery performance are among the most critical KPIs that customers track to optimize operational efficiency, enhance service levels, and drive continuous improvement.

OTIF is a key supply chain KPI that measures whether customers receive the correct quantity of goods on or before the committed or requested delivery date. It assesses overall delivery reliability by calculating the percentage of order items that are delivered both on time and in full.

By tracking OTIF, organizations can do the following:

- Identify gaps in fulfillment and logistics processes
- Improve customer satisfaction and service reliability
- Drive continuous supply chain optimization

There are standard data products related to sales orders and deliveries available in SAP Business Data Cloud, but OTIF domain content is not currently provided out of the box. As a result, OTIF needs to be developed as a custom analytical use case.

Figure 9.33 illustrates a very common use case that involves adopting standard SAP data products and leveraging custom data products to fulfill many different use cases.

Organizations can use these existing standard data products as a foundation to build OTIF models by combining relevant delivery and order fulfillment data. Using this approach, custom analytical models can be designed and custom dashboards can be created within SAP Business Data Cloud to accurately measure and monitor OTIF performance.

This flexibility allows businesses to tailor the KPI to their specific operational definitions and continuously refine it to drive supply chain optimization and improved service levels.

Artificial intelligence can play a significant role in improving OTIF performance by enabling predictive and prescriptive insights across the supply chain. By analyzing historical patterns, real-time data, and external signals, AI models can predict delayed shipments, identify potential inventory shortages, and highlight supply chain disruptions in advance.

Beyond prediction, AI can also recommend alternative actions, such as rerouting shipments, adjusting inventory allocation, or identifying substitute suppliers to mitigate risks. These proactive capabilities help organizations improve OTIF metrics, enhance delivery reliability, and maintain high service levels while optimizing supply chain operations. This represents a powerful closed-loop use case in which insights from analytics are seamlessly translated into action through AI agents, thus bridging the gap between data-driven insights and operational execution.

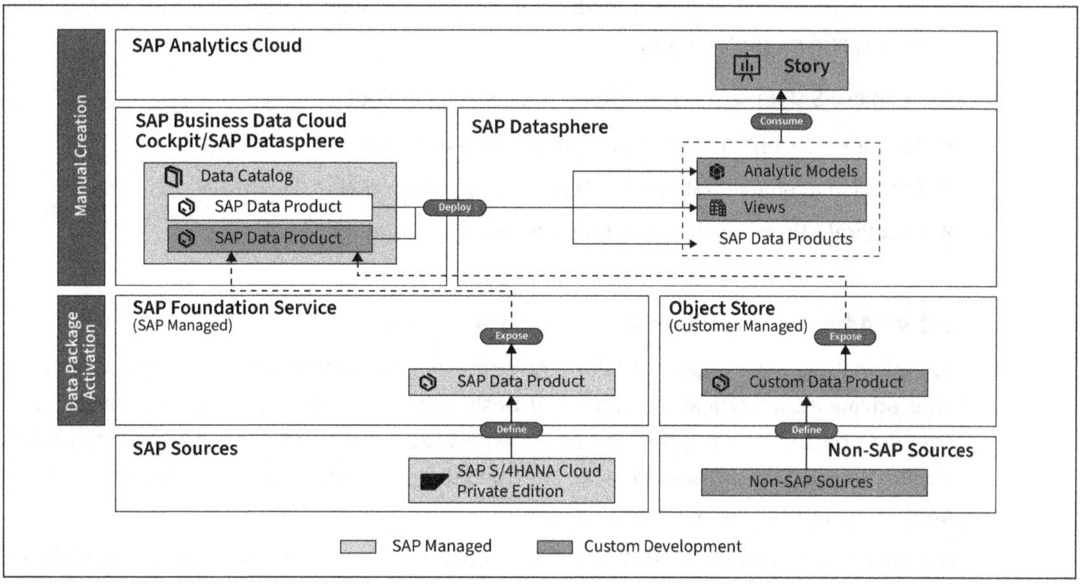

Figure 9.33: Custom Content

9.9.3 Total Spend

Total spend is one of the most critical KPIs for any organization because it directly impacts the following areas:

- Cash flow and liquidity
- Profit margins
- Working capital management
- Cost optimization strategies

You can access a unified, trusted view of total spend to eliminate data silos and reduce the effort required to integrate, cleanse, and manage spend data. A complete view of total spend typically includes four major categories:

- **Direct spend**
  Costs tied directly to production of goods or services, such as raw materials, components, and manufacturing inputs
- **Indirect spend**
  Operational expenses not directly tied to production, such as IT services, marketing, office supplies, and facilities

- **Contingent labor spend**
  Workforce-related costs not tied to permanent employees, managed via external vendors or platforms, such as contractors, freelancers, and temporary staff
- **Travel and expense spend**
  Employee-related expenses incurred during business activities, such as travel, lodging, meals, and reimbursements

You can use SAP Business Data Cloud to do the following:

- Access clean, enriched, and classified spend data out of the box
- Establish a reliable data foundation for analytics and AI
- Accelerate time to value with ready-to-use data products

9.9.4 Advanced Spend Intelligence Capabilities

Spend analysis and insights transform raw data into actionable business insights. They provide advanced analytical journeys and dashboards that help organizations understand spending patterns and opportunities. AI-generated insights are embedded directly within applications, enabling users to make informed decisions faster. Users can also interact with data using natural language through Joule.

Supplier enrichment with the Dun & Bradstreet integration improves supplier visibility and strengthens risk assessment capabilities. It provides deeper insights into supplier profiles, relationships, and business information. These enhanced insights help organizations make better sourcing decisions and improve supplier management practices.

AI-powered invoice classification uses advanced AI and ML models to automatically classify invoices with high accuracy. This reduces manual effort and minimizes classification errors. It also improves spend categorization and ensures greater consistency in reporting and analysis.

Organizations can gain real-time visibility into external workforce and services procurement through comprehensive monitoring and analytics. This enables them to track contingent labor spending and workforce utilization more effectively. They can also monitor vendor performance and identify workforce trends. These insights support data-driven workforce planning and resource management decisions

With the Spend Intelligence solution in SAP Business Data Cloud, organizations can achieve full transparency into their total spend across the enterprise. This comprehensive visibility helps businesses improve cash flow management and strengthen cost-control measures. By analyzing spending patterns and trends, organizations can identify savings opportunities and reduce spend leakage. The insights gained also support the development of stronger supplier and workforce strategies. In addition, Spend Intelligence accelerates AI-driven decision-making by providing timely, data-driven insights that enable faster and more informed business actions. The Spend Intelligence package within SAP Business Data Cloud provides organizations with a comprehensive solution to address fragmented and inconsistent spend data. This offering is designed to rapidly and efficiently unify, classify, and enrich data from both SAP and non-SAP sources, consolidating it into a cohesive semantic layer. As a result, organizations gain complete, enterprise-wide visibility into total spend within a single, centralized environment.

Built on SAP Business AI, the solution enables the automatic delivery of relevant insights and opportunities across key business workstreams. It supports faster and more informed decision-making by surfacing actionable intelligence at the right time.

The experience is designed to be intuitive, featuring detailed dashboards, advanced analytics, and natural language interaction capabilities through SAP's Joule copilot, which guides users and enhances productivity throughout the analytical process.

Figure 9.34 shows an example of Joule embedded in Spend Intelligence insights.

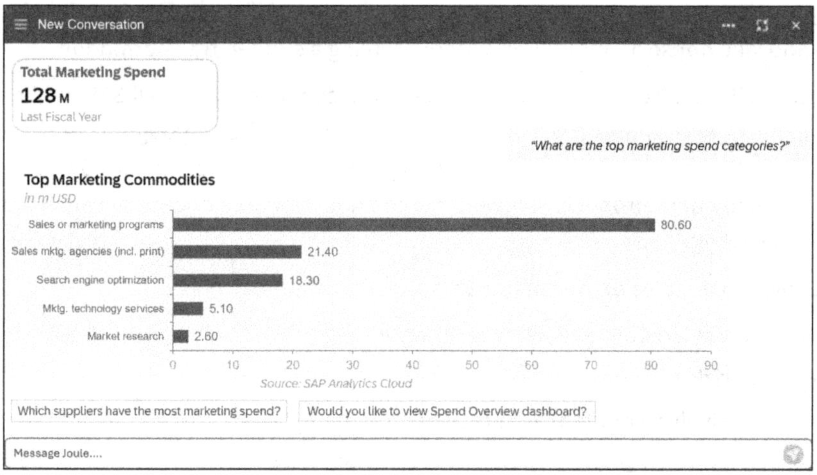

Figure 9.34: Joule Embedded in Spend Intelligence

In addition, the solution promotes collaboration by enabling the integration and sharing of spend data across functions such as finance, supply chain, human resources, and sales. This cross-functional visibility helps unlock deeper enterprise-level insights and supports more aligned decision-making.

The platform is also highly connected, with native integration across SAP's source-to-pay solutions. This integration simplifies access to relevant information and enables organizations to move seamlessly from insight to action, ultimately driving measurable business value.

9.9.5 SAP Business Warehouse in the Private Cloud

Many SAP customers have long-standing investments in SAP BW environments, often built and optimized over several years. These systems represent not only a significant financial investment but also accumulated technical debt and deeply embedded business logic.

SAP Business Data Cloud provides a pathway to protect and extend these investments while enabling a transition to modern data architecture.

The approach allows organizations to move existing SAP BW data models, queries, and transformations as is into an SAP-managed environment, covering the full stack, from infrastructure and hardware to application management. This minimizes disruption while preserving proven assets and established processes.

By leveraging existing SAP BW skill sets, organizations can avoid immediate large-scale replatforming efforts and instead adopt a phased, controlled transition strategy. This serves as a low-risk migration path, enabling teams to maintain continuity while gradually modernizing their data landscape.

Importantly, this transition strategy provides the flexibility to move to SAP Business Data Cloud at the company's own pace, avoiding costly and high-risk rebuilds. Organizations can incrementally adopt a data product–driven approach, aligning with modern principles such as data mesh and lake house architectures, without forcing a sudden transformation.

SAP Business Data Cloud also positions itself as a centralized platform for all SAP application data, simplifying access and governance across the enterprise. Through capabilities like the data product generator, existing SAP BW objects can be converted into SAP Business Data Cloud–ready data products, accelerating the adoption of cloud-native analytics, AI, and machine learning.

Overall, this approach enables organizations to do the following:

- Protect and maximize existing SAP BW investments
- Reduce migration risk and complexity
- Transition progressively to modern architecture
- Enable faster access to advanced analytics and AI capabilities

Once SAP BW data is made available on SAP Business Data Cloud, organizations can immediately unlock advanced capabilities, including AI- and ML-driven insights and agent-based and copiloted experiences powered by SAP Business AI.

Organizations can accelerate the modernization of their SAP BW environments by leveraging purpose-built capabilities within SAP Business Data Cloud.

SAP BW analytical workloads can be efficiently transitioned by transferring SAP BW query metadata through the query template generator, enabling continuity of existing reporting and analytics with minimal disruption.

SAP BW data models can be migrated and transformed using the data product generator, which converts legacy SAP BW objects into SAP Business Data Cloud–ready data products designed to support modern, agentic, and AI-driven use cases. You can enable self-service modeling capabilities for power users on top of the data products via modeling capabilities in SAP Business Data Cloud.

This approach allows enterprises to bridge legacy investments with next-generation innovation, turning historically siloed and rigid data models into flexible, reusable data products that support faster, smarter decision-making.

Figure 9.35 illustrates the step-by-step process of SAP BW modernization, from leveraging legacy investments to adopting standard data products and eventually retiring SAP BW.

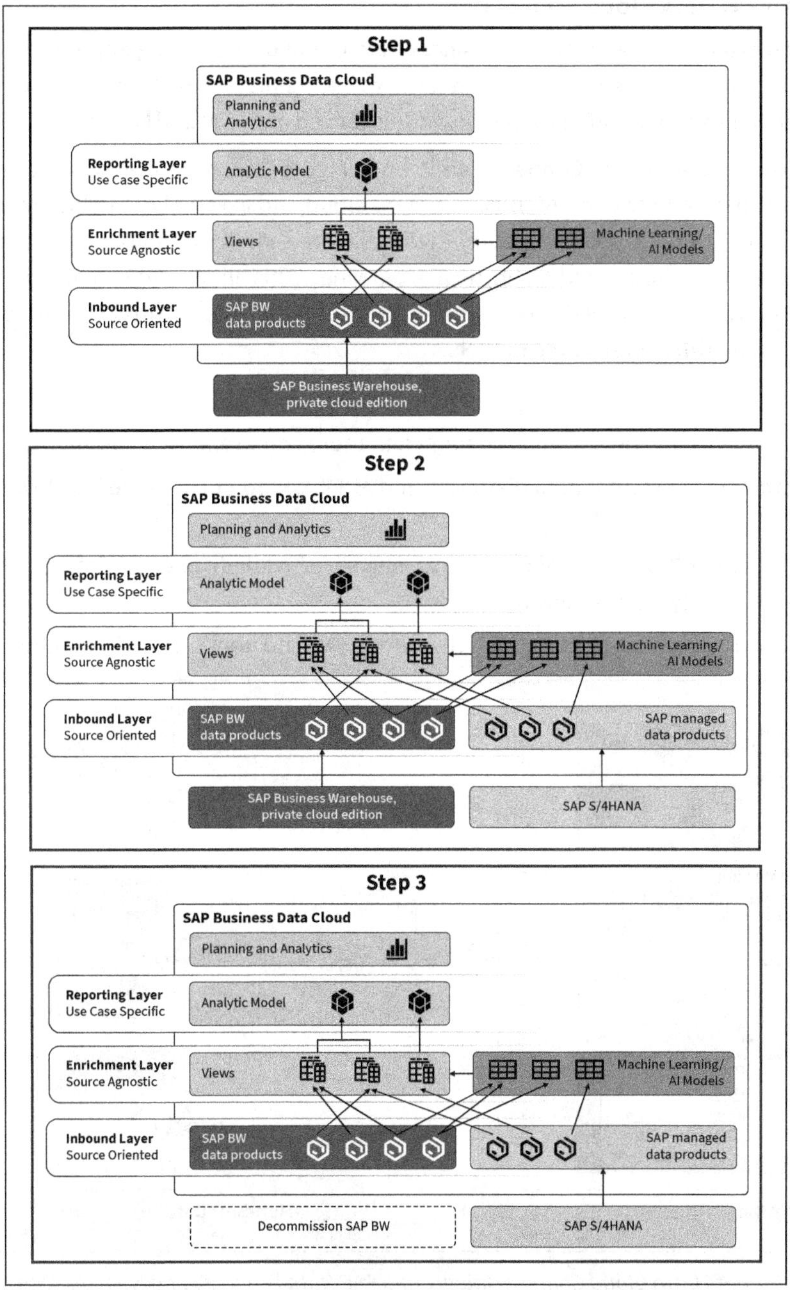

Figure 9.35: SAP BW Modernization to SAP Business Data Cloud

By bringing SAP BW into SAP Business Data Cloud, organizations can protect and fully leverage their existing investments while establishing a clear path toward modernization. This approach enables a gradual transformation of legacy SAP BW assets into modern, reusable data products, aligning with the emerging data product economy.

9.9.6 SAP Business Data Cloud Connect

Organizations can extend the value of SAP Business Data Cloud by leveraging SAP Business Data Cloud Connect to integrate with leading data engineering platforms. These platforms often host large volumes of non-SAP data and provide advanced native AI and ML capabilities.

Through SAP Business Data Cloud Connect, as shown in Figure 9.36, organizations enable bidirectional data sharing between SAP Business Data Cloud and external data platforms. This allows SAP data products and third-party datasets to be seamlessly combined into a unified data ecosystem without building brittle ETL pipelines. This allows data scientists to enrich trusted SAP records with external inputs (like Internet of Things [IoT] sensors or weather data) to train AI/ML forecasting models.

As a result:

- SAP and non-SAP data can be harmonized for richer, more comprehensive insights.
- Existing investments in external data platforms and AI/ML capabilities can be fully leveraged.
- Data products from SAP Business Data Cloud can be enriched with external datasets to support advanced analytics and new use cases.
- Organizations can operationalize insights across both SAP and non-SAP environments.

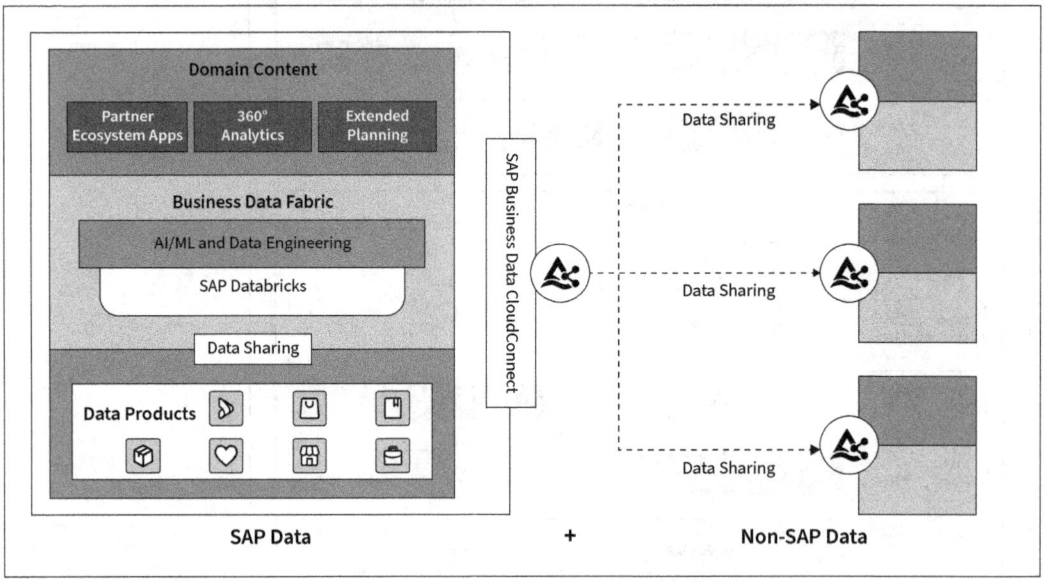

Figure 9.36: SAP Business Data Cloud Connect, Integrating SAP Data Products and Non-SAP Data

This integration approach enables a hybrid data strategy in which SAP Business Data Cloud serves as a trusted foundation for SAP data while remaining open and interoperable with the broader enterprise data landscape, ultimately driving more powerful, AI-driven outcomes.

9.10 Summary

SAP is advancing a comprehensive portfolio of integrated data and AI technologies, designed to unlock new business insights and seamlessly feed those insights back into operational processes and applications.

At the core of this vision is a foundational concept often described as the AI/data/application flywheel, which underpins the evolution toward the Autonomous Enterprise.

This flywheel operates as a continuous value loop: Business transactions generate high-value operational data. When this data is captured, governed, and structured effectively, it becomes a strategic asset. The data is then used to train and refine AI and ML models. These models power AI agents and intelligent automation, which can optimize business processes and decision-making. The resulting improvements in applications drive better outcomes and new transactions, restarting and accelerating the cycle.

This reinforcing cycle enables organizations to continuously improve and innovate, creating a self-optimizing system that becomes more intelligent over time.

A key implication of this model is that data strategy and AI strategy are inseparable. Without high-quality, well-managed data, AI initiatives cannot deliver meaningful value. Conversely, without AI, organizations cannot fully capitalize on the potential of their data.

By tightly integrating data, AI, and applications, SAP enables organizations to do the following:

- Embed intelligence directly into business processes
- Transition from reactive to proactive and autonomous operations
- Drive continuous optimization and innovation
- Deliver measurable business outcomes at scale

A critical success factor for any AI initiative is the quality and accessibility of the underlying data. Organizations cannot build effective AI systems without clean, well-governed, and easily accessible data. The data products economy is the foundation layer to gather curated, well-governed data, ready for AI to turn it into into real-world outcomes.

In the next chapter, you'll see how SAP S/4 HANA extensibility is developed by leveraging SAP BTP and agentic AI capabilities.

Chapter 10
Extensibility and Agent Development

This chapter brings the extensibility capabilities of SAP S/4HANA and SAP Business Technology Platform together with the emerging world of AI-powered agent development to show you how to build an intelligent, adaptive enterprise.

The speed of the modern world demands that your enterprise applications keep pace with changing business requirements. An enterprise that cannot adapt its processes to customer-specific rules, a finance team that cannot automate the daily grind of collections calls, or an operations group that cannot surface real-time risk signals in the receivables portfolio will always be working harder than necessary to achieve the same outcomes. SAP provides a layered extensibility framework on SAP S/4HANA and SAP Business Technology Platform (SAP BTP) that allows your organization to fill these gaps quickly, safely, and in a way that preserves the integrity of your core system so it can be upgraded without disruption. Paired with the AI agent capabilities of Joule Studio, these extensions become active, intelligent participants in your business processes.

SAP provides three primary extensibility layers, and selecting the right one for your requirement is the most important architectural decision you will make before development begins. These layers are as follows:

- **Key-user extensibility**
  For configuration-style extensions delivered through low-code tooling in SAP S/4HANA
- **Developer extensibility**
  For on-stack ABAP development using the ABAP RESTful application programming model in the ABAP development tools
- **Side-by-side extensibility on SAP BTP**
  For extensions that run outside SAP S/4HANA, built with SAP Cloud Application Programming Model, the ABAP RESTful application programming model on SAP BTP ABAP environment, or SAP Build

SAP recommends working your way down this list in order. That is, you should exhaust what is possible with key-user extensibility before moving to developer extensibility, and you should exhaust what is possible with on-stack developer extensibility before building a side-by-side extension on SAP BTP. This discipline keeps your SAP S/4HANA system as close to standard as possible and ensures that as many of your extensions as possible benefit from SAP's ongoing investment in its standard tooling.

In this chapter, we'll expand on the discussion from Chapter 8 with a deeper dive into extensibility topics. You'll learn how to extend SAP S/4HANA safely using key-user extensibility, developer extensibility, and side-by-side extensibility on SAP BTP in alignment with SAP's clean core standards. You'll explore SAP Cloud Application Programming Model and the ABAP RESTful application programming also model as the two premier frameworks for building enterprise-grade custom applications. The chapter then pivots to agent development using Joule Studio, and you will learn how to build, configure, and deploy custom Joule Skills and Joule Agents that can orchestrate actions across your SAP landscape.

10.1 In-App Extensibility for Key Users

As we introduced in Chapter 8, Section 8.1.1, key-user extensibility is SAP's low-code framework for making targeted customizations to SAP S/4HANA applications without writing ABAP code. It is delivered through a set of extensibility applications that are available to users who have been granted the key-user extensibility role collection. Key-user extensibility is the fastest path to extending SAP S/4HANA because its artifacts are automatically Tier A compliant, can be created and managed by business analysts with functional SAP knowledge, and do not require a transport to a development system. Extensions created through key-user extensibility tools are stored in a dedicated namespace that is separate from SAP's own namespace, ensuring they cannot accidentally overwrite standard objects.

The key-user extensibility toolbox includes several important apps:

- **Custom Fields**
  The Custom Fields app lets you add custom fields to standard SAP data objects. These fields can be appended to core data services (CDS) views and their corresponding OData APIs, making them available in standard SAP Fiori applications, reports, forms, and custom integration scenarios.
- **Custom Business Objects**
  The Custom Business Objects app lets you create entirely new data entities with their own database tables, CDS views, and SAP Fiori user interfaces for data entry and reporting all generated automatically from your definition.
- **Custom Logic**
  The Custom Logic app lets you implement business rules at defined extension points within standard SAP processes using a restricted ABAP syntax that enforces Tier A compliance. In addition, the Custom Analytical Queries application lets business analysts create report queries against standard and custom CDS views without writing code.
- **Custom CDS Views**
  With custom CDS views, you can create custom models inside SAP S/4HANA. With the custom CDS view tooling, you can create models by selecting existing primary data sources or create fully custom standalone models. You can define associations to existing models and

apply filters on the data or calculated fields or aggregations. Models can be used for analytical reports or transactional applications and can be configured to become external APIs.

10.1.1 Extending the Order-to-Cash Process

Consider the practical challenge of tracking customer payment behavior risk within the order-to-cash process. A standard SAP S/4HANA credit check evaluates a customer's credit limit and current exposure, but many organizations want to factor in a broader risk picture: how many invoices in the last 12 months were paid late, what is the average days-past-due for this customer, and whether this customer has a pattern of raising disputes to delay payment. This data exists inside SAP S/4HANA in the accounts receivable and dispute management applications, but the standard credit check does not surface it as a single risk indicator on the sales order.

Using key-user extensibility, a business analyst can create a custom field called *customer payment risk score* in the customer master object and append it to the relevant CDS views so that it appears in the customer fact sheet SAP Fiori app and in the sales order header. A custom logic rule can be implemented at the credit check extension point in the sales order process to read the payment risk score field and escalate the credit block to a senior credit analyst when the score exceeds a defined threshold, rather than triggering the standard automatic credit block workflow. The entire customization—the field, the logic, and the UI adaptation—can be built by a knowledgeable business analyst in a matter of hours and transported to production using the standard transport mechanism, with no developer involvement required.

This is the power of key-user extensibility in the order-to-cash context. The customization directly addresses a real business need, better early identification of customers who are likely to pay late, without touching a single line of SAP standard code. When SAP releases a new version of the credit management functionality, the customization is preserved because it is built on released extension points that SAP guarantees will remain stable across upgrades. Because the custom payment risk score field is appended to the standard CDS view and its OData API, it is immediately available for use in integration flows, Joule Agent skills, and analytics without any additional development work.

The Custom Business Objects app addresses a different class of requirement that is also common in the order-to-cash process. Many organizations need to track collections activities against open invoices: a log of every phone call made, every email sent, every promise to pay received, and every dispute raised during the collections process. This data does not belong in the standard dunning tables because it contains free-form notes, attachment references, and outcome codes that are specific to the organization's collections process. A key user can create a custom business object called *collections activity log* with fields for the invoice reference, the contact person, the date of contact, the contact method, the outcome, the promised payment date, and any notes. SAP will automatically generate the database table, the CDS view, the OData API, and a basic SAP Fiori application for the object. The collections team can begin logging activities immediately, and the data is immediately available for reporting, integration, and, as you will see in Section 10.5.2, consumption by a Joule collections agent.

10.1.2 Transporting Key-User Extensibility Artifacts

Key-user extensibility supports two transport modes.

Simple transport mode provides basic functionality for transporting extensions across the system landscape. It supports a single active business configuration transport and handles transport types automatically, with the option for individual instant transports when an urgent change needs to be moved to production.

Flexible transport mode provides fully flexible transport management, allowing you to create multiple business configuration transports in parallel and to define transport automations that align with your organization's release cadence.

For organizations running a structured SAP landscape with development, quality assurance, and production systems, flexible transport mode is the recommended approach as it aligns key-user extensibility transports with the same governance process used for developer extensibility and integration artifacts.

10.2 Developer Extensibility

When the requirements of a business process go beyond what key-user extensibility can deliver, SAP provides developer extensibility for on-stack ABAP development in SAP S/4HANA. Developer extensibility uses the ABAP development tools inside the Eclipse IDE as its development environment, and SAP's strong recommendation is that all developer extensibility development follow the *ABAP RESTful application programming model*, the modern ABAP development framework designed to help developers efficiently create SAP Fiori applications and their supporting backend data models, business logic, and APIs in a way that is both cloud-ready and Tier A compliant. SAP S/4HANA Cloud Private Edition customers who adhere to the ABAP RESTful application programming model gain the same upgrade safety guarantees as key-user extensibility, because the ABAP RESTful application programming model exposes only released ABAP APIs and extension points.

In the following sections, we'll take a closer look at the development model, walk through an example, and cover the supported Git-enabled change and transport system (gCTS) functionality.

10.2.1 The ABAP RESTful Application Programming Model

The ABAP RESTful application programming model was created from lessons learned across decades of ABAP development to give the next generation of SAP developers a structured, opinionated model for building transactional SAP Fiori applications. The model is built around the concept of *business objects*, representations of real-world enterprise entities such as a sales order, a purchase order, a customer, or a custom collections activity. Business objects in the ABAP RESTful application programming model have a hierarchical structure, with root nodes and child nodes connected by composition relationships. A sales order business object, for example, has a root sales order header node with a one-to-many composition with sales order item nodes, each of which may have a further composition with sales order item schedules.

The core components of an ABAP RESTful application programming model business object are the CDS data model, the behavior definition, the behavior implementation, the service definition, and the service binding. The CDS data model defines the structure of the business object: its fields, associations to related objects, and annotations that drive the SAP Fiori UI. The *behavior definition* specifies what operations the business object supports: which nodes can be created, read, updated, and deleted; which actions can be called; and which determinations and validations run during the object's lifecycle. The *behavior implementation* provides the actual ABAP code that executes the behaviors defined in the behavior definition. The *service definition* declares which CDS entities are exposed through the OData service. Finally, the *service binding* binds the service to a specific OData version and makes it accessible to SAP Fiori applications or integration consumers.

One of the most powerful features of the ABAP RESTful application programming model for the developer extensibility use case is the ability to extend existing SAP business objects using the ABAP RESTful application programming model's extension model. Rather than modifying the standard SAP business object code, a developer can create an extension of a standard business object that adds new fields, new validations, new determinations, or new actions to the object while the standard object remains untouched. This is the on-stack equivalent of the side-by-side decoupling philosophy: your extension code lives in your namespace, not SAP's namespace, and is therefore protected during upgrades.

10.2.2 Building a Credit Risk Scoring Service

To illustrate developer extensibility in the order-to-cash context, consider building a more sophisticated credit risk scoring service than what key-user extensibility alone can provide. Imagine that your organization wants to implement a proprietary credit risk score calculation that combines four inputs: the customer's payment history in SAP S/4HANA (retrieved from accounts receivable), the current open receivables balance relative to the credit limit, any active disputes in the SAP Dispute Management application, and an external credit bureau score retrieved via an API call to an external credit rating service. The calculation logic is complex enough to require developer involvement, but the result needs to be surfaced in the standard sales order process, in the customer fact sheet, and in a custom SAP Fiori risk dashboard.

Let's walk through the process flow:

1. A developer would begin by creating a CDS view that reads the relevant receivables data from the standard SAP S/4HANA CDS views exposed for the accounts receivable aging, the credit management objects, and the dispute management objects.
2. The developer creates an ABAP RESTful application programming model behavior implementation class in ABAP that implements the scoring algorithm, reading the four inputs, applying the weighting model, and returning a normalized score between 0 and 100 along with a risk category of low, medium, high, or critical.
3. The behavior definition for the credit risk score business object exposes a `ReadByAssociation` operation so that the score can be retrieved as an association from the standard customer CDS view.
4. A service binding exposes the credit risk score via an OData V4 API, making it available for consumption in SAP Fiori applications, integration flows, and Joule Agent skills.
5. The developer then uses the ABAP RESTful application programming model extension model to extend the standard customer business object and the standard sales order business object to include the credit risk score as a *virtual element*, a calculated field that is not stored in the database but is computed at runtime by calling the credit risk score service. This virtual element appears naturally in the standard customer fact sheet and sales order SAP Fiori applications alongside the other standard fields, requiring only a UI adaptation using key-user extensibility to position the field in the appropriate location on the screen.

The entire extension is built without modifying a single standard SAP object. The result is a credit risk scoring capability that gives the sales and credit teams a richer picture of customer risk the moment they are creating or approving a sales order, precisely where and when the information has the most business value.

10.2.3 Git-Enabled Transport Management for Developer Extensibility

Modern developer extensibility in SAP S/4HANA also supports gCTS, which allows developers to store and version ABAP development objects in Git repositories alongside other development artifacts, bringing the benefits of modern source control branching strategies, pull request reviews, rollback capability, and audit trails to ABAP development. When gCTS is enabled in a development system, ABAP objects are tracked in a Git repository and changes are moved between systems by checking out a specific branch or tag in the target system rather than by releasing transport requests in the traditional way. This approach aligns ABAP development with the DevOps practices that are standard in the broader SAP BTP development ecosystem, making it easier for organizations to adopt continuous integration and delivery practices across their entire development portfolio, both on stack and side by side.

10.3 Side-by-Side Extensibility

Side-by-side extensibility on SAP BTP, which we introduced in Chapter 8, Section 8.1.2, represents the most powerful tier of the extensibility framework. Where key user and developer extensibility both operate within the SAP S/4HANA system boundary, side-by-side extensions run entirely outside SAP S/4HANA on SAP BTP. This decoupled architecture provides several significant advantages (see Figure 10.1). Side-by-side extensions can combine data from multiple SAP systems, non-SAP systems, and external APIs into a single application or workflow, which is not possible with on-stack extensions. They can be built using any language or framework supported on SAP BTP: Node.js, Java, Python, ABAP, or any other container-based runtime, giving developers freedom of choice. They scale independently of the SAP S/4HANA system, so a high-volume batch process running on SAP BTP does not compete with transactional SAP users for system resources. Finally, because they communicate with SAP S/4HANA exclusively through released OData APIs and standard events, they are completely upgrade-safe.

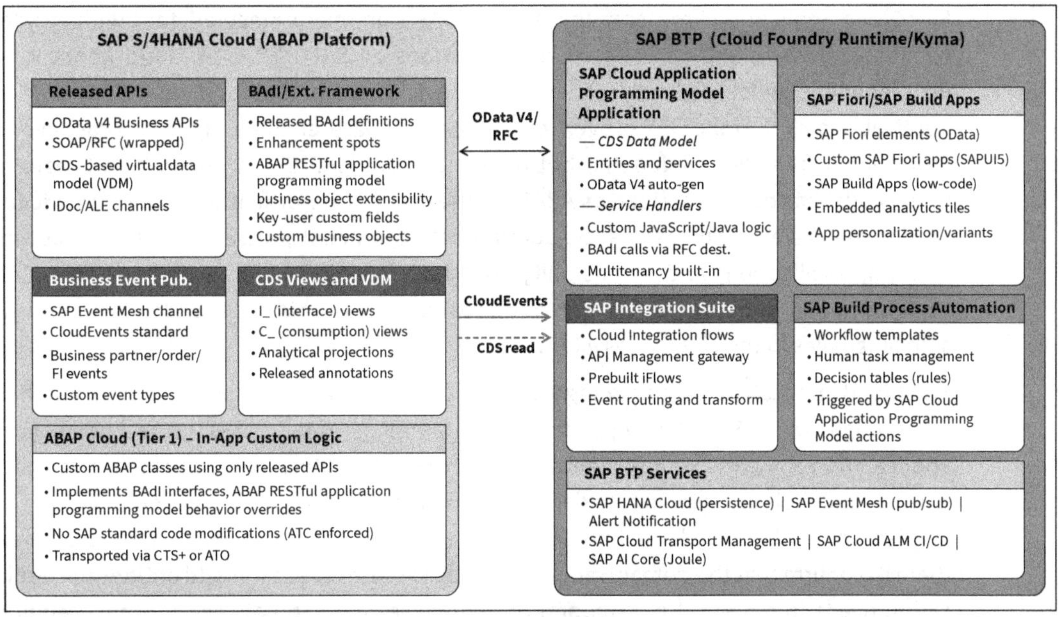

Figure 10.1: Developer Extensibility Options with Side-by-Side Development on SAP BTP

The most common patterns for side-by-side extensions in the order-to-cash context are as follows:

- **Preprocessing applications**
  Custom logic that evaluates or enriches data before it enters SAP S/4HANA, such as a customer onboarding portal that validates and cleanses customer master data before creating a business partner record via API.

- **Postprocessing applications**
  Logic that acts on data after it is created in SAP S/4HANA, such as an automated collections engine that reads open invoices from the FI-AR OData API and executes a configurable collections action sequence.
- **Analytical applications**
  Dashboards and reports that combine SAP S/4HANA data with data from other systems, such as a days sales outstanding (DSO) management dashboard that pulls open receivables from SAP S/4HANA, payment forecast data from SAP Cash Management, and customer engagement data from SAP Customer Experience (SAP CX).
- **Substitute applications**
  Custom SAP Fiori applications built on SAP BTP that replace or supplement a standard SAP S/4HANA SAP Fiori application with a richer user experience tailored to a specific user persona, such as a collector-specific workspace that consolidates all the information and actions a collections analyst needs into a single screen.

SAP Cloud Application Programming Model is SAP's premier framework for building side-by-side extensions and custom cloud applications on SAP BTP. SAP Cloud Application Programming Model is a framework of languages, libraries, and tools for building cloud applications that embodies proven enterprise software engineering best practices in a developer-friendly, opinionated-but-open structure. SAP Cloud Application Programming Model has been described as ABAP for the non-ABAP world; it offers the same focus on domain-driven design, semantic data modeling, and enterprise concerns like authorization, localization, and extensibility that have always characterized the best SAP development, but delivers it through Node.js and Java, the two most widely used server-side languages in the cloud ecosystem.

SAP Cloud Application Programming Model places its primary emphasis on domain modeling through CDS. In SAP Cloud Application Programming Model, CDS is the backbone of the application. You define your business entities, their attributes, their associations to other entities, and their behavior in CDS definition language files that read almost like structured English. These CDS models are human-readable enough that a business analyst can review them to confirm that the domain model correctly captures the real-world entities the application needs to represent a collaboration that closes the gap between business requirements and technical implementation before any application logic is written.

Consider building a DSO management application on SAP BTP as a side-by-side extension of SAP S/4HANA (see the end-to-end flow in Figure 10.2). The DSO Management application needs to serve the treasury and collections teams with a real-time view of the receivables portfolio, the ability to launch collections actions directly from the dashboard, and a workflow for managing dispute escalations that keeps the customer communication and the internal approval chain in one place. None of this is available as a single integrated experience in standard SAP S/4HANA; the data exists across multiple standard SAP Fiori applications, but no single standard application brings it all together with actionable workflows tailored to the collections team's daily routine.

The SAP Cloud Application Programming Model data model for the DSO Management application begins with the core entities the application needs to manage. The `Customer` entity represents the debtor; it includes the standard SAP S/4HANA business partner fields plus the custom credit risk score from the developer extensibility scenario described in Section 10.2. The `OpenInvoice` entity represents an open accounts receivable line item's invoice number, due date, original amount, open amount, currency, aging bucket, and payment terms. The `CollectionsCase` entity represents the collections team's working record for a customer: the assigned collector, the collections strategy in use, the next action date, the last contact date, and the current case status. The `DisputeCase` entity represents a formal dispute raised against an invoice: the disputed amount, the reason code, the responsible department, and the current approval status. Finally, the `CollectionsActivity` entity, which corresponds to the custom business object created via key-user extensibility in SAP S/4HANA, provides the chronological log of every action taken on a `CollectionsCase` entity.

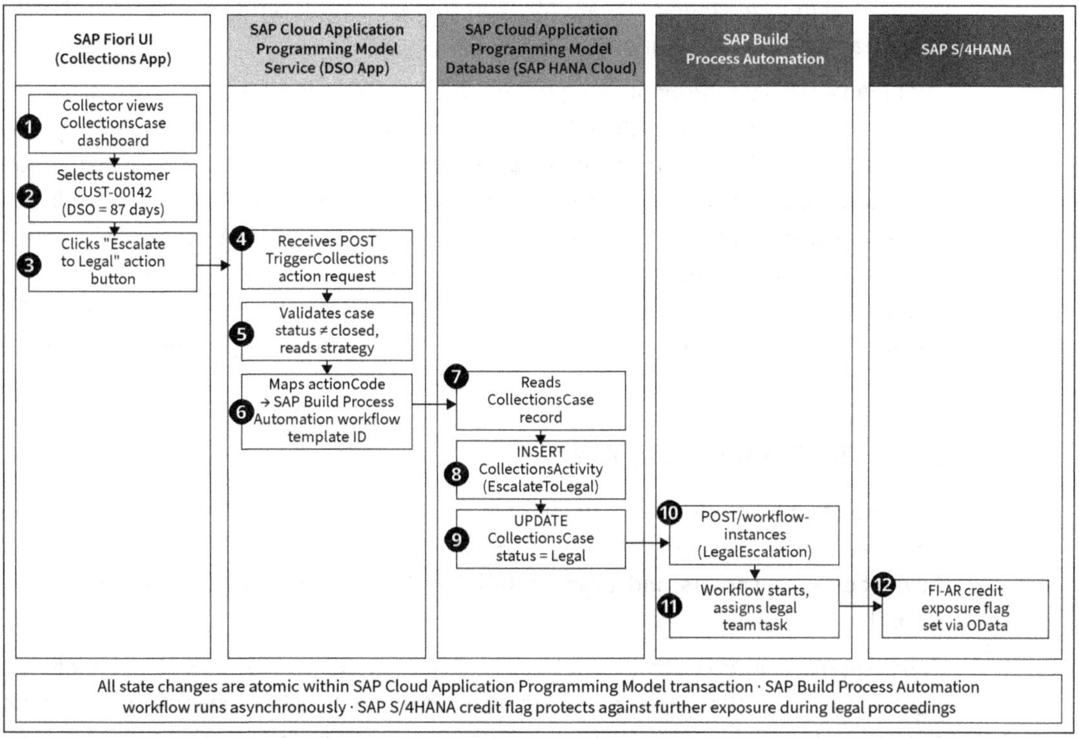

Figure 10.2: Example Flow from Custom SAP Fiori UI to SAP S/4HANA Backend

These entities are defined in CDS models, and their associations represent the real-world relationships between them: A `Customer` has many `OpenInvoices`, a `Customer` has many `CollectionsCases`, a `CollectionsCase` has many `CollectionsActivities`, and an `OpenInvoice` may have many `DisputeCases`. SAP Cloud Application Programming Model generates fully functional OData V4 APIs from these models automatically, so the SAP Fiori application development team can begin building the user interface before all the business logic is complete, using SAP Cloud Application Programming Model's built-in mock data capability to simulate the service responses.

The SAP Cloud Application Programming Model service layer defines the business logic that operates on these entities. The DSO Management application needs several service operations beyond basic create, read, update, and delete (CRUD). A `CalculateCustomerDSO` function reads all open invoices for a customer, calculates the weighted average days past due, and returns a DSO figure that the dashboard displays alongside the credit risk score. A `TriggerCollectionsAction` action takes a `CollectionsCase` identifier and an action code (`Email`, `Phone`, `FormalNotice`, or `EscalateToLegal`), and executes the appropriate workflow: calling the SAP Build Process Automation API to start the correct workflow template, logging the action in the `CollectionsActivity` entity, and updating the `NextActionDate` on the `CollectionsCase`. A `SubmitDisputeResolution` action takes a `DisputeCase` identifier and a resolution decision, updates the dispute record in SAP S/4HANA via the Dispute Management OData API, and triggers a notification workflow to the customer via the SAP CX integration.

SAP Cloud Application Programming Model's built-in support for multitenancy, extensibility, and authorization makes the DSO Management application enterprise-ready from the first line of code. The service definitions include authorization annotations that control which users can see which invoices based on their company code assignment in SAP S/4HANA, ensuring that a collector in the European region cannot accidentally view or action invoices belonging to the North American entity, for example. SAP Cloud Application Programming Model handles the enforcement of these rules in the service layer, so the SAP Fiori application developer does not need to implement security logic in the user interface.

10.4 The Extensibility Wizard

There are often many challenges involved in deciding how to extend an application in SAP S/4HANA. It is not always easy to identify the available extension options nor to determine the right extension points, and it can feel disjointed to navigate between SAP S/4HANA and SAP BTP.

The *extensibility wizard* consolidates both on-stack and side-by-side extension options into one centralized location. You can find this option in the dropdown menu under your user icon if you have been assigned the extensibility wizard role collection. The extensibility wizard provides the option to create either on-stack or side-by-side extensions with a guided experience; it considers the application the user has navigated to in SAP S/4HANA and updates the context accordingly. You will have the on-stack options of creating custom fields, adding custom logic, or creating a custom side-by-side process with SAP Build Process Automation.

Many processes in SAP S/4HANA use the core business workflow capability to enable preconfigured workflow scenarios, which can be enhanced with SAP S/4HANA's flexible workflow features that allow condition-based steps for approval, such as routing to an approver for a sales order based on a customer's payment history or usage of existing credit line. However, you will often need to go beyond these capabilities, and this is where the extensibility wizard in SAP BTP makes your life easier.

If you navigate to the sales order application, select the extensibility wizard, and choose **Process Extension**, you'll see a list of standard system events to choose from to trigger your custom workflow—for example, sales order created, deleted, or updated. Once you've selected an event, give your project a name and description; you'll be automatically brought into SAP Build, where you can start developing your custom extension workflow. Inside the SAP Build environment, you can interact with SAP Joule for Developers to assist you in creating your extension, reducing the time it takes to create your customization. The extensibility wizard makes it easy to bridge the gap between your standard applications and SAP BTP for adding your enterprise's custom workflow and logic to your processes and gives you the confidence that you are leveraging the right extension points in adherence to the highest standards of SAP's clean core principles.

10.5 Joule Skills and Agent Development

The extensibility capabilities described so far each represent a significant investment in extending the order-to-cash process. They automate data processing, surface richer information to users, and reduce manual reconciliation work. However, they still require users to navigate applications, interpret data, and decide what to do next. The next frontier of enterprise automation involves delegating more of that cognitive work to AI agents that can interpret the data, reason about the appropriate action to take, and either execute that action autonomously or present a recommendation to the user with the information needed to make a fast, well-informed decision.

Joule Studio, which we introduced in Chapter 7, Section 7.2.1, is the development environment within SAP Build for building these custom AI agents. Joule Studio provides no-code and low-code interfaces for creating Joule Skills and assembling them into Joule Agents that can be deployed across the SAP landscape. The skills and agents built in Joule Studio benefit from Joule's deep integration with SAP's business context; its understanding of SAP terminology, business process flows, and the semantic meaning of SAP data gives them a level of business intelligence that a generic AI agent built outside the SAP ecosystem would require enormous effort to replicate.

We'll explain the key parts of Joule Studio (skills, agents, and more) in the following sections.

10.5.1 Joule Skills

A *Joule Skill* is a deterministic, rule-based task that follows a predefined sequence to perform a specific business operation. Joule Skills are the atomic building blocks of Joule Agents. Each skill has three components: a trigger, a sequence of actions, and a response. The *trigger* has a natural language description that tells the Joule Agent when to invoke the skill; the agent uses this description to match user intent or incoming events to the appropriate skill. The *sequence of actions* defines what the skill does; it can call OData APIs, invoke SAP Build Process Automation workflows, read or write data from or into SAP BTP data stores, call external APIs, and branch conditionally based on data retrieved during execution. The *response* defines what the skill returns: either data that is passed to the next skill in a multistep agent workflow, or a response message delivered to the user through the Joule chat interface.

Skills should be narrow in scope. Each skill should do one thing well, with a clear and specific trigger description. This design principle is important not just for code quality reasons but because the Joule Agent's ability to correctly route a user's request to the right skill depends on the clarity and specificity of each skill's trigger description. A skill whose trigger description says "handle anything related to invoices" will conflict with other invoice-related skills and produce unreliable agent behavior. A skill whose trigger description says "retrieve the current open invoice balance and days overdue for a specific customer" is unambiguous and will be selected reliably whenever the relevant intent is detected.

10.5.2 The Joule Collections Agent

To make the order-to-cash extensibility story complete, consider building a Joule Collections Agent that assists the collections team with their daily work. This agent is deployed in the DSO Management SAP Fiori application as an embedded Joule chat interface, and it is also available to collections managers through the standard Joule interface in SAP Build Work Zone. Figure 10.3 presents a diagram for this agent.

We'll explore the Joule Collections Agent in more detail in the following sections.

Joule Collections Agent Skills

Figure 10.3 shows an architecture overview for the Joule Collections Agent, including its channels, runtime, skills, and data sources. The agent is equipped with five core skills:

- **`GetCustomerReceivablesSummary`**
  The `GetCustomerReceivablesSummary` skill takes a customer name or ID as input; calls the SAP S/4HANA FI-AR OData API and the SAP Cloud Application Programming Model DSO Management service to retrieve the customer's total open balance, aging breakdown, DSO figure, credit risk score, and the status of any open `CollectionsCases`; and returns a formatted summary to the user. A collector can ask Joule "what is the receivables summary for Contoso Manufacturing" and receive a complete picture of

the customer's financial position in seconds, without navigating between multiple SAP S/4HANA SAP Fiori applications.

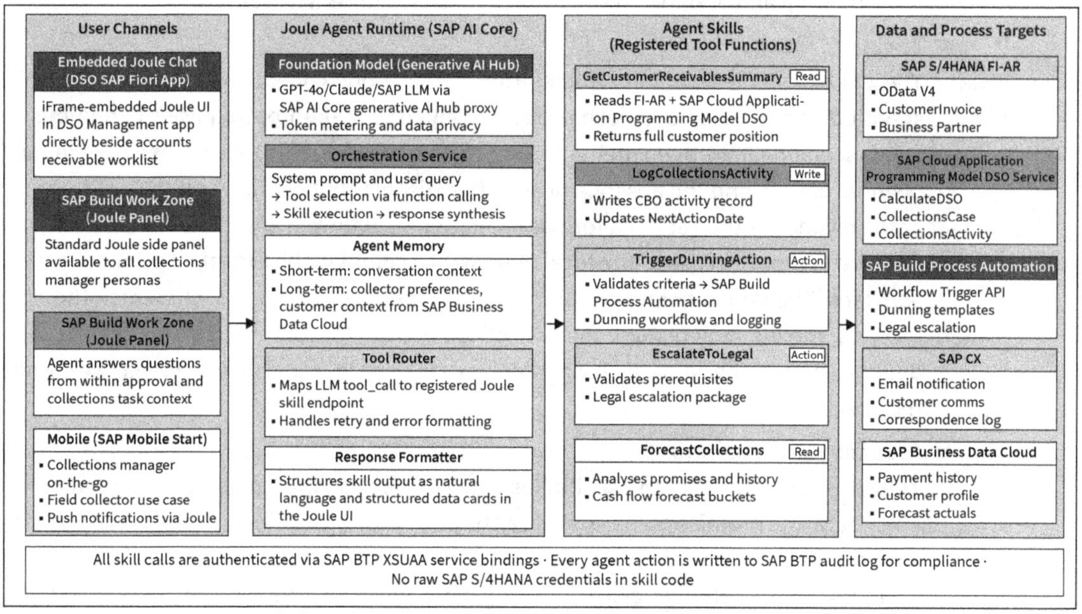

Figure 10.3: Reference Architecture Flow for Example Joule Collections Agents

- **`LogCollectionsActivity`**
  The `LogCollectionsActivity` skill accepts a customer ID, a contact method, a contact outcome, and free-form notes, and it writes a new `CollectionsActivity` record to the custom business object via its OData API. It also updates the `NextActionDate` on the open `CollectionsCase` based on the outcome code; if the customer promised to pay by a specific date, the skill captures the promise date and sets the next action date to one day before the promise date so the collector is reminded to follow up if the payment does not arrive. A collector can say "log a call with Contoso Manufacturing, spoke with accounts payable, they promised payment by June 30th," and Joule will capture and structure all that information correctly without the collector needing to navigate to a form and fill in each field manually.
- **`TriggerDunningAction`**
  The `TriggerDunningAction` skill takes a customer ID and an action type; validates that the customer meets the criteria for the requested action, checking the credit hold status, the open dispute count, and the last contact date to prevent inappropriate escalation; and then calls the SAP Build Process Automation API to start the corresponding dunning workflow. For a `Phone` action, it starts a workflow that routes the call task to the assigned collector with the relevant customer information preloaded. For a `FormalNotice` action, it starts a workflow that generates the formal payment demand letter using a template in

SAP Forms by Adobe, routes it for management approval, and sends it to the customer via the SAP CX email integration. The skill logs the triggered action in the `CollectionsActivity` log and updates the `CollectionsCase` status accordingly.

- **`EscalateToLegal`**
  The `EscalateToLegal` skill handles the most serious collections scenarios, in which standard dunning has been exhausted. It validates that all prerequisite conditions for legal escalation have been met: a minimum number of dunning letters sent, a minimum overdue period, and the absence of an active dispute that might affect the collectability of the balance. It then starts a formal escalation workflow in SAP Build Process Automation that routes the case to the legal team with a complete history of all prior collections activities. The skill generates a summary package of all relevant documents—the open invoices, the dunning history, the collections activity log, and the customer correspondence—and attaches it to the escalation workflow so that the legal team has everything they need to act.
- **`ForecastCollections`**
  The `ForecastCollections` skill takes a date range as input and uses the open receivables data, the promise-to-pay dates captured in previous collector calls, and the historical payment behavior of each customer to generate a collections forecast for the period. It returns a bucketed forecast showing how much of the current open balance is expected to be collected within the period, based on promises received and historical patterns. This forecast can be consumed by the treasury team to update their cash flow model without requiring manual data compilation from the collections team.

Joule Collections Agent Assembly

With the five skills defined, the Joule Collections Agent is assembled in Joule Studio by configuring the agent's context and capabilities. The agent is given a system prompt that defines its persona; it is a collections operations assistant with deep knowledge of the company's collections policies, the order-to-cash process, and the tools available to the collections team. It is given instructions on how to prioritize customer interactions, when to escalate versus when to continue standard dunning, and how to respond when it cannot fulfill a request because the necessary data is not available or a prerequisite condition has not been met.

The agent is connected to the five skills, and each skill's trigger description is reviewed to ensure it is specific enough for reliable routing. The agent is also connected to the DSO Management application's SAP Cloud Application Programming Model service as a data source so that it can answer general questions about the collections portfolio such as "which customers in the DACH region have the highest DSO this quarter" by querying the SAP Cloud Application Programming Model service directly via Joule's graph capability, without requiring a dedicated skill for every possible data question.

Testing in Joule Studio is conducted using a conversation simulator that allows the developer to interact with the agent in a test chat interface before deployment. The developer walks through representative collector scenarios, such as a morning review of the previous day's collections activity, a specific customer inquiry that requires a dunning action, or an escalation scenario to verify that the agent routes correctly to the appropriate skill; handles missing data gracefully; and returns responses that a collections team member would find genuinely useful rather than generic or confusing.

Joule Collections Agent Deployment and Governance

Once testing is complete, the Joule Collections Agent is deployed through the standard SAP Build deployment pipeline. The agent is published to the SAP Build Work Zone Joule interface and embedded in the DSO Management SAP Fiori application as a context-aware assistant. Users who have the collector role in SAP BTP's Identity and Authentication services are granted access to the agent, ensuring that the collections agent's capabilities, particularly the `TriggerDunningAction` and `EscalateToLegal` skills, are only accessible to authorized users.

Governance of the Joule Agent is handled through SAP Cloud ALM's integration monitoring capabilities. Every API call made by the Joule Agent's skills is logged as a business transaction, making it possible to audit the agent's activity, identify skill invocations that returned errors, and measure the agent's contribution to key metrics such as how many collections activities were logged through the agent versus manually, how many dunning actions were triggered, and whether the collections cases handled by the agent show better resolution times than those managed without agent assistance.

Continuous improvement of the agent is supported by the feedback mechanisms built into the Joule interface. Users can rate individual agent responses and provide freeform feedback when the agent's response was not helpful. These feedback signals are reviewed by the SAP BTP development team on a regular cadence and used to refine skill trigger descriptions, improve the agent's context prompt, and identify gaps in skill coverage where new skills should be added. This continuous improvement cycle means the Joule Collections Agent becomes progressively more valuable over time as it is refined based on real usage data from the collections team.

Building an extension or a Joule agent is not a one-time event. These assets have a lifecycle that spans initial development, quality assurance testing, production deployment, ongoing monitoring, periodic enhancement, and eventual retirement. SAP BTP provides a set of lifecycle management services that support this full lifecycle for both traditional code extensions and AI agent assets.

SAP Cloud Transport Management handles the movement of development artifacts like SAP Cloud Application Programming Model applications, Joule Agent configurations, SAP Build Process Automation workflows, and SAP Integration Suite integration flows from development to quality assurance to production in a controlled, auditable way. Each transport is tracked in SAP Cloud Transport Management, creating a complete audit trail of what was deployed, when, by whom, and whether it was approved through the required review process. SAP Cloud Transport Management can be integrated with SAP Cloud ALM to provide a top-down view of change activity across the entire SAP BTP landscape alongside changes in the SAP S/4HANA system itself, giving operations teams a single view of all changes that could affect the production environment.

SAP Continuous Integration and Delivery provides a predefined DevOps pipeline services for SAP Cloud Application Programming Model and other SAP BTP artifacts. When a developer commits a change to the SAP Cloud Application Programming Model DSO Management application's Git repository, the continuous integration/continuous delivery (CI/CD) pipeline automatically executes the unit test suite built with SAP Joule for Developers' test-generation assistance, runs a code quality check, and either approves the change for promotion to the quality system or returns it to the developer with a detailed test failure report. This "commit early, commit often" development practice supported by automated testing significantly reduces the volume of defects that reach the quality assurance team and shortens the overall release cycle.

For Joule Agents specifically, lifecycle management includes the conversation quality monitoring and feedback mechanisms discussed earlier, as well as a formal version control process for agent configurations. When a skill's trigger description is updated, when a new skill is added to an agent, or when the agent's system prompt is revised, these changes are version-controlled in SAP Build and can be rolled back to a previous version if the change produces unexpected behavior in production. This version control capability is essential for Joule Agent governance as the consequences of a poorly configured agent skill, such as a `TriggerDunningAction` skill that can be invoked without checking the prerequisite conditions, can have real financial and customer relationship consequences.

10.5.3 What's Ahead for Joule Studio

As of the time of writing, SAP has announced a major evolution in development with its new iteration of Joule Studio. Although not currently generally available at the time of writing, it is quite possible you are reading this book after the new Joule Studio has been released.

At its core, the new Joule Studio still has the same end goal of building, deploying, and managing custom agents. However, the new Joule Studio embraces the vision set forth with SAP Business AI platform to support the Autonomous Enterprise. Similar to SAP Build, Joule Studio will allow you to build AI agents and skills, full-stack applications and extensions, and workflows that extend SAP and non-SAP systems. However, Joule Studio will take an intent-based approach to development, integrated through the new Joule Work user experience. Users who are granted the roles to access Joule Studio will engage with Joule in a project-planning phase to scope out the requirements of their development. A Joule Agent will help guide users to assess the process steps needed in the scenario, the systems to communicate with, and the actions that the application, workflow, or agent should perform and, more importantly, should not perform. Once the planning is complete, you will be presented with a product requirements document (PRD). This document will help business users and IT validate early that the content that Joule is creating meets the expectations of your end users. From the PRD, Joule will then begin to build the specification, followed by coding the solution, testing scenarios, and then finally deploying the solution to the SAP-managed runtime.

Developers will have access to the code that Joule has created and can modify it in Joule Studio or adapt the code in the tooling they prefer, such as Claude Code or GitHub Copilot. Joule Studio will be integrated with SAP LeanIX and SAP Signavio so that agents can understand the existing agents and applications in your enterprise and the processes these agents and applications support. With this knowledge, Joule can easily decide where to integrate with an existing application so that it does not build duplicative content for functionality that already exists. When the agent or application is deployed, developers will gain insights into the agents' performance and execution, while SAP manages the deployment, scaling, and other related lifecycle management capabilities. The new Joule Studio represents an exciting leap forward for SAP to help your company realize the vision of the Autonomous Enterprise by bringing flexibility and governance together to create agents, agent skills, workflows, and applications grounded in your data and processes.

10.6 Custom AI Services and Development

SAP BTP provides the ability to connect to SAP and major third-party large language models (LLMs) via its SAP AI Core and SAP AI Launchpad services, which we introduced in Chapter 7, Section 7.4. These services on SAP BTP work closely together to help you bring custom AI scenarios to life in your organization and can be used to augment system extension, side-by-side applications, and custom Joule Agents. This section will look at how to leverage these services for your enterprise.

10.6.1 SAP AI Core

SAP AI Core, which we introduced in Chapter 7, is a Kubernetes-based, multitenant AI-execution environment delivered as a fully managed service on SAP BTP. Its primary purpose is to remove the infrastructure complexity that typically prevents enterprise teams from operationalizing machine learning (ML): provisioning GPU compute, managing containerized training pipelines, versioning model artifacts, scaling inference endpoints, and enforcing enterprise security and data residency requirements. SAP AI Core handles all of these as platform services, allowing data science and AI engineering teams to focus entirely on model development and business logic rather than on infrastructure management.

SAP AI Core ships with two particularly important platform capabilities that distinguish it from a generic Kubernetes service. The first is the AI API, a standardized REST API surface that abstracts the underlying orchestration and allows tooling, pipelines, and applications to interact with SAP AI Core through a stable, versioned interface regardless of what the underlying infrastructure evolves to. The second is the generative AI hub, which extends

SAP AI Core beyond custom model training into the consumption of LLMs. The generative AI hub provides a unified proxy layer through which an SAP BTP application can invoke foundation models from multiple providers such as GPT-4o from OpenAI, Claude models from Anthropic, Gemini from Google Vertex AI, and SAP's own Joule-aligned models through a single, harmonized API surface with centralized cost tracking, token usage metering, content safety filtering, and data privacy guarantees that prevent customer data from being used to train the underlying foundation models. This means an enterprise can consume best-of-breed LLMs for different use cases while maintaining a single governance and cost-management point on SAP BTP.

The following sections will take a closer look at SAP AI Launchpad, the interface for SAP AI Core, and the process to use SAP AI Core to extend SAP S/4HANA.

SAP AI Launchpad

SAP AI Launchpad is the operations and lifecycle management interface for SAP AI Core. Where SAP AI Core, the execution runtime, is the engine room, AI Launchpad is the control room. It provides an SAP Fiori–based web application for data scientists, ML engineers, and AI operations teams with a graphical interface to every capability exposed by the AI API. Through SAP AI Launchpad, a team can connect to one or more SAP AI Core instances across different SAP BTP subaccounts or hyperscalers, switch between them in a single browser session, and manage all AI assets from a unified view.

SAP AI Launchpad organizes its workspace around the same resource group concept used by SAP AI Core. Within a selected resource group, an operations engineer can browse all registered *scenarios*, which are named collections of executable pipelines and model configurations that represent a complete AI use case, such as Demand Forecasting or Invoice Anomaly Detection. Within a scenario, they can trigger new training executions, monitor execution statuses and logs in real time, compare metrics across runs in the experiment tracking view, register trained model artifacts, and promote a selected model version to a new deployment with a chosen replica count and compute resource profile. The deployments view shows all running inference endpoints, their health status, current request throughput, and latency percentiles, and it allows the operator to update the deployed model version or scale in response to higher volumes of request.

For generative AI use cases, SAP AI Launchpad provides the generative AI hub interface. This is a prompt-engineering workbench where developers can experiment with different foundation models, compare responses, craft and save system prompts, test retrieval-augmented generation (RAG) configurations, and evaluate output quality against defined test sets before committing a prompt or model configuration to a production deployment. The generative AI hub in SAP AI Launchpad also exposes the orchestration service, a pipeline that chains prompt construction, LLM invocation, content safety checks, and response grounding into a single callable endpoint, allowing application developers to consume a fully governed generative AI capability without implementing the orchestration logic themselves.

Extending SAP S/4HANA with SAP AI Core

The architectural pattern for extending SAP S/4HANA with SAP AI Core follows the same clean core principle that governs all SAP BTP–based extensions: no modification to the SAP S/4HANA application server, all AI logic running on SAP BTP, and SAP S/4HANA interacting with the AI capability through stable, versioned APIs or through SAP Integration Suite. This pattern enables a clear separation of upgrade risk: SAP upgrades SAP S/4HANA, the customer upgrades their AI models independently, and this allows AI capabilities to evolve on their own release cadence without dependency on SAP transport windows.

A typical extension pattern works as follows. An SAP Cloud Application Programming Model application running on SAP BTP, Cloud Foundry environment acts as the orchestration layer. It receives a trigger (a new sales order created event from SAP S/4HANA via SAP Integration Suite, advanced event mesh, a scheduled batch job, or a direct API call from an SAP Fiori application), enriches the incoming data by calling additional SAP S/4HANA OData APIs to gather context, assembles the inference request payload, calls the SAP AI Core deployment endpoint via an authenticated HTTP request using the AI API binding, receives the model prediction, applies any business rule postprocessing, and writes the result back to SAP S/4HANA via an OData `PATCH` or action call. The SAP S/4HANA application never needs to know that an AI model was involved; from its perspective, it received an API call that updated a field value, exactly as it would from any other SAP BTP extension.

Examples of SAP S/4HANA extension use cases powered by SAP AI Core include the following:

- **Intelligent credit risk scoring**
  A custom gradient-boosted model trained on SAP S/4HANA accounts receivable payment history, customer master data, and external credit bureau feeds predicts the probability of a customer invoice becoming overdue; the score is surfaced in the Collections Management app as a risk indicator alongside each open item.
- **Demand forecasting**
  A time-series model trained on SAP S/4HANA Sales history, enriched with weather, promotional, and macroeconomic signals, generates SKU-level demand forecasts that are written back to SAP S/4HANA as planned independent requirements, feeding the material requirements planning run without a planner manually adjusting every line.
- **Invoice anomaly detection**
  An unsupervised model trained on historical accounts payable posting patterns identifies invoices that deviate statistically from expected vendor behavior (unusual amount, atypical general ledger account combination, duplicate line pattern) and surfaces them as a priority review list in the accounts payable SAP Fiori inbox before payment is released. Figure 10.4 shows the end-to-end architecture for an invoice anomaly detection extension.

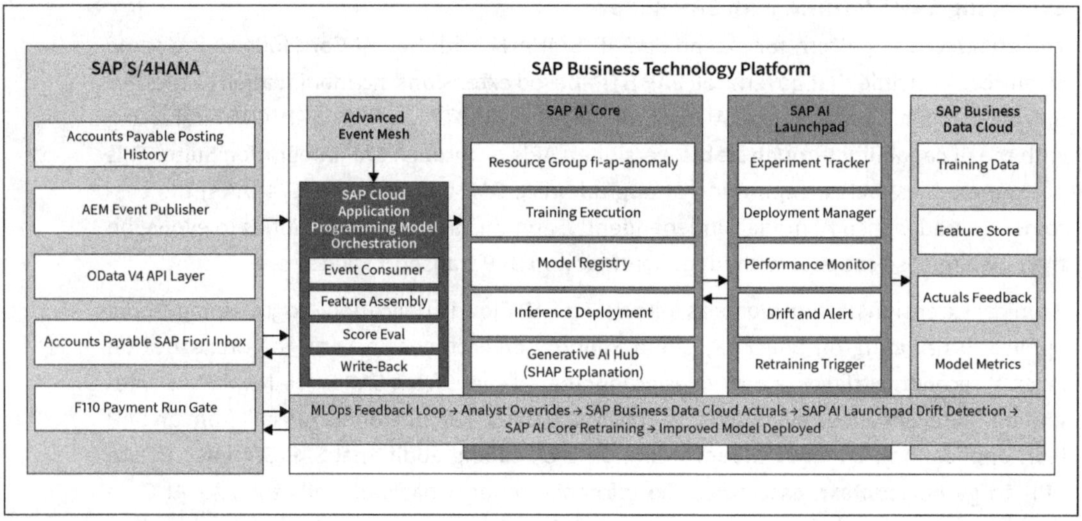

Figure 10.4: Example End-to-End Architecture for Invoice Anomaly Detection

10.6.2 Integration with SAP Business Data Cloud

SAP Business Data Cloud is SAP's unified data fabric platform that combines SAP Datasphere (governed data modeling and federation), SAP Analytics Cloud (planning and analytics), and deep integration with hyperscaler data platforms, particularly SAP's partnership with Databricks, into a single, governed data estate. We discussed SAP Business Data Cloud at length in Chapter 9.

The defining characteristic of SAP Business Data Cloud is that it provides a semantically harmonized, business-ready data layer over the raw transactional data held in SAP S/4HANA, SAP SuccessFactors, SAP Ariba, and other SAP applications. Rather than exposing raw database tables, SAP Business Data Cloud surfaces business entities such as customers, materials, sales orders, and cost centers in a form that is consistent in its terminology, currency, and time dimension across every source system.

The integration between SAP AI Core and SAP Business Data Cloud creates what SAP describes as *business-grounded AI*: AI models that are trained and served not on generic, uncontextualized data extracts but on the full, governed, semantically consistent enterprise data managed by SAP Business Data Cloud.

10.7 Summary

SAP provides a layered, comprehensive extensibility framework that allows your organization to fill functional gaps in the order-to-cash process and across the broader SAP Autonomous Suite in a way that is sustainable, upgrade-safe, and aligned with the clean core principles that protect your investment in SAP S/4HANA. Key-user extensibility delivers fast, business-analyst-led customizations for field additions, custom objects, and process rules that automatically achieve the highest tier of clean core compliance. Developer extensibility using the ABAP RESTful application programming model provides the power of full ABAP development for complex on-stack extensions while preserving upgrade safety through the discipline of using only released APIs and extension points. Side-by-side extensibility on SAP BTP using SAP Cloud Application Programming Model delivers the most powerful tier of customization applications that span multiple SAP and non-SAP systems, run at independent scale, and are completely decoupled from the SAP S/4HANA upgrade cycle.

The Joule Agent development capabilities of Joule Studio transform these extensions from passive data containers and static applications into active, intelligent participants in your business processes. The Joule Collections Agent built in this chapter illustrates how a set of well-defined skills, each narrow in scope, each connected to the rich data and process APIs exposed by the extensibility layers beneath it, can collectively deliver a capability that materially improves business outcomes, reducing DSO through earlier risk identification, faster collections response, and consistent policy execution.

The path from a raw extensibility requirement to a deployed, governed, and continuously improving custom capability is well-supported by SAP BTP's development, testing, transport, and monitoring services. SAP Cloud Transport Management, SAP Continuous Integration and Delivery, and SAP Cloud ALM together provide the operational discipline needed to run a high-quality extension and agent portfolio at enterprise scale, while SAP Joule for Developers, embedded throughout the SAP Build environment, ensures that the development teams building these capabilities have intelligent AI assistance at every step, from generating CDS models and service implementations to writing unit tests and refactoring legacy custom code to reach clean core compliance.

The organizations that invest in building these capabilities today and in building them correctly, with the governance and clean core discipline the framework demands, will be those that can adopt SAP's next generation of business AI capabilities the fastest. Every extension built on a released OData API, every event published to a managed event broker, and every Joule Skill with a clear and specific trigger description is another connection point through which the AI capabilities of the intelligent enterprise can flow.

In the next chapter, we'll switch gears to integration topics and dive into SAP Integration Suite.

Chapter 11
Integration Capabilities

We'll dive into integration capabilities in this chapter to showcase how SAP Integration Suite and SAP Integration Suite, advanced event mesh deliver flexible, intelligent, and future-ready integration capabilities that underpin the Autonomous Enterprise.

Integration acts as the nervous system of any enterprise. Now more than ever it is important that the applications that support your business operations communicate with each other, sharing data so that operations can run in near real time and react as changes happen across the organization.

This chapter will explore SAP's integration strategy for process integration. You will learn about SAP's integration strategy and the tools SAP delivers to support this strategy in SAP Integration Suite. This includes application programming interface (API) management, event-driven integration, AI-assisted integration, and master data integration.

11.1 Integration Strategy

Defining an integration strategy is a foundational step you need to perform during your Autonomous Enterprise transformation. The agents that will perform actions or surface insights for you and your fellow users will use these interfaces to communicate with your backend systems. As the number of agents, applications, systems, and users grows within your organization, it is important to provide a strategy to seamlessly connect the systems within your landscape so they can communicate with each other. Integration becomes the crucial connective tissue, acting as the central nervous system of your enterprise to allow your applications to send and receive data to each other and trigger actions and events when something happens across your enterprise. Ad hoc approaches to integration will create data silos over time within your enterprise, in both systems and teams. An integration strategy sets the foundation for allowing data to flow between your systems, both SAP and non-SAP, and to the broader ecosystem of external partners, suppliers, customers, regulatory agencies, and others to share information in as close to real time as possible. It is important to define an enterprise integration strategy to allow data to flow between systems; this allows developers to easily identify, monitor, and govern their interfaces and it future-proofs your landscape to make integrating new systems easier and faster.

In the following sections, we'll walk through the strategic imperatives that support a successful integration strategy. We'll also look at the key tools that support your integration strategy: SAP Integration Solution Advisory Methodology, SAP Business Accelerator Hub, and SAP Integration Suite.

11.1.1 Strategic Imperatives

SAP recommends that your organization establish an integration Center of Excellence to create and execute your strategy. SAP recommends five strategic imperatives for centering your integration Center of Excellence:

- **The platform**
  SAP recommends that your organization take a platform approach to integration. Integration can span many topic areas, from user authentication and authorization to data, user experience, and (the focus of this chapter) data and process integration. By having a platform in place to handle the different styles of integration, you will have the tools your organization needs to handle different styles and patterns of integration. A unified platform for integration creates a centralized location to handle, monitor and govern integration activities, making it easier for your organization to build new integrations, check the health of existing integrations, and modernize legacy integrations. This centralized approach allows you to easily respond to new and emerging integration needs within your organization. SAP has provided its integration tooling on the SAP Business Technology Platform (SAP BTP) so that organizations can easily use SAP's platform to rapidly integrate their SAP and non-SAP systems within their landscape.
- **Agility and scalability**
  As noted, a centralized platform creates a unified place to integrate your people, processes, and systems. By defining integration patterns and frameworks, you can create repeatable and reusable integration content to accelerate time to value for future integration needs. This agility comes from the foundation of having a defined API Management strategy. APIs act as the access point for your interfaces, business objects, and systems. Having a centralized point to register, monitor, and handle APIs throughout their lifecycle creates the foundation for an agile integration framework. These APIs will feed repeatable process integration patterns and allow your organization to meet future development needs. Scalability is the other side of agility; having a platform that can independently scale to process integration creates a stability pattern for your organization to grow and consume more data from more sources over time. Having an independent platform on SAP BTP allows for an environment that can scale to fit the growing needs of your enterprise while also meeting you where you are at today.
- **The citizen integrator**
  Business users in your organization without a background in development are known as *citizen integrators*. These citizen integrators often hold key functional knowledge for how a process needs to be integrated and where the integration points need to be placed between systems. They need low-code tooling to help model the business and process logic needed in an integration and allow them to engage with IT more effectively without

having to overcome the hurdle of learning a development language. When business and IT can collaborate effectively, you can create integrations faster that meet the needs of your organization without tedious rework.

- **Strong governance**
  Your organization needs to define a plan for managing integration paths and protocols. One of the key activities of an integration Center of Excellence is to define and document the rules around access control, change management, enhancements, system credentials, and data encryption. These governance rules become even more essential in the world of the AI agents who will call on these integration points to perform activities on behalf of your users. All too often, without the oversight of an integration Center of Excellence, this task is left up to IT developers, who are often only focused on the current task at hand and not the strategic direction of your organization. By defining an integration governance strategy, you can create a framework to enforce these standards while still meeting the changing needs of your organization.
- **Reporting and analytics**
  Do not forget about reporting when defining the goals of your integration Center of Excellence. Monitoring and reporting on the flow of data within your organization allows you to see how well your teams are adhering to the guidelines set by your integration Center of Excellence. Predictive intelligence can be used to suggest new integration patterns to improve the flow of data. Governance is never a one and done process; instead, it is an end state your organization will move to over time. With the emergence of new integration patterns for agent-to-agent communication and agent-to-backend-service communication, this reporting process allows your organization to easily shift and adapt to changing technology and world events.

11.1.2 SAP Integration Solution Advisory Methodology

SAP Integration Solution Advisory Methodology is available to support your adoption of an integration strategy. This is SAP's official methodology for classifying integration use cases into styles and domains, then mapping each style to the right integration technology. It replaces ad hoc tool selection with a repeatable, governed framework that works across cloud, on-premise, and hybrid landscapes.

SAP Integration Solution Advisory Methodology is broken into four key phases:

1. Identify the integration domains relevant for your organization. Integration domains are typical areas in a hybrid landscape where integration is needed, such as cloud-to-on-premise or cloud-to-cloud integration.
2. Identify the relevant integration styles and integration use case patterns.Integration styles and integration use case patterns comprise a set of typical, technology-agnostic integration use cases such as process integration, data integration, device integration, and so on.

3. Validate the key characteristics that are relevant for your integration architecture. Key characteristics are criteria that help enterprise architects and integration architects to map integration styles to the relevant capabilities of integration technologies.
4. Based on your settings, the system maps the selected set of integration domains, integration styles, and use case patterns to the best suitable integration technology.

The SAP Integration Solution Advisory Methodology is designed for use by all types of users, from enterprise architects to hands-on developers. Inside SAP's integration tooling are tools to create processes and questionaries to support adoption of SAP Integration Solution Advisory Methodology, tailored to your organization's needs.

There are five integration styles supported by SAP Integration Solution Advisory Methodology:

- Application-to-application (A2A)
- Business-to-business (B2B)
- Business-to-consumer (B2C) or API
- Analytics and data integration
- Internet of Things (IoT) and edge integration

11.1.3 SAP Business Accelerator Hub

SAP catalogs all its standard APIs, integrations, and events at SAP Business Accelerator Hub. This hub is not a product from SAP but a singular point of access to find important information about SAP systems and processes. Its primary function is to help developers, architects, and business analysts find all the material they need to extend and integrate SAP solutions. SAP Business Accelerator Hub catalogs all prepackaged SAP-to-SAP and SAP-to-non-SAP integrations that come out of the box with SAP Integration Suite, which we will discuss in further detail in Section 11.1.4. It also catalogs all standard interfaces for all SAP systems. SAP Business Accelerator Hub also provides sandbox environments for technical users to try out SAP APIs so that they can get hands-on practice calling modern SAP interfaces.

SAP Business Accelerator Hub has grown to catalog so much more than just APIs and integrations. Now, when you log onto SAP Business Accelerator Hub, you can view the following:

- SAP and partner APIs
- SAP Build content
- Integration content (integration flows, and value-mapping)
- Process blueprints
- Events
- CDS views
- Graph
- Open connectors
- Community content
- Adapters
- Analytics packages
- Transformation content
- Data products

Figure 11.1 outlines all the various content you can discover via SAP Business Accelerator Hub above and beyond the standard integrations and APIs available out of the box.

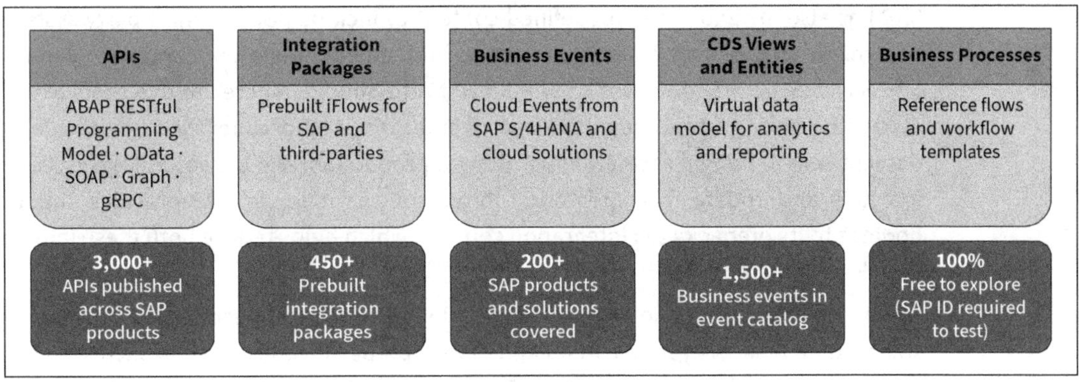

Figure 11.1: SAP Business Accelerator Hub: More Than Just SAP's API Catalog

With all this content on SAP Business Accelerator Hub, SAP has deployed a Joule Agent on the site to help you navigate. You can click the diamond icon in SAP Business Accelerator Hub and ask Joule your question to have it bring up the relevant content. For example, you can ask Joule to show you the standard integration content for integrating sales orders between SAP S/4HANA Cloud and Salesforce, and Joule will help navigate you to the out-of-the-box integration content provided by SAP.

11.1.4 SAP Integration Suite

SAP Integration Suite is a service on SAP BTP to help your organization meet its enterprise integration needs. It is an integration-platform-as-a-service (iPaaS) offering comprised of all the features you need to meet your organization's process integration needs. This offering was created by SAP to act as a single comprehensive solution to connect all your business solutions together seamlessly.

We'll provide an overview of SAP Integration Suite in the following sections, including the guiding principles, key capabilities, and monitoring features.

Core Principles

The development of SAP Integration Suite is steered by four guiding principles:

- **Predefined integration content**
  SAP has released over 1,800 prebuilt integration flows to connect SAP and non-SAP systems together. Predefined content is cataloged and viewable on SAP Business Accelerator Hub and in SAP Integration Suite so that integration developers can rapidly connect their business systems together. SAP understands the APIs that work between their business systems, such as the Sales Order APIs for the SAP S/4HANA and SAP Customer Experience (SAP CX) solutions, and as such has built predefined integration content for these

solutions. By leveraging out-of-the-box integration content, developers can rapidly accelerate their time to production.

SAP has also expanded the predefined content to include popular third-party systems that might exist in your enterprise, such as predefined integration content to connect SAP SuccessFactors to Kronos or SAP S/4HANA to Shopify. There are two costs associated with creating a new integration in your enterprise's landscape: The first is the development cost to bring a net-new integration to production, and the second is the cost to maintain and update that integration throughout its lifecycle. SAP provides periodic updates to its prepackaged integration content, which helps reduce both these integration development costs.

You might have a scenario in your organization in which you need to deviate from the standard. SAP allows you to extend and modify prepackaged content in such scenarios to fit the needs of your organization. For example, you might include custom fields in your sales order interface that would need to be accounted for in the integration; with SAP Integration Suite, you can adapt the prepacked content to fit your needs without having to spend the time building that integration from scratch. Modified standard content is flagged as such in the system and not overwritten when an update is released. Instead, the owner of the integration package is notified when an update is published to the standard content and can chose the best path for adopting the new features added to the standard content.

- **Open integration**
  SAP recognizes that it would be an anomaly if your organization was only running on SAP software and as such has made sure that SAP Integration Suite can integrate with the non-SAP systems in your enterprise. SAP does provide prepackaged content to integrate with non-SAP systems, as mentioned previously, but it would be impossible for SAP to deliver this content for every non-SAP system in the marketplace. Therefore, to facilitate such integrations, SAP Integration Suite provides a catalog of adapters for over 180 non-SAP systems. These adapters standardize access to non-SAP system APIs and create easy-to-reuse connectors where integration is needed with a third-party system.

  With these adapters in SAP Integration Suite, developers can easily connect to the systems they need for their integration without having to hop between multiple tools. The adapter catalog offers robust security measures such as secure authentication protocols, encrypted data transmission, and data masking. Where SAP does not have an adapter for a third-party system, SAP Integration Suite has technical adapters to connect via the protocol the non-SAP system communicates over, such as HTTPs, SFTP, AQMP, and OData.

- **Holistic integration**
  Integration is not just about connecting two systems together; it is about covering all aspects of an organization's integration needs, supporting integrations patterns for A2A, B2B, business-to-government (B2G), event-driven integration, API-led integration, and

legacy integration formats such as flat file transfer between systems. SAP Integration Suite provides tools for users no matter what integration pattern they need to support.

- **AI integration**
  SAP Integration Suite provides AI tooling to assist developers with creating integrations, and tooling for integration with artificial intelligence. The suite has its own Joule Agent that helps developers with day-to-day tasks tied to integration, reducing the manual effort needed to create a new custom integration or analyze the health of an existing integration. SAP Integration Suite also has tooling to integrate with AI protocols with an adapter for interacting with large language models and the ability to create and deploy custom Model Context Protocol (MCP) servers for custom AI scenarios. Both these topics will be explored in Section 11.4.

Key Capabilities

The key capabilities of SAP Integration Suite align with SAP's guiding development principles. Let's walk through them:

- **Cloud Integration**
  The Cloud Integration component of SAP Integration Suite is where you can copy, configure, and modify prebuilt integration flows as well as develop custom integration flows. The Cloud Integration environment is a no-code first designer where you can model the logic for your specific integration. The Cloud Integration capability is complimented by the additional features in SAP Integration Suite.
- **Open Connectors**
  The Open Connectors capability provides access to 180 third-party adapters for custom integrations.
- **API Management/Graph**
  The API Management and Graph capabilities allow you to create API proxies, monitor and manage the lifecycle of your enterprise APIs, and create an internal repository for your enterprise APIs that allows your developers to easily discover and consume the APIs across your landscape.
- **Integration Advisor/Trading Partner Management**
  Integration Advisor and Trading Partner Management are tools that use a predelivered SAP Knowledge Graph to assist with creating B2B integrations with your external trading partners. Data space integration compliments the Trading Partner Management capability by allowing your organization to exchange data with other participants in the same data space, such as Catena-X, based on the European Privacy Policy to enforce data sovereignty.
- **Integration Assessment**
  The Integration Assessment capability provides tooling to help you define your integration patterns and ensure consistent usage in line with SAP Integration Solution Advisory Methodology.

- **Migration Assessment**
  The Migration Assessment tool allows you to perform an assessment on your existing SAP Process Orchestration or SAP Process Integration content to evaluate what integration content can be migrated from SAP's legacy integration tools into SAP Integration Suite. The migration assessment also identifies content that is not ready to be brought to SAP Integration Suite and gives an estimate of the level of effort needed to make it ready.
- **Edge Integration Cell**
  The Edge Integration Cell capability provides a containerized runtime to deploy integrations in on-premise or private, managed cloud landscapes, giving you the flexibility to decide where you would like your integrations to run.
- **Event Mesh/advanced event mesh**
  SAP Integration Suite comes with the Event Mesh capability, which acts as a centralized message broker supporting publish and subscribe integration patterns. The Event Mesh capability has a companion SAP BTP service, *SAP Integration Suite, advanced event mesh*, which allows you to deploy a connected network of event brokers across the globe. We will take a deeper look at event-driven integration in Section 11.3.

SAP Integration Suite is a comprehensive platform offering all the capabilities you need to achieve your enterprise integration goals.

Integration Monitoring

SAP Integration Suite includes monitoring content for its capabilities. The bulk of the monitoring content in SAP Integration Suite is available via its **Monitoring** tab. Integrations, integration content, security artifacts, processing logs, message stores, queues, and access logs can all be centrally monitored from this view. Monitoring data from integration suite can be exported via standard-provided APIs from the tool and brought to tools like SAP Analytics Cloud for reporting dashboards or SAP Cloud ALM for centralized monitoring. The **Monitoring** tab in SAP Integration Suite can be customized to include tailored views for your specific needs.

As your landscape expands to include cloud, on-premise, and hybrid environments, it's important to have a top-down view in which you can monitor and raise alarms when issues arise with integration. Although SAP Integration Suite does include robust monitoring capabilities, integrations can fail outside its domain; for example, if the source system fails to send data to SAP Integration Suite, then the suite's monitoring capabilities will never be able to raise an alarm for your support employee. SAP supports holistic monitoring with SAP Cloud ALM. We'll discuss SAP Cloud ALM in more detail in Chapter 13, but for now, know that SAP Cloud ALM is the standard functionality for integration and exception monitoring. SAP Cloud ALM looks at the systems registered in its system landscape directory and the supported integration scenarios between these systems, then provides end-to-end monitoring of the data exchanged between these systems and integrations. The main features of integration and exception monitoring are as follows:

- End-to-end monitoring by correlating single integration artifacts with end-to-end message flows across components and across technology
- Embedded monitoring of integration-related exceptions
- Support of message flows processed by platforms like SAP Integration Suite as well as support of peer-to-peer interface technologies such as SAP Application Interface Framework messages as part of SAP S/4HANA Cloud
- Searching and tracking single messages based on exposed business context attributes like order number
- Embedded alerting to inform the integration owner in IT about critical issues, including drilldown into alert details
- Integration with operation automation to trigger operation flows context-sensitively

Integration and exception monitoring in SAP Cloud ALM can be enabled via a standard setup, which is a simple activation of standard monitoring content, and an expert setup, which allows for customizing the monitoring categories and alerts raised by SAP Cloud ALM. Once the setup is complete in SAP Cloud ALM, you can log into the tool and see high-level KPIs representing the status of your systems and integrations, see details of exceptions and alerts, and drill down into further details about the health of your integrations and any issues.

11.2 APIs and Integration Flows

APIs are the foundational components of building integrations. SAP supports a wide variety of API standards, from historical interfaces such as remote function calls (RFCs), Intermediate Documents (IDocs), and Business Application Programming Interfaces (BAPIs) to modern APIs such as OData, Simple Object Access Protocol (SOAP), Graph, and events.

SAP recommends, where possible, that you migrate from legacy API standards to adopt the modern APIs; it documents this practice in its clean core guidelines, outlined in Chapter 8, Section 8.1. Although not every classic SAP API has a successor modern API, SAP also has recommended frameworks for wrapping legacy APIs to be exposed as if they were modern APIs.

For non-cloud-enabled interfaces such as BAPIs and RFCs, SAP recommends creating a factory model wrapper class and releasing the wrapper to act as if it was a modern API with standard create, read, update, and delete (CRUD) functionality as defined in the legacy ABAP code. This approach allows you to modernize your legacy interface and easily adopt a fit-to-standard approach if SAP releases a successor API in the future.

For modern APIs, SAP has a large catalog of standard offerings in the form of OData and SOAP APIs that can be enabled out of the box. OData APIs align with the Representational State Transfer (REST) design for accessing and modifying data. They use the standard HTTP `GET`, `POST`, `PUT`, `PATCH`, and `DELETE` actions. The OData standard enhances the REST framework by providing capabilities to filter and manipulate the data by parameters passed via the URL. The SOAP APIs use a highly structured XML format for interacting with data. SOAP messages are encapsulated in an XML envelope that defines the data structure and how to process it, definitions of the data and data types inside the message, and finally the procedure calls and how to anticipate the response. OData APIs are delivered from SAP S/4HANA via the gateway server component; the gateway server handles the ingestion of data, authentication from the sending source, and the passing of data back to the SAP code processing it. SOAP interfaces are enabled via the SOAMANGER capabilities in SAP S/4HANA. OData interfaces are lightly coupled to the core SAP objects due to their deployment model on the gateway server and operate in a synchronous manner. They provide the foundation for the standard SAP Fiori applications to surface data to the user and create a good foundation for creating mobile applications. As such, they create integration points for integration flows that can run in a synchronous manner. SOAP interfaces, meanwhile, are tightly coupled to the core ERP platform, require the sharing of a formal WSDL file for integration, and operate in an asynchronous manner, making them useful for high-security integrations.

SAP Integration Suite's *API Management* capability helps you address your enterprise requirements for API security and governance, as shown in Figure 11.2. At its core, the API Management capability lets you publish, share, and manage the lifecycle of your APIs in a secure and scalable environment.

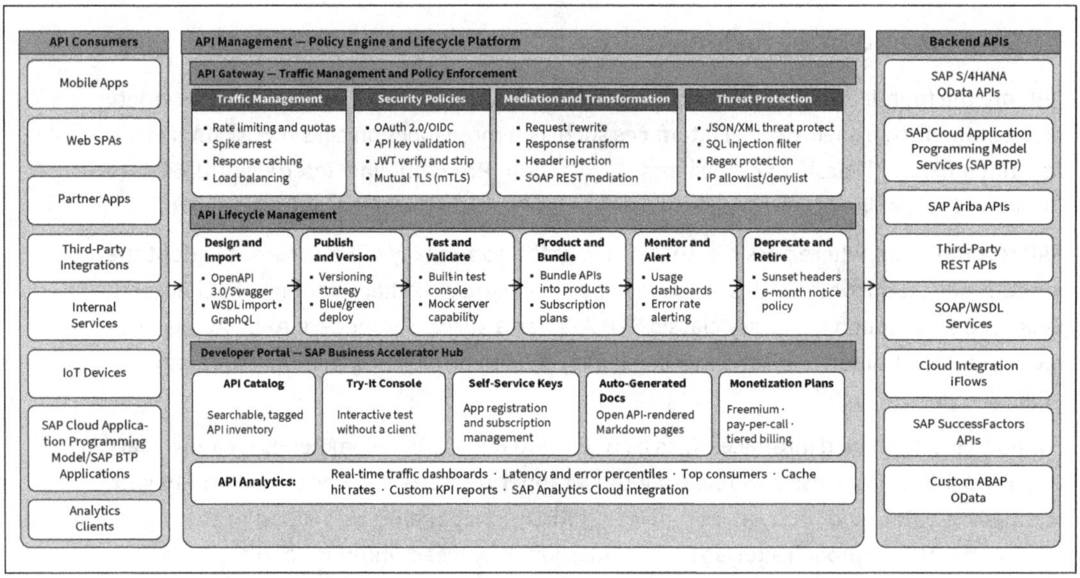

Figure 11.2: API Management Capability

API Management takes your core system APIs and allows you to create proxies, which can be harmonized against your own domain. By creating a proxy for your core OData, SOAP, or other APIs, you create a layer of additional security and monitoring. Proxies can be shared with internal or external groups, and you can hide the core system API behind a safe proxy address exposed in SAP Integration Suite. Between the proxy and the core system, API policies and rules can be set, such as security policies that protect against unauthorized access or threats.

Rules can be placed in the policy manager, such as traffic management rules that set how often an application can call an API so that an external system cannot act erroneously and put additional strain on the core system. APIs registered in API Management can be governed to manage the lifecycle, documentation, and updates to the API; these governed APIs then can be exposed in catalogs for developers to have self-service access to and consume if they have the necessary authorizations.

API Management also services insights and analytics; usage analytics tied the API, error logs, and consumers of the API can all be centrally monitored from the API Management platform. All public APIs that are visible in SAP Business Accelerator Hub can be found in SAP Integration Suite as well, listed under the **API** section of the **Discover** tab within SAP Integration Suite. APIs in the **Discover** tab can be copied into the API Management section of SAP Integration Suite, where they can be easily configured to point to your backend system.

Once an API is configured, you can create an API proxy with your desired governance rules and expose this API to the developer hub so it can be utilized by development teams within your organization. API proxies can be tested within SAP Integration Suite, where you can simulate API calls and responses, verify authentication mechanisms, and ensure that your created API policies execute as intended. APIs registered in API Management can be monetized as well. Within API Management, you can define a rate plan for monetizing your API and attach the rate plan to released API products; from there, application developers who consume that API can then view its cost and the billing details for their consumption of the API.

APIs provide the foundation for integrating systems together, but integrations are never as simple as passing data between system endpoints. For this reason, SAP Integration Suite offers its *Cloud Integration* capability. Figure 11.3 outlines the types of adapters that come out of the box, the transformation steps available, the monitoring capabilities, and the many systems that can be connected via SAP Integration Suite.

Similar to discovering standard APIs in the **Discover** section of SAP Integration Suite, you can find prebuilt SAP integrations in place as well. SAP has created over 1,800 prebuilt integration flows across its ERP and cloud line-of-business (LoB) solutions. These prebuilt integrations can be copied easily into your environment from the **Discover** tab and then configured and modified to meet your requirements.

The design tool in SAP Integration Suite is a low-code first designer where you can model the logic needed in your integration. Integrations are triggered by the *receiver adapters*, a mix of system-specific adapters via the Open Connectors capability and native system-specific adapters such as the adapters for SAP SuccessFactors and SAP Ariba. Supporting the system-specific adapters are adapters based on technical protocols such as HTTP/HTTPS, OData, SOAP, RFC, AQMP, SFTP, and other popular protocols. Between the integration adapters is where the integration flow logic resides. Through simple drag-and-drop controls, you can easily make message mappings between interfaces, perform transformations such as encrypting data or changing the format of data, and call out to external systems for message enrichment or data validation. SAP Integration Suite allows you to create robust models for your integration flow logic, creating looping routines, error routines, and conditional paths to handle any edge cases that might arise during an integration scenario. Custom code can also be embedded into SAP Integration Suite via script steps. Scripting is supported in two languages, JavaScript and Groovy; this functionality should be used when all low-code options have been exhausted.

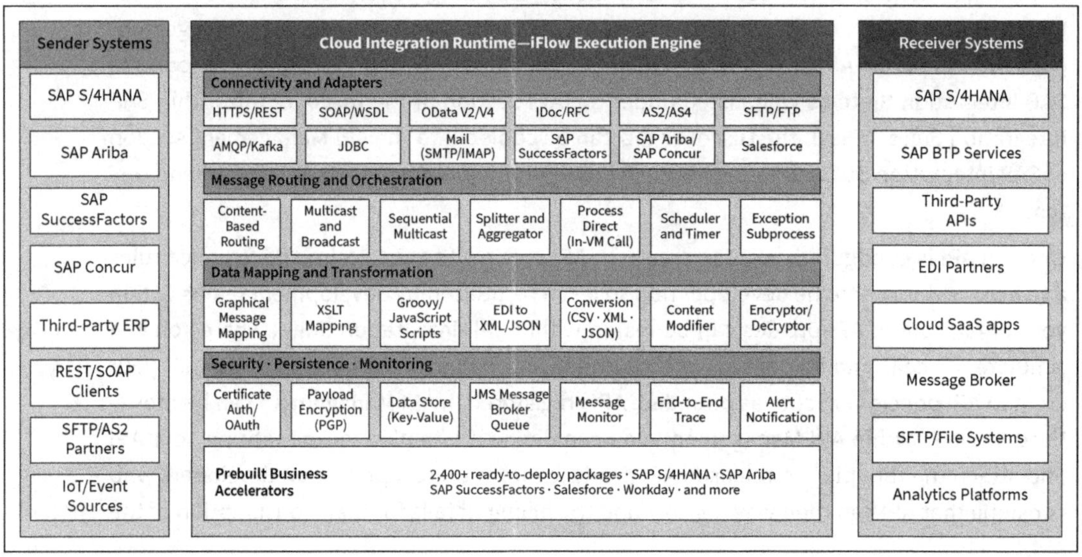

Figure 11.3: Key Capabilities of SAP Integration Suite's Cloud Integration Capability

Just like API Management, *integration flows* have a robust **Monitoring** tab to monitor the health of integrations. On the **Monitoring** tab in SAP Integration Suite, you can view logs and errors associated with deploying integrations and integrations currently running. If an integration goes into error, retry patterns can be established via SAP Integration Suite. If the message cannot be processed due to an error in the integration flow or the data sent, or if the receiver system is unreachable, SAP Integration Suite will log the error. In the **Monitoring** tab, you can analyze the root cause of the issue. Through configuration of the monitoring settings view in the low-code designer, you can determine the exact step in which the integration failed and view the logs associated with that failure. The **Monitoring** tab also catalogs other useful information, such as message consumption, which is filterable; security material, such as passwords and system certificates; locked development objects; and more. The **Monitoring** tab of SAP Integration Suite can be configured to filter for custom attributes, such as data inside the message of an integration. In addition, custom titles with KPIs related to integration activity can be configured on the **Monitoring** tab.

Note

Design guidelines also can be configured on the **Monitoring** tab of the SAP Integration Suite so that your development team can better adhere to the SAP Integration Solution Advisory Methodology principles mentioned in Section 11.1.2.

SAP Integration Suite has a special designer for B2B integrations delivered via its *Trade Partner Management* and *Integration Advisor* capabilities.

The Integration Advisor is an AI-powered tool to help you create message implementation guidelines (MIGs) and message mapping guidelines (MAGs). The first step of using the Integration Advisor is to set up a company profile, including unique identifiers for your organization and any suborganizations you may have. You can enrich this content by adding information about your business systems, security material for authenticating to your systems, and points of contact within your organization who are responsible for Electronic Data Interchange (EDI) and B2B integrations. You then perform the same activity with the relevant information from your trading partner to onboard them into Trade Partner Management. Once you have onboarded your organization and your trading partners, you can then, with the assistance of AI, create the interface specifications for your organization and your trading partners. SAP Integration Suite supports the OData, SOAP, and IDoc frameworks out of the box; it also supports major industry standards such as X12 and EDIFACT. Once you have created your MIGs for yourself and your trading partners, you can create your MAGs with the assistance of AI.

Once your MIGs and MAGs are created, it is time to put them to work in Trading Partner Management. You first create templates for your B2B scenarios, such as a trading partner sending a purchase order to your system via an EDI protocol such as X12. Once the template is created, you can adapt it to add partner-specific content, such as their unique mapping and MIG.

Now that all the setup is complete, you can deploy your integration and receive sales orders seamlessly from your trading partner via EDI. The Trading Partner Management tool will know how to identify your trading partner and will use the correct MIG and MAG to validate and transform the data before it is sent to your system. Messages received via Trading Partner Management receive an enhanced layer of monitoring, allowing you to view the message payload as it was received and after it was transformed, plus other supporting material such as the message disposition notification.

Figure 11.4 shows an example scenario of Trading Partner Management ingesting a sales order from a trading partner sending an X12 850 EDI message.

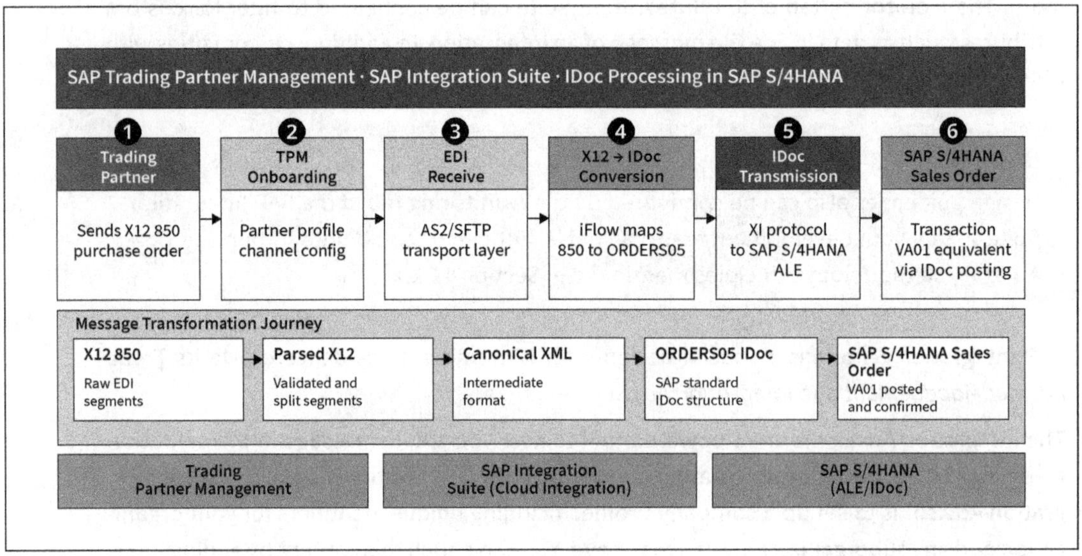

Figure 11.4: Example Scenario of SAP Integration Suite's Trading Partner Management Capability

As illustrated in Figure 11.4, you can use the Trading Partner Management capability in SAP Integration Suite to take in sales orders from your trading partners. In this scenario, your trading partner is sending you an order from their system via the X12 standard. Using the Trading Partner Management framework, you can create a partner flow that identifies the partner-specific mapping and interface specification to validate the message and correctly maps it into the format your system needs to accept the new sales order.

11.3 Event-Driven Integrations

Event-driven integration represents a modern pattern of integration between applications that can help reduce the time it takes to build new integrations and connect existing applications. Traditionally, applications were connected via point-to-point integrations, in which one application would either call a remote application or integration platform via an API call, or periodically drop a batch file of data. This required integration teams to develop new integrations every time a new system or functionality was onboarded in an organization, creating high lead times to connect systems together and growing technical debt for integration teams to maintain over time.

Event-driven integration supports a publish-and-subscribe pattern that allows a system to hand data off to an event broker, which in turn delivers the data to all subscribing applications. Events represent a change in state for a business application object. A simple example of this is a business partner being created, updated, or deleted in your SAP S/4HANA system. Rather than waiting for a nightly batch job to replicate that change, the event is published the moment the business partner record changes, and every subscribing system receives the update in near real time. Event-driven integration is an asynchronous pattern of integration that supports a multitude of integration scenarios, such as the following:

- **One-to-many distribution/many-to-one integrations**
  Cascading updates when an element is changed to multiple subscribing systems simultaneously
- **Shock absorber**
  Queueing high-volume transactions to distribute them to a subscriber at the rate they can consume without disrupting operations
- **High availability**
  Storing transactions while a system is experiencing an unexpected outage or going through a maintenance window, ensuring no data is lost

We'll explore the key capabilities for event-driven integration, Event Mesh and SAP Integration Suite, advanced event mesh, in the following sections.

11.3.1 Event Mesh and Advanced Event Mesh

SAP Integration Suite comes with the *Event Mesh* capability, which acts as a centralized message broker that supports the publish-and-subscribe pattern. Event Mesh provides a managed broker environment in which SAP Autonomous Suite systems and custom applications can publish and subscribe to events. SAP Autonomous Suite systems are preconfigured to connect to Event Mesh, making it straightforward to enable standard events and route them to the applications that need to consume them.

For organizations with more complex event-driven requirements, *SAP Integration Suite, advanced event mesh* expands upon the capabilities of Event Mesh. Advanced event mesh allows you to deploy a network of connected brokers to create a true mesh architecture for processing system events. This mesh architecture allows your organization to span event brokers across multiple geographies, ensuring that events are processed close to where your systems are deployed for optimal latency and resilience. Advanced event mesh is ideal for large enterprise organizations that need to process high volumes of events across globally distributed landscapes with the highest levels of availability.

Often the best scenarios for building extensions on SAP BTP are asynchronous data exchanges, and events provide a strong framework for passing data to extensions running on SAP BTP. Workflows and automations running on SAP Build Process Automation can be triggered by an event. SAP Cloud Application Programming Model and ABAP RESTful application programming model applications can both send events to and receive events from SAP Autonomous Suite applications. Custom Joule Agents can call events within Joule Skills to execute actions in an SAP Autonomous Suite system. By using Event Mesh in SAP Integration Suite or advanced event mesh, you can create resiliency with a circuit breaker design pattern to safely exchange data between multiple applications, ensuring that your enterprise continues to function even when individual systems experience disruptions.

We'll unpack the fundamentals of Event Mesh and advanced event mesh in the following sections.

Topics

To understand how to best leverage Event Mesh and advanced event mesh, first you should understand the concept of a *topic*. When a publisher sends a message, it addresses that message to a *topic string*, a hierarchical, slash-delimited path that describes the nature and origin of the event. A topic is not a stored resource that must be precreated; it is simply a string that the publisher specifies at the moment of sending. For example, a sales order created in SAP S/4HANA might be published to the `sap/s4hana/orders/created/1000/DE` topic, where the segments encode the source system, event type, company code, and country. This hierarchical structure is central to advanced event mesh's routing capability.

Subscribers do not need to know every topic in advance. Instead, they use wildcard subscriptions to express interest in a pattern of topics. Advanced event mesh supports two wildcard characters: The single-level wildcard (*) matches exactly one level in the topic hierarchy, and the multilevel wildcard (>) matches one or more levels from that point to the end of the topic string. A consumer that needs all order events from any company code in Germany, for example, could subscribe to `sap/s4hana/orders/*/*/DE`, while a consumer that needs every event published under the `sap/s4hana` namespace would subscribe to `sap/s4hana/>`. This topic-based publish-and-subscribe model means that new consumers can be added to the mesh without any change to the publisher, and new event types can be introduced without updating existing subscriber configurations.

Topic hierarchies also serve as the integration point between advanced event mesh and SAP event sources. SAP S/4HANA Cloud Public Edition publishes business events such as `BusinessPartner.Changed`, `SalesOrder.Created`, and `PurchaseOrder.Released` to the advanced event mesh using a structured topic namespace defined in SAP Event Catalog. Architects can browse this catalog within advanced event mesh's tooling, discover available event types, inspect their CloudEvents-compliant payloads, and immediately wire them to consumers or downstream queues.

Queues and Guaranteed Messaging

Where topics provide the publish-and-subscribe broadcast mechanism, queues provide the guaranteed, at-least-once delivery mechanism that enterprise integration requires. A *queue* is a named, durable endpoint stored on the advanced event mesh broker. Messages sent directly to a queue, or attached to a queue from a topic via a topic subscription, are persisted to broker storage and held until a consumer acknowledges receipt. If no consumer is connected, the broker holds the messages for the duration of the queue's configured message time to live. If a consumer connects and receives a message but fails before acknowledging it, the broker redelivers the message to the next available consumer, ensuring no message is silently lost.

Queues in advanced event mesh are exclusive or nonexclusive. An *exclusive* queue delivers all messages to a single active consumer at a time; when that consumer disconnects, the next registered consumer takes over without message loss, making this pattern suitable for ordered, single-consumer processing such as IDoc posting to SAP S/4HANA. A *nonexclusive* queue distributes messages in a round-robin fashion across all active consumers, enabling horizontal scaling of message processing. Adding more consumer instances increases throughput proportionally without any change to the publisher or the queue configuration as each instance competes for the next available message from the same queue.

Topic subscriptions attached to queues are the bridge between the broadcast world of topics and the guaranteed world of queues. A queue can have one or many topic subscriptions; any message published to a matching topic is automatically pulled into the queue and subject to the queue's durability and delivery guarantees. An example of this could be a dead letter queue that holds all messages that could not be successfully picked up by a system after all tries have been exhausted. This is the mechanism that underpins the shock absorber pattern: An upstream publisher fires events to a topic at whatever rate it chooses, and a downstream consumer reads from the queue at a controlled, sustainable rate, with the queue depth absorbing the difference between the two rates.

Event Bridge

The advanced event mesh's Event Bridge is one of the most strategically important capabilities in its portfolio. It provides native, configuration-driven connectivity between advanced event mesh and external messaging and event streaming services—specifically, Amazon SQS, Amazon SNS, Amazon EventBridge, Microsoft Azure Service Bus, Azure Event Hubs, Google Cloud Pub/Sub, and Apache Kafka. Rather than building and maintaining custom integration code to bridge between these services, architects configure a bridge in the advanced event mesh console by specifying the source, destination, topic or queue mappings, and credential references. Advanced event mesh handles the protocol translation, offset management, and error recovery automatically.

The Event Bridge operates bidirectionally. Events can flow from advanced event mesh into an external service; for example, an SAP S/4HANA business event published to advanced event mesh can be forwarded to an Amazon EventBridge event bus where AWS Lambda functions can react to it, or from an external service into advanced event mesh where downstream SAP Cloud Application Programming Model applications on SAP BTP or SAP Integration Suite integration flows consume it as a standard advanced event mesh queue message. This bidirectional capability means that advanced event mesh acts as the central event hub in a hybrid, multicloud architecture, normalizing event routing regardless of whether the producer or consumer runs on SAP BTP, AWS, Azure, Google Cloud Platform, or on-premise.

Event Bridge Use Case

A logistics partner publishes shipment status updates to Amazon SNS. The Event Bridge attracts those SNS messages into an advanced event mesh queue with a topic subscription. An SAP Cloud Application Programming Model application on SAP BTP consumes from the queue and updates the delivery status in SAP S/4HANA, all without the logistics partner needing to have any knowledge of SAP's event infrastructure.

For Apache Kafka specifically, the advanced event mesh Kafka Bridge supports both consumer and producer roles. Advanced event mesh can consume from one or more Kafka topics and republish the messages into the advanced event mesh under a mapped advanced event mesh topic hierarchy, or it can act as a Kafka producer, writing advanced event mesh messages to designated Kafka topics. The bridge manages Kafka consumer group offset commits, handles partition rebalancing transparently, and supports schema registry integration for Avro-encoded payloads. This makes advanced event mesh the natural integration point between Kafka-centric streaming architectures and SAP event-driven applications, without requiring SAP applications to take a direct Kafka client dependency.

Event Portal and Design-Time Governance

The Event Portal is the design-time and governance layer that sits alongside the runtime Event Mesh. It provides a centralized registry of event types, their schemas, topic addresses, and the applications that produce and consume them. Architects use the Event Portal to design the event-driven architecture before writing a line of code: they define application domains (bounded contexts for event ownership), create event objects with associated AsyncAPI or JSON Schema specifications, and model the relationships between producers, consumers, and events in a visual graph. The portal acts as the single source of truth for the event-driven application landscape.

The Event Portal's designer view allows architects to model which application publishes which event to which topic, and which applications subscribe to it. This model is not merely documentation; it can be used to generate AsyncAPI specifications that developers use as contracts when building publishers and consumers, and it integrates with the advanced event mesh runtime to validate that deployed applications are behaving according to their modeled intent. When a new event type is introduced, the portal captures the schema, the owning domain, and the lifecycle status (draft, released, deprecated, retired), giving governance teams the visibility to manage the event catalogue over time in the same disciplined way that API Management governs REST APIs.

11.3.2 Example: Advanced Event Mesh

If you are a retailer launching a flash sale, there can be unintended consequences for the underlying SAP S/4HANA system without the right integration architecture in place. This is a classic antipattern in systems integration, where an event meant to drive high traffic causes an adverse effect by putting too much stress on a system, causing poor performance or, in a worst case, bringing your system down so that end users cannot purchase the sale items. Within minutes of the promotion going live, multiple upstream channels converge simultaneously on the order management layer: the web storefront, which may be handling tens of thousands of concurrent browser sessions driven by marketing emails and social media; the in-store point-of-sale network, where cashiers and self-checkout lanes are processing a surge of basket completions triggered by the same promotional pricing; and the reseller channel, where partner platforms have been prenotified of the sale window and are submitting bulk order files or API calls the moment the price drops.

Each of these channels has its own throughput characteristics, its own retry logic, and its own tolerance for latency. Without a buffer between them and SAP S/4HANA, every order event hits the system directly, and the resulting spike can saturate the ABAP application server work processes, overwhelm the order creation functions, and degrade system performance precisely at the moment the business needs it most.

Advanced event mesh addresses this vulnerability through what architects call the *shock absorber pattern* (see Figure 11.5).

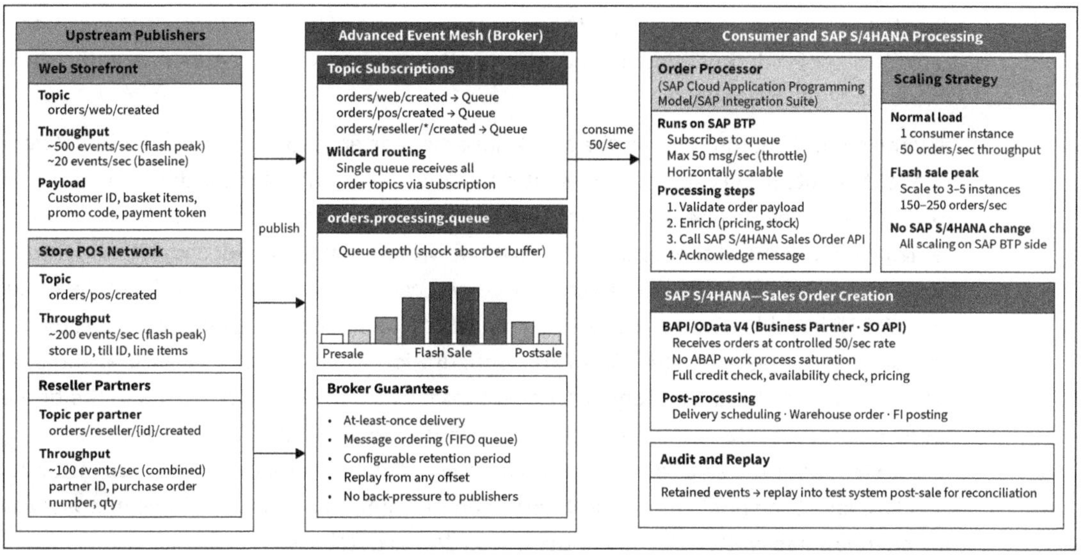

Figure 11.5: Advanced Event Mesh Acting as Shock Absorber to Resolve Flash Sale Integration Antipattern

Rather than allowing the three upstream channels to write sales orders directly into SAP S/4HANA, each channel publishes its order events to a dedicated topic on the Event Mesh. The web storefront publishes to `orders/web/created`, the POS network publishes to `orders/pos/created`, and each reseller partner publishes to its own name-spaced topic, such as `orders/reseller/{partnerId}/created`. Advanced event mesh guarantees delivery of every message regardless of what is happening downstream; if SAP S/4HANA is under load, the messages queue in the broker. The Event Mesh absorbs the spike in the same way a hydraulic shock absorber converts a sharp impact into a controlled, sustained force: The upstream channels experience no back pressure and no timeouts, while for the downstream consumer, an SAP Integration Suite flow or an SAP Cloud Application Programming Model service acting as the order processor reads from the broker at a rate that SAP S/4HANA can sustain. The asymmetry between the write rate and the read rate is exactly the point. The broker holds the difference.

The practical architecture places a dedicated queue, or a set of topic-subscribed queues, between the publisher topics and the SAP S/4HANA order creation consumer. The consumer is configured with a maximum message rate that reflects the throughput the ABAP order creation pipeline can handle under normal operating conditions—for example, 50 sales order creation calls per second. During the flash sale, the web storefront may be publishing at 500 events per second, the POS network at 200, and the reseller channel at a further 100. The advanced event mesh holds the backlog, preserving the message order and guaranteeing at-least-once delivery. The queue depth metric, visible in the advanced event mesh console, becomes the key operational indicator: A rising queue depth tells the operations team that the sale is performing ahead of forecast, while a falling depth tells them the backlog is clearing. Once the sale window closes and publishing rates drop back to baseline, the consumer works through the remaining queue at full pace until depth returns to zero. At no point during this sequence does SAP S/4HANA experience the raw traffic shape of the flash sale event itself.

Beyond smoothing throughput, the shock absorber pattern provides three additional operational benefits in this scenario. First, it decouples failure domains; if the web storefront experiences a deployment issue mid-sale and restarts, its unacknowledged messages remain safely in the broker and are redelivered without data loss. Second, it enables independent scaling; the consumer microservice processing orders can be horizontally scaled on SAP BTP by adding more instances reading from the same queue, increasing throughput without any change to the SAP S/4HANA configuration. Third, it creates a natural audit and replay capability; because every order event from every channel is persisted in the broker for its configured retention period, the operations team can replay the event stream into a test system after the sale to validate pricing logic, debug anomalies, or run reconciliation checks against the final SAP S/4HANA order count, something that would be impossible if the channels had written directly to the transactional database.

11.4 AI for Integration

AI is rapidly transforming how integration teams build and maintain enterprise integrations. SAP has embedded AI capabilities throughout SAP Integration Suite to reduce the manual effort required at every stage of the integration lifecycle, from initial design through to ongoing maintenance and monitoring. Let's look at what's available now and what's planned for the future.

11.4.1 AI-Assisted Integration Development

Within the Cloud Integration designer, SAP has introduced Joule-powered assistance to help integration developers build integration flows faster. Developers can describe their integration scenario in natural language and receive suggestions about the appropriate adapters, transformation steps, and routing logic needed to implement the scenario. This reduces the time a developer spends navigating the tooling and searching for the right components, allowing them to focus their effort on the business logic specific to their organization's requirements.

For data mapping, traditionally one of the most time-consuming activities in integration development, AI assistance can analyze the source and target message structures and suggest mapping rules between the two, dramatically decreasing the time it takes to complete a mapping before testing. AI-assisted mapping learns from the structural patterns of the data and the naming conventions of fields to suggest the most likely mappings, which integration developers can then review, adjust, and confirm. This brings the speed of low-code tooling to tasks that have historically required deep technical expertise to execute.

Beyond development, AI is also being applied to integration monitoring. SAP Integration Suite's monitoring capabilities are augmented with intelligent anomaly detection that can identify unusual patterns in integration message flows and raise alerts before a problem escalates. Rather than waiting for a critical failure to occur, your integration operations team can be proactively notified of degrading conditions such as rising error rates, unusual latency, or unexpected message volume changes. This shifts your integration operations team from a reactive posture to a proactive one, reducing the business impact of integration issues.

AI-powered root cause analysis is also available to help your support teams more quickly diagnose the source of integration failures. When an error occurs, the AI assistant can analyze the error message, the integration flow configuration, and the recent history of the failing interface to suggest the most likely causes and recommended resolution steps. This reduces the mean time to resolution for integration incidents and allows your team to maintain a healthier integration landscape with less manual effort.

The Integration Advisor capability within SAP Integration Suite uses a predelivered SAP Knowledge Graph to assist with creating B2B integrations. This graph contains definitions for hundreds of industry-standard message types and the typical mapping rules between them. When you are building a B2B integration, the Integration Advisor can automatically propose the message implementation guidelines and mapping templates based on the standard you are implementing, reducing a task that once took weeks of manual specification work down to hours.

This AI-powered capability within Integration Advisor continuously learns from the mapping work performed by users across the SAP ecosystem. When integration developers accept or refine the proposed mappings, SAP Knowledge Graph is updated, meaning that the suggestions provided to the next user performing a similar integration are more accurate and complete. This collaborative intelligence model makes SAP Integration Suite's B2B tooling increasingly powerful over time.

11.4.2 Future AI Enhancements

SAP is making a major investment into AI capabilities within SAP Integration Suite. At the time of writing, the features discussed here are the AI capabilities available in SAP Integration Suite, but many more are coming to the tool. SAP Integration Suite will have the ability to create and deploy custom Model Context Protocol (MCP) servers, for example. MCP is an open-source standard that allows AI agents and assistants to connect to external data sources such as APIs, databases, and other tools. Using an MCP server allows an agent to call the right tools to surface data back to the end user and reduces hallucinations. SAP Integration Suite's MCP server capabilities will extend its API Management capabilities, discussed earlier, and allow organizations to have a layer of governance over how an agent calls their critical backend systems. Other exciting AI features in the SAP Integration Suite roadmap include AI-assisted migration for customers using legacy SAP integration tooling, AI-assisted mapping generation, end-to-end integration flow generation with Joule, and enhanced AI-assisted error resolution.

11.5 SAP Master Data Integration

Master data is the foundation upon which your enterprise applications operate. Customer records, vendor records, product catalogs, organizational structures, and financial charts of accounts are all examples of master data that must remain consistent across the applications in your landscape. When master data is inconsistent between systems, it leads to processing errors, reporting inaccuracies, and costly manual reconciliation work. SAP Master Data Integration is a service on SAP BTP specifically focused on solving this challenge.

SAP Master Data Integration is a cloud service that acts as a central hub for distributing and synchronizing master data across your SAP landscape. It adheres to SAP One Domain Model, SAP's standardized semantic model for master data objects across all SAP Autonomous Suite applications. By aligning to a single domain model, SAP Master Data Integration ensures that master data exchanged between applications is semantically consistent, even when the applications themselves represent the same object in slightly different ways internally.

SAP Master Data Integration supports the initial load of master data into connected applications and keeps those applications in sync as master data changes over time. When a business partner record is updated in one system, that change is detected and distributed to all other connected systems that have subscribed to that business partner object. This eliminates the need for custom point-to-point master data replication integrations between individual applications, which are typically fragile and difficult to maintain.

SAP Master Data Integration supports the following key master data domains:

- Business partner
- Product
- Product group
- Company code
- Plant
- Cost center
- Profit center
- Purchasing organization
- Purchasing group
- Sales controlling object
- Project controlling object
- Fixed asset
- Symbolic general ledger account
- Bank
- Category
- Equipment
- Functional location

SAP Master Data Integration is most effective when paired with a master data governance strategy. Without governance, even the most technically capable integration platform will distribute and replicate data quality issues across your entire landscape, amplifying the problem rather than solving it. SAP Master Data Integration works in concert with SAP Master Data Governance (SAP MDG), which provides the workflow and approval processes to ensure that master data changes are reviewed and validated before they are distributed. Together, SAP MDG and SAP Master Data Integration form a comprehensive master data management solution that covers both the governance over who can change master data and the distribution of those approved changes to your landscape.

By implementing SAP Master Data Integration as part of your broader integration strategy, you reduce manual data entry across systems, minimize the risk of data inconsistencies causing downstream processing errors, and create a single reliable source of truth for the master data that your enterprise operations depend on.

11.6 Summary

Integration is the connective tissue of the Autonomous Enterprise. As the number of systems, applications, and AI agents in your landscape grows, a well-defined integration strategy becomes the foundation that allows your organization to scale without creating data silos or mounting technical debt. SAP provides a comprehensive set of integration services on SAP BTP to help your organization meet these challenges.

SAP Integration Solution Advisory Methodology gives your integration Center of Excellence a structured framework for defining the right integration technologies for your specific scenarios. SAP Business Accelerator Hub provides a central catalog of APIs, integrations, and events so your teams can quickly discover what is available before building from scratch. SAP Integration Suite delivers a single, holistic integration platform that covers process integration, API Management, B2B integration, event-driven integration, and legacy migration tooling, all enhanced with embedded AI that reduces the manual effort required to build and maintain your integrations.

Event-driven integration through Event Mesh and SAP Integration Suite, advanced event mesh gives your organization the ability to react to business changes in near real time, moving beyond the batch-oriented and point-to-point integration approaches of the past. The AI capabilities embedded throughout SAP Integration Suite accelerate development, improve monitoring, and reduce the burden on your operations team. Finally, SAP Master Data Integration ensures that the master data your applications depend on stays consistent and current across your entire landscape.

Together these tools give your organization the integration foundation it needs to support today's operations while positioning you to rapidly adopt the emerging patterns of AI-driven and event-driven enterprise integration that will define the intelligent enterprise of the future.

This concludes Part II of the book. In the next chapter, we'll begin the transformation journey of Part III with a discussion of how to build your business case.

PART III

Transformation and Execution

Chapter 12
Developing a Business Case for Your Autonomous Enterprise

A delay in a shipment, a missed invoice follow-up, a late maintenance decision: These are symptoms of a deeper structural gap between signal and action. That gap, the time between when a signal appears and when the enterprise responds, is where value is lost. The business case for the Autonomous Enterprise exists to measure and reduce that delay.

As discussed in the Preface, a container ship sits in Rotterdam, and it holds components needed in Stuttgart. Nothing dramatic happens: a missed docking slot, a customs processing delay, a crane crew running an hour late. The signals exist: port status updates, logistics schedules, customs backlogs. However, the signals sit in different systems, viewed by different teams, interpreted separately.

By the time someone connects the signals, production in Stuttgart is scheduled to start in 36 hours, without the components on hand. The shipment, which should have arrived with a comfortable margin, now arrives 18 hours late. Two weeks of production delays follow. The retailer in Chicago, planning for peak season inventory, faces a shortage. Finance revises earnings guidance three days after the moment that revision should have happened.

Nobody made a mistake; the data was accurate and the systems were working. The problem was timing. The gap between when the issue became visible in systems and when the organization could actually act on it was too large.

This is where the business case for the Autonomous Enterprise begins, not with technology, but with this gap. We'll chart the development of a business case in the following sections to see how AI and autonomous capabilities add value to organizations.

12.1 Where Is the Value Lost?

The Rotterdam delay is not unique to supply chain. The same pattern appears across finance, operations, and service delivery. It appears wherever signals must flow through manual processes before action occurs.

A customer's payment sits overdue. The invoice went out correctly; the customer received it. However, the receivables team, reviewing a batch of invoices each morning, does not identify this specific customer's pattern until it is already 10 days past due. By then, a phone call is needed. Negotiation happens, but the cash is delayed further. The CFO, tracking days sales outstanding (DSO), sees the metric move in the wrong direction and asks why. The answer comes a week later: "We had a late payment."

Inventory accumulates in a regional distribution center. The demand forecast showed steady consumption, but actual demand dropped sharply last week, not because information is missing but because the shift from forecast to reality happens in the weekly planning cycle. By the time the cycle runs, two days of overstocking have already occurred. Orders are placed, inventory ships, and the supplier has already manufactured what was not needed.

A maintenance schedule shows equipment operating within normal parameters. Vibration levels are increasing, but gradually. The data exists, logged every hour, but the analysis happens monthly. In Month 14, the equipment fails unexpectedly, and the unplanned downtime costs $500,000. Investigation reveals that deterioration was visible starting in Month 11, three months before failure. The signals were there.

These are not data problems, because the information exists. The issue is the delay between signal and action. In each case, better decisions that could have been made happen a day late, a week late, or a month late. The cost of that delay is what the business case must capture.

12.2 What Agents and Assistants Actually Change

The conventional approach to these problems is to hire more people: more collectors to watch receivables, more planners to monitor demand, more analysts to interpret equipment signals. This approach does not scale, however; it costs more, and it remains error-prone because human attention is finite.

Autonomous agents and assistants change the dynamic fundamentally. They do not make better decisions than humans, but they make decisions humans never get around to making.

Consider what happens in a traditional receivables process. A collector sits down at 9 a.m. and reviews the day's work. Open invoices are sorted by age. Emails from customers arrive throughout the day. The collector responds to what is urgent, tries to reach customers on phone, escalates issues that need management attention. By day's end, most of the collector's time has been consumed by reactive work. The truly at-risk accounts, the ones showing early signs of payment delay, never receive proactive outreach.

An agent operating in the same receivables function works differently. Overnight, before anyone arrives, the agent has analyzed payment patterns across all customers. It has identified which accounts showed changed behavior: slower recent payments, increased dispute frequency, credit limit utilization changes. It has classified disputes by type and complexity. It has routed simple disputes to standard responses. It has prepared context for complex ones.

When the collector arrives, the work is not to process all open invoices. The work is to intervene in the specific high-risk accounts where human judgment makes a difference. The agent handled the transactional sorting, and the collector handles the relationship complexity.

This is not efficiency in the traditional sense (doing the same work faster). This is effectiveness in a different sense. The collector can now focus on interventions that actually prevent payment delay rather than managing it after the fact. The organization can act hours or days earlier than before. The same headcount delivers fundamentally different outcomes.

This pattern repeats everywhere autonomous enterprise operates. Finance teams shift from invoice processing to exception handling, supply chain planners shift from data gathering to strategy refinement, maintenance technicians shift from reactive repair to predictive intervention: The work changes because the transactional burden is removed.

The business case rests on this shift, the value of changing the organizational focus, not removing people. This value is realized with the following outcome-oriented mindset, as highlighted in Figure 12.1:

1. Start with business outcomes (cash, margin, cycle time, resilience).
2. Design capabilities and processes backward from these outcomes.
3. Use systems and AI as accelerators, not anchors.

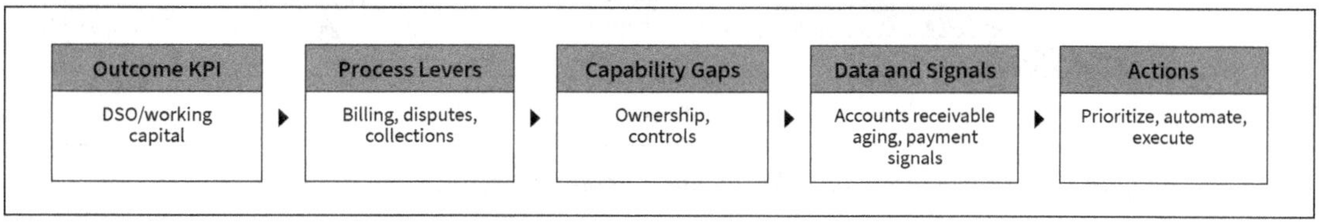

Figure 12.1: Outcome-Oriented Thinking

12.3 The Flywheel Model: How Autonomous Capabilities Amplify Traditional Value

The traditional flywheel that we introduced in Chapter 1, Section 1.2 works through four overlapping cycles: cloud operating model (faster upgrades, reduced IT burden), SAP Cloud ERP capabilities (optimized processes), SAP Business Data Cloud integration (unified data enables better insights), and continuous innovation adoption.

Each cycle reinforces the others. Better data enables better decisions, better decisions drive adoption, adoption creates cleaner data, and cleaner data enables more advanced analytics. Autonomous capabilities fundamentally accelerate this flywheel.

In the traditional model, data improves gradually through better processes. In the autonomous model, data improves rapidly because agents surface inconsistencies immediately and continuously. The difference compounds: Cleaner data means better predictions, better predictions mean faster action, and faster action means problems are solved before they cost money. The organization does not just move faster; it works differently because the foundation keeps improving without waiting for humans to figure it out.

That is where the real return comes from: not from doing the same work faster but from doing fundamentally different work because the baseline has improved. Without agents, each of these moves forward requires manual effort. With agents, each move forward happens automatically because the foundation improves continuously.

This is the multiplier effect. The traditional flywheel delivers value, while autonomous agents deliver multiplied value because they accelerate every turn of the wheel simultaneously. Key value areas of the flywheel model are highlighted in Figure 12.2.

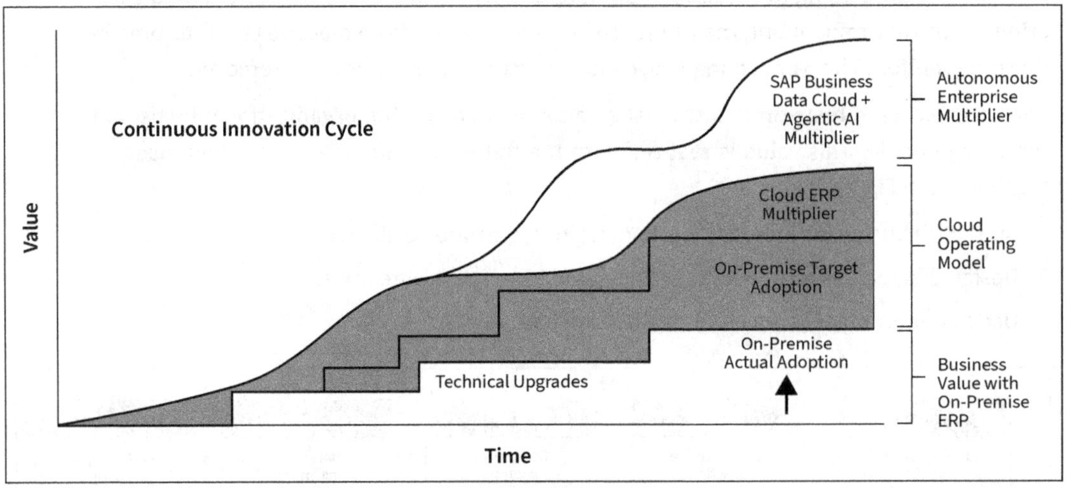

Figure 12.2: Flywheel Model: Value of Autonomous Enterprise

Key value areas of the flywheel model include the following:

- **Cloud operating model**
  Agents handle more use cases with less customization effort, reducing upgrade complexity faster. The benefit areas include improved IT productivity, reduction in total cost of ownership (TCO), reduction in upgrade efforts, and reduction in integration and compliance efforts.
- **SAP Cloud ERP capabilities**
  Agents embed into workflows, making processes more efficient without requiring reengineering.
- **SAP Business Data Cloud**
  Rather than relying on incremental data cleanup, agents continuously classify and improve data quality in the background. The greater value, however, lies in unlocking end-to-end insights from SAP Business Data Cloud, driving faster, more informed decisions. AI assistants and agents then operationalize those decisions by executing actions directly in core systems such as SAP S/4HANA.

- **Innovation adoption**
  With agents handling routine work, teams adopt new capabilities faster because they have bandwidth to learn.

The business case for Autonomous Enterprise is not that it adds a fourth wheel to the flywheel; instead, it accelerates all four wheels simultaneously, compounding the value each delivers.

12.4 Top-Down Value Versus Bottom-Up Value

A business case can be approached from two complementary perspectives. A top-down approach begins with the outcomes the organization is trying to achieve, such as improving working capital, increasing service levels, reducing operational costs, or accelerating decision-making. It asks how much value could be created if those business outcomes improved and establishes the potential opportunity at an enterprise level.

A bottom-up approach starts within the individual processes that contribute to those outcomes. It examines activities such as invoice processing, collections management, demand planning, procurement approvals, or customer service operations to identify inefficiencies, delays, manual effort, and sources of cost. These process-level improvements are then quantified and aggregated to estimate their contribution to broader business objectives.

Both perspectives are important. A top-down analysis ensures that the business case remains connected to strategic objectives, while a bottom-up analysis provides the operational evidence needed to support the projected benefits. Together, they create a business case that is both strategically relevant and grounded in measurable business improvements

Now let's discuss both approaches in the context of the working capital challenges that Maria the CFO was facing in Case Study 1, introduced in Chapter 1, Section 1.8.

12.4.1 Top-Down View: The Outcome That Matters

Begin with outcomes, not processes. What business metric must improve? For finance, it is often working capital. For supply chain, it is often inventory turns or on-time delivery. For service, it is uptime or response time. The choice depends on your strategy.

Work backward from that outcome. Per Case Study 1 in Chapter 1, the CFO concerned with working capital looks at DSO, which sits at 72 days. What does that number actually mean in cash terms?

A company with $1 billion in annual revenue has roughly $2.74 million in daily revenue. At 72 days DSO, approximately $197 million in cash is sitting in receivables that should have converted to payment.

If DSO could be reduced by just 15 percent—about 10 days—the released cash would be approximately $27 million. That cash is not new; it already exists. It is simply delayed.

This is the top-down target. Not *improve processes* but specifically, "Release $27 million in working capital by reducing DSO from 72 to 62 days." The business case begins here with a number the CFO already understands.

12.4.2 Bottom-Up View: Where the Cash Actually Gets Stuck

The $27 million target is real, but abstract. The bottom-up view shows where that money actually sits. Consider the following example.

Pull the detailed receivables data and analyze the open invoice items, and the pattern appears quickly: specific, measurable inefficiencies that delay cash. The following are the key bottom-up drivers for unlocking the working capital:

1. Overdue open items that should have been collected. Not disputed but simply not followed up on. These items sit in the queue waiting for action.
2. Disputes stuck in resolution, in the form of back and forth between customer service, finance, and operations. Resolution takes 15 days when the process could handle it in five. The customer payment waits.
3. Inconsistent collection activity. Some accounts have weekly follow-ups, while others sit 60 days untouched. The ones receiving attention move faster. The inconsistency is the problem.
4. Payment terms not optimized for reality. Customers who always pay on time could operate on tighter terms, while customers who are slow should have tighter terms but do not. Terms do not match behavior.

Each of these factors is specific, observable, and addressable. Together, they explain why DSO is at 72 days instead of something better.

Connecting Bottom-Up to Top-Down View

These inefficiencies drive toward the $27 million gap. They might not each be worth a specific dollar amount; the reality is interconnected. However, each one, when addressed, moves the needle.

Automated routing eliminates the queue. Dispute resolution that routes immediately shortens cycle time. Consistent follow-up ensures nothing falls through. Terms that match behavior accelerate payment. None of these is a silver bullet, but together, they shift the operating model from manual and reactive to continuous and coordinated. That is where the cash comes back.

The challenge with building a bottom-up business case is knowing where to look. Traditional approaches rely on interviews and spot checks. Teams describe what they think the process is, but the reality is often different.

Process management tools like SAP Signavio Process Insights reveal what actually happens versus what is documented. They show where work sits in queues.

SAP Signavio Process Insights runs directly on your SAP ERP or SAP S/4HANA system. Instead of asking teams how work happens, it observes how work actually happens. It maps every transaction, every workflow step, and every delay. It identifies where items sit in queue, where handoffs happen between systems, where the same work is done multiple ways, and where bottlenecks form.

The result is not a model or an estimate. It is the current operating reality. You see the exact percentage of items moving normally versus stuck, the financial impact of each delay, and the specific process steps that are causing the problem. This is the foundation for a defensible business case.

Let's examine a concrete example, as highlighted in Figure 12.3. SAP Signavio Process Discovery for a receivables workflow shows this specific pattern.

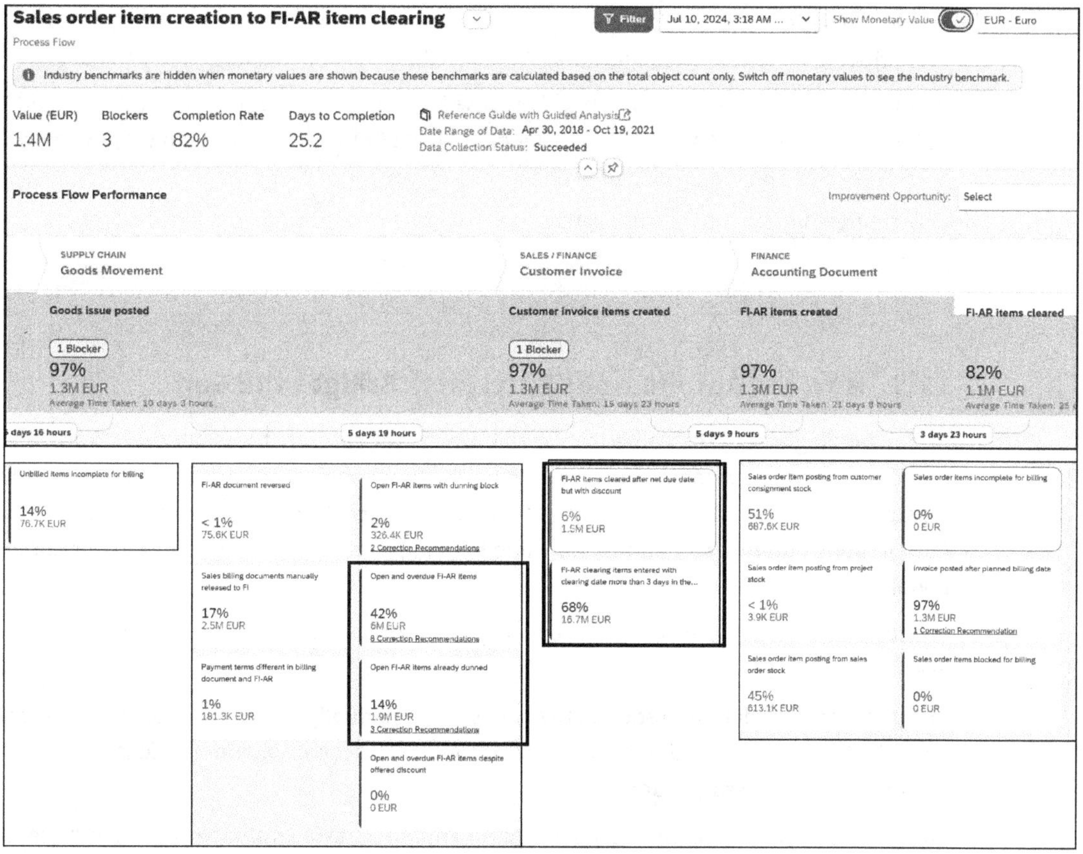

Figure 12.3: SAP Signavio for Bottom-Up Calculations

A total of 42 percent of accounts receivable items are open and overdue. That is 6 million EUR sitting beyond due date. Another 6 percent are cleared after the due date but with discount applied, which means that 1.5 million EUR in late payment discounts are being lost. And 68 percent of accounts receivable clearing items are entered with a clearing date more than three days in the past, for 16.7 million EUR in transactions for which the cash was received but the system was updated late.

Combined, these inefficiencies represent significant working capital trapped in the process. The workflow takes 25.2 days from sales order to accounts receivable item clearing. SAP Signavio identifies three specific blockers: payment terms inconsistency between billing and accounts receivable, invoices posted without proper blocking, and items stuck in incomplete status.

A combined 9.4 million EUR sits in receivables either overdue or paid late with discounts. This is not an estimation. This is what the system shows is happening. None of these are people problems; they are workflow design problems. The system creates the delay.

From Estimation to Evidence

Traditional business cases estimate: "We probably lose millions to late payments."

SAP Signavio findings are evidence: "We lose 9.4 million EUR to overdue and late-paid items. The workflow has eight identified blockers. Removing them would tighten the 25.2-day cycle and recover this cash."

The business case moves from aspirational to grounded in what your system is actually doing.

12.5 How the Autonomous Enterprise Bridges the Gap

Understanding where the value lives is one thing, but capturing it is another. Traditional approaches require organizational redesign: Hire more collectors, add collection analysts, create dispute resolution specialists. This approach costs money and takes time. The Autonomous Enterprise captures the value by removing transactional burden, not adding people.

Consider the first bottom-up issue: reducing overdue open items from 30 to 10 days average.

In a traditional model, this requires the collector to manually review all open invoices each day, identify which are aging beyond target, and send follow-up communications. This is routine work—necessary, but routine.

An agent operating autonomously does this overnight. It reviews all open invoices. It identifies which are aging beyond policy (e.g., more than 10 days unpaid). It checks customer payment history (Is this customer normally slow, or is this unusual?). It prepares a follow-up communication customized to the customer's situation. For routine customers with unusual delays, the agent sends an automatic follow-up. For customers with consistent slow payment patterns, the agent prepares the context for the collector to handle.

The collector arrives to find that routine follow-ups are already done. The focus is on customers who actually need a conversation.

As a result, overdue open items reduce from 30 days to 12 days without adding headcount. The same collector now handles more accounts with better targeting.

The second issue is reducing dispute resolution time. In a traditional model, disputes flow between teams—customer service, finance, operations—through email and meetings. Each handoff takes time.

An agent operates differently. When a dispute arrives, the agent classifies it: Is this a delivery issue, a billing issue, or a quality issue? Based on the classification, the agent routes it to the appropriate team. It attaches relevant supporting documentation. It flags any historical context (Has this customer disputed before? Is there a pattern?). For disputes that match standard resolution patterns, the agent proposes resolution. For complex disputes, the agent presents the full context to the human.

As a result, disputes that previously took 15 days to resolve now take eight days without additional headcount. The same people are working on preclassified, precontextualized issues instead of managing the routing.

The pattern repeats for each bottom-up issue. The agent removes the transactional sorting and routing work, while humans focus on the exceptions and relationships. The traditional efficiency story is "Do more with less": faster processing, shorter cycle times, more work per person. The organizational effectiveness story is different: "Focus on what actually matters."

When transactional work is removed, the organization's workforce changes fundamentally. Not in size, but in what they do. Collectors spend their day on collection conversations instead of data entry. Planners spend their day on strategy instead of data gathering. Finance teams spend their day on analysis instead of reconciliation. The same people become more effective because they are focused on the work that requires judgment.

This has cascading organizational benefits. People are more engaged when working on judgment calls rather than processing transactions. Turnover decreases, productivity increases, and quality improves because human attention is on what matters.

From a CFO perspective, this is compelling. You are not arguing for additional headcount but arguing for more effective use of existing headcount. The cost of autonomous capabilities is lower than the cost of hiring additional people. The retention benefit is higher and the quality improvement is measurable. For additional details on the processes and agents discussed in this section, refer to Chapter 5.

12.6 Building a Value-Based Roadmap

The business case should not stop at identifying the problem. It should sequence the solution so that each quarter delivers measurable value while building toward the larger outcome.

As shown in Figure 12.4, a typical roadmap looks like this:

- Q2 2026 focuses on the foundation: standardizing accounts receivable disputes and collections processes. SAP Signavio findings showed the workflow was inconsistent; this quarter aligns it. The expected value is 0.5 to 1 day DSO reduction. This is not transformational, but it is real and achievable in 12 weeks.
- Q3 2026 adds automation. Cash application automation eliminates the manual data entry that SAP Signavio showed was creating delays. SAP Business Data Cloud provides working capital intelligence so that teams see the metrics in real time. The expected value is a 0.3 to 0.7 day additional DSO reduction. Combined with Q2, we are now 0.8 to 1.7 days better.
- Q4 2026 brings the agents online. Embedded AI and predictive insights begin operating. Accounts receivable collections agents start handling routine follow-ups. The workflow that previously depended on manual daily reports now operates continuously. The expected value is one to two days additional DSO reduction.
- Q1 2027 focuses on optimization and rollout. The system is now mature enough to extend beyond the pilot. The value continues, with 1.5 to 2.5 days additional improvement as teams learn and refine.

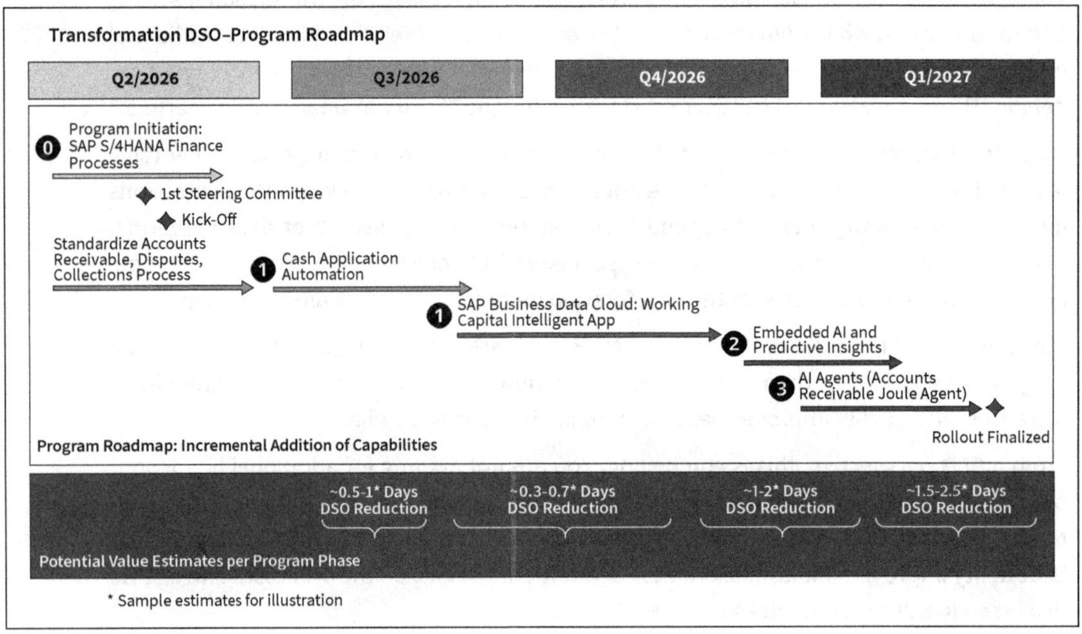

Figure 12.4: Outcomes-Based Roadmap

This is not a big bang transformation in which you invest heavily and hope for a return in 18 months; it is an incremental value realization. Each quarter delivers measurable benefit, and each quarter funds the next. Teams see progress, and leadership sees return on investment happening in real time.

When Q2 delivers 0.5 to 1 day of DSO improvement, that is 1.5 to 3 million EUR released immediately. Q3's additional 0.3 to 0.7 of a day of improvement is another 1 to 2 million EUR. By early Q4, you are 1.8 to 3.7 days better, with 2.5 to 5 million EUR in working capital recovered, and the organization has not waited 12 months to see it.

This is how transformation value is actually realized. Not in theory, but in quarters, with specific milestones and measurable outcomes for each step.

The Journey Starts Here

Q1 2027 is not the end. It is where continuous improvement begins. The baseline is now clean, the process is now standardized, and the data is now reliable. From this point forward, the organization can see opportunities that were invisible before.

New use cases emerge. The same process intelligence that improved receivables can be applied to payables, to procurement, to order-to-cash. The agents that learned to handle collections can be extended to handle disputes, to predict customer churn, to identify early payment opportunities.

The $27 million working capital improvement is not the destination but the foundation. Every improvement that follows builds on this one. The value does not stop in Q1 2027; it accelerates from there.

12.7 From Rotterdam to Outcomes

The Rotterdam story illustrates the fundamental constraint that autonomous enterprise addresses: the gap between signal and action. The gap exists everywhere. It costs money. The business case is the monetization of closing that gap.

The strongest business cases connect three things. They start with a visible outcome gap (DSO too high, inventory too high, equipment downtime too frequent). They trace that gap to specific process issues (overdue items, slow disputes, inconsistent collection). They show how removing transactional work changes what humans can focus on, making action faster and smarter, as shown in Figure 12.5.

The result is higher organizational effectiveness, not just lower cost. The same team, focused differently, delivers fundamentally better outcomes. The business case that shows this is the one that ends up funded because it shows something leaders already believe: People are the constraint, not processes. Removing transactional work and focusing on judgment calls makes organizations better.

The value-based roadmap then shows how to sequence that value expansion, year by year, building from high-confidence improvements to transformational ones. This is how Autonomous Enterprise stops being a project and starts being a capability. The value is not hidden; it is late. Most enterprises have the cash, the inventory, and the data. They just cannot act on it fast enough. By the time the signal gets through the systems and someone responds, the moment has passed: The payment is already overdue, the inventory already stocked, the equipment already failed.

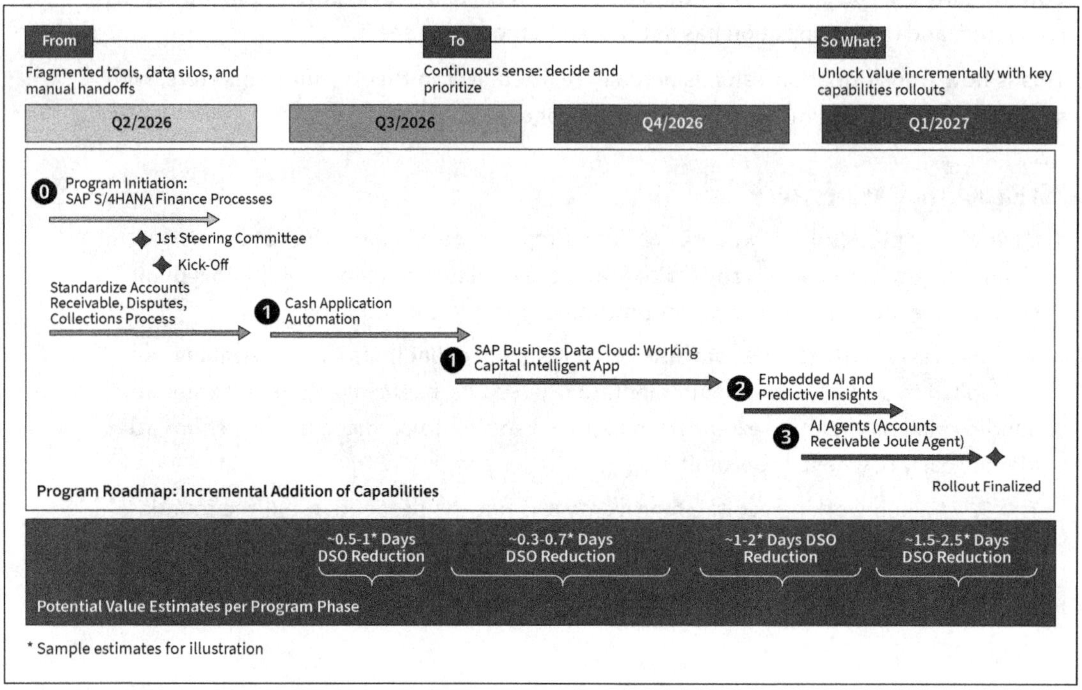

Figure 12.5: Business Case Narrative

The business case is about finding those delays, measuring what they cost, and fixing them. Not through more people or more dashboards, but through changing what work is actually done by humans versus what is handled automatically.

When that happens, the organization does not just work faster. It works at the right time. Decisions happen before problems compound, cash moves before it gets stuck, and inventory adjusts before it becomes waste. The $27 million was never lost; it was just sitting in the wrong place for too long. Fix the timing, and it shows up.

12.8 Summary

This chapter examined how to develop a business case for the Autonomous Enterprise by focusing on measurable business outcomes rather than technology initiatives. We explored how value can be identified by examining the gap between business objectives and current operational performance, and how that value can be quantified through both top-down and bottom-up methods.

The chapter introduced two complementary approaches to business case development. The top-down approach begins with business outcomes such as working capital improvement, service levels, operational efficiency, and financial performance. The bottom-up approach examines the individual processes that influence those outcomes, identifying delays, bottlenecks, manual activities, and inefficiencies that create measurable business impact. We also explored how process analysis tools such as SAP Signavio can provide the operational evidence needed to support business case assumptions and establish a baseline for improvement.

Finally, we showed how business value can be translated into an outcomes-based roadmap that delivers measurable benefits over time. By connecting strategic objectives to process-level improvements and sequencing initiatives around business outcomes, organizations can build a business case that is both financially credible and operationally achievable.

Understanding where value exists and how to quantify it is only part of the transformation journey. The next chapter examines the transformation toolchain and the capabilities, platforms, and methods that organizations use to move from business case development to execution.

Chapter 13
Leveraging Tools for Transformation

Thus far, we've discussed the capabilities within SAP's Autonomous Enterprise to support an agile business prepared for a future of rapid changes and continuous movement. We've painted the vision, but one of the key gaps in the technology space is how to actually make that vision come to life. That is the foundation of SAP's agent-led toolchain.

At this point in the book, hopefully you are excited about what the Autonomous Enterprise means for your organization, and the power available to innovate. The next question should be: How does our business actually accomplish this and transform to embrace this new reality? The digital landscape is rapidly evolving. Becoming masters at continuous change is no longer simply nice to achieve; it's a business imperative for staying relevant. The introduction of agentic AI broadly across businesses and industries is unprecedented, and the organizations that use this to embrace change wholly are the most likely winners in this scenario.

Recognizing the gap between desire and ability to change is what led to the inception of Business Transformation Management and the agent-led toolchain at SAP. The agent-led toolchain is a selection of tools that are embedded with AI and collaborate with assistants to create an interconnected understanding of processes, applications, data, and people to accelerate and lower the risk of transformation initiatives. These tools enable clarity into the current state, alignment of a future state, and the ability to continuously change and adapt. Transformation management is fundamentally about moving from thinking of change as a part of a project to a capability embedded in the fabric of your organization. These tools also align with the SAP Activate methodology that SAP recommends for transformation programs to simplify the move to the Autonomous Enterprise.

SAP developed this toolchain offering after undergoing its own transformation, moving from a primarily on-premise software company to a cloud-based software-as-a-service (SaaS) organization, and the tools now offered as the agent-led toolchain were critical to SAP's own internal change journey. SAP still uses enterprise-wide implementations of these applications today. These tools have come together through a series of acquisitions, partnerships, and continued innovation and investment to support critical capabilities of transformation, including change management, quality assurance, value engineering, and more, across the key dimensions of people, process, data, and applications (see Figure 13.1).

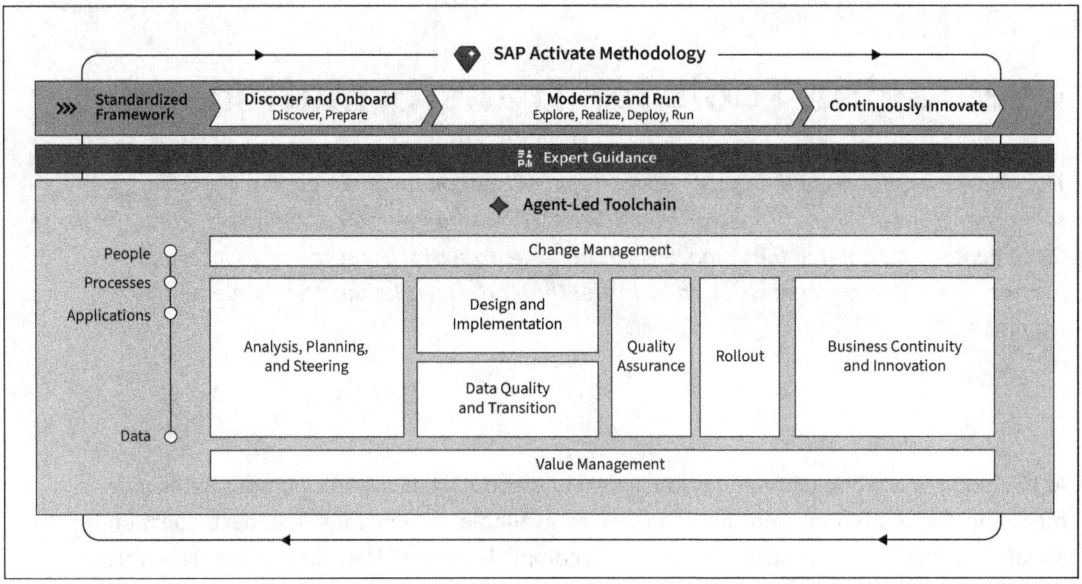

Figure 13.1: Key Dimensions of Transformation and Capabilities Supported Through SAP Activate Methodology

The key products we will introduce in this chapter as part of the agent-led toolchain are as follows:

- **SAP Signavio**
  Process management and mining for documenting, governing, and improving business processes
- **SAP LeanIX**
  Enterprise architecture management for visualizing and improving IT landscapes
- **SAP Cloud ALM**
  Application lifecycle management for requirements management and implementation delivery
- **Tricentis**
  Continuous automated testing for quality assurance
- **Syniti**
  Advanced data migration and management for data quality
- **OpenText**
  Content management for document management and access
- **WalkMe**
  Digital adoption for streamlined change enablement and conformance
- **Migration and modernization AI assistants**
  A collection of assistants that work in collaboration with this portfolio of tools for key tasks like test management, custom code analysis, and more

SAP Build and SAP Integration Suite are sometimes included as part of the toolchain as well, but they were already discussed in detail in Chapter 8. We will touch on how these components fit in with the other tools discussed in this chapter during the case study reviews.

After establishing what each of these tools are capable of alone, we'll then follow how Korindal's executives leverage the agent-led toolchain as they pivot your mindset to embrace the Autonomous Enterprise to reduce friction in your departments.

13.1 Process Optimization with SAP Signavio

The first tool we will explore in the agent led toolchain is SAP Signavio. SAP Signavio is a process management and optimization tool to help you understand, document, govern, and improve the processes within your organization. SAP Signavio is a modular solution, with two primary focus areas, process analytics and process management, with both collaboration and artificial intelligence embedded throughout the tool to act as a digital process consultant. Organizations that use SAP Signavio to identify process inefficiencies, design and implement solutions, and continuously change unlock an average of 21 million dollars in value. SAP Signavio can also be used to measure and communicate value potential and value realized as part of a transformation initiative.

On top of these capabilities is the Process Consulting Agent, a natural language entry point to interact with data within SAP Signavio. Here you can ask questions, synthesize data, and get actual recommendations for how to make process improvements. This is an accessible entry point for all users, even those without deep process backgrounds, to use the information within SAP Signavio. We will focus in this chapter on the underlying capabilities to build out this agentic AI experience.

In the following sections, we'll first describe analytics capabilities and then management capabilities; however, in practice, the right starting point will depend on each unique customer situation. In some cases, starting with analytics and leveraging plug and gain can accelerate a customer's process modeling journey. Other customers have well-documented processes and highly custom data, so starting with modeling might be a faster path to value. Because of the modular approach with SAP Signavio, the roadmap for when to use each component is flexible.

Note

It is important to note that though we focus here on SAP Signavio for SAP transformations, all the tools we'll discuss are system-agnostic and can be leveraged across different systems and end-to-end processes within your organization, bringing to life a full digital twin of your business.

13.1.1 Process Analytics in SAP Signavio

One of two major capability areas within SAP Signavio is how to analyze processes within an organization using operational data. SAP Signavio Process Intelligence uses process mining to visualize the operational truth of how a process runs. *Process mining* involves connecting into sources of data within the business and connecting to change logs to understand the actual complexity of business processes. Once you understand the process, you can use this data to surface trends, inefficiencies, and valuable areas of conformance and improvement.

SAP Signavio Process Intelligence can be operated in two ways. The first is rapid analytics options developed for SAP products, which involves a templated view of your process landscape across multiple performance indicators, using over 50 years of ERP experience to surface benchmarks and critical metrics. You can also go much deeper and customize the analysis to cross multiple systems and custom processes, and thus reflect the specifics of your organization. These capabilities work best in conjunction; it's harder to dig for treasure without a map, and the rapid analysis provides directional information to point to where the most value can be unlocked in a business. Additional capabilities use plug and gain to move the data easily between the two approaches in order to balance speed and complexity and highlight the most valuable areas to transform.

We'll discuss both approaches in more detail in the following sections.

Process Intelligence for SAP Products

To get started in SAP Signavio Process Intelligence, first imagine having visibility into all processes running in SAP, with full access to metrics based on all of SAP's business, industry, and process knowledge, and benchmarks and recommendations within 24 hours and with limited setup. That is the first step in the analytics suite. After connecting through SAP Business Technology Platform (SAP BTP), you are provided with a view of up to 140 SAP processes (the number populated is determined by your actual usage of SAP), hundreds of performance indicators, and correction and innovation recommendations to begin to determine ways to improve your system. This analysis is intended to provide critical insights into where the most pain is in your processes so that you can go deeper.

You can drill into processes in two major ways. First, end-to-end processes allow you to drill down by process area: source-to-pay, lead-to-cash, and so on. Lines of business (LoBs) can be used to isolate processes that address a specific improvement opportunity, like identifying automation opportunities or reducing maverick buying.

Process flows and process performance indicators provide data about how your processes are running today, as shown in Figure 13.2 for the accounts payable clearing process. Each process is measured for a specific date range, optimized to showcase a meaningful time frame for that process to be executed in and, where possible, comparing your metrics to those of industry peers. A high-level process flow showcases the critical steps to complete the process selected, how documents move between these process steps, and the time between each step. Below this, a wide variety of blockers and ways to measure the process are identified. Blockers are color-coded to help you identify where there is the most room for improvement. Every box is clickable to another level of detail; you can drill into internal benchmarking, filter the analysis to be more relevant or remove extraneous data, or see the complete list of data included in the data pull.

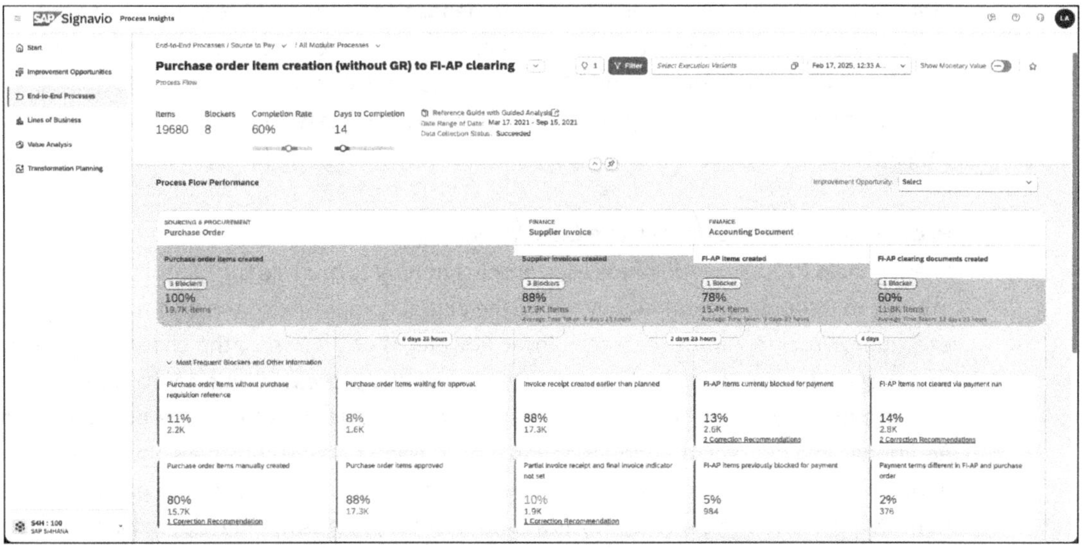

Figure 13.2: Example Process Flow from Process Insights, Including Key Metrics and Blockers

Many blockers are also tied to correction or innovation recommendations. *Correction recommendations* are tactical issues identified that can be fixed immediately in the source system. They are often related to old and open transactional data, poor master data, or underutilized automations. You can gain immense value from prioritizing and implementing these recommendations, each of which includes the exact steps to perform to execute the change. *Innovation recommendations* are broader and compare your usage of SAP today against all capabilities available from SAP, including SAP Business AI, SAP Fiori apps, and many other solution capabilities. This can help narrow down the scope for additional features needed from SAP and align the roadmap for continuous change to the most business value. These recommendations also serve as a useful tool as SAP rapidly releases new products, and they can help you understand what will be the most impactful for your organization.

Recommendations and metrics can also be tied to value cases to help prioritize improvement opportunities and build the business case to implement change. Value cases are calculated by identifying a metric to improve and determining a realistic improvement percentage. These can be tracked over time as new data is continuously added.

Prior to a transformation, you can also identify the most critical correction recommendations to execute prior to a system migration, usually tied to transactional data and current custom configurations. You can also tie all current SAP usage to the updated best practice documentation that is most relevant as a starting point for building a future state.

This is an excellent starting point in process analytics for companies running SAP. Connecting the ERP system to SAP Signavio is relatively easy and has extremely fast time to value, with only one to two days set aside to populate the data. It can provide a clear baseline of the processes being leveraged today and a way to measure change in critical performance indicators over time, as well as offer recommendations with rapid returns. The two drawbacks are (1) that though you can use filters to analyze the data, it is not configurable to specific customer situations, and (2) it is limited to SAP solutions. Both drawbacks are addressed by moving to enterprise-level SAP Signavio Process Intelligence and deeper process mining.

Process Intelligence Across the Enterprise

SAP Signavio Process Intelligence uses process mining to analyze the digital traces of each process in a source system and visualize the actual complexity of the end-to-end process execution. Process mining leverages three key pieces of data from the change logs of the source systems: a unique identifier to trace through the process, a timestamp, and an activity. With these, you can string together each unique variation of how the process is run. Additional attributes like supplier or customer, region, value, business unit, and so on enrich the change logs to help you derive impactful information from the process operations. Any system with a change log is a candidate for process mining, but most businesses start with highly transactional, high-volume processes like lead-to-cash or source-to-pay.

SAP Signavio offers a variety of value accelerators and data connectors across source systems to support companies that are beginning your process mining journey. These packages include key foundational templates for data extraction and identifying appropriate source tables, data transformation to translate system information into contextualized business terminology, and prebuilt dashboards around different value cases, including components like reducing rework or improving automation. SAP Signavio Process Intelligence is fully customizable, so most companies typically pick value accelerators as a starting point and add additional data, transformations, or dashboards to ensure the outputs are optimized for your unique business concerns.

For the SAP landscape, you can use *Plug and Gain*, an approach to rapidly analyze processes in the SAP landscape using full process mining, as a starting point to compare what you have already uncovered to what SAP recommends, running the process through a best run score as well as providing insight into custom transaction usage and how it impacts business performance. Prebuilt dashboards are designed to give meaningful insights as you move to the Autonomous Enterprise from legacy systems and to monitor and optimize your ongoing SAP usage.

As you create more unique situations or connect to more data sources, you can develop custom dashboards to fully analyze end-to-end business processes. First, there is a collection of process widgets to help visualize the different variants of the process. The most traditional process mining view is *Process Discovery*, where you can expand out all captured activities and paths to see the full complexity of all variant paths. We often call this a spaghetti diagram, and you can use it to visualize exactly how the process is actually executed, as shown in Figure 13.3. In addition to this view, there are additional widgets like Variant Explorer and Process Conformance that allow you to isolate specific variants of the process. Process Conformance also brings comparisons to process models, allowing you to compare the process that was designed, or a best practice model, to the actual execution of the process in the system.

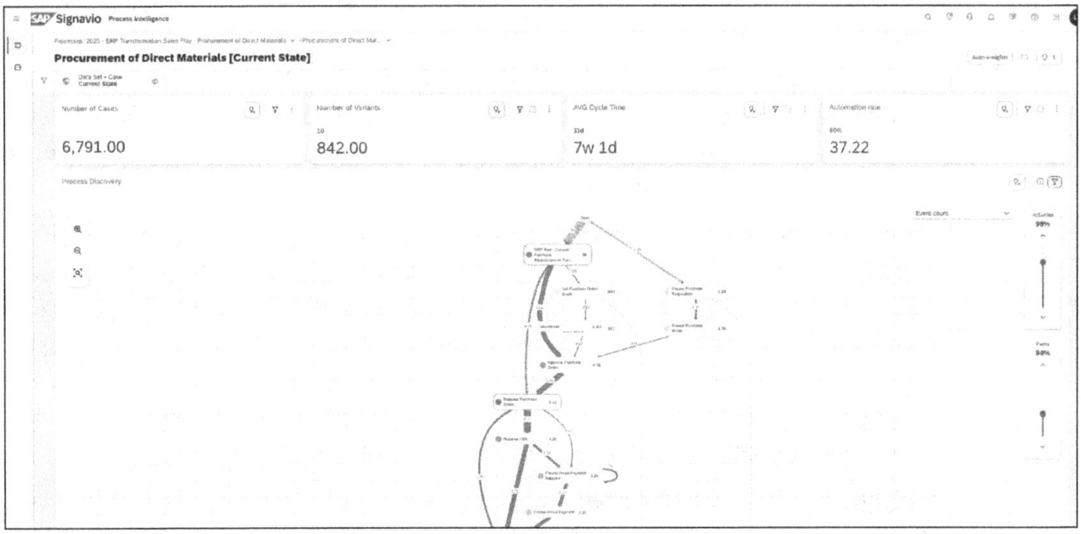

Figure 13.3: Example Process Discovery: Demonstrating Business Metrics and Complexity of Multiple Execution Paths in System

Additional widgets can be used to measure a variety of metrics. These can be created through out-of-the-box metrics provided in the value accelerators, drag-and-drop capabilities in Widget Builder, complex calculations and custom coding in the Signal Editor, or natural language conversation with the AI-powered Text to Widget. Building widgets through metrics, charts, graphs, and tables allows you to begin to isolate challenges within your organization and measure the impact that making specific changes could have on a program. Insights can be captured and shared across the team, as well as tied to initiatives in Transformation Manager (discussed in the next section) to collaborate on the most critical issues, assign owners, and track progress against key items.

This can be completed through manual analysis, but SAP Signavio also offers many ways to accelerate the process analysis. First, Auto Insights applies a strong correlation engine to surface widgets that highlight metrics affected by different dimensions. For example, it might highlight that there is a significant difference in fulfillment rate for a specific supplier or material used in the process. There is also the ability to use AI and prompt-specific metrics to surface correlations, such as by asking SAP Signavio to surface correlations around cycle time to identify patterns in the data. To take this one step further, AI-powered root cause analysis allows you to select a specific metric, specify how much you would like to improve it by, and set a series of attributes like date range, and the tool will analyze all possible correlations affecting that metric to provide the top contributors to good performance and the top detractors causing poor performance. This allows you to rapidly isolate where the issue might lie, be it a specific company code or region, a specific technology, or another avenue in which to explore improvements. This rapidly reduces the time to analyze your data and lets you move from identifying opportunities to implementing changes.

Executing the process improvements identified within SAP Signavio may take many forms, from retraining employees or leveraging a tool like WalkMe to support digital adoption, to making changes directly in the source system and fixing technical debt or issues, to building agents to support a process and full process reengineering. Within SAP Signavio, actions can be created around specific process triggers to generate email or chat notifications, trigger automation with SAP Build Process Automation (covered in Chapter 8) or other tools, or push information into APIs within the organization.

Data in SAP Signavio can be updated on a recurring basis with a direct connection to a source system. Many companies update your data daily for operational use and alignment. One benefit of this is the ability to measure the changes and highlight the value realized to the organization over time. This is also important as part of an ERP transformation program as it allows you to communicate the success by measuring performance before and after changing systems and thus justify the investment in the program. SAP Signavio's ability to measure operational performance is the ideal solution for achieving this.

For highly complex, interconnected processes, leveraging an object-based instead of case-based approach may be more valuable. Object-centric modeling and process networks let you look at the data in a different way and see how your end-to-end processes are interconnected in your organization. Although there are meaningful use cases in procurement or accounts receivable, for example, these processes are also interconnected to some extent, so depending on what case IDs you follow, you might not see the complete picture. Within SAP Signavio Process Intelligence, you can add business objects and case scopes to view the interdependencies between different processes. Looking at these processes as dependencies affecting one another helps break down silos within organizations and truly identify where in the business the real blockers exist.

One additional key component that lives within SAP Signavio Process Intelligence is the *Process Atom*, the smallest consumable piece of business context that exists in an organization. This might be something like "An order must go through two approvals if above $50,000" or "Whoever creates an order cannot approve an order." Historically, this type of business context and rule exists in lengthy company manuals, training documentation for employees to learn, or simply through organizational knowledge held by employees. Process atoms let you capture and organize all these business rules, and they can be created from process mining, derived from process models, or entered manually. They are governed in the same way process models are, and they are intended to be living components, so if data suggests the atoms are not accurate, updates and changes can be suggested automatically.

Process atoms are a key driver for enabling *Company Memory* in the organization, as well as feeding and monitoring AI agents. These bite-sized pieces of information are easy for agents to search against and ensure they are operating as expected within the organization, and you can then build agentic monitoring and conformance based on agents' adherence to atoms.

SAP Signavio Process Intelligence is the home base for all data-driven information about how processes are operating. It drives conformance and governance of these processes, and offers the correct insights to make meaningful organizational changes.

13.1.2 Process Management in SAP Signavio

SAP Signavio Process Modeler and SAP Signavio Collaboration Hub form a digital foundation for the designed processes in an organization. This acts as the source of truth for how a process is intended to be executed, and it is a valuable tool for onboarding, change management, governance, risk management, and compliance, enabling AI agents with business context and transformation to both understand the as-is process and design a more ideal to-be process. This work is classically completed by consultants as deliverables of transformation programs, but many of these become project artifacts. The goal with the process management tools within SAP Signavio is to use those artifacts as meaningful assets for a transformation and ongoing operationalization of processes as a discipline.

Collaboration Hub and Process Documentation

SAP Signavio Collaboration Hub is the starting point for most users within SAP Signavio. This home screen experience is configurable for different audiences and personas to manage access and how they interact with the solutions. The administrator of the system can create the click-through paths and configure any process hierarchy that makes sense for the organization or can get started by using out-of-the-box value accelerators to follow industry standards or SAP Best Practices as a starting point.

As users enter the view of a process model (see the procurement example in Figure 13.4), a few key pieces of information are visible. First, the process hierarchy indicates which process level the current process is at. Most organizations will document processes to Level 4 or 5, but there is no limit to the levels that can be supported. Revision history is also retained for all changes, including who completed the changes and when. This also allows for side-by-side comparison between different revisions of the process. The diagram itself includes swimlanes for who is executing the process and process tasks. SAP Signavio adheres to Business Process Modeling Notation (BPMN) 2.0 as its modeling standard. Additional diagram types like business decision diagrams, event-driven process chains, customer journey maps, organization charts, and others are also supported.

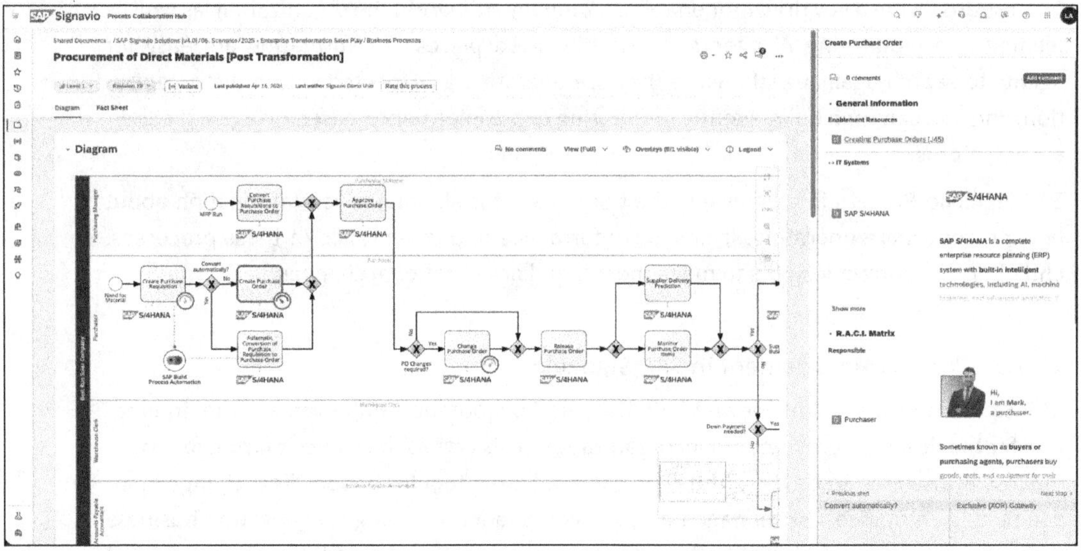

Figure 13.4: Example Process Diagram

Both the diagram and the tasks and elements within the diagram can be enhanced with attributes. Attributes are configurable to the organization, but the most common attributes customers start with are IT systems, roles, and risks and controls. Attributes add additional context and allow diagrams to be searchable and reportable by these fields. Attributes also can be visualized using overlays within the tool to enrich the process documentation.

Although not required, many attributes are tied to dictionary elements. Dictionary elements are reusable pieces of information stored in a central repository. When users use the dictionary, it ensures that all instances of, for example, a specific role like procurement manager are identical, and all usages of that item are linked together. Additional context for that component can be added or updated once through the dictionary and thus cascaded to each of the diagrams. The dictionary also can be populated through connectors and APIs. For IT systems, those leveraging the full agent-led toolchain can use your application fact sheets in SAP LeanIX to populate this dictionary element. The source of truth lives in SAP LeanIX, and SAP Signavio uses that information as needed to document where those IT systems are connected to a process. (For more on SAP LeanIX, see Section 13.2.)

Users can also leave comments on and add tasks to process maps. This real-time collaboration allows process modelers to collect feedback in the tool and encourages these process diagrams to be living assets within the organization, rather than static artifacts as part of a project deliverable.

Finally, process models can be enriched with data that comes from process intelligence. Live insights can be linked to widgets to display actual performance for respective processes and tasks alongside the process design. This is an excellent tool for reporting, tracking, and creating executive dashboards to communicate the intention of a process along with the actual execution all in one view.

Process Modeling

SAP Signavio has a robust process creation and reporting backend to enable process designers as they construct the process landscape in your organization.

To begin developing processes, there are several options for users. The AI-assisted SAP Signavio Process Modeler is the simplest starting point, via which you can use natural language, voice, images, standard operating procedures, and more to create BPMN 2.0–compliant process diagrams, add attributes, and iterate process models. Existing processes also can be imported from Visio or other tools in a BPMN format, and process models can be created in the design tool as well. Finally, process documentation can be imported from the Value Accelerator Library to access SAP Best Practices content and other partner- and industry-related processes as a starting point for design. Process models can also be updated manually with drag-and-drop capabilities, and SAP Signavio includes built-in checks for BPMN 2.0 compliance to drive process consistency.

SAP Signavio Process Modeler also includes a series of reports, like an IT system-usage diagram and a RACI (responsible, accountable, consulted, informed) chart that leverages the process attributes to generate these materials for change management and decision-making purposes. Users can also generate custom reports to better suit the needs of the organization.

Process Governance

As many users are engaging with a process, and this scales to a capability within the organization, it's important to have proper checks and balances in place to ensure consistency. SAP Signavio Process Governance is the built-in workflow tool that allows for embedded process approvals through the appropriate channels prior to publishing process changes. Instead of allowing users to publish new content directly, SAP Signavio can be enabled to push process changes through this approval process, which is configurable to fit whatever is required by the organization. This might require review by a manager, a process owner, a business owner, or a project manager. During each step of the approval, additional feedback or information can be collected through forms, and steps can be included to send emails and notifications to users or other teams or automatically publish new content when it completes the approval cycle. SAP Signavio Process Governance also supports time-based reviews for processes that require regular upkeep. As part of a transformation, this ensures all process content receives the required sign-offs prior to implementation, and all decision information can be captured within the comments in SAP Signavio Process Governance.

In addition to process approvals, SAP Signavio Process Governance can be initiated via a form or email. This is useful for collecting feedback and for performing other process-related tasks within the SAP Signavio platform.

Transformation Manager

SAP Signavio Process Transformation Manager is a solution within SAP Signavio that really ties things together for a program driving change within the organization. Within SAP Signavio Process Transformation Manager, the Transformation Advisor leverages AI and publicly available financial documentation to suggest areas of focus tied to metrics in SAP Signavio Process Intelligence. Programs and initiatives can be generated with AI to build out an entire project with tasks and activities to reach operational targets. An example of these tasks is provided in Figure 13.5, highlighting example initiatives for a procure-to-pay transformation, including its status, impact, and any related comments. Benchmarks are consolidated and can be tied to value cases for different insights, and all information collected across the different modules within SAP Signavio are streamlined into one view. SAP Signavio Process Transformation Manager is the orchestration layer to move from identifying areas of improvement to taking action in those programs.

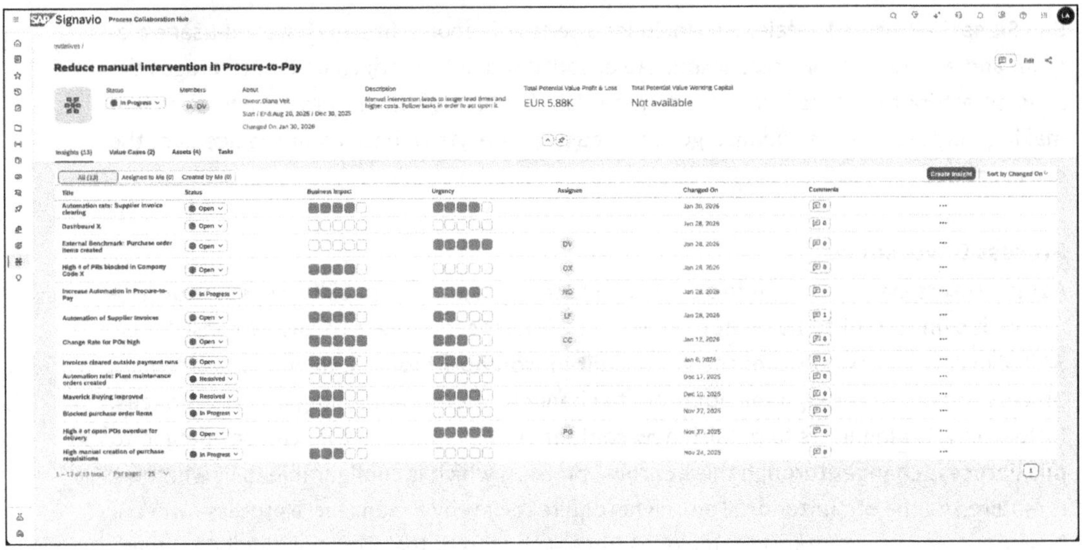

Figure 13.5: SAP Signavio Process Transformation Manager: Procure-to-Pay Example

13.2 Architecture Management with SAP LeanIX

SAP LeanIX is SAP's offering for enterprise application management within the agent-led toolchain. It is the market leader for enterprise architecture and supports customers in modernizing your IT landscape by providing transparency across the portfolio. Traditionally, IT organizations struggle to manage high complexity, concurrent projects, long implementation cycles, and high technical debt and maintenance costs, all issues that SAP LeanIX helps address. SAP LeanIX helps reduce IT spend, accelerate transformation, and reduce risk through three primary modules: SAP LeanIX Application Portfolio Management, SAP LeanIX Technology Risk and Compliance, and SAP LeanIX Architecture and Road Map Planning. SAP LeanIX embraces collaboration and transparency to provide organizations with usable data for your application portfolios and to bring together strategy and business capabilities for specific IT components.

To begin discussing SAP LeanIX, the metamodel concept (see Figure 13.6) is a critical foundation. SAP LeanIX includes an out-of-the-box metamodel that can be configured for specific scenarios, though most companies use it as is. It establishes the data collected in SAP LeanIX, sorted into consumable layers like strategy and application and data architecture. Data within SAP LeanIX is collected in fact sheets, and each fact sheet is defined by which component of the metamodel it addresses. The fact sheet for an application, which is the foundational piece of information within SAP LeanIX, will collect different key pieces of metadata than fact sheets for initiatives, for example. Different fact sheets can be collected, organized, and connected to represent the complete IT landscape, and they can branch out into different products and use cases.

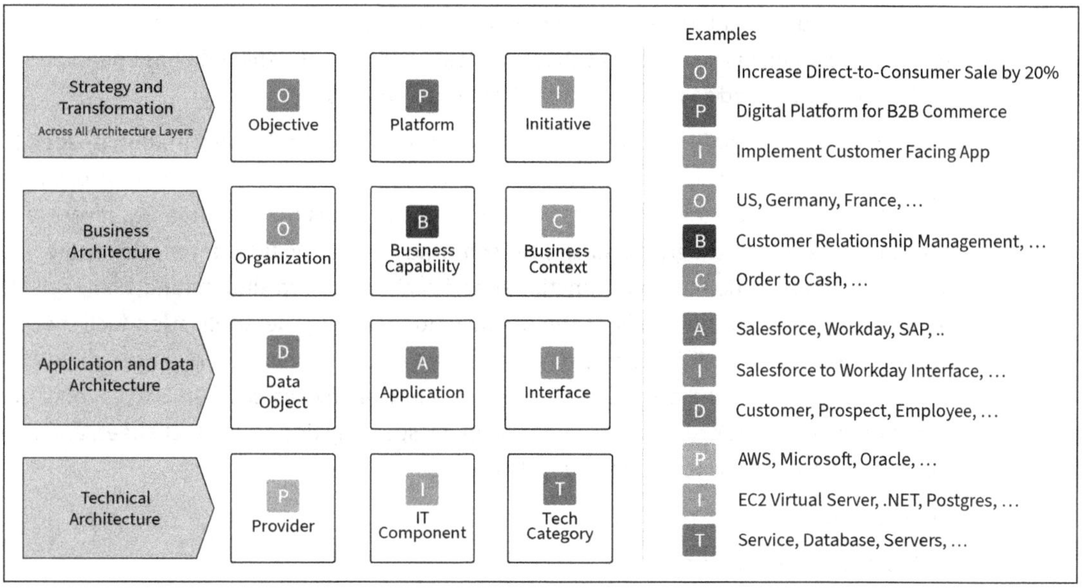

Figure 13.6: SAP LeanIX Metamodel, with Examples for Each Data Type

SAP LeanIX Application Portfolio Management is the first are within SAP LeanIX we will explore; it is the required SAP LeanIX foundation, with SAP LeanIX Technology Risk and Compliance and SAP LeanIX Architecture and Road Map Planning additional product lines to consider once the technology foundation is formed.

Now, let's take a closer look at each of the three SAP LeanIX modules.

13.2.1 SAP LeanIX Application Portfolio Management

SAP LeanIX Application Portfolio Management is a key starting point in SAP LeanIX to achieve some level of transparency in the IT landscape in your organization. In an ideal future state, this includes the entirety of the business, but many organizations start with one business unit, location, or other segment to start to add value quickly.

Once an organization establishes its metamodel and the fact sheet types it is starting with, the next step is getting data into SAP LeanIX. This can be done in a variety of ways, including leveraging AI-powered SaaS discovery, connecting to your configuration management database (CMDB) to import existing information, uploading from Excel, leveraging embedded AI capabilities and SAP landscape discovery to build inventory from connections to SAP Cloud ALM, and manual input.

Most companies will start with applications, business capabilities, and organizations to begin your inventory, as this allows for key reports to be generated and quickly add value. Data on fact sheets can be enhanced through surveys sent out to collect key information from identified experts on any given component and defined through subscriptions to each fact sheet, and they can be easily completed through teams, email, and other communication channels. Quality seals on fact sheets indicate how recently the data has been updated and can be configured to match organization requirements.

Fact sheets can be configured to provide an organization with the information needed for that fact sheet type, such as attributes, tags, and relations. Starting points for each fact sheet type are provided out of the box based on what fields most companies leverage. In the application fact sheet, for example, some key fields include descriptions, lifecycle stages, business criticality, and functional and technical fit to understand how well the application supports the needs of the organization. Linking the application fact sheet to related business capabilities shows how different applications support business goals. Having these few fields filled out supports a whole range of key use cases in SAP LeanIX. Tags are another way to capture key information in a visual way tied to the fact sheet, and relations offer another way to connect fact sheets.

There are three types of relations:

- **Hierarchical relations**
  These build out relationships between a parent and its children. This is great for creating clear structure, such as for regional breakdowns.
- **Generic relations**
  These versatile relationships connect any fact sheet types. They can be useful for required or required-by relationships and for establishing logical dependencies between fact sheet types.
- **Transitive relations**
  These relationships connect fact sheets indirectly and are not readily visible. For example, you could connect business capabilities together to show interdependencies between fact sheets of the same type.

Now that you have key information in SAP LeanIX, you can work through the first use cases for value creation.

Application Portfolio Assessment

Application portfolio assessment helps create transparency into the true, as-is state of an organization's applications. This is a key first step for any technology transformation, and that remains true as the starting point for an SAP S/4HANA Cloud transformation journey. You can use the outputs of the assessment to understand how the as-is applications support business goals, what functional and technical fit they serve, and where the applications are deployed across the organization. These can be analyzed via a variety of AI tools as well by using MCP servers to bring your data into any organizational interface. Multiple reports can be generated to understand the portfolio and begin to make decisions about where there are redundant applications, how to modernize, and what capabilities are the most critical to improve. In Figure 13.7 shows one example report that visualizes applications and how they support business capabilities. Further information is added here, including color-coding the apps by business criticality and adding your technical and functional fit in the bottom corners of the boxes. As you add more metadata to fact sheets, you can visualize these diagrams in more complex ways.

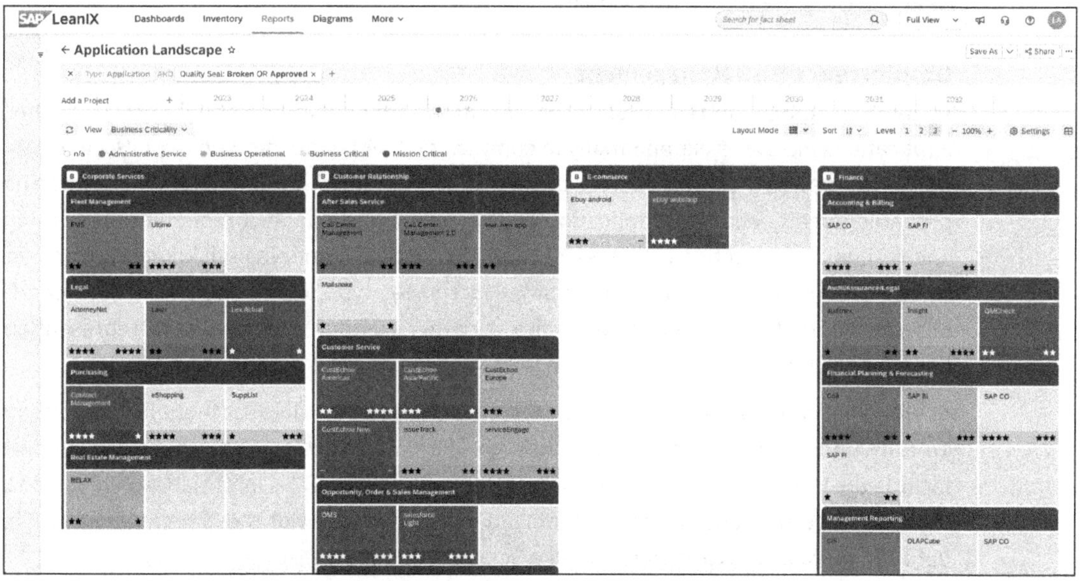

Figure 13.7: SAP LeanIX Application Portfolio Report

Application Rationalization

Application rationalization is the next use case that can be explored once you have key application portfolio information in SAP LeanIX. The goal of application rationalization is to reduce costs and streamline the current application portfolio to match key business outcomes like improving efficiency and reducing complexity. Within SAP LeanIX, you can use assessments like TIME (tolerate, invest, migrate, eliminate) and a variety of out-of-the-box reports to analyze IT spend and assess costs across a variety of dimensions. With this information, you can develop a roadmap for reviewing dependencies and decisions, and you can use SAP LeanIX to track the progress toward your goals and report the outcomes of the programs to reduce IT spend and redundant or noncompliant applications to executives.

Thinking about these two use cases in terms of an SAP transformation, these are critical first steps in developing the as-is landscape, evaluating both SAP and non-SAP systems and dependencies, and understanding the complexity of the upcoming transformation.

13.2.2 SAP LeanIX Technology Risk and Compliance

With a foundation of application portfolios, you can start to expand into more complex and beneficial use cases. The goal of technology risk and compliance is to enable your organization to understand the lifecycle, risk, and dependencies across your technology stack and make informed decisions about what risk to accept and what risk to address. Technology risk can cause major disruptions to an organization, including risk of system outages and obsolescence, additional risk of security incidents, added cost of ownership, and the additional cost of performing maintenance on systems that are outdated and potentially not optimized to support business growth.

Obsolescence Risk Management

As the technology landscape of an organization grows, managing the lifecycle of these applications increases exponentially in complexity. SAP LeanIX provides a platform to document and assess the impact across the tech landscape. At this point, your organization has probably already enabled application management and has at a minimum applications and business capabilities within SAP LeanIX. To enable technology risk assessments, you will also want IT Components as another fact sheet type within the tool. These can be imported directly from ServiceNow, Jira, or other custom integrations to represent the hardware, software, and configuration items used across the organization today.

As with application assessment, the next step is to enrich the data for these IT components. To fully understand the technology risk, the key attributes to add around the lifecycle include the URLs, support type, active support, and policies. This information can be added through surveys or automatically imported using AI and out-of-the-box capabilities. This data can then be used in other tools like ServiceNow to increase data quality.

Once this data is established, a plethora of reports are available to aggregate the risk and assess it against the business capabilities and the applications they support. This visualization exercise simplifies the understanding of the landscape, allows enterprise architects to recognize and prioritize more urgent and disruptive risks to business continuity, and allows you to make and communicate decisions and roadmaps that address the risks identified. Migration efforts can be tracked in roadmaps and monitored over time.

As part of a transformation project, understanding IT component risk and obsolescence helps to clarify what components of the integrations need to be upgraded, and visualize in the long term where additional change will be required. For certain systems and components, upgrading them during an SAP S/4HANA transformation will help reduce long-term risk and ensure the foundational building blocks underlying the systems are prepared for the future state.

13.2.3 SAP LeanIX Architecture and Road Map Planning

With key information now documented about the entire IT landscape, you can use all this data to build out robust transformation plans and execute against a future state technology roadmap. With SAP LeanIX Architecture and Road Map Planning, you can use templates and various scenarios to model a future state, understand all the interfaces and underlying component impacts of change, and effectively communicate and monitor the entire process. In Figure 13.8, we highlight the impact the changes will have on the overall portfolio, as well as how these impact the related business capabilities and the timeline for the project. These views are completely configurable to match how you want to communicate change in your organization.

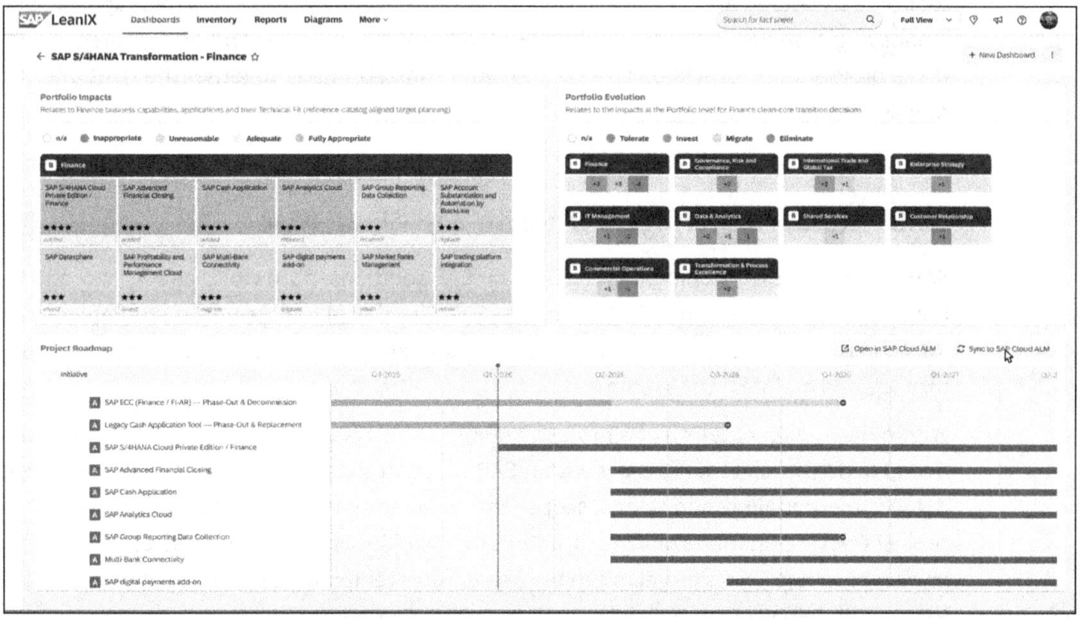

Figure 13.8: SAP LeanIX Architecture and Road Map Planning

First, you can set up initiatives within SAP LeanIX based around major transformation projects and align these with prebuilt templates for the transformation. This will import key milestones and information to consider. You can configure the transformation to understand the project impacts and include reports and dashboards to visualize elements of the current state landscape and dependencies. This is a particularly important step for ERP transformations to determine the impact the program has on additional systems and integrations across the organization.

Then, you can define the target state. As you start the transformation, you might want to compare multiple target states and evaluate which has the most potential to achieve the business results you desire. These what-if scenarios might be as high level as the impacts of a public versus private cloud setup for your organization, or as granular as specific module decisions. You can also leverage AI to make recommendations based on your landscape data. These visualizations allow your team to capture the critical information and make data-driven decisions together.

Once a transformation scenario is set, it can be used to build out the transformation roadmap with milestones and key impacts. There are many templates here to accelerate the build of these components. Projects can be tracked over time to ensure you remain on schedule, and as you complete components of the transformation, executing them in SAP LeanIX automatically updates your architecture to the future state you have designed.

Critically, transformation planning milestones and activities can be linked within other tools like SAP Cloud ALM to track key information between the tools. Progress against milestones might be documented within SAP Cloud ALM as the build is being executed, and all those project statuses will be reflected in SAP LeanIX, bringing together the right information within each tool.

13.3 Lifecycle Management with SAP Cloud ALM

SAP Cloud ALM is the deployment cockpit for SAP solutions, both cloud and hybrid, and the home base to manage your SAP solutions. SAP Cloud ALM handles both the implementation component of application lifecycle management and the operations after go-live to manage and maintain the systems.

Within the implementation of a project, SAP Cloud ALM is a self-contained tracker for what needs to be completed. A *project* within SAP Cloud ALM groups together tasks, test cases, defects, methodology and teams, scope, and requirements into one landing page, making project tracking seamless and giving additional visibility across teams to all elements of the implementation. These projects can be created with templates to easily apply the SAP Activate methodology or configured to match specific scenarios.

Within the domain of a project, one of the most critical elements of SAP Cloud ALM is operating as the source of truth for all design decisions and requirements. As functional design occurs through fit-to-standard workshops, SAP Cloud ALM is the digital source of all configuration decisions. SAP Cloud ALM can import business processes from SAP Signavio, as well as SAP Best Practices, to ensure all business context is captured throughout the design. Then you can build your requirements against the business processes and break these down into user stories and tasks that need to be completed. This complete view is shown in Figure 13.9, in which an accounts receivable process is imported through SAP Signavio, and the first requirements and user stories are tied directly to the process tasks. Governance of requirements and the decision-making process is built into the tool, and each task can be grouped into deliverables to measure meaningful outputs. Each deliverable can be tracked across the team, and all statuses are available, as well as the interdependencies between different tasks.

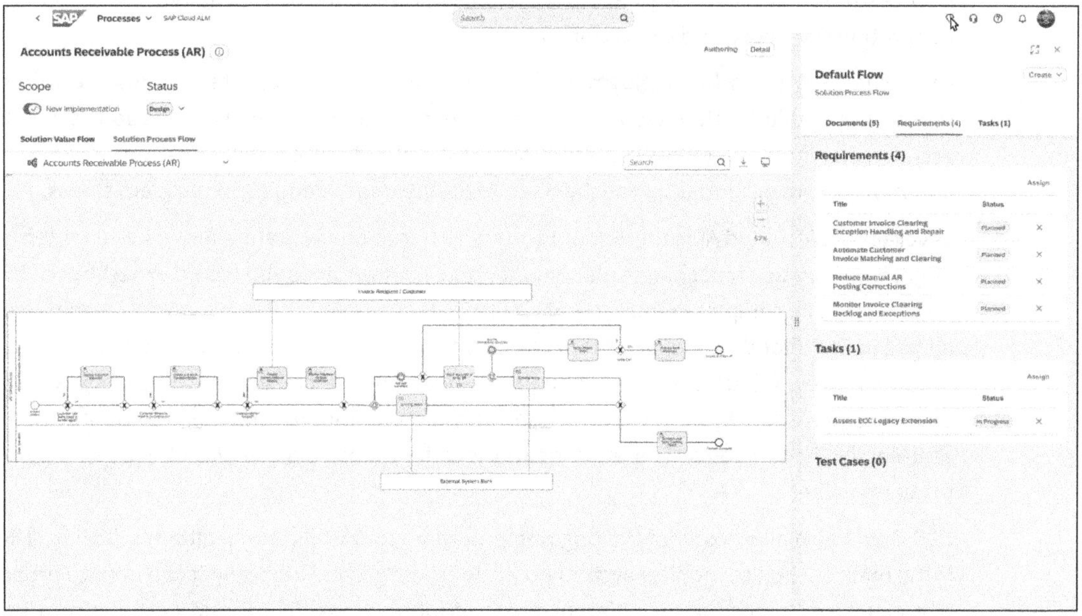

Figure 13.9: SAP Cloud ALM with Business Process Imported from SAP Signavio and Related Requirements

Once all requirements and deliverables are approved and assigned, this is the key place to track progress as well. Your build team can use SAP Cloud ALM to track your work and understand who is working on what. Project managers have visibility into any blockers in advance.

Requirements also can be tied to test cases, and manual testing is supported within SAP Cloud ALM directly. Each test case created in SAP Cloud ALM can be traced back to the original business process, providing traceability throughout the test phase. As test cases are executed, defects also can be tracked directly in the tool and monitored for resolution. Development of test cases is not limited to SAP solutions; it can be traced through the end-to-end process as well, including integrations between SAP and non-SAP solutions.

Testing can be further improved by leveraging additional solutions like Tricentis (Section 13.4) that specialize in automated testing. These integrate directly into SAP Cloud ALM, which continues to act as the orchestration and reporting layer and is the de facto single source of truth for all testing statuses as part of the implementation program.

As the solution is built and tested, the final step in the implementation program is to deploy the solutions. SAP Cloud ALM is the change engine to move features into different environments and track the deployment status of each requirement. Change managers can move from huge, big bang go-lives to incremental value and continuous feature delivery, giving much more agility to the program to begin delivering usable software. Change workflows are built into the tool, providing the appropriate framework for seamless management of new feature releases through a variety of mechanisms, including change and transport, adaptation transport, and cloud transport.

Overlaying key tasks for deployment from the design, build, and test teams are robust reporting capabilities that keep everyone on the project up-to-date with the status, blockers, defects, interdependencies, and more. This visibility streamlines the entire program and moves from outdated spreadsheets to real time understanding of program status.

Leveraging SAP Cloud ALM throughout the transformation program is also a key enabler for ongoing innovation and change management. Key deliverables you would expect out of the build phase like complete requirements, test cases coverage, and deployment mechanisms all live in SAP Cloud ALM in a digital ecosystem, which is now usable for ongoing transformation. As new features or capabilities are requested, you can assess the change impact and gaps more rapidly, recycle test cases and identify the net change, and deploy more seamlessly. This enables IT operations to move in a more agile way and deliver ongoing business value.

SAP Cloud ALM also is part of the ongoing program operations and health monitoring. This is the main cockpit to monitor system exceptions, usage, and jobs, and it can initiate proactive alerts and reporting for different types of interactions. SAP Cloud ALM is also a key home base to understand the full breadth of SAP usage and proactively monitor any performance gaps or issues. This extends into monitoring agent deployment in the SAP landscape and understanding any outages.

Overall, SAP Cloud ALM is the home base for deployment of SAP solutions throughout design, build, test, and ongoing monitoring. It provides organizations with a single source of truth and clear traceability from requirements through testing and go-live, and it enables agile continuous change for an organization on an ongoing basis. This source of truth is critical for the era of rapid change and for IT to meet the expectations of rapid and continuous innovation.

13.4 Testing Automation with Tricentis

Anyone who has undergone a system conversion knows testing can be one of the most painful steps. Executing testing, and designing, developing, executing, and communicating the fixes required with a build team when there are errors can be extremely time-consuming; as a manual process, it can be difficult to track and ensure complete coverage of the expected use cases. Testing is a point in a transformation at which the process, application management, and business outcomes converge, meaning there are a lot of perspectives to keep track of during this critical transformation step. This is where Tricentis comes in to support testing automation both during and after a transformation.

Tricentis integrates into the toolchain through SAP Cloud ALM with a complete platform for SAP application testing. The key components of this process include change impact analysis to understand what needs to be tested based on previous results and changes to the application; continuous test automation to ensure seamless regression testing with limited intervention; and performance testing to ensure no technical components or expected capabilities are lost. Tricentis can be used across SAP and non-SAP systems, with the focus here on testing as part of the SAP transformation.

Tricentis has innovated in the testing space by shifting from code-based testing to model-based testing, making it more intuitive to create, execute, and monitor testing progress. This also brings business users closer to test scenarios by making these scenarios easy to read and understand. Tricentis integrates into the toolchain through SAP Cloud ALM, allowing test cases to be built against the requirements and use cases documented. Tricentis allows you to create automations of test scripts that can run multiple times and to incorporate testing into third-party systems. As tests run, your status is synced back into SAP Cloud ALM for continuous monitoring and reporting purposes. Within Tricentis, it's also possible to drill down further into defects and understand where the process errors occurred, accelerating time to repair. Figure 13.10 demonstrates the same accounts receivable process highlighted throughout the examples, with the high-level business flows visualized for testing; each task has an indication of whether it has been automated for testing.

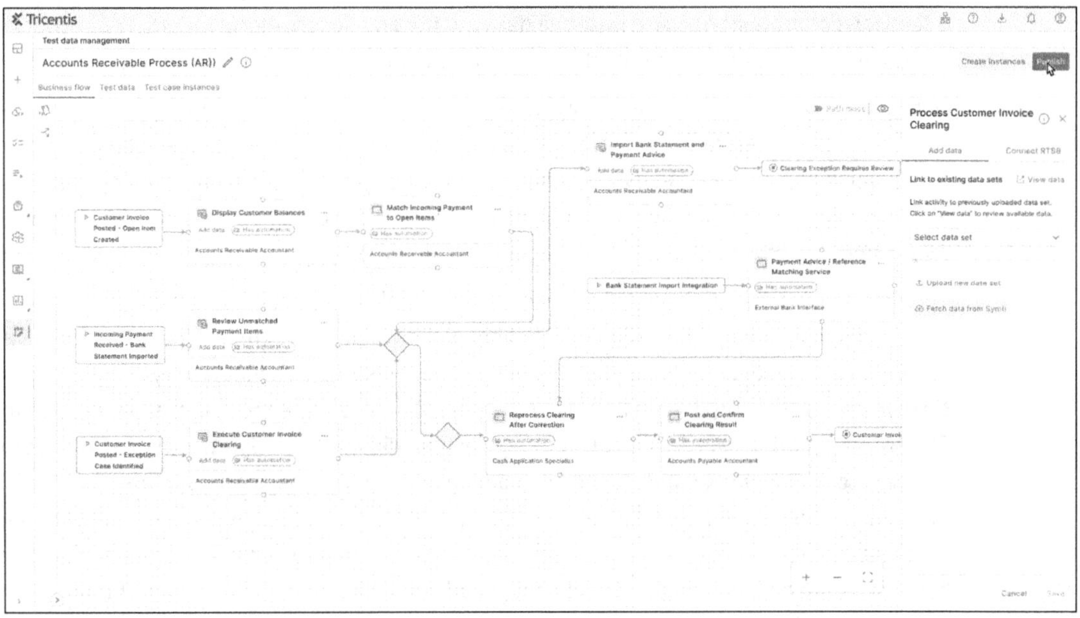

Figure 13.10: Tricentis Testing Coverage Linked to SAP Cloud ALM

After you complete all the initial testing setup during a transformation, you have an outstanding foundation for continuous testing after the transformation. Tricentis's change impact assessment can sync to read ongoing changes in the system from SAP Cloud ALM and automatically suggest which test scripts need to be run again, see where there are gaps, and execute the test automation for the new functionality. All the work from the transformation builds the muscle of change and establishes key capabilities that will benefit the organization on an ongoing basis, allowing for speed, agility, and scale as well as lowering the risk from all future changes made to the system.

13.5 Data Migration with Syniti

SAP Advanced Data Migration and Management by Syniti is part of the agent-led toolchain. It offers data quality, data cleansing, and data migration capabilities to your transformation efforts, leveraging AI and machine learning (ML) capabilities to migrate faster and smarter. Many organizations understand that leveraging your data is a strategic asset to transformation, but a high proportion of organizations report low to medium levels of sophistication in your data priorities. The data problem is not easy to solve. Data in organizations lives in a variety of systems, with varying levels of quality assurance applied to those systems. Moving any data from one system to another requires a deep understanding of both systems, along with the business context to make that data relevant and meaningful. Data migration and mapping can be one of the most difficult aspects of a system transformation, and that's why the solutions offered by Syniti not only help to accelerate the migration component but also improve data quality and access along the way, contributing to less risk and the ability to unlock value from the data already within your organization.

During a migration activity, the following key activities occur: understanding the data in the system today, understanding the quality and financial impact of any data issues, conducting cleanup and mapping activities, and ultimately migrating data to the new system. Ideally, the migration is what we call a boring go-live, in which there are no major issues to report because of a well-supported process during the project.

The first steps within the migration project are defining goals and leveraging them in the collaborative platform to connect to data assets as they are brought into the tool. This gives clear guidance and a vision for the rest of the activities and connects everything together. Next you need to bring data in as part of the migration. Here the focus is the SAP system for ERP transformation, but Syniti can connect to over 160 systems. Syniti is also a great complement to SAP Business Data Cloud and a home base to manage information from multiple data sources converging and adding context to that data. ML and AI capture data and automatically guide you to analyze it in your system today for both structure and quality based on both previous actions and best practices. A firm understanding of the quality in the system today is important to build a baseline for the effects of the transformation and to make key changes prior to moving data to the new system. There is also the opportunity to make small data changes today to add value immediately.

The next step is to start to design the mapping for the target system. In this step, you design, enrich, and transform the data to make it ready for migration. Business users can support establishing data policies and rules embedded within the tool, which are then applied automatically across systems as part of the mapping activities. The IT team takes over from here to profile the data that was collected and define rules for transformation into the target system. All decisions are documented directly in Syniti to ensure traceability. The mapping and transformation can be accelerated with automatically generated code that can be validated once it is transformed for the target system.

At this stage, the data is ready for migration. Syniti supports simulation of the load into the target system to conduct tests and checks for any data issues or irregularities that still exist. This provides confidence that the data is prepared and accurate prior to loading it into the new system. Once the simulation has passed all quality gates, the data is ready to be migrated into the target system automatically.

All of this is managed by leveraging advanced analytics and project management capabilities. Tasks can be divided among team members, forming sections of work that are easier to schedule and plan for. Syniti provides visibility into project timelines, completion rates, and more across these units with customizable views. This allows for granular reporting and the ability to react quickly to deviations from the plan.

After migration, Syniti has the foundational information to continue to manage data quality within the organization. The SAP data brought in during the migration can be enriched with data from other systems. Lessons learned from the migration can be applied to other system transformations within the business, and continuous monitoring of the data ensures that no new issues arise as users begin adopting new processes and that a high level of data quality is maintained. This digital record of data policies, procedures, and decisions sets a foundation of data excellence across the organization.

Beyond data migration, Syniti also integrates with Tricentis as a source for clean, available testing data. Syniti can sync with Tricentis to support identifying and managing data, ensuring realistic test scenarios but also leveraging data that matches the scenario defined. This partnership between the two tools further amplifies the benefits.

Note

We focused here on how data migration and management is useful for a large system transformation, but there are countless other applications of this as well. Mergers and divestitures is one example, in which you are combining or decoupling data from two different organizations; streamlining this process can reduce the time needed to complete these activities.

13.6 Content Management with OpenText

OpenText within a transformation program is the source of record for document management both in and outside of SAP, and it can be used to store and archive documents that are part of a process. Examples of documents that might be managed with OpenText include invoices, emails, spreadsheets, order forms, and more. Within OpenText, companies have a lower-cost option to store records and ensure compliance and risk management when migrating data to SAP S/4HANA.

The primary functionality of OpenText is to provide a folder-based repository, at a much lower cost, for storing documents. This can connect documents to business objects within SAP so they are easy to store and find and allow for quick annotations and edits as they are encountered in a process. For many companies, this use case is ideal for moving older data stored on premise that is required to be available from a regulatory and compliance perspective, but is not necessarily needed in day-to-day interactions within the future-state ERP system. This is critical for regulatory requirements, future audits, and more. Within OpenText, you can create a folder structure to find and manage documents and then load content into it from both SAP and non-SAP systems.

During the transformation project, you will need to move old transaction documents into an archive. OpenText offers a solution to maintain these records without having to move this data directly into SAP Autonomous Suite. This makes storage easier and means the actual data cutover is done on a lean database, without outdated and cumbersome records involved. Both SAP Signavio and Syniti can help identify documents that can be archived prior to a transformation project.

OpenText also supports document management for non-SAP systems. As part of the application rationalization activities conducted in SAP LeanIX, you may identify systems that will be decommissioned as part of the transformation, either because they are redundant or because the capabilities they support today will be supported by SAP in the future. Any critical documentation that is found today in those systems can be migrated and stored in OpenText for retention and future access.

Leveraging OpenText also promotes collaboration and transparency by having a single source of truth for documents across the organization. For any future migrations and upgrades, the structure is already in place for document management and storage.

13.7 Digital Adoption with WalkMe

No matter how great the processes and systems you design and deliver as part of a transformation, they are only truly as effective as the people using them. Over the last decade, the importance of people, change, and adoption has gained significant traction and is now considered an integral part of the program, and WalkMe is the leading digital adoption platform that accelerates that user journey even further. WalkMe sits on top of the applications in your organization, both SAP and non-SAP, and has two key components to drive transformation success: user analytics and metrics, and embedded user guidance, combined with underlying AI to understand software the way humans do. These capabilities in tandem, along with the rest of the toolchain, support faster adoption of a new application. This focus on getting users to use the new system in the way it is intended is one of the most important steps in realizing value from the new solutions, as well as reducing risk and compliance concerns.

WalkMe follows a fundamental data-action-experience methodology to deliver these improved adoption activities to consumers. Some of the key use cases in which WalkMe can help include improved onboarding, reduced need for support tickets, increased productivity, and, what we will focus on in this section, feature adoption and change management for new technology solutions.

13.7.1 Data and Analytics in WalkMe

As we've talked about with many of these tools, the first step in solving some of these challenges is understanding where the problems exist and how much reward is there for solving them. WalkMe offers a variety of data-driven capabilities to better quantify user adoption issues within the organization and develop a plan for how to close the gaps.

The first tool is WalkMe Discovery, which can build a view of the web-based applications leveraged by users to complete a business process. From this view, it's possible to find data like user count, how frequently users access the tools, exposure to WalkMe, and penetration. There are also capabilities to add in license optimization to understand the cost to the organization and other factors tied to the allocation of the software throughout the organization. This gives increased visibility into applications with high user interaction, and it augments data from SAP LeanIX for an application rationalization exercise to identify applications that are not being fully leveraged. There are also capabilities to identify where AI is being used to interact with these applications.

To gain an additional perspective about tool adoption, digital experience analytics gives more user- and task-level interaction data. Here, you can capture information about how end users are engaging with the system and on-screen guidance, where they get stuck, where they drop off or have to do rework, and how much time they spend on specific tasks. This is critical information to identify where there is the most friction and then identify actions to reduce this for end users. Workflow insights also shows where people become stuck or stalled and which interactions occur. Think of this like double-clicking into the user experience from the process analytics gathered in SAP Signavio. It gives additional insight into the root causes behind key process steps having a longer turnaround time, and it offers more granular details about the screen-level interactions driving friction.

The analysis is also supported by AI tools in UI intelligence to continuously monitor process analytics and help suggest the best actions to improve. UI intelligence focuses on data-heavy activities, getting down to the form level and helping identify fields and data entry types that are cumbersome, forgotten, entered incorrectly, frequently corrected, and more. This gives ideas for how to reduce time and risk in data-related tasks within the workflow.

All this data can be collected and visualized out of the box, and there are customizable dashboards available for different purposes, from an executive audience to change prioritization discussions. This gives everyone the information they need to act on the data.

As noted earlier, this data is complementary to the type of data you might collect about your applications in SAP LeanIX or your processes in SAP Signavio. Taking process-level data as an example, SAP Signavio gives end-to-end process-performance metrics, focusing on key system activities, while WalkMe can provide an additional detailed perspective to understand exactly how many clicks it takes for a user to complete a task or where they have to manually enter a lot of information, making that task take longer. We explore how these tools all come together in Section 13.8.

Adoption data becomes critical as part of a transformation to support both change planning and prioritization, showing where users are going to experience pain points in the new process and measuring success and planning for continuous improvement post go-live. Getting user-level data about tool adoption allows you to report and communicate success and to realign and pivot where needed to ensure users understand and are leveraging updated applications and processes. This feeds into the loop of continuous innovation and helps continue to train users.

13.7.2 Platform and Experience in WalkMe

Providing built-in system guidance to end users is what WalkMe is primarily known for, supporting better adoption of software solutions. This goes to the action and experience part of the methodology: what you can provide end users, in the moment of change or friction, to ensure the desired outcome in a seamless way. WalkMe is like having an extensive change network available in your organization. Before a digital adoption tool, you may have relied on classroom or self-paced training and augmented that with a group of super users, champions available to help continuously train your colleagues on a new tool and become experts. WalkMe enables every single user in the organization to work alongside a super user, offering digital guidance for every step of a new process as they learn and referenceable material at any time in the future.

As you build workflows in WalkMe, the primary starting point is the *solutions gallery*, providing prebuilt out-of-the-box content for common processes and systems, which you can then configure to match your organization's policies and processes. In-app guidance can offer the following capabilities to direct users to the next best step:

- **Smart walkthroughs with branching**
  Guided analysis in the application for real-time guidance through processes that may take multiple paths
- **Smart tips**
  Give contextual information for users to validate your input
- **Shoutouts**
  Get users' attention and deliver time-sensitive information when it matters most
- **Launchers**
  Provide contextual access to the assets users need when they need them

- **Automations**
  Decrease user interactions and speed up manual processes by automating repetitive workflows for users, autofilling information across applications

These capabilities come together to provide clear information to end users about how to run the processes and offer guidance as processes and screens change. One of the most beneficial components of WalkMe is that these in-app directions can also leverage a user's contextual knowledge to provide the most optimal experience. A first-time user might receive a welcome banner and a deeper walk through of all the things to look for on a screen, while a user that accesses a page regularly may only receive new feature updates but can still access the walkthrough if they forget a step at any time. Someone who hasn't used the application in a while may receive targeted interactions or reminders about policy-related steps. This helps drive compliance, a user experience that adapts to usage, and better adoption of underutilized features.

As you roll out walkthroughs, you will also want to analyze your data and see how effective the walkthroughs are. If you notice trends about specific user types or segments, you can customize the experience for those users to receive more specific tips and guidance. Feedback is also built into WalkMe, and users can self-identify points of friction or clarity needed.

The action bar (see Figure 13.11) is an additional capability within WalkMe that guides and prompts users across applications. It semantically understands the page the user is on and provides information like relevant training, next best actions, and the available tools, information, AI agents, and more that can help guide users. The action bar is a familiar interface that can be accessed anywhere and seamlessly provides information across systems.

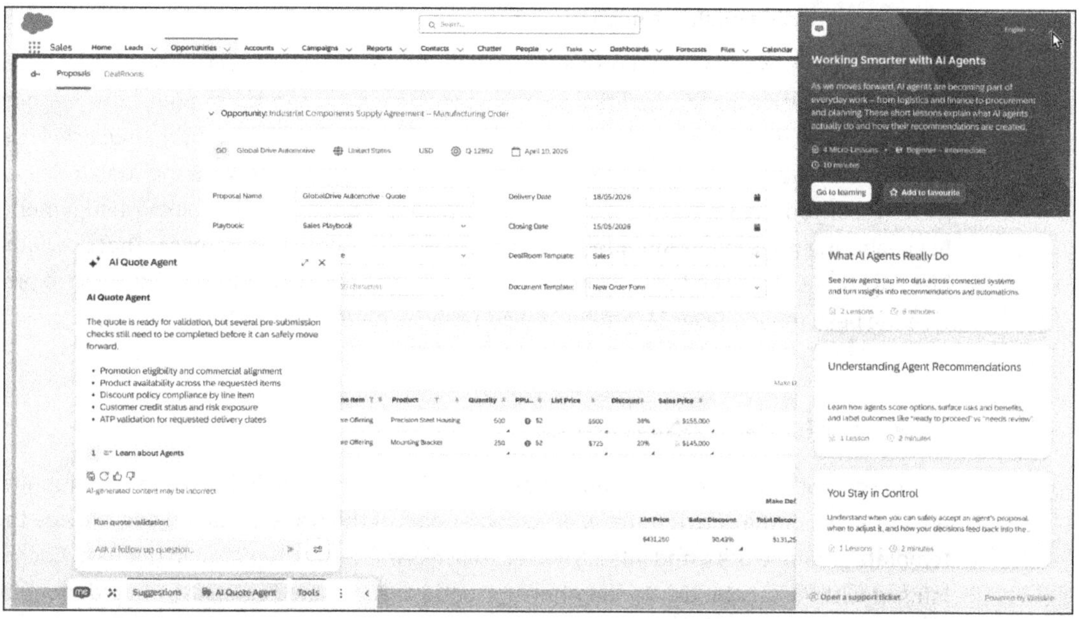

Figure 13.11: WalkMe Action Bar

Generative AI is embedded throughout the digital adoption experience. Three key types of AI are available to provide contextualized information at the moment of need: pinned, proactive, and on demand. Pinned AI exists mostly as part of Smart Walkthroughs, providing suggestions for free-form text entry, searching backend sources for specific codes that are needed for data entry, and serving other use cases. This provides users with information at specific points of the process. Proactive AI works as part of the action bar, scanning the page and highlighting actions and open items, connecting to other systems, and even bringing in content from Joule Skills to provide next best actions. This AI type prompts users for what in your workflow is the highest priority. On-demand AI exists as a side panel, offering free-form text search across the enterprise policies, processes, and workflows to answer specific questions.

In addition to the core digital adoption platform, Learning Arc is WalkMe's capability for traditional learning content. This AI-enabled capability is where trainings, simulations, learnings, and certifications can be designed to augment the adoption journey. This doesn't become a learning management system as a source of truth for organization-wide training but is instead a design studio to build meaningful learning content. Learning can be designed in varying lengths and formats and accessed by users from the learning portal. WalkMe suggests learning content based on user roles and what applications they are accessing, ensuring access to content to build the skills needed to execute work.

Creators can build this content from scratch, import content from other platforms, or leverage AI by describing training needs and using that to create baseline training content and training outlines. It can also suggest quiz questions and structure based on the training format and rigor required. This central source of content streamlines where training is created and updated for continuous improvement. It also aids in reviewing data and understanding where new training might be needed.

Learning becomes even more powerful when you connect it to data within the digital adoption and walkthrough capabilities in the tool. Now you can determine how learning improves business outcomes. You can go from pull learning, in which users must seek the training they want, to push learning, in which users are guided toward the most relevant learnings in your flow of work. This might be anything from new policy acknowledgements to specific microlearnings intended to correct user behavior in the moment. These capabilities together unlock the full power of digital user adoption and enable the business to provide the best, most relevant learning and adoption experiences to users.

13.8 Toolchain Convergence

Alone, each of these tools adds immense value. They form a digital platform and foundation for many of the activities to be executed as part of the transformation and provide the templates, frameworks, and infrastructure to perform them better than in the past. Each is infused with AI capabilities to help transformations run more seamlessly. In older digital transformations, process design might have involved weeks of workshops and thousands of sticky notes to determine an as-is state, to be followed by weeks of defining the to-be processes after functional analysis. Requirements and test cases were created, tracked, and executed in endless spreadsheets. Training was done in classrooms, in a traditional presentation format. What the tools gathered here have done is give you a toolbox full of the best-in-class digital versions of each of the activities that have to be completed to execute a successful transformation, and that on its own is powerful and will accelerate these activities and reduce the time, cost, and risk of the transformation. However, the true value of these tools is unlocked when they begin to become interconnected. What if you could easily access an approved business process as you write out technical requirements, and have all the context from the business team to design related test cases? What if data for the digital adoption of the delivered product could help inform and prioritize your roadmap for what to fix and improve?

One of the most common challenges companies talk about for digital transformations is alignment across teams and communication. The agent-led toolchain helps to alleviate that by bringing the right information and the right tools to the right people at the right time across all dimensions of a transformation, making aligning people, process, applications, and data easier.

In the following sections, we'll explore the critical integrations between the tools already discussed as part of the agent-led toolchain, as well as revisit the Korindal case studies to understand where and how these tools benefit companies both during and after a large-scale digital transformation.

13.8.1 SAP Transformation

For this section, let's follow the case study organization, Korindal, as it transforms the key use cases in its business, improving DSO and on-time, in-full (OTIF) delivery rates during an SAP transformation project. We'll explore this as the company upgrades its ERP system to SAP Autonomous Suite, and we'll discuss leveraging the SAP Activate methodology as a standardized framework for modernization. There are many approaches to transformation, ranging from a purely technical migration without making business process changes to a full greenfield implementation in which SAP is implemented using best practices as extensively as possible, but for this example we will assume the organization is following a combination approach, transforming key business processes to use new capabilities while maintaining critical differentiators that support the business. We will also assume that this organization is currently leveraging SAP ERP. If moving from a different ERP into SAP, the methodology and steps would be similar, but without the ability to use some of the rapid analytics for understanding the current state. There is also potentially less value in focusing on the current state when moving between two tools as the level of transformation will be high, and many customers will limit the work in this step to focus on the future state design with new capabilities.

Discover and Onboard

At the beginning of the transformation, the most critical component is understanding where you are today so that you can determine how to get to the desired future state. The critical tools to start with to understand the current state are SAP Signavio and SAP LeanIX for visibility into the as-is process and technology landscape. It is important to have a foundation in what is happening today, develop technology and change roadmaps to best define the future state and achieve organizational goals, and start to make decisions about what is going well and what needs to change as part of the transformation.

Understanding Current State Process Performance

The first step is understanding process performance today. Determining the baseline process performance will allow the team to understand which opportunities offer the most value, where the business is performing well today and how to minimize disruptions, and prioritize an implementation roadmap. Thinking long term, this creates data and information to later understand and communicate the value delivered through implementing SAP S/4HANA and a foundation in process excellence to fully leverage AI.

You can start the journey with SAP Signavio Process Transformation Manager, in which you can document key initiatives you are hoping to accomplish, including the goals of improving DSO and OTIF rates. Within SAP Signavio Process Transformation Manager, you can tie initiatives to value drivers so that it's easy to organize and quantify your goals. You can also return here frequently to capture and document insights as you explore data and opportunities and assign owners to these projects. This creates accountability and visibility and allows you to link all relevant information together in a central repository.

For an organization currently running SAP ERP, you can use SAP Signavio Process Intelligence to rapidly scan the ERP system, offering a data-driven approach to discovering what is occurring within the system today. This will help you collect and identify all the processes being executed in SAP ERP today and offer a plethora of metrics to see performance across the business. If you have a decentralized system and multiple instances, you can bring all that information into one tool as you ideally start to consolidate systems as well. You can hone in on key initiatives and sort data by how to improve these value drivers. You may discover your organization is well below the industry benchmark for customer invoice clearing as well as open and overdue accounts receivable items. You can correlate these to innovation recommendations that use both SAP S/4HANA Cloud capabilities that will improve the experience and SAP Business AI functions that you can add to the project roadmap to have an impact on these metrics.

At this point, you can also identify key transactional data to clean up and master data issues to resolve before transferring to a new system with specific correction recommendations. This is a great starting point prior to leveraging Syniti later to make the data migration as seamless as possible. Having a foundation of clean, usable data enables the Autonomous Enterprise to operate at scale.

Once you have a good understanding of the high-level process metrics and the process landscape, you can go deeper to get a clearer view of process execution. Here, you might start to analyze order management processes and how they relate to OTIF metric.

For Korindal, the *best-run score*, comparing its current operations to those SAP recommends as the most optimal paths, is quite low. In addition, supply chain-related processes are highly customized, using a variety of custom transaction codes to execute the same delivery tasks. You can benchmark this information internally to understand the impact on business performance and highlight where there are high volumes of poor performance. These are insights that Korindal can capture and tie to future state functionality across the Autonomous Enterprise to improve. There are some standard transaction codes that should give both better performance and a more streamlined process, and the company adds that to the roadmap. One other point to note is that based on the high proportion of the workforce leveraging this customization today, the change impact will be high. This is something to document as a high-value use case for WalkMe as the company readies for go-live.

Now that you have a clear view of the data around how processes run, you can create a repository of process documentation to communicate the current state, design the future state, understand change impacts, and provide a source of truth for business decisions during the actual build. The example organization will use the data from process mining to reverse-engineer the critical business steps that are executed and decide on the top variants to use in the BPMN process model. A team will work with process experts to determine the roles associated with the tasks.

This is a great starting point, but most processes do not occur solely in SAP system tasks. At this point, you can work with business process subject matter experts to fill in out-of-system tasks, those for which the process touches other systems, and tie this all together by capturing knowledge from on-the-ground expertise. Here, you can use the AI-assisted SAP Signavio Process Modeler to provide native language inputs to make these changes in a conversational flow.

These activities can be repeated for each LoB and key process area. As an outcome, the organization now has visibility into process performance in its current ERP system, a list of improvement initiatives and insights to prioritize for the transformation, and a process landscape showing what the as-is process architecture looks like. Without a tool like SAP Signavio, this work can take months to execute and can pull business stakeholders away for extended workshops to capture process knowledge. Leveraging a data-driven approach gives richer insights into the process execution with a fraction of the time and effort.

Understanding Current State Technology Landscape

In parallel with documenting current-state processes, Korindal uses SAP LeanIX to get the same level of granularity of details for its current-state technology landscape. Again, this foundational information is key to understand where the company needs to change, see impacts on things like data elements and integrations, and have a baseline to communicate the ultimate impact on the organization.

As the company gets started defining its current-state technology landscape, the first things to identify are business capabilities and applications. The company can start with templates for capabilities, and Korindal's applications can be captured through AI SaaS discovery, current state documentation in a CMDB, or even from images of the landscape. SAP LeanIX has discovery governance tools to ensure there is no duplicate data, and the reference catalog is complete upon upload.

At this point, the company can take some time to customize and enrich the gathered data. This ensures all applications are supporting the correct business capabilities. You can use tools like surveys to collect business user information around functional fit, and you might continue to add information like technical components, licensing costs, integrations, interdependencies, and more to visualize the as-is landscape. As part of the SAP upgrade, you will especially want to capture information about data inflows to SAP and from SAP to other systems to understand and document that impact and see where it is possible to reduce friction. Other views to consider include analyzing the adherence to a clean core strategy in the current state and using that to compare to the future state, again communicating the improvement gained from moving to SAP S/4HANA.

As fact sheets are filled out to the appropriate level to analyze the as-is state, you can also create dashboards to showcase how the current applications support your business capabilities and the complexity of the landscape in scope. This helps reduce redundant applications and identify where new capabilities can help streamline the landscape.

The last step within SAP LeanIX will be to establish the connection to SAP Signavio. During the transformation and beyond, SAP Signavio is the source of truth for processes, while SAP LeanIX is the source of truth for applications and business capabilities. Through the out-of-the-box integration, you can populate the dictionary in SAP Signavio with the applications and capabilities from SAP LeanIX. This means you now have a live connection in SAP Signavio to the source of truth for applications, and you can use that exact same list as you identify what applications are supporting the business processes. It then takes a small effort to link each of the as-is processes already defined to the appropriate applications they are executed in. As you add these connections through the dictionary, the process information will sync back to SAP LeanIX as a business context fact sheet, and the relationship is automatically established. This gives an additional, critical way to view the as-is landscape, sorted by which applications are needed to support each business process, and it ensures you have the appropriate coverage and change model ongoing.

Additional Steps in Discover

Beyond the key steps of setting up the as-is process and technology landscape, this is also a good time to establish the setup and administration for the other tools in the toolchain and to set the baseline projects up in SAP Cloud ALM. These tasks may overlap with the next preparation phase as well, depending on your project timeline and scope.

Within SAP Cloud ALM, you can establish a project from a template associated with RISE with SAP and customize it to your scenario by defining the scope, team, roles, timeline, milestones, deliverables, and other fields that are relevant to provide the structure needed to execute the project. As you prepare the project, you can also establish the sync between SAP Cloud ALM and SAP LeanIX. With this integration, you can push the project timeline and milestones into SAP LeanIX and sync the SAP Cloud ALM project with initiative fact sheets in SAP LeanIX. This allows for another single source of truth for project activities, and a change in SAP Cloud ALM will be automatically reflected in SAP LeanIX. While you set up the integration to SAP LeanIX, it is also a good time to connect SAP Cloud ALM and SAP Signavio; you will need this connection to use the future-state process models in SAP Cloud ALM as part of the build cycle.

The last activity in the prepare phase that may or may not apply is to use any data available from WalkMe to further build out initiatives and insights. Especially if your organization uses WalkMe today, this can provide additional task-level data to further understand any pain points in the process, drop offs between tasks, and difficulties for users moving between systems.

Now, as an outcome of this part of the SAP Activate methodology, we have the following:

1. A clear list of initiatives and insights about processes today. This is something you can use to prioritize your project, maintain accountability, and visualize the opportunity you are trying to improve.
2. A built-out current-state process hierarchy, describing end-to-end processes in scope, including system and nonsystem tasks, and how these are supported by your organization's applications.
3. Data about how the processes are operating today, providing a baseline for the transformation.
4. A current-state technology landscape, including SAP and non-SAP systems and key integrations, how they support the business hierarchy, and what processes they are linked with.
5. A starting-point project in SAP Cloud ALM.
6. Core integrations built between SAP Signavio and SAP LeanIX, SAP Signavio and SAP Cloud ALM, and SAP LeanIX and SAP Cloud ALM.

You now have the foundation to continue to the prepare phase. Again, these tasks are critical for any transformation, and the agent-led toolchain establishes a digital, interconnected source of truth, executed and built with the best-in-class platforms to manage each component of the transformation.

Prepare for the Transformation

The focus of the first step is to understand the current state of the organization. What do we do today? How are we performing? What opportunity exists? What happens outside of SAP?

The goal of the prepare phase is to really establish the operating model for how these tools will be used throughout the transformation. These activities could also occur before those in the previous section, or simultaneously with the discover activities, but ensuring proper governance, collaboration models, and enablement and awareness are in place will support the adoption of the toolchain through the transformation. This also begins to change hearts and minds in the organization to establish change capabilities as you move toward a state of continuous innovation.

This is full project-planning time. As you develop the scope and project timelines, from a transformation perspective, the key activities here include the following:

- Effectively communicate the scope of tooling to be used as part of the transformation program. Ensure proper training and enablement is provided so that the team is ready to get started. Make the most of SAP's prebuilt training wherever possible.
- Establishing process governance and an appropriate approval workflow in SAP Signavio, including a routing system for processes, clear ownership, and consistent guidelines for reviewing and approving process flow diagrams as part of the definition-to-be.
- Fully document architecture decisions and targets within SAP LeanIX and the architecture review process as the state-to-be is being defined.
- Establish the SAP Cloud ALM setup and ensure project milestones are aligned with the overall program timeline.
- Begin establishing a change management program and communications plan, and identify an adoption strategy with WalkMe at the core of digital adoption after go-live.

Modernize and Explore the Future State

At this point, you have fully established the baseline. There are ways of working in place to use the toolchain effectively. You have data on what is and is not working from a process and IT landscape perspective, and you've established goals and initiatives for this project. This is when things become fun: How do you design a future state to improve on what you are doing today? This is when you get to explore everything being offered in the new software, challenge how you do things today, and decide what really matters to fully realize the benefits of becoming an Autonomous Enterprise.

In this section, we'll look at this through again improving DSO and OTIF by transforming your technology. Additional common improvement goals that we'll highlight in this section include establishing a clean core and applying best practices wherever possible to fully embrace all future innovations.

Building a Future-State Process Design

The first step for understanding the future state is building the process flows and determining how to work in the future. Through the discovery phase, you have a prioritized list of where transformation will bring the most value and understand the scope of and goals for the transformation. There are several options for each of the process flows you can gather for future-state design. First, you can apply best practices from SAP. These can be imported directly into SAP Signavio from the Value Accelerator Library, or you can use Joule to help find the most relevant content for how you operate. You can also leverage your subject matter experts and AI tools like the AI-assisted SAP Signavio Process Modeler to provide context for what your dream process would look like. A key consideration at this point is also what SAP Autonomous Suite functionality you are going to embed into your processes. You know from your discovery in SAP Signavio Process Insights that you have many SAP Fiori apps, SAP Business AI, SAP Build Automation, and more to consider, and you can use your data about what transactions occur in SAP today to identify the most relevant and valuable. Here you can identify a Sales Order Fulfillment SAP Fiori app, for example, that you want to include in your design to support your goal to improve OTIF.

At this stage, you also use data you collected from your process analysis to determine where you can minimize customizations within your system as you move toward more standard practices. As you analyze the best-run scores and business benchmarking for the finance area, you might determine that by performing more steps in an automated fashion and bringing more of your process into your ERP system, you can improve your disputes process. You bring this knowledge forward to the business to build the future state design that includes updated technology, controls, prompts, and automation together to streamline your organization's finance processes.

You can use process modeling to visualize the future state and identify your end-to-end business processes, including all the steps that will be performed in SAP, identifying out-of-system tasks, handoffs, and more. The goal is to align with one global template, but if there are legal or regulatory requirements based on country, business unit, product, or other relevant dimensions, localized variants can be captured in variant management.

Leveraging SAP Signavio, it is possible to do a lot of this documentation prior to workshops, coming up with starting point designs, incorporating best practices, and more. You can also socialize updated process documentation digitally in advance of hosting design workshops and collect feedback as comments directly in the process. This iterative, collaborative approach allows the process design team to maintain a centralized repository and align on lower-stakes decisions offline as prework. As you bring teams together to complete design sessions, you are now leveraging that time to align on big rocks and decisions that require in-person discussions and cross-team collaboration to make work. This saves time and decision fatigue in meetings and allows for focused and productive sessions.

The outcome of your design sessions will be a repository of process models documenting key design decisions, new features and functionality, and updated processes. At this point, you've considered functional changes that cloud ERP brings to the table, determined where you need to transform to fully leverage AI in your future state, and been thoughtful about roles and responsibilities, handoffs, risks, and technical complexity. Here you might implement formal governance processes and approvals to document the appropriate checks and balances and confirm alignment.

Now, you can use SAP Signavio and these future-state processes as a critical input to your change management plans. With your as-is execution data in hand from SAP Signavio Process Intelligence, you can compare your future-state design to identify major change impacts, create training plans, and prioritize your implementation plans for WalkMe around major system changes.

Throughout this process, you should continue to link your decisions back to the transformation initiatives you defined in SAP Signavio Process Transformation Manager. Have you made decisions in support of your goals? In this case, you know you've redesigned your finance process to conduct more steps in your ERP system, as you are introducing new applications within your supply chain processes in support of improving OTIF rates. This is critical to measure the success of your transformation project.

Defining User Stories and Requirements

As you finalize your process design and obtain the required business approval in SAP Signavio, the next step is to sync this to your functional team in SAP Cloud ALM. Once processes are loaded into SAP Cloud ALM, your functional design team can now leverage all the business context that has been agreed to as they build out the user stories and requirements for development. This streamlines knowledge transfer and brings together the right information to the right teams at the right time.

Now requirements can be linked directly to the related process flow. As requirements are written, SAP Cloud ALM also uses AI to enhance requirement documentation and applies best practices across the team, encouraging additional layers of standardization.

Your documented requirements are essential to build the solution and are the foundation to create test cases; they are also an additional input to your change management plans. Once you have documented requirements, your testing team can start to develop a plan for automated test cases in Tricentis, data requirements, and more.

Building a Future-State Enterprise Architecture

Now that you have a thorough understanding of the future-state business landscape, it is important to also map out the future-state application landscape. This serves several purposes: ensuring you understand how all critical business capabilities will be supported in the future, visualizing and documenting the future state, understanding interdependencies with the larger technology landscape including critical integrations, creating reports to ensure your technology goals are being met (like maintaining clean core where possible), and mapping the roadmap for key milestones and implementation strategies (business units, regions, modules, etc.).

To accomplish these goals, you can turn to SAP LeanIX, and specifically SAP LeanIX Architecture and Road Map Planning. You will start by documenting your transformations against the initiatives you built out earlier. You can either use the SAP Reference Solution Architecture recommendations, which is the preferred approach for most transformation development, or you can create transformations manually. Leveraging SAP Reference Solution Architecture has the added benefits of being quicker to deploy as its recommendations are prebuilt and you are already incorporating best practices. Manual transformations require more effort but may be necessary for parts of the architecture deviating from the standard.

To import the future-state capabilities, you can select the relevant business capabilities and map these to the scoped processes that are affected by your transformation scope. SAP LeanIX will automatically suggest the applications within SAP that align with the business capabilities identified, and the team can select the solutions decided upon to add to the transformation project and build the transformation. For any items that cannot be determined through this approach, the manual approach gives additional flexibility to align to your specific organizational needs.

Once the capabilities are mapped, you can develop reports to visualize your future state architecture. One view would be mapping applications to the business capabilities and leveraging the color scale to indicate technologies being phased in, being phased out, and remaining active. You can leverage the timeline as you link target implementation or end of life dates to understand how this changes at different points. You also can create visualizations for integrations that need to be added or removed within your landscape to identify and prioritize integrations to external systems.

Finally, you can use the roadmap report to see how the landscape changes over time, including interdependences and major milestones. Remember, milestones can be linked from SAP Cloud ALM so that your projects are aligned between systems. You can view your roadmap sorted by different initiatives, capabilities, organizations, and other visualizations.

SAP LeanIX allows you to create a visual representation of the impact of your transformation program on how you run your business, and how this connects within the broader technical landscape.

Deploy, Realize, and Run the New Transformation

At this point, you've done quite a lot of groundwork to make your transformation successful. You have a baseline of how you operated previously, as well as clear transformation goals and initiatives. You have a designed and approved future state, linked to both functional requirements and your target enterprise architecture. You have approvals and buy in from business and IT, and leadership as needed, a plan built out to accomplish this, and an expected value case for the improvements you are making. All this prep work leads to where this starts to make an impact: building and running the new applications.

You will primarily orchestrate this phase through SAP Cloud ALM, the tool of choice for deployment, providing end-to-end documentation of software changes and building audit trails and integration with deployment providers for transparency. In the previous step, you defined requirements in SAP Cloud ALM based on your business processes. Your next step in SAP Cloud ALM is to align those requirements to implementation activities and assign owners. You can link these together to visualize your plan for deploying software and prioritize different tasks and dependencies to be completed. This gives a critical project management capability to track the progress of features, trace them to requirements, and understand how all the pieces of the project fit together. This fits with a continuous delivery model of design, build, and test to iteratively reach the solution.

As activities are tracked and developed through the build, you also build out your test cases. These can be built and tracked directly in SAP Cloud ALM for manual testing, or you can leverage your integration with Tricentis for automated testing. Automated testing offers the advantage of being repeatable at scale, and reusable post transformation as needed. Within Tricentis, you build out your test strategy and coverage, monitor failures, and iteratively move between build and test. As you conduct test cases, the status syncs back to SAP Cloud ALM for full transparency.

Parallel to the building and testing of the solution, your migration teams can build and test your strategy to move data and documents to your new implementation. Syniti is the orchestration layer to understand the data in your current ERP system, conduct data cleanup activities, map the data to the new system, and conduct practice loads prior to the full cutover. The data gathered through Syniti can also be connected to Tricentis to optimize building test cases so they better reflect actual business data. This helps make data cutover prior to go-live as seamless as possible.

The last consideration here is document management and defining a strategy for what components in your ERP system can be archived and moved into OpenText, and what you need to carry forward into the new ERP system. This can be accomplished by reviewing data in Syniti and SAP Signavio and aligning with regulatory and compliance to understand what documents you need to bring forward for business operations and what documents you can move into your archiving tool to reduce cost of storage.

At this point, you now have your updated modules ready to go. Depending on the implementation strategy, this might be done in waves or all at once.

For the case study, the business is ready to go with core finance functionality, as well as the additional business apps in Korindal's supply chain that are expected to improve the OTIF rates. As the company implements this software, it will also publish the updated processes and execute transformations in SAP LeanIX, ensuring that these remain the source of truth for how the business operates over time.

This is when change management becomes essential. Change should be a key consideration throughout the entire transformation program, and you incorporate it during each phase. Ideally, you have cohesive communications plans, leadership buy-in and support, well-articulated metrics and goals you are hoping to achieve, user communities for support, and, most importantly, the right adoption tools. For your cloud ERP implementation, you've identified the workflows with the most change using SAP Signavio and built out adoption paths in WalkMe, ready for your new users to log in. It does not matter what improvements and innovations you have made in the software if your users are not prepared to use the tool as designed.

Finally, you've arrived at your go-live date. Your test plans have all passed, with all new modules in production in the cloud. Your data migration was successful, and you've brought in the information you need to move forward in the new system, with thoughtful data cleanup and remapping based on any data structure change. As your first users log in, they are greeted with prompts from WalkMe that help them navigate the new interfaces and execute your work in the new system. Essentially, each user in SAP has a super user sitting next to them, helping them understand where to click, what information to fill out, and even when they need to navigate to other systems to complete the end-to-end process. Users also have access to view the full process documentation in SAP Signavio, so they can understand the bigger picture and where your work fits into the larger flow.

It takes several weeks to stabilize the system, ensuring there are no major gaps in technical or functional requirements, no security issues, and any other major changes required. After several weeks, you can start collecting data around your performance to understand if you are directionally meeting your goals. One of the first things you may notice is that your adoption is going well, with many users turning off the more detailed views of WalkMe as they learn the new workflows and moving to just quick tips for fewer tasks. You can also start to measure conformance to the new process in SAP Signavio, seeing much higher rates of individuals leveraging best practices within the tool.

Most critically, you can measure how you were performing in SAP ERP previously and the major metrics you want to measure in SAP S/4HANA. Your CFO and CSCO now have dashboards in SAP Signavio where they can keep a pulse on live metrics to drive your business. They can see with confidence that the metrics that matter to them, DSO and OTIF, have reduced as part of the transformation program; even more valuable, they have visibility into exactly where, when, and why these metrics might change over time, with a new ability to course-correct quickly and to explain deviations with ease and confidence.

Now that you have reached the end of your transformation, you can reflect on how the toolchain impacted your journey. Fundamentally, the key tasks you've always done as part of a transformation journey are still being executed, but instead of doing this work manually and in a decentralized fashion, the toolchain establishes a best-in-class digital foundation for these activities. Besides making the entire process easier and faster, it also ensures you can reuse all this work for continuous improvement efforts, as discussed in the next section.

13.8.2 Continuous Improvement Effort

During the SAP transformation and bringing these tools across teams and business units, the ideal foundation has been set to build a true practice of continuous improvement. Now the team has all the tools in place and have learned the critical skills to make this part of the engine to make the business run better all the time. In the age of cloud software, change is not part of a migration, or part of annual or even quarterly releases, but something that is ongoing on a day-to-day basis. The technologies in the organization will continuously be bringing forward new features and updates, and the ways people work evolve rapidly in today's technology landscape. However, the hard work is done, and at this point the transformation is not part of any project but a capability built across the business.

This brings us to the third use case in this book: the CIO dilemma experienced by David at Korindal. As a CIO, setting up the AI-enabled toolchain is going to be a critical enabler to shift how IT delivers business value continuously and at scale. Let's explore using the toolchain to identify, prioritize, and implement use cases outside of a full system implementation.

Let's evolve the finance example. Between SAP Signavio, SAP LeanIX, and WalkMe, you have a very strong understanding of a digital twin of how your organization runs. From SAP Signavio, you can keep a pulse on your process metrics and performance and quickly detect deviations from the process you have defined. Within SAP LeanIX, you get full visibility into your IT landscape and can understand any upcoming changes to your application architecture. From WalkMe, you can get to user- and task-level data to see where users are commonly frustrated, stuck, or spending a lot of time executing tasks. Both your business and IT departments can access this level of visibility, turning process improvement into an organization-wide initiative.

As you analyze the data, you discover that your turnaround time on disputes has increased dramatically over the past quarter. As you investigate this metric in more detail, you might discover that there is additional rework needed for certain customers that is driving increased cycle times. You can also see this reflected at the task level in WalkMe, with users dropping off during some data reconciliation processes. Using SAP Signavio Process Insights, you can investigate what features within SAP could be useful to help improve your collections process. Based on your transaction code usage, the tool suggests a new Accounts Receivable AI Agent that can be installed from SAP Business AI, and your IT team gets to work on this.

First, you determine how this AI will change the process, and you develop your future-state accounts receivable process with this new AI agent installed and execute any defined governance and approvals. You can identify change gaps and how you need to restructure and retrain resources to work with this new software. This new process, along with data from SAP Signavio Process Intelligence and *process atoms*, consumable pieces of the process like dispute thresholds and approval requirements, can all be used to make your AI agent more effective from day one.

You also can review how this fits into your broader AI landscape within SAP LeanIX. As you build out your transformation, you can add this new AI capability into SAP AI Agent Hub and ensure you don't have other AI solutions that would already fill this gap. Your AI Agent

Radar allows you to ensure you are using the most compliant underlying models defined in your organization's AI governance planning. You can also map out your transformation and track milestones.

You again sync your processes into SAP Cloud ALM so that you can build requirements and test cases and track your implementation. Because you already have a strong foundation of automated test cases in Tricentis, you only have to make a few tweaks to effectively complete end-to-end testing, streamlining your implementation. You also communicate your change plan to your users and do some light retraining for human-in-the-loop interventions and update your WalkMe prompts to reflect the changes.

Now you're ready to deploy and track agentic performance in Signavio to understand if you achieved the goal of continuing to improve your finance execution. You moved what would previously be bundled into a larger release, and take weeks to months of research, redesign, building, testing, and deployment, into a streamlined execution engine. You now know exactly where to find the data, root cause analysis, and solutions you need, and you have the tools and foundations to execute against that rapidly. The CIO can now move from operating as a service to the business to becoming an innovation partner, identifying where technology can solve business execution gaps and delivering that quickly and with compliance to organizational governance.

This is the future of transformation. Once your organization is in the cloud, the rate of innovation is exponential, limited only by how adept the team is at identifying and executing improvements. This solution highlighted adding features, but other solutions might include retraining, adjusting adoption workflows, moving pieces of the process, retiring or migrating old or outdated IT systems, and more. Organizations that have built these capabilities across all departments are at an advantage as they can move fast and with precision.

13.9 Summary

In this chapter, you explored the multitude of tools within the AI-enabled toolchain, first from the functionality standpoint of each tool and how it added value across key transformation activities like process design, adoption, testing, and others, and then with a look at how these benefits are multiplied by connecting the right data and capabilities between these tools, creating an interconnected machine that drives transformation initiatives. These tools add value both during the digital transformation program and in the long term with ongoing innovation. You also explored how the agent-led toolchain supports all three key use cases from this book, helping your CFO and CSCO improve your operations and giving the CIO the digital platform to become a true business enabler.

This chapter focused on the use case for an SAP transformation program, but it's important to note that the tools discussed here, except for SAP Cloud ALM, are system-agnostic. SAP Signavio, SAP LeanIX, WalkMe, Tricentis, Syniti, and OpenText can all be used across any digital transformation initiative following the same steps: Understand the current state, design the future state, build, and then monitor and continuously improve. In the long term, these solutions can represent the actuality of how the entire organization is structured and run, across all business units, products, underlying software, and more, extending the benefits of building these capabilities.

In the next chapter, we'll focus more specifically on AI through the lens of governance and responsibility.

Chapter 14
AI Governance and Responsible AI

Responsible and compliant artificial intelligence secures the operational state of the agentic enterprise through rigorous ethical and legal guardrails. It ensures every autonomous action is grounded in business context, maintaining absolute transparency and traceability across SAP Autonomous Suite. By prioritizing human oversight and regional sovereignty, this governed approach allows organizations to scale intelligence with confidence.

When autonomous agents are empowered to act on behalf of the organization, trust becomes the primary currency of the digital enterprise. The question is no longer what artificial intelligence (AI) can do, but whether the organization can govern, secure, and audit what it does. SAP Business AI is built on this premise: Responsibility, transparency, and human-centered oversight are architectural requirements, not afterthoughts.

This chapter examines the frameworks and technical guardrails that transform ethical intent into operational reality. For the modern leader, governance is not a brake on innovation; it is the accelerator that allows the organization to move with confidence. The deep integration of security protocols, regional data sovereignty, and human-centered design into the SAP Business AI architecture means that breakthroughs do not come at the expense of integrity.

Context is the differentiator. In a generic environment, AI can suffer from hallucinations or produce answers that are technically plausible but irrelevant to the business. Within SAP Business Data Cloud, every model is grounded in the semantic reality of specific business processes, master data, and historical relationships. SAP Knowledge Graph acts as both an accuracy catalyst and a security guardrail, ensuring that an agent understands not just the command it receives but the organizational boundaries within which it must operate.

In the mature agentic model, the system evolves from a passive system of record into an active, self-governing partner. It operates autonomously for routine tasks, such as a procurement agent that detects a minor supply discrepancy and reallocates stock based on preapproved logic, and it becomes highly agentic for complex exceptions, surfacing these to human experts with full contextual support. The software shoulders the operational load while high-stakes judgment remains a human prerogative.

Four pillars structure this chapter. Ethical AI guidelines establish the guiding principles for proportionality, safety, and fairness. Privacy and auditability details how the data architecture maintains corporate data sanctity while providing regulatory transparency. Sovereign AI and global compliance explain how the architecture adapts to regional data laws and the

EU AI Act. Finally, autonomous suite resilience and the secure core cover the defense-in-depth methodology that protects business-critical systems from emerging generative AI threats.

Figure 14.1 highlights the framework via which SAP translates high-level ethical principles into actionable business results through a robust governance model. It demonstrates the transition from the *as-is* state of fragmented governance to a unified *to-be* state in which intelligence is natively governed and compliant by design.

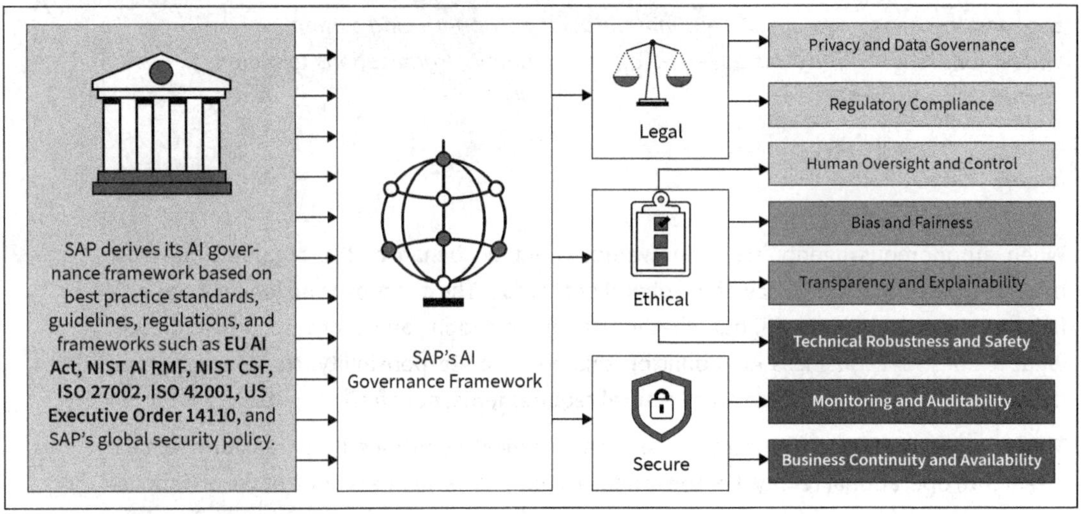

Figure 14.1: Pillars of Responsible AI

14.1 Ethical AI Guidelines: Principles into Action

For SAP, ethical AI is not an optional layer added at the end of the development process; it is the operating system that governs how SAP Business AI is conceived, built, deployed, and overseen across the enterprise landscape. The SAP Global AI Ethics policy is grounded in the company's Global Human Rights Commitment Statement and aligned with UNESCO's Recommendation on the Ethics of Artificial Intelligence, giving SAP a value framework that is both internationally recognized and legally relevant for the design and use of AI systems. This policy applies to every AI system SAP develops, integrates, sells, or uses internally and to all actors involved in the AI system lifecycle, from initial design and data collection through deployment, monitoring, and retirement.

Within this framework, SAP positions SAP Business AI around three strategic attributes:

- *Relevance* comes from deeply embedding AI into end-to-end business processes rather than treating it as a detached analytics layer.

- *Reliability* stems from grounding models in high-quality, semantically rich enterprise data.
- *Responsibility* is expressed through binding ethical, security, and privacy requirements that govern every AI feature and supplier across the portfolio.

Ethical AI, in other words, is codified in policies, handbooks, product standards, and lifecycle checklists that together constitute a management system rather than a statement of intent. The following sections detail how SAP translates these principles into practice: first through 10 guiding principles adapted from UNESCO, then through the business AI lifecycle that operationalizes ethics at every stage of development, and finally through the ecosystem governance model that extends accountability to partners and regulators.

14.1.1 SAP's 10 Ethical AI Guiding Principles

At the core of this ethical posture are 10 guiding principles that SAP has adapted from UNESCO and translated into operational requirements. They articulate, in a compact form, what SAP expects of every AI system and every actor in the AI value chain. Let's walk through them:

1. **Proportionality and do no harm**
   AI systems must be necessary and appropriate for a clearly defined purpose, and never deployed where risks exceed benefits. SAP requires explicit impact assessments and prohibits use cases that could cause disproportionate harm.
2. **Safety and security**
   AI systems must be robust against failures, adversarial attacks, and unintended behaviors throughout their lifecycle. SAP mandates continuous security testing and fail-safe mechanisms at every stage.
3. **Fairness and nondiscrimination**
   AI must be designed to prevent reproduction or amplification of existing biases. SAP requires proactive bias testing across datasets, models, and outputs, with documented mitigation strategies.

> **Case Study Example**
> For Korindal, this directly shapes how the accounts receivable agents are designed: Collection tactics are calibrated by customer payment history rather than applied uniformly, preventing unequal treatment of customers experiencing temporary hardship.

4. **Sustainability**
   AI should contribute positively to environmental and social sustainability. SAP applies energy-efficiency requirements to model training and inference, embedding sustainability metrics into deployment decisions.

5. **Right to privacy and data protection**
   AI systems must comply with General Data Protection Regulation (GDPR) and equivalent regional frameworks, respecting data minimization, purpose limitation, and individual rights throughout.
6. **Human oversight and determination**
   AI must remain subject to appropriate human oversight. SAP operationalizes this through human-in-the-loop, human-on-the-loop, and human-in-command governance patterns matched to each use case's risk profile.
7. **Transparency and explainability**
   AI capabilities, intended use, and limitations must be clearly communicated. SAP requires AI to be visibly identifiable in user interfaces with documented confidence levels for outputs.
8. **Responsibility and accountability**
   Ethical and legal responsibility rests with human actors. SAP assigns explicit accountability by role across the AI lifecycle through impact assessments and compliance checks.
9. **Awareness and literacy**
   Organizations must invest in AI literacy for employees, customers, and stakeholders. SAP supports this through open learning resources, SAP TechEd, and digital literacy initiatives.
10. **Multistakeholder governance and collaboration**
    Ethical AI requires ongoing dialogue across disciplines and jurisdictions. SAP embodies this through its AI Ethics Advisory Panel, participation in the EU AI Pact, and governance processes that invite external feedback.

14.1.2 Operationalizing Ethics: The Business AI Lifecycle and Risk Assessment

Rather than leaving these principles as static declarations, SAP operationalizes them through the SAP AI Ethics Handbook (available at *https://www.sap.com/products/artificial-intelligence/ai-ethics.html?t*) and the SAP Business AI lifecycle. Every AI initiative begins with an ideation phase in which use case owners consult the SAP Global AI Ethics policy, identify affected users and stakeholders, and complete an AI Use Case Impact Assessment to surface potential harms, privacy challenges, or disproportionate effects on vulnerable groups. Red-line scenarios such as human surveillance that abuses rights, deanonymization of already anonymized data, manipulative systems that erode autonomy or democratic debate, social scoring, and deliberate environmental damage are explicitly prohibited and must be abandoned as soon as they are identified.

Figure 14.2 illustrates a five-stage lifecycle for delivering embedded AI into business operations. It shows how use cases move from ideation and feasibility experiments through development and integration into business processes, culminating in operational deployment of AI functions at scale.

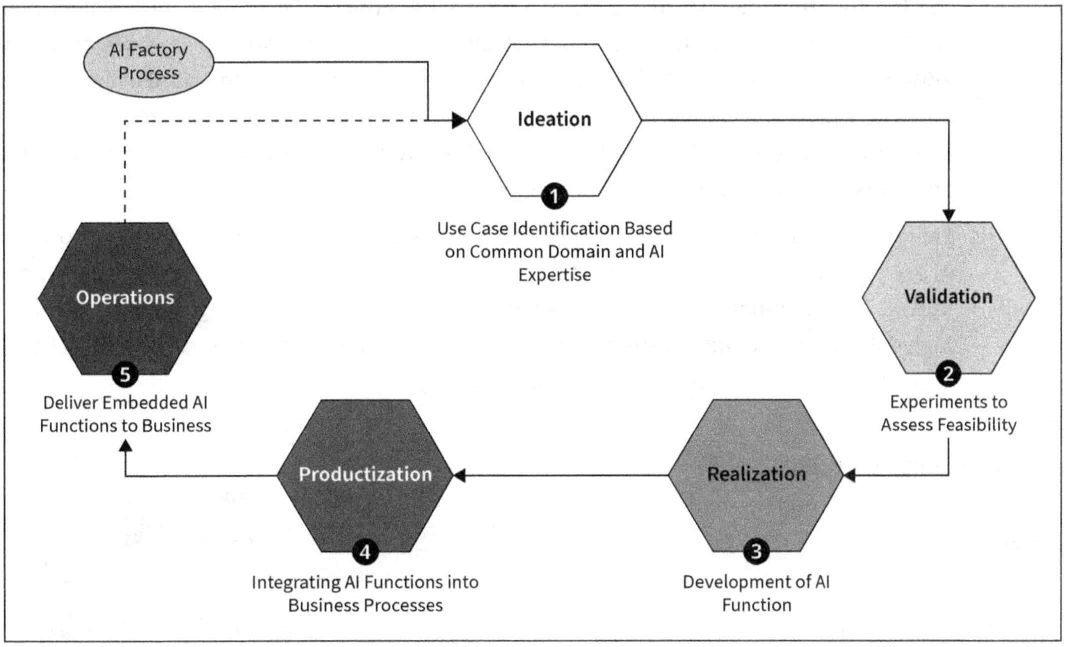

Figure 14.2: SAP Business AI Lifecycle

Use cases that process personal data, involve largely automated decision-making, or could have significant consequences for individuals, especially in domains like HR, healthcare, critical infrastructure, or democratic processes, are classified as high risk and routed to the AI Ethics Steering Committee for dedicated review before development can proceed. This structured classification into standard, high-risk, and red-line categories makes the ethical status of each AI feature explicit and determines which checkpoints, approvals, or corrective measures are required.

As ideas move into validation, realization, productization, and operations, the same principles reappear as concrete design and governance tasks. Teams test models for fairness and robustness, implement content filtering and data masking for generative models, document capabilities and limitations, and design user experiences that both signal the presence of AI and preserve human agency. Each release triggers self-assessments and, where necessary, renewed impact assessments, recognizing that risk profiles change when scope, data, or models evolve. In operations, AI or large language model (LLM) ops engineers monitor performance, bias, and safety events, while product teams track environmental metrics and consider model updates that reduce energy and resource consumption without sacrificing accuracy.

14.1.3 Ecosystem Governance and Global Regulatory Alignment

The ethical AI story at SAP is inseparable from its governance and ecosystem strategy. A formal structure of roles and bodies—the AI Ethics Steering Committee, the AI Ethics Office, the Trustworthy AI workstream, and the external AI Ethics Advisory Panel—provides oversight and escalation paths, ensuring that difficult trade-offs and high-risk use cases are addressed at the right level of seniority. The Steering Committee, composed of senior leaders from across the company, oversees SAP's AI ethics approach and approves or rejects high-risk use cases, while the advisory panel of external experts offers independent perspectives on policies and specific decisions.

Figure 14.3 illustrates SAP's AI risk assessment and classification process, showing how new AI use cases are evaluated step by step. Through a series of questions about personal data, impact on individuals, automation, and sector, each use case is categorized as standard, high risk, or red line, with corresponding outcomes such as business continuation, AI Ethics Steering Committee review, or mandatory stoppage.

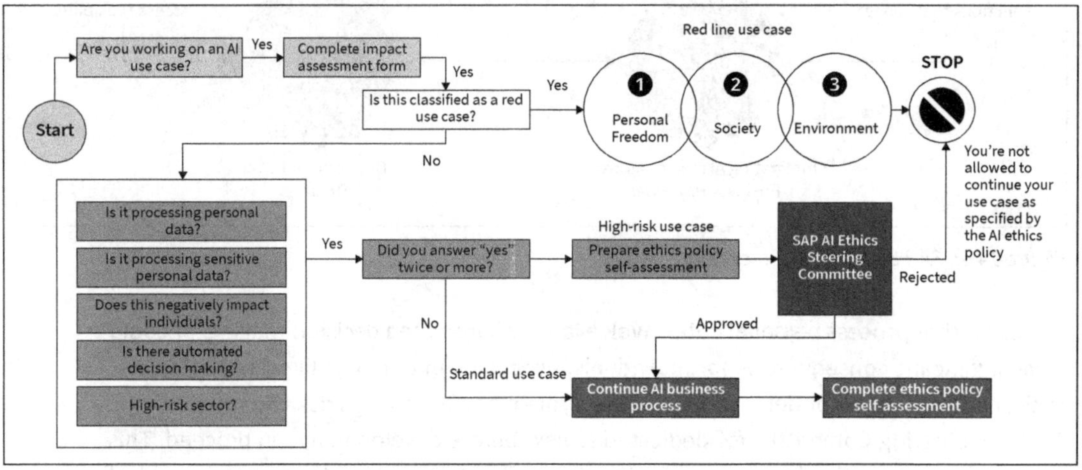

Figure 14.3: AI Risk Assessment and Classification Process

Operationally, the AI Ethics Office coordinates the governance bodies, maintains and updates the AI Ethics Policy and guiding principles, and serves as the first point of contact for AI ethics questions, including concerns raised through mechanisms like the Speak Out at SAP tool (*https://sap.integrityline.com*). The Trustworthy AI workstream, an internal cross-functional program within SAP's product engineering organization that drives responsible AI implementation, and the roles defined in the SAP Business AI lifecycle, such as use case owners, AI data scientists, AI engineers, designers, AI/LLM ops engineers, and product developers, carry explicit responsibilities for embedding ethics requirements into everyday work, from data handling and model design to UX writing and production monitoring.

Crucially, SAP treats partners and suppliers as part of a *chain of responsibility*. Partner applications built on SAP Business Technology Platform (SAP BTP) and the generative AI hub, as well as inbound third-party models, undergo structured onboarding, risk assessment, and in many cases certification programs that test adherence to responsible AI and product standards before integration or resale. This ensures that the ethical posture of SAP Business AI is preserved across the AI value chain, not just within SAP-built components.

SAP's AI governance framework aligns with the EU AI Act, the National Institute of Standards and Technology (NIST) AI Risk Management Framework, International Organization for Standardization (ISO) 27002, and ISO/International Electrotechnical Commission (IEC) 42001. The company is implementing the EU AI Act's prohibitions, participating in the EU AI Pact, and running a board-supervised classification program for its AI systems. Externally, the World Benchmarking Alliance identified SAP as meeting all assessed ethical AI dimensions, and the ISO/IEC 42001 certification covers Joule, SAP AI Core, and SAP AI Launchpad.

SAP's ethical AI guidelines define what kinds of AI SAP will never build, which use cases demand heightened oversight, how customer data may be used, and how human beings remain in command. This ethical scaffolding makes it possible to move fast with confidence.

14.2 Privacy and Auditability: The Data Sanctuary

The architecture is built on a fundamental principle: Although context is essential for an agent to be effective, that context must never leave the governed perimeter in an unprotected state. Privacy is enforced through a multilayered strategy of technical guardrails.

In a standard consumer AI model, data is often used for training, leading to significant risks of data leakage or "memorization" of sensitive information. Within the SAP Business AI framework, a definitive line is drawn: Customer data is never used to train the underlying foundational models. By utilizing SAP's AI Foundation and SAP Business Data Cloud, the enterprise ensures that its data remains its own, isolated within a secure environment that respects the rigorous standards of global data protection regulations.

The operationalization of privacy within the agentic core relies on a multilayered strategy within the following technical guardrails. These guardrails act as an automated filtration system, ensuring that sensitive information is never exposed to external LLMs in an unencrypted or identifiable state:

- **Secure data masking and anonymization**
  Before any data is transmitted to a generative model via SAP AI Core, it undergoes a rigorous masking process. Personally identifiable information (PII) is identified and replaced with synthetic tokens or anonymized markers. This ensures that the model can still process business logic such as calculating a budget variance or drafting an HR policy, but without ever seeing the actual names or specific identifiers of the individuals involved.
- **Tenant isolation and secure multitenancy**
  In a cloud environment, the risk of cross-tenant data leakage is a primary concern. SAP maintains strict logical isolation between different customer environments. Even when multiple organizations utilize the same underlying AI infrastructure, their data remains

within their specific *data sanctuary*, preventing any possibility of data from one organization influencing the outputs or models of another.

- **Content filtering and PII detection**
  Every prompt sent to a model and every response received by the agent passes through a real-time content filter. This system is designed to detect and block the accidental inclusion of sensitive data types, such as credit card numbers, social security identifiers, or specific proprietary product codes. This proactive layer ensures that even if a user inadvertently includes sensitive data in a query, the system automatically intervenes to protect the organization's privacy.

Figure 14.4 presents SAP's *privacy and security by design and default* concept for cloud applications with embedded SAP Business AI. It links contractual assurances in the SAP data processing agreement and SAP's technical and organizational measures to concrete controls in the application layer, such as authentication, encryption, anonymization, logging, lifecycle management, and data transparency features.

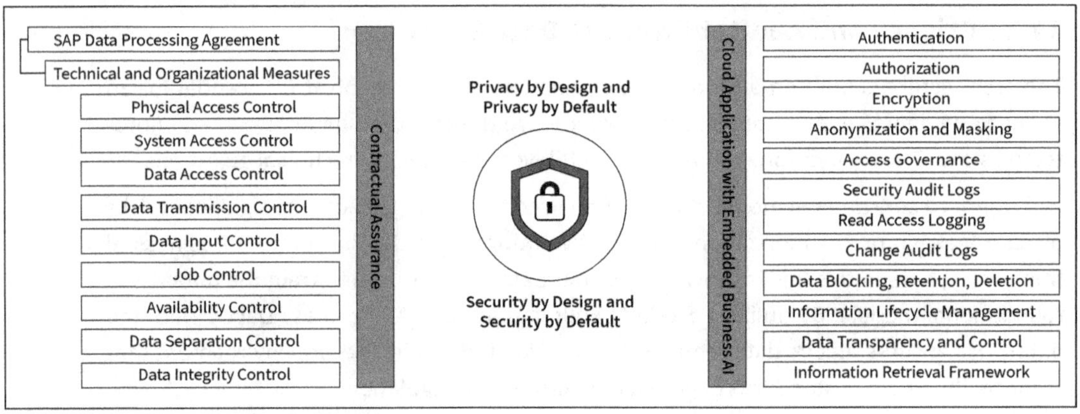

Figure 14.4: Privacy and Security by Design and Default

Privacy alone is insufficient for the responsible enterprise; it must be coupled with absolute auditability. In a world where agents act on behalf of a company, an organization must be able to reconstruct every decision and every action for regulatory, legal, or internal review. Auditability is the mechanism that transforms black box AI into a transparent and accountable business partner.

SAP ensures this level of transparency through a comprehensive audit trail architecture. Every interaction between an agent and a business system is logged, creating a chronological record of the intent, the data utilized, and the outcome. This is not merely a record of what happened, but an explanation of why it happened. Through advanced explainability features, the system can surface the specific business documents and data points that influenced an agentic decision. For instance, if a procurement agent recommends a specific vendor, the audit trail can highlight the historical performance data and contract terms that led to that specific recommendation.

Case Study Example

This proved critical for Korindal's compliance team, who could trace every agent action across their finance workflows back to the originating authorization context during their first postdeployment audit.

This level of auditability is essential for meeting the stringent requirements of the European Union AI Act and other global regulatory frameworks. By providing a clear, immutable record of AI operations, organizations can successfully navigate audits and demonstrate that their autonomous systems are operating within the boundaries of established corporate policies and legal mandates.

The final dimension of the data sanctuary is respect for regional sovereignty. As data laws become increasingly fragmented across the globe, the intelligent enterprise must be able to adapt its data processing to match local jurisdictional requirements. SAP addresses this through its global network of data centers and the implementation of EU Access and other regional residency protocols.

This means that for organizations operating in highly regulated regions, the processing of AI tasks, including the masking of data and the generation of logs, occurs within specified geographic boundaries. This ensures that sensitive data never leaves its primary jurisdiction, even when interacting with global foundational models. By combining this regional control with deep technical security, the SAP Business AI architecture provides a secure, private, and fully auditable environment where the agentic enterprise can flourish without compromise.

14.3 Sovereign AI and Global Compliance

As AI becomes the engine of the global economy, the regulatory landscape is shifting from general data protection to specific, high-stakes AI governance. For the global enterprise, the challenge is no longer just managing data but ensuring that autonomous systems comply with a fragmented world of regional laws and jurisdictional requirements. SAP's strategy for sovereign AI is designed to solve this complexity, allowing organizations to maintain the scale of a global cloud while respecting the boundaries of local sovereignty. This is not merely about where data is stored, but about where it is processed, who can access it, and which legal frameworks govern its use.

The following subsections examine the two primary governance dimensions: regulatory alignment with the EU AI Act and related frameworks, and the ISO/IEC 42001 certification that establishes a global standard for AI management systems.

14.3.1 Navigating the Regulatory Frontier and the EU AI Act

The most significant shift in global compliance is the EU AI Act, the world's first comprehensive horizontal legal framework for artificial intelligence. Unlike previous regulations that focused primarily on data privacy, the EU AI Act categorizes AI systems based on risk, ranging from unacceptable to high risk and low risk. For SAP customers, this means that every agentic interaction must be evaluated against these legal tiers to ensure that the technology does not infringe upon fundamental rights or create systemic vulnerabilities.

SAP has proactively aligned its SAP Business AI architecture with these rigorous requirements. By acting as a secure bridge between the provider of the foundation model and the enterprise deployer, SAP ensures that the responsibilities of the AI value chain are clearly distributed. This includes providing the comprehensive documentation, transparency reports, and technical guardrails required for high-risk use cases, such as those found in human resources or critical infrastructure management. Compliance is treated as a native feature of SAP Autonomous Suite, ensuring that as new implementing acts and national laws emerge, the underlying core remains automatically aligned with the latest legal mandates.

To support the requirement for digital sovereignty, SAP utilizes its global network of data centers to provide localized processing environments. A cornerstone of this strategy is the EU Access protocol. For customers with stringent sovereignty needs, AI features embedded within SAP cloud services, including the generative capabilities of Joule, adhere to specific data processing agreements that ensure data remains within the European Economic Area.

This regional governance extends to how data is handled by third-party models. Through the generative AI hub, data is masked and filtered within a company's specified jurisdiction before any request is sent to a global foundation model. Furthermore, SAP ensures that its AI infrastructure is available in key regional hubs, such as EU10, EU11, US10, and AP10. This allows organizations to choose the deployment zone that best matches their regulatory obligations, providing a sovereign cloud experience where the intelligence is local, but the innovation is global.

14.3.2 ISO/IEC 42001 and the Gold Standard for Governance

Beyond regional laws, global compliance requires a standardized approach to organizational management. SAP has achieved the ISO/IEC 42001 certification, which is the premier international standard for AI management systems. This certification provides a verified, globally recognized framework for developing, deploying, and governing AI systems responsibly. It comprehensively covers the internal governance of the Joule platform, SAP AI Core, and SAP AI Launchpad.

By adhering to this standard, SAP provides customers with an audit-ready environment. When a company is asked by a regulator or an internal auditor to prove that its autonomous agents are safe and ethical, the ISO/IEC 42001 certification serves as a powerful testament to the maturity of the organization's governance. It ensures that risk management, data quality, and system transparency are not just theoretical goals, but documented and certified operational realities.

A critical aspect of the sovereign AI strategy is the clear definition of roles. In the agentic era, SAP can act as a provider, a deployer, or an integrator depending on the specific use case. The SAP Business AI framework simplifies this complex ecosystem for the customer by providing a robust contractual assurance layer. Whether an organization is using a prebuilt Joule Skill or building its own custom agent on the generative AI hub, the SAP cloud services agreement provides the definitive legal basis for data processing and the exercise of user rights.

14.4 Autonomous Suite Resilience and the Secure Core

Equally critical to the responsible enterprise is the technical architecture required to maintain the stability and integrity of core business operations. As AI moves from the periphery into the center of applications like SAP S/4HANA and SAP SuccessFactors, it must adhere to the same rigorous security standards that govern financial transactions and human resources records. In the agentic era, resilience is defined by the ability of a system to remain functional, secure, and accurate even when subjected to the unique stresses and vulnerabilities introduced by LLMs. SAP approaches this challenge through a comprehensive strategy that protects the secure core while enabling safe, intelligent extensibility.

The following sections examine three layers of this defense: the threat mitigation architecture that addresses emerging generative AI vulnerabilities, the generative AI hub as a hardened processing enclave for all model interactions, and the Joule security architecture that governs the conversational interface.

14.4.1 The Defense Architecture and Emerging Threat Mitigation

Traditional enterprise defenses are necessary but insufficient for the generative era. The secure core of SAP Autonomous Suite relies on robust authentication, authorization, and auditability. However, natural language interfaces blur the boundaries between user input, system commands, and generated output. SAP enforces strict separation between core application logic and the generative extensibility layer, ensuring that even if a model behaves unpredictably, fundamental business processes remain insulated. The architecture limits the blast radius of any AI failure; hallucinations or adversarial inputs cannot corrupt master data or trigger unauthorized workflows. Figure 14.5 illustrates this layered defense model.

SAP aligns its defensive architecture with the OWASP Top Ten vulnerabilities for LLMs. The agentic enterprise must be hardened against *prompt injection*, in which malicious instructions are disguised within legitimate queries, and against data exfiltration attempts that exploit conversational interfaces. SAP addresses these issues through input validation, output filtering, and strict separation of instruction and data channels.

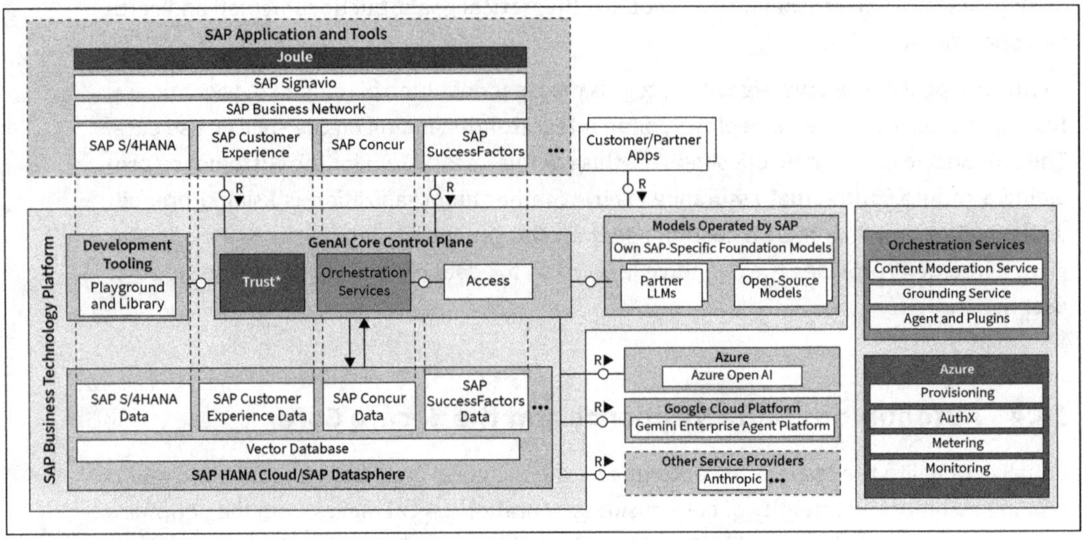

Figure 14.5: LLM Security with Generative AI Hub

Critical vulnerabilities also arise from insecure output handling and excessive agency. SAP mitigates these through validated zero-trust output processing, explicit role-based authorization for every plugin interaction, and strict input validation with rate limiting and aggressive data sanitization.

14.4.2 The Generative AI Hub as a Hardened Processing Enclave

A cornerstone of this defensive strategy is the generative AI hub within SAP AI Core. This environment functions as a resilient and segregated processing enclave that facilitates communication between SAP Autonomous Suite and various foundational models while strictly enforcing data privacy through a process known as *orchestrating prompt privacy*.

The generative AI hub's security rests on non-negotiable principles: The prompt is authorized and directed only to approved models that comply with regional compliance requirements. A strict zero-copy architecture ensures data sovereignty, and every interaction is logged for audit purposes.

Figure 14.6 shows a generative AI reference architecture on SAP BTP, spanning provider and subscriber subaccounts. It depicts how end user software-as-a-service (SaaS) applications connect via HTTPS to UI components and services, which in turn use the generative AI hub (SAP AI Core and SAP AI Launchpad), infrastructure services, and external foundation models like Azure OpenAI and Aleph Alpha, all within a multitenant SaaS setup.

Before data leaves the corporate perimeter, the system applies data masking to natural language input, identifying and replacing PII with synthetic tokens. This ensures the model processes the business logic without ever seeing actual names, account numbers, or proprietary details.

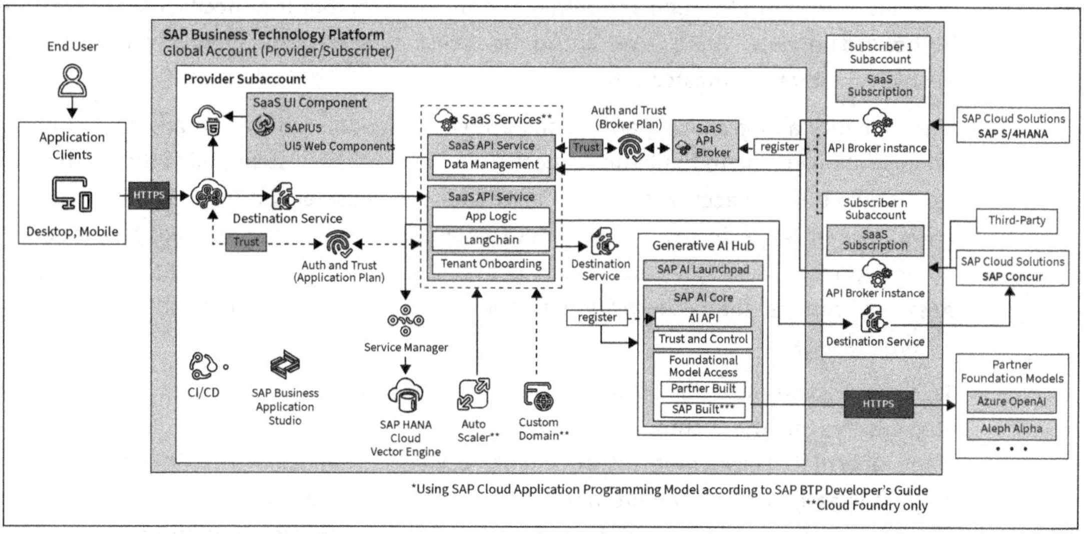

Figure 14.6: Generative AI Reference Architecture

Once the foundation model generates a response, the hub continues its defensive orchestration. It applies rigorous output filtering to the returning payload, ensuring no malicious instructions or inappropriate content can reenter SAP BTP. The system then performs a demasking process, where it rehydrates the data by replacing the synthetic tokens with the original secure information before presenting the result to the user.

The entire orchestration operates under a strict zero-retention policy. Third-party model providers are contractually and technically prohibited from retaining prompts or using them to train foundation models. Proprietary organizational knowledge is never absorbed into a public algorithm. This ensures the Autonomous Suite remains a resilient sanctuary for corporate data.

14.4.3 Securing the Digital Assistant: The Joule Architecture

As Joule becomes the primary conversational interface for the enterprise, securing this specific interaction point is paramount. The Joule security architecture is designed to ensure that the ease of natural language querying does not bypass established corporate data governance or introduce new vectors for unauthorized data access.

The first line of defense is robust identity and access management integrated directly with SAP Cloud Identity Services. Before Joule even processes a request, the user must be authenticated through secure single sign-on protocols. More importantly, Joule operates on a strict principle of *principal propagation*. This means the assistant inherits the exact role-based access control permissions of the user directly from the underlying backend systems, whether that is SAP SuccessFactors, SAP S/4HANA, or SAP Customer Experience. When an employee asks Joule a question, the system does not use a generic service account; it searches through and retrieves only the data that the specific user is explicitly authorized to view. There is no backdoor access or elevated system privilege that can expose unauthorized master data.

Once identity and granular permissions are verified, Joule utilizes an advanced prompt orchestration engine to fulfill the request. This is where the system builds the secure context required for an accurate answer. Instead of relying solely on the pretrained, general knowledge of an external model, Joule leverages retrieval-augmented generation to ground its responses. By securely tapping into SAP Knowledge Graph and connected business applications, the orchestration engine retrieves specific, authorized, and real-time enterprise data.

This retrieved business data is then injected into the prompt as factual context. This critical step ensures that the eventual response is highly accurate, factually grounded in the corporate reality of that exact moment, and highly resistant to hallucinations. Once this localized context is gathered and the prompt is securely constructed, Joule routes the structured query into the generative AI hub for processing. This seamless integration ensures that the user experiences the fluid intelligence of a conversational agent, while the enterprise maintains the ironclad security of its foundational systems.

14.5 Agent Lifecycle Management

There is a governance problem that every organization scaling AI will hit, and most hit it sooner than expected. Individual agents work well: A collections agent resolves disputes, a sourcing agent identifies alternative suppliers, a planning agent rebalances inventory. Each one delivers value in its domain. But when you have 50 agents running, then 150, then 500, across finance, procurement, supply chain, and HR, you face a fundamentally different challenge. How do you know what is running, whether it is delivering value, and when it should be retired? Figure 14.7 shows how SAP AI Agent Hub manages agents across their full lifecycle, from planning and building through discovery, monitoring, governance, and eventual decommissioning.

This is not hypothetical. By mid-2026, SAP AI Agent Hub had discovered over 100,000 AI agents across 150 customer environments (*http://s-prs.co/v633514*). Many of those agents were unknown to the teams nominally responsible for them. Shadow agents, redundant agents, agents accessing data they should not have (a pattern Korindal's IT team encountered when auditing their own AI deployment), agents consuming compute without delivering measurable outcomes: The pattern mirrors what happened with SaaS sprawl a decade ago, except that agents can act, not just store data, and so the governance stakes are higher.

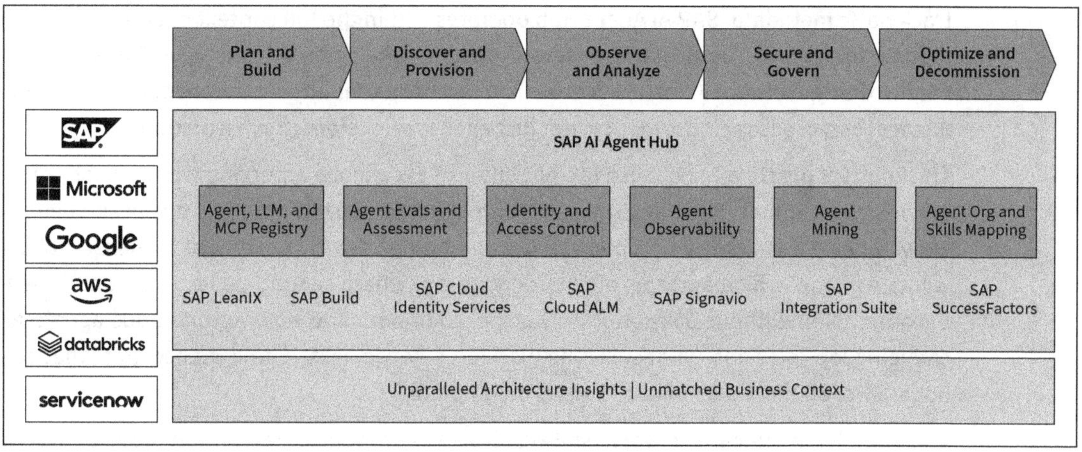

Figure 14.7: Agent Lifecycle

SAP addresses this through SAP AI Agent Hub, which manages agents across their full lifecycle in the following phases. Think of it as application lifecycle management, but purpose-built for autonomous AI assets:

1. **Plan and build**
   An agent should not exist without a reason tied to business architecture. Before anything reaches production, SAP AI Agent Hub integrates with SAP LeanIX for enterprise architecture context and SAP Signavio for process intelligence. This means that every proposed agent maps to a defined business capability, a specific process step, and a known set of application and data dependencies. The question is not "Can we build this agent?" but "Does this agent align with our transformation strategy, and does it duplicate something we already have?"
2. **Discover and provision**
   You cannot govern what you cannot see. The generative AI hub creates a comprehensive inventory of every agent in the landscape: SAP-delivered agents, custom agents built in Joule Studio, third-party agents accessed through Model Context Protocol (MCP) servers. Each is registered with its purpose, owner, scope, and permissions. This is the foundation. Without a complete registry, governance is guesswork.
3. **Observe and analyze**
   Once agents are operational, telemetry flows through SAP Cloud ALM and the SAP Cloud SDK. The hub surfaces execution traces, decision paths, response times, error rates, and cost per agent. This data answers the questions that matter: Is the collections agent actually reducing days sales outstanding? Is the sourcing agent finding better suppliers or just cycling through the same list? Observation without business context is just monitoring, but observation mapped to process outcomes is governance.

What makes this framework distinct from competing approaches is the depth of context available at each phase. Most AI governance tools operate as registries: they list agents and track basic metadata. SAP AI Agent Hub operates within the full context of enterprise architecture (from SAP LeanIX) and business process intelligence (from SAP Signavio). This means that governance decisions are informed by actual application dependencies, data lineage, and process conformance, not just agent-level telemetry in isolation.

This matters practically because the AI Center of Excellence emerging within most organizations serves multiple constituencies with different questions. The business wants to know where AI is improving operations and whether the improvement is measurable. IT wants to know which systems agents depend on, where redundancies exist, and how to decommission without disruption. HR wants to understand how autonomous agents are changing job roles and what skills the workforce needs next. A governance layer that cannot answer all three sets of questions is incomplete.

The underlying principle is worth stating directly: Autonomous execution requires proportional governance. Every agent activated, every scope expanded, every new domain automated increases the governance surface. The organizations that scale successfully are not the ones that build the most agents. They are the ones that govern what they build with the same discipline they apply to financial controls, data privacy, and application security. SAP AI Agent Hub provides the mechanism, but the discipline is organizational.

14.6 Summary

The transition to an agentic future is defined not by technological capability alone but by governed trust. Ethics, security, and compliance form the essential scaffolding that allows organizations to scale autonomous operations with confidence. SAP's 10 guiding principles, enforced through the SAP Business AI lifecycle and the AI Ethics Steering Committee, ensure that AI amplifies human judgment rather than replacing it.

This ethical foundation is matched by legal alignment with the EU AI Act, ISO/IEC 42001 certification, and localized processing through EU Access, and by the technical resilience of SAP Autonomous Suite. The generative AI hub's defense-in-depth methodology, the zero-retention policy, and Joule's role-based security architecture together ensure that proprietary data remains insulated from public models while conversational interfaces operate within strict authorization boundaries.

In the agentic era, governance is the accelerator. When AI is rooted in enterprise reality and guided by a secure, ethical core, it becomes a transparent, reliable, and powerful ally. The enterprise that masters these dimensions does not just automate; it secures organizational trust and empowers the human insight that shapes the future of the business.

The next chapter translates this governance foundation into a practical implementation roadmap, showing how organizations activate these capabilities progressively, build the internal competencies to sustain them, and measure the business outcomes that justify continued investment in responsible, autonomous operations.

Chapter 15
Operationalizing Your Vision

The previous chapters laid out what the Autonomous Enterprise is. This chapter shows how to build it. We will walk through the sequence of decisions, including deployment approach, instance strategy, readiness assessment, roadmap sequencing, and value governance, that translate vision into operations.

Most enterprises that decide to pursue the Autonomous Enterprise have clarity on what they want to achieve: Release trapped working capital. Improve order fulfillment speed. Reduce the time it takes to respond to an exception before it becomes a failure. The vision is usually clear enough. What breaks most transformations is not the clarity of the destination but the absence of a credible, detailed path from here to there. This chapter is about that path. It covers four elements that turn the Autonomous Enterprise vision from strategic intent into operational reality: the deployment approach you choose, the way you structure your systems to scale, the tools you use to assess readiness, and the governance framework that tracks whether you are delivering value as you go.

The four people in this chapter are the same ones you have met before: Maria, the CFO of Korindal, trying to close a $400 million working capital gap; James, running supply chain, managing a 14 percent fulfillment miss that costs $30 million per year; David, heading IT, watching his release calendar fill faster than it empties; and one new voice, the program executive who has to connect all three problems into a single transformation plan that does not destroy the business in the process of improving it.

By the end of this chapter, you will understand how these elements work together. You will see how Korindal's three separate problems roll into a single, multiyear roadmap through which value emerges continuously, humans and agents work side by side, and every quarter produces measurable progress.

15.1 Deployment Strategy and Instance Planning: Choosing Your Starting Architecture

The first structural decision any organization faces is simple: how to move from its current systems to the cloud-based, artificial intelligence (AI)–enabled architecture without disrupting the business in the process.

Three deployment approaches dominate this decision, each making different trade-offs between speed, risk, and freedom to reshape operations:

- **Greenfield: Starting clean**

  A greenfield deployment builds a new SAP S/4HANA system from the ground up, using SAP Best Practices content as a starting point. Rather than inheriting existing configurations and customizations, the organization designs processes aligned to standard capabilities and modern platform features such as automation, embedded analytics, and AI.

  The advantage is a meaningful reset. Legacy complexity, redundant custom code, and inconsistent processes can be removed, enabling a cleaner and more scalable operating model. The trade-off is effort and change impact. Greenfield programs often require significant time for process design, testing, and organizational readiness, and they demand a willingness to operate differently, not just on a new system.

  Greenfield is most appropriate when customization debt is high, when process transformation is a strategic goal, and when the organization can absorb significant change.

- **Brownfield: Carrying the past forward**

  A brownfield (system conversion) approach upgrades an existing SAP system to SAP S/4HANA in place. Data, configurations, and custom code are largely retained, though custom code must be assessed and adapted for compatibility. The result is a system that remains familiar to users while benefiting from modern infrastructure, improved performance, and the SAP Fiori user experience. Business disruption is minimized, and transition timelines are often shorter than for more transformational approaches.

  The trade-off is limited transformation. Existing processes, whether efficient or not, are preserved, and heavy customization can make it harder to fully adopt standard innovations such as advanced automation or AI-driven capabilities.

  Brownfield is well suited for organizations with stable, effective processes or where minimizing disruption is a priority.

- **Selective data transition: Taking what works**

  Selective data transition is a hybrid approach that combines characteristics of both the greenfield and brownfield options. It enables organizations to selectively reuse existing processes, configurations, and data while redesigning others. Rather than moving everything or starting from scratch, this approach allows organizations to define which business units, processes, and datasets are migrated. This includes selectively transferring master data and relevant historical data while leaving behind obsolete information.

  This flexibility enables phased deployments, such as by region or function, thereby reducing the risk of a single, big bang go-live and supporting a more controlled transformation. The trade-off is complexity. Managing selective migration decisions, ensuring process and data consistency, and coordinating phased go-lives require strong integration architecture and disciplined program governance. Decisions about what to retain, redesign, or exclude must be clearly defined up front, as changes later in the program are costly.

  This approach is particularly effective for large, multidivision organizations in which a single migration approach does not fit all parts of the business.

As highlighted in the Chapter 1 case study, Korindal's landscape is a collection of regional platforms built over years of acquisition, with limited integration and no unified data model. In that context, the choice of deployment strategy is less about preference and more about what the organization can realistically execute. A full greenfield approach would deliver the cleanest outcome but would require replacing multiple ERP backbones simultaneously, a level of disruption that Korindal may not be currently positioned to absorb. A brownfield conversion, while lower risk, would largely preserve the fragmentation that is already preventing Maria from achieving a consolidated view of receivables and cash across the group.

Selective data transition therefore becomes a credible path. Rather than forcing a single transformation model across the entire enterprise, it allows Korindal to establish a clean finance core while selectively retaining and integrating regional systems where they continue to serve operational needs. The decision ultimately comes down to three factors: the level of customization in existing systems, the organization's capacity to absorb change, and the urgency of building a unified data foundation. In Korindal's case, these conditions point toward a selective approach as a practical way to begin resolving the constraints surfaced in Chapter 1.

The deployment approach answers, How do we get from here to there? The *instance strategy* answers a different question: Once there, how do we structure the landscape so that it can continue to evolve without becoming harder to change over time?

An *instance* is a single production ERP system. There are four key instance patterns, as highlighted in Figure 15.1:

- **Single global instance**
  Single global instances work when an organization is willing to operate on a single shared platform. All processes run from end to end without cross-system integration. Local variants such as company codes, country tax configurations, master data extensions, and the like remain possible within the instance. The real risks are that every change, upgrade, or remediation affects the entire enterprise simultaneously and that global standardization pressure can suppress legitimate local needs.
- **Line-of-business instances**
  Line-of-business (LoB) instances deploy a separate instance per functional domain, such as finance, procurement, order management, or manufacturing, where each instance serves its respective function globally. This gives functional teams the autonomy to change and deploy independently. The cost is integration: Every functional handoff (a confirmed purchase order flowing into finance, an order settling across systems) requires an interface, and there is no single, in-system view of end-to-end processes.
- **Regional instances**
  Regional instances are common when regulatory, tax, or language complexity varies significantly. One system per region means that you can comply with local requirements. It also means building integration layers to roll up to global finance and global reporting.

- **Template/hybrid model**

  A template model means that one master instance is used as a configuration baseline from which other instances are cloned and rolled out. Each derived instance starts identically to the template and then receives local additions. The governance challenge is keeping those derivatives aligned as the template evolves; if local instances accumulate custom code and diverge, you lose the ability to roll out template upgrades without expensive reharmonization.

  A hybrid model means deliberately mixing instance patterns, such as running a single global finance instance for financial consolidation while allowing regional or LoB instances for supply chain or order management. The governance challenge here is different: It is about deciding which functions belong on which instance type and managing the integration between them.

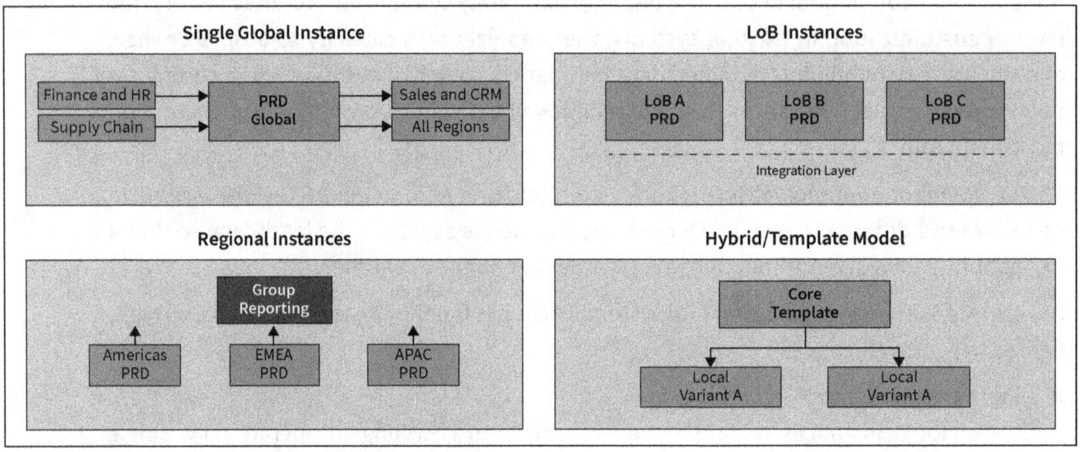

Figure 15.1: Instance Strategy Patterns

For Korindal, the instance strategy centers on establishing a global finance core as the single system of record, supported by a template-driven model across regions. A new cloud-based global ERP instance provides the standardized finance layer, including group reporting and control. Around this, regional systems continue to operate where required for local processes and compliance, progressively aligned to the global template over time. The existing European system remains in place initially, integrated into the global layer for consolidated financial reporting.

This approach reflects the underlying structure of the business: a need for centralized financial control combined with regional operational flexibility, enabled through a template that balances standardization with local variation.

15.2 Organization Readiness Assessment

Once the deployment approach and instance strategy have been established, programs encounter a different kind of problem. The question is no longer what architecture should look like, but whether the organization can actually execute it.

This is where most transformations begin to diverge from plans, not because the target state was wrong, but because the conditions required to deliver it were never fully understood. The purpose of organization readiness assessment is not to measure maturity or compare against peers. It is to establish whether there is enough clarity to proceed without introducing avoidable delay, cost, and rework during execution.

In practice, baseline readiness comes down to six areas that consistently determine whether a program accelerates or stalls. These are not technical assessments. They are a leadership alignment mechanism, a way to establish whether decisions will move at speed, whether ownership will be clear, and whether value will compound or leak away.

Figure 15.2 shows the six dimensions: scope and value, non-negotiables, capabilities and processes, data systems and integration, governance, and change and adoption. Each one is tightly connected. Weaknesses in one will surface later through another, usually when the cost of resolving it is the highest.

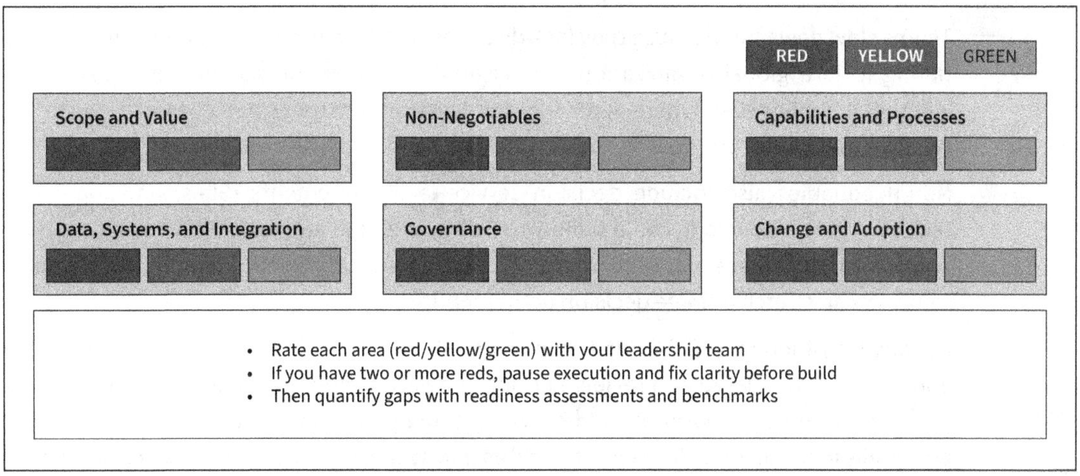

Figure 15.2: Organization Readiness Baseline

We'll walk through these six dimensions in more detail in the following sections.

15.2.1 Scope and Value

Scope and value is where the first level of clarity is required. Programs often begin with broad statements of intent, such as *transform finance globally*, but execution depends on specific outcomes. It must be clear what is in scope, what is not, which metrics are expected to move, and who is accountable for those outcomes.

When scope is vague and value is undefined upfront, decisions drift during delivery and value erodes without a single obvious failure point. A CFO discovers at go-live that the program was supposed to improve days sales outstanding (DSO), but no one documented a baseline, no one owns the metric, and the finance organization has already moved on to other priorities.

The key is asking three questions early and answering them with precision: What is explicitly in and out? Which outcomes or KPIs must move, and what are their current baselines and targets? Who owns value realization from end to end?

At Korindal, Maria (the CFO) made this explicit from the start: DSO reduction is owned by her and measured monthly, with specific milestones tied to each phase of the roadmap. That clarity changed everything downstream. Every design decision could be tested against that outcome.

15.2.2 Non-Negotiables

Non-negotiables establish the boundaries within which those decisions are made. These typically include regulatory constraints, timeline commitments, constraints on customization, and architectural choices that have already been fixed. When these are not explicit, every design decision becomes open for reconsideration.

Teams slow down not because they lack direction, but because they are constantly renegotiating it. In a global organization, one region's requirement becomes a debate about whether it applies everywhere. A payment processing constraint becomes a thread in governance that blocks progress for weeks because the boundaries were never clear.

Non-negotiables also include decision service-level agreements (SLAs) and mandatory technical gates. How long can a build versus buy decision stay open? Five business days? That constraint forces clarity. Without it, consensus requirements and multistakeholder sign-offs can stretch a single decision across months.

For any transition to SAP S/4HANA from SAP ERP, SAP Readiness Check is non-negotiable and must be completed and reviewed prior to design freeze. Its findings on simplification items, custom code footprint, add-on compatibility, and integration inventory become guardrails for clean core decisions, the extension strategy, and the technical critical path.

SAP Readiness Check provides an initial assessment to help customers prepare for key transformation scenarios including SAP ERP to SAP S/4HANA conversion, SAP S/4HANA upgrades, and SAP Business Warehouse (SAP BW) to SAP BW/4HANA transitions. The analysis identifies relevant simplification items, custom code impacts, add-on compatibility, integration considerations, data volume and sizing requirements, and recommended remediation activities to support project planning and scope definition. Refer to SAP Notes 2913617 and 3059197 for SAP S/4HANA conversion and SAP S/4HANA upgrades, respectively, to understand the latest prerequisites, data collection steps, and execution guidance.

The practical output is a small set of guardrails, such as *no custom code on core transactional processes* or *customer master data standards apply globally*. Certain regional compliance requirements are non-negotiable; others are additive. When these are written down and agreed upon by leadership, the work moves faster.

15.2.3 Capabilities and Processes

Capabilities and processes introduce a different type of reality. What the organization believes its processes look like is rarely how work is actually performed. Variations, manual interventions, and region-specific adaptations accumulate over time, particularly in organizations that have grown through acquisition. These are not anomalies. They are mechanisms the business relies on to function.

Organizations can adopt different levels of process analysis depending on their transformation objectives. For a high-level assessment of a selected few business processes, SAP Signavio Process Insights, discovery edition provides rapid visibility into process performance issues, improvement opportunities, and SAP S/4HANA innovation recommendations. For more comprehensive analysis, SAP Signavio Process Insights (full edition) delivers broad process monitoring, benchmarking, and continuous improvement capabilities across multiple end-to-end business processes and performance indicators. For the deepest level of insight, SAP Signavio Process Intelligence leverages process mining to uncover root causes, process variants, and execution bottlenecks, making it particularly valuable for process reengineering and large-scale business transformation initiatives. Collectively, these solutions help organizations identify where AI-enabled process automation and intelligent workflows can deliver the greatest business value and maximize the transformation return on investment (ROI). Refer to Chapter 13 for more details on these process analysis options, their capabilities, and recommended use cases.

Process readiness also identifies where standardization will deliver the most value. If one region processes invoices consistently and another has built 50 local variations, that difference exists because someone solved a real business problem at some point. Simply eliminating it is a good way to break something that was working. The key is understanding what is driving each variation, not assuming it is unnecessary.

15.2.4 Data, Systems, and Integration

Data, systems, and integration move the focus from process to information. The question here is not whether data exists, but whether it can be trusted and used consistently. In most multisystem landscapes, business entities are represented multiple times with different identifiers and attributes. A customer, for example, can exist in separate forms across regional systems, each carrying partial information.

This becomes visible the moment systems are examined together. If a customer master has three different representations, then credit decisions, collections prioritization, and reporting outcomes will vary depending on which system is used. The same customer appears as three different entities, each with different outstanding balances, different credit limits, and different payment histories.

SAP LeanIX provides enterprise architecture management capabilities in three offerings: SAP LeanIX Application Portfolio Management, SAP LeanIX Architecture and Road Map Planning, and SAP LeanIX Technology Risk and Compliance. Using SAP landscape discovery, organizations can view their SAP applications and SAP Business Technology Platform (SAP BTP) services to establish a baseline of what is running, where it is running, and who owns it. A primary value driver for SAP LeanIX is reducing expenditure on redundant SaaS applications by identifying where the portfolio contains duplication, where applications carry technology or compliance risk, and where consolidation decisions are warranted. The outcome is an application estate that can be reasoned about: what to keep, what to retire, and what integrations the estate currently depends upon. For organizations moving from landscape discovery into integration planning, SAP offers the SAP LeanIX integration setup assessment as an advisory service.

For transitions from SAP ERP, SAP Readiness Check quantifies data model changes, custom code remediation, add-on compatibility, and integration touchpoints. In SAP ERP to SAP S/4HANA transitions, running this check surfaces exactly these inconsistencies, alongside custom code dependencies and data issues that were not visible at the planning stage.

Technology readiness builds on these data and system findings. The issue is not whether the organization has modern platforms but whether the landscape can support the way processes need to run. In many environments, integration is still batch-driven, interfaces are partially documented, and data is not available at the point where decisions need to be made.

To address this, organizations can perform an integration readiness assessment using the Integration Assessment, based on SAP Integration Solution Advisory Methodology, which we introduced in Chapter 11, Section 11.1.2. This assessment inventories current APIs, interfaces, middleware, event sources, and integration patterns while evaluating opportunities to adopt API-led and event-driven architectures using SAP Integration Suite and SAP BTP. Such an assessment helps determine whether the landscape can support near-real-time data availability for AI agents, intelligent applications, and autonomous workflows. Stable, governed APIs enable access to trusted business information at the point of decision, while event-driven integration patterns can significantly reduce dependency on traditional batch-oriented processing. Equally important, the assessment highlights the governance, observability, and traceability requirements needed to support autonomous operations, capturing what triggered an action, what data was used, what policies or rules were applied, and what outcome was produced.

Note

For additional guidance, refer to the following resources:

- SAP Integration Solution Advisory Methodology: *http://s-prs.co/v633515*
- Integration Assessment: *http://s-prs.co/v633516*

Collectively, these assessments help answer a broader question: Can the technology landscape support autonomous operations? Technology readiness is determined not by whether an organization has deployed SAP S/4HANA, SAP BTP, or other modern platforms, but whether the underlying architecture can support intelligent, cross-system decision making. Autonomous workflows depend on trusted data, stable APIs, well-governed integrations, and timely access to business information. Every automated decision may require information from multiple systems, making integration quality and data availability foundational capabilities rather than implementation details. Organizations that rely heavily on batch-based integration, fragmented interfaces, or inconsistent data models often discover that these constraints limit their ability to deploy AI-enabled processes and autonomous agents at scale. Equally important are governance and traceability: understanding what triggered an action, what information was used, what policies were applied, and what decision was produced. Together, SAP LeanIX, SAP Readiness Check, the Integration Assessment, and SAP Integration Suite on SAP BTP provide complementary perspectives on application portfolio readiness, ERP system readiness, and integration architecture readiness. These assessments help organizations assess whether the technical foundation for AI-enabled processes and autonomous workflows is already in place or still needs to be built.

15.2.5 Governance

Governance determines how issues are resolved when they arise, and how quickly. The quality of governance is visible not in documents but in outcomes: Do decisions that should take five business days actually take five business days? Do design reviews not resurrect decisions already made? Are architecture and security embedded early, not retrofitted late?

When ownership is unclear or decision rights are distributed without clear escalation paths, programs stall. When ownership is clear, they move.

The governance structure in Figure 15.3 illustrates this principle. The pyramid is not complicated. Each level has a specific function and cadence: The board sets the strategic direction quarterly. Executive sponsors ensure alignment monthly. The steering committee resolves cross-functional dependencies monthly. The architectural review board validates technical decisions semimonthly. The design authority board owns the detailed solution design semimonthly. Transformation management coordinates execution weekly. Finally, value assurance runs continuously.

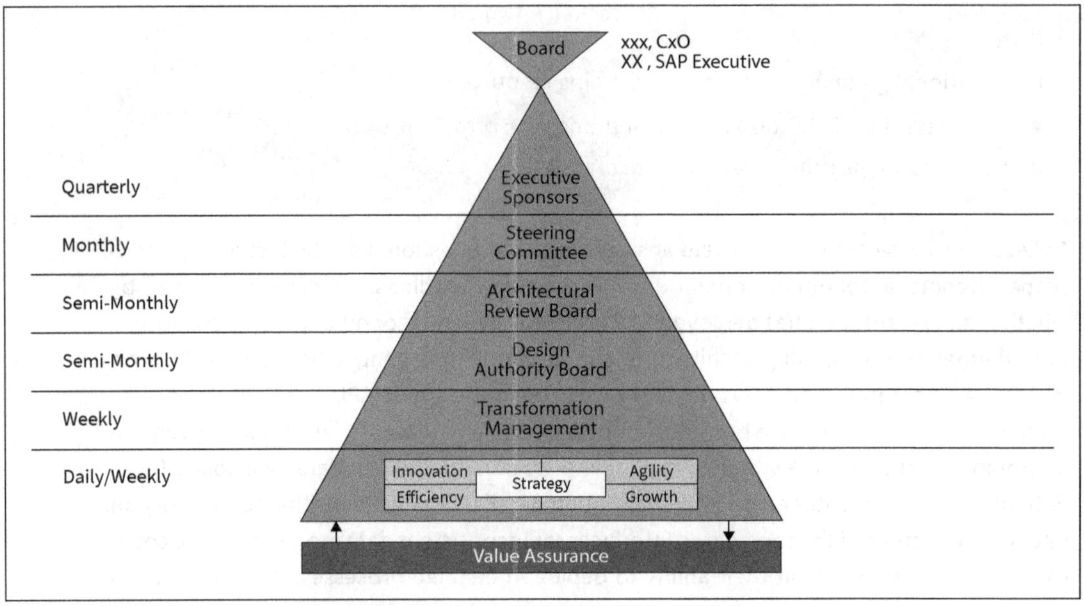

Figure 15.3: Example Governance Structure for Transformation Program

The structure works because it is explicit. Everyone knows which issues go to which level, and escalation paths are defined. A question about platform modernization does not bounce between committees; it goes to the design authority. An architectural constraint that affects multiple workstreams does not get revisited. It is settled at the architectural review board level and carried forward. Cross-functional trade-offs are resolved at the steering committee level, not debated at every working session.

The governance structure described previously focuses on transformation execution. As organizations introduce AI-enabled processes, assistants, and agents, governance responsibilities expand beyond program management and solution design. New questions emerge regarding accountability, operational boundaries, and decision authority. Governance must establish who approves autonomous capabilities for production use, what rules constrain their behavior, and how decisions can be reviewed and explained.

Three governance dimensions become increasingly important as organizations move from recommendation-based systems toward greater levels of autonomy:

- **Agent lifecycle governance**
  Organizations need a formal process for determining when a capability is ready to move from experimentation into production operations. Readiness criteria typically include business acceptance, performance thresholds, operational controls, governance maturity, and validation of expected outcomes. The objective is to establish confidence that the capability can operate consistently within the business environment.

- **Policy and constraint governance**
  Autonomous processes must operate within clearly defined business boundaries. For example, a collections process may be permitted to send reminders automatically but require human approval before offering significant discounts or changing payment terms. These policies must be owned by the relevant business functions, formally approved, version controlled, and consistently enforced across all deployments.
- **Observability and audit governance**
  As organizations delegate more operational activities to intelligent systems, traceability becomes essential. Business and audit teams must be able to determine what triggered an action, what information was used, which policies were applied, who approved the capability, and what outcome resulted. This level of transparency is necessary not only for compliance purposes but also for building organizational trust and business acceptance.

As levels of autonomy increase, these governance responsibilities naturally become embedded within existing governance bodies. The Architectural Review Board validates the technical controls, integration standards, and operational boundaries within which capabilities operate. The Design Authority Board defines the policies, approval requirements, and business constraints that govern behavior. Together, these governance functions help ensure that increasing autonomy remains aligned with business objectives, compliance requirements, and organizational risk tolerance.

The result is speed. Issues move through the structure, decisions stick, and teams execute rather than relitigate.

15.2.6 Change and Adoption

Every transformation alters how work is performed: sometimes in subtle ways, sometimes in fundamental ones. Manual effort shifts toward exception handling. Responsibilities move across roles. Process ownership becomes more explicit. These changes do not happen automatically because systems change.

The practical question is whether the organization has acknowledged and prepared for these shifts. In organizations where change is treated as a communication activity rather than an operating model adjustment, adoption tends to lag behind technical completion. Systems go live, but behaviors remain unchanged, and expected outcomes are delayed.

Role-based change planning is different from change communication. It identifies which roles change or are repurposed, how those changes will be introduced, when training aligns to real deployment waves, and how the organization will hear concerns and respond. Collection specialists shift from manually processing 90 percent of routine exceptions to handling 100 percent of tough, judgment-based exceptions. The work becomes higher value and lower repetition, but it is different work. Organizations that are organizationally ready accept this premise and retrain people into the higher-value work. Refer to Chapter 16 for more information on best practices for change management.

15.2.7 The Baseline Assessment Framework

In practice, a readiness assessment is used to establish all six dimensions before detailed design work begins. The work focuses on validating assumptions against actual system behavior and operational reality. It typically involves a combination of structured leadership discussion and direct system analysis using tools such as SAP Signavio Process Insights, discovery edition; SAP Signavio Process Intelligence (process mining); SAP LeanIX; and SAP Readiness Check.

The assessment is a leadership gate, not a scoring exercise. Each dimension is rated as red, yellow, or green by the executive team. Specific red conditions include the following:

- **Data/systems and integration**
  Red if the SAP Readiness Check is not completed or shows unresolved high-impact simplification items or add-on conflicts
- **Capabilities and processes**
  Red if baseline KPIs/PPIs are missing for in-scope processes
- **Scope and value**
  Red if metrics lack baselines, targets are not quantified, or ownership is undefined
- **Governance**
  Red if decision rights are unclear or escalation paths are undefined
- **Change and adoption**
  Red if role impact analysis is incomplete or training is not sequenced to deployment waves
- **Non-negotiables**
  Red if guardrails, decision SLAs, or mandatory gates (e.g., customization, readiness check completion) are not agreed upon by leadership

If two or more dimensions are red, execution pauses and clarity is established before build begins. Yellow dimensions are addressed through the initial design phase, and green dimensions proceed.

Across all six areas, a consistent pattern emerges: Issues that are visible early but left unresolved become constraints later. At that point, programs face a choice between delay and compromise. Most choose compromise, introducing workarounds that allow progress but limit value. The result is a system that works, but not in the way originally intended.

For large transformations, the readiness work takes 8 to 12 weeks. Gathering a reliable view across processes, systems, and regions is not immediate. However, the time spent early is typically recovered during execution, where fewer surprises occur, where fewer critical decisions need to be revisited, and where value compounds rather than leaks.

What the assessment ultimately provides is not a maturity ranking or competitive benchmark. It provides clarity on what will enable execution and what will constrain it. Programs that start with that clarity move with fewer interruptions and deliver outcomes more consistently. Those that do not encounter the same issues later, when the cost of resolving them is significantly higher.

With readiness established across all six dimensions and governance gates in place, the organization is now prepared to define the target architecture that will support the journey. The question shifts from whether the organization can execute to what the organization is actually building toward.

15.3 From Readiness to Target Architecture

Readiness assessments establish whether an organization is prepared to execute a transformation. They do not, however, define the architecture toward which the transformation is moving. Once the deployment strategy, instance strategy, process readiness, data readiness, and governance have been established, organizations require a clear target architecture that aligns technology modernization with business outcomes.

Throughout this book, we have described the Autonomous Enterprise as the combination of SAP Cloud ERP, LoB applications, SAP Business Data Cloud, SAP BTP, and intelligent capabilities operating as a connected system. The SAP AI-native North Star architecture provides an alternative view of that same destination, organizing those capabilities into four architectural layers: user experience, processes, business context, and platform services.

The SAP AI-native North Star architecture provides this target-state view. Rather than treating applications, data, integrations, and intelligent capabilities as separate initiatives, it organizes them into a unified operating model consisting of four interconnected layers—the user experience layer, process layer, foundation layer, and platform layer—as shown in Figure 15.4. Together, these layers provide the architecture through which users, business processes, enterprise data, and operational platforms interact as a connected system.

The user experience layer provides a unified engagement model through which users interact with business capabilities. The process layer connects applications, workflows, integrations, and agents into end-to-end business processes. The foundation layer provides business context through enterprise data, knowledge, analytics, and semantic models. The platform layer provides runtime, governance, observability, security, and lifecycle management capabilities that allow solutions to operate consistently and at scale.

For transformation programs, this architecture serves as a directional guide rather than an implementation blueprint. Organizations do not deploy the four layers independently. Instead, capabilities introduced during each deployment wave gradually contribute to the target state. Process standardization contributes to the process layer. Data harmonization contributes to the foundation layer. Integration modernization contributes to both the process and platform layers. Governance investments strengthen the platform layer. Over time, these initiatives converge into a more connected and adaptive operating environment.

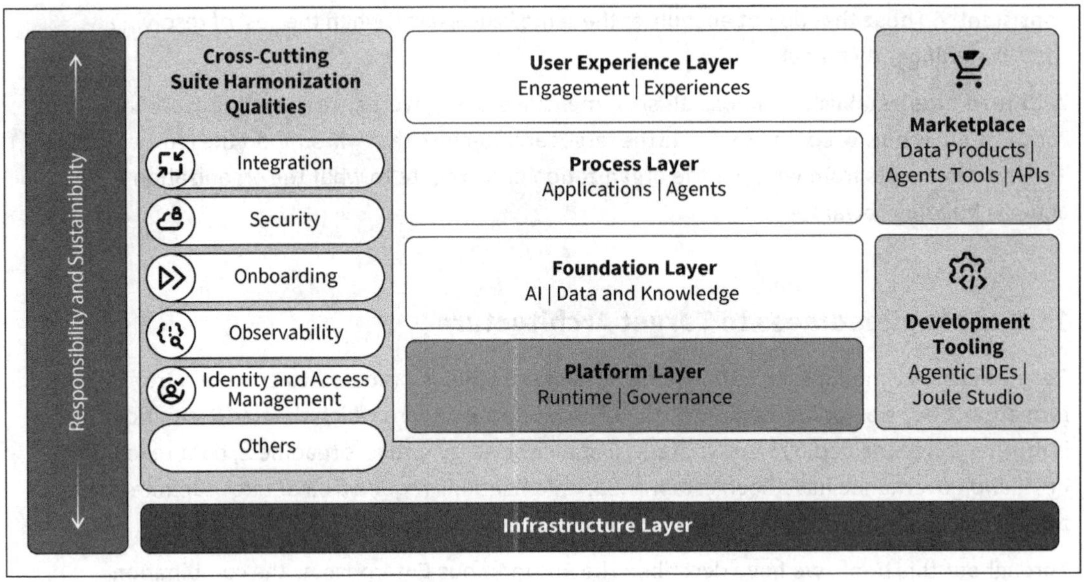

Figure 15.4: Building Blocks of SAP AI-Native North Star Architecture

The North Star architecture defines the destination. The question that follows is not whether to build toward it, but how to sequence the journey so that each step delivers measurable business value and builds organizational capability for the next phase. That sequencing is the roadmap—the mechanism that transforms architecture from aspiration into outcomes.

Note

To learn more about the North Star architecture, visit *http://s-prs.co/v633504*.

For comprehensive guidance on implementing the North Star architecture through SAP LeanIX-based landscape discovery and SAP Cloud ALM-enabled execution, see Chapter 13. Together, these capabilities provide a structured approach for translating architectural vision into measurable outcomes and tangible business value.

15.4 Multiyear Roadmap

The roadmap sequences how capabilities mature over time. It is not a list of projects. It is a governance mechanism that connects strategy, execution, and value realization through three overlapping phases: crawl (foundation), walk (optimization), and run (autonomy).

Figure 15.5 highlights an example of a multiyear transformation roadmap. The following are its key phases:

- **Plan and Design (Q2–Q3 2026)**
  Readiness assessment, process standardization, master data reconciliation, AI governance framework.
- **Crawl (Q3 2026–Q1 2027)**
  Foundation infrastructure live. Systems connected via event-driven APIs.
- **Walk (Q1–Q4 2028)**
  Automation of routine processes. Collections and planning workflows accelerate. AI agents begin limited autonomy (handle routine exceptions within defined guardrails; humans oversee).
- **Run (Q4 2028 onward)**
  Agents operate autonomously on routine work. Humans shift to exceptions and strategy. Continuous learning and adaptation.

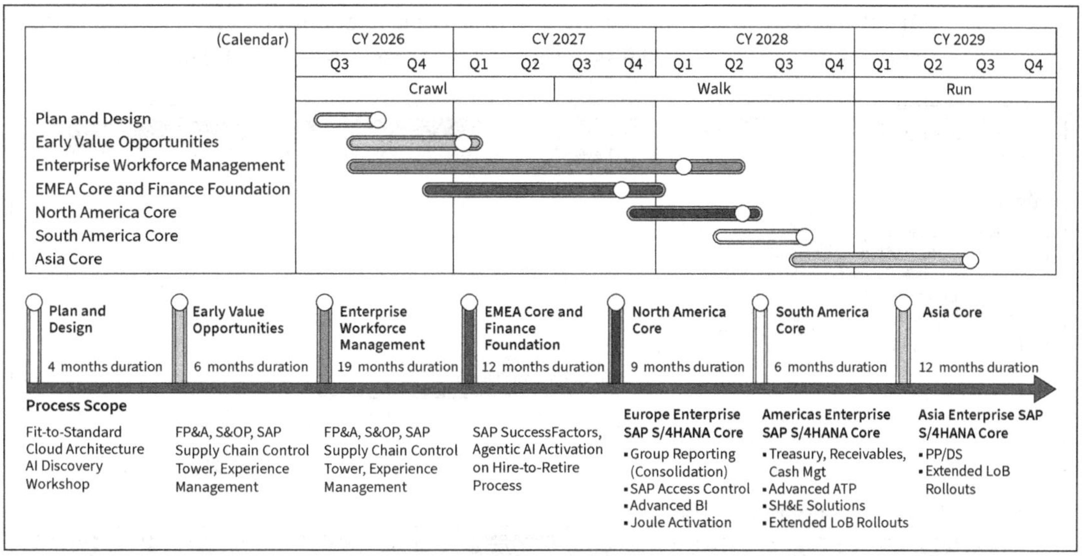

Figure 15.5: Broader Multiyear Transformation Program: Example

Three regional SAP S/4HANA cores deploy sequentially: EMEA, NA, and Asia. Progression to each subsequent region depends on achieving measurable business value from the previous deployment, rather than merely completing the implementation project. Dependency is explicit. Finance DSO improvements in the first phase enable collections automation in the next phase. Supply chain demand sensing in the first phase enables planning and agent autonomy in Phase 2. Value compounds because each phase enables the next.

Now let's zoom into one outcome thread (DSO) to show quarterly progression through that same timeline. Every outcome in Figure 15.5 has the same structure—dependencies, value gates, AI agent activation—but they progress at different speeds because they have different prerequisites and business readiness points.

Figure 15.6 shows what *outcome-driven* means. Start with the business outcome, to reduce DSO by two to four days. Work backward through the phases:

1. **Foundation**
   Standardize accounts receivable/disputes/collections. Lock baseline. Real-time visibility. AI agents assist (recommend, flag); humans decide. Value: 0.5–1.0 days.
2. **Optimize**
   Automate invoice matching. Collections agent activates with limited autonomy (handles routine dunning, discount offers); specialists handle disputes. Value: 0.3–0.7 days.
3. **Insights**
   Machine learning predicts payment risk. SAP Business Data Cloud unifies data. Agent authority expands as confidence and governance mature. Value: 0.5–0.7 days.
4. **Autonomous**
   Agent operates within policy. Humans focus on relationships and exceptions. Agent autonomy expands as business acceptance increases. Value: 0.7–1.0 days.

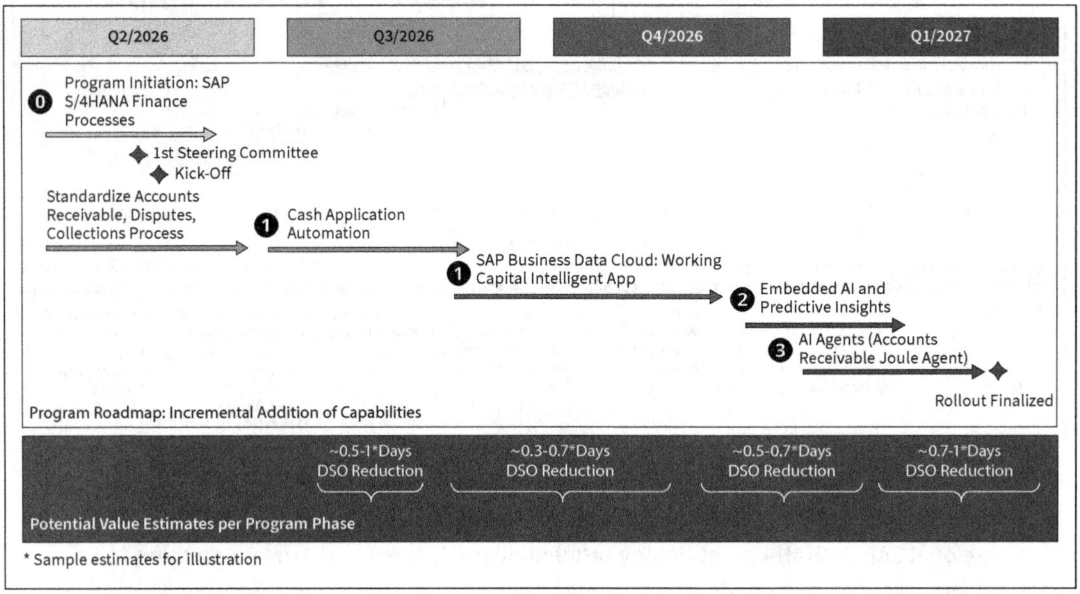

Figure 15.6: Outcome-Based Roadmap: Working Capital, Days Sales Outstanding

The outcome-oriented mindset is all about starting with the business outcome, not the technology. A project-driven roadmap says: "Implement ERP, build a data lake, deploy AI agents." An outcome-driven roadmap says: "Reduce DSO from 55 to 45 days. Here are the sequence of capabilities and the AI agent activation points that will make it real."

AI agents are not activated because the technology exists. They are activated when the business has achieved sufficient process standardization, data quality, governance maturity, and organizational trust to delegate decisions safely. Autonomy emerges from business readiness supported by technology, not technology readiness alone. This is the discipline that makes Autonomous Enterprise real.

The roadmap defines the phases. Within those phases, the question becomes: Which AI capabilities should be activated in each phase? This is not a rhetorical question. The appropriate AI approach varies depending on whether the business problem involves structured data, unstructured content, pattern recognition, or coordinated decision-making across multiple systems. Understanding these distinctions allows organizations to sequence not just when capabilities activate, but which capabilities activate at each stage. The following section explores various AI approaches—each with distinct strengths, requirements, and ideal scenarios—to help you make that decision with clarity.

15.5 The AI Golden Path: From Business Intent to Deployment

A roadmap defines what capabilities will be delivered and when. Organizations must also determine how those capabilities progress from business concepts into production solutions. This is particularly important for capabilities that combine applications, workflows, analytics, integrations, and agents.

SAP's AI Golden Path provides a structured approach for moving from business intent to deployed solutions. Rather than beginning with technology components, it starts with a clearly defined business objective and progressively refines that objective into requirements, specifications, implementation artifacts, and deployed capabilities. This creates a consistent delivery framework that connects business outcomes with technical execution.

The process begins with business intent. Examples include reducing DSO, improving on-time delivery performance, accelerating onboarding, or improving demand planning accuracy. These objectives are translated into business requirements and solution specifications that define the processes, data, controls, integrations, and governance needed to support the desired outcome. The resulting specifications are then implemented as applications, workflows, analytics, integrations, and agents before being deployed into production environments.

An important characteristic of the AI Golden Path is that business context remains central throughout the lifecycle. Process knowledge, enterprise architecture standards, application landscapes, data products, governance requirements, and operational constraints are incorporated during design rather than after deployment. This helps ensure that capabilities are aligned with enterprise standards and can be operated consistently at scale.

The specific AI technologies that operationalize this path depend on four factors: the structure of available data, response-time requirements, governance maturity, and organizational readiness. Different business objectives require different technologies, and the same objective may require multiple technologies depending on which phase of the roadmap an organization is in.

For Korindal, this means the objective of improving working capital is not translated directly into a collections agent. The organization first establishes process standards, data quality, governance policies, and integration readiness. Only then are automation and autonomous capabilities introduced. The result is a progression in which technology adoption follows business readiness rather than preceding it.

The comparison in Table 15.1 shows when each AI approach is most appropriate within the AI Golden Path.

Characteristic	Relational Foundation Models	Classic Machine Learning	LLMs and RAG	AI Agents
Data type	Structured data (tabular/relational with numerical, textual, categorical fields)	Structured data (tabular/relational, mostly numerical or categorical)	Unstructured or semistructured data (text, documents, conversations, code)	Multimodal: structured data, unstructured content, tool outputs, external system states
Problem type	Classification, regression, scoring tasks	Classification, regression, time series, anomaly detection, clustering, ranking	Language understanding, generation, transformation, semantic reasoning	Multistep reasoning, decision-making, planning, task orchestration across multiple tools
Output/ task type	Label, score, numeric value with clearly defined output	Label, score, numeric value, anomaly flag, rank, forecast	Generated text, extracted entities, transformations, embeddings, semantic matches	Action sequences, decisions with state, multiturn interactions, orchestrated workflows
Interpretability	High; transparent predictions with in-context examples	High; can be tuned and explained; feature importance visible	Medium; reasoning is partially opaque (black box generation)	Medium to high; decision flows are traceable but reasoning chain can be complex

Table 15.1: Selecting the Appropriate AI Approach

Characteristic	Relational Foundation Models	Classic Machine Learning	LLMs and RAG	AI Agents
Control and flexibility	High; flexible to changing data models and new tasks without retraining	Maximum control over model properties; requires ongoing model lifecycle management	Flexible; no retraining; semantic reasoning without custom engineering	Fine-grained control over flows, tools, rules, memory, error handling, and reflection steps
When to use	Structured business data, predictive tasks, when you want instant accuracy without training	Time series, anomaly detection, clustering, when latency <200ms is critical or data locality required	Unstructured text, semantic reasoning, knowledge extraction, conversational interfaces	Multistep workflows, cross-system coordination, goal-oriented task execution, autonomous operation
Use case examples	▪ Customer churn prediction ▪ Credit risk scoring ▪ Payment delay prediction ▪ Sales order completion ▪ Offer recommendations	▪ Demand forecasting ▪ Anomaly detection in sensors ▪ Search ranking ▪ Large catalog recommendations ▪ Time series prediction	▪ Document summarization and translation ▪ Information extraction ▪ Conversational Q&A ▪ Code generation ▪ Semantic search	▪ Collections workflow automation ▪ Supply chain optimization ▪ IT ticket routing and resolution ▪ Dispute resolution ▪ Process automation
Ideal scenario	You have clean, structured business data and need predictive insights	You need maximum control, lowest latency, or must handle time series/clustering problems	You work primarily with text/unstructured data and need flexible semantic reasoning	You need autonomous systems that reason across multiple tools, maintain state, and adapt over time

Table 15.1: Selecting the Appropriate AI Approach (Cont.)

Table 15.1 provides a quick reference for comparing the four primary AI modalities. Each has distinct strengths, operational characteristics, and ideal deployment scenarios. Let's elaborate on when and how each approach fits within the Golden Path, using Korindal's specific challenges to illustrate the decision logic:

- **Relational foundation models**
  Relational foundation models (e.g., SAP-RPT-1) excel in the foundation and optimize phases, when organizations need instant predictions on structured business data. Their primary advantage is that they can leverage pretrained business representations and operate directly on enterprise business data with significantly less task-specific training and feature engineering than traditional machine learning approaches. At Korindal, SAP-RPT-1 could predict payment risk by analyzing historical receivables data, providing collections specialists with risk scores on every customer interaction. This makes it ideal for rapid value realization early in the roadmap.
- **Classic machine learning**
  Classic machine learning (ML) is the default choice for time-series forecasting, anomaly detection, and complex pattern recognition. Supply chain demand forecasting at Korindal uses classic ML because forecasting benefits from model tuning, feature engineering, and the ability to incorporate domain knowledge about seasonality, supplier lead times, and regional variations. ML trades operational simplicity for predictive accuracy when that trade is worth making.
- **Large language models (LLMs) and retrieval-augmented generation (RAG)**
  LLMs with RAG become critical when business processes involve unstructured data: customer communications, contract terms, technical documentation, or support tickets. In collections, disputes often arrive as free-form customer emails. LLMs+RAG classifies disputes by category, sentiment, and risk—without retraining—because these technologies excel at semantic understanding without custom engineering. Collections specialists then route disputes based on AI-extracted categories, accelerating triage without requiring manual data cleanup.
- **AI agents**
  AI agents represent the highest maturity of the AI Golden Path. They coordinate decisions across multiple AI models, enforce business policies, maintain state, and adapt based on outcomes. Agents activate only after foundation and optimize phases have delivered measurable value and organizational trust is established. A collections agent in the run phase doesn't replace the organization's existing decision logic; it extends it. The agent enforces policies (maximum discount authority, escalation triggers, compliance rules) and orchestrates actions across systems, but always within guardrails that the steering committee defines and can revoke immediately.

Note
To learn more about SAP's AI Golden Path, visit *https://architecture.learning.sap.com/docs/ai-golden-path.*

Important: Check Before Building

Before designing a custom agent for your business objective, consult Chapter 5 to identify existing agents within the SAP Autonomous Suite (finance, supply chain, spend, HCM, customer experience). Many common business processes—collections automation, demand planning, procurement workflows—already have prebuilt agents ready for deployment. Custom agents should only be built when your use case falls outside existing domain coverage. Starting with what already exists accelerates time-to-value and reduces implementation risk.

Delivering a capability into production does not automatically create business value. The final step is ensuring that deployed capabilities produce measurable operational outcomes and that those outcomes are continuously monitored over time. This is the role of value realization, which will be discussed in the next section.

15.6 Value Realization Plan: Measuring and Governing the Outcomes

Deploying systems and activating AI agents is not the same as delivering business value. The difference is the gap between go-live and outcome. Systems go live and teams celebrate, but if DSO doesn't move, OTIF doesn't improve, and release velocity doesn't accelerate, then the program has succeeded in deployment but stopped short of impact. Systems went live, but the business did not change. Value realization is the discipline that prevents this. It is the mechanism that connects deployment activity to business impact and holds the program accountable for outcome delivery, not output completion.

Most transformations measure themselves by project milestones: go-live dates, user adoption rates, training completion. These are outputs. They tell you whether the program is executed, but they do not tell you whether the business has changed. The accountability gap emerges because outputs and outcomes have different timelines and owners. A system goes live in Q2 (output, owned by IT), but DSO doesn't move until Q3, when collections teams adopt the new process (outcome, owned by finance). If the program stops measuring after go-live, it never knows whether the outcome materialized.

Value realization fills this gap. It establishes what outcomes the program is trying to achieve, measures them constantly after deployment, and holds leadership accountable for moving the needle, not just deploying the system.

Value management runs in parallel with execution, as highlighted by an example framework in Figure 15.7.

It does not start after go-live. It starts in Week 2 of the program, during planning, before any system deployment begins. Baselines must be locked and ownership must be clear before execution starts. You cannot measure whether DSO improved if you do not know what the baseline was. You cannot hold Maria accountable for DSO if her role and decision authority are not defined.

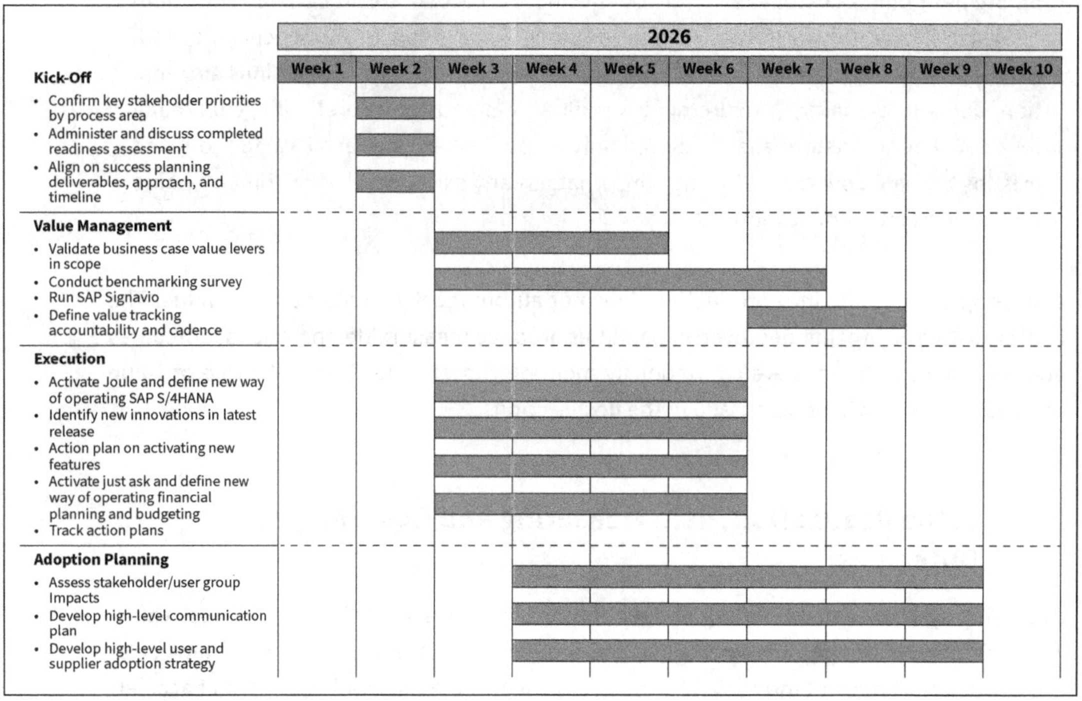

Figure 15.7: Value Management Governance

The value management workstream includes three key building blocks:

1. **Baseline establishment**
   Measure the current state accurately. How many days is DSO today? Not "around 18 days," but the exact number across all regions. How many collections exceptions are there per day? What percentage is resolved on first contact? What is the cost per exception? These numbers become the reference point for all future measurements.
2. **Ownership clarity**
   Assign a single owner to each outcome. Not a committee, not shared ownership, but one person who owns DSO, for example, and makes decisions about collections process changes; one person who owns OTIF and decides on supply chain prioritization; one person who owns release velocity. These owners commit publicly to targets and report monthly on progress.
3. **Measurement infrastructure**
   Build the dashboards and reports that track needles in real time: daily DSO trending, weekly collections metrics, monthly OTIF by region, quarterly release velocity. Without this infrastructure, programs rely on retrospective reporting ("Did we hit our target?") instead of real-time monitoring ("Are we on track to hit it? If not, why not?").

Figure 15.8 shows value realization in practice. It's a progression, not a one-time measure at go-live.

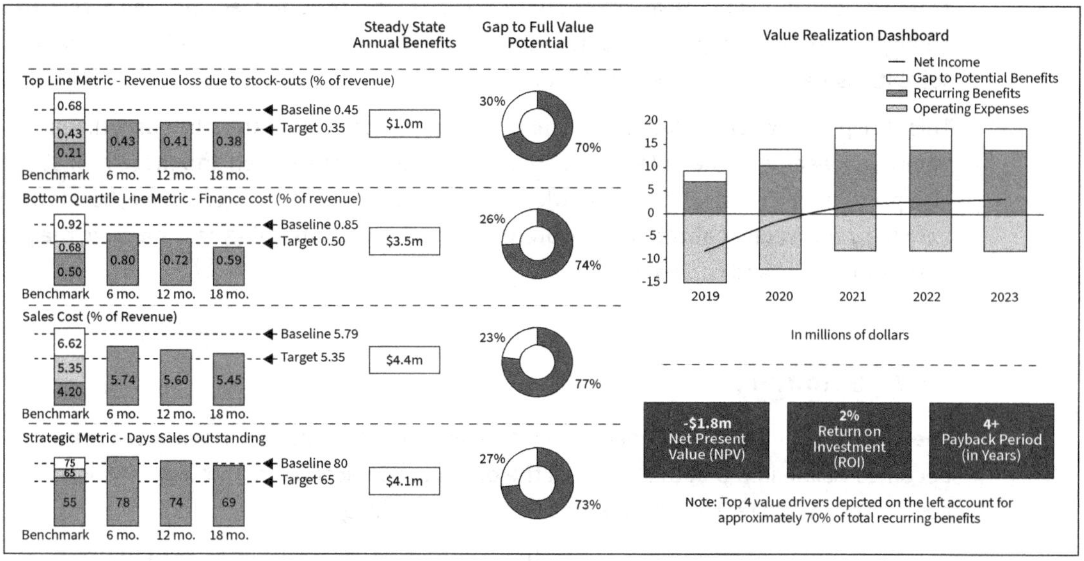

Figure 15.8: Value Tracking Dashboard

Baseline is locked at the start of the program. Revenue loss from stockouts is 0.45 percent of revenue. Cost per collections exception is $15. Release cycle time is four months. These are the starting points.

Then benchmark progression is measured on a regular basis. At six months, revenue loss from stockouts improves to 0.43 percent (a minor improvement as processes stabilize). At 12 months, 0.41 percent. At 18 months, 0.38 percent. The needle is moving, but slower than the target. This triggers an investigation: Are process changes working? Is adoption tracking? Is the target realistic?

The critical insight is that value does not appear instantly at go-live. It accumulates over quarters and years as adoption takes hold, processes stabilize, and AI agents learn from outcomes. Programs that expect value in Q2 and measure it through Q4 are disappointed. Programs that track value across 18 months see the real trajectory and make informed decisions about continuing an investment or redirecting effort.

The most common mistake is assuming that when an AI agent goes live, value materializes automatically. It does not. An agent can be deployed correctly, perform as designed, and still deliver zero business value if the conditions around it are not ready. A collections agent that routes disputes to specialists may work perfectly, but if specialists do not change how they handle disputes, then DSO does not move. A planning agent that recommends production adjustments may work perfectly, but if planners ignore recommendations because they do not trust the model, then OTIF does not improve.

Value realization measurement forces this reality into the open. When the collections agent is live but DSO has not moved after 30 days, the program stops and diagnoses why. Is agent accuracy low? Are specialists overriding decisions? Is adoption tracking? Is there a data quality issue? The program does not move to the next phase and hope that Phase 3 will solve it. Instead, it fixes the Phase 2 blocker.

This discipline, which includes pausing when value does not materialize and diagnosing before proceeding, is the difference between programs that succeed and those that stall. It also prevents the most common postdeployment problem: "The system is live, but nothing has changed." Value realization measurement ensures that if nothing has changed, the program knows why and can adjust.

15.7 Summary

The Rotterdam story that opened the Preface was about the gap between when something becomes visible in the data and when the organization acts on it. That gap is where value gets left on the table. This chapter has been about closing it.

The Autonomous Enterprise is not a technology upgrade. It is an operating model change. It requires a deployment strategy that is realistic about your starting point and your capacity for change. It requires instance planning that lets you scale autonomy without creating technical and operational gridlock. It requires a readiness assessment that is honest about where you are and what you need to change. It requires a multiyear roadmap that sequences capability maturation and delivers value continuously rather than in one big bang. It requires value realization governance that actually measures whether you are delivering value and corrects course when reality diverges from plan. And it requires an organizational commitment to genuinely changing how work is done rather than installing technology and expecting everything to change automatically.

Maria's $400 million working capital gap does not disappear overnight; it closes through a series of incremental improvements: faster invoice processing, better exception handling, earlier collection intervention, more sophisticated predictions, and finally autonomous agents that handle routine collection work continuously. Year after year, DSO improves and cash is released. By Year 3, the gap that looked insurmountable has been substantially closed because the organization has fundamentally changed how it makes decisions about receivables.

James's 14 percent OTIF miss does not disappear because Korindal implements an AI agent. It disappears because the company changes how supply chain planning works. It moves from batch cycles to continuous sensing. It introduces agents that can optimize within constraints. It changes how planners spend their time, from running analysis to making judgment calls on what to do with the analysis the system provides. OTIF improves quarter by quarter as the operating model matures.

David's release calendar that fills faster than it empties does not clear because Korindal modernizes its infrastructure. It clears because the company introduces a platform architecture that lets change happen without touching the core. It invests in integration and API consistency so that systems can evolve independently. It shifts IT's focus from maintaining technical debt to enabling business innovation. Delivery velocity improves because the architecture and the governance finally align.

This is not easy, but it is achievable through disciplined execution and an honest assessment of reality. The Autonomous Enterprise vision is not about technology. It is about organizations that can perceive what is happening in their operations, decide what to do about it, and act continuously, at scale, with increasing autonomy, while keeping the humans in the loop for decisions that require human judgment. This chapter has been about how to build exactly that.

By the end of the journey, the gap between signal and action closes. The working capital flows more freely. The orders are fulfilled on time. The systems change at the speed the business requires. The organization discovers that autonomy was never about removing humans from the process. It was about removing the time-consuming, repetitive parts so that humans could focus on what they actually do best: make judgment calls that drive strategy and relationships.

That discovery, that autonomy amplifies human value instead of replacing it, is what makes the vision real, and that is the path this chapter has laid out.

Technology enables transformation, but people deliver it. In the next chapter, we explore how organizational change management turns new capabilities into lasting business outcomes.

Chapter 16
Change Management

The previous chapter provided the strategic blueprint (deployment approach, instance strategy, readiness, roadmap, value measurement). This chapter ensures people adopt it through structured change management, behavior change, and sustained skill development.

When systems go live and teams celebrate, change management is often considered complete. In reality, it is just beginning. The Autonomous Enterprise requires a different caliber of change work than previous enterprise resource planning (ERP) programs because the nature of work itself changes. This chapter examines why, and how to prepare your organization for it.

We'll discuss how to develop that capability through structured change management, AI fluency building, and learning architecture. Strategy without execution becomes a white paper; execution without change management becomes a system that works but changes nothing about how the business operates.

16.1 Why Change Management Is Crucial

Figure 16.1 shows six implementation phases (discover, prepare, explore, realize, deploy, run), with organizational change management activities mapped to each. It also illustrates how change management is woven through the entire program lifecycle, not just at go-live.

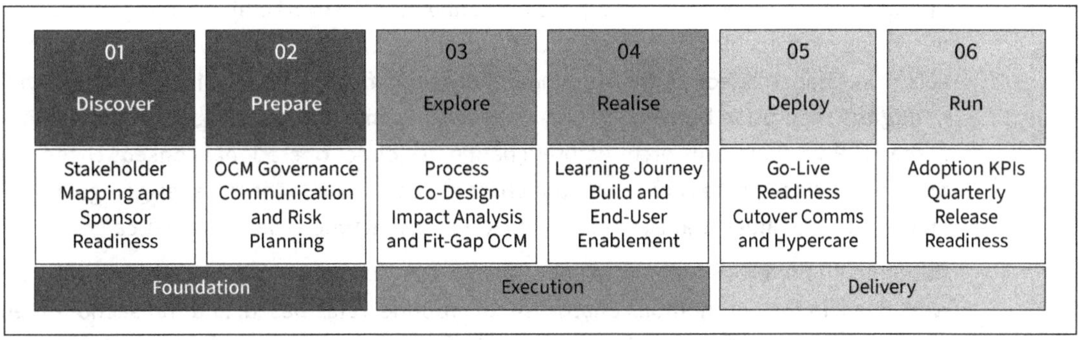

Figure 16.1: Organizational Change Management Phases

In the following sections, we'll explain why organizational change management activities are critical in your transition to the Autonomous Enterprise.

16.1.1 The Problem: Technology Does Not Transform Organizations

Figure 16.1 illustrates a critical insight: Change management is not a go-live event. It runs through the entire program lifecycle, from initial discovery through steady-state operations. The figure shows how organizational change management (OCM) activities are embedded in each of the six SAP Activate phases, starting with discovery work, where the organization assesses what must change; continuing through realization and deployment, where adoption happens; and extending into the run phase, where sustainment becomes the operating discipline. This continuous thread is what separates transformations that stick from those that revert.

Technology does not transform organizations; people do. Systems can be installed flawlessly, dashboards can be built perfectly, and AI agents can perform as designed, but if people do not change how they work, the organization does not transform. This distinction matters most in the Autonomous Enterprise era because the change is not incremental. It is structural.

In previous ERP programs, change was often about learning new screens and new processes. A finance controller still did reconciliation work, just on different software. A buyer still negotiated terms, just in a different system. In the Autonomous Enterprise, roles themselves change. Finance controllers shift from manual reconciliation to exception handling and analysis. Buyers shift from negotiating to validating machine-generated sourcing recommendations. Planners shift from running demand forecasts to interpreting recommendations and making trade-off decisions. Collections specialists shift from processing routine exceptions to handling judgment-based customer interactions.

This is an identity problem, not a process training problem. People have to renegotiate what their job is, how they create value, and where their expertise matters.

16.1.2 What Actually Changes

When a customer organization embarks on an SAP S/4HANA Cloud implementation or an upgrade to the Autonomous Enterprise operating model, several things change simultaneously. The most visible changes are process changes. SAP S/4HANA Cloud is built on a Universal Journal architecture that fundamentally simplifies the financial data model, eliminating the separation between financial and management accounting that characterized earlier ERP versions. This architectural change has direct operational consequences: The month-end close process changes; the chart of accounts may need redesigning; and the analytical capabilities that executives can access in real time expand dramatically.

But beneath the process changes lies a deeper transformation: a change in the nature of work itself. In the Autonomous Enterprise, the routine, rules-based, and transaction-intensive aspects of knowledge work are progressively automated. This is good news for productivity, but it requires a fundamental renegotiation of professional identity and a significant expansion of analytical and decision-making skills.

This pattern repeats across every functional domain. In procurement, intelligent sourcing changes the buyer's role from negotiator to algorithm auditor. In human resources (HR), AI recommendations reshape how managers handle performance and development. In finance, reconciliation automation shifts work from clerical to analytical. In supply chain, demand-sensing agents change planning from forecasting to scenario management. In collections, agent-assisted exception handling moves specialists from processing to relationship management.

16.1.3 The Adoption Gap

Most programs measure adoption via training completion rates and system usage metrics, but these are insufficient. The real measure is behavioral change: Are people using AI-generated insights? Are they trusting agent recommendations? Are they working differently, or just doing the same work on a new system? Without deliberate change management, organizations default to old behaviors: A finance analyst receives an AI-generated reconciliation but reconstructs it manually "just to be sure." A planner receives a demand forecast from an agent but runs their own analysis. A buyer receives a sourcing recommendation but goes to trusted suppliers anyway. The system works, but the business does not change, and value does not materialize.

In previous waves of ERP, change management was important, but in the Autonomous Enterprise era, it is the determinant of success. When the critical mass of the organization trusts system-driven recommendations and adopts new workflows, value compounds. When only a small minority do, value stalls. There is no middle ground where the program succeeds by half-measures.

This is why change management cannot be a project workstream. It is an ongoing operating discipline that runs in parallel with system deployment and extends well past go-live. Leadership commitment must be sustained, measurement must be continuous, learning must be reinforced, and the organization must genuinely renegotiate how work is done.

16.2 Structured Change Management

Effective change management in autonomous enterprise programs operates in three overlapping phases: readiness, adoption, and sustainment. Each phase has specific objectives, activities, and governance.

ADKAR® (awareness, desire, knowledge, ability, reinforcement), a registered trademark of Prosci Inc., is the foundational model underlying all change work (*http://s-prs.co/v633517*). This model recognizes that change is a progression, not a switch. Awareness (people understand what is changing) must precede desire (people want to change). Knowledge (people know how to change) requires training and learning journeys. Ability (people can actually do the new work) takes practice and coaching. Reinforcement (people sustain the new behaviors) requires ongoing leadership attention and measurement. Skipping a stage, like jumping from awareness to knowledge without building desire, or deploying without reinforcement, causes adoption to stall. This framework prevents programs from assuming that training alone is sufficient.

We'll dive into these phases of progression in more detail in the following sections.

16.2.1 Phase 1: Readiness

Chapter 15 addressed organizational readiness assessment as a planning discipline before execution begins (assessing six dimensions: scope, non-negotiables, capabilities, data/systems, governance, and change adoption). The readiness phase in change management is operationally different. It covers the first few months of the implementation lifecycle, when an organization must be explicitly prepared for what will change and why. Both assessments are required and complementary.

Readiness is not about awareness. It is about honest assessment of what will change and why, followed by explicit preparation. This phase establishes the foundation that all subsequent adoption will rest upon. Organizations that rush readiness consistently encounter resistance and adoption delays six months into the program when assumptions prove incorrect.

Stakeholder impact mapping identifies specifically which roles will change and how. Not just "finance will change," but specific, illustrative targets from a representative deployment: Invoice matching may be 90 percent or more automated; collections specialists may handle a fraction of their current manual volume directly. This specificity forces the organization to confront reality and prepare accordingly. The impact map should identify not just headcount changes but role evolution: What new skills will be needed? What decisions will shift to different people? What will become obsolete? When a role fundamentally shifts, people need to understand that evolution months before it happens, not weeks.

Resistance identification is a critical readiness activity. Assume resistance exists and identify where it will emerge. Finance leaders who built their reputation on heroic month-end closes may resist automation that eliminates the crisis management that once defined their value. Buyers with deep supplier relationships may distrust algorithms. Long-tenured specialists may fear that automation makes their expertise irrelevant. Identify these pockets early through candid conversations, not surveys. A direct conversation with a finance director about what they will do when the month-end close is automated is more valuable than a readiness assessment questionnaire.

Leadership commitment must be visible and sustained. Change succeeds when leadership is present week after week.

Case Study Example

Maria the CFO spends two hours a week reviewing adoption metrics and asking, "Why didn't this team adopt the new process?" James the CSCO sits in planning meetings watching agents work and asks, "Do you trust this recommendation?" David the CIO attends collections team standups and observes how specialists interact with the agent. Leaders being present communicates that all of this matters.

Role redesign happens before training starts. This is different from process design, which decides how work flows. Role redesign decides what the job is: Write new job descriptions, define new decision authorities, establish new reporting lines if needed. Collections specialists move from processing exceptions to managing customer relationships. Their performance metrics change, their career paths change, and their team structure changes. This must be explicit and documented before training starts so that people understand the magnitude of the shift and can prepare mentally and professionally.

OCM governance structures the readiness work. At minimum, governance includes a steering committee that meets monthly to review readiness assessments, approve the impact map, and clear blockers. An OCM workstream lead reports to the steering committee and coordinates stakeholder engagement, communication planning, and training design. If readiness governance is ad hoc or delegated entirely to HR, the program will encounter adoption surprises.

The readiness assessment itself should be grounded in measurable criteria, not sentiment. Does leadership understand the outcome targets and their role in achieving them? Can the finance controller articulate what will change about their job? Can the supply chain planner name the key decisions that will shift to AI agents? Do buyers understand which parts of the negotiation become automated and which remain human? An assessment that answers these questions with clarity moves forward. An assessment in which leaders cannot articulate what will change pauses and repeats until they can. This discipline prevents the "leadership said yes to the program but did not understand what it meant" problem that often derails adoption.

16.2.2 Phase 2: Adoption

Adoption runs in parallel with system deployment and continues past go-live. This is the longest and most critical phase because it determines whether the organization will actually change or whether it will adapt the system to existing behaviors.

Phased rollout with feedback loops prevents training everyone on day one. You can train in waves tied to deployment phases. The first wave is for foundation processes (real-time visibility, dashboards). Measure adoption for two to three weeks, identify obstacles, and fix them. Train for the second wave. Repeat. This iterative approach allows the program to diagnose and resolve adoption blockers quickly rather than discovering widespread resistance after full deployment. An obstacle for Wave 1 might be "the dashboard is confusing" or "nobody understands what this metric means." Fix it for Wave 2. Small corrections early prevent systemic problems later.

Guided practice replaces instruction. Training should not be "here is how to use the system." It should be "here is a real scenario from your work. Now decide: Do you trust the system recommendation or do you override it? Why?" Scenario-based learning creates judgment rather than procedure. People learn to think about AI-generated intelligence, not memorize screens. After training, a collections specialist should be able to explain their reasoning for accepting or rejecting an agent recommendation; that is judgment. If they can only remember the process, that is procedure, and it will not survive when the process changes.

Communities of practice emerge naturally if given space and air. Collections specialists learn from each other about how to interpret agent recommendations faster than formal training can teach. Planners share how they validate demand forecasts. Finance teams discuss reconciliation exceptions and what patterns mean. These communities become the real learning mechanism, supplementing formal training. Organizations that try to suppress communities of practice in favor of "consistent training" encounter adoption stalling because the practical learning that people need is not being shared.

Manager coaching is often overlooked, but it is critical. Managers need to understand what changed, but also why they need to hold their teams accountable for adopting it. A manager who does not understand why the new process matters will not push adoption when people resist. Manager coaching should include the business case for the change, how to recognize adoption, what obstacles to escalate, and what behaviors to reinforce. This is different from training; it is leadership preparation.

Escalation for resistance goes to line management, not to HR. When individuals resist change, escalate to their manager and one level above: "Your team member is not adopting the new process. Here is what we see. What is blocking them? How can we help?" This keeps change management visible to leadership and prevents it from becoming a project workstream issue. If escalation goes to HR, the issue becomes "why is this person resisting change" instead of "what organizational barrier is preventing adoption."

Adoption gates operationalize the overall discipline. A *gate* is a decision point at which specific metrics determine whether the program proceeds to the next phase. Gates should be grounded in business outcomes and technical readiness. For example, a Collections Adoption Gate might specify that the first-contact resolution rate must be at or above a defined target and that collections team satisfaction must be neutral or positive. If all conditions are met, the program proceeds to expand to remaining regions. If any is unmet, the program pauses and diagnosis begins.

Discipline is crucial because it prevents programs from proceeding with weak adoption foundations. A program might hit a high agent-acceptance rate and declare victory, moving forward while assuming the remainder will fall in line. More often, those who have not adopted remain resistant, and the weak foundation becomes a constraint when the program needs to scale.

Adoption metrics infrastructure matters more than the metrics themselves. A program that tracks adoption metrics manually on spreadsheets loses real-time visibility. Metrics that sit in dashboards no one checks weekly become historical artifacts. Effective adoption metrics infrastructure should include the following:

1. Automated data collection from system logs (how often does each user accept vs. override recommendations)
2. Weekly reporting delivered to adoption leaders and workstream leads
3. Escalation triggers when metrics drop below thresholds
4. Root cause investigation protocol when escalations occur
5. Cross-functional review forums where metric owners explain performance and propose adjustments

These infrastructure steps ensure adoption does not drift silently.

16.2.3 Phase 3: Sustainment

Sustainment is the often-forgotten phase in which change either sticks or reverts. Most programs measure success at go-live and declare victory. Sustainment is where that victory is either consolidated or lost.

Change management is a continuous cycle, not a linear program with a finish line. The assessment of readiness (what new capability is coming?) feeds into communication (why does this matter?), which enables people through learning and certification, which supports deployment, which is measured through adoption metrics, which informs iteration for the next cycle. When SAP releases new capabilities, organizations must assess readiness, communicate, enable, deploy, measure, and iterate. Without this disciplined cycle, improvements plateau and adoption stalls. Organizations that treat change as continuous (cycling through readiness, adoption, and sustainment repeatedly) maintain momentum. Those that declare victory after go-live and stop the cycle watch adoption decline.

Continuous learning recognizes that technology evolves. Agent models improve, processes are redesigned based on operational learning, new capabilities are released quarterly. Learning does not end at go-live. It should include monthly training updates on new capabilities, quarterly forums for teams to share how they are working differently, annual skills assessments to catch skill degradation. When organizations stop learning after go-live, they lose months of operational learning that could be captured and spread. Teams that develop new workarounds or discover better ways to interpret agent recommendations should be sharing that learning, not siloing it.

Reinforcement from leadership prevents backsliding. Maria still reviews days sales outstanding (DSO) metrics, still attends steering committees, still asks, “Are we adopting the new process or reverting?" This sustained visibility prevents backsliding. Organizations that declare the change program over once systems go live are surprised six months later when usage drops and old behaviors emerge. Leadership visibility is the counterweight that prevents that reversion.

Measurement of business outcomes matters more than adoption metrics alone. Not just "90 percent of users trained," but determine the business outcomes: Has DSO moved or stalled? Has on-time, in-full (OTIF) delivery improved or declined? Has release velocity accelerated? If outcomes are not materializing despite high adoption, investigate. Is adoption real or performative? Are people using systems but reverting to old decisions? Is the agent model wrong? Is there an organizational constraint preventing the change from delivering value? This investigation should change the program.

16.2.4 Adoption Patterns Across Functional Domains

Change management approaches differ by functional area because the nature of work, the skills required, and the resistance patterns are fundamentally different. Finance change differs from supply chain change, which differs from IT change. Let's explore how:

- **Finance domain change**
  Finance transformations often focus on moving from manual reconciliation to exception handling and analytics. Collections teams move from exception processing to relationship management. The adoption challenge in finance is often skepticism: "Will the system really match these complicated items?" "What if something goes wrong?" Finance leadership typically has high risk aversion, so resistance comes from wanting to keep manual review and audit trail processes that are familiar. Change management in finance must build confidence through gradual automation (foundation phase with visibility only, not decisions) and require sign-off from audit or compliance that new matching logic is acceptable. Finance change also requires explicit communication about what data quality issues will be discovered during transition and how they will be addressed. Many finance teams encounter surprise data quality findings that derail adoption if they are not mentally prepared for diagnosis and cleanup work.
- **Supply chain domain change**
  Supply chain transformations often focus on moving from demand forecasting to demand sensing and scenario planning. Planners move from running forecasts to interpreting agent recommendations and making business trade-offs. The adoption challenge in supply chain is often trust: "Will the algorithm account for what I know about this customer?" "What if the forecast is wrong?" Supply chain professionals often have deep domain knowledge and relationships with customers and suppliers. They see the agent as potentially naive. Change management in supply chain must involve planners deeply in model development and validation. When planners help train the model on historical data and understand the feature importance, they are more likely to trust it going forward. Pilot programs in which the agent forecasts are compared side by side with planner forecasts for six weeks before switching to agent-only mode significantly

increase adoption. Planners see that the model captures patterns they would have missed and handles routine adjustments automatically.

- **IT domain change**
  IT transformations often focus on moving from a project-based change release to a continuous flow release and increasing deployment frequency materially. IT teams move from planning major releases to managing continuous integration and deployment pipelines. The adoption challenge in IT is often throughput anxiety: "If we deploy twice a week, we will break production." "Our compliance team will not allow this velocity." IT change management must address both capability gaps (many IT teams have never done continuous deployment) and governance concerns (compliance, security, change advisory board processes). Starting with a dedicated low-risk product or application that can deploy frequently, building confidence, and then expanding to higher-risk systems works well. Also, establishing clear approval criteria upfront such as automated test coverage percentage, security scanning results, and performance thresholds prevents the change advisory board from becoming a bottleneck. When approval is based on objective criteria met by automation, governance strengthens instead of blocking change.

16.3 Measurement and Diagnosis: When Adoption Stalls

The most important capability in change management is not communication or training. It is the ability to diagnose why adoption is not happening and fix it before proceeding to the next phase. This requires measurement discipline and root cause analysis.

Adoption stalling looks like this: The system is live, training completion is at 90 percent, system login rates are at 85 percent, and yet nothing has changed. DSO collections is still 72 days, OTIF is still 86 percent, and the release cycle is still four months. The technology worked, but the people did not change.

Root cause diagnosis starts with specific metrics. If agent acceptance is 40 percent when it should be 80 percent, the next question is why. Let's take an example of a company that implemented a few AI agents, such as collection and demand sensing agents. There are five possible categories of root cause:

- **Model accuracy issue**
  The agent recommendations are wrong: The collections agent recommends calling customers who pay consistently and have zero payment risk, or the demand sensing agent predicts a demand increase when demand is actually declining. When model accuracy is the root cause, the fix is retraining the model or adjusting the feature engineering. Organizations should pause adoption progression and invest the time to fix the model.

- **User understanding issue**
  The system outputs are confusing or people do not understand how to interpret them. Collections specialists do not know what it means when the agent flags a customer as *high priority*. Planners do not understand which parts of a forecast are based on history

and which are based on external data. When understanding is the issue, the fix is targeted training, better system documentation, or clearer labeling of outputs.

- **Organizational readiness issue**
  The organization has a structural barrier preventing adoption. The collections department still has a multiperson approval process for every dispute, so agent recommendations are being systematically overridden. Finance has a policy that all reconciliation items must be manually reviewed, preventing autonomous matching. Buyers report to a director who does not believe in algorithms and actively tells them to ignore recommendations. When readiness is the issue, the fix requires leadership intervention to remove the barrier.
- **Capability gap issue**
  People do not have the skills to use the output productively. Finance analysts do not know how to interpret reconciliation exception patterns; planners do not know how to make trade-off decisions with the new demand sensing data; collections specialists do not know how to handle complex disputes that the agent flags. When capability is the issue, the fix is targeted training on the specific gap.
- **Confidence issue**
  People understand the system and have the skills, but they do not trust it enough to act on recommendations. They accept a small percentage of recommendations "to be safe" but override most of them. When confidence is the issue, the fix is often time-based: Exposure to accurate recommendations, visible business results, and peer adoption accelerate confidence building.

Building confidence requires sustained visibility into what the agent is recommending, tracking accuracy over time, and celebrating early wins. When a collections specialist sees that following the agent's recommendation for one month results in higher first-contact resolution and lower cost per exception, confidence builds. When they see peers successfully handling complex scenarios using agent recommendations, trust accelerates. Confidence-building programs include the following:

1. Weekly metrics showing agent accuracy on past recommendations
2. Monthly success stories shared across teams (what worked, why, and the outcome)
3. Peer learning forums where specialists teach each other how to work effectively with agents
4. Leadership recognition for specialists who show strong adoption and good judgment

A change program that measures weekly and takes root cause diagnosis seriously will identify these issues in weeks and fix them, not months. The program that does not measure closely will discover the adoption failure a few months later during the post-go-live business review. By that time, adoption patterns are entrenched, specialist frustration is high, and changing direction becomes costly in both time and organizational morale.

16.4 AI Fluency and Certification

The Autonomous Enterprise requires a new baseline competency: the ability to work with AI-generated intelligence rather than replace it entirely or ignore it entirely. This is not "AI skills" in the data science sense. It is judgment about when to trust recommendations and when to override them. It is understanding what data feeds an agent so you know whether to trust its output. It is pattern recognition: "This recommendation looks wrong because the context has changed."

AI literacy for all does not mean that everyone needs to build AI models. However, everyone needs to understand what an AI recommendation represents. What data does it use? How was it trained? What are its known failure modes? A finance controller does not need to code a machine learning model, but they need to understand that a reconciliation recommendation is only as good as the data that trained it. If the data included a period of unusual activity that skews the patterns, then the controller needs to recognize when that unusual data is influencing recommendations.

The AI fluency framework should define competency tiers. Level 1 (all users) includes understanding what an AI recommendation represents and recognizing when context has changed. Level 2 (decision-makers) includes assessing recommendation accuracy, knowing how to validate it, and understanding when to override it. Level 3 (technical staff) includes understanding data inputs, model retraining, and performance monitoring. This tiered approach prevents overtraining and targets depth where it matters most.

Role-specific scenarios tie AI literacy to actual work. Collections specialists need different AI literacy than planners. A scenario-based certification for collections might be: "The system recommends calling this customer now. The data shows they pay in 5 to 10 days. Last month they paid on Day 3. Last week there was a payment dispute that was resolved. Should you call?" This builds judgment about context and patterns. A planner scenario might be: "Demand forecast is up 20 percent versus last month. Last month we had an unexpected order from a large customer that dropped off. Should you adjust production?" This builds judgment about signal versus noise.

The assessment process for AI fluency validates judgment, not knowledge retention. A collections specialist passes if they can articulate why they made a specific decision about an agent recommendation as part of their real work, not if they correctly answer multiple-choice questions about the system. Did they have sound reasoning? Can they explain it? Have they shown the ability to override when appropriate and accept when warranted? These are the markers of genuine AI fluency. Someone who follows the agent blindly has not developed fluency. Neither has someone who overrides everything on principle.

Continuous validation recognizes that AI models drift over time. A demand-sensing model trained on prepandemic data produces wrong recommendations in a supply shortage. Certification is not a one-time activity; it is ongoing: quarterly recertification on new scenarios, annual skills assessments, tracking where model performance has changed. When an organization discovers that an agent's recommendations are systematically wrong because the training data is stale, everyone should know why and what changed, not just complain that the agent is broken.

Certification should measure whether someone can interpret what an agent recommendation means, identify when that recommendation might be wrong, make the right decision (trust or override) with evidence, and explain their decision to their manager. Certification is not a test score. It is demonstrated capability. A collections specialist who has handled 100 disputes through the new agent-assisted process, correctly overriding recommendations five times and explaining why, has demonstrated AI fluency. That person has learned more than someone who passed a certification exam.

16.5 Learning Journeys and Consultant Enablement

Consultant competence and customer adoption rise together or fall together. If consultants do not understand the Autonomous Enterprise operating model, they will teach customers old behaviors. If they do, they accelerate adoption significantly.

In the following sections, we'll explore two teaching methods to accelerate adoption: learning journeys and customer enablement.

16.5.1 Learning Journeys: Structured Paths

Learning journeys are curated learning paths tied to roles and deployment phases. A Collections Specialist Journey in Month 1 might cover an introduction to AI-assisted collections: what agents do and do not do, and how recommendations are generated. Month 2 covers working with agent recommendations: interpreting output and validating accuracy. Month 3 is scenario-based practice with real cases. Month 4 covers advanced judgment on when to override. Month 5 is continuous learning on new agent capabilities and model improvements.

Learning journeys are not generic. They are specific to an organization's deployment approach, its industry, its baseline capability, and the systems it is running. A company using a greenfield approach builds different competency than one performing a brownfield conversion. A financial services company needs different scenarios than a manufacturing company. A team that has never used real-time analytics needs different learning than a team that has. A customer with legacy systems integration challenges needs more learning on data quality and matching rules than a customer starting clean.

The journey must also account for role-specific learning gaps. A collections specialist who has worked in that role for 15 years has deep knowledge of customer relationships and dispute patterns, but may have no experience with data analysis or algorithm validation. The learning journey should build on existing strengths while filling specific gaps. Month 1 is not "learn about AI"; that is vague. Month 1 is "here is how the collections agent uses the customer payment history to prioritize who to call. You have 15 years of pattern recognition.

This is where your judgment is still irreplaceable." Build confidence through recognition of existing expertise, then expand into new areas.

Governance of learning journeys includes a learning council that meets quarterly to review completion rates, assessment pass rates, and feedback. Are people moving through the journey on schedule? Are assessment pass rates low, suggesting the content is not landing? Are people finding the scenarios realistic? This council does not just track compliance; it gathers feedback and adjusts the journey for the next cohort. A learning journey that is static becomes stale.

Learning journey effectiveness requires measurement beyond completion rates. Let's explore an example to track the effectiveness of a training:

1. **Assessment pass rates by topic**
   If a significant share of collections specialists fail the payment priority assessment, that topic needs redesign.
2. **Time to proficiency**
   This measures how long it takes before a specialist is making independent decisions with the new process.
3. **Mistake rates over time**
   As people progress through the journey, do mistakes decrease?
4. **Trusted teams**
   Manager confidence in team capability makes managers feel their teams are ready for expanded autonomy.
5. **Business outcome improvement correlation**
   Do specialists who complete the journey show better DSO improvement than those still early in it?

Such measurements ensure learning journeys are not just training completion exercises but genuine capability builders.

Role transitions within journeys deserve explicit attention. A collections specialist who moves from the journey's Month 2 (understanding output) to Month 4 (making override decisions) is taking on new responsibility and risk. That transition should include the following:

1. An explicit conversation with their manager about the skill progression
2. Recognition of the new capabilities they are developing
3. Adjustment of performance metrics to reflect expanded scope
4. Access to experts who can coach them through complex edge cases

Without explicit transitions, people progress through journeys invisibly, and their expanded capabilities go unrecognized and underutilized.

Learning journeys should also include exit criteria or graduate pathways. A specialist who completes all five months should have a clear career path: Advance to lead specialist, move to a different functional area, join the adoption coaching team, or continue as a subject matter expert. Without these pathways, the best learners become frustrated and exit before their expertise can compound. The organization invests in their development, they reach proficiency, and they leave because there is nowhere to go.

16.5.2 Consultant Enablement Programs

Consultants are the transfer mechanism for change. If consultants understand the Autonomous Enterprise mindset and the customer's specific context, they teach customers to change. If they understand only product features, they teach customers to adapt the system to old behaviors.

Consultant learning requirements include deep understanding of the operating model shift, not just feature functionality. The consultant needs to understand that a finance controller's job is changing, not just that there is a new reconciliation screen. They need to understand why a buyer's role shifts from negotiator to algorithm auditor. They need to understand the business case for each change so they can explain to customers why this matters.

Context knowledge is critical. What is the company's deployment approach? What is its instance strategy? What readiness gaps did it identify? What is its organizational capacity for change? Different companies need different consultant approaches even if they are running the same system. A company with strong change management capability might move fast. One managing multiple concurrent programs might need more support. One with weak data governance needs more focus on data quality.

Real scenario practice with mentoring is the core of consultant enablement. Not classroom training but real customer scenarios from the field, analyzed with subject matter experts. What worked? What failed? Why? When did the company adopt and when did it resist? Learning from real experiences accelerates consultant competence far more than generic training.

Consultant competency frameworks define what consultants must know at each stage. Entry-level consultants know the SAP S/4HANA features and basic processes. Mid-level consultants understand operating model shifts and can coach businesses on capability progression. Senior consultants have navigated multiple deployments and can read early warning signs of an adoption stalling or organizational resistance. Promotion through the framework is tied to demonstrated competency, not years of service. A consultant with two years of experience on five implementations may have more judgment than one with four years on two implementations.

Ongoing enablement includes regular updates on new capabilities and model improvements, quarterly forums where consultants share what is working and what is not in customer deployments, and annual recertification on industry-specific scenarios and emerging practices. When consultants stop learning after their initial enablement, their advice stales. Quarterly forums surface these changes and help consultants adjust. Annual recertification ensures that consultants stay grounded in current practice rather than reminiscing about how things used to work.

Incentives for consultants should reward adoption outcomes in company deployments, not just project hours. If a consultant delivers a system on time but the company does not adopt the new ways of working, the consultant has not succeeded. A consultant who is willing to extend an engagement to help a company through an adoption delay has succeeded. This incentive shift changes consultant behavior from "get the project done" to "get the company to change."

Consultant sustainability requires ongoing investment. Many consulting organizations treat initial enablement as a one-time event: onboard the consultant, certify them, deploy them. Then they wonder why consultant quality decays over time. The reality is that consultant competency naturally stales without reinforcement. Product capabilities evolve; customer contexts change; new scenarios emerge that earlier cohorts did not encounter. A consultant who was highly effective two years ago but has not updated their knowledge in six months is now potentially giving well-intentioned advice based on outdated understanding.

Sustaining consultant effectiveness requires the following:

- Mandatory regular learning hours on new product capabilities
- Participation in quarterly forums where consultants share customer scenarios and discuss what worked and what did not
- Annual recertification through scenario-based assessment in which consultants must demonstrate they can diagnose real customer problems and recommend appropriate approaches
- Mentorship relationships in which senior consultants coach emerging consultants on judgment and company engagement
- Periodic deployments where consultants embed themselves in organizations to reconnect with real operational challenges

This investment is expensive, but the cost of poor consultant advice is higher. Programs that stall or go wrong often trace back to consultant guidance that was well-intentioned but based on outdated or incomplete understanding.

Consultant specialization also matters. Some consultants thrive on broad enterprise transformations. Others excel in specific domains such as finance, supply chain, or IT. Rather than treating all consultants as interchangeable, organizations should support specialization. A consultant with five finance transformations under their belt understands the nuances of finance change that a generalist does not. They know how to work with controllers who are skeptical of automation. They understand the regulatory and compliance anxieties that surface during finance transformations. They are more valuable for finance work than for supply chain work. Consultant career paths should reflect this, offering advancement through specialization, not just seniority or hours billed.

16.6 Summary

Technology does not transform organizations; change management does. The Autonomous Enterprise requires a caliber of change work that is different in kind, not just degree. It is not about training people on new screens. It is about helping people renegotiate their professional identity, their decision-making authority, and what makes them valuable in an organization where routine work is increasingly automated.

This requires sustained leadership commitment that is visible and ongoing past go-live, not one-time ceremonial events. Consistent allocation of leadership attention to adoption metrics, behavior change, and business outcome realization matters. When a CFO stops asking about adoption metrics in Month 6, adoption starts declining. When a CSCO stops attending planning meetings to observe agent work, people start reverting to old planning approaches. Leadership visibility is the primary counterbalance to natural organizational inertia.

Structured change programs must move through readiness, adoption, and sustainment phases, an operating discipline that runs as long as the technology is in use. Each phase has specific objectives and governance, and each phase builds on the previous one. Programs that skip readiness encounter adoption surprises. Programs that skip sustainment watch adoption decline after go-live. The three-phase model is not an option but a structural requirement for change to stick.

Within each phase, governance structures matter. Readiness needs impact mapping and role redesign decisions that are explicit and documented. Adoption needs weekly metrics and adoption gates that trigger investigation when targets are missed. Sustainment needs continuous learning programs and quarterly forums where people share how they are adapting to new capabilities. This governance does not happen accidentally. It happens because a named person is accountable for it and reports monthly to leadership.

AI fluency must be built into the baseline competency model. Not just for data scientists, but for everyone whose work touches intelligent systems. Judgment about when to trust machine-generated recommendations and when to override them is as important as knowing how to use the system. When an organization treats AI fluency as optional, it creates a tier system: Some people learn to work with AI while others default to old approaches. Value compounds for the first group and stalls for the second.

Consultant accountability must be tied to adoption outcomes, not just project delivery. Consultants are the transfer mechanism. When they understand the operating model shift, companies learn it. When they do not, companies default to old behaviors. If a consultant is measured only on go-live date and training completion, they have no incentive to stay engaged through adoption. If they are measured on whether companies have actually changed how they work 12 months after go-live, they will invest in adoption and push back on timeline pressures that would compromise change.

Continuous measurement of behavior change and business outcome delivery is essential. Not just training completion, not just system usage, but whether people are actually making different decisions and whether the business is actually changing as a result. A dashboard showing that 85 percent of users are trained and 92 percent log in daily tells you nothing about whether DSO is moving or OTIF is improving. A dashboard showing DSO trend, OTIF trend, adoption by role, and first-contact resolution rate tells you whether the transformation is working. Measure what matters.

By the end of the transformation, the working capital flows more freely, orders are fulfilled on time, and systems change at the speed the business requires, but only if the people in the organization have genuinely changed how they work. That is the work of change management. Technology is just the enabler.

The relationship between technology and change is asymmetric. Technology enables change but does not guarantee it. An AI agent that correctly predicts payment risk is worthless if collections specialists ignore the predictions. A dashboard showing real-time inventory is useless if planners do not trust the data or know how to act on it. A system that routes disputes to the right specialist is pointless if that specialist uses the same manual workarounds they used before. The technology can be perfect, and the transformation can still fail if people do not change.

Conversely, people acting within outdated systems encounter frustration and constraints that limit what they can achieve. A specialist who wants to work with company context but whose system only shows transaction history is blocked by technology constraints. A planner who wants to make faster decisions is slowed by batch-based integration that delivers data weekly instead of hourly. Technology constraints are real, but an absence of technology constraints does not guarantee change. Organizations discover this painful truth regularly: They implement a new system, and the business operates differently for approximately two weeks until people revert to what they know and understand.

This means that change management is not optional. It is the connective tissue between technology capability and business outcome. It is the discipline that translates "the system can do this" into "the organization is doing this." It is the work that turns a good system into a catalyst for transformation.

The Autonomous Enterprise is built on this understanding. It assumes that technology will evolve: Capabilities will improve; new AI agent functionalities will be released. The organization's ability to adopt those capabilities and translate them into business outcomes will determine whether the enterprise becomes more autonomous or simply more expensive. That translation happens through the structured change management, AI fluency building, learning architecture, and consultant enablement described in this chapter.

The journey does not end with adoption. It ends when autonomous ways of working become part of how the enterprise operates every day. The final chapter shows how to make that shift from transactions to autonomy.

Chapter 17

From Analog Transactions to Autonomous Value

The Autonomous Enterprise is more than a collection of technologies; it is a different way of operating. This final chapter illustrates how applications, data, and AI come together to create that new operating model.

The story of enterprise software over the past 40 years has been a story of acceleration. Transactions that once took weeks now take seconds. Visibility that once came quarterly through closing processes now arrives in real time. Tasks that took entire back-office departments to accomplish now run in the cloud on platforms that scale without human intervention.

The Autonomous Enterprise is not the endpoint of that journey, but it is a fundamental shift in direction, and it is built on a specific convergence of three layers that have never before operated in concert at enterprise scale.

17.1 The Three Layers of the Autonomous Enterprise

Let's recall the three layers of the Autonomous Enterprise from Chapter 1. The first layer is the applications layer: SAP S/4HANA Cloud and line-of-business (LoB) applications running on a clean core, standardized and integrated, providing the execution platform through which the business actually operates. SAP S/4HANA Cloud for finance and supply chain, SAP Integrated Business Planning for Supply Chain (SAP IBP) for demand and supply, SAP Ariba for procurement: These are not new systems, but they are the operational backbone of the enterprise, now running cloud-native with the process standardization that autonomous operations require.

The second layer is the platform foundation: SAP Business AI Platform, SAP's enterprise AI foundation that brings together SAP Business Technology Platform (SAP BTP), SAP Business Data Cloud, and SAP Signavio. This is not a single product so much as a unified architecture. SAP BTP provides the development and runtime environment for AI applications and the agent infrastructure. SAP Business Data Cloud provides the data fabric that connects SAP application data, external signals, and partner networks into a coherent context that AI agents can reason over. SAP Signavio provides the process intelligence layer that gives agents an understanding of business workflows and process context, not just raw data records. Together, SAP Business AI Platform unifies data, process context, and governance into the trusted foundation that makes AI accurate, secure, and scalable at enterprise scale. Without this foundation, agents operate on fragments. With it, they operate on truth.

The third layer is the intelligence layer, which brings AI directly into operations. Joule Assistants are domain-level orchestrators embedded in each business domain, coordinating purpose-built Joule Agents to execute end-to-end workflows. Autonomous Finance, Autonomous Supply Chain, Autonomous Spend, Autonomous HCM, and Autonomous CX are the domains through which SAP Autonomous Suite delivers its value, each with specialized, role-based Joule Assistants and the agents they coordinate. Sitting at the human interface of this architecture is Joule Work, the dynamic AI workspace. Joule Work, based on each person's intent, surfaces the right insights, automates routine work, and coordinates AI agents across the enterprise so leaders and operators can focus on decisions that move the business forward. This layer does not replace the execution platform or the SAP Business AI Platform foundation. It acts on them, which includes perceiving patterns, recommending actions, making decisions within guardrails, and coordinating responses across domains. Together, these three layers create something no single layer could achieve: an enterprise that can sense what is happening, decide what to do about it, and act continuously rather than in batches.

This is the architecture of the Autonomous Enterprise. The journey to it does not follow a fixed calendar but a maturity progression, the four phases that every organization traverses at its own pace determined not by the availability of the technology but by the readiness of the organization. The phased model in this chapter describes maturity progression, not a simultaneous product release. Many of the Phase 1 and Phase 2 capabilities are available across SAP's platform today. Phase 3 and Phase 4 describe what organizations achieve as they build data readiness, trust, and autonomous authorization progressively. Specific capabilities and integrations continue to evolve. The phases are a map of the journey, not a checklist of features available at a single point in time.

To understand what this journey produces in practice, we return to Korindal Consumer Group, a company that started its journey with three familiar problems and ended it with something it had never had before: an operation that learns.

17.2 Where Korindal Started: One Layer, Not Three

Recall Maria's working capital problem. Korindal's regional systems maintained separate accounts receivable processes running on multiple ERP systems, but without the process standardization that makes automation possible. Collections staff performed manual exception handling. Disputes took weeks to resolve. Days sales outstanding (DSO) was running 18 days above target, not because the applications were wrong but because the data flowing through them was fragmented across regions, and no intelligence layer existed to act on that data continuously. At Korindal's revenue scale, eighteen days of DSO above target represented roughly $400 million in cash the company had earned and not yet collected. The execution platform existed, but the data fabric and the AI layer did not. The components of the solution were there, such as reconciliation rules, payment history, and dispute classifications, but they were scattered across departments, spreadsheets, and individual expertise.

James faced a parallel problem in supply chain. Multiple ERP systems managed order management and inventory execution, while legacy supply chain planning systems ran demand and supply planning. However, the two ran on separate cadences, so the planning cycle was disconnected from the ERP execution layer. On-time, in-full (OTIF) delivery was down by 14 percent not because the applications were incapable but because the signal, including a demand shift visible in POS data, in shipment patterns, in customer ordering behavior, and so on, arrived too slowly and was acted on too late. The planner was always chasing reality. A data fabric that unified those signals in real time, and an AI layer that could detect and act on them continuously, would change the equation entirely.

David's release problem was different in nature but identical in structure. Legacy application lifecycle systems provided the change and release management infrastructure, but the governance process layered on top of it was human-committee-based: months of planning, approval cycles, and risk review before any deployment. The application was not the constraint; the process around it was. An AI system could continuously assess deployment requirements by asking the same governance questions as a review committee would, which could reduce the process from weeks to hours while still maintaining control.

These three problems had a common cause: The enterprise had one layer, applications, but not the other two. Data was fragmented, not unified. Intelligence was human, not augmented. The gap between knowing something and acting on it was structural because the architecture required human effort to bridge it. This is where the Autonomous Enterprise begins, not by replacing the applications layer but by adding the data and AI layers that allow the applications to operate at their full potential.

The journey from this fragmented starting point to autonomous operations follows four phases of maturity. The phases are not measured in years; rather, they are stages of organizational and technical readiness that progress at different speeds depending on where an enterprise starts. For Korindal, starting from a fragmented multi-ERP landscape, all four phases were part of its transformation story. For organizations already running SAP S/4HANA Cloud with an active data fabric, the first phase is already complete. The intelligence can be activated from that foundation without delay.

17.2.1 Phase 1: The Foundation: Standardize, Connect, and First Intelligence

Korindal's first challenge was architectural. The enterprise could not add an intelligence layer to processes that were not standardized or a data fabric to systems that did not share a common model. The foundation had to come first.

The migration to SAP S/4HANA Cloud brought Korindal's fragmented regional operations into a single, standardized execution platform. The European core, which was previously running SAP ERP with 15 years of accumulated customization, moved to SAP S/4HANA Cloud's clean core architecture. The North American operation, running on an entirely separate ERP system with its own chart of accounts, moved to the same platform. The Asia Pacific acquisition followed. For the first time in Korindal's history, finance, supply chain, and procurement ran on a consistent process model with consistent data definitions across all regions. SAP IBP replaced the fragmented planning systems, SAP Ariba unified procurement, and SAP Cloud ALM provided the structured release management infrastructure for the program itself and all future deployments.

The moment SAP S/4HANA Cloud went live in the first region, something changed that had never been possible before: SAP Business Data Cloud could begin connecting. SAP Business Data Cloud began assembling a unified view of accounts receivable, demand signals, and working capital across all live regions. Maria's finance team received the first consolidated receivables dashboard across regions that ran on the same data model. It was a partial picture since the migration was still running, but it was real, current, and did not depend on a month-end consolidation. That alone changed the conversation. The problem was no longer invisible.

Before the full migration was complete, Korindal activated its first AI agent. In the European region, which was already live on SAP S/4HANA Cloud, the Collection Insights Agent was deployed in recommendation mode within the Accounts Receivable Assistant. It did something immediately valuable: It analyzed payment history, behavioral patterns, and account status across the European receivables portfolio and produced a daily prioritized contact list for the collections team. The team received actionable, ranked intelligence rather than a raw aging report. Ninety percent of the time, the recommendation aligned with what the specialist would have done anyway. Ten percent of the time, the specialist overrode it and the system registered each override as a signal that made the next recommendation more accurate.

This first agent produced two things simultaneously: a measurable reduction in DSO in the pilot region, and a growing performance record that would later authorize autonomous operation. Every recommendation that was accurate, every override that was documented, every exception that was logged became training signal and governance evidence for the phases that followed.

For organizations that have already completed this foundation phase, including running SAP S/4HANA Cloud with SAP Business Data Cloud, Phase 1 is done. The intelligence layer can be activated from that foundation without waiting for further infrastructure work. The pace of the journey ahead is determined by how quickly the organization is ready to adopt new ways of working, not by technology readiness.

17.2.2 Phase 2: Intelligence Activated: Agents as Advisors

With SAP S/4HANA Cloud live across all major regions and SAP Business Data Cloud fully operational, Korindal entered Phase 2, in which it activated the full intelligence layer in recommendation mode. This is the phase in which agents become genuinely useful and start surfacing patterns, providing context, and building the organizational trust that autonomous operation requires later.

In Joule Work, Korindal's finance and supply chain teams now had a workspace where intent became action. The Accounts Receivable Assistant became active across all regions. The Planning Assistant began working alongside SAP IBP's AI-assisted demand sensing. The governance infrastructure for agents was established through SAP AI Agent Hub, the centralized capability for discovering, inventorying, and governing all deployed agents before their scope expanded.

For Maria, Phase 2 was transformative in a way that had not been possible in Phase 1. SAP Business Data Cloud now unified receivables data across all regions in real time, replacing month-end consolidation with a live view of performance. The Accounts Receivable Assistant helped Maria prioritize collections activities, improve receivables visibility, and act earlier on at-risk accounts. Orchestrating agents such as the Collection Account Preparation Agent, Collection Insights Agent, and Collections Email Outreach Agent reduced manual effort and enabled a more proactive approach to collections management. The Billing Assistant streamlined billing operations and improved billing accuracy. Leveraging agents such as the Billing Creation Agent, Billing Adjustment Agent, and Billing Anomaly Agent helped identify exceptions earlier and reduce time spent on routine billing activities.

Together, these assistants automated repetitive work, improved visibility across the order-to-cash process, and allowed finance teams to focus on customer engagement, exception management, and decisions that required business judgment. The performance record the agents built in this phase was as important as the operational outcomes. By the end of Phase 2, DSO had reduced by approximately four days against target, releasing around $90 million in previously trapped cash through agents that were still operating in recommendation mode.

James activated the Planning Assistant within Korindal's SAP IBP environment. SAP IBP's AI-assisted demand sensing capability began generating demand signal updates far more frequently than the previous monthly planning cycle, drawing on short-term transactional data, including shipments, customer order patterns, and inventory movements, and near-term signals from the supply network. Where the monthly planning run had produced a single forecast revision once every 30 days, the AI capability identified meaningful pattern shifts within the planning cycle and surfaced them to planners continuously.

The Planning Assistant translated demand-sensing outputs into planning context that James and his team could act on immediately. Behind the scenes, it coordinated capabilities and agents such as the Lead Time Agent, Exception and Scenario Management Agent,

Inventory Investment Allocation Agent, and Demand Fulfillment Agent to continuously evaluate the implications of changing demand patterns across the network.

Rather than waiting for the next monthly planning cycle and reviewing a large replanning output, James's team received timely insights when demand began to diverge from plan. The Planning Assistant highlighted emerging demand shifts, potential inventory exposure, capacity constraints, and fulfillment risks, providing planners with the context needed to evaluate trade-offs and determine the appropriate response.

Planners could review the underlying demand signals, assess the recommendations generated by the supporting agents, and either act on them or override them based on market conditions, customer commitments, or business knowledge that was not reflected in the planning models. OTIF improvement in Phase 2 was modest; the underlying cadence of planning had not yet changed, and planners still made every final decision. However, those decisions were now informed by more current signals, better visibility into supply and demand imbalances, and earlier identification of potential shortages or fulfillment risks. The time between signal detection and planner response began to shrink. That narrowing of the gap between insight and action became the foundation for the more autonomous planning capabilities introduced in Phase 3.

David's IT organization used Phase 2 to make governance more consistent and transparent. Instead of relying primarily on review meetings and individual judgment, deployment approvals were based on clearly defined criteria such as test coverage, security results, performance checks, and configuration validation.

At the same time, the move to a cloud operating model changed how IT teams spent their time. Many routine infrastructure activities such as patching, backups, system maintenance, and technical upgrades became more standardized and less time-consuming. This allowed David's team to focus less on keeping systems running and more on preparing the business to take advantage of new SAP capabilities. SAP Cloud ALM helped automate the collection and evaluation of deployment evidence, giving teams a clearer view of whether a release was ready to proceed and where gaps remained. Readiness assessments, upgrade planning activities, and release preparation became more structured, helping teams understand the impact of new functionality, identify required actions, and plan adoption more effectively.

As a result, discussions shifted away from debating opinions and toward resolving specific issues that had been identified. Routine changes moved through the approval process more efficiently, while IT teams were able to dedicate more attention to evaluating new release capabilities, reducing custom code where appropriate, and preparing the business for innovation. The improvement came not from people working faster, but from having clear criteria, better visibility into compliance, automated evidence gathering, and more time to focus on value-adding activities. As a result, Korindal's average time from change request to approved deployment shortened by approximately 30 percent during Phase 2, not through automation of approvals but through better evidence, clearer criteria, and less time spent in preparation. It was a foundation, but not yet a transformation.

Phase 2 established three foundations that enabled the next stages of the transformation. First, it showed that the organization could successfully incorporate assistants and agents into everyday work without disrupting existing processes. Employees remained in control of decisions while benefiting from better information, recommendations, and automation of routine activities.

Second, it provided greater visibility into where recommendations delivered value and where they required refinement. This allowed teams to measure outcomes, build confidence in the new capabilities, and establish a performance baseline for assistants and agents across different business processes.

Third, it put the governance, operating, and technology foundations in place to support broader automation. With clearer controls, defined approval criteria, and centralized management of assistants and agents, the organization was better prepared to adopt the more autonomous capabilities introduced in phase 3.

17.2.3 Phase 3: Autonomous Within Guardrails: Agents as Decision-Makers

Phase 3 marked a shift in how work was performed across Korindal. The assistants did not suddenly gain new capabilities; instead, the organization became confident enough to allow assistants to coordinate and execute routine activities within clearly defined business rules and governance controls. The focus was no longer on whether assistants could provide recommendations but on where assistants could be trusted to act, while employees concentrated on the situations that required experience, judgment, and relationship management. The assistants became the primary coordination layer, orchestrating specialized agents across finance, supply chain, and IT workflows.

For Maria's organization, the Accounts Receivable Assistant became the central way collections activities were managed. Working through agents such as the Collection Insights Agent, Collection Account Preparation Agent, and Collections Email Outreach Agent, the assistant continuously prioritized accounts, prepared outreach activities, and coordinated routine collection processes. Collection specialists no longer spent most of their time reviewing reports, gathering account information, or preparing standard communications. Their attention shifted to strategic customers, dispute resolution, and complex collection situations where human expertise added the greatest value. As routine activities were handled more consistently and more quickly, DSO improved by approximately six days from the Phase 2 baseline, moving significantly closer to the organization's working capital targets. The Billing Assistant further improved the order-to-cash process by coordinating billing creation, billing adjustments, and anomaly detection activities, helping reduce delays and improve billing accuracy. At the same time, the Governance Assistant extended its scope to continuous controls monitoring, replacing the quarterly internal audit cycle with real-time assurance. Maria's compliance team moved from preparing for audits to managing work from a live controls dashboard.

In the supply chain organization, the Planning Assistant evolved from providing planning insights to coordinating routine planning activities across stable areas of the business. Working through agents such as the Inventory Investment Allocation Agent, Exception and Scenario Management Agent, Demand Fulfillment Agent, and Supply Issue Resolution Agent, the assistant continuously monitored inventory levels, capacity constraints, component availability, and replenishment needs. Planners were no longer required to spend most of their time managing routine operational decisions. Instead, they focused on supplier negotiations, constrained materials, new product introductions, and strategic trade-offs between inventory, cost, and service levels. Planning cycles became shorter and more responsive, and OTIF improved to approximately 91 percent as the organization responded faster to changing conditions.

David's IT organization experienced a similar transformation. The cloud operating model reduced the effort required for routine infrastructure activities such as patching, backups, technical upgrades, and platform maintenance. Governance controls and SAP Cloud ALM provided greater transparency and consistency across release management. Instead of spending most of their time keeping systems running, IT teams focused on evaluating emerging capabilities, identifying new assistants and agents that could deliver business value, and working with business leaders to prioritize adoption. Release planning increasingly centered on business outcomes rather than technical readiness alone. New capabilities could be assessed, piloted, and activated more quickly because governance, testing, and operational foundations were already in place.

As a result, release cycles accelerated significantly. Routine changes that met established governance criteria moved through the process with less manual oversight, helping reduce delivery timelines by roughly half compared to earlier phases. The flywheel effect became visible. Every successful deployment increased confidence in the operating model, making it easier to adopt additional assistants and agents, which in turn created new opportunities for automation and business improvement.

By the end of Phase 3, the organization's original constraints had materially improved. DSO had been reduced by approximately 10 days from the original baseline, with most of that improvement coming from Phase 3's autonomous operations. OTIF had increased to about 91 percent. Release cycles were roughly twice as fast as they had been at the start of the transformation. More importantly, finance teams, planners, and IT specialists were spending less time on routine execution and more time on customer outcomes, operational optimization, innovation, and business growth. The organization had demonstrated that assistants could coordinate work at scale within defined guardrails while maintaining governance, transparency, and accountability. That confidence became the foundation for Phase 4.

As assistants took on a larger role in coordinating work, governance did not disappear; it changed.

In the earlier phases, employees reviewed recommendations and made the final decisions. By Phase 3, many routine activities were being coordinated by assistants within clearly defined business rules. The governance focus therefore shifted from reviewing individual actions to defining, monitoring, and continuously improving the rules that guided those actions. Korindal learned that successful autonomy required three capabilities. First, business users needed visibility into why recommendations and actions were being taken. Second, governance controls had to be established before assistants operated at scale rather than after issues occurred. Third, employees needed the ability to intervene whenever business conditions changed or an outcome required additional review.

These capabilities had already been established through the governance foundations, operating model, and controls introduced in Phase 2. As a result, assistants could operate with greater autonomy while remaining transparent, accountable, and aligned to business objectives.

17.2.4 Phase 4: Coordinated Autonomy: The Flywheel in Motion

Phase 4 represented a shift from optimizing individual functions to optimizing the enterprise as a whole. In the earlier phases, the Accounts Receivable Assistant, Billing Assistant, Planning Assistant, and Governance Assistant each improved outcomes within their respective domains. In Phase 4, they began operating with a shared business context built on SAP Business Data Cloud. Information that had previously remained within departmental boundaries became available across functions, creating a more connected view of customers, products, suppliers, inventory, cash flow, and operational performance.

For Maria's finance organization, this meant that collection priorities were no longer evaluated solely through receivables information. The Accounts Receivable Assistant could consider broader business context, including customer order patterns, service performance, and operational disruptions. At the same time, the Billing Assistant helped ensure that billing and dispute activities were coordinated with changing business conditions. Decisions were no longer made using isolated departmental data. They were made using a shared view of the business.

The same evolution occurred in supply chain operations. The Planning Assistant continued coordinating activities through capabilities such as the Inventory Investment Allocation Agent, Exception and Scenario Management Agent, Lead Time Agent, and Demand Fulfillment Agent. However, planning decisions now reflected a broader understanding of enterprise conditions. Planners could evaluate supply, demand, customer commitments, and financial implications using a common set of business signals rather than separate functional reports.

The largest change occurred within David's IT organization. By this stage of the transformation, the cloud operating model had significantly reduced the effort spent on infrastructure operations, patching, backups, and technical upgrades. IT teams increasingly focused on business outcomes rather than technical maintenance. Working with leaders across finance and supply chain, they evaluated new assistants and agents as they became available, identified high-value use cases, and accelerated business adoption. The SAP LeanIX enterprise architecture capability helped simplify the discovery and assessment of new capabilities, while governance processes ensured that new assistants could be introduced in a controlled and repeatable manner.

A supply disruption demonstrated the value of this operating model: When sourcing conditions changed unexpectedly, finance, supply chain, procurement, and IT no longer worked through separate analyses and sequential decision cycles. The Planning Assistant highlighted inventory and supply impacts. Finance assistants evaluated potential customer and cash flow implications. IT teams were able to rapidly assess whether new capabilities or process changes could support the response. Business leaders reviewed a coordinated view of the issue and focused their attention on selecting a course of action rather than assembling information from multiple systems.

This created a fundamentally different way of operating. The greatest value no longer came from automating individual tasks but came from connecting decisions across the enterprise. Finance, supply chain, and IT worked from the same business context, responded to changes more quickly, and aligned around common outcomes. What began in Phase 1 as isolated productivity improvements had become an enterprise-wide decision environment in which assistants coordinated work across business functions and helped the organization move from reactive operations to proactive management.

By Phase 4, Korindal had become a different enterprise.

DSO was within two to three days of target, representing more than 15 days of improvement from the original baseline. Hundreds of millions of dollars in working capital had been released back into the business. Collection teams spent less time on routine processing and more time on customer relationships, dispute resolution, and strategic account management.

Supply chain performance improved as well. OTIF reached 94 percent, an improvement of eight percentage points from the original baseline. Planning teams focused on exceptions, supplier relationships, new product introductions, and operational risks rather than routine planning activities. Decisions were made earlier, with better information and greater visibility across the supply chain.

IT experienced an equally significant transformation. Release cycles that once took months were measured in weeks. The cloud operating model reduced the effort required for infrastructure and operational maintenance, allowing IT teams to focus on business outcomes. New assistants and agents were evaluated, piloted, and deployed in partnership with business leaders, creating a continuous cycle of innovation and adoption.

The most important outcome, however, was not reflected in any single metric: Korindal no longer operated through long delays between recognizing a problem and responding to it. Finance, supply chain, and IT worked from a shared understanding of the business and could coordinate action much more quickly when conditions changed. Human expertise remained essential, leaders still made strategic decisions, and specialists still managed complex situations. What changed was that people spent less time gathering information, coordinating routine work, and reacting to issues after the fact.

Korindal had not simply improved DSO, OTIF, or release velocity. It had developed the ability to sense change earlier, respond faster, and continuously adapt. That capability became a lasting advantage, and the true measure of its transformation.

17.3 The Starting Point: What to Do Now

The four-phase journey from Korindal's fragmented starting point to a connected and adaptive operating model may feel ambitious. For most organizations, the more immediate questions are where they are today and what steps they should take next.

The first step is to assess the foundation honestly. Organizations that are still operating across fragmented processes, disconnected data, and multiple system landscapes should focus first on establishing a common digital core with SAP S/4HANA Cloud and a unified data foundation through SAP Business Data Cloud. Assistants can only be as effective as the information available to them. If business processes and data remain fragmented, assistants will simply operate within the same constraints that people face today.

The second step is to introduce assistants in an advisory role before expanding their responsibilities. Whether deploying the Accounts Receivable Assistant, Billing Assistant, Planning Assistant, or Governance Assistant, the initial objective should be to improve decision-making rather than to automate decisions immediately. Allow business users to review recommendations, validate outcomes, and build confidence in the operating model. Every exception, override, and unexpected outcome provides valuable insight into where processes, business rules, or assistant behavior should be refined.

The third step is to establish governance before introducing greater autonomy. Organizations should define business rules, escalation paths, approval requirements, and accountability models early in the journey. For every activity that may eventually be coordinated autonomously, business leaders should determine what conditions qualify for autonomous handling, who retains oversight, how exceptions are managed, and when human intervention is required. Organizations that build governance into the operating model from the beginning are far more likely to scale successfully. SAP AI Agent Hub can provide a centralized inventory and governance layer for deployed assistants and agents, helping organizations maintain visibility, ownership, and control as adoption expands.

Equally important is investing in SAP Business Data Cloud as a strategic capability rather than a technical integration project. A shared business context does not emerge automatically. It requires trusted data, integrated processes, and a consistent view of customers, products, suppliers, inventory, cash flow, and operational performance. The greatest value is realized when assistants operate from a common business context rather than isolated functional datasets.

Organizations should also embrace the cloud operating model as part of their transformation. Too often, companies deploy new capabilities while continuing to operate IT as though infrastructure management remains the primary objective. The real opportunity comes when technical teams spend less time on patching, system maintenance, backups, and upgrades, and more time helping the business adopt innovation. SAP Cloud ALM helps establish operational discipline through release management, testing, monitoring, and lifecycle governance. For larger organizations, SAP LeanIX can provide the architectural visibility needed to understand how assistants, applications, processes, and business capabilities fit together as the landscape evolves. As SAP Business AI introduces new assistants and agents, IT teams should focus on evaluating business value, piloting capabilities with business stakeholders, and accelerating adoption across finance, supply chain, procurement, and customer operations.

Finally, governance capabilities should be established before scale is reached. As assistants become embedded across business processes, organizations need visibility into what has been deployed, how performance is being measured, and where oversight is required. Performance monitoring, auditability, business ownership, and intervention mechanisms should be designed early while the environment remains manageable.

The lesson from Korindal is straightforward: Success does not come from deploying more assistants or agents; it comes from establishing the right foundation with SAP S/4HANA Cloud, SAP IBP, SAP Business Data Cloud, SAP Business AI, SAP Cloud ALM, and SAP AI Agent Hub and then combining those capabilities with strong governance, clear business ownership, and a cloud operating model focused on outcomes. Organizations that build the foundation first, establish trust second, and expand autonomy third are far more likely to realize lasting value from their transformation journey.

17.4 The Moment in History

The Autonomous Enterprise is becoming possible because three technology foundations are reaching maturity at the same time and are finally beginning to work together rather than in isolation.

The first is the execution layer. SAP business applications such as SAP S/4HANA Cloud, SAP IBP, SAP Ariba, SAP SuccessFactors, and SAP Customer Experience provide the systems through which core business processes are executed. Process standardization, clean core principles, and cloud operating models make automation and continuous innovation possible at enterprise scale.

The second is the business network and data foundation. SAP Business Network extends business processes beyond enterprise boundaries to suppliers, logistics providers, partners, and customers. SAP Business Data Cloud connects business application data, external signals, and partner network information into a shared business context. Without this layer, assistants and agents operate on isolated data. With it, they can operate with a more complete and trusted view of their customers, products, suppliers, inventory, financial performance, and operational activity. This is the layer that enables decisions to move beyond individual functions and become enterprise-wide.

The third is the intelligence layer. SAP Business AI provides the foundation for assistants and agents that help coordinate work across business processes. Joule Assistants and Joule Agents bring operational intelligence into finance, supply chain, procurement, customer engagement, and human resources, while Joule Work provides the experience through which people interact with these capabilities. Rather than replacing business processes, the intelligence layer helps organizations execute those processes faster, more consistently, and with greater awareness of changing business conditions.

Running across all three layers is a governance and process foundation. SAP Signavio provides process intelligence and visibility into how work actually flows through the organization. SAP Cloud ALM provides lifecycle management, testing, monitoring, release management, and operational discipline. SAP AI Agent Hub provides centralized governance and visibility for deployed assistants and agents. Together, these capabilities help ensure that business processes, automation, and autonomous operations remain transparent, measurable, and aligned to enterprise objectives.

Individually, each layer provides value: Business applications improve operational execution. Business networks extend collaboration across enterprise boundaries. SAP Business Data Cloud improves visibility and context. SAP Signavio provides process understanding. SAP Business AI improves decision-making and execution. SAP Cloud ALM and SAP AI Agent Hub provide governance and operational control. Together, they create something no one piece could achieve alone: an enterprise that can sense what is happening, understand what it means, coordinate decisions across functions, and act continuously rather than through periodic review cycles.

17.5 Summary

Organizations are no longer lacking access to technology. The challenge is sequencing the journey correctly. Many organizations attempt to deploy assistants and agents before establishing a common execution foundation, trusted business data, and governance processes. In practice, intelligence amplifies whatever foundation is underneath it. If processes are fragmented and data is inconsistent, technology simply accelerates those limitations.

The organizations that will create the greatest value over the next decade will not necessarily be those that deploy the most assistants or agents. They will be the organizations that build and integrate all these layers into a coherent operating model. They will establish governance before they need it, invest in business process discipline, create a trusted data foundation, and adopt a cloud operating model that allows innovation to scale continuously.

Korindal's journey from a fragmented enterprise struggling with disconnected processes, delayed decisions, and organizational friction to a coordinated and adaptive enterprise is not fundamentally a technology story. It is an operating model story.

The Autonomous Enterprise is the outcome of bringing together applications, networks, data, intelligence, process governance, and operational discipline into a single system that continuously learns, adapts, and improves. It's not a technology to deploy, but a way of operating to build.

The Authors

Nitin Singh is the managing principal of architecture advisory at SAP America, helping Fortune 500 companies translate strategy into transformation outcomes. He leads a team of enterprise architects focused on simplifying technology landscapes, accelerating cloud adoption, and moving organizations from AI-assisted to AI-executed decisions.

With 21 years of experience guiding large-scale SAP transformations, his work centers on this book's core question: how enterprises move from automating processes to operating autonomously. Nitin is known for his pragmatic approach to clean core and enterprise architecture and an unwavering focus on measurable business value over technical perfection. His philosophy is that the best architecture is invisible. Success is measured not by the platform you build, but by how much faster the business moves.

Nitin is the author of two SAP PRESS books on SAP S/4HANA and SAP Signavio, a frequent speaker at SAP and ASUG events, and is based in Newtown Square, Pennsylvania.

Jason Porterfield leads SAP America's Americas Architecture Advisory practice, guiding strategic enterprise transformations across the region. An almost 30-year SAP ecosystem veteran, he has served as a customer, partner, and SAP executive—bringing a 360-degree perspective to modernizing ERP and accelerating outcomes. As enterprise architecture leader, he mentors teams that help organizations design their path to the cloud and realize measurable value. He is based in the Kansas City area and remains an active voice in the SAP enterprise architecture community.

Srini Devarapalli is an AI GTM strategy and advisory lead guiding enterprises on adopting SAP Business AI with a clear strategy, value roadmap, and responsible agentic and embedded AI capabilities. Prior to this role, he spent more than 16 years in management and technology consulting, leading large-scale transformations as an enterprise architect with broad experience in strategy, process design, and systems implementation across multiple industries. He has been a keynote speaker at SAP Insider and ASUG events, sharing insights on AI strategy and enterprise transformation.

Joffy Mathew is a senior director in SAP Business Data Cloud product management. He has 28 years of global SAP experience and is a seasoned strategist and results-oriented professional with demonstrated success leading high-profile, complex initiatives to develop strategy and delivery of enterprise data management and analytics solutions. At SAP, Joffy is focused on formulating business-driven data strategies as part of the product, enabling core technology solutions for intelligent and sustainable enterprises to improve data strategy and analytics. As a trusted advisor to a diverse list of strategic SAP clients across multiple industries, Joffy is a regular speaker at ASUG conferences and events. He conducts regular webinars on data strategy and digitization.

Cooper Walsh is a senior solution advisor for SAP BTP. He has worked at SAP for 10 years, focusing exclusively on SAP BTP. He has helped guide customers on their journey to SAP's modern business suite and understands the critical role that SAP BTP plays in helping customers integrate their solutions and adopt SAP's clean core framework. With the recent advances in generative AI, he has also developed a strong focus on how customers can adopt an AI strategy as part of their strategic roadmap, and on how SAP BTP can support AI for their business operations. Cooper is based out of SAP's Palo Alto location and supports its west enterprise customers with experience in high-tech, retail, consumer products, life sciences, agriculture, automotive, and other industries.

Lyssa Aruda is a senior solution advisor at SAP with more than a decade of experience leading large-scale enterprise transformations. She began her career in consulting, partnering with global organizations to drive process and organizational change, implement complex software solutions, and deliver operations at scale. In recent years, Lyssa has specialized in process optimization and process mining, helping organizations turn operational data into actionable insights and measurable business outcomes. Known for bridging technical depth with functional expertise, she works closely with customers to translate strategy into execution and ensure value realization across the transformation lifecycle.

Index

A

B

C

D

G

H

I

J

Q

R

S